Grundlehren der mathematischen Wissenschaften 348

A Series of Comprehensive Studies in Mathematics

Series editors

M. Berger   P. de la Harpe   N.J. Hitchin
A. Kupiainen   G. Lebeau   F.-H. Lin
S. Mori   B.C. Ngô   M. Ratner   D. Serre
N.J.A. Sloane   A.M. Vershik   M. Waldschmidt

Editor-in-Chief

A. Chenciner   J. Coates   S.R.S. Varadhan

For further volumes:
www.springer.com/series/138

Dominique Bakry · Ivan Gentil · Michel Ledoux

Analysis and Geometry of Markov Diffusion Operators

BOWLING GREEN STATE
UNIVERSITY LIBRARIES

Dominique Bakry
Institut de Mathématiques de Toulouse
Université de Toulouse
   and Institut Universitaire de France
Toulouse, France

Michel Ledoux
Institut de Mathématiques de Toulouse
Université de Toulouse
   and Institut Universitaire de France
Toulouse, France

Ivan Gentil
Institut Camille Jordan
Université Claude Bernard Lyon 1
Villeurbanne, France

ISSN 0072-7830   ISSN 2196-9701 (electronic)
Grundlehren der mathematischen Wissenschaften
ISBN 978-3-319-00226-2   ISBN 978-3-319-00227-9 (eBook)
DOI 10.1007/978-3-319-00227-9
Springer Cham Heidelberg New York Dordrecht London

Library of Congress Control Number: 2013952461

Mathematics Subject Classification: 39B62, 39B72, 47D07, 53C21

© Springer International Publishing Switzerland 2014

This work is subject to copyright. All rights are reserved by the Publisher, whether the whole or part of the material is concerned, specifically the rights of translation, reprinting, reuse of illustrations, recitation, broadcasting, reproduction on microfilms or in any other physical way, and transmission or information storage and retrieval, electronic adaptation, computer software, or by similar or dissimilar methodology now known or hereafter developed. Exempted from this legal reservation are brief excerpts in connection with reviews or scholarly analysis or material supplied specifically for the purpose of being entered and executed on a computer system, for exclusive use by the purchaser of the work. Duplication of this publication or parts thereof is permitted only under the provisions of the Copyright Law of the Publisher's location, in its current version, and permission for use must always be obtained from Springer. Permissions for use may be obtained through RightsLink at the Copyright Clearance Center. Violations are liable to prosecution under the respective Copyright Law.

The use of general descriptive names, registered names, trademarks, service marks, etc. in this publication does not imply, even in the absence of a specific statement, that such names are exempt from the relevant protective laws and regulations and therefore free for general use.

While the advice and information in this book are believed to be true and accurate at the date of publication, neither the authors nor the editors nor the publisher can accept any legal responsibility for any errors or omissions that may be made. The publisher makes no warranty, express or implied, with respect to the material contained herein.

Printed on acid-free paper

Springer is part of Springer Science+Business Media (www.springer.com)

How far can you go with the Cauchy-Schwarz inequality and integration by parts?

To Leonard Gross

Preface

Semigroups of operators on a Banach space provide very general models and tools in the analysis of time evolution phenomena and dynamical systems. They have a long history in mathematics and have been studied in a number of settings, from functional analysis and mathematical physics to probability theory, Riemannian geometry, Lie groups, analysis of algorithms, and elsewhere.

The part of semigroup theory investigated in this book deals with Markov diffusion semigroups and their infinitesimal generators, which naturally arise as solutions of stochastic differential equations and partial differential equations. As such, the topic covers a large body of mathematics ranging from probability theory and partial differential equations to functional analysis and differential geometry for operators or processes on manifolds. Within these frameworks, research interests have grown over the years, now encompassing a wide variety of questions such as regularity and smoothing properties of differential operators, Sobolev-type estimates, heat kernel bounds, non-explosion properties, convergence to equilibrium, existence and regularity of solutions of stochastic differential equations, martingale problems, stochastic calculus of variations and so on.

This book is more precisely focused on the concrete interplay between the analytic, probabilistic and geometric aspects of Markov diffusion semigroups and generators involved in convergence to equilibrium, spectral bounds, functional inequalities and various bounds on solutions of evolution equations linked to geometric properties of the underlying structure.

One prototypical example at this interface is simply the standard heat semigroup $(P_t)_{t \geq 0}$ on the Euclidean space $\mathbb{R}^n$ whose Gaussian kernel

$$u = u(t,x) = p_t(x) = \frac{1}{(4\pi t)^{n/2}} e^{-|x|^2/4t}, \quad t > 0, \ x \in \mathbb{R}^n,$$

is a fundamental solution of the heat equation

$$\partial_t u = \Delta u, \quad u(0,x) = \delta_0,$$

for the standard Laplace operator Δ, thus characterized as the infinitesimal generator of the semigroup $(P_t)_{t \geq 0}$.

From the probabilistic viewpoint, the family of kernels $p_t(x)$, $t > 0$, $x \in \mathbb{R}^n$, represents the transition probabilities of a standard Brownian motion $(B_t)_{t \geq 0}$ as

$$\mathbb{E}\big(f(x + B_{2t})\big) = \int_{\mathbb{R}^n} f(y)\, p_t(x - y)\, dy = P_t f(x), \quad t > 0, \ x \in \mathbb{R}^n,$$

for all bounded measurable functions $f : \mathbb{R}^n \to \mathbb{R}$.

The third aspect investigated in this work is geometric, and perhaps less immediately apparent than the analytic and probabilistic aspects. It aims to interpret, in some sense, the commutation of derivation and action of the semigroup as a curvature condition. For the standard Euclidean semigroup example above, the commutation $\nabla P_t f = P_t(\nabla f)$ will express a zero curvature, although this corresponds not only to the curvature of Euclidean space as a Riemannian manifold but rather to the curvature of Euclidean space equipped with the Lebesgue measure, invariant under the heat flow $(P_t)_{t \geq 0}$, and the bilinear operator $\Gamma(f, g) = \nabla f \cdot \nabla g$.

In order to carry out the investigation along these lines, the exposition emphasizes the basic structure of a Markov Triple[1] (E, μ, Γ) consisting of a (measurable) state space E, a carré du champ operator Γ and a measure μ invariant under the dynamics induced by Γ. The notion of a carré du champ operator Γ associated with a Markov semigroup $(P_t)_{t \geq 0}$ with infinitesimal generator L given (on a suitable algebra $\mathcal{A}$ of functions on E) by

$$\Gamma(f, g) = \frac{1}{2}\big[L(fg) - f L g - g L f\big],$$

will be a central tool of investigation, the associated Γ-calculus providing, at least at a formal level, a kind of algebraic framework encircling the relevant properties and results.

These analytic, stochastic and geometric features form the basis of the investigation undertaken in this book, describing Markov semigroups via their infinitesimal generators as solutions of second order differential operators and their probabilistic representations as Markov processes, and analyzing them with respect to curvature properties. The investigation is limited to symmetric (reversible in the Markovian terminology) semigroups, although various ideas and techniques go beyond this framework. We also restrict our attention to the diffusion setting, that is when the carré du champ operator is a derivation operator in its two arguments, even in those cases where the result could be extended to a more general setting. These restrictions rule out many interesting fields of applications (discrete Markov chains, models of statistical mechanics, most of the analysis of algorithms of interest in optimization theory or approximations of partial differential equations, for example), but allow us to concentrate on central features in the analysis of semigroups, in the same way that ordinary differential equations are in general easier to handle than discrete sequences. Even within the field of symmetric diffusion semigroups, we have not tried

---

[1] The terminology "Markov triple" should not, of course, be confused with solutions of the Markov Diophantine equation $x^2 + y^2 + z^2 = 3xyz$!

to cover all the possible interesting cases. In order to keep the monograph within a reasonable size, we have had to omit, among other things, the specific analysis related to hypoelliptic diffusions, special features of diffusions on Lie groups, and many interesting developments arising from infinite interacting particle systems.

In addition, although we have largely been motivated by the analysis of the behavior of diffusion processes (that is, solutions of time homogeneous stochastic differential equations), rather than concentrating on the probabilistic aspects of the subject, such as almost sure convergence of functionals of the trajectories of the underlying Markov processes, recurrence or transience, we instead chose to translate most of the features of interest into functional analytic properties of the Markov structure (E, μ, Γ) under investigation.

Heat kernel bounds, functional inequalities and their applications to convergence to equilibrium and geometric features of Markov operators are among the main topics of interest developed in this monograph. A particular emphasis is placed on families of inequalities relating, on a Markov Triple (E, μ, Γ), functionals of functions $f : E \to \mathbb{R}$ to the energy induced by the invariant measure μ and the carré du champ operator Γ,

$$\mathcal{E}(f, f) = \int_E \Gamma(f, f) d\mu.$$

Typical functionals are the variance, entropy or $\mathbb{L}^p$-norms leading to the main functional inequalities of interest, the Poincaré or spectral gap inequality, the logarithmic Sobolev inequality and the Sobolev inequality. A particular goal is to establish such families of inequalities under suitable curvature conditions which may be described by the carré du champ operator Γ and its iterated Γ_2 operator.

Similar inequalities are investigated at the level of the underlying semigroup $(P_t)_{t \geq 0}$ for the heat kernel measures, comparing $P_t(\varphi(f))$ (for some $\varphi : \mathbb{R} \to \mathbb{R}$) to $P_t(\Gamma(f, f))$ or $\Gamma(P_t f, P_t f)$, which give rise to heat kernel bounds. With this task in mind, we will develop the main powerful tool of heat flow monotonicity, or semigroup interpolation, with numerous illustrative applications and strong intuitive content. To illustrate the principle, as a wink towards what is to come, let us briefly present here a heat flow proof of the classical Hölder inequality which is very much in the spirit of this book. In particular, the reduction to a quadratic bound is typical of the arguments developed in this work. Let f, g be suitable (strictly) positive functions on $\mathbb{R}^n$ and $\theta \in (0, 1)$. For fixed $t > 0$, consider, at any point (omitted), the interpolation

$$\Lambda(s) = P_s\left(e^{\theta \log P_{t-s} f + (1-\theta) \log P_{t-s} g}\right), \quad s \in [0, t],$$

where $(P_t)_{t \geq 0}$ is the standard heat semigroup on $\mathbb{R}^n$ as recalled above. Together with the heat equation $\partial_s P_s = \Delta P_s = P_s \Delta$, the derivative in s of Λ is given by

$$\Lambda'(s) = P_s\left(\Delta(e^H) - e^H\left[\theta e^{-F} \Delta(e^F) + (1-\theta) e^{-G} \Delta(e^G)\right]\right)$$

where $F = \log P_{t-s} f$, $G = \log P_{t-s} g$ and $H = \theta F + (1-\theta)G$. Now by standard calculus,

$$e^{-H}\Delta(e^H) - \left[\theta e^{-F}\Delta(e^F) + (1-\theta)e^{-G}\Delta(e^G)\right]$$
$$= |\nabla H|^2 - \theta|\nabla F|^2 - (1-\theta)|\nabla G|^2$$

which is negative by convexity of the square function. Hence $\Lambda(s)$, $s \in [0, t]$, is decreasing, and thus

$$\Lambda(t) = P_t\left(f^\theta g^{1-\theta}\right) \leq (P_t f)^\theta (P_t g)^{1-\theta} = \Lambda(0).$$

Normalizing by $t^{n/2}$ and letting t tend to infinity yields Hölder's inequality for the Lebesgue measure. Actually, the same argument may be performed at the level of a Markov semigroup with invariant finite discrete measure, thus yielding Hölder's inequality for arbitrary measures.

While functional inequalities and their related applications are an important focal point, they also give us the opportunity to discuss a number of issues related to examples and properties of Markov semigroups and operators. One objective of this work is thus also to present the basic tools and ideas revolving around Markov semigroups and to illustrate their usefulness in different contexts.

The monograph comprises three main parts.

The first part, covering Chaps. 1 to 3, presents some of the main features, properties and examples of Markov diffusion semigroups and operators as considered in this work. In a somewhat informal but intuitive way, Chap. 1 introduces Markov semigroups, their infinitesimal generators and associated Markov processes, stochastic differential equations and diffusion semigroups. It also describes a few of the standard operations and techniques while working with semigroups. Chapter 2 develops in detail a number of central geometric models which will serve as references for later developments, namely the heat semigroups and Laplacians on the flat Euclidean space, the sphere and the hyperbolic space. Sturm-Liouville operators on the line, and some of the most relevant examples (Ornstein-Uhlenbeck, Laguerre and Jacobi), are also presented therein. On the basis of these preliminary observations and examples, Chap. 3 then tries to describe a general framework of investigation. While it would not be appropriate to try to cover in a unique formal mould all the cases of interest, it is nevertheless useful to emphasize the basic properties and tools in order to easily and suitably develop the Γ-calculus. In particular, it is necessary to describe with some care the various classes and algebras of functions that we shall be dealing with and to show their relevance in the classical smooth settings. Note that while infinite-dimensional models would require further care in this abstract formalism, the methods and principles emphasized throughout this work are similarly relevant for them. Taking the more classical picture as granted, Chap. 3 may be skipped at first reading (or limited to the summary Sect. 3.4).

Part II, forming the core of the text, includes Chaps. 4 to 6 and covers the three main functional inequalities of interest, Poincaré or spectral gap inequalities, logarithmic Sobolev inequalities and Sobolev inequalities. For each family, some basic properties and tools are detailed, in tight connection with the reference examples of Chap. 2 and their geometric properties. Stability, perturbation and comparison properties, characterization in dimension one, concentration bounds and convergence to equilibrium are thus addressed for each family. The discussion then distinguishes between inequalities for the heat kernel measures (local) and for the invariant measure (global) which are analyzed and established under curvature hypotheses. Chapter 4 is thus devoted to Poincaré or spectral gap inequalities, closely related to spectral decompositions. Chapter 5 deals with logarithmic Sobolev inequalities, emphasized as the natural substitute for classical Sobolev-type inequalities in infinite dimension, and their equivalent hypercontractive smoothing properties. Sobolev inequalities form a main family of interest for which Chap. 6 provides a number of equivalent descriptions (entropy-energy, Nash or Gagliardo-Nirenberg inequalities) and associated heat kernel bounds. A significant proportion of this chapter is devoted to the rich geometric content of Sobolev inequalities, their conformal invariance, and the curvature-dimension conditions.

On the basis of the main functional inequalities of Part II, Part III, consisting of Chaps. 7 to 9, addresses several variations, extensions and related topics of interest. Chapter 7 deals with general families of functional inequalities, each of them having their own interest and usefulness. The exposition mainly emphasizes entropy-energy (on the model of logarithmic Sobolev inequalities) and Nash-type inequalities. In addition, the tightness of functional inequalities is studied by employing the tool of weak Poincaré inequalities. Chapter 8 is an equivalent description of the various families of inequalities for functions presented so far in terms of sets and capacities for which co-area formulas provide the suitable link. The second part of this chapter is concerned with isoperimetric-type inequalities for which semigroup tools again prove most useful. Chapter 9 briefly presents some of the recent important developments in optimal transportation in connection with the semigroup and Γ-calculus, including in particular a discussion of the relationships between functional and transportation cost inequalities (in a smooth Riemannian setting).

The last part of the monograph consists of three appendices, on semigroups of operators on a Banach space, elements of stochastic calculus and the basics of differential and Riemannian geometry. At the interface between analysis, probability and geometry, these appendices aim to possibly supplement the reader's knowledge depending on his own background. They are not strictly necessary for the comprehension of the core of the text, but may serve as a support for the more specialized parts. It should be mentioned, however, that the last two sections of the third appendix on the basics of Riemannian geometry actually contain material on the Γ-calculus (in a Riemannian context) which will be used in a critical way in some parts of the book.

This book has been designed to be both an introduction to the subject, intended to be accessible to non-specialists, and an exposition of both basic and more advanced results of the theory of Markov diffusion semigroups and operators. Indeed we chose to concentrate on those points where we felt that the techniques and ideas are central and may be used in a wider context, even though we have not attempted to reach the widest generality. Every chapter starts at a level which is elementary for the notions developed in it, but may evolve to more specialized topics which in general may be skipped at first reading. It should be stressed that the level of exposition throughout the book fairly non-uniform, sometimes putting emphasis on facts or results which may appear as obvious or classical for some readers while developing at the same time more sophisticated issues. This choice is motivated by our desire to make the text accessible to readers with different backgrounds, and also by our aim to provide tools and methods to access more difficult parts of the theory or to be applied in different contexts. This delicate balance is not always reached but we nevertheless hope that the chosen style of exposition is helpful.

The monograph is intended for students and researchers interested in the modern aspects of Markov diffusion semigroups and operators and their connections with analytic functional inequalities, probabilistic convergence to equilibrium and geometric curvature. Selected chapters may be used for advanced courses on the topic. Readers who wish to get a flavor of Markov semigroups and their applications should concentrate on Part I (with the exception of Chap. 3) and Part II. Via an appropriate selection of topics, Part III tries to synthesise the developments of the last decade. The book demands from the reader only a reasonable knowledge of basic functional analysis, measure theory and probability theory. It is also expected that it may be read in a non-linear way, although the various chapters are not completely independent. The reader not familiar with the main themes (analysis, probability and geometry) will find some of the basic material collected together in the appendices.

Each Chapter is divided into Sections, often themselves divided in Sub-Sections. Section 1.8 is the eighth section in Chap. 1. Theorem 4.6.2 indicates a theorem in Chap. 4, Sect. 4.6, and (3.2.2) is a formula in Sect. 3.2. An item of a given chapter is also referred to in other chapters by the page on which it appears. There are no references to articles or books within the exposition of a given chapter. The Sections "Notes and References" at the end of each chapter briefly describe some historical developments with pointers to the literature. The references are far from exhaustive and in fact are rather limited. There is no claim for completeness and we apologize for omissions and errors. For books and monographs, we have tried to present the references in historical order with respect to original editions (although the links point toward the latest editions).

This book began its life in the form of lectures presented by the first author at Saint-Louis du Sénégal in April 2009. He thanks the organizers of this school for the opportunity to give this course and the participants for their interest. This work presents results and developments which have emerged during the last three decades. Over the years, we have benefited from the vision, expertise and help of a number of friends and colleagues, among them M. Arnaudon, F. Barthe, W. Beckner, S. Bobkov, F. Bolley, C. Borell, E. Carlen, G. Carron, P. Cattiaux, D. Chafaï, D. Cordero-Erausquin, T. Coulhon, J. Demange, J. Dolbeault, K. D. Elworthy, M. Émery, A. Farina, P. Fougères, N. Gozlan, L. Gross, A. Guillin, E. Hebey, B. Helffer, A. Joulin, C. Léonard, X. D. Li, P. Maheux, F. Malrieu, L. Miclo, E. Milman, B. Nazaret, V. H. Nguyen, Z.-M. Qian, M.-K. von Renesse, C. Roberto, M. de la Salle, L. Saloff-Coste, K.-T. Sturm, C. Villani, F.-Y. Wang, L. Wu and B. Zegarliński. We wish to thank them for their helpful remarks and constant support.

F. Bolley, S. Campese and C. Léonard went through parts of the manuscript at several stages of the preparation, and we warmly thank them for all their corrections and comments that helped to improve the exposition.

We sincerely thank the Springer Editors C. Byrne and M. Reizakis and the production staff for a great editing process.

We apologize for all the errors, and invite the readers to report any remarks, mistakes and misprints. A list of errata and comments will be maintained online.

Lyon, Toulouse
June 2013

Dominique Bakry
Ivan Gentil
Michel Ledoux

Basic Conventions

Here are some classical and basic conventions used throughout the book.

$\mathbb{N}$ is the set of integers $\{0, 1, 2, \ldots\}$. The set of real numbers is denoted by $\mathbb{R}$. Functions (on some state space E) are always real-valued. Points in $\mathbb{R}$ are usually denoted by x (if $\mathbb{R}$ is the underlying state space) or by r.

An element $r \in \mathbb{R}$ is positive if $r \geq 0$, strictly positive if $r > 0$, negative if $r \leq 0$ and strictly negative if $r < 0$. Moreover, $\mathbb{R}_+ = [0, \infty)$ is the set of positive real numbers while $(0, \infty)$ denotes the set of strictly positive numbers. For $r, s \in \mathbb{R}$, $r \wedge s = \min(r, s)$ and $r \vee s = \max(r, s)$. We agree that $0 \log 0 = 0$.

In the same way (and somewhat against the current), a positive (respectively negative) function f (on E) is such that $f(x) \geq 0$ (respectively $f(x) \leq 0$) for every $x \in E$. The function is strictly positive or strictly negative whenever the inequalities are strict. Similarly, an increasing (respectively decreasing) function f on $\mathbb{R}$ or some interval of $\mathbb{R}$ satisfies $f(x) \leq f(y)$ (respectively $f(x) \geq f(y)$) for every $x \leq y$. The function f is said to be strictly increasing or strictly decreasing whenever the preceding inequalities are strict. A function is monotone if it is increasing or decreasing.

Points in $\mathbb{R}^n$ are denoted by $x = (x_1, \ldots, x_n) = (x_i)_{1 \leq i \leq n}$ (or sometimes $x = (x^1, \ldots, x^n) = (x^i)_{1 \leq i \leq n}$ depending on the geometric context). The scalar product and Euclidean norm in $\mathbb{R}^n$ are given by

$$x \cdot y = \sum_{i=1}^n x_i\, y_i, \quad |x| = (x \cdot x)^{1/2} = \left(\sum_{i=1}^n x_i^2 \right)^{1/2}.$$

The notation $|\cdot|$ is used throughout to denote the Euclidean norm of vectors and of tensors.

The constant function equal to 1 on a state space E is denoted by $\mathbb{1}$. If $A \subset E$, $\mathbb{1}_A$ is the characteristic or indicator function of A.

All measures on a measurable space $(E, \mathcal{F})$ considered here are positive measures. Positive (measurable) functions on $(E, \mathcal{F})$ may take the value $+\infty$. If μ is a (positive) measure on $(E, \mathcal{F})$, and if f is a function on E which is integrable with respect to μ, its integral with respect to μ is denoted by $\int_E f\, d\mu$ or $\int_E f(x) d\mu(x)$, or sometimes as $\int_E f(x)\mu(dx)$. The Lebesgue measure on the Borel sets of $\mathbb{R}^n$ is denoted by dx. If B is a Borel set in $\mathbb{R}^n$, its Lebesgue measure is sometimes denoted by $\text{vol}_n(B)$.

The terminology "change of variables" is used in the broad sense of changing a variable x into $h(x)$ and a function f into $\psi(f)$. "Chain rule" is understood more in an algebraic sense when h and ψ are polynomials.

The notations are supposed to be reasonably stable throughout the monograph. Further definitions and conventions will be given in the text when they are needed. A list of symbols and notations with the corresponding reference pages is given on p. 523.

Contents

Part I Markov Semigroups, Basics and Examples

1 Markov Semigroups 3
  1.1 Markov Processes and Associated Semigroups 7
  1.2 Markov Semigroups, Invariant Measures and Kernels 9
  1.3 Chapman-Kolmogorov Equations 16
  1.4 Infinitesimal Generators and Carré du Champ Operators 18
  1.5 Fokker-Planck Equations 23
  1.6 Symmetric Markov Semigroups 24
  1.7 Dirichlet Forms and Spectral Decompositions 29
  1.8 Ergodicity 32
  1.9 Markov Chains 33
  1.10 Stochastic Differential Equations and Diffusion Processes 38
  1.11 Diffusion Semigroups and Operators 42
  1.12 Ellipticity and Hypo-ellipticity 49
  1.13 Domains 52
  1.14 Summary of Hypotheses (Markov Semigroup) 53
  1.15 Working with Markov Semigroups 56
  1.16 Curvature-Dimension Condition 70
  1.17 Notes and References 74

2 Model Examples 77
  2.1 Euclidean Heat Semigroup 78
  2.2 Spherical Heat Semigroup 81
  2.3 Hyperbolic Heat Semigroup 88
  2.4 The Heat Semigroup on a Half-Line and the Bessel Semigroup . 92
  2.5 The Heat Semigroup on the Circle and on a Bounded Interval . . 96
  2.6 Sturm-Liouville Semigroups on an Interval 97
  2.7 Diffusion Semigroups Associated with Orthogonal Polynomials . 102
  2.8 Notes and References 118

3	**Symmetric Markov Diffusion Operators**	119
	3.1 Markov Triples	120
	3.2 Second Order Differential Operators on a Manifold	137
	3.3 Heart of Darkness	151
	3.4 Summary of Hypotheses (Markov Triple)	168
	3.5 Notes and References	173

Part II   Three Model Functional Inequalities

4	**Poincaré Inequalities**	177
	4.1 The Example of the Ornstein-Uhlenbeck Semigroup	178
	4.2 Poincaré Inequalities	181
	4.3 Tensorization of Poincaré Inequalities	185
	4.4 The Example of the Exponential Measure, and Exponential Integrability	187
	4.5 Poincaré Inequalities on the Real Line	193
	4.6 The Lyapunov Function Method	201
	4.7 Local Poincaré Inequalities	206
	4.8 Poincaré Inequalities Under a Curvature-Dimension Condition	211
	4.9 Brascamp-Lieb Inequalities	215
	4.10 Further Spectral Inequalities	220
	4.11 Notes and References	230
5	**Logarithmic Sobolev Inequalities**	235
	5.1 Logarithmic Sobolev Inequalities	236
	5.2 Entropy Decay and Hypercontractivity	243
	5.3 Integrability of Eigenvectors	250
	5.4 Logarithmic Sobolev Inequalities and Exponential Integrability	252
	5.5 Local Logarithmic Sobolev Inequalities	257
	5.6 Infinite-Dimensional Harnack Inequalities	265
	5.7 Logarithmic Sobolev Inequalities Under a Curvature-Dimension Condition	268
	5.8 Notes and References	273
6	**Sobolev Inequalities**	277
	6.1 Sobolev Inequalities on the Model Spaces	278
	6.2 Sobolev and Related Inequalities	279
	6.3 Ultracontractivity and Heat Kernel Bounds	286
	6.4 Ultracontractivity and Compact Embeddings	290
	6.5 Tensorization of Sobolev Inequalities	291
	6.6 Sobolev Inequalities and Lipschitz Functions	293
	6.7 Local Sobolev Inequalities	296
	6.8 Sobolev Inequalities Under a Curvature-Dimension Condition	305
	6.9 Conformal Invariance of Sobolev Inequalities	313
	6.10 Gagliardo-Nirenberg Inequalities	323
	6.11 Fast Diffusion Equations and Sobolev Inequalities	330
	6.12 Notes and References	340

Part III Related Functional, Isoperimetric and Transportation Inequalities

7 Generalized Functional Inequalities 347
   7.1 Inequalities Between Entropy and Energy 348
   7.2 Off-diagonal Heat Kernel Bounds 355
   7.3 Examples 362
   7.4 Beyond Nash Inequalities 364
   7.5 Weak Poincaré Inequalities 373
   7.6 Further Families of Functional Inequalities 382
   7.7 Summary for the Model Example μ_α 386
   7.8 Notes and References 387

8 Capacity and Isoperimetric-Type Inequalities 391
   8.1 Capacity Inequalities and Co-area Formulas 392
   8.2 Capacity and Sobolev Inequalities 396
   8.3 Capacity and Poincaré and Logarithmic Sobolev Inequalities .. 399
   8.4 Capacity and Further Functional Inequalities 403
   8.5 Gaussian Isoperimetric-Type Inequalities Under a Curvature Condition 411
   8.6 Harnack Inequalities Revisited 421
   8.7 From Concentration to Isoperimetry 425
   8.8 Notes and References 429

9 Optimal Transportation and Functional Inequalities 433
   9.1 Optimal Transportation 434
   9.2 Transportation Cost Inequalities 438
   9.3 Transportation Proofs of Functional Inequalities 442
   9.4 Hamilton-Jacobi Equations 451
   9.5 Hypercontractivity of Solutions of Hamilton-Jacobi Equations .. 454
   9.6 Transportation Cost and Logarithmic Sobolev Inequalities 458
   9.7 Heat Flow Contraction in Wasserstein Space 462
   9.8 Curvature of Metric Measure Spaces 464
   9.9 Notes and References 466

Appendices

Appendix A Semigroups of Bounded Operators on a Banach Space .. 473
   A.1 The Hille-Yosida Theory 473
   A.2 Symmetric Operators 475
   A.3 Friedrichs Extension of Positive Operators 477
   A.4 Spectral Decompositions 478
   A.5 Essentially Self-adjoint Operators 481
   A.6 Compact and Hilbert-Schmidt Operators 483
   A.7 Notes and References 485

Appendix B Elements of Stochastic Calculus 487
    B.1   Brownian Motion and Stochastic Integrals 487
    B.2   The Itô Formula 491
    B.3   Stochastic Differential Equations 493
    B.4   Diffusion Processes 495
    B.5   Notes and References 498

Appendix C Basic Notions in Differential and Riemannian Geometry   499
    C.1   Differentiable Manifolds 500
    C.2   Some Elementary Euclidean Geometry 502
    C.3   Basic Notions in Riemannian Geometry 504
    C.4   Riemannian Distance 509
    C.5   The Riemannian Γ and Γ_2 Operators 511
    C.6   Curvature-Dimension Conditions 513
    C.7   Notes and References 518

Afterword 521
    Chicken "Gaston Gérard" 521

Notation and List of Symbols 523

Bibliography 527

Index 547

Part I
Markov Semigroups, Basics and Examples

Chapter 1
Markov Semigroups

This chapter is introductory and descriptive. It aims to introduce, somewhat informally, some of the basic ideas and concepts in the investigation of Markov semigroups, operators and processes, at the interface between analysis, partial differential equations, probability theory and geometry, loosely jumping from one area to another. Readers only familiar with one, or two, or even none of these fields, will find the necessary background in the various appendices.

The chapter is not intended to be read linearly, but should be used as a *vade mecum* of ideas and tools on the topic of Markov semigroups. Each section introduces and sheds light on one or several aspects of the investigation of Markov semigroups, aiming at this point to develop intuition rather than giving precise definitions and hypotheses. Pointers to more precise descriptions and examples appearing in subsequent chapters of the book are provided. In particular, some of the main features of the analysis will be illustrated by model examples in Chap. 2, while Chap. 3 will develop the complete formalism describing the suitable environment in which the full theory may easily be developed. For the reader's convenience, a first set of precise assumptions concerning state spaces, measures, semigroups etc., is nevertheless put forward here in Sect. 1.14.

There are many ways to address Markov semigroups and generators. In this chapter, we investigate their elementary properties starting with the assumption that we already know the semigroup itself which then leads to the definition of the so-called infinitesimal generator of the semigroup, which in return entirely describes the latter. In Chap. 3, conversely, we begin with the generator, or rather its carré du champ operator, given on some suitable class of functions, and use it to determine properties of the semigroup. This setting then describes the formalism in which the monograph will evolve.

Generally speaking, a semigroup $\mathbf{P} = (P_t)_{t \geq 0}$ is a family of operators acting on some suitable function space with the semigroup property $P_t \circ P_s = P_{t+s}$, $t, s \geq 0$, $P_0 = \text{Id}$. Such families naturally arise in numerous settings, and describe evolution equations which may be investigated from various viewpoints. In particular, these semigroups appear in the probabilistic context describing the family of laws of Markov processes $(X_t)_{t \geq 0}$ living on a measurable space E, the fundamental relation being given by the conditional expectation representation

$$P_t f(x) = \mathbb{E}\big(f(X_t) \mid X_0 = x\big)$$

for $t \geq 0$, $x \in E$ and $f : E \to \mathbb{R}$ a suitable measurable function.

In order to illustrate, and introduce the reader to, these investigations, let us describe the simplest example, namely the heat or Brownian semigroup on the Euclidean space $\mathbb{R}^n$. This example may be introduced in many ways, depending on the culture and purpose. According to the three main aspects of this monograph, we briefly describe it from the analytic, probabilistic and geometric viewpoints.

First we start with the family of Gaussian kernels

$$p_t(x) = \frac{1}{(4\pi t)^{n/2}} e^{-|x|^2/4t}, \quad t > 0, \; x \in \mathbb{R}^n.$$

Formally, we may already agree that $p_0(x)$ may be considered as the Dirac mass at 0. The normalization is chosen so that $p_t(x)$ are probability densities with respect to the Lebesgue measure (denoted dx), that is $\int_{\mathbb{R}^n} p_t(x) dx = 1$. It is a classical result, and easy to see, that p_t solves the (parabolic) heat equation

$$\partial_t p_t = \Delta p_t$$

where Δ is the standard Laplacian on $\mathbb{R}^n$.

From these kernels, one may define operators P_t, $t \geq 0$, sending suitable functions $f : \mathbb{R}^n \to \mathbb{R}$ to new functions $P_t(f) = P_t f$ defined by

$$P_t f(x) = \int_{\mathbb{R}^n} f(y) \, p_t(x, y) dy, \quad t > 0, \; x \in \mathbb{R}^n,$$

where $p_t(x, y) = p_t(x - y)$, $(x, y) \in \mathbb{R}^n \times \mathbb{R}^n$. This definition applies for example to any function f which is bounded and measurable, but the condition that f is bounded may obviously be relaxed provided $P_t f$ still makes sense. P_0 is the identity operator, $P_0 f = f$. By standard convolution, these operators satisfy the semigroup property $P_t \circ P_s = P_{t+s}$, $t, s \geq 0$, and thus form a so-called semigroup (moreover it is Markov, or mass preserving, since $P_t(\mathbb{1}) = \mathbb{1}$). On smooth functions $f : \mathbb{R}^n \to \mathbb{R}$,

$$\partial_t P_t f = \Delta P_t f = P_t(\Delta f).$$

In other words, $u(x, t) = P_t f(x)$, $t \geq 0$, $x \in \mathbb{R}^n$, solves the partial differential (heat) equation $\partial_t u = \Delta u$ with initial condition $u(x, 0) = f(x)$. This initial condition has however to be interpreted carefully. It is certainly true that, for any $x \in \mathbb{R}^n$, $\lim_{t \to 0} u(x, t) = f(x)$ provided f is continuous and bounded, but has to be understood in a weaker sense when f is only measurable.

Accordingly, the Laplacian Δ is called the infinitesimal generator of the (heat) semigroup $(P_t)_{t \geq 0}$. This semigroup is one simple prototypical example which anticipates several developments in this monograph. The heat equation $\partial_t u = \Delta u$ is the basic parabolic partial differential equation linking the first order time derivative to

1 Markov Semigroups

second order space derivatives, highlighting the central importance of second order differential operators of interest in this monograph.

On the probabilistic side, the densities p_t describe the transition probabilities of a random process $(B_t)_{t \geq 0}$ in $\mathbb{R}^n$ called Brownian motion. More precisely, $(B_t)_{t \geq 0}$ is a (continuous) random process with values in $\mathbb{R}^n$ such that $B_0 = 0$ (almost surely) and the increments

$$B_{t_1}, B_{t_2} - B_{t_1}, \ldots, B_{t_k} - B_{t_{k-1}}, \quad 0 \leq t_1 < \cdots < t_k,$$

are independent centered Gaussian vectors in $\mathbb{R}^n$ with respective covariance matrices $t_1 \, \mathrm{Id}, (t_2 - t_1) \, \mathrm{Id}, \ldots, (t_k - t_{k-1}) \, \mathrm{Id}$ where Id is the identity matrix. Note that this n-dimensional Brownian motion may be written as $B_t = (B_t^1, \ldots, B_t^n)$, $t \geq 0$, where $(B_t^i)_{t \geq 0}$, $i = 1, \ldots, n$, are independent one-dimensional Brownian motions. From this point of view, there is not much more to n-dimensional Brownian motion than in the one-dimensional case although dimension could (and will) play a fundamental role in many functional inequalities developed later.

By construction, the distribution of $B_t - B_s$, $t > s \geq 0$, is given by

$$\mathbb{E}\big(f(B_t - B_s)\big) = \int_{\mathbb{R}^n} f\left(\frac{y}{\sqrt{2}}\right) p_{t-s}(y) dy$$

(for f bounded and measurable, for example). In terms of the semigroup $(P_t)_{t \geq 0}$, the preceding may also be written as

$$\mathbb{E}\big(f(x + B_{2t})\big) = \mathbb{E}\big(f(x + \sqrt{2}\, B_t)\big) = P_t f(x), \quad t \geq 0, \ x \in \mathbb{R}^n.$$

The process $(x + B_t)_{t \geq 0}$ is (by definition) a Brownian motion starting from $x \in \mathbb{R}^n$. In this interpretation, the semigroup property is a consequence of the fact that B_{t+s} ($t, s \geq 0$) may be written as the sum of two independent Gaussian vectors $(B_{t+s} - B_s) + B_s$ with respective covariance matrices $t\,\mathrm{Id}$ and $s\,\mathrm{Id}$. This observation also illustrates the Markov property of the process $(B_t)_{t \geq 0}$, which states that starting from B_s, the process $(B_{t+s})_{t \geq 0}$ is again a Brownian motion.

The somewhat annoying $\sqrt{2}$ factor in the preceding formulas justifies the probabilistic convention of working with $\frac{1}{2}\Delta$ instead of Δ as the infinitesimal generator. However we avoid the factor $\frac{1}{2}$ throughout and mainly work with the generator Δ. Working then with stochastic calculus involving Brownian motion, we use from time to time the process $(B_{2t})_{t \geq 0}$.

The geometric content of the heat or Brownian semigroup model is less immediately perceptible since it is developed on the flat Euclidean space $\mathbb{R}^n$. However, analogous heat equations and Brownian processes may be considered on different state spaces, such as the sphere or the hyperbolic space. The geometry of the underlying state space, or rather of the Laplace operator acting on it, is then closely connected with various properties of the corresponding Markov semigroups and processes. In particular, curvature properties play an important role in this analysis, as we shall see. On the flat space $\mathbb{R}^n$, the (null) curvature is reflected on the heat semigroup $(P_t)_{t \geq 0}$ by the commutation property $\nabla(P_t f) = P_t(\nabla f)$ between the action of the gradient operator ∇ and the semigroup P_t (on smooth functions $f : \mathbb{R}^n \to \mathbb{R}$). By Jensen's inequality, this commutation produces the gradient bound

$$\big|\nabla(P_t f)\big| \leq P_t\big(|\nabla f|\big).$$

In fact, as will be developed extensively, this commutation inequality corresponds to a functional curvature property of the Laplace operator Δ. The analysis of similar relationships or gradient bounds in more general settings under certain curvature conditions will reveal important information about the semigroups, and in turn will shed light on one of our main objects of study, the functional inequalities.

The fundamental example of the heat or Brownian semigroup in Euclidean space may therefore be analyzed in different ways, via partial differential equations, probability theory or geometry. These three modes of analysis will be central to our investigation of Markov semigroups and operators. The elementary properties of the heat semigroup just presented may be generalized and deepened in many different directions. As announced in the introduction, our investigation shall be restricted to the family of diffusion (symmetric) Markov semigroups for which the generator (the Laplace operator in the example of the heat semigroup) is a second order differential operator satisfying the change of variables formula.

In this chapter we shall make some fairly informal remarks on various aspects of the analysis of Markov semigroups and operators, using the Euclidean heat semigroup as a focal example, in anticipation of future developments. We also introduce some basic concepts, some of them being further developed in the appendices. Readers are encouraged to complement this overview with relevant references from the literature, a selection of which can be found in the Notes and References Sect. 1.17 at the end of the chapter.

Let us outline the contents of this chapter. The first sections describe some of the basic elements in the general description of Markov semigroups, operators and processes, dealing with state space, the Markov property, invariant measures, infinitesimal generators and carré du champ operators. The associated Chapman-Kolmogorov and Fokker-Plank equations are discussed next, connecting with probabilistic Markov processes on one side and partial differential equations on the other. Sections 1.6 and 1.7 are concerned with symmetric semigroups, Dirichlet forms and spectral decompositions, while Sect. 1.8 introduces ergodicity. Markov chains on finite or countable state spaces are then briefly presented as a simple illustration of the previous concepts (although the study of Markov semigroups and operators will be restricted in this book to the so-called diffusion semigroups, thus excluding such discrete models). Probabilistic intuition is further developed on the basis of stochastic differential equations and diffusion processes in Sect. 1.10 which illustrates the main diffusion hypothesis central to the investigation. Diffusion semigroups and their domains are discussed next, together with ellipticity and hypo-ellipticity. At this stage, a first summary of the various hypotheses from the semigroup viewpoint is outlined in Sect. 1.14 (Chap. 3 will provide a complete description of the formalism and hypotheses). Section 1.15 then presents a variety of basic tools and operations on semigroups and their infinitesimal generators, which are of constant use and help in applications. The last section introduces the concept of curvature and dimension of a Markov (diffusion) operator which will play a crucial role throughout, in particular in the study of functional inequalities.

1.1 Markov Processes and Associated Semigroups

Although one main object of investigation will be the study of Markov semigroups by themselves, the reader should keep in mind that their study is mainly motivated by the description of laws of Markov processes. The close relationships between Markov semigroups and Markov processes will be a constant theme throughout this work. Indeed, some very important features of Markov semigroups become clear from the probabilistic interpretation, and many aspects of the behavior of Markov processes follow from a careful analysis of the associated semigroups and infinitesimal generators.

In this interplay, the very first example of a Markov process, briefly discussed above, is that of Brownian motion with associated heat semigroup and infinitesimal generator the Laplacian Δ. This example took place in the standard Euclidean space $\mathbb{R}^n$ (equipped with its Borel σ-field). In general, a Markov process $(X_t)_{t \geq 0}$ is a family of random variables constructed on some probability space $(\Omega, \Sigma, \mathbb{P})$ with values in some set E, equipped with a σ-field $\mathcal{F}$, thus defining a measurable space $(E, \mathcal{F})$. The set E need not be topological, however $\mathcal{F}$ cannot in general be any σ-field.

Indeed, some measurable spaces are better than others, namely those for which the *measure decomposition theorem* applies. This theorem states that whenever μ is a probability measure on the product σ-field $\mathcal{F} \otimes \mathcal{F}$ on $E \times E$, and if μ_1 denotes its projection on the first coordinate, then μ can be decomposed as $\mu(dx, dy) = k(x, dy)\mu_1(dx)$ for some kernel $k(x, dy)$ (that is, a probability measure depending in a measurable way on the variable x). This decomposition theorem is true for a rather large class of measurable spaces, but nevertheless it is not always true. For example, it holds when E is a so-called Polish space (that is a separable topological space, the topology of which may be defined by a complete metric) equipped with its Borel σ-field. This example of course includes the case of $\mathbb{R}^n$, or any open set in $\mathbb{R}^n$, with the Borel σ-field. In the following, by a *good measurable space* we shall mean a measurable space $(E, \mathcal{F})$ for which the measure decomposition theorem applies, and for which there exists a countable family generating the σ-algebra $\mathcal{F}$. Whenever there is a reference measure on $(E, \mathcal{F})$, the latter requirement may be relaxed to the weaker condition that the countable family generates the σ-algebra only up to sets of measure 0. In particular, all $\mathbb{L}^p$-spaces, $1 \leq p < \infty$, on $(E, \mathcal{F})$ (equipped with a σ-finite measure) are separable under this assumption. Functions on a measure space are understood as equivalence classes of (real-valued) functions with respect to the relation of almost everywhere equality, and as such equalities and inequalities between functions always hold almost everywhere. (See Sect. 1.14 for precise definitions and properties.)

Throughout, we shall always work with good measurable spaces. This level of generality is justified by the numerous examples of applications and illustrations of the use of Markov evolutions in mathematical physics and biology, for example, dealing in particular with processes on spaces with an infinite number of coordinates. For such spaces, the setting of locally compact spaces is clearly not sufficient. Although we do not really discuss such infinite-dimensional models in this book, it is important to choose the correct framework to start with.

Markov processes are characterized by the basic Markov property. Let $(X_t^x)_{t\geq 0}$ be a measurable process on a probability space $(\Omega, \Sigma, \mathbb{P})$ starting at time $t=0$ from $x \in E$. When the initial point x is fixed or clear from the context, $(X_t^x)_{t\geq 0}$ is denoted more simply by $(X_t)_{t\geq 0}$. For a given starting point $x \in E$, denote by $\mathcal{F}_t = \sigma(X_u\,;\,u \leq t)$, $t \geq 0$, the natural filtration of $(X_t)_{t\geq 0}$. The *Markov property* then indicates that for $t > s$, the law of X_t given $\mathcal{F}_s$ is the law of X_t given X_s, as well as the law of X_{t-s} given X_0, the latter property reflecting the fact that the Markov process is time homogeneous, which is the unique case that will be considered here. The process $(X_t)_{t\geq 0}$ (starting at x) is then said to be a *Markov process*. One prototypical example is Brownian motion (with values in $E = \mathbb{R}^n$) for which the independence of the increments immediately ensures the Markov property.

In a more analytical language, the Markov property may be described on the finite-dimensional distributions of the process $(X_t)_{t\geq 0}$. Namely, if $p_t(x, dy)$ is the probability kernel describing the distribution of X_t starting from x, the law of the pair (X_{t_1}, X_{t_2}), $t_1 \leq t_2$, is obtained after conditioning by $\mathcal{F}_{t_1}$ as $p_{t_1}(x, dy_1)\, p_{t_2-t_1}(y_1, dy_2)$. Iterating the procedure, the law of the sample $(X_{t_1}, \ldots, X_{t_k})$, $0 < t_1 \leq \cdots \leq t_k$, given that $X_0 = x$, is

$$p_{t_1}(x, dy_1)\, p_{t_2-t_1}(y_1, dy_2) \cdots p_{t_k - t_{k-1}}(y_{k-1}, dy_k).$$

The description of the laws of the various k-uples $(X_{t_1}, \ldots, X_{t_k})$, $0 < t_1 \leq \cdots \leq t_k$, is certainly not enough all the properties of the Markov process itself. For example, it does not allow the description of the law of the random variable X_T where T is the first (random) time when the process $(X_t)_{t\geq 0}$ enters some set A. For such tasks, it is in general necessary to request additional assumptions such as the fact that $(X_t)_{t\geq 0}$ lives in a topological space and has regular paths (continuous, or at least right-continuous). For the purpose of our investigation, the study of finite-dimensional marginal laws will suffice, so we shall mainly concentrate on them.

Given such a Markov process (or family of processes) $\{X_t^x\,;\,t \geq 0, x \in E\}$, define the associated Markov semigroup $(P_t)_{t\geq 0}$ on suitable measurable functions $f : E \to \mathbb{R}$ by the conditional formula

$$P_t f(x) = \mathbb{E}\big(f(X_t^x)\big) = \mathbb{E}\big(f(X_t) \mid X_0 = x\big), \quad t \geq 0,\ x \in E. \tag{1.1.1}$$

In particular $P_0 f = f$. The Markov property then indicates that, for every suitably integrable function $f : E \to \mathbb{R}$, and every $t, s \geq 0$,

$$\begin{aligned}
P_{t+s} f(x) = \mathbb{E}\big(f(X_{t+s}^x)\big) &= \mathbb{E}\big(\mathbb{E}\big(f(X_{t+s}^x) \mid \mathcal{F}_t\big)\big) \\
&= \mathbb{E}\big(\mathbb{E}\big(f(X_{t+s}^x) \mid X_t^x\big)\big) \\
&= \mathbb{E}\big(P_s f(X_t^x)\big) = P_t(P_s f)(x).
\end{aligned}$$

1.2 Markov Semigroups, Invariant Measures and Kernels

In other words, $P_{t+s} = P_t \circ P_s$, $t, s \geq 0$, which is the fundamental property of Markov semigroups $(P_t)_{t \geq 0}$.

Note that by duality, those semigroups $(P_t)_{t \geq 0}$ also act on the set of measures ν on $\mathcal{F}$ via the formula

$$\int_E P_t f \, d\nu = \int_E f \, d(P_t^* \nu) \tag{1.1.2}$$

(for every $t \geq 0$ and suitable function $f : E \to \mathbb{R}$), introducing the dual semigroup $(P_t^*)_{t \geq 0}$. Formally speaking, and in relation with the probabilistic interpretation of Markov semigroups in terms of Markov processes, when ν is the law of X_0, $P_t^* \nu$ is the law of the variable X_t.

As will be discussed at length in subsequent sections and chapters, such a Markov semigroup $(P_t)_{t \geq 0}$ may be entirely described by an operator L, called the infinitesimal generator of $(P_t)_{t \geq 0}$, which occurs in many examples as a differential operator (such as the standard Laplacian in the case of Brownian motion). Furthermore, although no topology on the set E is required, as mentioned earlier, it is quite important in practice that the trajectories $t \mapsto X_t$ are not too irregular. This regularity is seen through the action of good functions and there is usually a nice family $\mathcal{A}$ of functions $f : E \to \mathbb{R}$ which is rich enough (the domain or a core of it, as defined later) such that, for any $f \in \mathcal{A}$, the processes $t \mapsto f(X_t), t \geq 0$, are almost surely right-continuous with left-limits (càdlàg in the French abbreviation). We shall not need this type of abstract construction. In the best cases, these processes are actually almost surely continuous, corresponding to the so-called diffusion case (presented in Sect. 1.11 below). This is in particular the case when the Markov processes are given as solutions of stochastic differential equations as described in Sect. 1.10.

1.2 Markov Semigroups, Invariant Measures and Kernels

This section introduces the basic notions of Markov semigroup and invariant measure and, still in an informal way, some of the general principles governing their analysis. The framework will be made progressively more precise in subsequent sections. The basics for the construction and properties of abstract semigroups are presented in Appendix A.

1.2.1 Markov Semigroups, Invariant Measures

The fundamental object of investigation consists of a family $\mathbf{P} = (P_t)_{t \geq 0}$ of operators defined on some set of real-valued measurable functions on $(E, \mathcal{F})$. The following list describes the first definitions and properties of such an object:

(i) For every $t \geq 0$, P_t is a linear operator sending bounded measurable functions on $(E, \mathcal{F})$ to bounded measurable functions.

(ii) $P_0 = \mathrm{Id}$, the identity operator (*initial condition*).
(iii) $P_t(\mathbb{1}) = \mathbb{1}$, where $\mathbb{1}$ is the constant function equal to 1 (*mass conservation*).
(iv) If $f \geq 0$, then $P_t f \geq 0$ (*positivity preserving*).
(v) For every $t, s \geq 0$, $P_{t+s} = P_t \circ P_s$ (*semigroup property*).

If $f : E \to \mathbb{R}$ is bounded measurable, we write $P_t f = P_t(f)$. Properties (iii) and (iv) are often also described as "P_t is a Markov operator", giving rise to the terminology. Note that, as a consequence of the positivity and mass conservation properties, by Jensen's inequality, for every convex function $\phi : \mathbb{R} \to \mathbb{R}$, every $t \geq 0$ and every, say bounded, measurable function f on E,

$$P_t(\phi(f)) \geq \phi(P_t f). \tag{1.2.1}$$

In particular, for the functions $\phi(r) = |r|^p$, $r \in \mathbb{R}$, $1 \leq p \leq \infty$, $P_t(|f|^p) \geq |P_t f|^p$.

The preceding five conditions actually (almost) characterize what will be called a Markov semigroup (the semigroup property being described by the last property (v)). What is missing is a continuity property at $t = 0$ which will turn out to be important, but requires a new notion to be fully appreciated, namely the notion of *invariant* (or *stationary*) measure.

Definition 1.2.1 (Invariant measure) Given a family $\mathbf{P} = (P_t)_{t \geq 0}$ on a measurable space $(E, \mathcal{F})$ as before, a (positive) σ-finite measure μ on $(E, \mathcal{F})$ is said to be invariant for $\mathbf{P}$ if for every bounded positive measurable function $f : E \to \mathbb{R}$ and every $t \geq 0$,

$$\int_E P_t f \, d\mu = \int_E f \, d\mu. \tag{1.2.2}$$

In other words, according to (1.1.2), μ is invariant if $P_t^* \mu = \mu$ for every $t \geq 0$.

Note that, by (1.2.1), if a measure μ is invariant for $\mathbf{P}$, then, for every $t \geq 0$, P_t is a contraction on the bounded functions in $\mathbb{L}^p(\mu)$ for any $1 \leq p \leq \infty$. As is traditional, $\mathbb{L}^p(\mu) = \mathbb{L}^p(E, \mu)$ with associated norm $\|\cdot\|_p$ (sometimes also denoted $\|\cdot\|_{\mathbb{L}^p(\mu)}$), $1 \leq p \leq \infty$, denote the Lebesgue spaces on the measure space $(E, \mathcal{F}, \mu)$. By density, P_t may thus be extended to a bounded operator on $\mathbb{L}^p(\mu)$ for all $1 \leq p \leq \infty$, and even a contraction in the sense that $\|P_t\|_{p,p} \leq 1$ where $\|P\|_{p,p}$ denotes the operator norm of P in $\mathbb{L}^p(\mu)$. In particular, the invariance definition (1.2.2) extends to any $f \in \mathbb{L}^1(\mu)$.

The existence of such an invariant measure is not always guaranteed, but most semigroups of interest do have an invariant measure. The uniqueness question, or whether the invariant measure is finite or not, should then be further investigated. These questions already arise for Markov chains on countable spaces (see Sect. 1.9). Very often, an invariant measure may be constructed by choosing any reasonable initial probability measure μ_0 and by considering any weak limit of $\frac{1}{t} \int_0^t P_s^* \mu_0 ds$ as $t \to \infty$ (where we recall the dual semigroup $(P_t^*)_{t \geq 0}$ described in (1.1.2)). However, this method only works in general when the invariant measure is finite, and must be adapted in the general case. A typical example is the heat or Brownian semigroup

1.2 Markov Semigroups, Invariant Measures and Kernels

on $\mathbb{R}^n$ for which the invariant measure is given (up to multiplication by a constant) by the Lebesgue measure (which may be obtained from the heat semigroup through the normalization $(4\pi t)^{n/2} P_t^* \mu_0$ as $t \to \infty$ for any reasonable initial probability measure μ_0).

The invariant measure is in general only defined up to a multiplicative constant. When it is finite, a canonical choice is to normalize it into a probability measure. It is then usually unique. However, there is no canonical normalization when the measure is infinite. The analysis of Markov semigroups with a finite (normalized) invariant measure is usually easier than the general case. This issue is similar to the one for Markov chains on countable state spaces, for which the existence of a finite invariant measure automatically ensures that the associated Markov process is recurrent (see Sect. 1.6 below).

The invariant measure, at least when it is a probability measure, has a clear probabilistic interpretation on the associated Markov process $\{X_t^x ; t \geq 0, x \in E\}$. If the process starts at time $t = 0$ with initial distribution μ (that is, the law of X_0 is μ), then it keeps this distribution at each time t since, by the Markov property, for any (say, bounded) measurable function $f : E \to \mathbb{R}$,

$$\mathbb{E}(f(X_t)) = \mathbb{E}(\mathbb{E}(f(X_t) \mid X_0))$$
$$= \mathbb{E}(P_t f(X_0)) = \int_E P_t f \, d\mu = \int_E f \, d\mu = \mathbb{E}(f(X_0)). \quad (1.2.3)$$

This property justifies the terminology of stationary measure for μ in a probabilistic context. The same interpretation remains true in general if one accepts the idea of random variables with distributions having infinite mass.

Summarizing the construction of a Markov semigroup at this point, provided μ is a σ-finite invariant measure for $\mathbf{P} = (P_t)_{t \geq 0}$, the latter consists of a family of operators bounded in $\mathbb{L}^p(\mu)$ satisfying the properties (i)–(v). The last condition required to deal with a Markov semigroup is the following continuity property:

(vi) For every $f \in \mathbb{L}^2(\mu)$, $P_t f$ converges to f in $\mathbb{L}^2(\mu)$ as $t \to 0$ (*continuity property*).

Since $(P_t)_{t \geq 0}$ is a contraction and a semigroup, the continuity property (vi) expresses that $t \mapsto P_t f$ is continuous in $\mathbb{L}^2(\mu)$ on $\mathbb{R}_+$. We could have required convergence in $\mathbb{L}^p(\mu)$ for any $1 \leq p < \infty$, although for simplicity we stick to $p = 2$. This property will almost always be satisfied in the applications (however not for $p = \infty$).

The continuity property (vi) actually reflects the regularity properties of the associated Markov process $\{X_t^x ; t \geq 0, x \in E\}$ through the representation (1.1.1). Indeed, assume that $\mathcal{A}$ is a dense subset of bounded functions in $\mathbb{L}^2(\mu)$ such that, for any $f \in \mathcal{A}$, the process $t \mapsto f(X_t)$ has right-continuous paths for any initial value $X_0 = x$ and, for any $t_0 > 0$, $\sup_{t \leq t_0} |P_t f| \in \mathbb{L}^2(\mu)$. This is for example the case for bounded functions as soon as μ is a probability measure, and for any $f \in \mathbb{L}^2(\mu)$ if μ is reversible—see Lemma 1.6.2. Then, for such functions, and for any $x \in E$,

$\lim_{t \to 0} P_t f(x) = f(x)$ and $\lim_{t \to 0} P_t f = f$ in $\mathbb{L}^2(\mu)$ (a property extended then to any $f \in \mathbb{L}^2(\mu)$ since P_t is a contraction).

On the basis of the above properties (i)–(vi), the following statement summarizes the definition of a Markov semigroup with invariant measure μ.

Definition 1.2.2 (Markov semigroup) A family of operators $\mathbf{P} = (P_t)_{t \geq 0}$ defined on the bounded measurable functions on a state space $(E, \mathcal{F})$ with invariant σ-finite measure μ satisfying the properties (i)–(vi) is called a Markov semigroup of operators.

Accordingly, the Markov semigroup $\mathbf{P} = (P_t)_{t \geq 0}$ defines a semigroup of contractions on $\mathbb{L}^2(\mu)$ in the sense of Appendix A.

Very often property (iii) may be relaxed to $P_t(\mathbb{1}) \leq \mathbb{1}$. This condition appears naturally when solving Schrödinger-type equations with negative potential (see Sect. 1.15.6 below), but also when the underlying state space is not complete (see Sect. 2.5, p. 96, or Sect. 3.2.2, p. 141). In such cases, $\mathbf{P} = (P_t)_{t \geq 0}$ will be called a *sub-Markov* semigroup. Sub-Markov semigroups are only considered in this book when the need arises.

1.2.2 Kernels

Markov operators $P_t, t \geq 0$, as given in Definition 1.2.2, may be represented by (probability) *kernels* corresponding to the transition probabilities of the associated Markov process (and this is the way Markov semigroups are usually given). Namely, for every bounded measurable function $f : E \to \mathbb{R}$,

$$P_t f(x) = \int_E f(y)\, p_t(x, dy), \quad t \geq 0, \ x \in E, \qquad (1.2.4)$$

where $p_t(x, dy)$ is, for every $t \geq 0$, a probability kernel (that is, for every $x \in E$, $p_t(x, \cdot)$ is a probability measure, and for every measurable set $A \in \mathcal{F}$, $x \mapsto p_t(x, A)$ is measurable). Recall that such a representation formula is actually understood for μ-almost every x in E, as are all identities between functions throughout this work. The distribution at time t of the underlying Markov process X_t^x starting at x is thus given by the probability $p_t(x, \cdot)$.

From the representation (1.2.4), the operators P_t, $t \geq 0$, can be extended to positive measurable functions, possibly with infinite values. Often, inequalities involving $P_t f$ are accordingly stated for bounded or positive measurable functions f.

The representation (1.2.4) requires good measurable spaces and the measure decomposition theorem and is established in the next proposition (more for self-consistency, since again Markov semigroups associated with Markov processes are usually given explicitly in terms of kernels).

1.2 Markov Semigroups, Invariant Measures and Kernels

Proposition 1.2.3 (Kernel representation) *Given a good measure space $(E, \mathcal{F}, \mu)$ (for example a Polish space), with μ σ-finite, let P be a linear operator sending positive measurable functions to positive measurable functions. Then, if $P(\mathbb{1}) = \mathbb{1}$ and if P is bounded on $\mathbb{L}^1(\mu)$, P may be represented by a probability kernel $p(x, A)$, $x \in E$, $A \in \mathcal{F}$, in the sense that for every bounded or positive measurable function f on E and (μ-almost) every $x \in E$,*

$$Pf(x) = \int_E f(y)\, p(x, dy).$$

Proof We use a result from measure theory, holding on good measurable spaces, known as the bi-measure theorem. This theorem states that whenever $(E, \mathcal{F})$ is a good measurable space, if there is a map $(A, B) \mapsto \nu(A, B)$ from $\mathcal{F} \times \mathcal{F}$ into $\mathbb{R}_+$ such that for every $A, B \in \mathcal{F}$, the maps $C \mapsto \nu(A, C)$ and $C \mapsto \nu(C, B)$ are measures on $\mathcal{F}$, then there exists a measure ν_1 on the product space $E \times E$ equipped with the product σ-field $\mathcal{F} \otimes \mathcal{F}$ such that for every $(A, B) \in \mathcal{F} \times \mathcal{F}$, $\nu(A, B) = \nu_1(A \times B)$.

On the basis of this result, we establish the proposition when μ is a probability measure (the case when μ is σ-finite then easily follows). Set, for $(A, B) \in \mathcal{F} \times \mathcal{F}$,

$$\nu(A, B) = \int_E \mathbb{1}_A P(\mathbb{1}_B) d\mu,$$

where $\mathbb{1}_A$ (respectively $\mathbb{1}_B$) is the characteristic function of the set A (respectively B). It is a bi-measure. To prove this, it is enough to check that whenever $(A_k)_{k \in \mathbb{N}}$ is a decreasing sequence of sets in $\mathcal{F}$ with $\bigcap_k A_k = \emptyset$, then $\nu(A_k, B)$ converges to 0 as $k \to \infty$, and similarly for the second argument with a corresponding sequence $(B_k)_{k \in \mathbb{N}}$ decreasing to the empty set. For the sequence $(A_k)_{k \in \mathbb{N}}$, it suffices to apply the dominated convergence Theorem to the sequence $\mathbb{1}_{A_k} P(\mathbb{1}_B)$, $k \in \mathbb{N}$. For the sequence $(B_k)_{k \in \mathbb{N}}$, observe that since P is continuous in $\mathbb{L}^1(\mu)$, $P(\mathbb{1}_{B_k})$ converges to 0 in $\mathbb{L}^1(\mu)$, and hence, there exists a subsequence converging to 0 μ-almost everywhere. Again by the dominated convergence Theorem, $\nu(A, B_k) \to 0$ along this subsequence, and since $(\nu(A, B_k))_{k \in \mathbb{N}}$ is decreasing, $\nu(A, B_k) \to 0$.

The bi-measure theorem thus applies to ν, and therefore there exists a (probability) measure ν_1 on the σ-field $\mathcal{F} \otimes \mathcal{F}$ such that

$$\nu_1(A \times B) = \int_E \mathbb{1}_A P(\mathbb{1}_B) d\mu$$

for every $A, B \in \mathcal{F}$. Since $P(\mathbb{1}) = \mathbb{1}$, the first marginal of ν_1 is μ and ν_1 may be decomposed by the measure decomposition theorem as

$$\nu_1(dx, dy) = p(x, dy)\mu(dx).$$

It is then easily checked that the kernel $p(x, dy)$ fulfills the required conditions. The proposition is established. □

Very often, the family of kernels $p_t(x, dy)$ have densities with respect to a reference measure (often also the invariant measure).

Definition 1.2.4 (Density kernel) A Markov semigroup $\mathbf{P} = (P_t)_{t \geq 0}$ on $(E, \mathcal{F})$ is said to admit density kernels with respect to a reference σ-finite measure m on $\mathcal{F}$ if there exists for every $t > 0$ a positive measurable function $p_t(x, y)$ defined on $E \times E$ (up to a set of $m \otimes m$-measure 0) such that, for every bounded or positive measurable function $f : E \to \mathbb{R}$ and (m-almost) every $x \in E$,

$$P_t f(x) = \int_E f(y) \, p_t(x, y) dm(y).$$

In this case, $\int_E p_t(x, y) dm(y) = 1$ for (m-almost) every $x \in E$ (reflecting the fact that $P_t(\mathbb{1}) = \mathbb{1}$).

In order for $P_t f$ to make sense for any $f \in \mathbb{L}^2(m)$ in this definition, it is in general required that for all $t > 0$ and (m-almost) every $x \in E$, $\int_E p_t(x, y)^2 dm(y) < \infty$. In this case, $(P_t)_{t \geq 0}$ is said to admit $\mathbb{L}^2$-density kernels with respect to m. As already mentioned, the reference measure will often be chosen to be the invariant measure μ.

In the context of Definition 1.2.4, many of the properties of the Markov semigroup $(P_t)_{t \geq 0}$ may be translated into properties of the density kernels (for example the Fokker-Planck equation described in Sect. 1.5 or reversibility in Sect. 1.6). However, it is not always easy to check that density kernels exist, which usually requires additional knowledge on the semigroup. Beyond the case of Hilbert-Schmidt operators (see Appendix A, Sect. A.6, p. 483) where an explicit expression for this kernel may be given (see, for example, (1.7.3) below), the main task would be to first prove that the semigroup is a bounded operator from $\mathbb{L}^1(m)$ into $\mathbb{L}^\infty(m)$. The following proposition, which will illustrated in the study of heat kernel bounds in Chaps. 6 and 7, is a useful result in this regard.

Proposition 1.2.5 (Existence of density kernel) *Let $(E, \mathcal{F}, m)$ be a good measure space with m σ-finite. Let P be a linear operator mapping $\mathbb{L}^1(m)$ into $\mathbb{L}^\infty(m)$ with operator norm $\|P\|_{1,\infty} \leq M$. Then, there exists a measurable function (kernel density) $p(x, y)$ on $E \times E$ with $|p(x, y)| \leq M$ for $m \otimes m$-almost every $(x, y) \in E \times E$ such that, for any $f \in \mathbb{L}^1(m)$ and (m-almost) every $x \in E$,*

$$P f(x) = \int_E f(y) \, p(x, y) dm(y).$$

Proof We only prove the result in the case when m is a probability measure, the general case being easily adapted by σ-finiteness of m. For a good measurable space, there exists a sequence $(A_\ell)_{\ell \in \mathbb{N}}$ of measurable sets which generates the σ-field $\mathcal{F}$ up to sets of m-measure 0. For each $\ell \in \mathbb{N}$, let $\mathcal{F}_\ell$ be the σ-field generated by $A_0, \ldots, A_\ell$. It is also generated by a partition $(B_1^\ell, \ldots, B_{j_\ell}^\ell)$ of E from which the

1.2 Markov Semigroups, Invariant Measures and Kernels

sets with m-measure 0 have been removed. For $f \in \mathbb{L}^1(m)$, define $P_\ell f$ as the conditional expectation $\mathbb{E}(Pf \mid \mathcal{F}_\ell)$ of Pf with respect to $\mathcal{F}_\ell$. The operator P_ℓ is bounded from $\mathbb{L}^1(m)$ into $\mathbb{L}^\infty(m)$ with norm M and may be represented as

$$P_\ell f(x) = \sum_{j=1}^{j_\ell} Q_j^\ell f \, \mathbb{1}_{B_j^\ell}(x)$$

where

$$Q_j^\ell f = \frac{1}{m(B_j^\ell)} \int_{B_j^\ell} Pf \, dm.$$

The Q_j^ℓ's, $j = 1, \ldots, j_\ell$, are linear operators, bounded on $\mathbb{L}^1(m)$, with norm bounded above by M. By duality,

$$Q_j^\ell f = \int_E f(y) \, q_j^\ell(y) dm(y)$$

for some bounded (by M) measurable functions q_j^ℓ, $j = 1, \ldots, j_\ell$. Restricted to functions which are $\mathcal{F}_\ell$-measurable, the functions q_j^ℓ are linear combinations of $\mathbb{1}_{B_j^\ell}$, $j = 1, \ldots, j_\ell$. Thus, there exists a kernel $p_\ell = p_\ell(x, y)$ which is $\mathcal{F}_\ell \otimes \mathcal{F}_\ell$-measurable, bounded by M, and such that for any integrable function $f : E \to \mathbb{R}$ measurable with respect to $\mathcal{F}_\ell$ and any $x \in E$,

$$P_\ell f(x) = \int_E f(y) \, p_\ell(x, y) dm(y).$$

In other words, p_ℓ is the kernel of the operator $\mathbb{E}(P(\mathbb{E}(f \mid \mathcal{F}_\ell)) \mid \mathcal{F}_\ell)$. It is then not hard to see that p_ℓ, $\ell \in \mathbb{N}$, is an $m \otimes m$-martingale with respect to the increasing family of σ-fields $\mathcal{F}_\ell \otimes \mathcal{F}_\ell$, $\ell \in \mathbb{N}$. Since it is bounded, by elementary martingale theory, it converges ($m \otimes m$-almost everywhere) to a measurable function $p(x, y)$, $(x, y) \in E \times E$, bounded by M, which is easily seen to be a kernel of the operator P. The proof is complete. □

As a further general fact, it is worth mentioning that as soon as a Markov operator P on a probability space $(E, \mathcal{F}, \mu)$ is bounded from above by a constant $C < 2$, then $P^2 = P \circ P$ is also bounded from below. This is the content of the following simple proposition.

Proposition 1.2.6 *Let P be a Markov operator (thus satisfying properties (iii) and (iv) of the definition of a Markov semigroup) with density kernel $p(x, y)$, $(x, y) \in E \times E$, with respect to an invariant probability measure μ on $(E, \mathcal{F})$. Then, as soon as $\|P\|_{1,\infty} \leq 1 + \varepsilon$ for some $0 \leq \varepsilon < 1$, for every positive function f in $\mathbb{L}^1(\mu)$,*

$$P^2 f \geq \left(1 - \sqrt{\varepsilon}\right)^2 (1+\varepsilon) \int_E f \, d\mu.$$

In particular, the density kernel $p^2(x, y)$ of P^2 is bounded from below by $(1 - \sqrt{\varepsilon})^2(1 + \varepsilon)$.

Proof Assume by homogeneity that $\int_E f d\mu = 1$. Setting $g = Pf$, $g \leq 1 + \varepsilon$ and $\int_E g d\mu = 1$ since μ is invariant for P. For any $0 < a < 1 + \varepsilon$, $P^2 f = P(g) \geq a P(\mathbb{1}_{\{g \geq a\}})$. But

$$P(\mathbb{1}_{\{g \geq a\}}) = 1 - P(\mathbb{1}_{\{g < a\}}) \geq 1 - (1 + \varepsilon)\mu(g < a)$$

and by Markov's inequality,

$$\mu(g < a) = \mu(1 + \varepsilon - g > 1 + \varepsilon - a)$$
$$\leq \frac{1}{1 + \varepsilon - a} \int_E (1 + \varepsilon - g) d\mu = \frac{\varepsilon}{1 + \varepsilon - a}.$$

Therefore,

$$P^2 f \geq \frac{a(1 - \varepsilon^2 - a)}{1 + \varepsilon - a}$$

and it remains to choose the optimal $a = (1 + \varepsilon)(1 - \sqrt{\varepsilon})$. The proof is complete. □

It is easy to see that the condition $\varepsilon < 1$ is necessary in the above statement: consider for example on $E = \{0, 1\}$ the Markov operator P given by $Pf(0) = f(1)$ and $Pf(1) = f(0)$, invariant with respect to the uniform measure μ, for which $\varepsilon = 1$ and $P^2 = \text{Id}$. For this P, the kernel $p^2(x, y)$ with respect to μ is such that $p^2(0, 1) = p^2(1, 0) = 0$.

1.3 Chapman-Kolmogorov Equations

Section 1.1 describes the basic principle used to construct a Markov semigroup from a Markov process. Now conversely, given a Markov semigroup, the construction of a Markov process associated to it relies on the Chapman-Kolmogorov equation which expresses the semigroup property from a probabilistic point of view. As constructed in the preceding sections, kernels and density kernels are only defined almost everywhere with respect to the underlying measure, and identities between them are understood in this sense below.

Let $\mathbf{P} = (P_t)_{t \geq 0}$ be a Markov semigroup on $\mathbb{L}^2(\mu)$ according to Definition 1.2.2. The semigroup property $P_t \circ P_s = P_{t+s}$ translates to the kernels $p_t(x, dy)$ of the representation (1.2.4) via the composition property, for all $t, s \geq 0$, $x \in E$,

$$p_{t+s}(x, dy) = \int_{z \in E} p_t(z, dy) p_s(x, dz). \quad (1.3.1)$$

Such a relation is called a *Chapman-Kolmogorov equation*.

1.3 Chapman-Kolmogorov Equations

When the kernels admit densities $p_t(x, y)$, $t > 0$, $(x, y) \in E \times E$, with respect to a reference measure m on $(E, \mathcal{F})$ as in Definition 1.2.4, the latter equation expresses that, for all $t, s > 0$, $(x, y) \in E \times E$,

$$p_{t+s}(x, y) = \int_E p_t(z, y) \, p_s(x, z) dm(z). \tag{1.3.2}$$

Now, as announced, from the Chapman-Kolmogorov equation (1.3.1), one may construct, starting from any point $x \in E$, a Markov process $(X_t)_{t \geq 0}$ on E by specifying the distribution of $(X_{t_1}, \ldots, X_{t_k})$, $0 \leq t_1 \leq \cdots \leq t_k$, as

$$\mathbb{E}\big(f(X_{t_1}, \ldots, X_{t_k})\big) = \int_{E \times \cdots \times E} f(y_1, \ldots, y_k) \, p_{t_k - t_{k-1}}(y_{k-1}, dy_k) \\ \times p_{t_{k-1} - t_{k-2}}(y_{k-2}, dy_{k-1}) \cdots p_{t_1}(x, dy_1) \tag{1.3.3}$$

for every, say, positive or bounded measurable function f on the product space $E \times \cdots \times E$. When the distribution of such a k-dimensional vector $(X_{t_1}, \ldots, X_{t_k})$ is specified, it determines the distributions of each extracted lower dimensional vector (for example, the law of (X_{t_1}, X_{t_3}) can be deduced from that of $(X_{t_1}, X_{t_2}, X_{t_3})$). It is thus necessary that the system of finite-dimensional distributions be compatible, which is precisely the content of the Chapman-Kolmogorov equation.

Equation (1.3.3) describes the law of the process $(X_t)_{t \geq 0}$ starting at $X_0 = x$. If the initial distribution of X_0 is given by some other measure ν, then the law of $(X_0, X_{t_1}, \ldots, X_{t_k})$ is given by

$$\mathbb{E}\big(f(X_0, X_{t_1}, \ldots, X_{t_k})\big) = \int_{E \times \cdots \times E} f(y_0, y_1, \ldots, y_k) \, p_{t_k - t_{k-1}}(y_{k-1}, dy_k) \cdots \\ \times p_{t_1}(y_0, dy_1) \nu(dy_0). \tag{1.3.4}$$

Given a Markov semigroup $\mathbf{P} = (P_t)_{t \geq 0}$, we therefore have candidates for the distribution of a Markov process $(X_t)_{t \geq 0}$ starting at a point $x \in E$. This entails a correspondence between Markov processes and Markov semigroups (even without the continuity condition (vi)). However, as already mentioned, the preceding finite-dimensional description is not enough in general to characterize the full law of the Markov process and its regularity properties. Usually, is does not even describe the joint laws of the processes starting from different initial values x and y, as is the case for example when solving stochastic differential equations. In fact, the process often lives on a topological space, and one then looks for a process such that the trajectories (the random maps $t \mapsto X_t(\omega)$ for $\omega \in \Omega$) are continuous, or right-continuous with left limits (càdlàg). It requires extra work to construct these laws as probability measures on the space of continuous (or càdlàg) maps on $\mathbb{R}_+$ with values in the state space E. This question will not be addressed in this book, since most of the time the concrete processes will be explicitly given, for example as solutions of stochastic differential equations (see Sect. 1.10).

In the following, we will indistinctly speak of a Markov semigroup $\mathbf{P} = (P_t)_{t \geq 0}$ with state space E and of its associated Markov process, or family of Markov processes, $\mathbf{X} = \{X_t^x \, ; \, t \geq 0, x \in E\}$. When the starting point is implicit from the context or fixed, we often simply denote the Markov process by $(X_t)_{t \geq 0}$.

1.4 Infinitesimal Generators and Carré du Champ Operators

This section introduces infinitesimal generators and carré du champ operators of Markov semigroups which are central objects of interest encoding numerous features and properties.

1.4.1 Infinitesimal Generators

A Markov semigroup $\mathbf{P} = (P_t)_{t \geq 0}$ as defined in Definition 1.2.2 is driven by an operator called the infinitesimal generator of the Markov semigroup. According to Appendix A, it may actually be introduced in a general framework.

A family of bounded linear operators $(P_t)_{t \geq 0}$ on a Banach space $\mathcal{B}$ with the semigroup ($P_t \circ P_s = P_{t+s}, t, s \geq 0$) and continuity ($P_t f \to f$ for every $f \in \mathcal{B}$ as $t \to 0$) properties is called a semigroup of (bounded) linear operators (on the Banach space $\mathcal{B}$). In the preceding setting of Markov (contraction) semigroups, the Banach space $\mathcal{B}$ will always be the Hilbert space $\mathbb{L}^2(\mu)$, but we could similarly work (adapting (vi) of the definition of Markov semigroup) with $\mathbb{L}^p(\mu)$, $1 \leq p < \infty$, or the space of bounded continuous functions (provided E is topological), or any space on which the family of operators $(P_t)_{t \geq 0}$ acts naturally.

In this setting, the Hille-Yosida theory (cf. Sect. A.1, p. 473) indicates that there is a dense linear subspace of $\mathcal{B}$, called the *domain* $\mathcal{D}$ of the semigroup $(P_t)_{t \geq 0}$, on which the derivative at $t = 0$ of P_t exists in $\mathcal{B}$. The operator that maps $f \in \mathcal{D}$ to this derivative $\mathrm{L} f \in \mathcal{B}$ of $P_t f$ at $t = 0$ is a linear (usually unbounded) operator, called the *infinitesimal generator* of the semigroup $(P_t)_{t \geq 0}$, denoted L. The domain $\mathcal{D}$ is also called the domain of L, denoted $\mathcal{D}(\mathrm{L})$, and depends on the underlying Banach space $\mathcal{B}$. Actually, the generator L and its domain $\mathcal{D}(\mathrm{L})$ completely characterize the semigroup $(P_t)_{t \geq 0}$ acting on $\mathcal{B}$.

Applied to a Markov semigroup $\mathbf{P} = (P_t)_{t \geq 0}$ as presented in Definition 1.2.2 with $\mathcal{B} = \mathbb{L}^2(\mu)$ the preceding yields the following definition.

Definition 1.4.1 (Infinitesimal generator)   Let $\mathbf{P} = (P_t)_{t \geq 0}$ be a Markov semigroup with state space $(E, \mathcal{F})$ and invariant measure μ. The infinitesimal generator L of $\mathbf{P}$ in $\mathcal{B} = \mathbb{L}^2(\mu)$ is called the Markov generator of the semigroup $\mathbf{P} = (P_t)_{t \geq 0}$ with $\mathbb{L}^2(\mu)$-domain $\mathcal{D}(\mathrm{L})$.

When dealing with $\mathcal{B} = \mathbb{L}^p(\mu)$, $1 \leq p < \infty$, in Definition 1.4.1, we sometimes speak of the $\mathbb{L}^p(\mu)$-domain of L. In view of the connection with the Markov process

1.4 Infinitesimal Generators and Carré du Champ Operators

X associated with the semigroup **P**, the generator L will also be called the Markov generator of **X**.

The linearity of the operators P_t, $t \geq 0$, together with the semigroup property, shows that L is the derivative of P_t at any time $t > 0$. Namely, for $t, s > 0$,

$$\frac{1}{s}[P_{t+s} - P_t] = P_t\left(\frac{1}{s}[P_s - \mathrm{Id}]\right) = \left(\frac{1}{s}[P_s - \mathrm{Id}]\right)P_t.$$

Letting $s \to 0$ then yields

$$\partial_t P_t = L P_t = P_t L. \qquad (1.4.1)$$

While traditionally reserved for Laplace operators Δ, by extension $(P_t)_{t \geq 0}$ will often be called the *heat semigroup* or *heat flow* with respect to the generator L (thus solving the *heat equation* (1.4.1)).

Note that replacing L with cL for some $c > 0$ amounts to the time change $t \mapsto ct$. This seemingly insignificant transformation actually allows us to freely normalize generators. More sophisticated transformations, changing L to $c(x)L$ where $c(x)$ is a strictly positive function, will be considered below in Sect. 1.15.2.

In the illustrations, it is usually difficult, but also often useless, to exactly describe the domain $\mathcal{D}(L)$ of the generator L. It will be enough to determine a so-called *core* $\mathcal{D}_0$, that is a dense subspace with respect to the topology of the domain in the sense that for every element $f \in \mathcal{D}(L)$, there is a sequence $(f_k)_{k \in \mathbb{N}}$ in the core $\mathcal{D}_0$ converging to f and such that Lf_k converges to Lf. The question of the description of the domain, or a core of it, is a delicate question which we will return to later (see Sect. 1.13). We refer the reader to Appendix A for further examples. In most examples of interest, and in particular in this monograph, the operator L is a differential operator (for example the usual Laplace operator Δ for the heat or Brownian semigroup on $\mathbb{R}^n$). The statement that a function f is in the domain $\mathcal{D}(L)$ then hides a number of properties. A first property would be that the function f should be smooth enough so that the action of the generator (for example in the distributional sense) gives rise to a function Lf in $\mathbb{L}^2(\mu)$. As a second property, the function should be well-behaved at infinity (or at the boundary of the state space) so that integration by parts formulas (hidden in the symmetry property introduced in Sect. 1.6 and at the heart of the analysis in this book) can be used.

The reader should also pay attention to the fact that in Definition 1.4.1 the semigroup $\mathbf{P} = (P_t)_{t \geq 0}$ is given and that the infinitesimal generator L (and its domain) is determined from it. One of the main difficulties in the analysis of semigroups is that in general we would like to work in the other direction, that is, starting from the knowledge of a Markov generator L on a set $\mathcal{A}_0$ of "nice" functions (for example, when L is a differential operator, the set of smooth compactly supported functions). This amounts to proving that this set of functions is a core in the domain, but for a semigroup not yet described. Whether this knowledge is enough to determine the semigroup $(P_t)_{t \geq 0}$ is a delicate matter, related in the symmetric case to the question of essential self-adjointness (see Appendix A, Sect. A.5, p. 481, and Chap. 3, Definition 3.1.10, p. 134). The question of determining when some unbounded operator

L given on such a class $\mathcal{A}_0$ may be extended to a larger domain in order to become the infinitesimal generator of a Markov semigroup will be addressed in Chap. 3 together with the necessary tools to achieve this goal in practice. In the core of the text, some efficient criteria to achieve the latter property in concrete examples will be presented (see for instance Proposition 2.4.1, p. 95, Proposition 2.6.1, p. 102, Proposition 3.2.1, p. 142, or Proposition A.5.4, p. 482.)

With respect to abstract semigroups of operators, the Markov semigroups $\mathbf{P} = (P_t)_{t \geq 0}$ considered in this work share some further properties described by Definition 1.2.2. In particular, they act on function spaces, satisfy the fundamental stability property of preserving the unit function, and admit an invariant measure. These properties may be translated (equivalently) on the generator L of the semigroup $\mathbf{P}$. If the constant $\mathbb{1}$ function is in $\mathbb{L}^2(\mu)$, then it is in the domain $\mathcal{D}(L)$ and $L(\mathbb{1}) = 0$. This is the case when μ is finite. When the measure μ is infinite, $\mathbb{1}$ cannot be in the domain, and it is not enough in general to know that $L(\mathbb{1}) = 0$ in order to conclude that conversely $P_t(\mathbb{1}) = \mathbb{1}$ for every $t \geq 0$. Further conditions are necessary (such as boundary conditions), or the generator should itself satisfy suitable properties. Concerning the invariant measure, differentiating at $t = 0$ the identity $\int_E P_t f d\mu = \int_E f d\mu$ (which is justified as soon as Lf is bounded for example, or by Lemma 1.6.2 below in the reversible case) shows that if f is in the $\mathbb{L}^1(\mu)$-domain, then

$$\int_E L f \, d\mu = 0.$$

Conversely, if this condition holds on a set of functions f which is dense in the $\mathbb{L}^1(\mu)$-domain of L, then the measure μ will be invariant for $\mathbf{P}$.

1.4.2 Carré du Champ Operators

To a Markov semigroup $\mathbf{P} = (P_t)_{t \geq 0}$ and its infinitesimal generator L, with $(\mathbb{L}^2(\mu)$-$)$ domain $\mathcal{D}(L)$, is naturally associated a bilinear form. Assume that we are given a vector subspace $\mathcal{A}$ of the domain $\mathcal{D}(L)$ such that for every pair (f, g) of functions in $\mathcal{A}$, the product fg is in the domain $\mathcal{D}(L)$ ($\mathcal{A}$ is an algebra).

Definition 1.4.2 (Carré du champ operator) The bilinear map

$$\Gamma(f, g) = \frac{1}{2} \big[L(fg) - f \, Lg - g \, Lf \big]$$

defined for every $(f, g) \in \mathcal{A} \times \mathcal{A}$ is called the carré du champ operator of the Markov generator L.

The definition of the carré du champ operator is clearly subordinate to the algebra $\mathcal{A}$. Such a class $\mathcal{A}$ will be mostly natural in a given context (for example smooth functions on a manifold) and will be carefully described and discussed in Chap. 3.

1.4 Infinitesimal Generators and Carré du Champ Operators

The carré du champ operator Γ (which is so named for reasons which will be made clear below) will play a crucial role throughout this work. Numerous examples will be made discussed in this monograph, but it may already be useful to mention the simple example of the Laplacian $L = \Delta$ on $\mathbb{R}^n$ giving rise to the standard carré du champ operator $\Gamma(f, g) = \nabla f \cdot \nabla g$ (the usual scalar product of the gradients of f and g) for smooth functions f, g on $\mathbb{R}^n$.

If $f \in \mathcal{A}$, since $P_t(f^2) \geq (P_t f)^2$ for every $t \geq 0$ by (1.2.1), in the limit as $t \to 0$, $L(f^2) \geq 2f L f$. It follows that the carré du champ operator is positive on $\mathcal{A}$ in the sense that

$$\Gamma(f, f) \geq 0, \quad f \in \mathcal{A}. \tag{1.4.2}$$

This is a fundamental property of Markov semigroups. To somewhat lighten the notation, we set $\Gamma(f) = \Gamma(f, f)$, $f \in \mathcal{A}$. Observe also that by bilinearity, (1.4.2) immediately yields the Cauchy-Schwarz inequality

$$\Gamma(f, g)^2 \leq \Gamma(f)\Gamma(g), \quad (f, g) \in \mathcal{A} \times \mathcal{A}. \tag{1.4.3}$$

As already mentioned, the operator L is usually only determined on a core of the domain, however it completely characterizes the semigroup $\mathbf{P} = (P_t)_{t \geq 0}$. One of the main difficulties is to translate the properties of L to the semigroup $\mathbf{P}$. A first instance is positivity. As emphasized above, the positivity of the carré du champ operator (1.4.2) is one of the main properties which will be carefully described in examples of interest. Positivity may actually be addressed more generally. If $\phi : \mathbb{R} \to \mathbb{R}$ is a convex function, and f is a function in the domain $\mathcal{D}(L)$ such that $\phi(f)$ also belongs to $\mathcal{D}(L)$, then Jensen's inequality (1.2.1) together with the definition of L shows that

$$L\phi(f) \geq \phi'(f) L f. \tag{1.4.4}$$

The fact that $\Gamma(f) = \Gamma(f, f) \geq 0$ is precisely this property for the convex function $\phi(r) = r^2$, $r \in \mathbb{R}$. This property, together with the invariant measure μ, actually establishes (at least formally) the positivity preserving property of the semigroup $(P_t)_{t \geq 0}$. Indeed, if $f \geq 0$, setting $\Lambda(t) = \int_E \phi(P_t f) d\mu$, $t \geq 0$, and differentiating together with (1.4.4) yields that

$$\Lambda'(t) = \int_E \phi'(P_t f) L P_t f \, d\mu \leq \int_E L\phi(P_t f) d\mu.$$

But now, if $\phi(P_t f)$ is in the domain $\mathcal{D}(L)$, then the integral on the right-hand side of the preceding inequality is zero since μ is invariant. Therefore Λ is decreasing so that

$$\int_E \phi(P_t f) d\mu \leq \int_E \phi(f) d\mu$$

for every $t \geq 0$. In particular, for the convex function $\phi(r) = |r|$, and with $f \geq 0$,

$$\int_E |P_t f| d\mu \leq \int_E |f| d\mu = \int_E f \, d\mu = \int_E P_t f \, d\mu. \tag{1.4.5}$$

Therefore, $P_t f \geq 0$ μ-almost everywhere whenever $f \geq 0$, showing that P_t preserves positivity for every $t \geq 0$.

Now of course the preceding argument relies on (1.4.4) rather than only the positivity of Γ. However, it is not hard to see that the positivity of Γ implies (1.4.4) when ϕ is a convex polynomial. Extending this to more general ϕ's requires further hypotheses. But there should be no surprise that in any given example, as soon as Γ is positive in the sense of (1.4.2), then (1.4.4) is also valid. In particular, for diffusion semigroups as discussed later in Sect. 1.11, the diffusion property

$$L\phi(f) = \phi'(f)Lf + \phi''(f)\Gamma(f)$$

(for smooth $\phi : \mathbb{R} \to \mathbb{R}$) immediately yields (1.4.4) from the positivity property (1.4.2).

1.4.3 Martingale Problem

To conclude this section, we illustrate the notion of infinitesimal generators from the probabilistic viewpoint. Given a Markov semigroup $\mathbf{P} = (P_t)_{t \geq 0}$, the generator L of $\mathbf{P}$ may be used to describe the law of the associated Markov process $\mathbf{X} = \{X_t^x ; t \geq 0, x \in E\}$. In the good cases (that is when there exists a countable dense subset of the domain $\mathcal{D}(L)$ of L), the process $(X_t)_{t \geq 0}$ may be chosen in such a way that for any f in the domain $\mathcal{D}(L)$, the map $t \mapsto f(X_t)$ is right-continuous and

$$M_t^f = f(X_t) - \int_0^t Lf(X_s)ds, \quad t \geq 0, \quad (1.4.6)$$

is a local martingale. This type of abstract construction will not be necessary in what follows, but it might be useful to know that the Markov process $\mathbf{X}$ (as a probability measure on the space of right-continuous functions with values in E) may be obtained directly from the operator L. This construction amounts to solving the so-called *martingale problem* associated with L, that is defining a probability law (or a family of them) on the space of functions $t \mapsto X_t$ from $\mathbb{R}_+$ with values in E such that for any function f in the domain, the expression M_t^f given in (1.4.6) is a martingale. When the process is given as a solution of a stochastic differential equation as described in Sect. 1.10, this martingale property will appear as a by-product of Itô's formula (see Appendix B, Sect. B.4, p. 495). Observe that, when the preceding local martingale is a true martingale (for example when it is uniformly integrable, or better bounded), then taking expectation in (1.4.6) yields $\mathbb{E}(M_t^f) = \mathbb{E}(M_0^f)$ and therefore the identity

$$P_t f(x) = f(x) + \int_0^t P_s Lf(x)ds \quad (1.4.7)$$

(for every $t \geq 0$ and $x \in E$).

1.5 Fokker-Planck Equations

This picture may be further extended to smooth functions $F(t, x)$ depending on $t \geq 0$ and $x \in E$ under suitable conditions such that

$$F(t, X_t) - \int_0^t (\partial_s + L) F(s, X_s) \, ds, \quad t \geq 0, \tag{1.4.8}$$

is also a (local) martingale. In particular, when choosing $F(t, x) = P_{T-t} f(x)$ for f bounded and $t \in [0, T]$, $F(t, X_t)$ is a martingale on $[0, T]$. Since $(\partial_s + L) F(s, X_s) = 0$, taking expectations at $t = T$, one recovers that

$$P_T f(x) = \mathbb{E}(f(X_T) \mid X_0 = x).$$

From the abstract semigroup viewpoint, the stochastic representation of the martingale problem (1.4.7) indicates that, for every $t \geq 0$,

$$P_t = \mathrm{Id} + \int_0^t P_s L \, ds. \tag{1.4.9}$$

The second formulation (1.4.8) on the other hand is concerned with the heat equation $\partial_t G = LG$ ($G(0, x) = f(x)$). To formally solve it, it suffices to consider the exponential of the operator L as

$$Q_t f = e^{tL} f = \sum_{k \geq 0} \frac{t^k}{k!} L^k f \tag{1.4.10}$$

which clearly satisfies $\partial_t Q_t = LQ_t$, and thus $P_t = Q_t = e^{tL}$. Hence, the first formulation (1.4.7) expresses that $\partial_t P_t = P_t L$ while the second (1.4.8) indicates that $\partial_t P_t = LP_t$. The fact that the operator L commutes with P_t actually tells us that P_t is in a sense a function of L. This statement will be given a precise meaning in various examples below, but this basic observation can serve as a guide throughout this monograph. Note, however, that it is quite impossible, given the representation (1.4.10) of $P_t = e^{tL}$, to analyze properties such as positivity preservation, which is central to the analysis of Markov semigroups.

1.5 Fokker-Planck Equations

This short section describes a dual point of view on the Chapman-Kolmogorov equation, which is more commonly found in the partial differential equations literature.

Given a Markov semigroup $\mathbf{P} = (P_t)_{t \geq 0}$ with infinitesimal generator L as in the preceding sections, by (1.2.4) the operators $P_t, t \geq 0$, may be represented by Markov kernels. Usually, they admit densities $p_t(x, y), t > 0, (x, y) \in E \times E$, with respect to some reference measure m (for example the Lebesgue measure), not necessarily the invariant measure, as described in Definition 1.2.4, so that

$$P_t f(x) = \int_E f(y) p_t(x, y) dm(y), \quad t > 0, \, x \in E,$$

on a suitable class of functions f on E. Since L is the generator of the semigroup $\mathbf{P} = (P_t)_{t \geq 0}$, the function $p_t(x, y)$, $t > 0$, $x \in E$, is a solution of the heat equation with respect to L

$$\partial_t p_t(x, y) = \mathrm{L}_x p_t(x, y), \quad p_0(x, y) dm(y) = \delta_x, \tag{1.5.1}$$

where L_x denotes the operator L acting on the x variable. This expresses that

$$\partial_t P_t f = \mathrm{L} P_t f.$$

But one may similarly consider the dual equation, called the *Fokker-Planck equation* (or Kolmogorov forward equation), for $t > 0$,

$$\partial_t p_t(x, y) = \mathrm{L}_y^* p_t(x, y), \tag{1.5.2}$$

where L^* is the *adjoint* of L with respect to the reference measure m in the sense that

$$\int_E f \mathrm{L}^* g \, dm = \int_E g \mathrm{L} f \, dm$$

for suitable functions f, g (see Sect. A.2, p. 475). The latter identity (1.5.2) then expresses that

$$\partial_t P_t f = P_t \mathrm{L} f.$$

This dual Fokker-Planck equation may be obtained, at least formally, admitting that derivation in $t > 0$ commutes with the integral in y, via the identities

$$\int_E \partial_t p_t(x, y) f(y) dm(y) = \int_E p_t(x, y) \mathrm{L} f(y) dm(y)$$
$$= \int_E f(y) \mathrm{L}_y^* p_t(x, y) dm(y)$$

which are valid on a suitable class of functions f, giving rise to (1.5.2).

Care has to be taken with the fact that the heat and Fokker-Planck equations do not play the same role. When writing $\partial_t P_t f = P_t \mathrm{L} f$, the function f is implicitly in the domain of L (which usually requires that f is sufficiently smooth). However the equation $\partial_t P_t f = \mathrm{L} P_t f$ can in general be applied for irregular functions (only bounded and measurable, for example) since the operator P_t will usually be smoothing, mapping irregular functions to regular ones (see Sect. 1.12). More details will be provided in Chap. 3. One of the by-products of the tools developed in this work in order to measure the convergence to equilibrium will be to quantify the smoothing properties of the operators P_t.

1.6 Symmetric Markov Semigroups

One of the central themes of this work will be to detail the many useful properties of the important class of symmetric Markov semigroups and their associated reversible

1.6 Symmetric Markov Semigroups

Markov processes. As before, denote by $\mathbf{P} = (P_t)_{t \geq 0}$ a Markov semigroup with state space $(E, \mathcal{F})$, invariant measure μ and infinitesimal generator L with $\mathbb{L}^2(\mu)$-domain $\mathcal{D}(L)$. The associated Markov process is denoted by $\mathbf{X} = \{X_t^x ; t \geq 0, x \in E\}$.

1.6.1 Symmetric Markov Semigroups, Reversible Measures

Symmetry of a Markov semigroup, or *reversibility* of the invariant measure, are defined as follows.

Definition 1.6.1 (Symmetric Markov Semigroup) The Markov semigroup $\mathbf{P} = (P_t)_{t \geq 0}$ is said to be symmetric with respect to the invariant measure μ, or μ is reversible for $\mathbf{P}$, if for all functions $f, g \in \mathbb{L}^2(\mu)$ and all $t \geq 0$,

$$\int_E f \, P_t g \, d\mu = \int_E g \, P_t f \, d\mu. \qquad (1.6.1)$$

If the semigroup $\mathbf{P} = (P_t)_{t \geq 0}$ admits density kernels $p_t(x, y)$, $t > 0$, $(x, y) \in E \times E$, with respect to μ, then p_t is a symmetric function on $E \times E$.

If μ is a probability measure, the symmetry property for $f = \mathbb{1}$ shows that μ is invariant. The converse is not true in general. Under reasonable minimal assumptions, an invariant measure is unique. Such a measure may or may not be reversible. Hence Markov semigroups are divided into two classes, according to whether the invariant measure is reversible or not. In the fundamental examples studied in Chap. 2 we will learn how to distinguish between these two classes.

From a probabilistic point of view, the name reversible refers to reversibility in time of the associated Markov process whenever the initial law is the invariant measure. Indeed, from (1.2.3), we know that if the process starts from the invariant distribution μ, then it keeps the same distribution for any time. Moreover, if the measure is reversible, and if the initial distribution of X_0 is μ, then for any $t > 0$ and any partition $0 \leq t_1 \leq \cdots \leq t_k \leq t$ of the time interval $[0, t]$, the law of $(X_0, X_{t_1}, \ldots, X_{t_k}, X_t)$ is the same as the law of $(X_t, X_{t-t_1}, \ldots, X_{t-t_k}, X_0)$. This is easily seen from (1.3.4) when the initial measure ν is μ since in this case the measure $p_t(x, dy)\mu(dx)$ is symmetric in (x, y), and then, by an immediate induction, the measures

$$p_{t_k - t_{k-1}}(y_{k-1}, dy_k) \, p_{t_{k-1} - t_{k-2}}(y_{k-2}, dy_{k-1}) \cdots p_{t_1}(y_0, dy_1) \, \mu(dy_0)$$

are invariant under the change $(y_0, \ldots, y_k) \mapsto (y_k, \ldots, y_0)$. Therefore, for any $t > 0$, the law of the process $(X_s)_{0 \leq s \leq t}$ is the same as the law of the process $(X_{t-s})_{0 \leq s \leq t}$. Hence, the law of the Markov process is "reversible in time".

In terms of the generator L of the semigroup $\mathbf{P}$, the reversibility property may be expressed as

$$\int_E f \, \mathrm{L} g \, d\mu = \int_E g \, \mathrm{L} f \, d\mu \qquad (1.6.2)$$

on suitable functions f, g in the domain $\mathcal{D}(L)$, as can be checked by differentiating the identity (1.6.1) at $t = 0$. The operator L is thus a symmetric operator, however it is unbounded on $\mathbb{L}^2(\mu)$.

As discussed in Appendix A, there is a fundamental difference in infinite-dimensional Hilbert spaces, namely that a symmetric operator is not necessarily self-adjoint (see there for a summary of the fundamental properties of symmetric and self-adjoint operators). For a symmetric semigroup $(P_t)_{t \geq 0}$, the generator L defined on its domain $\mathcal{D}(L)$ is always self-adjoint. Nevertheless, as already emphasized and precisely described later in Chap. 3, we are in general given some symmetric operator L on a class of functions (such as differential operators defined on smooth compactly supported functions) for which the semigroup has yet to be constructed. Those symmetric operators considered in this book always admit at least one self-adjoint extension, that is the domain $\mathcal{D}(L)$ may always be extended in such a way that L becomes a self-adjoint operator, which in turn entirely describes the semigroup $(P_t)_{t \geq 0}$ (see Sect. 3.1, p. 120). Whether or not this extension is unique is related, as mentioned earlier, to the question of the uniqueness of the semigroup $(P_t)_{t \geq 0}$ associated to the description of L on $\mathcal{D}(L)$ and to the essential self-adjointness property (Definition 3.1.10, p. 134). For example, on the interval $(-1, +1)$, there are many semigroups $(P_t)_{t \geq 0}$ for which the generator coincides with the Laplace operator $Lf = f''$ acting on functions f which are smooth and compactly supported in $(-1, +1)$, which defines a symmetric operator for the Lebesgue measure dx. In a probabilistic context, two such examples are the semigroup associated with the Brownian motion killed at the boundary and the Brownian motion reflected at the boundary. These examples correspond to two different self-adjoint extensions of the operator, although the space of functions which are smooth and compactly supported in $(-1, +1)$ is dense in $\mathbb{L}^2(dx)$ (see Sects. 2.4, 2.5 and 2.6, pp. 92, 96 and 97).

The following is a simple useful application of martingale theory to the analysis of symmetric Markov semigroups which illustrates the interplay between probability and analysis. The following lemma is in particular useful as mentioned earlier in the study of the regularity properties of the associated Markov processes. The proof relies on maximal inequalities for martingales. The result itself will not really be used later and should be considered more as an illustration of the power of probabilistic techniques in this context.

Lemma 1.6.2 (Rota's Lemma) *Let* $\mathbf{P} = (P_t)_{t \geq 0}$ *be a Markov semigroup symmetric with respect to* μ. *Then, for any* $1 < p < \infty$, *there exists a* $C_p > 0$ *such that for any measurable function* $f : E \to \mathbb{R}$,

$$\left\| \sup_{s \geq 0} P_s f \right\|_p \leq C_p \|f\|_p.$$

Proof We sketch the proof in the case where μ is a probability measure (the general case has to be adapted with some care). It suffices to assume that f is positive and, by approximation, bounded. Let $(X_t)_{t \geq 0}$ be the Markov process with initial

1.6 Symmetric Markov Semigroups

distribution μ associated with $(P_t)_{t\geq 0}$. For $T > 0$, consider the (positive) martingale $M_t = P_{T-t}(f)(X_t)$, $t \in [0, T]$ (cf. Sect. 1.4.3). By Doob's maximal inequality,

$$\mathbb{E}\Big(\sup_{0\leq t\leq T} M_t^p\Big) \leq C_p^p \, \mathbb{E}(M_T^p)$$

for some constant $C_p > 0$. But $M_T = f(X_T)$ and since (by invariance) the law of X_T is μ, $\mathbb{E}(M_T^p) = \|f\|_p^p$. Furthermore, conditional expectation is a contraction in $\mathbb{L}^p(\mathbb{P})$, so that

$$\mathbb{E}\Big(\mathbb{E}\Big(\sup_{0\leq t\leq T} M_t \mid X_T\Big)^p\Big) \leq C_p^p \, \|f\|_p^p .$$

Now reversibility ensures that the law of (X_t, X_T) is the same as that of (X_{T-t}, X_0) so that $\mathbb{E}(M_t \mid X_T) = P_{2(T-t)} f(X_T)$. It follows that

$$\mathbb{E}\Big(\sup_{0\leq t\leq T} M_t \mid X_T\Big) \geq \sup_{0\leq t\leq T} \mathbb{E}(M_t \mid X_T) = \sup_{0\leq t\leq T} P_{2(T-t)} f(X_T).$$

In conclusion

$$\Big\| \sup_{0\leq t\leq 2T} P_t f \Big\|_p \leq C_p \|f\|_p$$

and it remains to let T go to infinity to get the result. □

1.6.2 Markov Triple

When the measure μ is symmetric, the generator L and the semigroup $\mathbf{P} = (P_t)_{t\geq 0}$ are completely described by the measure μ and the carré du champ operator Γ from Definition 1.4.2. Indeed, for all functions f, g in the algebra $\mathcal{A} \subset \mathcal{D}(L)$ on which Γ is defined,

$$\int_E \Gamma(f,g) d\mu = - \int_E f \, Lg \, d\mu . \tag{1.6.3}$$

This basic formula is immediate by integration of the definition (1.4.2) of $\Gamma(f, g)$ together with the invariance property $\int_E L f d\mu = 0$ and is actually an integration by parts formula as will be seen explicitly in the case of diffusion processes. It will turn out to be a fundamental tool in the analysis of Markov semigroups. (Note that in the non-symmetric case, with μ invariant, the above integration by parts formula (1.6.3) does not hold for all f and g.) Since Γ is bilinear and positive ((1.4.2)), for every function f for which Lf is well-defined,

$$\int_E f(-Lf) d\mu \geq 0.$$

The operator $-L$ is therefore a positive operator, and for such positive operators, the theory of symmetric and self-adjoint operators is much easier, as presented in Appendix A.

As the carré du champ operator Γ and the measure μ completely determine the symmetric Markov generator L, we will mostly work throughout this monograph with what will be called a *Markov Triple* (E, μ, Γ) consisting of a (σ-finite) measure μ on a state space $(E, \mathcal{F})$ and a carré du champ operator Γ (on some suitable class $\mathcal{A}$ of functions on E). This framework will be carefully presented in Chap. 3. The adoption of this framework is particularly justified in our investigation of functional inequalities in later chapters. As a standard example, consider the Lebesgue measure on the Borel sets of $\mathbb{R}^n$ and the carré du champ operator $\Gamma(f, g) = \nabla f \cdot \nabla g$ for smooth functions f, g on $\mathbb{R}^n$. Further basic examples will be discussed in this and the next chapter.

1.6.3 Heat Equations

In the last part of this section, we collect a few observations on the Chapman-Kolmogorov and Fokker-Planck heat equations in the symmetric context.

We first briefly revisit the Chapman-Kolmogorov and Fokker-Planck equations for a symmetric Markov semigroup $(P_t)_{t \geq 0}$, and draw some useful conclusions. Thus let μ be reversible for L. The operators P_t, $t \geq 0$, are then symmetric in $\mathbb{L}^2(\mu)$ and often admit kernel densities with respect to μ, denoted $p_t(x, y)$, $t > 0$, $(x, y) \in E \times E$, which are symmetric in (x, y).

As discussed in Appendix A, Sect. A.2, p. 475, while symmetry and self-adjointness might not coincide for unbounded operators, the generator L of a symmetric semigroup is self-adjoint, that is $L = L^*$ where the adjoint operator L^* is computed with respect to the invariant measure μ. Now, from the heat and Fokker-Planck equations (1.5.1) and (1.5.2),

$$\partial_t p_t(x, y) = L_x p_t(x, y) = L_y p_t(x, y), \quad t > 0.$$

In particular

$$L_x p_t(x, y) = L_y p_t(x, y).$$

Recall that the notations L_x and L_y are to emphasize the action on the respective variables x and y of the kernel $p_t(x, y)$. This is however not a simple consequence of the symmetry property. For example, on $\mathbb{R}^2$, the function $h(x, y) = x^4 + y^4$ is symmetric in (x, y), but $\partial_x^2 h \neq \partial_y^2 h$. On the other hand, on $\mathbb{R}^n$, all smooth functions F of $|x - y|^2$ satisfy $\Delta_x F = \Delta_y F$. This will be the case for the heat kernels of the standard Laplacian.

The second illustration is related to the trace of the semigroup. Namely, the Chapman-Kolmogorov equation (1.3.2) indicates that the scalar product in $\mathbb{L}^2(\mu)$

of the functions $z \mapsto p_s(x,z)$ and $z \mapsto p_t(z,y)$ is $p_{t+s}(x,y)$. In particular, by the Cauchy-Schwarz inequality, for all $t,s > 0$ and $(x,y) \in E \times E$,

$$p_{t+s}(x,y)^2 \leq p_{2t}(x,x) p_{2s}(y,y).$$

Note that this inequality, when applied for $t = s$ (and after the change $t \mapsto \frac{t}{2}$), indicates that, for every $t > 0$, the function $p_t(x,y)$ of (x,y) achieves its maximum on the diagonal $x = y$. Moreover, in the symmetric case, and provided the following expression makes sense, one has

$$\int_E p_{2t}(x,x) d\mu(x) = \int_E \int_E p_t(x,y)^2 d\mu(x) d\mu(y). \tag{1.6.4}$$

When the latter is finite, the operator P_t is Hilbert-Schmidt (see Sect. A.6, p. 483, for more information about Hilbert-Schmidt operators). In terms of the associated Markov process $\mathbf{X} = \{X_t^x ; t \geq 0, x \in E\}$, $p_t(x,x)$ is the probability that X_t returns at time $t > 0$ to its initial point x (as for the Brownian bridge for example). Of course, this naive representation is formal since in general this probability vanishes. This basic formula however links the distribution of the so-called bridge process to the trace of the semigroup.

1.7 Dirichlet Forms and Spectral Decompositions

In the context of symmetric Markov semigroups, the carré du champ operator immediately gives rise to the notion of a *Dirichlet form*. Dirichlet forms play an important role in the construction of semigroups (see Chap. 3), and the functional inequalities studied throughout this work are also closely related to this object. Dirichlet forms may be considered in non-symmetric frameworks but here we restrict the exposition to the symmetric case.

1.7.1 Dirichlet Forms

For a symmetric Markov semigroup $\mathbf{P} = (P_t)_{t \geq 0}$ with infinitesimal generator L, reversible measure μ and carré du champ operator Γ on a class $\mathcal{A}$ of functions on E, consider the (symmetric) bilinear operator

$$\mathcal{E}(f,g) = \int_E \Gamma(f,g) d\mu, \quad (f,g) \in \mathcal{A} \times \mathcal{A}.$$

By the integration by parts formula (1.6.3),

$$\mathcal{E}(f,g) = \int_E \Gamma(f,g) d\mu = -\int_E f \mathrm{L} g \, d\mu \tag{1.7.1}$$

for any $f,g \in \mathcal{A}$.

The bilinear form $\mathcal{E}$ may actually be defined on a class of functions larger than $\mathcal{A}$. To this end, observe first that from (1.7.1), $\mathcal{E}(f) = \mathcal{E}(f,f)$ may be defined for any function $f \in \mathcal{D}(L)$. Also (and this type of argument will be used constantly throughout this book), for any $f \in \mathbb{L}^2(\mu)$ and any $t > 0$,

$$\partial_t \int_E (P_t f)^2 d\mu = 2 \int_E P_t f\, LP_t f\, d\mu = -2\mathcal{E}(P_t f)$$

and

$$\partial_t \mathcal{E}(P_t f) = -\partial_t \int_E P_t f\, LP_t f\, d\mu$$

$$= -\int_E (LP_t)^2 d\mu - \int_E P_t f\, L^2 P_t f\, d\mu \qquad (1.7.2)$$

$$= -2 \int_E (LP_t f)^2 d\mu.$$

In particular, $\mathcal{E}(P_t f)$ is a decreasing function of t and

$$\int_E f^2 d\mu - \int_E (P_t f)^2 d\mu = 2 \int_0^t \mathcal{E}(P_s f) ds \geq 2t\, \mathcal{E}(P_t f).$$

Changing t into $\frac{t}{2}$, this inequality shows that the quantity

$$\frac{1}{t}\left[\int_E f^2 d\mu - \int_E (P_{t/2} f)^2 d\mu\right]$$

is decreasing in $t > 0$. But the latter may be rewritten by the reversibility property as $\frac{1}{t}\int_E f(f - P_t f) d\mu$, and therefore, for $f \in \mathcal{D}(L)$, it converges to $\mathcal{E}(f)$ when t tends to 0.

This observation allows us to extend the definition of $\mathcal{E}(f)$ to a wider class of functions, defining $\mathcal{D}(\mathcal{E})$ as the set of functions $f \in \mathbb{L}^2(\mu)$ for which the quantity

$$\frac{1}{t}\left[\int_E f^2 d\mu - \int_E (P_{t/2} f)^2 d\mu\right] = \frac{1}{t}\int_E f(f - P_t f) d\mu$$

has a finite (decreasing) limit as t decreases to 0, and $\mathcal{E}(f)$ to be this limit for $f \in \mathcal{D}(\mathcal{E})$.

Definition 1.7.1 (Dirichlet form) For a symmetric Markov semigroup $\mathbf{P} = (P_t)_{t \geq 0}$ with reversible measure μ, the energy $\mathcal{E}(f)$ is defined as the limit as $t \to 0$ of

$$\frac{1}{t}\int_E f(f - P_t f) d\mu$$

for all functions $f \in \mathbb{L}^2(\mu)$ for which this limit exists, defining in this way the domain $\mathcal{D}(\mathcal{E})$. The Dirichlet form $\mathcal{E}(f,g)$ is defined by polarization for f and g in the Dirichlet domain $\mathcal{D}(\mathcal{E})$, and $\mathcal{E}(f) = \mathcal{E}(f,f)$.

1.7 Dirichlet Forms and Spectral Decompositions

If $f, g \in \mathcal{D}(L)$,

$$\mathcal{E}(f, g) = \int_E f(-Lg)d\mu = \int_E g(-Lf)d\mu,$$

while if f, g belong to the class $\mathcal{A}$ on which Γ is defined,

$$\mathcal{E}(f, g) = \int_E \Gamma(f, g)d\mu.$$

Hence $\mathcal{D}(L) \subset \mathcal{D}(\mathcal{E}) \subset \mathbb{L}^2(\mu)$. These objects will be further investigated in Chap. 3 where in particular $\mathcal{D}(\mathcal{E})$ will be constructed first and then $\mathcal{D}(L)$ will be extracted from it.

1.7.2 Spectral Decompositions

For symmetric semigroups, the spectral decomposition of the generator L is a very efficient tool for analyzing properties of functions and operators. We refer to Sect. A.4, p. 478, for abstract definitions and to Proposition 3.1.6, p. 131, for concrete applications in the setting of Markov semigroups. In particular, for any function $f \in \mathbb{L}^2(\mu)$ and any $t > 0$, the function $P_t f$ always belongs to the domain $\mathcal{D}(L)$, and also to $\mathcal{D}(L^k)$ for any integer $k \in \mathbb{N}$ (by induction, $f \in \mathcal{D}(L^k)$ if $f \in \mathcal{D}(L^{k-1})$ and $L^{k-1} f \in \mathcal{D}(L)$—see Remark 3.1.7, p. 132). In particular, this would not be the case for a non-symmetric semigroup.

We next briefly comment on the spectral decomposition of a Markov operator L in the simple instance when there is a countable basis of eigenvectors for the operator L. Assume therefore that $(e_k)_{k \in \mathbb{N}}$ is a Hilbertian basis of $\mathbb{L}^2(\mu)$ consisting of eigenfunctions of L with corresponding sequence of eigenvalues $(\lambda_k)_{k \in \mathbb{N}}$. Hence

$$-Le_k = \lambda_k e_k, \quad k \in \mathbb{N}.$$

In this case, by the integration by parts formula (1.6.3), for every integer k,

$$\mathcal{E}(e_k) = \int_E \Gamma(e_k, e_k)d\mu = -\int_E e_k L e_k \, d\mu = \lambda_k \int_E e_k^2 \, d\mu = \lambda_k.$$

As a fundamental consequence, the eigenvalues of a Markov generator are always negative (justifying the choice of sign in the definition $-Le_k = \lambda_k e_k$). This expresses in another way that $-L$ is positive. The operator $-L$ is sometimes called the *opposite* operator (to L).

Since (formally) $P_t = e^{tL}$, $t \geq 0$, for a function $f \in \mathbb{L}^2(\mu)$ decomposed as $f = \sum_{k \in \mathbb{N}} f_k e_k$ in the basis $(e_k)_{k \in \mathbb{N}}$,

$$P_t f = \sum_{k \in \mathbb{N}} e^{-\lambda_k t} f_k e_k.$$

Then setting, for $t > 0$ and $(x, y) \in E \times E$,

$$p_t(x, y) = \sum_{k \in \mathbb{N}} e^{-\lambda_k t} e_k(x) e_k(y), \tag{1.7.3}$$

shows that

$$P_t f(x) = \int_E f(y) \, p_t(x, y) d\mu(y)$$

in accordance with the kernel representation (1.2.4). That the density kernels $p_t(x, y)$ thus defined are positive is however not easy to see directly, even in simple examples. From (1.6.4), the trace formula

$$\int_E p_t(x, x) d\mu(x) = \sum_{k \in \mathbb{N}} e^{-t\lambda_k}, \quad t > 0, \tag{1.7.4}$$

follows. Hence, information on the functions $p_t(x, x)$ yields estimates on the spectrum of L. Of course, there are numerous examples where the operator admits densities without being Hilbert-Schmidt (the standard Euclidean heat semigroup being the simplest example).

Contrary to the finite-dimensional case, in infinite dimension, symmetric or self-adjoint operators L do not always admit an orthonormal basis of eigenvectors. This is replaced in the general setting by a spectral decomposition. As described in Sect. A.4, p. 478, the spectrum, which is the set of $\lambda \in \mathbb{R}$ such that $L - \lambda \text{Id}$ is not invertible, is actually divided into two parts, the discrete spectrum, which corresponds to isolated eigenvalues associated with a finite-dimensional eigenspace, and the essential spectrum. For example, the Laplace operator on $\mathbb{R}^n$, Markov generator of the heat or Brownian semigroup, has an empty discrete spectrum. In ideal settings, as described later in Chap. 4, the essential spectrum is entirely determined by the behavior of L at infinity. This is in particular the case for Persson operators, which are developed in Sect. 4.10.3, p. 227.

1.8 Ergodicity

In probability theory, ergodic properties usually relate to the long time behavior. In the context of Markov processes $\{X_t^x; t \geq 0, x \in E\}$, and when the invariant measure μ is a probability measure, it is in general expected that quantities such as

$$\frac{1}{t} \int_0^t f(X_s^x) ds$$

converge almost surely as $t \to \infty$ to $\int_E f d\mu$, whatever the starting point x and for suitable functions f. Although such results may in general be deduced from the analysis developed in this monograph (see for example Sect. 1.15.9), we adopt here a less ambitious approach. *Ergodicity* in our context will relate to the convergence

1.9 Markov Chains

of the law of X_t^x to the invariant probability measure μ as $t \to \infty$. In other words, for suitable functions $f : E \to \mathbb{R}$, $P_t f(x)$ converges to $\int_E f d\mu$ for any $x \in E$. (See the precise Definition 3.1.11, p. 135.)

From both the point of view of probability and analysis, these ergodic properties may be seen as "convergence to equilibrium", the equilibrium being the invariant measure (whenever finite). The analysis of convergence to equilibrium plays an important role in many fields, and many of the methods and functional inequalities developed in this monograph will lead to (precise) quantitative bounds on the convergence to equilibrium.

Determining the class of functions f such that $P_t f(x)$ converges, and the sense in which it converges, strongly depends on the context. According to the spectral decomposition of self-adjoint operators as presented in Sect. A.4, p. 478, (see also Sect. 1.7.2), the easiest way to understand ergodicity is to look at $\mathbb{L}^2(\mu)$-convergence. In the context of symmetric Markov semigroups for which the infinitesimal generators L are self-adjoint, the associated semigroups $\mathbf{P} = (P_t)_{t \geq 0}$ are actually $\mathbb{L}^2(\mu)$-ergodic. Indeed, the spectrum of L lies in $(-\infty, 0]$ and for any $f \in \mathbb{L}^2(\mu)$, $P_t f$ converges to $P_\infty f$ in $\mathbb{L}^2(\mu)$ as $t \to \infty$ where P_∞ is the orthogonal projection onto the space associated with the eigenvalue 0, called the invariant space. This eigenspace consists of the solutions of the equation $Lf = 0$ ($f \in \mathcal{D}(L) \subset \mathbb{L}^2(\mu)$). In particular, since $\int_E (-Lf) f d\mu = \mathcal{E}(f) = \int_E \Gamma(f) d\mu$, and since $\Gamma(f) \geq 0$, the invariant functions satisfy $\Gamma(f) = 0$ (μ-almost everywhere). Now, in many cases, these invariant functions are constant. For example, in the diffusion case (Sect. 1.11), $\Gamma(f)$ stands for the square of the length of ∇f. A similar description is available in the discrete Markov chain setting (Sect. 1.9). This property is related to connexity of the state space as developed below in Sect. 3.3, Definition 3.3.7, p. 151, the precise relationship between ergodicity and connexity being described in Proposition 3.3.10, p. 157.

Once we know that invariant functions are constant, there are only two possibilities: either $\mu(E) < \infty$, and assuming that μ is a probability measure, $P_\infty f = \int_E f d\mu$, or $\mu(E) = \infty$, and since 0 is the only constant in $\mathbb{L}^2(\mu)$, $P_\infty f = 0$. These two situations are represented by the Ornstein-Uhlenbeck semigroup and the standard heat semigroup in $\mathbb{R}^n$ (see Sect. 2.7.1, p. 103, and Sect. 2.1, p. 78).

Convergence to equilibrium therefore reflects the fact that $P_t f$ converges in $\mathbb{L}^2(\mu)$ to $P_\infty(f)$, which is the projection on the space of invariant functions. A similar phenomenon can be found in other settings. For example, in differential geometry, one may also construct semigroups on p-forms (which are no longer functions, or functions of a certain type on a larger space) for which the spectrum is still negative, and for which the invariant space may carry some non-constant or non-zero terms, the so-called space of harmonic p-forms. The non-existence of non-vanishing invariant forms in this context yields important topological information.

1.9 Markov Chains

Although this monograph is concerned with a continuous (diffusion) setting as presented in the next section (that is, diffusion semigroups and Markov processes with continuous paths), it is of interest to briefly test and motivate the preceding abstract definitions and properties of Markov semigroups and their infinitesimal generators on the concrete example of a Markov chain on a finite or countable state space. Furthermore, many of the difficulties that will be encountered in general may already be described in the context of Markov chains on a countable state space. Indeed, in this simple setting, generators appear as matrices L for which the associated semigroup is nothing else than $e^{t\text{L}}$, that is the usual exponential of matrices. In particular, it is natural in this context to start from the Markov generator and define from it the associated semigroup. Most of the properties of general Markov semigroups and generators have an analogue in this context, and this setting therefore provides a rich source of ideas and basic counterexamples. Moreover, both the finite and countable models also have great importance when one considers numerical estimates. For example, when solving a partial differential equation such as the heat equation, discretization of space is often considered. When discretizing differential operators like those described in Sect. 1.11 below, the resulting operators are generators of Markov processes on finite or countable spaces.

In the following, we (briefly) review for the example of Markov chains on a finite or countable state space some of the objects and properties considered in the abstract framework of the previous sections. We refer the reader to general references on Markov chains for a more complete picture.

1.9.1 Finite Markov Chains

We begin with one of the simplest cases, a finite state space. Thus given a finite set E (equipped with the σ-field $\mathcal{F}$ of all its subsets), consider, as domain of any Markov generator L, the vector space of all real-valued functions on E (with dimension the cardinality of E, say N). On the basis of the functions $f_x = \mathbb{1}_x$, $x \in E$, a Markov generator L can be represented as a matrix $(\text{L}(x, y))_{(x,y) \in E \times E}$.

In this context, the (Markov) property $\text{L}(\mathbb{1}) = 0$ expresses that, for every $x \in E$, $\sum_{y \in E} \text{L}(x, y) = 0$. The carré du champ operator is given by

$$\Gamma(f)(x) = \sum_{y \in E} \text{L}(x, y) \big[f(x) - f(y)\big]^2, \quad x \in E.$$

Hence, the carré du champ operator is positive if and only if all off-diagonal terms of the matrix L are positive. A square matrix $\text{L} = \text{L}(x, y)$, $(x, y) \in E \times E$, with the two properties $\sum_{y \in E} \text{L}(x, y) = 0$ for all $x \in E$, and $\text{L}(x, y) \geq 0$ for all $x \neq y$, is called a Markov generator on the finite state space E.

The easiest way to produce such a Markov generator is to consider a stochastic (or Markov) matrix P and to set $\text{L} = \lambda(P - \text{Id})$ for some $\lambda > 0$. The matrix $P = (P(x, y))_{(x,y) \in E \times E}$ actually represents the transition probabilities of a Markov chain jumping at integer times from x to y with probability $P(x, y)$. The continuous time Markov process associated with P is obtained from the chain by replacing the times between the jumps by independent exponential variables with parameter λ.

1.9 Markov Chains

It is an elementary exercise to check that, for a square matrix L, the matrix e^{tL}, $t \geq 0$, has all its entries strictly positive if and only if the off-diagonal entries of L are positive. This exponential then describes the semigroup $P_t = e^{tL}$, $t \geq 0$, with infinitesimal generator L.

In this finite state space context, for a convex function $\phi : \mathbb{R} \to \mathbb{R}$, Jensen's inequality $L\phi(f) \geq \phi'(f) L f$ (1.4.4) expresses that, at any given point $x \in E$,

$$\sum_{y \in E} L(x, y) [\phi(f(y)) - \phi(f(x))] \geq \sum_{y \in E} L(x, y) \phi'(f(x)) [(f(y) - f(x)]$$

which is a clear consequence of the convexity of ϕ.

A (finite) measure μ is invariant with respect to L if and only if, for all $y \in E$,

$$\sum_{y \in E} \mu(x) L(x, y) = 0.$$

Such a measure always exists (it is an eigenvector of the transpose of L, with eigenvalue 0 and positive coordinates). It is unique as soon as the matrix L is irreducible, that is, for any pair of points (x, y) in $E \times E$, there exists a path in E, $(x = x_0, x_1, \ldots, x_\ell = y)$, such that for every $i = 0, \ldots, \ell - 1$, $L(x_i, x_{i+1}) > 0$. For the measure μ to be reversible, the condition is that

$$\mu(x) L(x, y) = \mu(y) L(y, x) \qquad (1.9.1)$$

for all $(x, y) \in E \times E$ (also known as the detailed balance condition). Summing over x this identity shows that μ is invariant, so that in this particular case, reversibility implies invariance. Observe that when the matrix L is symmetric in the ordinary sense, then the uniform measure on E is reversible for L. Moreover, when L is irreducible, $\mu(x) > 0$ for every $x \in E$, and the matrix

$$R(x, y) = \sqrt{\mu(x)} L(x, y) \frac{1}{\sqrt{\mu(y)}}, \quad (x, y) \in E \times E,$$

is symmetric in the usual sense. It can then be diagonalized with respect to an orthonormal basis, which means that there is a basis of eigenvectors of L which is orthonormal in $\mathbb{L}^2(\mu)$. The eigenvalues are therefore negative, and if e_k, λ_k, $k = 0, \ldots, N$, denote respectively the eigenvectors and eigenvalues of $-L$, then, for every k,

$$e^{tL}(e_k) = e^{-\lambda_k t} e_k.$$

If the vectors e_k are normalized in $\mathbb{L}^2(\mu)$, that is, if $\sum_{x \in E} e_k^2(x) \mu(x) = 1$, then

$$\sum_{k=0}^{N} e^{-\lambda_k t} e_k(x) e_k(y)$$

is positive for every pair (x, y) in $E \times E$ and every $t \geq 0$ (this is not obvious).

The invariant measure μ is thus viewed as a positive eigenvector of the transposed matrix of L. The fact that the transposed matrix of L admits a positive eigenvector is not specific to those Markov generators. It is a general property of matrices with positive entries (see the above transformation $\widehat{L}$ yielding a trivial change on the spectrum without changing the eigenvectors). Such matrices always admit an eigenvector with positive entries. It corresponds to the eigenvalue with maximal modulus, and this eigenvalue is real and positive. Moreover, provided the matrix is irreducible in the sense described above, this eigenvector is unique up to scaling with strictly positive coordinates. These are known as the *Perron-Frobenius* eigenvector and eigenvalue.

The matrices with positive entries (outside the diagonal) may serve as a toy model for generators of positivity preserving semigroups (without the Markov property $P_t(\mathbb{1}) = \mathbb{1}$). The analogue of the Perron-Frobenius eigenvector is then a positive eigenvector (known as the ground state in the physics literature), and may be used for example to remove a potential term from a generator (see for example Sects. 1.15.8 and 1.15.6 below and Proposition 4.5.3, p. 198).

1.9.2 Countable Markov Chains

Beyond the simplest case of finite spaces, the next case to consider is when the space is countable. It can thus be assumed that $E = \mathbb{N}$ but we tend to prefer the abstract notation E below, which is particularly justified when one has to take into account the geometric aspects of the state space E. For example, random walks on $\mathbb{Z}$ and $\mathbb{Z}^n$ do not have the same behavior concerning recurrence and transience (that is, whether or not the corresponding Markov process tends to infinity with time).

In the context of a countable state space E, a Markov semigroup $\mathbf{P} = (P_t)_{t \geq 0}$ is described by an "infinite matrix" of positive kernels $(p_t(x, y))_{(x,y) \in E \times E}$, $t \geq 0$, such that for all $t \geq 0$ and $x \in E$, and any positive function f on E,

$$P_t f(x) = \sum_{y \in E} f(y) \, p_t(x, y).$$

As in the finite case, for any $x \in E$, $p_t(x, y)$ is a probability measure on E, that is, $p_t(x, y) \geq 0$ for all $y \in E$ and $\sum_{y \in E} p_t(x, y) = 1$. Then the generator L is also given by an infinite matrix $(L(x, y))_{(x,y) \in E \times E}$ as

$$L(x, y) = \partial_t p_t(x, y)_{|t=0}, \quad (x, y) \in E \times E,$$

and, for any, say finitely supported, function f on E,

$$Lf(x) = \sum_{y \in E} L(x, y) f(y).$$

1.9 Markov Chains

The heat equation reads

$$\partial_t p_t(x, y) = \sum_{y \in E} L(x, y) p_t(x, y), \quad p_0(x, y) = \delta_x(y).$$

Solving such an equation requires some properties of the matrix L. Among them, note that whenever $x \neq y$, $L(x, y) \geq 0$, and, for every $x \in E$, $\sum_{y \in E} L(x, y) = 0$.

As in the case of a finite space, the carré du champ operator is given by

$$\Gamma(f)(x) = \sum_{y \in E} L(x, y) [f(x) - f(y)]^2, \quad x \in E$$

(on, again, say finitely supported functions), and an invariant measure μ for L is characterized by

$$\mu(y) = \sum_{x \in E} \mu(x) L(x, y)$$

for every $y \in E$. Again, finding an invariant measure μ requires the solution of an infinite-dimensional eigenvector problem. The measure μ is reversible if

$$\mu(x) L(x, y) = \mu(y) L(y, x)$$

for all $(x, y) \in E \times E$ (detailed balance property). Hence, in order for μ to be reversible the ratios $\frac{L(x,y)}{L(y,x)}$ must be expressible in the form $\frac{R(x)}{R(y)}$ for some function R. A reversible measure may then be easily defined directly from the generator L, fixing the value at one point $x_0 \in E$, say $\mu(x_0) = 1$, and setting $\mu(x) = \frac{L(x_0, x)}{L(x, x_0)}$, $x \in E$. If this measure is finite, it can be normalized into a probability measure. An invariant measure will exist and is unique in the recurrent case. When the underlying Markov chain is transient, there may exist many invariant measures.

It is also of interest to describe the paths of the Markov process (or family of Markov processes) $\mathbf{X} = \{X_t^x ; t \geq 0, x \in E\}$ with generator L in these finite or countable contexts. Working directly in the countable state space setting, we introduce the Markov matrix

$$K(x, y) = -\frac{L(x, y)}{L(x, x)}, \quad x \neq y, \quad K(x, x) = 0.$$

We therefore exclude the possibility that $L(x, x) = 0$ for any x (such a point would be a trap point, meaning that when the process arrives at this point, it stays there forever). Look first at the discrete time Markov chain $(\widehat{X}_k)_{k \in \mathbb{N}}$ with matrix K as Markov kernel, that is the sequence of random variables with $\widehat{X}_0 = x$ such that, for all $k \in \mathbb{N}$ and $(x, y) \in E \times E$,

$$\mathbb{P}(\widehat{X}_{k+1} = y \mid \widehat{X}_k = x) = K(x, y).$$

This process is known as the underlying Markov chain of the process $\mathbf{X}$. Then the associated Markov process $\mathbf{X}$ may described in the following terms. For each $k \in \mathbb{N}$, if T_k is the k-th jump time of $(X_t)_{t \geq 0}$, the sequence $(X_{T_k})_{k \in \mathbb{N}}$ is a Markov chain which follows the same law as $(\widehat{X}_k)_{k \in \mathbb{N}}$. The interval $T_{k+1} - T_k$ between the jump follows an exponential law with parameter $-\mathrm{L}(X_{T_k}, X_{T_k})$, independently of what happened before T_k. In other words, when the process arrives at some point $x \in E$, it waits for an exponential time with parameter $-\mathrm{L}(x,x)$, and then jumps to y with probability $K(x,y)$.

This rough description hides some serious difficulties. It may happen that the underlying Markov chain $(\widehat{X}_k)_{k \in \mathbb{N}}$ is transient (that is, goes to infinity when $k \to \infty$) and that the jumps are so quick (when $\mathrm{L}(x,x)$ goes to infinity) that the process so described goes to infinity in finite time. The actual construction of the process given the data contained in the matrix L requires further restrictions. We refer the reader to the standard literature for a complete account of the topic.

These discrete models may often be thought of as approximations for diffusion processes. As a basic example, consider the Markov process on $\mathbb{Z}/N \subset \mathbb{R}$, which jumps from x to $x \pm \frac{1}{N}$ at exponential times with parameter N^2. Its generator may be described as

$$\mathrm{L}_N f(x) = N^2 \left[f\left(x + \frac{1}{N}\right) + f\left(x - \frac{1}{N}\right) - 2f(x) \right],$$

which clearly converges to $\mathrm{L} f(x) = f''(x)$ when N goes to infinity (at least along smooth functions). In a naive way, this reflects Donsker's theorem which describes convergence of a suitable renormalization of the random walk on $\mathbb{Z}$ to Brownian motion.

1.10 Stochastic Differential Equations and Diffusion Processes

Diffusion semigroups are the main objects of investigation in this work, and will be introduced in Sect. 1.11. They occur quite naturally in the probabilistic context of stochastic differential equations and we present here some of the intuition behind this description. The reader is referred to Appendix B for the necessary stochastic calculus background and more precisely to Sect. B.4, p. 495, for the necessary material on diffusion processes.

1.10.1 Diffusion Processes

In this framework, we begin with a stochastic differential equation, in $\mathbb{R}$ or $\mathbb{R}^n$,

$$dX_t^x = \sigma(X_t^x) dB_t + b(X_t^x) dt, \quad X_0^x = x, \tag{1.10.1}$$

where $(B_t)_{t \geq 0}$ is a standard Brownian motion (on $\mathbb{R}$ or $\mathbb{R}^n$, respectively) and σ and b are smooth functions with bounded derivatives. More precisely, say on $\mathbb{R}^n$,

1.10 Stochastic Differential Equations and Diffusion Processes

σ is an $n \times n$ matrix and b a vector. A process $\mathbf{X} = \{X_t^x ; t \geq 0, x \in \mathbb{R}^n\}$ which is a solution of (1.10.1) is usually called a *diffusion process* (more accurately, since the initial point x is a variable, it is a family of diffusion processes). For a given smooth function $f : \mathbb{R}^n \to \mathbb{R}$, at least C^2, Itô's formula (Theorem B.2.1, p. 491) indicates that

$$df(X_t^x) = \sum_{i,j=1}^n \sigma_i^j(X_t^x)\partial_j f(X_t^x)dB_t^i + Lf(X_t^x)dt$$

$$= dM_t^f + Lf(X_t^x)dt, \quad X_0^x = x,$$

where $(M_t^f)_{t \geq 0}$ is a local martingale (given as a stochastic integral) such that $M_0^f = 0$ and

$$Lf = \frac{1}{2} \sum_{i,j=1}^n \sum_{k=1}^n \sigma_k^i \sigma_k^j \partial_{ij}^2 f + \sum_{i=1}^n b^i \partial_i f. \qquad (1.10.2)$$

In other words, $(M_t^f)_{t \geq 0}$ solves the martingale problem (1.4.6) for L. Actually $(M_t^f)_{t \geq 0}$ is a true martingale as soon as f and Lf are bounded functions, and in this case, taking expectations, for all $t \geq 0$ and $x \in \mathbb{R}^n$,

$$\mathbb{E}(f(X_t^x)) = f(x) + \int_0^t \mathbb{E}(Lf(X_s^x))ds. \qquad (1.10.3)$$

In terms of the associated Markov semigroup

$$P_t f(x) = \mathbb{E}(f(X_t^x)) = \mathbb{E}(f(X_t^x) \mid X_0 = x), \quad t \geq 0, \ x \in \mathbb{R}^n,$$

so that (1.10.3) leads to the fundamental formula (1.4.9)

$$P_t f(x) = f(x) + \int_0^t P_s Lf(x) ds, \quad t \geq 0, \ x \in \mathbb{R}^n.$$

We therefore recover here in a concrete way the characterization of the law of the process $\mathbf{X}$ in terms of a martingale problem (1.4.6) discussed in Sect. 1.4.3. The second description (1.4.8) of the martingale problem may similarly be recast in terms of stochastic calculus. Namely, observe, again from Itô's formula, that if $F(t, x)$ is a (smooth) function of the two variables $t \geq 0$ and $x \in \mathbb{R}^n$, then

$$dF(t, X_t^x) = dM_t^F + (\partial_t + L)F(t, X_t^x)dt, \quad X_0^x = x$$

(where L is acting on the space variable x). In other words, if $\partial_t F = LF$ (and $F(0, x) = f(x)$), then $F(T - t, X_t^x)$ is a martingale on the time interval $t \in [0, T]$ which is equal to $F(T, x)$ at $t = 0$ and to $f(X_T^x)$ at time $t = T$. Taking expectations,

$$F(x, T) = \mathbb{E}(F(0, X_T^x)) = \mathbb{E}(f(X_T^x)).$$

Hence if $F(t, x)$ is a solution of the partial differential (heat) equation $\partial_t F = LF$ with $F(0, x) = f(x)$, then $F(t, x) = P_t f(x)$, $t \geq 0$, $x \in \mathbb{R}^n$. Note that here the operator L (that is, the coefficients of the original stochastic differential equation (1.10.1)) does not depend on t. The case of an equation with time dependence would actually lead to similar consequences, however it is somewhat more difficult to express.

Therefore understanding the distributions of X_t^x, $t \geq 0$, $x \in \mathbb{R}^n$, reduces to the study of the solutions of heat equations of the type $\partial_t F = LF$. To this end, let us have a closer look at the second order differential operator L of (1.10.2). It may be written as

$$Lf = \frac{1}{2} \sum_{i,j=1}^{n} g^{ij}(x) \partial_{ij}^2 f + \sum_{i=1}^{n} b^i(x) \partial_i f \tag{1.10.4}$$

where $\mathfrak{g} = (g^{ij})_{1 \leq i,j \leq n} = (g^{ij}(x))_{1 \leq i,j \leq n}$ denotes the symmetric matrix $\sigma \sigma^*$ and $b = (b^i)_{1 \leq i \leq n} = (b^i(x))_{1 \leq i \leq n}$. For a matrix $\mathfrak{g}$ to be expressible in this form, it is necessary that, at every point x, it is symmetric and positive, that is

$$\mathfrak{g}(V, V) = \sum_{i,j=1}^{n} g^{ij}(x) V_i V_j \geq 0 \quad \text{for all } V = (V_i)_{1 \leq i \leq n} \in \mathbb{R}^n.$$

We refer to the latter property of the operator L of (1.10.4) as semi-ellipticity. Ellipticity on the other hand usually refers to semi-ellipticity together with non-degeneracy of $\mathfrak{g}$, that is, it must be strictly positive or positive-definite. A further discussion of these notions appears in Sect. 1.12 below. Every symmetric positive matrix $\mathfrak{g}$ can be diagonalized with respect to an orthonormal basis, and admits a symmetric square root such that $\sigma \sigma^* = \sigma^2 = g$. The fact that this can be achieved continuously on x requires some care, but in general σ may be chosen with at least the same regularity as $\mathfrak{g}$ as long as $\mathfrak{g}$ is invertible (elliptic case). Even when $\mathfrak{g}$ is degenerate, one may choose a square root σ to be at least locally Lipschitz, provided $\mathfrak{g}$ is at least $\mathcal{C}^2$. We will generally not be concerned with such regularity issues in a general setting since the operators L will usually be given explicitly in the form $\mathfrak{g} = \sigma \sigma^*$.

Once L is given, that is the matrix $\mathfrak{g} = \mathfrak{g}(x)$ and the vector $b = b(x)$, one can construct the associated Markov process $\mathbf{X} = \{X_t^x ; t \geq 0, x \in \mathbb{R}^n\}$ described above by solving a stochastic differential equation, and then find the solution to the heat equation $\partial_t F = LF$ with the help of the distribution of X_t^x. There is therefore a close connection between solutions of such heat equations and laws of diffusion processes.

Note finally that the operator L does not fully describe the process $\mathbf{X} = \{X_t^x ; t \geq 0, x \in \mathbb{R}^n\}$ since there is always a choice of the square root of the matrix $\mathfrak{g}$. However, it is possible to completely describe from L the distribution of X_t^x for every t as a martingale problem (1.4.6), the formulation of which only relies on L (and not on the choice of σ).

1.10.2 Hörmander Form

Very often, the operator L is given in the so-called *Hörmander form*. Recall (cf. Appendix C, Sect. C.1, p. 500) that a *vector field* on $\mathbb{R}^n$ or on an open set of $\mathbb{R}^n$ is a first order operator $f \mapsto Zf$ which may be written as

$$Zf(x) = \sum_{i=1}^{n} Z^i(x)\, \partial_i f(x).$$

The coefficients $Z^i(x)$ are the coordinates of Z in the base $(\partial_i)_{1 \le i \le n}$. If Z is a vector field, then $Z^2 = Z \circ Z$ is a second order differential operator. Then, given a collection $(Z_j)_{0 \le j \le d}$ of vector fields (d does not need to be the dimension of the space, it may be smaller or larger), the operator

$$L = \frac{1}{2} \sum_{j=1}^{d} Z_j^2 + Z_0 \qquad (1.10.5)$$

is a second order differential operator, semi-elliptic with no 0-order term. For example, the ordinary Laplacian on $\mathbb{R}^n$ is given in this form, with $d = n$ and $Z_j = \partial_j$ for $1 \le j \le n$ and $Z_0 = 0$. At least locally, an operator L given by (1.10.4) may always be represented in this Hörmander form, so that there is no loss of generality in restricting to operators of this form. This Hörmander form is particularly useful when analyzing the hypoelliptic properties of L (see Sect. 1.12 below), however it is not unique.

For such an operator L in Hörmander form (1.10.5), to construct the associated Markov process, say on $\mathbb{R}^n$, $\mathbf{X} = \{X_t^x\,;\, t \ge 0,\, x \in \mathbb{R}^n\}$ with generator L, it is enough to choose d independent Brownian motions $(B_t^j)_{t \ge 0}$ in $\mathbb{R}^n$, $1 \le j \le d$, and to solve the stochastic differential equation

$$dX_t^x = \sum_{j=1}^{d} Z_j(X_t^x) \circ dB_t^j + Z_0(X_t^x)dt, \quad X_0^x = x. \qquad (1.10.6)$$

Here, the notation $Z_j(X_t^x) \circ dB_t^{(j)}$ means that the standard Itô integral has been replaced by the Stratonovich integral (Sect. B.4, p. 495, in Appendix B). The previous notation has to be understood component-wise, that is, for any coordinate $X_t^{x,i}$, $1 \le i \le n$, of X_t^x,

$$dX_t^{x,i} = \sum_{j=1}^{d} Z_j^i(X_t^x) \circ dB_t^j + Z_0^i(X_t^x)dt.$$

This change from Itô's integral to Stratonovich's integral may be viewed as a mere change of notation, but it has the distinct advantage of having a simpler chain rule formula than the Itô integral, and therefore leads to a direct representation of L in the desired form. The link between this stochastic differential equation and the associated Markov semigroup as a martingale problem (1.4.6) can be seen, as before, by checking that $df(X_t^x) = dM_t^f + Lf(X_t^x)dt$ where M_t^f is a martingale, and this is again the result of a direct application of Itô's formula for Stratonovich integrals.

1.11 Diffusion Semigroups and Operators

Markov semigroups associated with solutions of stochastic differential equations as presented in the previous section are prototypical examples of the so-called diffusion semigroups. (Symmetric) Markov diffusion semigroups and operators are the central objects of analysis in this monograph. In this section, we do not deal with the most general family of diffusion operators and only present some basic examples of interest. In particular, we only loosely describe the classes of functions on which the various operators are acting. It should be pointed out that, with respect to the previous probabilistic interpretation, we work here, and throughout the book, with the generator L rather than $\frac{1}{2}$L.

1.11.1 Diffusion Operators

To introduce the family of diffusion semigroups and operators (discussed in further detail in Chap. 3), we begin with $E = \mathbb{R}^n$ or an open subset of $\mathbb{R}^n$ as underlying state space. Working on such a state space, smooth functions (at least C^2 but often C^∞) with compact support in E are in the domains of the diffusion operators L under investigation. To determine whether smooth functions form a core (in other words whether it is enough to know L on these functions in order to describe the associated Markov semigroup $\mathbf{P} = (P_t)_{t \geq 0}$) actually involves the intrinsic metric and the boundary of E as will be discussed below (see Sect. 3.2.2, p. 141, and also Sect. A.5, p. 481, for a general discussion when the operator L is symmetric).

On smooth functions f, thus on $E = \mathbb{R}^n$ or some open subset of $\mathbb{R}^n$, a Markov diffusion operator L is a second order differential operator of the form

$$Lf = \sum_{i,j=1}^n g^{ij} \partial_{ij}^2 f + \sum_{i=1}^n b^i \partial_i f \qquad (1.11.1)$$

where $\mathfrak{g} = \mathfrak{g}(x) = (g^{ij}(x))_{1 \leq i,j \leq n}$ and $b = b(x) = (b^i(x))_{1 \leq i \leq n}$ are smooth, respectively $n \times n$ symmetric matrix-valued and $\mathbb{R}^n$-valued, functions of $x \in E$. The fact that L does not involve constant terms expresses the fact that $L(\mathbb{1}) = 0$. The carré du champ operator Γ (cf. Definition 1.4.2) takes the form, on smooth functions f, g,

$$\Gamma(f,g) = \sum_{i,j=1}^n g^{ij} \partial_i f \, \partial_j g. \qquad (1.11.2)$$

1.11 Diffusion Semigroups and Operators

That Γ is positive expresses the semi-ellipticity property of the generator L. This also explains why we limit ourselves to second order differential operators since a differential operator of higher order can never have a positive carré du champ operator.

When the operator L is given in the Hörmander form (1.10.5) (without the factor $\frac{1}{2}$), the carré du champ operator is given by

$$\Gamma(f,g) = \sum_{j=1}^{d} Z_j f Z_j g,$$

from which positivity is immediate. It is also plain from this expression that L is elliptic as soon as, at any point x, the vector space spanned by the vectors $Z_j(x)$, $1 \leq j \leq d$, is $\mathbb{R}^n$. This requires in particular $d \geq n$.

The continuous setting of the previous examples allows for a crucial diffusion, or change of variables, property emphasized in the following basic definition. At this point, it is more a property of the differential operators (1.11.1) but it will make sense in a general framework (emphasized in Chap. 3) for Markov generators L acting on some class $\mathcal{A}$ of (smooth) functions. For such an operator, recall the carré du champ operator Γ of Definition 1.4.2, as well as the shorthand notation $\Gamma(f) = \Gamma(f, f)$.

Definition 1.11.1 (Diffusion operator) An operator L, with carré du champ operator Γ, is said to be a diffusion operator (or generator) if

$$L\psi(f) = \psi'(f) L f + \psi''(f) \Gamma(f) \tag{1.11.3}$$

for every $\psi : \mathbb{R} \to \mathbb{R}$ of class at least $\mathcal{C}^2$ and every suitably smooth function f.

The operator L is also said to satisfy the diffusion property or change of variables formula. In the examples of second order differential operators L considered above, one may for example choose sufficiently smooth functions f in this definition. Precise assumptions and descriptions of relevant classes of functions will be addressed later in Chap. 3.

There is a similar, equivalent, change of variables formula for functions of several variables,

$$L\big(\Psi(f_1, \ldots, f_k)\big) = \sum_{i=1}^{k} \partial_i \Psi(f_1, \ldots, f_k) L f_i$$
$$+ \sum_{i,j=1}^{k} \partial_{ij}^2 \Psi(f_1, \ldots, f_k) \Gamma(f_i, f_j) \tag{1.11.4}$$

for $\Psi : \mathbb{R}^k \to \mathbb{R}$ of class $\mathcal{C}^2$ (actually $\mathcal{C}^\infty$ will mainly be used in the core of the book), and $f_1, \ldots, f_k$ smooth. Formally, the diffusion property expresses that L is a second order differential operator, or equivalently that Γ is, in each argument, a first order differential operator. For example,

$$\Gamma(\psi(f), g) = \psi'(f)\Gamma(f, g) \qquad (1.11.5)$$

for smooth f and g, and similarly for families of functions.

From a purely algebraic point of view, if the operator L and its carré du champ operator Γ are defined on an algebra $\mathcal{A}$ of functions included in the domain $\mathcal{D}(L)$, the formula (1.11.4) for a polynomial function Ψ boils down to the chain rule formula

$$\Gamma(fg, h) = f\Gamma(g, h) + g\Gamma(f, h)$$

for any triple (f, g, h) of elements of $\mathcal{A}$, which precisely translates the fact that, in each of its arguments, Γ is a first order differential operator. Extending this to the wider class of smooth functions Ψ generally requires further assumptions (see Remark 3.1.4).

It may be observed that (1.11.4) applied for a system of coordinates $(f_1, \ldots, f_n)$ yields precisely the expression (1.11.1) of L in this system with $g^{ij} = \Gamma(f_i, f_j)$ and $b^i = Lf_i$, $1 \leq i, j \leq n$, written in those coordinates.

For diffusion operators, given a (smooth) function h, the operator $f \mapsto \Gamma(h, f)$ is a first order differential operator, that is a vector field. In a smooth context, it is denoted ∇h. With this notation,

$$\nabla h(f) = \sum_{i,j=1}^{n} g^{ij} \partial_j h \, \partial_i f.$$

Alternatively, $\nabla h(f) = \sum_{i=1}^{n} Z^i \partial_i f$ where $Z^i = \sum_{j=1}^{n} g^{ij} \partial_j h$, $1 \leq i \leq n$.

The diffusion property of a given Markov generator L is closely related to the continuity properties of the sample paths of the associated Markov process $\mathbf{X} = \{X_t^x; t \geq 0, x \in \mathbb{R}^n\}$. As discussed in Sect. 1.1, given any Markov semigroup, the associated Markov process may be constructed so that for every "nice" function f (in the domain of L), the real-valued process $(f(X_t))_{t \geq 0}$ is right-continuous with left-limits. What is hidden behind the diffusion property is that these processes $(f(X_t))_{t \geq 0}$ are continuous in this case. The proof of this claim requires some work in stochastic calculus (with respect to non-continuous processes). At least, the diffusion property implies a kind of continuity in measure for functions of the domain as described in Theorem 3.1.16, p. 136.

Observe that the diffusion operators L for which the second order terms are all zero (that is, for which the carré du champ operator is identically zero) are far from being uninteresting. They correspond to vector fields, that is, first order differential operators. For such an operator, the associated semigroup $\mathbf{P} = (P_t)_{t \geq 0}$ is of the form $P_t f(x) = f(X_t^x)$ where X_t^x is the path following the vector field starting at x. It is the analogue of the Markov processes described by solutions of stochastic differential equations, but deterministic. If the vector field is of the form $Zf(x) = \sum_{i=1}^{n} Z^i(x) \partial_i f(x)$, then X_t^x is a point in $\mathbb{R}^n$ whose coordinates are the solution of the differential system $X_0 = x$, $dX_t^i = Z^i(X_t)dt$. The law of the process at time t is a Dirac mass at X_t^x and randomness has disappeared. Randomness (or diffusion) arises precisely from the second order terms in the differential operator.

1.11.2 Invariant Measures of Differential Operators

Given a diffusion operator L as in (1.11.1), let us now describe its invariant measure when it admits a (smooth, strictly positive) density w with respect to the Lebesgue measure dx (on $E = \mathbb{R}^n$ or an open subset of $\mathbb{R}^n$). This density w is given by the equation $L^* w = 0$ where L^* is the adjoint of L with respect to the Lebesgue measure which may be explicitly described, on smooth functions f, by

$$L^* f = \sum_{i,j=1}^n \partial_{ij}^2 (g^{ij} f) - \sum_{i=1}^n \partial_i (b^i f). \tag{1.11.6}$$

For a different reference measure, it is necessary to take the adjoint with respect to this new measure (for example, when the diffusion matrix is not the identity, there is a better choice than the Lebesgue measure as will be seen below). The invariant measure will usually be given in the further developments, and its existence or finiteness will be clear from the context.

In general, the adjoint operator L^* of (1.11.6) is a second order differential operator, but including a non-zero constant term (usually called the potential in quantum mechanics and in the study of Schrödinger equations). In the simplest instances $L = \Delta + Z$ where Δ is the usual Laplacian on $\mathbb{R}^n$ and $Z = (Z^i)_{1 \leq i \leq n}$ is a vector field, with the Lebesgue measure as reference measure. Hence, by (1.11.6), on smooth functions f, the adjoint operator L^* with respect to the Lebesgue measure is

$$L^* f = \Delta f - Zf - \mathrm{div}(Z) f \tag{1.11.7}$$

where $\mathrm{div}(Z) = \sum_{i=1}^n \partial_i Z^i$. Indeed, given smooth compactly supported functions f, g (on $E = \mathbb{R}^n$ or an open subset of $\mathbb{R}^n$), $\int_E f \Delta g \, dx = \int_E g \Delta f \, dx$ whereas, by integration by parts, for every $1 \leq i \leq n$,

$$\int_E f Z^i \partial_i g \, dx = - \int_E \partial_i (Z^i f) g \, dx = - \int_E g Z^i \partial_i f \, dx - \int_E g \partial_i Z^i f \, dx$$

thus justifying the above form of L^*.

Note that it is often easier to describe the evolution of the density with respect to the (an) invariant measure since then there is no potential term in the adjoint of L. Indeed, if μ is invariant for L and if L^* denotes the adjoint of L with respect to μ, for every compactly supported function f in the domain $\mathcal{D}(L)$ and every function g in $\mathcal{D}(L)$, $\int_E f L^* g \, d\mu = \int_E Lf \, g \, d\mu$. Therefore, if $g = \mathbb{1}$, by invariance of μ, $\int_E f L^*(\mathbb{1}) d\mu = 0$ so that $L^*(\mathbb{1}) = 0$ and L^* does not include constant terms.

1.11.3 Reversible Measures of Differential Operators

In the context of second order differential operators L of the form (1.11.1), invariant measures μ are not easy to identify in general. For example, on an open subset of $\mathbb{R}^n$, the density w of μ with respect to the Lebesgue measure is a solution of $L^* w = 0$. But when the measure μ is reversible, things are a lot easier. For example, on an open set in $\mathbb{R}^n$, or on a manifold given any local system of coordinates, an operator given in the form

$$\mathrm{L} f = \frac{1}{w} \sum_{i,j=1}^n \partial_i \big(w g^{ij} \partial_j f \big), \qquad (1.11.8)$$

where w is smooth and strictly positive, is clearly (by integration by parts) symmetric in $\mathbb{L}^2(\mu)$ where $d\mu = w dx$ is the measure with density w with respect to the Lebesgue measure. This is indeed the general case when the reversible measure has a non-vanishing density w, and therefore operators of the form

$$\mathrm{L} f = \sum_{i,j=1}^n g^{ij} \partial^2_{ij} f + \sum_{j=1}^n \bigg(\sum_{i=1}^n \partial_i g^{ij} + g^{ij} \partial_i (\log w) \bigg) \partial_j f$$

are exactly those operators with second order terms given by the matrix (g^{ij}) and reversible measure $d\mu = w dx$. Therefore, the knowledge of (g^{ij}) and (b^i) in the representation (1.11.1) allows for a direct identification of the invariant measure $d\mu = w dx$ when it is reversible. In this case, for f, g smooth and compactly supported,

$$\int_{\mathbb{R}^n} f \,\mathrm{L} g\, d\mu = - \int_{\mathbb{R}^n} \sum_{i,j=1}^n g^{ij} \partial_i f \, \partial_i g \, d\mu = - \int_{\mathbb{R}^n} \Gamma(f, g) d\mu. \qquad (1.11.9)$$

When the matrix $\mathfrak{g} = (g^{ij})$ is everywhere non-degenerate (elliptic case), there is a more canonical way of decomposing the operator L of (1.11.1). Namely, instead of using the Lebesgue measure as a reference measure to compute densities, it is preferable in general to employ the *Riemannian measure* $\mu_\mathfrak{g}$ with density $w_\mathfrak{g} = \det(\mathfrak{g})^{-1/2}$ with respect to the Lebesgue measure. In this case, one introduces the Laplace-Beltrami operator which has the same second order terms as L and $\mu_\mathfrak{g}$ as reversible measure: in a local system of coordinates,

$$\Delta_\mathfrak{g} = \frac{1}{w_\mathfrak{g}} \sum_{i,j=1}^n \partial_i \big(w_\mathfrak{g}\, g^{ij} \partial_j \big)$$

and then $\mathrm{L} = \Delta_\mathfrak{g} + Z$ where Z is a vector field. This operator $\Delta_\mathfrak{g}$, called the *Laplace-Beltrami operator*, or *Laplacian*, associated with the (co-)metric $\mathfrak{g} = \mathfrak{g}(x) = (g^{ij}(x))_{1 \leq i,j \leq n}$ is presented in Appendix C, Sect. C.3, p. 504, in a Riemannian framework. With respect to the Riemannian framework, it is therefore

1.11 Diffusion Semigroups and Operators

the co-metric g which naturally enters into the investigation of Markov diffusion operators. We thus privilege the co-metric g in the notation. As explained there, the decomposition $L = \Delta_g + Z$ has the advantage of being preserved under a change of coordinates.

Under this decomposition

$$L = \Delta_g + Z,$$

L is symmetric with respect to a measure $d\mu = w d\mu_g$ as soon as $Zf = \Gamma(\log w, f)$. In other words, if $Zf = \sum_{i=1}^n Z^i \partial_i f$,

$$Z^i = \sum_{j=1}^n g^{ij} \partial_j(\log w), \quad 1 \leq i \leq n.$$

The operator $f \mapsto \Gamma(\log w, f)$ is often identified with $\nabla \log w \cdot \nabla f$, and in this case Z is called a *gradient vector field*. The invariant reversible measure is then typically written as

$$d\mu = e^{-W} d\mu_g$$

with $w = e^{-W}$ and μ_g the Riemannian measure. The coordinate representation (1.11.8) takes the form

$$L = \Delta_g - \Gamma(W, \cdot) = \Delta_g - \nabla W \cdot \nabla \quad (1.11.10)$$

(see Sect. C.5, p. 511, for further details). Both the coordinate representation of Γ (1.11.2) and the integration by parts formula (1.11.9) are similar in this context. The latter decomposition of L (the canonical decomposition) is most useful when computing iterated carré du champ operators in view of curvature-dimension conditions (see Sect. 1.16 below) since it is mainly with the language of Riemannian geometry that computations are made understandable.

When dealing with an operator $L = \Delta_g - \nabla W \cdot \nabla$ symmetric with respect to $d\mu = e^{-W} d\mu_g$ on a Riemannian manifold (M, g) with Riemannian measure μ_g, we sometimes speak of M as a *weighted Riemannian manifold*. The basic example of a weighted measure $d\mu = e^{-W} dx$ on $\mathbb{R}^n$ is already of significant interest.

In summary, it is easier to describe the invariant measure in the reversible case, since it is then given locally, while in the general case, it is obtained from a differential equation. For symmetric diffusions, the carré du champ operator Γ determines the second order terms of the generator L while the invariant measure μ determines the first order terms. (For comparison, note that such a characterization is not available in the discrete case, since for example on finite state spaces, the operator Γ already describes the generator L.)

1.11.4 Alternative Geometric Representations of Generators

In the last part of this section, we briefly discuss variations and a few operations on the preceding representations.

Adding a new vector field $\widehat{Z}$ to an operator $\mathrm{L} = \Delta_\mathfrak{g} - \nabla W \cdot \nabla$ as above, it may happen that the measure $d\mu = e^{-W} d\mu_\mathfrak{g}$ is still an invariant measure although no longer reversible, that is the new operator $\widehat{\mathrm{L}} = \mathrm{L} + \widehat{Z}$ is no longer symmetric in $\mathbb{L}^2(\mu)$. For this to happen, it is enough that for smooth compactly supported functions f, $\int_E \widehat{Z} f d\mu = 0$, or in other words that the adjoint $(\widehat{Z})^*$ of $\widehat{Z}$ in $\mathbb{L}^2(\mu)$ satisfies $(\widehat{Z})^*(\mathbb{1}) = 0$. In a local system of coordinates for which $\widehat{Z} f = \sum_{i=1}^n \widehat{Z}^i \partial_i f$, an easy computation shows that

$$(\widehat{Z})^* f = -\sum_{i=1}^n \widehat{Z}^i \partial_i f - \left(\sum_{i=1}^n \partial_i \widehat{Z}^i + \sum_{i=1}^n \widehat{Z}^i \partial_i (\log \rho) \right) f$$

where ρ is the density of μ with respect to the Lebesgue measure dx (while $w = e^{-W}$ is the density with respect to the Riemannian measure $\mu_\mathfrak{g}$). The quantity

$$\mathrm{div}_\mu(\widehat{Z}) = \sum_{i=1}^n \partial_i \widehat{Z}^i + \sum_{i=1}^n \widehat{Z}^i \partial_i (\log \rho) = \sum_{i=1}^n \partial_i \widehat{Z}^i + \widehat{Z}(\log \rho)$$

then plays the role of the divergence (and actually is the usual divergence when μ is the Lebesgue measure itself, that is $\rho = 1$). Then, the adjoint $\widehat{\mathrm{L}}^*$ of $\widehat{\mathrm{L}} = \mathrm{L} + \widehat{Z}$ in $\mathbb{L}^2(\mu)$ may be written

$$\widehat{\mathrm{L}}^* = \Delta_\mathfrak{g} - \nabla W - \widehat{Z} - \mathrm{div}_\mu(\widehat{Z}),$$

a description similar to (1.11.7) in the Euclidean case. Hence, the measure μ is invariant precisely when $\mathrm{div}_\mu(\widehat{Z}) = 0$, and then the adjoint $\widehat{\mathrm{L}}^*$ of $\widehat{\mathrm{L}} = \mathrm{L} + \widehat{Z}$ in $\mathbb{L}^2(\mu)$ is $\mathrm{L} - \widehat{Z}$.

There is still another most useful representation of $\mathrm{L} = \Delta_\mathfrak{g} - \nabla W \cdot \nabla$. We again begin with the Euclidean case $E = \mathbb{R}^n$ where $\mathfrak{g}$ is the usual Euclidean metric and $d\mu = e^{-W} dx$. For any smooth compactly supported function f on $\mathbb{R}^n$, consider the gradient operator $df = (\partial_i f)_{1 \leq i \leq n} = \nabla f$, which is a vector of smooth compactly supported functions. Its adjoint ∇^* is defined on smooth compactly supported vector-valued functions V in such a way that, for any smooth compactly supported function f,

$$\int_E (\nabla^* V) f \, d\mu = \int_E V \cdot \nabla f \, d\mu.$$

If $V = (V_i)_{1 \leq i \leq n}$, then

$$\nabla^* V = -\sum_{i=1}^n \partial_i V_i + \sum_{i=1}^n V_i \partial_i W$$

so that

$$\mathrm{L} = -\nabla^* \nabla. \tag{1.11.11}$$

This representation immediately extends to the general Riemannian case, where ∇f is replaced by the 1-form df given in a local system of coordinates by

1.12 Ellipticity and Hypo-ellipticity

$(\partial_i f)_{1 \leq i \leq n}$. The scalar product of 1-forms w and w' is then given in this system of coordinates by $w \cdot w' = \sum_{i,j=1}^n g^{ij} w_i w'_j$, so that again $L = -d^*d$, which may be seen as an extension of (1.11.11).

When the operator is elliptic, df may be replaced by its gradient ∇f which is now a vector field, where in a local system of coordinates,

$$\nabla^i f = \sum_{j=1}^n g^{ij} \partial_j f, \quad 1 \leq i \leq n.$$

The scalar product on vectors V and $\widehat{V}$ is defined as $V \cdot \widehat{V} = \sum_{i,j=1}^n g_{ij} V^i \widehat{V}^j$ where (g_{ij}) is the inverse matrix of (g^{ij}). The adjoint ∇^* is then defined by its action on a vector field Z as

$$\int_E \nabla^* Z f \, d\mu = \int_E Z \cdot \nabla f \, d\mu$$

for any smooth compactly supported function f, and we still have $L = -\nabla^*\nabla$, which also extends (1.11.11) to the Riemannian case. We refer to Appendix C for further details.

We conclude this section with a brief comment on integration by parts in the context of diffusion operators. Namely, observe that when the measure μ is only invariant, and only for diffusion operators, the integration by parts formula (1.6.3) still holds when $g = \psi(f)$. Indeed, setting $\psi = \Psi'$, invariance yields

$$0 = \int_E L\Psi(f) d\mu = \int_E \psi(f) Lf \, d\mu + \int_E \psi'(f) \Gamma(f) d\mu,$$

and thus, with $g = \psi'(f)$,

$$\int_E \Gamma(f, g) d\mu = \int_E \psi'(f) \Gamma(f) d\mu = -\int_E g \, Lf \, d\mu.$$

In particular, this observation explains why, in dimension one (for the so-called Sturm-Liouville operators, Sect. 2.6, p. 97), the analysis will be easier and invariant measures will automatically be reversible.

1.12 Ellipticity and Hypo-ellipticity

It has already been mentioned in Sect. 1.10 that ellipticity or hypo-ellipticity play an important role in the theory. In particular, under these conditions, solving a differential (heat) equation of the form $\partial_t u = Lu$ with respect to a differential operator L with initial condition $u(0, x) = f(x)$ has regularizing properties: even when f is only measurable, the solution $u(t, x)$ is smooth for $t > 0$.

Hypo-ellipticity is a notion intermediate between semi-ellipticity and ellipticity. Recall first the notions of semi-ellipticity and ellipticity for a diffusion operator L of the form (1.11.1)

$$Lf = \sum_{i,j=1}^{n} g^{ij} \partial_{ij}^2 f + \sum_{i=1}^{n} b^i \partial_i f$$

where $\mathfrak{g} = \mathfrak{g}(x) = (g^{ij}(x))_{1 \leq i,j \leq n}$ and $b = b(x) = (b^i(x))_{1 \leq i \leq n}$. The operator L is said to be *semi-elliptic* if the matrix $\mathfrak{g}$ is, at every point x, symmetric and positive,

$$\mathfrak{g}(V, V) = \sum_{i,j=1}^{n} g^{ij}(x) V_i V_j \geq 0 \quad \text{for all } V = (V_i)_{1 \leq i \leq n} \in \mathbb{R}^n. \tag{1.12.1}$$

Ellipticity refers to semi-ellipticity together with non-degeneracy of $\mathfrak{g}$, that is strictly positive or positive-definite, in the sense that

$$\text{if } \mathfrak{g}(V, V) = 0, \quad \text{then } V = 0. \tag{1.12.2}$$

We sometimes speak of *uniform ellipticity* whenever there exists a $c > 0$ such that

$$\mathfrak{g}(V, V) \geq c |V|^2 \quad \text{for all } V \in \mathbb{R}^n. \tag{1.12.3}$$

Semi-elliptic operators in particular include first order differential operators (vector fields) for which the matrix (g^{ij}) is zero everywhere (the associated carré du champ operator $\Gamma(f)$ vanishing for every f). Such a first order differential operator would never have regularizing properties as may be seen, for instance, from the example $L = \partial_x$ on $E = \mathbb{R}$ for which the solution of $\partial_t u = Lu$ with initial condition $u(0, x) = f(x)$ is just $u(t, x) = f(x + t)$.

However, when the operator L is symmetric with respect to some measure μ (or more precisely self-adjoint), the solution $u(t, x) = P_t f(x)$ is always smooth in some weak $\mathbb{L}^2(\mu)$ sense, that is in the domain of L^k for any integer $k \geq 1$ (p. 31). This property may be deduced from spectral analysis (see Remark 3.1.7, p. 132). But this global result is in general not enough to provide precise information on smoothness properties of the solution in the usual sense. Moreover, when solving any heat equation on $u(t, x)$ of the form

$$\partial_t u = Lu, \quad u(0, x) = u_0, \tag{1.12.4}$$

boundary conditions have to be imposed to ensure some kind of uniqueness (see, for example, in dimension one, Sects. 2.4, p. 92, 2.5, p. 96, and 2.6, p. 97). The simplest conditions (which however produce in general only sub-Markov semigroups) are Dirichlet boundary conditions which amount to assuming that $u(t, x)$ is 0 at the boundary of the domain for any $t > 0$, or in probabilistic terms to consider the associated Markov process killed at the boundary. Again, such restrictions imply that it is impossible to solve the heat equation (1.12.4) locally.

To be more precise, if L is defined, say, on an open set $\mathcal{O} \subset \mathbb{R}^n$, and if we consider a solution of the equation in $\mathcal{O}_1 \subset \mathcal{O}$ with the initial condition u_0 say compactly supported in $\mathcal{O}_1$, the solution is not unique: any kind of boundary condition on any

1.12 Ellipticity and Hypo-ellipticity

intermediate set $\mathcal{O}_1 \subset \mathcal{O}_2 \subset \mathcal{O}$ gives rise to a different solution in $\mathcal{O}_1$. The solution of (1.12.4) in $\mathcal{O}_1$ actually strongly depends on the behavior of L outside $\mathcal{O}_1$.

Moreover, smoothness of u_0 and of the coefficients g^{ij} and b^i in $\mathcal{O}_1$ does not necessarily ensure smoothness of the solution u in $\mathcal{O}_1$. In the degenerate case $(g^{ij}) = 0$, it is easy to construct, in dimension two for example, smooth coefficients b^i on $\mathcal{O}_1 = (-1, +1)^2$ and a solution $u(t, x)$ in $\mathcal{O}_1$ with $u(0, x)$ smooth and compactly supported in $\mathcal{O}_1$ such that $u(t, x)$ is not even continuous for $t = 1$ (the price to pay is that the b^i's are not continuous far from $\mathcal{O}_1$). However, when solving a stochastic differential equation in $\mathcal{O}$, if the coefficients are smooth and if the resulting process does not explode in $\mathcal{O}$, then the resulting law of X_t^x smoothly depends on $x \in \mathcal{O}$, and as a result $P_t u_0$ is smooth for any compactly supported function u_0 in $\mathcal{O}$ (see Sect. B.4, p. 495). But again this is a global result, requiring knowledge of the operator L everywhere where the underlying process lives.

Therefore, there is very little hope of obtaining in a general setting local properties of the solution of (1.12.4). This is the place where a more precise analysis, namely ellipticity or hypo-ellipticity, has to come into play.

As an introduction to this more refined analysis, we consider here only operators L of the type (1.11.1) with smooth coefficients (g^{ij}) and (b^i) on an open set $\mathcal{O} \subset \mathbb{R}^n$ (or an open set in a manifold). Then, as already mentioned earlier (see (1.12.2)), we say that L is elliptic in $\mathcal{O}$ if the matrix $(g^{ij}(x))$ is positive at any point $x \in \mathcal{O}$. In this situation, any solution on $[0, s] \times \mathcal{O}$ of the equation $\partial_t u = \mathrm{L} u$ is $\mathcal{C}^\infty$ on $(0, s) \times \mathcal{O}$, whatever the initial condition u_0, which may even only be measurable.

In the theory of partial differential equations, a linear operator P defined on some open set $\mathcal{O}$ is said to be *hypo-elliptic* if any u such that $Pu \in \mathcal{C}^\infty$ is itself $\mathcal{C}^\infty$. Classical regularity results ensures that if L is elliptic on $\mathcal{O}$, then it is hypo-elliptic. Furthermore $P = \mathrm{L} - \partial_t$ is hypo-elliptic on $(0, \infty) \times \mathcal{O}$ (parabolic hypo-ellipticity).

Hörmander's theory actually allows us to go beyond the elliptic case and to give a precise criterion for hypo-ellipticity in terms of the operator. Assume that L is defined on $\mathcal{O}$ in Hörmander form (1.10.5) as

$$\mathrm{L} = \sum_{j=1}^d Z_j^2 + Z_0$$

where the Z_j's, $0 \leq j \leq d$, are smooth vector fields (which may always be achieved at least locally). Consider then, at any point $x \in \mathcal{O}$, the vector space $\mathcal{V}_p$ generated by the vector fields Z_j and their commutators up to some fixed order $p \geq 1$, that is

$$\mathcal{V}_1 = \mathrm{span}\{Z_j,\, 0 \leq j \leq d\},$$
$$\mathcal{V}_2 = \mathrm{span}\{Z_j, [Z_j, Z_k],\, 0 \leq j, k \leq d\},$$
$$\ldots$$
$$\mathcal{V}_p = \mathrm{span}\{\mathcal{V}_{p-1} \cup \{[Z_j, V],\, V \in \mathcal{V}_{p-1},\, 0 \leq j \leq d\}\}.$$

Then, one main conclusion of the Hörmander theory is that if there exists an integer p such that, for any $x \in \mathcal{O}$, $\mathcal{V}_p = \mathbb{R}^n$, the operator L is hypo-elliptic on $\mathcal{O}$.

When going from L to the parabolic form $P = L - \partial_t$, Z_0 has to be replaced by $Z_0 - \partial_t$. Reserving Z_0 to span the time direction, the following statement holds.

Proposition 1.12.1 *For* L *in Hörmander form on* $\mathcal{O} \subset \mathbb{R}^n$ *as above, set* $\widehat{\mathcal{V}}_1 = \mathrm{span}\{Z_j, 1 \leq j \leq d\}$, *and, for any* $p > 1$,

$$\widehat{\mathcal{V}}_p = \mathrm{span}\{\widehat{\mathcal{V}}_{p-1} \cup \{[Z_j, V], V \in \widehat{\mathcal{V}}_{p-1}, 0 \leq j \leq d\}\}.$$

Then, if there exists an integer p *such that for any* $x \in \mathcal{O}$, $\widehat{\mathcal{V}}_p = \mathbb{R}^n$, *the operator* $L - \partial_t$ *is hypo-elliptic on* $(0, \infty) \times \mathcal{O}$. *In other words, under this condition, any solution of* $\partial_t u = Lu$ *on* $[0, s] \times \mathcal{O}$ *is smooth on* $(0, s) \times \mathcal{O}$, *whatever the behavior of* L *outside* $\mathcal{O}$.

The main difference with the "non-parabolic Hörmander Theorem" is that Z_0 is no longer used to span $\mathcal{V}_1$. For example, on $\mathbb{R}^2$, $L = \partial_x^2 + \partial_y$ satisfies Hörmander's condition but does not fulfill the conditions of Proposition 1.12.1 whereas $L = \partial_x^2 + x \partial_y$ does. In probabilistic terms, under the Hörmander condition of Proposition 1.12.1, the laws of the underlying process X_t^x, $t \geq 0$, have smooth densities for all $x \in \mathcal{O}$. We will come back to these notions and issues in Chap. 3, in particular in Sect. 3.2, p. 137.

1.13 Domains

This section briefly discusses a few observations and hypotheses about domains of Markov semigroups and their infinitesimal generators as they arose in the preceding sections. It is only aimed at giving a flavor of the necessary framework and will be completed in Chap. 3, Sect. 3.3, p. 151, in which precise hypotheses on the various classes of functions necessary to develop the Markov semigroup calculus are carefully discussed.

Throughout this chapter, we are dealing, on some state space $(E, \mathcal{F})$, with a symmetric Markov semigroup $\mathbf{P} = (P_t)_{t \geq 0}$ with infinitesimal generator L with $L^2(\mu)$-domain $\mathcal{D}(L)$ where μ is the invariant and reversible measure. As discussed earlier in Sect. 1.4, to define conversely the semigroup $\mathbf{P}$ from its generator L, it is necessary to know L on a core of its domain $\mathcal{D}(L)$. Usually however, the generator is only given on some set $\mathcal{A}_0$ of functions. Some simple criteria to ensure that a dense vector subspace of functions is a core (in particular in the symmetric case of most interest) are actually available. In Sect. 3.2, p. 137, we show for example that on a complete Riemannian manifold (in particular, on open sets in Euclidean spaces with a suitably chosen metric), $\mathcal{C}^\infty$ functions with compact support always form a core for operators of the form $L = \Delta_\mathfrak{g} - \nabla W \cdot \nabla$ for some smooth potential W on the manifold. Further examples will be discussed later. In general, when dealing with second order operators L with smooth coefficients on an open set in $\mathbb{R}^n$ or on a manifold as given by (1.11.1), every solution of the (heat) equation $\partial_t f = Lf$ will be smooth (as is the case when solving stochastic differential equations, as already mentioned in Sect. 1.12). The solution is actually as smooth as the initial data. If the operator is smooth and elliptic (or only hypo-elliptic), then even the regularity on the initial data is not required, as discussed in the previous section. We could then work on the class of C^∞ functions, but unfortunately it is not included in $\mathbb{L}^2(\mu)$, unless the manifold is compact. These observations lead us to consider an extended algebra $\mathcal{A} \supset \mathcal{A}_0$ on which the operator L is defined as an extension together with the associated carré du champ operator Γ, regardless of integrability properties. In the regular instances, this class $\mathcal{A}$ will be the class of smooth C^∞ functions.

Given a vector space $\mathcal{A}_0$ of real-valued functions f on E, a variety of hypotheses may be considered in order to conveniently work with **P** and L on $\mathcal{A}_0$. A general set of conditions is that $\mathcal{A}_0$ is contained in all $\mathbb{L}^p(\mu)$-spaces, $1 \leq p < \infty$, is dense in the domain $\mathcal{D}(L)$ of L (that is, is a core of the domain), and is stable under products and compositions with C^∞ functions vanishing at 0. In particular, it is an algebra. In addition, it may be imposed that whenever $f, g \in \mathcal{A}_0$, then $f P_t g \in \mathcal{A}_0$ for every $t \geq 0$, or even that $\mathcal{A}_0$ is stable under the action of P_t. Observe that under the latter assumption, $\mathcal{A}_0$ is automatically a core of the domain. The first hypothesis is formulated to include the example of the set $\mathcal{A}_0$ of smooth compactly supported functions on a complete manifold, while the second covers compact manifolds (but also non-compact instances such as, for example, the space of Schwartz functions for the Laplacian on $\mathbb{R}^n$). The algebra $\mathcal{A}$ on which the generator and its carré du champ operator may naturally be considered, corresponding to the class of smooth functions, may be constructed as an extension of $\mathcal{A}_0$.

In some situations, it is not necessary to assume stability by composition with smooth functions (as for the Ornstein-Uhlenbeck semigroup if one wishes to work only with polynomials, cf. Sect. 2.7.1, p. 103). Actually, there are numerous intermediate hypotheses in order to suitably cover one example or another, and it would be rather difficult to cover all the cases in this way. We often choose to work with the maximal hypothesis (of stability by the semigroup), leaving to the reader to adapt proofs and arguments to more specific or complicated cases. Section 3.3, p. 151, supplies the necessary arguments to fully justify this reduction, and describes in more detail the different technical conditions required on a set of functions on which L is given in order to ensure that the described extension can be achieved.

1.14 Summary of Hypotheses (Markov Semigroup)

At this stage of the exposition, summarizing some of the objects and ideas put forward so far, we collect a first set of hypotheses on the datum consisting of a Markov semigroup and its infinitesimal generator which will be in force throughout the book. Recall that Chap. 3 develops the complete framework in which the results of this monograph will be presented, emphasizing the carré du champ operator and the invariant measure, and demonstrating consistency between the two viewpoints.

H1. The *state space* is a measurable space $(E, \mathcal{F})$. All (positive) measures on $(E, \mathcal{F})$ are assumed to be σ-finite. Given a (σ-finite) measure μ on $(E, \mathcal{F})$,

$(E, \mathcal{F}, \mu)$ is assumed to be a *good measure space* in the sense that there is a countable family of sets which generates $\mathcal{F}$ (up to sets of μ-measure 0), and that both the measure decomposition theorem and the bi-measure theorem apply (cf. Sect. 1.2). To make things simpler, and since it is unlikely that the reader would ever need more, it may actually always be assumed that $\mathcal{F}$ is the completion with respect to the measure μ of the Borel σ-field for a topology on E which makes E a complete separable metric space (a so-called Polish space). However, the topology will not be used. Every $\mathbb{L}^p(\mu)$-space $(1 \leq p < \infty)$ is therefore separable. Functions, always real-valued, are classes of functions for the μ-almost everywhere equality, and equalities and inequalities such as $f \leq g$ are always understood to hold μ-almost everywhere. (See p. 7.)

H2. On a good measure space $(E, \mathcal{F}, \mu)$, a *Markov semigroup* $\mathbf{P} = (P_t)_{t \geq 0}$ is a family of operators with the following properties (see pp. 9 and 11):

(i) For every $t \geq 0$, P_t is a linear operator sending bounded measurable functions on $(E, \mathcal{F})$ into bounded measurable functions.
(ii) $P_0 = \text{Id}$, the identity operator (*initial condition*).
(iii) $P_t(\mathbb{1}) = \mathbb{1}$, where $\mathbb{1}$ is the constant function equal to 1 (*mass conservation*).
(iv) If $f \geq 0$, then $P_t f \geq 0$ (*positivity preserving*).
(v) For every $t, s \geq 0$, $P_{t+s} = P_t \circ P_s$ (*semigroup property*).
(vi) For every $f \in \mathbb{L}^2(\mu)$, $P_t f$ converges to f in $\mathbb{L}^2(\mu)$ as $t \to 0$ (*continuity property*).
(vii) For every p, $1 \leq p < \infty$, the operators P_t, $t \geq 0$, extend to every $\mathbb{L}^p(\mu)$ as bounded (contraction) operators.

H3. The measure μ is *invariant* for the semigroup $\mathbf{P} = (P_t)_{t \geq 0}$ in the sense that for any $f \in \mathbb{L}^1(\mu)$, $\int_E P_t f d\mu = \int_E f d\mu$ (Definition 1.2.1).

(vii) is then automatic and the operators P_t, $t \geq 0$, are contractions on every $\mathbb{L}^p(\mu)$, $1 \leq p \leq \infty$

H4. The operators P_t, $t \geq 0$, are represented by *Markov kernels*

$$P_t f(x) = \int_E f(y) p_t(x, dy), \quad x \in E,$$

where $p_t(x, dy)$, $t \geq 0$, is a family of probability kernels on E (that is, for fixed $x \in E$, a probability measure on $(E, \mathcal{F})$ such that for any $A \in \mathcal{F}$, $x \mapsto p_t(x, A)$ is measurable) (see p. 12).

H5. The *Markov process* (or family of processes) $\mathbf{X} = \{X_t^x ; t \geq 0, x \in E\}$ on a probability space $(\Omega, \Sigma, \mathbb{P})$ associated with the Markov semigroup $\mathbf{P} = (P_t)_{t \geq 0}$ satisfies, on bounded or positive measurable functions $f : E \to \mathbb{R}$, the conditional formula

$$P_t f(x) = \mathbb{E}\big(f(X_t^x)\big) = \mathbb{E}\big(f(X_t) \mid X_0 = x\big), \quad t \geq 0, \ x \in E.$$

The semigroup determines the finite-dimensional distributions of the processes $\{X_t^x ; t \geq 0\}$, $x \in E$, by the Chapman-Kolmogorov equations (Sect. 1.3).

1.14 Summary of Hypotheses (Markov Semigroup)

H6. The Markov semigroup $\mathbf{P} = (P_t)_{t \geq 0}$ is *symmetric* with respect to the measure μ, or μ is *reversible* for $\mathbf{P} = (P_t)_{t \geq 0}$, in the sense that for any pair (f, g) of functions in $\mathbb{L}^2(\mu)$,

$$\int_E f P_t g \, d\mu = \int_E g P_t f \, d\mu$$

(Definition 1.6.1).

H7. The *domain* $\mathcal{D}(L)$ of the *infinitesimal generator* L of the semigroup $\mathbf{P} = (P_t)_{t \geq 0}$ is the (linear) space of functions $f \in \mathbb{L}^2(\mu)$ such that the limit

$$\lim_{t \to 0} \frac{1}{t}(P_t f - f)$$

exists in $\mathbb{L}^2(\mu)$. This limit, denoted by Lf, defines the infinitesimal (Markov) generator (or operator) L. The domain $\mathcal{D}(L)$ is always a dense subspace of $\mathbb{L}^2(\mu)$ (cf. Definition 1.4.1). On this domain, the semigroup $(P_t)_{t \geq 0}$ solves the *heat equation* (with respect to L),

$$\partial_t P_t = L P_t = P_t L.$$

For a symmetric semigroup, the generator L is self-adjoint on its domain.

H8. The domain $\mathcal{D}(\mathcal{E})$ of the *Dirichlet form* $\mathcal{E}$ is the space of functions f in $\mathbb{L}^2(\mu)$ such that the limit

$$\lim_{t \to 0} \frac{1}{t} \int_E f(f - P_t f) d\mu$$

exists. This limit, denoted by $\mathcal{E}(f, f) = \mathcal{E}(f)$, defines the Dirichlet form $\mathcal{E}$ of the semigroup $\mathbf{P} = (P_t)_{t \geq 0}$ with domain $\mathcal{D}(\mathcal{E})$. When $f \in \mathcal{D}(L)$,

$$\mathcal{E}(f) = \int_E f(-Lf) d\mu$$

and $\mathcal{D}(L) \subset \mathcal{D}(\mathcal{E}) \subset \mathbb{L}^2(\mu)$ (cf. Sect. 1.7).

H9. The *carré du champ operator* Γ associated with the Markov generator L is a bilinear operator defined on an algebra $\mathcal{A}$ of functions included in the domain $\mathcal{D}(L)$ by

$$\Gamma(f, g) = \frac{1}{2}\big[L(fg) - f Lg - g Lf\big], \quad (f, g) \in \mathcal{A} \times \mathcal{A}$$

(Definition 1.4.2). One always has $\Gamma(f) = \Gamma(f, f) \geq 0$ and

$$\int_E \Gamma(f) d\mu = -\int_E f L f \, d\mu = \mathcal{E}(f).$$

Various hypotheses may hold on the vector space $\mathcal{A}$ of functions on which this carré du champ operator is defined (see Sect. 1.13).

H10. The generator L satisfies the *diffusion property* which states that for any family $(f_1, \ldots, f_k)$ of functions in $\mathcal{A}$ on which the carré du champ operator Γ is defined, and any smooth function $\Psi : \mathbb{R}^k \to \mathbb{R}$ such that $\Psi(f_1, \ldots, f_k)$ belongs to $\mathcal{D}(L)$,

$$L\big(\Psi(f_1, \ldots, f_k)\big) = \sum_{i=1}^{k} \partial_i \Psi(f) L f_i + \sum_{i,j=1}^{k} \partial_{ij}^2 \Psi(f) \Gamma(f_i, f_j).$$

Equivalently, the carré du champ operator Γ is, in each argument, a first order differential operator. The diffusion property translates the fact that the trajectories of the associated Markov process $\mathbf{X} = \{X_t^x \, ; t \geq 0, x \in E\}$ are in some sense continuous (cf. Sect. 1.11).

1.15 Working with Markov Semigroups

This section reviews a variety of basic operations which may be performed with Markov semigroups and generators. These operations are mostly presented on some specific, but important, illustrative examples which may then be adapted to more general settings if necessary. Depending on the operation, the various sections emphasize more the analytic, probabilistic or geometric aspect.

According to the preceding summary, we have a symmetric Markov semigroup $\mathbf{P} = (P_t)_{t \geq 0}$ on a state space $(E, \mathcal{F})$, with invariant and reversible measure μ and infinitesimal generator L with $\mathbb{L}^2(\mu)$-domain $\mathcal{D}(L)$. The semigroup and its generator yield the associated carré du champ operator Γ and Dirichlet form $\mathcal{E}$. As usual, $\mathbf{X} = \{X_t^x \, ; t \geq 0, x \in E\}$ denotes the associated Markov process for the probabilistic interpretation. If necessary, we use some, or all, of the hypotheses alluded to at the end of Sect. 1.13 on some suitable subset $\mathcal{A}_0 \subset \mathcal{D}(L)$ of "nice" functions, although these will usually be clear from the given context (here we retain our somewhat informal level of discussion).

1.15.1 Change of Coordinates

Given a Markov semigroup $(P_t)_{t \geq 0}$, or its associated Markov process $(X_t)_{t \geq 0}$, on a state space E, a bi-measurable bijection B between E and another state space E' yields a Markov process $(B(X_t))_{t \geq 0}$ on E', with Markov semigroup

$$P'_t f = \big(P_t(f \circ B)\big) \circ B^{-1}, \quad t \geq 0,$$

not really different from the original one. The properties of semigroups which we would like to investigate should actually be invariant under such transformations. There are indeed intrinsic quantities in a given Markov generator which are independent of the choice of such representations. Properties of semigroups are often

1.15 Working with Markov Semigroups

easier to detect in a given system of coordinates, so that it is important to perform such changes of coordinates. For example, when working on $\mathbb{R}^n$, or some open set in $\mathbb{R}^n$, one may consider diffeomorphisms B (that is, bijective differentiable maps with differentiable inverses). Such a diffeomorphism is classically called a change of variables. Consider its action on a diffusion generator

$$\mathrm{L}f = \sum_{i,j=1}^{n} g^{ij}(x)\partial^2_{ij} f + \sum_{i=1}^{n} b^i(x)\partial_i f$$

as in (1.11.1). To describe the image generator $\mathrm{L}'f = (\mathrm{L}(f \circ B)) \circ B^{-1}$, denote by $(y_1,\ldots,y_n) = B(x_1,\ldots,x_n)$ the coordinates of B. These functions are the new coordinates with respect to which the operator will be given and

$$\mathrm{L}'f = \sum_{i,j=1}^{n} \widehat{g}^{ij}(y)\partial^2_{ij} f + \sum_{i=1}^{n} \widehat{b}^i(y)\partial_i f$$

where

$$\widehat{g}^{ij}(y) = \sum_{k,\ell=1}^{n} J^i_k(y) J^j_\ell(y) g^{k\ell}(B^{-1}(y))$$

with J^i_j the Jacobian of the transformation

$$J^i_j(y) = \left(\frac{\partial y^i}{\partial x^j}\right)(B^{-1}(y)).$$

The formula for $\widehat{b}$ is however a bit more complicated since it involves both b and the derivatives of the g^{ij}'s. There are expressions which are better behaved than others (such as the matrix $\mathfrak{g} = (g^{ij}(x))$ or the vector fields, and more generally tensors, see below) under a change of variables. It is important here to notice that, from this point of view, the first order part $\sum_{i=1}^{n} b^i \partial_i$ of L is not a vector field. However, in the decomposition (1.11.10) $\mathrm{L} = \Delta_\mathfrak{g} + Z$, the term Z is a vector field. What does not behave properly in this expression is the second order term $\sum_{i,j=1}^{n} g^{ij}(x)\partial^2_{ij}$, which may appear as $\sum_{i,j=1}^{n} \widehat{g}^{ij}(y)\partial^2_{ij} + \sum_{i=1}^{n} \widehat{b}^i \partial_i$ after a change of coordinates. For example, if $\Delta = \partial^2_x + \partial^2_y$ is the usual Laplacian on $\mathbb{R}^2$ for which $\mathfrak{g} = \mathrm{Id}$ and $b=0$, in polar coordinates it takes the form

$$\partial^2_r + \frac{1}{r^2}\partial^2_\theta + \frac{1}{r}\partial_r.$$

In particular $b=0$ in the first form but $b \neq 0$ in the second, so that neither the vector b, nor the first order operator $\sum_{i=1}^{n} b^i \partial_i$, are intrinsic.

As already mentioned, many ideas in the study of diffusion generators are taken from Riemannian geometry, in which precisely invariants under the action of changes of coordinates are investigated. This is in particular the case for elliptic operators, for which the diffusion matrix $\mathfrak{g} = \mathfrak{g}(x) = (g^{ij}(x))_{1 \leq i,j \leq n}$ is, at every point x, symmetric positive-definite (see (1.12.2)). A Riemannian metric is thus naturally associated with such an elliptic operator, and there is a way to work with and to identify invariants under changes of coordinates in this geometry. The natural framework for differential operators and invariants under changes of coordinates is the setting of differentiable manifolds. The reader will find in Appendix C the basic geometric notions necessary for the further developments.

Moreover, quantities invariant under a change of coordinates, called intrinsic, will be of use in the analysis of semigroups. Among these quantities, some are easier to manipulate, as mentioned above, namely tensors (cf. Sect. C.3, p. 504, for a complete description of tensor quantities). In a change of coordinates, a tensor is multiplied by the Jacobian matrix as much as there are top indices in the tensor and by its inverse as much as there are bottom indices. In the definition of a diffusion operator, $\mathfrak{g} = (g^{ij})_{1 \leq i,j \leq n}$ is a tensor (twice covariant) but $b = (b^i)_{1 \leq i \leq n}$ is not.

1.15.2 Time Change

The time change from t to ct ($c > 0$) on a Markov semigroup $\mathbf{P} = (P_t)_{t \geq 0}$ clearly corresponds to a change of infinitesimal generator from L to cL. We investigate here changes of L to $c(x)$L for some function $c(x)$ that also correspond to time changes, although of more complicated forms, but which have a natural probabilistic interpretation.

Observe to start with that only positive functions $c(x)$ have to be considered since the carré du champ operator associated to cL is $c\Gamma$, which has to be positive. For the discussion here, we further restrict to strictly positive bounded functions $0 < a \leq c(x) \leq b < \infty$ and to diffusion operators L (so that the associated Markov processes $\mathbf{X}$ are sample continuous). While perhaps not the most interesting situation, once the basic principle is well understood in this case, one can easily adapt it to the general case.

Now, given the Markov process $(X_t)_{t \geq 0}$ (on some probability space $(\Omega, \Sigma, \mathbb{P})$) with generator L, starting at some (fixed) point $x \in E$, set for every $t \geq 0$,

$$C_t = \int_0^t \frac{1}{c(X_s)} \, ds.$$

This is an increasing continuous and differentiable process, with derivative $C_t' = \frac{1}{c(X_t)}$. For each fixed $\omega \in \Omega$, $C_t = C_t(\omega)$ is a bijective map from $\mathbb{R}_+$ into $\mathbb{R}_+$. Its (generalized) inverse map A_t, $t \geq 0$, is given by

$$A_t = \inf \{ s > 0 \, ; \, C_s \geq t \}.$$

It is easily checked that, for every $t \geq 0$,

$$A'_t = \frac{1}{C'_{A_t}} = c(X_{A_t}).$$

Since the process $(C_t)_{t \geq 0}$ is adapted to the natural filtration $(\mathcal{F}_t)_{t \geq 0}$ of $(X_t)_{t \geq 0}$, A_t is, for every $t \geq 0$, a stopping time of this filtration. The process $\widehat{X}_t = X_{A_t}$, $t \geq 0$, is a Markov process (due to the strong Markov property applied to the stopping times A_t). For any function f in the domain of L, by Itô's formula,

$$df(X_t) = dM_t^f + \mathrm{L}f(X_t)dt,$$

and, by composition,

$$df(X_{A_t}) = A'_t \big[dM_{A_t}^f + \mathrm{L}f(X_{A_t})dt\big].$$

The latter may be rewritten as

$$df(\widehat{X}_t) = d\widehat{M}_t^f + c(\widehat{X}_t)\mathrm{L}f(\widehat{X}_t)dt.$$

Here $(M_t^f)_{t \geq 0}$ is a martingale for the filtration $(\mathcal{F}_t)_{t \geq 0}$, and $(\widehat{M}_t^f)_{t \geq 0}$ is a martingale for the filtration $(\mathcal{F}_{A_t})_{t \geq 0}$. Hence, $(\widehat{X}_t)_{t \geq 0}$ is a Markov process for the generator $c(x)\mathrm{L}$. Note that when c is constant, $A_t = ct$ and we are back to the simple linear change of time.

1.15.3 Products

This operation will often be referred to in this book as the tensorization procedure. Given two independent families of Markov processes $\mathbf{X} = \{X_t^x \, ; t \geq 0, x \in E\}$ and $\mathbf{Y} = \{Y_t^y \, ; t \geq 0, y \in F\}$ on possibly different state spaces E and F, the pair (X_t^x, Y_t^y), $t \geq 0$, $(x, y) \in E \times F$, is again a Markov process on the product space $E \times F$. If $\mathbf{P} = (P_t)_{t \geq 0}$ and $\mathbf{Q} = (Q_t)_{t \geq 0}$ denote their respective semigroups, the product semigroup $\mathbf{R} = \mathbf{P} \otimes \mathbf{Q}$ is given by the tensor products $R_t = P_t \otimes Q_t$, $t \geq 0$. More precisely, the product semigroup $(R_t)_{t \geq 0}$ acts on products of functions $h(x, y) = f(x)g(y)$, $f : E \to \mathbb{R}$, $g : F \to \mathbb{R}$, as

$$R_t h(x, y) = P_t f(x) Q_t g(y), \quad t \geq 0, \ x \in E, \ y \in F.$$

The semigroup $(R_t)_{t \geq 0}$ may also be defined via the kernels of the representation (1.2.4) as

$$r_t\big((x, y), (dz, du)\big) = p_t(x, dz) \otimes q_t(y, du)$$

for every $t \geq 0$ and $(x, y) \in E \times F$ (with the obvious notation). In case the kernels of $(P_t)_{t \geq 0}$ and $(Q_t)_{t \geq 0}$ admit respective densities $p_t(x, z)$ and $q_t(y, u)$, $t > 0$,

$(x, z) \in E \times E$, $(y, u) \in F \times F$, with respect to measures m and n on E and F following Definition 1.2.4, then the product semigroup $(R_t)_{t \geq 0}$ has kernel densities $p_t(x, z) q_t(y, u)$ with respect to the product measure $m(dz) \otimes n(du)$.

The invariant measure of the product semigroup is the product of the invariant measures. It is reversible as soon as each factor is. The generator of $(R_t)_{t \geq 0}$ is the sum of the two generators. That is, if L and M are the respective generators of $(P_t)_{t \geq 0}$ and $(Q_t)_{t \geq 0}$, the generator $L \oplus M$ acts on a function $h(x, y)$ as $L_x h + M_y h$ where $L_x h$ denotes the action on the x variable, y being fixed, and conversely for $M_y h$. In particular, the Laplacian of $\mathbb{R}^n$ is the sum of the Laplacians on the coordinates, and the (heat or Brownian) semigroup on $\mathbb{R}^n$ is the product of the one-dimensional semigroups along each coordinate (and Brownian motion on $\mathbb{R}^n$ is an n-tuple of independent one-dimensional Brownian motions).

This book will often be concerned with tensorization properties of various functional inequalities, and will study how those functional inequalities behave under this product procedure. This is an important issue when dealing with the behavior of $\mathbf{P}^{\otimes n}$ for large n, which arises naturally in many areas such as statistical mechanics, infinite particle systems, probability theory, statistics and non-linear evolution equations.

1.15.4 Quotients, Commuting Vector Fields and Projections

Quite often, given a semigroup $\mathbf{P} = (P_t)_{t \geq 0}$ on a state space E with Markov generator L and associated Markov process $\mathbf{X} = \{X_t^x ; t \geq 0, x \in E\}$, there is a map $h : E \to F$ such that for every $t \geq 0$ and every bounded measurable function $g : F \to \mathbb{R}$, $P_t(g \circ h)$ is a function of h, call it $(Q_t g) \circ h$. In this case, the family of operators $\mathbf{Q} = (Q_t)_{t \geq 0}$ is again a Markov semigroup, with state space F, and the push-forward process $(h(X_t))_{t \geq 0}$ is a Markov process with semigroup $\mathbf{Q}$. In order for this property to hold, it is enough that $L(g \circ h) = (\widehat{L}g) \circ h$ for sufficiently many suitable functions $g : F \to \mathbb{R}$ where $\widehat{L}$ is an operator on F. The semigroup $\mathbf{Q} = (Q_t)_{t \geq 0}$ then has Markov generator $\widehat{L}$. This new semigroup $\mathbf{Q}$ and generator $\widehat{L}$ may then be considered as the *quotient* semigroup and generator of $\mathbf{P} = (P_t)_{t \geq 0}$ and L respectively via the projection map h. In this setting, if the semigroup $\mathbf{P}$ has invariant measure μ, then the image semigroup $\mathbf{Q}$ has invariant measure ν where ν is the image measure of μ induced by the map h. This procedure may sometimes be used to identify image measures, in particular when the operator $\widehat{L}$ is easy to identify.

Such a situation occurs in particular for diffusion operators when h is real-valued. In this case, it suffices that Lh be a function of h, as well as $\Gamma(h)$. Indeed, if $Lh = b(h)$ and $\Gamma(h) = a(h)$, then by the diffusion property (1.11.3) and for a sufficiently regular function $g : \mathbb{R} \to \mathbb{R}$,

$$L(g \circ h) = a(h) g''(h) + b(h) g'(h)$$

1.15 Working with Markov Semigroups

so that

$$\widehat{L}g = a(x)g'' + b(x)g'.$$

The same method applies when h is vector-valued via a multi-dimensional change of variables formula.

Another situation which often occurs is where there are vector fields Z_j, $1 \leq j \leq k$, commuting with L, and such that a function f of the domain $\mathcal{D}(L)$ is of the form $g \circ h$ if and only if $Z_j f = 0$, $1 \leq j \leq k$. The space F is then viewed the quotient of E by the action of these vector fields and the function h is the projection on this quotient space. When there is just one such vector field Z, F is then simply a parametrization of the integral curves of this vector field. If $Z_j = Z$ commutes with L, then Z also commutes with P_t for every $t \geq 0$, and if $Zf = 0$, then $ZP_t f = 0$. Indeed, taking the derivative in $s \in [0, t]$ of $P_s Z(P_{t-s} f)$ yields

$$P_s(LZ - ZL)P_{t-s} f = 0.$$

Therefore, identifying the value at 0 with the value at t, $P_t Z f = Z P_t f$. Hence the semigroup $(P_t)_{t \geq 0}$ preserves the functions vanishing on the vector field Z. Moreover, if Z is a vector field as above, let $(T_t)_{t \geq 0}$ be the flow solution of the differential equation $\partial_t T_t f = Z T_t f$ (called the exponential of the vector field Z). The flow $(T_t)_{t \geq 0}$ is simply the semigroup—in fact here a group, since there is no need for the restriction $t \geq 0$—with generator Z. Now, if Z with exponential $(T_t)_{t \geq 0}$ commutes with the generator L with semigroup $(P_t)_{t \geq 0}$, then P_t and T_t commute for every $t \geq 0$ (simply repeat the previous procedure by taking the derivative of $T_u P_t T_{s-u} f$ for $u \in [0, s]$).

As an illustration, a (smooth) function f in $\mathbb{R}^n$ is radial if and only if for all infinitesimal rotations $\Theta_{ij} = x^i \partial_j - x^j \partial_i$, $1 \leq i, j \leq n$, $\Theta_{ij} f = 0$. It is clear that these vector fields Θ_{ij} commute with the (standard) Laplace operator Δ on $\mathbb{R}^n$. Thus the heat semigroup $(P_t)_{t \geq 0}$ preserves radial functions, and a new semigroup may then be constructed by projection on the quotient space $\mathbb{R}_+$, called the Bessel semigroup (see Sect. 2.4.2, p. 94). Similarly, f only depends on $(x_1, \ldots, x_{n-1})$ if and only if $\partial_n f = 0$, and the image of the Brownian motion in $\mathbb{R}^n$ is again a Markov process in $\mathbb{R}^{n-1}$ (in this case simply Brownian motion in $\mathbb{R}^{n-1}$). This principle will be further illustrated in the next chapter in the case of the heat or Brownian semigroup in Euclidean space, but also on the sphere or on the hyperbolic space in which the Laplace operator admits a number of symmetries (see Sects. 2.2 and 2.3, pp. 81 and 88).

Another way to form a semigroup on some image space F can be illustrated using the carré du champ operator Γ and the reversible measure μ of the semigroup $(P_t)_{t \geq 0}$ (that is, on the Markov Triple structure (E, μ, Γ), which will be emphasized later). Namely, let ν be the image measure of μ by a measurable map $h : E \to F$, and define a new carré du champ operator on functions g on F by

$$\Gamma_1(g)(y) = \mathbb{E}\big(\Gamma\big(g(h)\big) \mid h = y\big), \quad y \in F,$$

the conditional expectation being taken under the law μ. (It is possible to give a meaning to this expression even when μ is not a probability measure via a decomposition of μ with respect to ν.) It is easily checked that Γ_1 is indeed a carré du champ operator, and if Γ is of diffusion-type, then so is Γ_1. In contrast to the previous procedure, this construction does not require any special properties of the function h with respect to L. However, the resulting generator is in general difficult to compute explicitly. This approach will be used repeatedly throughout this work when dealing with a Lipschitz map $h : E \to \mathbb{R}$ to transfer functional inequalities or concentration tail estimates of the space (E, μ) onto $\mathbb{R}$ (equipped with the image measure), and will be the basis for most tail estimates for Lipschitz functions discussed in the following chapters.

1.15.5 Adding a Vector Field

Adding a vector field is a simple analytic operation which however, from the probabilistic point of view, is somewhat delicate. It takes the form of the so-called *Girsanov transformation*. The basic principle is outlined here only for diffusion processes, and even for simplicity for processes on $E = \mathbb{R}^n$ with generator given in the Hörmander form (1.10.5)

$$L = \frac{1}{2} \sum_{j=1}^{d} Z_j^2 + Z_0$$

for vector fields $Z_0, Z_1, \ldots, Z_d$ (with the probabilistic convention). As we have seen in Sect. 1.10.2, the associated Markov process $\mathbf{X} = \{X_t^x ; t \geq 0, x \in \mathbb{R}^n\}$ may be described in this case via the stochastic differential equation (1.10.6).

We would like to add to L a vector field Z, for simplicity of the form

$$Z = \frac{1}{2} \sum_{j=1}^{d} a^j(x) Z_j$$

where the a_j's, $1 \leq j \leq d$, are smooth and bounded functions (beware that Z_0 is excluded in the previous sum). The new operator is $\widehat{L} = L + Z$, and the task is to compare the semigroup $(\widehat{P}_t)_{t \geq 0}$ with generator $\widehat{L}$ with the semigroup $(P_t)_{t \geq 0}$ with generator L. From a probabilistic viewpoint, the idea is to consider the law of the process $(X_t)_{t \geq 0}$ as a measure on the space of continuous functions on $\mathbb{R}_+$ with values in $\mathbb{R}^n$. Then, the law of the process $(\widehat{X})_{t \geq 0}$ with generator $\widehat{L}$ has a density with respect to the law of $(X_t)_{t \geq 0}$ given, at each time t, by

$$N_t = \exp\left(\int_0^t \sum_{j=1}^{d} a^j(X_s) dB_s^j - \frac{1}{2} \int_0^t \sum_{j=1}^{d} (a^j)^2(X_s) ds \right),$$

1.15 Working with Markov Semigroups

where $(B^j)_{1 \leq j \leq d}$ are the independent Brownian motions in $\mathbb{R}^n$ which appear in the stochastic differential equation (1.10.6). For a (sketch of a) proof, observe that the process $(N_t)_{t \geq 0}$ thus defined is a martingale by Itô's formula and that

$$d(f(X_t)N_t) = f(X_t)\,dN_t + N_t \bigg[dM_t^f + Lf(X_t)dt$$

$$+ \frac{1}{2} \sum_{i,j=1}^{d} \partial_j f(X_t) a^i(X_t) Z_i^j\, dt \bigg].$$

Therefore,

$$\mathbb{E}_x \bigg(N_t \bigg[f(X_t) - \int_0^t \widehat{L}f(X_t)dt \bigg] \bigg) = f(x)$$

where the notation $\mathbb{E}_x$ denotes the expectation conditional on $X_0 = x$. This shows that multiplying the law of X_t by the density N_t solves the martingale problem (1.4.6) associated to $\widehat{L}$. In fact, defining for every suitable function f, every $t \geq 0$ and $x \in \mathbb{R}^n$,

$$\widehat{P}_t f(x) = \mathbb{E}_x(f(X_t)N_t),$$

we have $\partial_t \widehat{P}_t f = \widehat{P}_t \widehat{L} f$ so that $(\widehat{P}_t)_{t \geq 0}$ is indeed the semigroup with generator $\widehat{L}$. There are of course many different ways to reach this conclusion. Observe that in order for this conclusion to hold that the vector field Z be a linear combination of the vector fields Z_j, $1 \leq j \leq d$. This condition is not restrictive when the operator L is elliptic, but becomes a serious restriction when it is not.

1.15.6 Adding a Potential

It is often the case that one has to look at differential operators of the form

$$\widehat{L}f = Lf + Vf,$$

where L is the generator of a Markov semigroup and Vf is multiplication by some non-constant function V. This happens in particular when solving Schrödinger-type equations. Such functions V are then called *potentials*. The heat equation $\partial_t f = \widehat{L}f$, and the associated semigroup $(\widehat{P}_t)_{t \geq 0}$, are then similarly of interest. The latter semigroup is still positivity preserving, but no longer satisfies $\widehat{P}_t(\mathbb{1}) = \mathbb{1}$. This is easy to understand in the simplest example of finite spaces, since then the matrix $L(x, y)$ is replaced by the matrix $\widehat{L}(x, y)$ which is equal to $L(x, y)$ when $x \neq y$ while $\widehat{L}(x, x) = L(x, x) + V(x)$.

As in the preceding sections, the aim is to compare the semigroup $(\widehat{P}_t)_{t \geq 0}$ with generator $\widehat{L}$ with the semigroup $(P_t)_{t \geq 0}$ with generator L. When V is constant, we have $\widehat{P}_t = e^{tV} P_t$, $t \geq 0$. Hence $\widehat{P}_t(\mathbb{1}) \leq \mathbb{1}$ (that is $(\widehat{P}_t)_{t \geq 0}$ is sub-Markov) as soon as

$V \leq 0$ (and this will remain the case for every non-constant potential $V \leq 0$). These sub-Markov semigroups share a lot of common features with the semigroups killed at the boundary such as those described later in Sect. 2.4, p. 92, and in Sect. 2.6, p. 97. For a general (but bounded) potential V, the easiest way to represent $(\widehat{P}_t)_{t \geq 0}$ starting from $(P_t)_{t \geq 0}$ is to use the same trick as for the Girsanov transformation, that is to describe $(\widehat{P}_t)_{t \geq 0}$ from the law of the Markov process $(X_t)_{t \geq 0}$ with generator L. This is done this time with the help of the *Feynman-Kac formula*. Working again for simplicity on $E = \mathbb{R}^n$, if

$$A_t = \exp\left(\int_0^t V(X_s)ds\right), \quad t \geq 0,$$

then, for $t \geq 0$ and $x \in \mathbb{R}^n$,

$$\widehat{P}_t f(x) = \mathbb{E}_x\big(f(X_t)A_t\big).$$

This is much easier to see than for the Girsanov transformation, since in this case

$$d\big(f(X_t)A_t\big) = A_t\big[dM_t^f + Lf(X_t)dt\big] + f(X_t)A_t V(X_t)dt.$$

Therefore, for every $t \geq 0$ and $x \in \mathbb{R}^n$,

$$\widehat{P}_t f(x) = f(x) + \mathbb{E}_x\left(\int_0^t A_s \widehat{L}(X_s)ds\right),$$

and hence $\partial_t \widehat{P}_t f = \widehat{P}_t \widehat{L} f$.

If the starting operator L is symmetric with respect to some measure μ, so is the new operator $\widehat{L}$. But this time, even for a probability measure μ, the constant function $\mathbb{1}$ is no longer an eigenvector. If $V \leq 0$, then the spectrum still lies in $(-\infty, 0)$. Very often, it is in fact in $(-\infty, \lambda_0)$ for some $\lambda_0 < 0$. When this lower bound of the spectrum of $-\widehat{L}$ is in fact an eigenvalue, it corresponds to (at least) one eigenvector U_0 which is positive. This eigenvector is called the *fundamental state* (or *ground state*) of the system (corresponding to the Perron-Frobenius eigenvector in the finite Markov chain setting of Sect. 1.9). When it is strictly positive everywhere, then the transformation of $\widehat{P}_t$, $t \geq 0$, into

$$R_t f = e^{-\lambda_0 t}\frac{1}{U_0}\widehat{P}_t(U_0 f), \quad t \geq 0,$$

satisfies $R_t(\mathbb{1}) = \mathbb{1}$, and therefore defines a new Markov semigroup (often called the ground state transform of the previous one). Hence, the study of such Markov semigroups may be reduced to the study of Markov semigroups provided that the ground state U_0 may be identified, which is not an easy task in general (see also Sect. 1.15.8 below for more details).

1.15.7 *Removing a Gradient Field*

Removing a gradient field $\Gamma(W, \cdot)$ $(= \nabla W \cdot \nabla)$ is not just adding the vector field with opposite sign. The transformation described here is based on the same principle as the h-transform and turns out to be very useful. Thus assume that L_1 is a diffusion generator written in the form

$$L_1 f = L f - \Gamma(W, f)$$

on suitable functions f, where L is another generator with carré du champ operator Γ. It is easily seen that L_1 also has Γ as carré du champ operator. Applying L_1 to $g = e^{W/2} f$ yields $L_1 g = e^{W/2}(L f - W_1 f)$ with

$$W_1 = -\frac{LW}{2} + \frac{\Gamma(W)}{4}.$$

In other words, denoting by M multiplication by $e^{W/2}$,

$$M^{-1} L_1 M = L - W_1.$$

The two operators L_1 and $L - W_1$ are conjugate to each other. In particular, for the respective associated semigroups $(P_t)_{t \geq 0}$ and $(Q_t)_{t \geq 0}$,

$$Q_t f = e^{-W/2} P_t(e^{W/2} f), \qquad P_t f = e^{W/2} Q_t(e^{-W/2} f), \quad t \geq 0.$$

Moreover, if P_t has kernel densities $p_t(x, y)$, $t > 0$, $(x, y) \in E \times E$, with respect to some measure m, that is

$$P_t f(x) = \int_E f(y) \, p_t(x, y) dm(y),$$

then Q_t has a kernel density with respect to m given by

$$e^{-(W(x) - W(y))/2} p_t(x, y), \quad (x, y) \in E \times E.$$

Furthermore, if L is symmetric with respect to μ, then L_1 is symmetric with respect to $d\mu_1 = e^{-W} d\mu$, and the map $f \mapsto e^{W/2} f$ is an isometry between $\mathbb{L}^2(\mu_1)$ and $\mathbb{L}^2(\mu)$. In this case, the operators L_1 and $L - W_1$ have the same spectral properties and if the spectrum of one of them is discrete, so is the spectrum of the other, with the same eigenvalues, the eigenvectors being in correspondence via the above map.

Very often, the latter may be applied in $\mathbb{R}^n$ with $L_1 = \Delta - \nabla W \cdot \nabla$. This operator shares the same spectral properties as the Schrödinger operator $\Delta - W_1$. The analysis of such operators is quite well-known, in particular there is a lot of classical analysis concerning their spectral properties, all of which carries over to the study of the operators $\Delta - \nabla W \cdot \nabla$.

1.15.8 h-Transform

This subsection deals with another current transformation of a Markov semigroup known as the *h-transform* in potential theory. Given a Markov semigroup $\mathbf{P} = (P_t)_{t \geq 0}$ on a state space E, let $h : E \to \mathbb{R}$ be a strictly positive measurable function. Under appropriate conditions on h, consider the new semigroup given by

$$P_t^h f = \frac{1}{h} P_t(fh), \quad t \geq 0,$$

on a suitable class of functions f. If P_t is represented by a kernel $p_t(x, dy)$ as in (1.2.4), then P_t^h is represented by the kernel

$$\frac{h(y)}{h(x)} p_t(x, dy).$$

If $(P_t)_{t \geq 0}$ has generator L, then $(P_t^h)_{t \geq 0}$ has generator

$$\mathrm{L}^h f = \frac{1}{h} \mathrm{L}(hf).$$

As an illustration, the h-transform can be used to describe the Fokker-Planck operator (see Sect. 1.5) in the symmetric setting. For example, if $\mathrm{L} = \Delta_\mathfrak{g} - \nabla W \cdot \nabla$ on a weighted Riemannian manifold with Laplace-Beltrami operator $\Delta_\mathfrak{g}$ and Riemannian measure $\mu_\mathfrak{g}$, then for $h = e^W$, $\mathrm{L}^h = \mathrm{L}^*$ where L^* is the adjoint of L in $\mathbb{L}^2(\mu_\mathfrak{g})$.

But this is not the main use of the h-transform. The aim is to continue working with the generator of a Markov semigroup, and L^h is actually such a generator if h is *harmonic* in the sense that $\mathrm{L}h = 0$. In this case, and provided L is a diffusion operator according to Sect. 1.11,

$$\mathrm{L}^h f = \mathrm{L} f + 2 \Gamma(\log h, f)$$

for every suitable function f.

There is a similar representation with strictly positive solutions of $\mathrm{L}h = \lambda h$ ($\lambda \in \mathbb{R}$) in which case $\mathrm{L}^h f = \mathrm{L} f + 2 \Gamma(\log h, f) + \lambda f$ corresponding to the Markov semigroup

$$P_t^h = e^{-\lambda t} \frac{1}{h} P_t(hf), \quad t \geq 0.$$

In general, it is difficult to construct such strictly positive harmonic functions on the whole space (for example on $\mathbb{R}^n$, any strictly positive harmonic function is constant when $\mathrm{L} = \Delta$), but it may be possible when restricted to suitable open sets (for example on $\mathbb{R}$, the function $h(x) = x$ on $(0, \infty)$). Now, for some classes of harmonic functions on a subset A of the state space (the minimal strictly positive ones), this new semigroup corresponds to the Markov process conditioned to stay in A. The associated process is known as the *h-process* associated to the original Markov process with semigroup **P**.

It is beyond the scope of this short section to develop the full theory of h-processes (which is indeed quite complicated and for which we refer the reader to the appropriate literature). However, we may at least give a flavor of it by comparison with Markov chains (cf. Sect. 1.9). Indeed, consider a Markov chain $(X_k)_{k \in \mathbb{N}}$ (on a finite set E, say) with transition matrix $P = (P(x, y))_{(x,y) \in E \times E}$, and assume for simplicity that for any $(x, y) \in E \times E$, $P(x, y) > 0$. The quantity $P(x, y)$ represents the probability that the chain will jump from x to y. Now restrict the matrix $P(x, y)$ to $(x, y) \in A \times A$ where $A \subset E$, and denote by P^A this restricted square matrix. The (unique strictly positive) Perron-Frobenius eigenvector U of P^A satisfies

$$\sum_{y \in A} P^A(x, y) U(y) = \lambda U(x)$$

for all $x \in A$, corresponding to the (real) eigenvalue λ with maximum modulus. Then, consider the new Markov matrix on $A \times A$ defined by

$$Q^A(x, y) = \frac{1}{\lambda U(x)} P^A(x, y) U(y), \quad (x, y) \in A \times A.$$

Given $x \in A$, for $k < \ell$, consider the law of $(X_1, \ldots, X_k)$ conditioned to the event $\{X_i \in A, 1 \leq i \leq \ell\}$. This is not the law of a Markov chain, but it becomes so if $\ell \to \infty$. In the limit, this law is the law of the Markov chain with transition matrix Q^A. This appears as a consequence of some elementary (but rather tedious) direct computations, which, in this case, can be made explicit from the matrix P. The h-transform is then simply the transfer of this principle to the continuous time case, in the larger setting of Markov semigroups. Notice that in this case the h-transform is the same operation as the ground phase state transformation described in Sect. 1.15.6.

1.15.9 Subordination

Subordination is another way to produce new semigroups from a given one, through time averaging. Let $(P_t)_{t \geq 0}$ be a Markov semigroup on E with generator L. For any probability measure ν on $\mathbb{R}_+$, define a new Markov operator by setting

$$P_\nu = \int_0^\infty P_t \, d\nu(t).$$

The integral may be considered as an integral operator, or more naively as acting on bounded measurable functions $f : E \to \mathbb{R}$ as $P_\nu f = \int_0^\infty P_t f \, d\nu(t)$.

From a probabilistic point of view, if $(X_t)_{t \geq 0}$ is the Markov process associated with the semigroup $(P_t)_{t \geq 0}$, consider a random variable T with values in $\mathbb{R}_+$ with law ν, independent of the process $(X_t)_{t \geq 0}$. Then P_ν is the law of the random variable X_T in the sense that for all bounded measurable functions f on E, and all $x \in E$,

$$P_\nu f(x) = \mathbb{E}\big(f(X_T) \mid X_0 = x\big).$$

From the definition, it is clear that P_ν is positivity preserving and satisfies $P_\nu(\mathbb{1}) = \mathbb{1}$ (ν is a probability measure). When P_t for $t > 0$ has a kernel density $p_t(x, y)$, $(x, y) \in E \times E$, with respect to a measure m, then P_ν has the kernel density $\int_0^\infty p_t(x,y) d\nu(t)$ as soon as this integral converges. Formally, $P_\nu = G_\nu(-L)$ where

$$G_\nu(\lambda) = \int_0^\infty e^{-\lambda t} d\nu(t), \quad \lambda \in \mathbb{R}.$$

For example, if $d\nu(t) = \alpha e^{-\alpha t} dt$ for some $\alpha > 0$, the resulting operator G_ν is known as the *resolvent* $G_\alpha = \frac{\alpha}{\alpha - L}$ of L (see Sect. A.1, p. 473).

The preceding construction may actually be pushed further to define a new semigroup $(Q_s)_{s \geq 0}$ of operators via a family ν_s, $s \geq 0$, of probability measures. For this to happen, it is enough that the family ν_s, $s \geq 0$, satisfies, for any $\lambda \in \mathbb{R}_+$

$$\int_0^\infty e^{-\lambda t} d\nu_s(t) = e^{-s\psi(\lambda)}$$

(for some positive function ψ on $\mathbb{R}_+$). Such families of measures ν_s, $s \geq 0$, on $\mathbb{R}_+$ are called *subordinators*. The semigroup $(Q_s)_{s \geq 0}$ then has generator $-\psi(-L)$. In general, the admissible functions ψ are described as

$$\psi(\lambda) = d\lambda + \int_0^\infty \big(1 - e^{-\lambda x}\big) d\Pi(x), \quad \lambda \in \mathbb{R}_+,$$

where $d \in \mathbb{R}_+$ and Π is a measure with support in $\mathbb{R}_+$ such that $\int_0^\infty (1 \wedge x) d\Pi(x) < \infty$.

The main example of a subordinator (known as the $\frac{1}{2}$-stable subordinator) is described by the function $\psi(\lambda) = \lambda^{1/2}$ giving rise to the family

$$d\nu_s^{1/2}(t) = \frac{1}{2\sqrt{\pi}} \frac{s}{t^{3/2}} e^{-s^2/4t} dt, \quad s \geq 0, \tag{1.15.1}$$

corresponding to $d = 0$ and $d\Pi(x) = C x^{-3/2} dx$. (In general, α-stable subordinators satisfy $\psi(\lambda) = \lambda^\alpha$ for $\alpha \in (0, 1)$ and $d\Pi(x) = C_\alpha x^{-1-\alpha} dx$.) In this case, since $Q_s = e^{-s\sqrt{-L}}$, for a given f, the function $H(s, x) = Q_s f(x)$, $s \geq 0$, $x \in E$, is a solution of the equation

$$\big(\partial_s^2 + L\big) H = 0, \quad H(0, x) = f.$$

Therefore, while $(P_t)_{t \geq 0}$ describes the solution of the parabolic heat equation associated to the generator L, the semigroup $(Q_s)_{s \geq 0}$ is the solution of the extended elliptic equation $\partial_s^2 + L = 0$ on $\mathbb{R}_+ \times E$ (elliptic at least when L itself is elliptic).

1.15 Working with Markov Semigroups

Again, the preceding operation has a clear probabilistic interpretation. The measure $v_s^{1/2}$ arising in (1.15.1) is precisely the law of the hitting time $T = \inf\{t > 0; B_t = 0\}$ for a one-dimensional Brownian motion $(B_t)_{t \geq 0}$ starting from $B_0 = s > 0$. Indeed, it is enough to observe that, for $\alpha > 0$, the function $F(t,x) = e^{-\sqrt{\alpha} x - \alpha t}$, $t \geq 0$, $x \in \mathbb{R}$, satisfies $(\partial_t + \partial_x^2)F = 0$ and is bounded on $\mathbb{R}_+$. Therefore $F(t, B_t)$ is a martingale on $[0, T]$ such that $\mathbb{E}(F(T, B_T)) = \mathbb{E}(F(0, s))$ (since $B_0 = s$) from which it follows that

$$\mathbb{E}_s(e^{-\alpha T}) = e^{-\sqrt{\alpha} s}$$

which is the known Laplace transform of the probability measure $v_s^{1/2}$.

The preceding representation of the solutions of the elliptic equation $(\partial_s^2 + L)H = 0$ from the semigroup $(Q_s)_{s \geq 0}$ and the law of some random variable on $\mathbb{R}_+$ may be easily extended to more general equations of the form

$$(\partial_s^2 + \alpha(s)\partial_s + L)H = 0, \quad H(0,x) = f,$$

with the help of the hitting time $\inf\{t > 0, \widehat{B}_t = 0\}$ where $(\widehat{B}_t)_{t \geq 0}$ is the diffusion on $\mathbb{R}_+$ with generator $\partial_t^2 + \alpha(t)\partial_t$ and initial value $\widehat{B}_0 = s$. At least for H bounded, one may consider to this end the martingale $H(\widehat{B}_t, X_t)$, $t \geq 0$, with $(X_t)_{t \geq 0}$ the Markov process with generator L assumed to be independent of $(\widehat{B}_t)_{t \geq 0}$. Then, by the martingale property, we use the fact that $\mathbb{E}_{(s,0)}(H(\widehat{B}_0, X_0)) = \mathbb{E}_{(s,0)}(H(\widehat{B}_T, X_T))$. Of course, this construction yields a new semigroup only when α is constant. For example, later in Sect. 2.2, p. 81, a slight modification of the $\frac{1}{2}$-stable subordinator will be considered, defined for $\alpha > 0$ by $dv_s^{\alpha,1/2}(t) = e^{\alpha s - \alpha^2 t} dv_s(t)$ and satisfying

$$\int_0^\infty e^{-\lambda t} dv_s^{\alpha,1/2}(t) = e^{-s(\sqrt{\lambda + \alpha^2} - \alpha)}, \quad \lambda \in \mathbb{R}_+. \quad (1.15.2)$$

The resulting semigroup is solution of

$$(\partial_s^2 - 2\alpha\partial_s + L)H = 0, \quad H(0,x) = f.$$

To conclude this section, we note that the subordination procedure may be developed for lots of different representations of the operator L. These usually make sense only in some specific situations or on some restricted classes of functions. For example, the Riesz potentials $(-L)^{-\alpha}$ may be represented for $\alpha > 0$ as

$$\gamma_\alpha^{-1} \int_0^\infty t^{\alpha-1} P_t \, dt$$

where $\gamma_\alpha = \int_0^\infty t^{\alpha-1} e^{-t} dt$. This operator may be bounded from some $\mathbb{L}^p$-space into another $\mathbb{L}^q$-space as is the case, for example, for the heat semigroup, but is only defined on the space of mean-zero functions in the finite measure case in general. Although these representations should be handled with care in general, they provide useful norm bounds and positivity preserving properties of the associated operators.

Of particular interest is the value $\alpha = 1$ corresponding to another useful representation in this direction, which is the potential

$$(-L)^{-1} = \int_0^\infty P_t \, dt.$$

This integral does not make sense in general although it may be defined under minimal reasonable conditions. Furthermore, it has a clear meaning in specific instances. For example, as will be discussed later in Chap. 4, as soon as the invariant measure μ of $(P_t)_{t \geq 0}$ is finite (and normalized into a probability measure) and satisfies a Poincaré inequality, it holds that for some constant $\lambda > 0$,

$$\|P_t f\|_2 \leq e^{-\lambda t} \|f\|_2, \quad t \geq 0,$$

on the set $\mathbb{L}_0^2(\mu)$ of functions $f \in \mathbb{L}^2(\mu)$ with $\int_E f \, d\mu = 0$. By invariance of μ, P_t maps $\mathbb{L}_0^2(\mu)$ into itself, so that in this context $(-L)^{-1}$ is well-defined as a bounded operator on $\mathbb{L}_0^2(\mu)$. It produces for any function $f \in \mathbb{L}_0^2(\mu)$ a unique function $g \in \mathbb{L}_0^2(\mu) \cap \mathcal{D}(L)$ such that $Lg = -f$.

These observations are actually the key to ergodicity as outlined in Sect. 1.8. Indeed, given $f \in \mathbb{L}_0^2(\mu)$ with g such that $Lg = -f$, the martingale representation (1.4.6) shows that

$$g(X_t^x) - g(X_0) = M_t^g + \int_0^t f(X_s) \, ds, \quad t \geq 0.$$

With the help of martingale theory to handle $\frac{1}{t} M_t^g$, it may then be deduced that $\frac{1}{t} \int_0^t f(X_s) \, ds$ converges to 0 (μ-almost surely) as $t \to \infty$.

1.16 Curvature-Dimension Condition

This last section introduces the notion of a curvature-dimension condition through the iterated carré du champ operator Γ_2 which will play a central role in the investigation of functional inequalities associated to Markov semigroups and operators. It describes the geometric aspect of this investigation. We only give here the main flavor of this basic idea which will be developed in subsequent chapters. The geometric Riemannian background at the origin of the concept is presented in Sects. C.5 and C.6, p. 511 and p. 513 of Appendix C.

1.16.1 Γ_2 Operator

Following the previous developments, consider a symmetric Markov semigroup $\mathbf{P} = (P_t)_{t \geq 0}$ with invariant reversible measure μ and infinitesimal generator L. Re-

1.16 Curvature-Dimension Condition

call from Definition 1.4.2 the carré du champ operator Γ defined on a suitable algebra $\mathcal{A}$ of functions in the $\mathbb{L}^2(\mu)$-domain $\mathcal{D}(L)$ of L by

$$\Gamma(f,g) = \frac{1}{2}\bigl[L(fg) - fLg - gLf\bigr], \quad (f,g) \in \mathcal{A} \times \mathcal{A}.$$

The idea (taken from a Riemannian viewpoint, as explained below) is to formally repeat this definition, replacing the product operation by Γ, to define a new operator Γ_2 (iterated carré du champ operator) as

$$\Gamma_2(f,g) = \frac{1}{2}\bigl[L\Gamma(f,g) - \Gamma(f,Lg) - \Gamma(Lf,g)\bigr], \quad (1.16.1)$$

for any pair (f,g) of functions such that the various terms on the right-hand side are well-defined. As for the carré du champ operator Γ, we often write more simply $\Gamma_2(f) = \Gamma_2(f,f)$.

By symmetry of L with respect to the measure μ, and as long as f, Lf, g, Lg and $\Gamma(f,g)$ are in the $\mathbb{L}^2(\mu)$-domain of L, and if moreover $\Gamma(f,g)$ is in the $\mathbb{L}^1(\mu)$-domain, the integration by parts formula for the Γ_2 operator reads as

$$\int_E \Gamma_2(f,g) d\mu = \int_E (Lf)(Lg) d\mu. \quad (1.16.2)$$

It might also be useful to record at this stage that the diffusion property for L or Γ (Definition 1.11.1) leads to a change of variables formula for the Γ_2 operator. For example, using (1.11.3) and (1.11.5), if $\psi : \mathbb{R} \to \mathbb{R}$ is smooth enough, an elementary calculation yields that

$$\Gamma_2\bigl(\psi(f)\bigr) = \psi'(f)^2 \Gamma_2(f) + \psi'(f)\psi''(f) \Gamma\bigl(f,\Gamma(f)\bigr) + \psi''(f)^2 \Gamma(f)^2. \quad (1.16.3)$$

Note that this formula is presented in (C.6.7), p. 516, as a consequence of the standard differential calculus rules in differentiable manifolds whereas the diffusion property from Definition 1.11.1 emphasizes here a more intrinsic and efficient calculus, called Γ-*calculus* (or Γ_2-calculus). The power of this calculus will be demonstrated at length throughout this work.

Some care has to be taken in the definition of Γ and Γ_2 concerning the choice of the class $\mathcal{A}$ of functions. This issue will be discussed in Chap. 3. For the examples considered in this short introduction, say on $E = \mathbb{R}^n$ or on a manifold, $\mathcal{A}$ may be taken for simplicity to be the class of smooth functions with compact support. If Δ is the standard Laplacian on $\mathbb{R}^n$, and if $f : \mathbb{R}^n \to \mathbb{R}$ is smooth, then

$$\Gamma(f) = |\nabla f|^2 \quad \text{and} \quad \Gamma_2(f) = |\nabla\nabla f|^2$$

(where $\nabla\nabla f = (\partial_{ij} f)_{1 \leq i,j \leq n}$ is the Hessian of f). More generally, as developed in Sect. C.6, p. 513, and in a Riemannian language, if $L = \Delta_g - \nabla W \cdot \nabla$ as described in Sect. 1.11.3, then $(\Gamma(f) = |\nabla f|^2$ and)

$$\Gamma_2(f) = |\nabla\nabla f|^2 + \mathrm{Ric}(L)(\nabla f, \nabla f) \quad (1.16.4)$$

where Ric(L) is a symmetric tensor defined from the Ricci tensor $\text{Ric}_\mathfrak{g}$ of the Riemannian (co-) metric $\mathfrak{g}$ by $\text{Ric}(L) = \text{Ric}_\mathfrak{g} + \nabla\nabla W$. On $\mathbb{R}^n$ with the flat Euclidean metric, and for the usual Laplacian Δ, this would simply be $\text{Ric}(L) = \nabla\nabla W$ and thus

$$\Gamma_2(f) = |\nabla\nabla f|^2 + \nabla\nabla W(\nabla f, \nabla f) \tag{1.16.5}$$

on smooth functions $f : \mathbb{R}^n \to \mathbb{R}$.

1.16.2 Curvature-Dimension Inequalities

As is clear from these examples, unlike the carré du champ operator Γ, the new operator Γ_2 is not always positive. For Laplacians $\Delta_\mathfrak{g}$ on Riemannian manifolds $(M, \mathfrak{g})$, the Γ_2 operator is an expression of the Bochner-Lichnerowicz formula (1.16.4) (cf. Theorem C.3.3, p. 509) and is positive if (and only if) the Ricci curvature $\text{Ric}_\mathfrak{g}$ of the manifold is positive. More precisely, for an elliptic operator L, there is always a function $\rho(x)$ such that for every smooth function f and at every point x in the space, $\Gamma_2(f) \geq \rho(x)\Gamma(f)$. For a Laplacian on a Riemannian manifold, the best possible function $\rho(x)$ in such an inequality is precisely the infimum of the Ricci tensor (that is, at any point, the smallest eigenvalue of some symmetric matrix evaluated from the coefficients of L). For non-elliptic operators, there is no such function in general.

Many differential inequalities may be seen as consequences of an inequality of the form

$$\Gamma_2(f) \geq \rho\,\Gamma(f) \tag{1.16.6}$$

for some $\rho \in \mathbb{R}$ and all f's in $\mathcal{A}$ (or in the respective domains of Γ and Γ_2). Such an inequality will be called a *curvature* condition $CD(\rho, \infty)$. To explain the meaning of the second parameter ∞, we turn to a more general definition.

Definition 1.16.1 (Curvature-dimension condition) A diffusion operator L is said to satisfy the curvature-dimension condition $CD(\rho, n)$, for $\rho \in \mathbb{R}$ and $n \in [1, \infty]$, if for every function $f : E \to \mathbb{R}$ in a sufficiently rich class $\mathcal{A}$,

$$\Gamma_2(f) \geq \rho\,\Gamma(f) + \frac{1}{n}(Lf)^2$$

(μ-almost everywhere).

The preceding definition is actually subordinate to the chosen class $\mathcal{A}$ of functions. Detailed and useful conditions on the class $\mathcal{A}$ for which such a definition should be considered will be deeply investigated in Chap. 3.

As discussed in Sect. C.6, p. 513 (to which we refer for a more detailed discussion), for a Laplacian on a Riemannian manifold, the curvature-dimension condition $CD(\rho, n)$ holds if and only if the Ricci curvature is bounded from below by ρ and

1.16 Curvature-Dimension Condition

n is greater than or equal to the (topological) dimension of the manifold (that is, the example from which the inequality actually takes its name). For example, the usual Laplacian Δ on $\mathbb{R}^n$ satisfies the condition $CD(0, n)$ (since $|\nabla\nabla f|^2 \geq \frac{1}{n}(\Delta f)^2$). But this is not necessarily true for other operators. For example, the Ornstein-Uhlenbeck operator investigated in Sect. 2.7.1, p. 103, satisfies $CD(1, \infty)$ on any (finite-dimensional) state space, but does not satisfy $CD(\rho, n)$ for any finite n.

As presented in Sect. 1.11.3, a general elliptic differential operator L on a manifold with dimension n is uniquely decomposed as $L = \Delta_\mathfrak{g} + Z$ where $\Delta_\mathfrak{g}$ is the Laplacian associated to a Riemannian (co-) metric $\mathfrak{g}$ and Z is a vector field. Then L satisfies a curvature-dimension condition $CD(\rho, m)$ if and only if $m \geq n$ and, setting $\nabla_S Z$ to be the symmetric covariant derivative of Z in the metric $\mathfrak{g}$ (that is the symmetrized tensor ∇Z),

$$\text{Ric}_\mathfrak{g} - \nabla_S Z \geq \rho\,\mathfrak{g} + \frac{1}{m-n} Z \otimes Z. \tag{1.16.7}$$

Note in particular that $m = n$ only when $Z = 0$ so that, in this sense, Laplacians are operators with minimal dimension among all elliptic operators. Note that here the curvature condition $CD(\rho, \infty)$ boils down to $\text{Ric}_\mathfrak{g} - \nabla_S Z \geq \rho\,\mathfrak{g}$. In particular, when $Z = -\nabla W \cdot \nabla$, as will normally be the case throughout this book, (1.16.7) reads as

$$\text{Ric}_\mathfrak{g} + \nabla\nabla W \geq \rho\,\mathfrak{g}. \tag{1.16.8}$$

In particular, under this condition, a weighted Riemannian manifold $(M, \mathfrak{g})$ equipped with the measure $d\mu = e^{-W} d\mu_\mathfrak{g}$ will sometimes be said to be of curvature bounded from below by ρ. On the flat manifold $\mathbb{R}^n$ for which $\text{Ric} = 0$, the latter simply amounts to a convexity condition on the potential W.

Curvature-dimension conditions behave nicely under the product procedure described in Sect. 1.15.3. Indeed, if two operators L_i on respective spaces E_i with associated carré du champ operators Γ_i, $i = 1, 2$, satisfy the curvature-dimension conditions $CD(\rho_i, n_i)$, $i = 1, 2$, then the operator $L_1 \oplus L_2$ on the product space $E_1 \times E_2$ with associated carré du champ operator $\Gamma_1 \oplus \Gamma_2$ satisfies the $CD(\rho, n)$ condition with $\rho = \min(\rho_1, \rho_2)$ and $n = n_1 + n_2$. This may be seen directly from the definition of the product carré du champ operator, or in a more naive way when both spaces are manifolds in a local system of coordinates through the computations developed in Sect. C.5, p. 511. Note that this tensorization property is not restricted to the reversible case. This observation in particular fully justifies the name "dimension" for the parameter n.

Finally, note that the curvature-dimension condition $CD(\rho, n)$ takes a simple form in dimension one for operators defined on an interval by $Lf = f'' - a(x)f'$. In this case, $CD(\rho, n)$ is equivalent to the differential inequality on a given by

$$a' \geq \rho + \frac{a^2}{n-1}. \tag{1.16.9}$$

Such one-dimensional operators will be illustrated in model examples in Chap. 2. Owing to their simplicity, they will prove to be useful when testing our intuition regarding similar convexity properties for more general operators on more general spaces.

1.17 Notes and References

This section collects some general references concerning various aspects of the study of Markov semigroups, generators and processes, which was mostly developed in the second half of the 20th century. At this stage, only a few general references (with subjective choices) are included. More precise references to specific results and properties will be pointed out within the main text. Since this chapter only surveys and briefly presents a few specific aspects of the general theory of Markov semigroups and processes, we refer the reader to these references, and more, for a complete account of this vast subject. Recall that Chap. 3 will set up a consistent framework in which the results of this monograph may be developed.

General references on semigroups of operators are [95, 96, 100, 143, 154, 168, 169, 241, 244, 264, 355, 356, 444]. Applications of semigroups of operators to partial differential equations are developed in [174, 205, 344, 392, 393, 419]. The standard main references [203, 247, 248] provide complete accounts on elliptic and hypo-elliptic partial differential equations of the second order relevant to the topics developed here. Elementary martingale theory may be found, for example, in [152, 289, 350, 358, 363, 440] (cf. also Appendix B). The interplay between potential theory, martingales and Markov processes is developed in the monographs [50, 72, 132, 142, 152, 153, 164, 170, 178, 252, 255, 307, 362, 363, 393, 418]. For a point of view related to Dirichlet forms, see [91, 189, 190, 294, 398].

Early developments of the theory in the spirit and at the root of this monograph can be found in the lecture notes [26]. G. Royer's introduction to logarithmic Sobolev inequalities [372] presents basic and relevant material on Markov semigroups and processes along the lines of this investigation. The book [431] by F.-Y. Wang is a comprehensive investigation of Markov operators and functional inequalities with some overlap with the content of this book.

The outline on Markov processes and semigroups of Sects. 1.1, 1.2, 1.3 and 1.6 follows the preceding general references. Standard expositions on probabilities and measures on general measurable spaces are [67, 83]. The measure decomposition theorem in good measurable spaces, as well as Propositions 1.2.3 and 1.2.5, may be found in [151] (see also [150]). (Actually, a statement such as for example Proposition 1.2.5 does not necessarily require the good measurable space framework provided suitable martingale convergence theorems are available.)

The history of the carré du champ operator Γ of the infinitesimal generator of a Markov semigroup (Sect. 1.4) is discussed in [153], with a particular reference to [366].

The Fokker-Planck point of view of Sect. 1.5 is emphasized mostly in the partial differential equation literature with the Lebesgue measure as reference measure.

The basic notion of Dirichlet form presented in Sect. 1.7, gradually put forward in various parts of mathematics, is taken here from the standard reference [189] by

1.17 Notes and References

M. Fukushima (see also [91, 190, 294, 431]). Rota's Lemma (Lemma 1.6.2) may be found in [196].

Ergodic properties refer to a variety of behaviors as time goes to infinity and only a very specific feature is examined in Sect. 1.8 and throughout this book.

Elements on Markov chains (Sect. 1.9) on finite and countable spaces may be found in [309, 333, 357].

Complete accounts on stochastic differential equations and diffusion processes and semigroups, as outlined in Sects. 1.10 and 1.11, are [50, 152, 252, 255, 263, 350, 358]. Martingale problems and the interplay between analysis and probability theory is exposed in [256, 393].

Hypo-ellipticity is one major topic of the Hörmander theory [247, 248] briefly discussed in Sect. 1.12.

The relevant aspects on domains of infinitesimal operators are briefly presented in Sect. 1.13 following the standard material and references on the subject (see above and also Appendix A and Chap. 3).

Section 1.15 gathers a variety of tools in the investigation of Markov semigroups, generators and processes. Some of them are part of the folklore and not always explicitly stated in standard references. Girsanov (Sect. 1.15.5) and Feynman-Kac (Sect. 1.15.6) formulas are presented in standard references on stochastic calculus such as [252, 350, 358, 393]. Potential theory and h-transforms as mentioned in Sect. 1.15.8 are deeply investigated in [164] (see also [153]) where the reader will find a comprehensive account of the interplay between potential theory and probability theory. More on subordinators (Sect. 1.15.9) may be found in [15, 64, 65, 379] and in the references therein. The basics on Riesz potentials may be found for example in [164, 342, 388].

The Γ_2 operator and the associated notion of a curvature-dimension condition were introduced in the early contributions [24, 36] on the basis of the Bochner-Lichnerowicz formula in Riemannian geometry towards the study of logarithmic Sobolev inequalities (see Chap. 5) and Riesz transforms (cf. [26]). Curvature or curvature-dimension conditions in terms of the Γ_2 operator are sometimes referred to as the "Γ_2 criterion".

Chapter 2
Model Examples

This chapter is devoted to some basic model examples, which will serve as a guide throughout this book. These model examples will also provide an opportunity to illustrate some of the ideas, definitions and properties of Markov semigroups, operators and processes introduced in the first chapter. Moreover, they will help to set up the framework for the investigation of more general Markov semigroups and generators as will be achieved in the next chapter.

The two simplest examples of diffusion semigroups and generators (at least among those for which the semigroup is explicitly known) are the (Euclidean) heat semigroup and the Ornstein-Uhlenbeck semigroup, with associated Brownian motion and Ornstein-Uhlenbeck process. There are of course other fundamental examples, as models or references for comparison, for which there is however in general no explicit formulas for the semigroups so that they are only described in terms of their generators. We present here some of these models, with a special focus on the underlying Laplacian or diffusion generator. With respect to Chap. 1, most of the Markov semigroups presented in this chapter will indeed be introduced by their generators (defined on classes of smooth functions). The examples considered here will actually present an opportunity to discuss the existence and uniqueness of (symmetric) semigroups with given generators (emphasized in the preceding chapter as the essential self-adjointness issue). Complete justifications are developed in the next chapter together with the description of the relevant classes of functions.

The chapter starts with the three geometric models of the heat semigroup on the Euclidean space, the sphere and the hyperbolic space. For each model, we carefully describe the geometric framework and present the natural Laplacian giving rise to the associated heat, or Brownian, semigroup. Sections 2.4 and 2.5 discuss the heat semigroup with Neumann, Dirichlet or periodic conditions on $\mathbb{R}$ or on an interval of $\mathbb{R}$. More general Sturm-Liouville operators on an interval of the real line are examined next, and are illustrated in Sect 2.7 by the examples of the Ornstein-Uhlenbeck or Hermite, Laguerre and Jacobi operators. These important examples are the three models of one-dimensional diffusion operators which may be diagonalized with respect to a basis of orthogonal polynomials, thus leading to deep analytic connections. Following the first chapter, the exposition often combines the various analytic, probabilistic and geometric viewpoints.

2.1 Euclidean Heat Semigroup

From the operator-theoretic viewpoint developed in the first chapter, the heat or Brownian semigroup in the Euclidean space $\mathbb{R}^n$ is the Markov diffusion semigroup with infinitesimal generator the usual Laplacian $\Delta = \Delta_{\mathbb{R}^n}$ on $\mathbb{R}^n$ defined on smooth functions $f : \mathbb{R}^n \to \mathbb{R}$ by

$$\Delta f = \Delta_{\mathbb{R}^n} f = \sum_{i=1}^{n} \partial_i^2 f$$

and with invariant and reversible measure the Lebesgue measure dx. The carré du champ operator is simply given by

$$\Gamma(f, g) = \nabla f \cdot \nabla g = \sum_{i=1}^{n} \partial_i f \, \partial_i g$$

for smooth functions f, g on $\mathbb{R}^n$, and in particular $\Gamma(f) = |\nabla f|^2$ is the standard Euclidean length of the gradient of f. A convenient function algebra to use here is, for example, the class of smooth ($\mathcal{C}^\infty$) compactly supported functions.

The associated Markov process is the standard Brownian motion $\mathbf{B} = (B_t)_{t \geq 0}$ in $\mathbb{R}^n$. It consists of n independent standard real Brownian motions $B_t = (B_t^1, \ldots, B_t^n)$, $t \geq 0$, starting at $x = (x_1, \ldots, x_n) \in \mathbb{R}^n$. The distribution of B_t at time t is the law of a Gaussian vector centered at x with covariance matrix t Id. $\mathbf{B} = (B_t)_{t \geq 0}$ actually corresponds to the more probabilistic normalization which works with $\frac{1}{2}\Delta$. However, we mostly deal with Δ later and therefore with $(\widetilde{B}_t)_{t \geq 0} = (B_{2t})_{t \geq 0}$ as associated Markov process. Observe that this change from t to $2t$ in the formulas may lead to some confusion when comparing with the standard literature on the subject.

The Laplace operator Δ with the Lebesgue measure as invariant and symmetric reference measure defines a Markov semigroup $\mathbf{P} = (P_t)_{t \geq 0}$ with kernel densities

$$p_t(x, y) = \frac{1}{(4\pi t)^{n/2}} e^{-|x-y|^2/4t}, \quad t > 0, \, (x, y) \in \mathbb{R}^n \times \mathbb{R}^n. \tag{2.1.1}$$

In other words, for every, say, bounded measurable function $f : \mathbb{R}^n \to \mathbb{R}$,

$$P_t f(x) = \int_{\mathbb{R}^n} f(y) \, p_t(x, y) dy, \quad t > 0, \, x \in \mathbb{R}^n.$$

These kernels classically solve the (parabolic) heat equation $\partial_t p_t = \Delta p_t$ (where Δ acts either on x or y). The semigroup $(P_t)_{t \geq 0}$ may via described in probabilistic

2.1 Euclidean Heat Semigroup

terms via the representation formula

$$P_t f(x) = \mathbb{E}_x\big(f(\widetilde{B}_t)\big) = \mathbb{E}\big(f(x + \sqrt{2t}\, G)\big), \quad t \geq 0,\ x \in \mathbb{R}^n, \qquad (2.1.2)$$

where G is a standard Gaussian vector in $\mathbb{R}^n$ with distribution the standard Gaussian probability measure $(2\pi)^{-n/2} e^{-|x|^2/2} dx$.

As such, the semigroup $(P_t)_{t \geq 0}$ is a Markov semigroup in the sense of (i)–(v) of Definition 1.2.2, p. 12. To check the continuity assumption (vi), that is, for any $f \in \mathbb{L}^2(dx)$, $P_t f \to f$ in $\mathbb{L}^2(dx)$ as $t \to 0$, start with a smooth and compactly supported function f and extend the result by density.

The semigroup $\mathbf{P} = (P_t)_{t \geq 0}$ is thus called the *heat or Brownian semigroup* on $\mathbb{R}^n$. It is one of the few examples for which an explicit description of the transition probabilities is available. The Chapman-Kolmogorov equation (1.3.2), p. 17, appears as a consequence of the fact that the sum of two independent Gaussian vector in $\mathbb{R}^n$ with respective covariance matrices $2t\,\mathrm{Id}$ and $2s\,\mathrm{Id}$ is a Gaussian vector with covariance $2(t+s)\,\mathrm{Id}$. Similarly, the dual Fokker-Planck equation (1.5.2), p. 24, is easily checked with the observation that, for $R = |x-y|^2$, $\Delta R = 2n$ and $\Gamma(R) = 4R$ (on either x or y).

Note that the constant function $\mathbb{1}$ is not in the $\mathbb{L}^2(dx)$-domain of Δ since not integrable with respect to the Lebesgue measure dx. As observed in Sect. 1.4, p. 18, the equation $P_t(\mathbb{1}) = \mathbb{1}$ is not a direct consequence of the fact that $\Delta(\mathbb{1}) = 0$. We come back to this issue later when dealing with completion and boundary values.

Among the further useful observations on the Euclidean heat semigroup $(P_t)_{t \geq 0}$, observe that the operators P_t are not Hilbert-Schmidt. Indeed, for every $t > 0$ and $x \in \mathbb{R}^n$, $\int_{\mathbb{R}^n} p_t(x,y)^2 dy = \frac{1}{(8\pi t)^{n/2}}$ and hence

$$\int_{\mathbb{R}^n} \int_{\mathbb{R}^n} p_t(x,y)^2 dx dy = \infty$$

(see Sect. A.6, p. 483). This is again a consequence of the fact that the reversible measure (the Lebesgue measure) is infinite. Actually, there are no square integrable eigenvectors of the Laplacian Δ. Indeed, the eigenvectors are typically of the form $y \mapsto e^{ix \cdot y}$, which are never integrable.

On the geometric side, one can immediately verify that the Γ_2 operator, (1.16.1), p. 71, of the standard Laplacian Δ on $\mathbb{R}^n$ is given on smooth functions $f : \mathbb{R}^n \to \mathbb{R}$ by

$$\Gamma_2(f) = \Gamma_2(f,f) = |\nabla\nabla f|^2 = \sum_{i,j=1}^{n} (\partial_{ij}^2 f)^2.$$

Since by the Cauchy-Schwarz inequality $\sum_{i,j=1}^{n}(\partial_{ij} f)^2 \geq \frac{1}{n}(\sum_{i=1}^{n} \partial_i^2 f)^2$, for every smooth f,

$$\Gamma_2(f) \geq \frac{1}{n}(\Delta f)^2$$

so that the Laplace operator Δ on $\mathbb{R}^n$ satisfies the curvature-dimension condition $CD(0,n)$ of Definition 1.16.1, p. 72 (with optimal values).

Using this example, we next illustrate some of the semigroup tools and properties described in Chap. 1. We start with the commutation relations with suitable vector fields as described in Sect. 1.15.4, p. 60. Brownian motion $\mathbf{B} = (B_t)_{t \geq 0}$ is clearly translation invariant. The process starting at x is the process starting at 0 translated from x. This property actually illustrates the commutation of the Laplacian with translations. A translation in the direction $u \in \mathbb{R}^n$ is expressed on functions as $f(x) \mapsto T_t f(x) = f(x + tu)$, $t \geq 0$, $x \in \mathbb{R}^n$. As $t \geq 0$ evolves, the family $(T_t f)_{t \geq 0}$ is nothing else but the solution of the differential equation $\partial_t T_t f = Z T_t f$ where Z is the constant vector field $Zf = \sum_{i=1}^n u_i \partial_i f$. The vector field Z clearly commutes with Δ, and thus in the heat semigroup $\mathbf{P} = (P_t)_{t \geq 0}$ we have $P_t(T_s f) = T_s P_t f$ for all choices of t and s (≥ 0).

A similar observation may be applied to rotations. The Laplacian Δ indeed also commutes with the operators $\Theta_{ij} = x_i \partial_j - x_j \partial_i$, $1 \leq i, j \leq n$. In particular, this commutation justifies the invariance by rotation of the law of Brownian motion $\mathbf{B} = (B_t)_{t \geq 0}$ when $x = 0$. Namely, if r_{ij}^t denotes the rotation of angle t in the 2-plane (e_i, e_j), then for any f and $x \in \mathbb{R}^n$, $f(r_{ij}^t x) = R_t f(x)$ where $R_t = e^{t \Theta_{ij}}$ is the (semi-) group generated by the vector field Θ_{ij}. Since Θ_{ij} vanishes at the origin, and hence 0 is left-invariant under R_t, the commutation between Δ and Θ_{ij} ensures that for any bounded measurable function f, $P_s R_t f(0) = R_t P_s f(0) = P_s f(0)$, justifying the rotational invariance of Brownian motion. Of course, there is no need to use such a complicated argument to observe the obvious fact that the (Gaussian) law of B_t is rotation invariant when the origin is 0. This presentation however emphasizes how to use similar arguments in more involved instances, when for example the law of the associated Markov process is not explicitly known.

The law of the Brownian motion $\mathbf{B} = (B_t)_{t \geq 0}$ has of course several other remarkable properties. In particular, for every $t \geq 0$, the law of B_t starting from 0 is also the law of $t^{1/2} B_1$. This again is a commutation property, this time with the vector field $Df(x) = \sum_{i=1}^n x_i \partial_i f(x)$, the exponential of which being the dilation semigroup $f(x) \mapsto D_t f(x) = f(e^t x)$ for which

$$[\Delta, D] = \Delta D - D\Delta = 2\Delta.$$

By the same argument as above, $P_s D_t = D_t P_{e^{2t} s}$ for every $s, t \geq 0$ which indeed leads to the dilation property after noticing that $D_t f(0) = f(0)$ for every t.

Finally, the heat kernels used to solve the heat equation lead through subordination to solutions to other equations (Sect. 1.15.9, p. 67). For example, the $\frac{1}{2}$-stable subordinator given in (1.15.1), p. 68, leads to the Cauchy kernels $q_t(x, y)$, $t > 0$, $(x, y) \in \mathbb{R}^n \times \mathbb{R}^n$, given by

$$q_t(x, y) = \int_0^\infty p_s(x, y) dv_t^{1/2}(s) = C_n \frac{t}{(t^2 + |x - y|^2)^{(n+1)/2}}$$

(where $C_n > 0$ is the suitable normalizing constant). For any bounded measurable function f on $\mathbb{R}^n$, the Cauchy kernels yield the *harmonic extension* $H(t, x)$ of f on

2.2 Spherical Heat Semigroup

$\mathbb{R}_+ \times \mathbb{R}^n$, that is the solution of

$$\bigl(\partial_t^2 + \Delta\bigr)H = 0, \quad H(0,x) = f(x),$$

via the representation $H(t,x) = \int_{\mathbb{R}^n} f(y) q_t(x,y) dy$.

2.2 Spherical Heat Semigroup

This section is devoted to the analogue of the heat semigroup on the standard sphere $\mathbb{S}^n$ (in $\mathbb{R}^{n+1}$). This is one most useful model spaces of a compact manifold without boundary, moreover of strictly positive (constant) curvature. It is closely related to the Jacobi semigroups described below in Sect. 2.7.4.

To get a clear picture, it is important to first provide suitable geometric descriptions of the state space in order to address the various objects under investigation, such as the generator and carré du champ operator. The *spherical Laplacian*, denoted by $\Delta_{\mathbb{S}^n}$ below, may be defined as the Laplace-Beltrami operator on the Riemannian manifold $\mathbb{S}^n$ as introduced in Sect. C.3, p. 504. It is invariant and symmetric with respect to the unique probability measure on $\mathbb{S}^n$, denoted σ_n, which is invariant under rotations in $\mathbb{R}^n$. As an embedded (in $\mathbb{R}^{n+1}$) manifold, representations of the sphere $\mathbb{S}^n$ in explicit charts may actually be provided. We present two main representations, the orthogonal projection representation and the stereographic projection representation. For each representation, we provide an explicit description of the spherical Laplacian, the invariant measure and the associated carré du champ operator. Below, we use the standard unit sphere $\mathbb{S}^n$ in $\mathbb{R}^{n+1}$, but simple scaling arguments cover spheres of arbitrary radius. Recall also that, according to the exposition in Appendix C, emphasis is placed on the (Riemannian) co-metric rather than the usual metric.

2.2.1 Orthogonal Projection Representation

The *orthogonal projection representation* of a point in $\mathbb{S}^n \subset \mathbb{R}^{n+1}$, $n \geq 1$, amounts to symmetrically cutting the sphere by the horizontal hyperplane $\mathbb{R}^n$, and associating to a given point $x = (x_1, \ldots, x_n)$ in the open unit ball B of this hyperplane the point

$$\bigl(x_1, \ldots, x_n, \sqrt{1 - |x|^2}\,\bigr)$$

on the sphere $\mathbb{S}^n$ in $\mathbb{R}^{n+1}$. As usual, $|x|^2 = \sum_{i=1}^n x_i^2$ is the square of the Euclidean norm in $\mathbb{R}^n$. The unit ball B of $\mathbb{R}^n$ is therefore a chart of the upper half-sphere (in the language of Sect. C.1, p. 500). This chart is obtained by considering the orthogonal projection from $\mathbb{R}^{n+1}$ onto $\mathbb{R}^n$. In $\mathbb{R}^{n+1}$, consider now a vector U tangent to the sphere at a point x. Via this projection, the vector U is transformed into a vector $\widehat{U}$ of

$\mathbb{R}^n$ with coordinates $(\widehat{U}^i)_{1 \leq i \leq n}$. Similarly, for every vector $\widehat{U}$ of $\mathbb{R}^n$, and any $x \in \mathbb{S}^n$, $\widehat{U}$ may be seen as the projection of a vector tangent to the sphere at x. However, the norm $|U|$ of U in $\mathbb{R}^{n+1}$ is not the norm of $\widehat{U}$ in $\mathbb{R}^n$, but a quantity related to some Euclidean metric $G = G(x) = (g_{ij}(x))_{1 \leq i,j \leq n}$ by $|U|^2 = \sum_{i,j=1}^n g_{ij}(x) \widehat{U}^i \widehat{U}^j$ where

$$g_{ij}(x) = \frac{x_i x_j}{1 - |x|^2} + \delta_{ij}, \quad 1 \leq i, j \leq n.$$

The dual metric (or co-metric, cf. Sect. C.3, p. 504) of interest here, given by the inverse matrix $\mathfrak{g}(x)$ of $(g_{ij})(x)$, is, at any point $x \in \mathbb{S}^n \subset \mathbb{R}^{n+1}$ in these local coordinates,

$$g^{ij}(x) = \delta^{ij} - x^i x^j, \quad 1 \leq i, j \leq n.$$

At this level, we switch to upper indices in agreement with the notation of Riemannian geometry (cf. Appendix C). (In the present setting, x^i is at the same time the coordinate and the differential $\frac{1}{2} \partial_i(|x|^2)$. Therefore the choice of notation x_i or x^i is irrelevant and depends on the context.) It is easily seen that the latter defines a strictly positive metric on B. The Ricci tensor Ric^{ij} of this metric is equal to $(n-1) g^{ij}$. This metric is thus of constant Ricci curvature (equal to $n-1$). Actually, the full Riemann curvature tensor is constant (cf. Sect. C.3, p. 504). At this point, the Laplacian is only defined in the upper half-space. In order to extend the Laplacian to the whole space one might consider a similar chart for the lower half-sphere, but this would still exclude the equator (i.e., in this system of coordinates, the boundary of the unit ball). However, it is not necessary to use charts to define the Laplacian: any chart (in particular, projecting onto any hyperplane) gives the same operator via a change of coordinates.

Let us now describe more precisely the Laplacian and its associated carré du champ operator and invariant measure in the preceding system of coordinates $x = (x^1, \ldots, x^n) \in B$. In this chart, the carré du champ operator of the spherical Laplacian $\Delta_{\mathbb{S}^n}$ of smooth functions $f = f(x)$ and $g = g(x)$ is given by

$$\Gamma(f, g) = \sum_{i,j=1}^n \left(\delta^{ij} - x^i x^j \right) \partial_i f \, \partial_j g.$$

The invariant and reversible measure in this system of coordinates is

$$d\mu(x) = c_n \left(1 - |x|^2 \right)^{-1/2} dx = w(x) dx$$

(where dx is the Lebesgue measure on the unit ball B of $\mathbb{R}^n$ and $c_n > 0$ is the normalization constant which ensures that μ is a probability measure). This measure μ is therefore, up to some constant, the image of σ_n under the orthogonal projection from $\mathbb{S}^n$ onto B. The Laplacian $\Delta_{\mathbb{S}^n}$ itself in this chart takes the form (compare (1.11.8), p. 46)

$$\Delta_{\mathbb{S}^n} = \frac{1}{w} \sum_{i,j=1}^n \partial_i \left(w g^{ij} \partial_j \right) = \sum_{i,j=1}^n \left(\delta^{ij} - x^i x^j \right) \partial^2_{ij} - n \sum_{i=1}^n x^i \partial_i. \quad (2.2.1)$$

2.2 Spherical Heat Semigroup

Having computed $\Delta_{\mathbb{S}^n}$ in this local system of coordinates, observe that if we consider x^i and x^j as the restrictions to the sphere $\mathbb{S}^n$ of the coordinates x^i and x^j ($1 \leq i, j \leq n$) in the ambient space $\mathbb{R}^{n+1}$, that is the restriction of two linear forms corresponding to two unit orthogonal vectors, then

$$\Delta_{\mathbb{S}^n}(x^i) = -nx^i, \qquad \Gamma(x^i, x^j) = \delta^{ij} - x^i x^j.$$

Via the chain rule formula, if f is the restriction to the sphere of a smooth function $f(x^1, \ldots, x^{n+1})$ defined in $\mathbb{R}^{n+1}$, then $\Delta_{\mathbb{S}^n} f$ is the restriction to the sphere of the quantity

$$\sum_{i,j=1}^{n+1} (\delta^{ij} - x^i x^j) \partial^2_{ij} f - n \sum_{i=1}^{n+1} x^i \partial_i f. \tag{2.2.2}$$

In other words, (2.2.1) is in fact valid in $\mathbb{R}^{n+1}$, and not only in the local system of coordinates. Hence, for explicit computations, it is not necessary to replace one of the coordinates, say x^{n+1} by $\pm\sqrt{1 - \sum_{i=1}^{n} (x^i)^2}$.

From the representation (2.2.2) of $\Delta_{\mathbb{S}^n}$ as an operator on $\mathbb{R}^{n+1}$, it is not immediate that the associated Markov process lives on the unit sphere. However, the characterization of $\Delta_{\mathbb{S}^n}$ may be changed slightly into an operator L in $\mathbb{R}^{n+1}$ satisfying

$$L(x^i) = -nx^i, \qquad \Gamma(x^i, x^j) = \delta^{ij} |x|^2 - x^i x^j, \quad 1 \leq i, j \leq n+1.$$

This operator obviously coincides with $\Delta_{\mathbb{S}^n}$ on the unit sphere. Moreover, it may be observed that for this new operator $L(|x|^2) = 0$ and $\Gamma(|x|^2) = 0$. Then, it is quite immediate that for the diffusion process $(X^x_t)_{t \geq 0}$ with generator L, $|X^x_t|^2$ is constant, and therefore that the process stays forever on the sphere it started from.

Both the carré du champ operator Γ and the invariant measure μ of the spherical Laplacian are invariant under the action of the rotations of $\mathbb{R}^{n+1}$. This is easily seen for rotations with vertical axis. Consider the first order differential operators in $\mathbb{R}^{n+1}$, $\Theta_{ij} = x^i \partial_j - x^j \partial_i$, $1 \leq i, j \leq n+1$, already introduced in Sect. 2.1. Then $[\Delta_{\mathbb{S}^n}, \Theta_{ij}] = 0$. Indeed, a smooth function f is rotationally invariant if and only if $\Theta_{ij} f = 0$ for every i, j. As already mentioned in the preceding section, the vector field Θ_{ij} generates a (semi)-group of rotations in the plane (e^i, e^j). Since every rotation is a composition of planar rotations in orthogonal planes, the claim follows.

2.2.2 Stereographic Projection Representation

A second representation of interest is the so-called *stereographic projection* of $\mathbb{S}^n$ in $\mathbb{R}^{n+1}$ on the hyperplane $\mathbb{R}^n$. Denote by N the north-pole (that is the point in $\mathbb{R}^{n+1}$ with coordinates $(0, \ldots, 0, 1)$), and for every $x \in \mathbb{S}^n$, $x \neq N$, consider the line joining N to x. It cuts the horizontal hyperplane at $T(x) \in \mathbb{R}^n$. The stereographic map $x \mapsto T(x)$ is a chart of $\mathbb{S}^n \setminus N$ on $\mathbb{R}^n$. In this chart, the carré du champ operator takes the form

$$\Gamma(f) = \frac{1}{4}(1+|x|^2)^2 |\nabla f|^2$$

where $|\nabla f|^2$ is the usual carré du champ operator associated with the Euclidean Laplacian of a smooth function f. This formula can be proved by a direct computation of the Jacobian matrix of the stereographic projection. The invariant measure in this system of coordinates is written as $d\mu(x) = c_n (1+|x|^2)^{-n} dx$ where dx is Lebesgue measure on $\mathbb{R}^n$ and $c_n > 0$ is the normalization constant so that μ is a probability measure. Once again, μ is the image of σ_n under the stereographic projection T. The latter measure μ is sometimes called a Cauchy measure, in analogy with the one-dimensional case. In the stereographic system of coordinates, the Laplace operator on $\mathbb{S}^n$ becomes

$$\Delta_{\mathbb{S}^n} = \frac{(1+|x|^2)^2}{4}\Delta - \frac{n-2}{2}(1+|x|^2)\sum_{i=1}^n x^i \partial_i \qquad (2.2.3)$$

where Δ is the Euclidean Laplacian in $\mathbb{R}^n$. We refer the reader to Sect. 2.3 for more details about the stereographic projection, viewed there as the restriction to the sphere of an inversion in $\mathbb{R}^{n+1}$.

The preceding orthogonal (2.2.1) and stereographic (2.2.3) projection representations of the Laplacian on the sphere $\mathbb{S}^n$, regarded as a sub-manifold of $\mathbb{R}^{n+1}$, do keep the same metric and are thus equivalent. One of the interesting properties of the stereographic projection is that, in this chart, the Euclidean and spherical metrics are proportional, in other words conformally equivalent. Generally speaking, a *conformal map* from a Riemannian manifold M into itself is a map under which the metric $\mathfrak{g}$ of M is transformed to $c(x)\mathfrak{g}$ for some strictly positive function $c(x)$. Two metrics $\mathfrak{g}_1$ and $\mathfrak{g}_2$ on a given manifold M are conformally equivalent when $\mathfrak{g}_2(x) = c(x)\mathfrak{g}_1(x)$ for some strictly positive function $c(x)$. Conformal maps and conformally equivalent metrics will play a crucial role in the study of Sobolev inequalities on Euclidean space and on the sphere (cf. Sect. 6.9, p. 313). Furthermore, from a more analytical point of view, uniform ellipticity (that is, the existence of a $c > 0$ such that $(g^{ij}) \geq c(\delta^{ij})$ in the sense of symmetric matrices, cf. (1.12.3), p. 50) is not satisfied in the orthogonal projection representation at the boundary of the unit ball (the equator of the sphere) whereas it is satisfied there in the stereographic representation. Conversely, the reverse inequality $(g^{ij}) \leq c(\delta^{ij})$ holds in the projection representation in the neighborhood of the south-pole but does not hold in the stereographic projection. Therefore, care must be taken over such properties which are not invariant by changes of coordinates, and thus are not intrinsic.

There are still many other ways to consider the Laplace operator on the sphere $\mathbb{S}^n$ and we briefly describe below some further examples (these, however, are not really used later).

First, for $\mathbb{S}^n$ embedded into $\mathbb{R}^{n+1}$ in the usual way, extend any smooth function f on $\mathbb{S}^n$ to a function $\widehat{f}$ defined on a neighborhood of $\mathbb{S}^n$ in $\mathbb{R}^{n+1}$, which is indepen-

2.2 Spherical Heat Semigroup

dent of the radius by setting $\widehat{f}(x) = f(\frac{x}{|x|})$. Then, if $\widehat{h} = \Delta \widehat{f}$ for the usual Laplacian in $\mathbb{R}^{n+1}$, $h = \Delta_{\mathbb{S}^n} f$ where h is the restriction of $\widehat{h}$ to $\mathbb{S}^n$.

The connection between the usual Laplacian on $\mathbb{R}^{n+1}$ and the spherical Laplacian $\Delta_{\mathbb{S}^n}$ may also be clarified in polar coordinates. Indeed, if a point $x \in \mathbb{R}^{n+1} \setminus \{0\}$ is parametrized by (r, θ) where $r = |x|$ is the Euclidean norm and $\theta = \frac{x}{|x|} \in \mathbb{S}^n$, then the Euclidean Laplacian $\Delta = \Delta_{\mathbb{R}^{n+1}}$ may be written as

$$\Delta_{\mathbb{R}^{n+1}} = \partial_r^2 + \frac{n}{r} \partial_r + \frac{1}{r^2} \Delta_{\mathbb{S}^n}. \qquad (2.2.4)$$

There is still another representation, more intrinsic in view of Lie group actions. Consider the sphere $\mathbb{S}^n$ as the quotient space $SO(n+1)/SO(n)$, where $SO(n+1)$ is the special orthogonal group in $\mathbb{R}^{n+1}$ ($SO(n)$ is then regarded as the subgroup of $SO(n+1)$ which leaves the point $(1, 0, \ldots, 0)$ invariant). Recall the infinitesimal rotations in $\mathbb{R}^{n+1}$, $\Theta_{ij} = x^i \partial_j - x^j \partial_i$, $1 \leq i, j \leq n+1$. These vector fields preserve functions which are independent of the radius (since they commute with the operator $\sum_{i=1}^{n+1} x^i \partial_i$). Now, for a function f on $\mathbb{S}^n$ given as the restriction to $\mathbb{S}^n$ of a smooth function defined in a neighborhood of $\mathbb{S}^n$ in $\mathbb{R}^{n+1}$, the operator $\Delta_{\mathbb{S}^n}$ may be represented as

$$\Delta_{\mathbb{S}^n} f = \sum_{1 \leq i < j \leq n+1} \Theta_{ij}^2 f. \qquad (2.2.5)$$

In this representation, observe that, as for the usual Laplace operator on $\mathbb{R}^n$, $\Delta_{\mathbb{S}^n}$ is given as the sum of squares of vector fields which commute with it, although they do not commute with each other. Moreover, many more vectors than the dimension of the space have to be used. This is an example of a Casimir operator on a homogeneous space. Such operators play a fundamental role in the analysis of compact Lie groups.

2.2.3 Spherical Heat Kernel

According to the general theory presented in Chap. 3, the spherical Laplacian, considered for example on the class of smooth ($\mathcal{C}^\infty$) functions on $\mathbb{S}^n$, defines the generator of the so-called *spherical heat or Brownian semigroup* $(P_t)_{t \geq 0}$ on the sphere $\mathbb{S}^n$. The Markov process associated with this operator is called the *spherical Brownian motion*. The spherical heat semigroup admits kernel densities $p_t(x, y)$, $t > 0$, $(x, y) \in \mathbb{S}^n \times \mathbb{S}^n$, with respect to the invariant measure σ_n (cf. Definition 1.2.4, p. 14). Since the semigroup commutes with rotations on the sphere, and therefore with any rotation R, the heat kernels satisfy $p_t(x, y) = p_t(Rx, Ry)$ ($t > 0$). But for any two pairs (x, y) and (x', y') of points on the sphere, there exists a rotation R such that $(x', y') = (Rx, Ry)$ if and only if $x \cdot y = x' \cdot y'$. Using the intrinsic distance on the sphere defined from the Laplacian (see Sect. C.4, p. 509, or (3.3.9), p. 166), which in this case is $d(x, y) = \arccos(x \cdot y)$, $(x, y) \in \mathbb{S}^n \times \mathbb{S}^n$, $p_t(x, y)$ may be expressed as a function of the distance $d(x, y)$ (and of t). Unfortunately, no explicit value for this function is easy to handle, but it may be expressed in terms of the heat kernels for Jacobi operators (see Sect. 2.7.4 below).

The harmonic analysis of the spherical Laplacian $\Delta_{\mathbb{S}^n}$ is classical. The eigenvectors of $\Delta_{\mathbb{S}^n}$ are the restrictions to the sphere $\mathbb{S}^n$ of the harmonic polynomials (in $\mathbb{R}^{n+1}$) homogeneous of degree k, and the eigenvalues, of $-\Delta_{\mathbb{S}^n}$, are $\lambda_k = k(k+n-1)$, $k \in \mathbb{N}$. In particular, the eigenvectors U associated with the first non-trivial eigenvalue n are the restrictions to the sphere of the linear maps in the ambient Euclidean space $\mathbb{R}^{n+1}$. Actually, they satisfy $\nabla \nabla U = -U \, \mathrm{Id}$ (here and in what follows $\mathrm{Id} = (g^{ij})$) and the eigenvalue n is obtained after taking the trace.

The polar coordinate representation (2.2.4) of the Euclidean Laplace operator from the spherical representation actually leads to a representation of harmonic functions in the unit ball with given boundary value f on the sphere via subordination (cf. Sect. 1.15.9, p. 67). Indeed, setting $r = e^{-t}$ in (2.2.4), the Laplace operator in $\mathbb{R}^{n+1}$ is given by

$$e^{2t} \big(\partial_t^2 - (n-1)\partial_t + \Delta_{\mathbb{S}^n} \big).$$

Using the subordinator given in (1.15.2), p. 69, with $\alpha = \frac{n-1}{2}$, the harmonic function H in the ball with boundary value f in polar coordinates $(r = e^{-t}, x \in \mathbb{S}^n)$ takes the form

$$H(e^{-t}, x) = \int_{\mathbb{S}^n} f(y) q_t(x, y) d\sigma_n(y)$$

where

$$q_t(x, y) = \int_0^\infty p_s(x, y) \, dv_t^{(n-1)/2, 1/2}(s), \quad t > 0, \ (x, y) \in \mathbb{S}^n \times \mathbb{S}^n.$$

While there are no simple expressions for heat kernels on spheres, the latter kernel $q_t(x, y)$ is on the other hand quite simple and is given by the celebrated *Poisson formula*. Indeed, it is classical that the harmonic extension H to the unit ball in $\mathbb{R}^{n+1}$ of a function $f : \mathbb{S}^n \to \mathbb{R}$ may be represented as

$$H(z) = \int_{\mathbb{S}^n} \frac{1 - |z|^2}{|z - y|^{n+1}} f(y) d\sigma_n(y). \tag{2.2.6}$$

This formula is quite easy to check. Observe first that the map

$$z \mapsto \frac{1 - |z|^2}{|z - y|^{n+1}}$$

is harmonic in the open unit ball B. Furthermore, for any $z \in B$, the measure $v(z, dy) = \frac{1-|z|^2}{|z-y|^{n+1}} d\sigma_n(y)$ is a probability measure on $\mathbb{S}^n$. This is a consequence of the harmonic property since the function $m(z) = \int_{\mathbb{S}^n} v(z, dy)$ is harmonic in B, constant on any sphere of radius $r < 1$ (due to the rotation invariance of σ_n) and satisfies $m(0) = 1$, so that the constant value on spheres of radius r is 1. It remains

2.2 Spherical Heat Semigroup

to observe that as $z \to z_0 \in \mathbb{S}^n$, $\nu(z, dy)$ converges to the Dirac mass at z_0. Then, the function $\int_{\mathbb{S}^n} f(y) \nu(z, dy)$ is harmonic on the ball, and for f continuous on $\mathbb{S}^n$, converges to $f(z_0)$ when z converges to $z_0 \in \mathbb{S}^n$. This is exactly what is expected from the Poisson kernel, proving (2.2.6).

According to the Poisson formula, in polar coordinates ($r = e^{-t}x$, $x \in \mathbb{S}^n$), the density kernel $q_t(x, y)$ with respect to σ_n of the measure $\nu(z, dy)$ may be written

$$q_t(x, y) = \frac{1 - e^{-2t}}{|e^{-t}x - y|^{n+1}}.$$

Writing the harmonic extension of f as $Q_t f$, the subordination representation expresses that, as a semigroup,

$$Q_t = \exp\left(-t\left(\sqrt{-\Delta_{\mathbb{S}^n} + \frac{(n-1)^2}{4}} - \frac{n-1}{2}\right)\right), \quad t \geq 0.$$

Applying Q_t to an eigenvector U_k on the sphere with eigenvalue $\lambda_k = k(k+n-1)$, we get $Q_t U_k = e^{-kt} U_k$ since

$$\sqrt{\lambda_k + \frac{(n-1)^2}{4}} - \frac{n-1}{2} = k.$$

Hence $Q_t U_k = r^k U_k$ (since $r = e^{-t}$) and we recover that U_k is indeed the restriction to the sphere of a harmonic polynomial of degree k in $\mathbb{R}^{n+1}$.

The representation (2.2.6) of the Poisson kernel will be most useful for example when dealing with Jacobi operators and polynomials (see Sect. 2.7.4 below). In the next section, similar formulas will be considered for the hyperbolic Laplacian (see (2.3.2)), in particular with recurrence identities in the dimension. Actually, a similar recurrence (2.7.15) holds in the sphere case, as will be developed with the tools of Jacobi operators.

2.2.4 Curvature-Dimension Condition

In the last part of this section, we describe the geometric and curvature features of the sphere and its Laplacian. As the sphere $\mathbb{S}^n$ of dimension n is of constant curvature $n - 1$, the Laplace operator $\Delta_{\mathbb{S}^n}$ satisfies the curvature-dimension condition $CD(n-1, n)$

$$\Gamma_2(f) \geq (n-1)\Gamma(f) + \frac{1}{n}(\Delta_{\mathbb{S}^n} f)^2$$

(for all f, say, in the algebra $\mathcal{A}$ of smooth functions) from Definition 1.16.1, p. 72, with optimal values (cf. Sect. C.6, p. 513). Such a curvature-dimension condition is similar to those of the one-dimensional Jacobi operators investigated in Sect. 2.7.4

below which may actually be viewed as projections on a diameter of the spherical Laplacian. As discussed there, curvature-dimension conditions for one-dimensional generators are analyzed via (1.16.9), p. 73.

The status of n as a dimension in the curvature-dimension condition $CD(n-1, n)$ may however be somewhat further analyzed. Starting from the projection representation (2.2.2) of $\Delta_{\mathbb{S}^n}$ (defined on the unit ball), projecting it on a vector subspace of dimension $n_0 < n$ (that is, letting $\Delta_{\mathbb{S}^n}$ act on functions depending only on the n_0 first coordinates) yields the same operator but this time on the unit ball of dimension n_0. In particular, this operator may thus be written as $\Delta_{\mathbb{S}^{n_0}} - (n-n_0)\sum_{i=1}^{n_0} x^i \partial_i$ where $\Delta_{\mathbb{S}^{n_0}}$ is the spherical Laplacian of dimension n_0 (in orthogonal projection). It is readily checked that, in these coordinates,

$$-\sum_{i=1}^{n_0} x^i \partial_i f = \Gamma\left(\log \sqrt{1-|x|^2}, f\right).$$

But now, the function $U = \sqrt{1-|x|^2}$ is nothing else than the restriction to the half-upper sphere of the first coordinate. It is therefore an eigenfunction associated to the first eigenvalue, so that the operator $\Delta_{\mathbb{S}^{n_0}}$ can be decomposed as

$$\Delta_{\mathbb{S}^{n_0}} + (n-n_0)\nabla \log U$$

where $\nabla\nabla_{\mathbb{S}^{n_0}} U = -U \operatorname{Id}$. In particular, with $Z = (n-n_0)\nabla \log U$,

$$\operatorname{Ric} - \nabla_S Z - \frac{1}{n-n_0} Z \otimes Z = (n-1)\operatorname{Id}.$$

Comparing with (1.16.7), p. 73, observe that there is equality between tensors. Thus, here we have found a fundamental example of operators satisfying the curvature-dimension condition $CD(n-1, n)$ in an optimal way but for which $n > n_0$ is not the topological dimension of the state space. The latter operator will be further analyzed in Sect. 6.9, p. 313, in connection with Sobolev-type inequalities.

2.3 Hyperbolic Heat Semigroup

The third model example is the heat semigroup on hyperbolic space. In the core of this book, this model will not be used as often as the preceding models in Euclidean and spherical spaces. So its description here will be somewhat more sketchy. We start again with a geometric description of the underlying state space. Various representations of the hyperbolic metric are available and here we emphasize two of them.

2.3 Hyperbolic Heat Semigroup

2.3.1 Upper Half-Space Representation

The first one is the *upper half-space representation* given by

$$g^{ij} = (x^n)^2 \delta^{ij}, \quad 1 \leq i, j \leq n,$$

on $E = \mathbb{R}^{n-1} \times (0, \infty)$. As in the preceding section, we use here upper indices in accordance with the Riemannian geometry convention, putting emphasis on the co-metric. It is not necessary here to change charts since this one covers at once the entire manifold. The resulting manifold, called the *hyperbolic space* $\mathbb{H}^n$ (in the upper half-space representation), is not compact. In this representation, the carré du champ operator may be written as

$$\Gamma(f) = (x^n)^2 |\nabla f|^2$$

and the invariant reversible measure is $d\mu(x) = (x^n)^{-n} dx$ (beware of the notation here, x^n is the n-th coordinate and $(x^n)^{-n}$ denotes its $-n$ power). Here, as usual, $|\nabla f|^2$ is the carré du champ operator of a smooth function f on $\mathbb{R}^n$ for the usual Laplacian $\Delta_{\mathbb{R}^n}$ and dx is the Lebesgue measure. The associated *hyperbolic Laplacian*, denoted $\Delta_{\mathbb{H}^n}$, is given in this representation by

$$\Delta_{\mathbb{H}^n} = (x^n)^2 \Delta_{\mathbb{R}^n} - (n-2) x^n \partial_n. \tag{2.3.1}$$

It should be pointed out that the hyperbolic metric degenerates at the boundary $\{x^n = 0\}$ of the upper half-space $E = \mathbb{R}^{n-1} \times (0, \infty)$. However, the description (2.3.1) is enough to define a unique symmetric semigroup with infinitesimal generator $\Delta_{\mathbb{H}^n}$. Following the developments in Chap. 3, for an elliptic operator on a manifold, knowledge of its action on the class of smooth compactly supported functions is enough to describe a unique symmetric semigroup with this operator as infinitesimal generator (this is the issue of essential self-adjointness) as soon as the manifold is complete (see Proposition 3.2.1, p. 142, and Corollary 3.2.2, p. 143). Now the distance on E induced by $\Delta_{\mathbb{H}^n}$ (see Sect. C.4, p. 509) is the standard hyperbolic Riemannian metric on E, and this metric is complete (the boundary is at an infinite distance with respect to it). Therefore, the operator $\Delta_{\mathbb{H}^n}$ is essentially self-adjoint on E and is the infinitesimal generator of a unique symmetric semigroup called the *hyperbolic heat or Brownian semigroup*. Moreover the operator satisfies the curvature-dimension condition $CD(-(n-1), n)$ (see below), hence this unique semigroup is indeed Markov.

From a probabilistic viewpoint, the heat semigroup on $\mathbb{H}^n$ is the Markov semigroup (up to the probabilistic normalization of the Laplacian) associated with the process solving the stochastic differential equation

$$dX_t^i = X_t^n \circ dB_t^i - (n-1) X_t^n dt, \quad 1 \leq i \leq n,$$

(in Stratonovich form). It is easily seen that this process does not reach the boundary $\{x^n = 0\}$ in finite time and therefore that the semigroup is indeed Markov. The Markov process $(X_t)_{t \geq 0}$ thus constructed is called *hyperbolic Brownian motion*.

The hyperbolic heat semigroup admits density kernels $p_t(x, y)$, $t > 0$, $(x, y) \in \mathbb{H}^n \times \mathbb{H}^n$, with respect to the invariant measure. They are expressed in terms of the intrinsic distance associated with hyperbolic Laplacian $\Delta_{\mathbb{H}^n}$ (see Sect. C.4, p. 509 or (3.3.9), p. 166). In the upper half-space representation $E = \mathbb{R}^{n-1} \times (0, \infty)$, the distance between (x_1, y_1) and (x_2, y_2) is given by

$$\cosh^{-1}\left(\frac{|x_1 - x_2|^2 + y_1^2 + y_2^2}{2 y_1 y_2}\right).$$

The density kernels $p_t(x, y)$ in dimension n are then expressed as functions $k_n(t, d(x, y))$ of the distance, where $k_n(t, d)$ may be defined by induction on $n \geq 2$ by

$$k_2(t, d) = \frac{\sqrt{2}}{(4\pi t)^{3/2}} e^{-t/4} \int_d^\infty \frac{s e^{-s^2/4t}}{(\cosh(s) - \cosh(d))^{1/2}} ds$$

$$k_3(t, d) = \frac{1}{(4\pi t)^{3/2}} \frac{d}{\sinh(d)} \exp\left(-t - \frac{d^2}{4t}\right)$$

and

$$k_{n+2}(t, d) = \frac{e^{-nt}}{2\pi \sinh(d)} \partial_d k_n(t, d). \tag{2.3.2}$$

The expressions take a different form according as n is even or odd (and are simpler for odd n's) while becoming increasingly more complicated as n increases. A similar recurrence formula for the heat kernels on spheres will be given below in (2.7.15) after the appropriate analysis of the corresponding Jacobi operators and their expansions in Jacobi orthogonal polynomials.

As for the sphere, the hyperbolic metric is conformally equivalent to the Euclidean metric. But now the invariant measure is infinite. The Riemann curvature tensor of the metric is constant, as is the Ricci curvature, which is equal to $-(n-1)$. (We have actually described in these three sections the only three metrics and spaces with this property, the Euclidean space, the sphere and the hyperbolic space.) It may be checked directly from the definition of $\Delta_{\mathbb{H}^n}$ that it satisfies a curvature-dimension condition $CD(-(n-1), n)$ in the sense of Definition 1.16.1, p. 72.

The Laplace operator $\Delta_{\mathbb{H}^n}$ on hyperbolic space is invariant under rotations around the axis of the last coordinate vector e_n, as well as under translations parallel to the hyperplane $\{x^n = 0\}$. According to Sect. 1.15.4, p. 60, the associated Markov semigroup leaves invariant the set of functions depending only on the last coordinate, and gives rise along this coordinate to a one-dimensional semigroup on $(0, \infty)$ with generator $x^2 \partial_x^2 - (n-2) x \partial_x$. After a change of variable setting $x = e^y$, the latter operator takes the simpler form $\partial_y^2 - (n-1) \partial_y$. In particular $\Gamma(y) = 1$, the function y (which may be seen as the distance from infinity) having gradient 1. By (1.16.9), p. 73, this one-dimensional operator still satisfies the $CD(-(n-1), n)$ condition, and moreover there is equality in the differential inequality (1.16.9), which characterizes this condition in dimension one. Therefore,

2.3 Hyperbolic Heat Semigroup

the geometric properties of the Laplace operator of hyperbolic space may be recovered from one-dimensional projections. Moreover, from the latter description, the function $h(x) = (x^n)^{n-1}$ satisfies $\Delta_{\mathbb{H}^n} h = 0$ and therefore defines a harmonic positive function. The measure $h d\mu$ is invariant (although not reversible) for $\Delta_{\mathbb{H}^n}$, and in particular there exists in this way at least one invariant measure different from the reversible one (many such invariant measures exist, this is just one of them).

2.3.2 Open Ball Representation

Like the sphere $\mathbb{S}^n$, the hyperbolic space $\mathbb{H}^n$ admits another representation on the open unit ball B of $\mathbb{R}^n$. In order to describe this, we return to the stereographic projection, but this time from $\mathbb{S}^{n-1}$ to the hyperplane $\{x^n = 0\}$ identified with $\mathbb{R}^{n-1}$. This transformation is an inversion. In the Euclidean space $\mathbb{R}^n$, the *inversion* with center $x_0 \in \mathbb{R}^n$ and radius $r > 0$ is the map which associates to each $x \neq x_0$ its inversion $x' \in \mathbb{R}^n$ defined by the condition that $x - x_0$ and $x' - x_0$ are proportional and $|x - x_0||x' - x_0| = r^2$. Analytically, it is given by

$$\varphi_{x_0, r} : x \mapsto x_0 + r^2 \frac{x - x_0}{|x - x_0|^2}.$$

The sphere with center x_0 and radius r is clearly stable under this transformation (which is equal to the identity on it). This sphere is called the sphere of the inversion $\varphi_{x_0, r}$. It is only defined for $x \neq x_0$ and it is an involution ($\varphi_{x_0, r}^2 = \mathrm{Id}$). Actually, the inversion $\varphi_{x_0, r}$ sends every sphere not containing x_0 to sphere, and every sphere containing x_0 to a hyperplane not containing x_0. Similarly, it transforms every hyperplane containing x_0 into a hyperplane containing x_0, and every hyperplane not containing x_0 into a sphere. Moreover, $\varphi_{x_0, r}$ preserves all the spheres orthogonal to the sphere of inversion (two spheres are orthogonal if at any intersection point x, the radii joining x to the centers of the spheres are orthogonal).

Now, the stereographic projection with pole N (north-pole) is actually the restriction to the unit sphere $\mathbb{S}^{n-1}$ of an inversion with center N and radius $\sqrt{2}$ (this inversion clearly preserves the intersection of the unit sphere with the horizontal hyperplane). But it also sends the upper half-space $\{x_n > 0\}$ onto the unit ball of $\mathbb{R}^n$. Via this transformation, the hyperbolic Laplacian $\Delta_{\mathbb{H}^n}$ of (2.3.1) becomes an operator on the open unit ball B with carré du champ operator

$$\Gamma(f) = \frac{1}{4}\left(1 - |x|^2\right)^2 |\nabla f|^2$$

and reversible measure $c_n(1 - |x|^2)^{-n} dx$. Actually, the inversions are conformal maps of the Euclidean space, and it is not so surprising that the image under an inversion of a metric conformally equivalent to the Euclidean metric is again conformally equivalent to the Euclidean metric. In this system of coordinates, the operator

$\Delta_{\mathbb{H}^n}$ becomes the operator defined on B as

$$\Delta_{\mathbb{H}^n} = \frac{1}{4}\left(1-|x|^2\right)^2 \Delta_{\mathbb{R}^n} + \frac{n-2}{2}\left(1-|x|^2\right)\sum_{i=1}^{n} x^i \partial_i. \qquad (2.3.3)$$

This representation is quite similar to the stereographic representation (2.2.3) of the spherical Laplacian $\Delta_{\mathbb{S}^n}$.

The latter representation (2.3.3) of the hyperbolic Laplacian gives rise to projections rather similar to the first representation on the upper half-space. Indeed, in the previous form, the Laplacian $\Delta_{\mathbb{H}^n}$ commutes with rotations centered at the origin. Therefore, following Sect. 1.15.4, p. 60, it preserves radial functions. Setting $|x| = \frac{\sinh(y/2)}{\cosh(y/2)}$, the operator acting on radial functions becomes the one-dimensional operator

$$\partial_y^2 + (n-1)\frac{\cosh(y)}{\sinh(y)}\,\partial_y.$$

Again $\Gamma(y) = 1$, and y is in fact the distance (in the hyperbolic metric) to the center of the unit ball. The latter operator is therefore $\Delta_{\mathbb{H}^n}$ acting on functions depending only on this distance. Using once more (1.16.9), p. 73, this one-dimensional operator satisfies the curvature-dimension condition $CD(-(n-1), n)$, with equality in the differential inequality (1.16.9) characterizing this condition.

In the unit ball representation, it may be further observed that $\Delta_{\mathbb{H}^n}$ is also invariant under rotations preserving the center of the unit ball. Going back and forth between the upper half-space representation and the open unit ball representation of $\mathbb{H}^n$, we may therefore construct many such invariant transformations (indeed the group of isometries of the hyperbolic space). Moreover, the harmonic function $h(x) = (x^n)^{n-1}$ described in the upper half-space may be seen in the ball as a harmonic function vanishing on the boundary except at one point. Using rotations, to any point on the boundary of the ball may be associated a harmonic strictly positive function, and therefore a new invariant measure. In this context, inversions assume the role that is played by rotations in the case of the sphere. In the unit ball representation of $\mathbb{H}^n$, the inversions with respect to spheres orthogonal to the unit sphere are transformations from the unit ball into itself. These inversions are isometries of the hyperbolic space, and the Laplacian $\Delta_{\mathbb{H}^n}$ is invariant under these inversions. It is of interest to look for the vector fields on the ball corresponding to this invariance which commute with $\Delta_{\mathbb{H}^n}$.

2.4 The Heat Semigroup on a Half-Line and the Bessel Semigroup

Before addressing more general Sturm-Liouville operators and semigroups in Sect. 2.6, we briefly discuss in this and the next section the heat semigroup on a half-line and on a bounded interval of the real line. These examples will provide an

2.4 The Heat Semigroup on a Half-Line and the Bessel Semigroup

opportunity to introduce the Neumann and Dirichlet boundary conditions, which are discussed more generally in Sect. 2.6, to address the issue of self-adjointness and to provide intuitive probabilistic descriptions in terms of reflected and killed Brownian motion.

2.4.1 The Heat Semigroup on a Half-Line

Consider the operator $Lf = f''$ on $(0, \infty)$ acting on the set $\mathcal{C}_c^\infty(0, \infty)$ of smooth and compactly supported functions f on $(0, \infty)$. As this operator is symmetric with respect to the Lebesgue measure dx, we may look for a symmetric semigroup for which $\mathcal{C}_c^\infty(0, \infty)$ is included in the domain of its generator and for which this generator coincides with L on this class of functions. Note that while we considered L on $(0, \infty)$, we will see that the associated Markov process may in fact live in $\mathbb{R}_+ = [0, \infty)$ and that we will have to consider the natural state space on which the semigroup lives to be $[0, \infty)$ instead of $(0, \infty)$. Whether we regard the space state to be $(0, \infty)$ or $[0, \infty)$ is more a matter of taste, and in any case the boundary behavior will have to be examined. This observation is relevant for most examples studied here and in the next sections.

A semigroup with generator L in this setting is not unique. Indeed, a bounded measurable function f defined on $(0, \infty)$ may be extended in at least two different ways to the whole real line $\mathbb{R}$. It may actually be extended to a symmetric function $\hat{f}$ (that is $\hat{f}(-x) = \hat{f}(x)$) or to an anti-symmetric function $\check{f}$ (that is $\check{f}(-x) = -\check{f}(x)$), its value at 0 will not matter. Then, if $(P_t)_{t \geq 0}$ is the heat semigroup on $\mathbb{R}$, $P_t \hat{f}$ is symmetric, while $P_t \check{f}$ is anti-symmetric. Setting $P_t^N f = P_t \hat{f}$ and $P_t^D f = P_t \check{f}$, taking the restriction to $\mathbb{R}_+ = [0, \infty)$ yields two different semigroups $(P_t^N)_{t \geq 0}$ and $(P_t^D)_{t \geq 0}$ on $\mathbb{R}_+$. It is easily seen that both semigroups are symmetric with respect to the Lebesgue measure, and that they are positivity preserving. $(P_t^N)_{t \geq 0}$ and $(P_t^D)_{t \geq 0}$ admit simple kernel densities via the standard heat kernel (2.1.1) (on the line) given, for $t > 0$ and $(x, y) \in (0, \infty) \times (0, \infty)$, by

$$p_t^N(x, y) = \frac{1}{2}\big[p_t(x, y) + p_t(x, -y)\big]$$

and

$$p_t^D(x, y) = \frac{1}{2}\big[p_t(x, y) - p_t(x, -y)\big].$$

The semigroup $(P_t^N)_{t \geq 0}$ is Markov while $(P_t^D)_{t \geq 0}$ is only sub-Markov ($P_t^D(\mathbb{1}) \leq \mathbb{1}$). For any function f on $(0, \infty)$, both semigroups $u(t, x) = P_t^{N,D} f$ solve the heat equation $\partial_t u = Lu$ on $\mathbb{R}_+ \times (0, \infty)$ and thus have L as infinitesimal generator on $\mathcal{C}_c^\infty(0, \infty)$. In conclusion, $\mathcal{C}_c^\infty(0, \infty)$ is not a core of the domain of L, or equivalently L is not essentially self-adjoint on this set (cf. Sect. A.5, p. 481, Sect. 1.6, p. 24, and Sect. 1.12, p. 49).

For any bounded measurable function f on $(0, \infty)$, and for any $t > 0$, $P_t^N f$ is a smooth function, symmetric, so that its derivative vanishes at $x = 0$. The semigroup $(P_t^N)_{t \geq 0}$ is called the semigroup with infinitesimal generator L and *Neumann boundary conditions* (derivatives vanish at the boundary). On the other hand, $P_t^D f$ is also smooth, anti-symmetric and therefore vanishes at $x = 0$. The semigroup $(P_t^D)_{t \geq 0}$ is then called the semigroup with infinitesimal generator L and *Dirichlet boundary conditions* (functions vanish at the boundary).

The Neumann and Dirichlet semigroups have clear probabilistic descriptions in terms of the associated Brownian motion. Letting $\{\widetilde{B}_t^x ; t \geq 0, x \in \mathbb{R}\}$ be Brownian motion with speed 2 on the line, it is easily shown that for every suitable function $f : (0, \infty) \to \mathbb{R}$, and every $t \geq 0$, $x \in \mathbb{R}_+$,

$$P_t^N f(x) = \mathbb{E}\big(f\big(|\widetilde{B}_t^x|\big)\big).$$

The family $(P_t^N)_{t \geq 0}$ is the semigroup of Brownian motion *reflected at $x = 0$*. On the other hand,

$$P_t^D f(x) = \mathbb{E}\big(f\big(\widetilde{B}_t^x\big)\mathbb{1}_{\{t < T\}}\big)$$

where T is the first time $s \geq 0$ for which $\widetilde{B}_s^x = 0$ (that is $\widetilde{B}_s = -x$). The latter assertion is actually not so immediate. In order to see why this is, observe that $F(t, x) = P_t^D f(x)$, $t \geq 0$, $x \in \mathbb{R}$, satisfies $\partial_t F = \partial_x^2 F$ and vanishes at $x = 0$. Then, by Ito's formula (Theorem B.2.1, p. 491), for any $t > 0$,

$$M_s = F\big(t - s \wedge t, \widetilde{B}_{s \wedge t}^x\big), \quad 0 \leq s \leq t,$$

is a martingale, and hence $\mathbb{E}(M_{T \wedge t}) = \mathbb{E}(M_0)$. Now, by definition of the stopping time T, if $T \leq t$ then $F(t - T, \widetilde{B}_T^x) = 0$ while when $T > t$, $F(0, \widetilde{B}_t^x) = f(\widetilde{B}_t^x)$ from which the conclusion follows. Therefore $(P_t^D)_{t \geq 0}$ is the semigroup of Brownian motion *killed at $x = 0$*.

2.4.2 The Bessel Semigroup

In the probabilistic realm, another operation on standard Brownian motion is the h-transform as described in Sect. 1.15.8, p. 66, which yields the law of Brownian motion conditioned to stay ever positive. The positive harmonic function h on the half-line $\mathbb{R}_+$ which vanishes at $x = 0$ is just $h(x) = x$. Performing the corresponding h-transform turns the generator $Lf = f''$ into the new generator

$$Lf(x) = f''(x) + \frac{2}{x} f'(x)$$

(for smooth functions f on $(0, \infty)$). This operator belongs to the family of so-called *Bessel semigroups*, associated with *Bessel processes* with parameter 2. These processes naturally appear when looking at radial parts of Brownian motion in dimension n. Indeed, since the standard Euclidean Laplacian commutes with rotations, it

2.4 The Heat Semigroup on a Half-Line and the Bessel Semigroup

preserves the class of radial functions. As described in Sect. 1.15.4, p. 60, if r denotes the function $|x|$, then $\Delta(r^2) = 2n$ while $\Gamma(r^2) = 4r^2$. Hence, for any smooth function f compactly supported in $(0, \infty)$

$$\Delta f(r) = f''(r) + \frac{n-1}{r} f'(r).$$

(Compare with (2.2.4).) The image operator on $(0, \infty)$ is thus $\mathrm{L} f = f'' + \frac{n-1}{x} f'$. The family of operators

$$\mathrm{L}_{B_\kappa} f = f'' + \frac{\kappa}{x} f'$$

on $(0, \infty)$ is known as the family of *Bessel operators* with parameter κ. Here we restrict our attention to the case $\kappa \geq 0$. The associated reversible measure is $x^\kappa dx$ on $(0, \infty)$. By means of the harmonic function $h_\kappa(x) = x^{1-\kappa}$ ($= \log x$ when $\kappa = 1$), it is easily seen that the Markov process with generator L_{B_κ} starting from $x > 0$ never reaches the boundary 0 as soon as $\kappa \geq 1$. A more precise explanation of why this is the case will be given in Sect. 2.6 below.

Now, one may naively think that if a process does not reach the boundary, then its generator is essentially self-adjoint, since different symmetric extensions correspond to various boundary behaviors of the process. The above example of L_{B_2} shows that this is not the case. Indeed, performing the h-transform procedure with $h(x) = x$ for the semigroups $(P_t^N)_{t \geq 0}$ and $(P_t^D)_{t \geq 0}$ of respectively reflected and killed Brownian motions on $(0, \infty)$ yields two semigroups with generator L_{B_2} on $(0, \infty)$, while the process driven by the associated stochastic differential equation does not reach the boundary. (Observe that the h-transform of the killed Brownian motion produces a Markov semigroup, while the transform of the reflected Brownian motion produces a semigroup $(P_t)_{t \geq 0}$ which only satisfies $P_t(\mathbb{1}) \geq \mathbb{1}$.) Moreover, the h-transform of L_{B_κ} yields the Bessel operator $\mathrm{L}_{B_{2-\kappa}}$. Hence, following Proposition 2.4.1 below, L_{B_κ} is essentially self-adjoint as soon as $\kappa > 2$. Since h-transforms preserve such properties, L_{B_κ} is also self-adjoint in the range $\kappa < 0$. In this setting, the associated process has a drift which pushes it to the boundary so strongly that, once it has reached 0, there is no way to return to the open set $(0, \infty)$. This kind of duality between L_{B_κ} and $\mathrm{L}_{B_{2-\kappa}}$ will also be observed for Laguerre semigroups (Sect. 2.7.3) with similar conclusions.

The value which appears as a limiting case for essential self-adjointness for $\kappa \geq 0$ is thus $\kappa = 2$. The following, somewhat more general, Proposition 2.4.1 provides a useful criterion to ensure that a given generator L on the half-line is essentially self-adjoint, that is the space $\mathcal{C}_c^\infty(0, \infty)$ of $\mathcal{C}^\infty$ compactly supported functions in $(0, \infty)$ is dense in the domain $\mathcal{D}(\mathrm{L})$ of L. As announced, it shows in particular that, as soon as $\kappa > 2$, the Bessel generator L_{B_κ} defined on $\mathcal{C}_c^\infty(0, \infty)$ is essentially self-adjoint.

Proposition 2.4.1 *Let* $\mathrm{L} f = f'' + a(x) f'$ *be defined on* $\mathcal{C}_c^\infty(0, \infty)$, *where a is a smooth function on* $(0, \infty)$. *Then the operator* L *is symmetric with respect to the measure* $d\mu = e^A dx$ *where* $A' = a$. *Moreover, as soon as there exist two constants* $c, C > 0$ *such that* $a'(x) + \frac{a^2(x)}{2} \geq \frac{c}{x^2} - C$, $x > 0$, *then* L *is essentially self-adjoint.*

Proof We briefly outline the arguments. The fact that L is symmetric with respect to $d\mu = e^A dx$ is immediate (see Sect. 2.6). Remove then the gradient in L according to the technique described in Sect. 1.15.7, p. 65. The problem is reduced to proving that if $K = \frac{a'}{2} + \frac{a^2}{4}$ then the operator $L_1 f = f'' - Kf$ is essentially self-adjoint on $(0, \infty)$ with respect to the Lebesgue measure. To this end, according to Proposition A.5.3, p. 482, it is enough to show that for some $\lambda \in \mathbb{R}$, the equation $f'' = (\lambda + K)f$ (understood in the distributional sense) has no solution in $\mathbb{L}^2(dx) = \mathbb{L}^2((0, \infty), dx)$ except 0. By the hypothesis, λ may be chosen so that $\lambda + K > K_0(x)$ for some $\varepsilon > 0$ where $K_0(x) = \frac{\varepsilon}{x^2}$. Any solution f on $(0, \infty)$ of $f'' = (K + \lambda)f$ is as smooth as K. Assuming that f is not identically 0, up to a sign change, let $f(x_0) > 0$ for some $x_0 > 0$. Now, if $f'(x_0) > 0$, it is easy to see from $f'' \geq K_0(x)f$ that f is increasing on (x_0, ∞), and is therefore convex on this interval. Being convex it grows at least linearly at infinity and therefore is not in $\mathbb{L}^2(dx)$. On the other hand, if $f'(x_0) < 0$, from standard arguments, f is bounded from below by the solution f_0 of $f_0'' = K_0 f_0$ which has the same value and same derivative at x_0. To verify that f is not in $\mathbb{L}^2(dx)$, it is therefore enough to show that f_0^2 is not integrable near 0. But the solutions of $f_0'' = K_0 f_0$ are linear combinations of x^{α_1} and x^{α_2} where α_1 and α_2 are solutions of $\alpha(\alpha - 1) = \varepsilon$. Since $f_0'(x_0) < 0$, f_0 behaves like $\beta x^{-\alpha_1}$ near the origin, with $\beta > 0$ and $-2\alpha_1 = 1 + \sqrt{1 + 4\varepsilon}$. The conclusion then easily follows and the proposition is established. □

2.5 The Heat Semigroup on the Circle and on a Bounded Interval

Considering the unit circle $\mathbb{S}^1 = \{z \in \mathbb{C} ; |z| = 1\}$ via the parametrization $z = e^{2i\pi x}$, a function on $\mathbb{S}^1$ is simply a periodic function on $\mathbb{R}$ with period 1. A function $\tilde{f}$ on $\mathbb{S}^1$ is thus identified with the 1-periodic function $f(x) = \tilde{f}(e^{2i\pi x})$. While periodic functions are never integrable with respect to the Lebesgue measure on $\mathbb{R}$, the Lebesgue measure on $\mathbb{S}^1$ may be introduced by $\int_{\mathbb{S}^1} \tilde{f}(z)dz = \int_0^1 f(x)dx$. If f and g are smooth and 1-periodic, then

$$\int_0^1 f''g\, dx = \int_0^1 fg''dx = -\int_0^1 f'g'dx$$

since via integration by parts the boundary terms vanish.

On the real line, the standard heat semigroup $(P_t)_{t \geq 0}$ commutes with translations. Therefore, if f is periodic with period 1, so is $P_t f$ for every $t \geq 0$, and it is also smooth when $t > 0$. Hence, the standard heat semigroup acting on periodic functions induces a Markov semigroup $(P_t^p)_{t \geq 0}$ on $\mathbb{S}^1$. To describe it more precisely, note first that for any periodic function f, one has by definition $\partial_t P_t^p f = (P_t^p f)''$. Then, this semigroup has the Lebesgue measure on $\mathbb{S}^1$ as invariant measure. All these considerations show that the above semigroup $(P_t^p)_{t \geq 0}$ on $\mathbb{S}^1$ is just the heat semigroup with infinitesimal generator $Lf = f''$ acting on (smooth) periodic functions. Observe that $(P_t^p)_{t \geq 0}$ commutes with rotations (which are nothing else but translations on the real line via the preceding identification).

Any function f on $[0, 1]$ may extended by periodicity to a new function $\hat{f}$ on the whole real line $\mathbb{R}$. The restriction to $[0, 1]$ of the action $P_t \hat{f}$ of the standard heat semigroup on the line is precisely $P_t^p f$. This construction yields a useful representation of the kernel densities p_t^p of the semigroup $(P_t^p)_{t \geq 0}$ on $\mathbb{S}^1$ as

$$p_t^p(x, y) = \sum_{k \in \mathbb{Z}} p_t(x, y+k), \quad t > 0, \ (x, y) \in [0, 1] \times [0, 1].$$

Unfortunately, this representation as a sum of a series does not admit a closed form.

The preceding analysis may be pushed a bit further on the basis of the properties of the standard heat semigroup $(P_t)_{t \geq 0}$ on the real line to further illustrate the Neumann and Dirichlet boundary conditions emphasized in the previous section. Indeed, any bounded measurable function f on $[0, 1]$ may be extended by symmetry to $[-1, +1]$, and then by periodicity with period 2. The resulting function $\hat{f}$ is invariant under the symmetries about 0 and 1. For every $t \geq 0$, $P_t \hat{f}$ is a function on $\mathbb{R}$ which shares the same symmetries. Since a smooth function on $\mathbb{R}$ which is symmetric under those symmetries has zero derivatives at 0 and 1, the resulting semigroup on $[0, 1]$ corresponds to the Neumann boundary conditions. In the same way, any function f defined on $[0, 1]$ may be extended by anti-symmetry at 0 and then by periodicity. The resulting function $\check{f}$ is anti-symmetric about $x = 1$, and so is $P_t \check{f}$. The associated semigroup on $[0, 1]$ then corresponds to the Dirichlet boundary conditions.

2.6 Sturm-Liouville Semigroups on an Interval

In this section, we turn to the analysis of the family of diffusion operators on an interval (bounded or unbounded) of the real line. This study extends, in the same language, the previous examples of the heat semigroups corresponding to Neumann, Dirichlet or periodic boundary conditions. The investigation of this family is quite accessible, but nevertheless already indicates some of the main difficulties in the study of general diffusion operators and semigroups.

2.6.1 Sturm-Liouville Operators

To fix the ideas, we begin with the case of a bounded interval of the real line $\mathbb{R}$, say $[-1, +1]$. Choose a generator L with smooth coefficients of the form

$$Lf = af'' + bf' \tag{2.6.1}$$

where a and b are smooth on $[-1, +1]$. A natural class of functions f on which L acts is the family of smooth functions on $[-1, +1]$ (equivalently the restrictions to $[-1, +1]$ of smooth functions defined on a neighborhood of $[-1, +1]$). Although it is of interest and often necessary to consider coefficients which are unbounded at the boundary of the interval (as for example in Sect. 2.4 or Sect. 2.7.4), we assume here for simplicity of exposition that a and b are bounded on the whole closed interval $[-1, +1]$. An immediate computation shows that the carré du champ operator Γ associated with the generator L of (2.6.1) is given, on smooth functions f and g, by

$$\Gamma(f, g) = a f' g'.$$

Since $\Gamma(f) = \Gamma(f, f) = a f'^2$ is always positive, we have to choose $a \geq 0$. To simplify the matter, we assume furthermore that $a > 0$ on $[-1, +1]$, so that the operator L is elliptic (Sect. 1.12, p. 49).

To investigate the operator L of (2.6.1), in order to simplify the expression we first perform a change of variables. To this end, set $a = \alpha^2$ where $\alpha > 0$, and look for $y(x)$ such that $dy = \frac{dx}{\alpha(x)}$. In other words, $\frac{d}{dy} = \alpha \frac{d}{dx}$. Hence, every function $f(x)$ is of the form $g(y(x))$, and $f'(x) = \frac{g'(y)}{\alpha(x)}$. Therefore,

$$\frac{d^2}{dy^2} = \alpha^2 \frac{d^2}{dx^2} - \frac{\alpha'}{\alpha} \cdot \frac{d}{dx}.$$

In the new variable y, the operator L thus takes the form $\partial_y^2 + c(y) \partial_y$ and the carré du champ operator is the usual $\Gamma(f) = f'^2$. Observe also that if α vanishes at one of the boundaries of $[-1, +1]$, the new interval on which varies may be unbounded (although we assumed here for simplicity that this is not the case). For further purposes, note that this change of variables is actually only possible in dimension one (since there is only one possible metric up to a multiplicative function). Another feature of dimension one is that the invariant measure is always reversible as well as explicit (since every vector field is a gradient).

After this change of variable, we may again rescale the state space as to interval $[-1, +1]$ and switch back to the x variable, so that the operator now takes the form

$$\mathrm{L}f = f'' + c(x) f' \qquad (2.6.2)$$

where c is a smooth function on $[-1, +1]$. In this form, the invariant measure is given, up to a multiplicative constant, by its density

$$w(x) = \exp\left(\int_{x_0}^{x} c(y) dy\right),$$

the initial point x_0 being a normalization constant for the underlying Markov process. Denote by μ the probability measure with (normalized) density w with respect to the Lebesgue measure (if the coefficients are all bounded, then the invariant measure is indeed finite).

2.6.2 Boundary Conditions

The description of the domain $\mathcal{D}(L)$ of the Sturm-Liouville operator L (2.6.2) is a more delicate question. In contrast to what happens when the drift term $c(x)$ is infinite at the boundary, as in Proposition 2.4.1, the space of smooth compactly supported functions (in the interval $[-1, +1]$) is never dense in the domain when $c(x)$ is bounded, and knowing how the generator behaves in the interior of the interval is not enough to fully describe the associated semigroup. As for the examples of the two last sections, it is necessary to describe the boundary behavior. If the operator has to be symmetric, for all functions f, g in some domain $\mathcal{A}_0$ to be made precise,

$$\int_{-1}^{+1} f\, \text{L} g\, d\mu = \int_{-1}^{+1} g\, \text{L} f\, d\mu$$

(where we recall that $d\mu = w dx$). But for bounded (smooth) functions f and g, the integration by parts formula indicates that

$$\int_{-1}^{+1} f\, \text{L} g\, d\mu = -\int_{-1}^{+1} \Gamma(f, g)\, d\mu + \left[fg'w \right]_{-1}^{+1}$$

and

$$0 = \int_{-1}^{+1} (f\, \text{L} g - g\, \text{L} f) d\mu = \left[(fg' - f'g) w \right]_{-1}^{+1}.$$

It is thus necessary to choose a domain $\mathcal{A}_0$ on which the boundary terms vanish for functions $f, g \in \mathcal{A}_0$. This determines the so-called "boundary conditions". As already observed for the heat semigroup on the half-line in Sect. 2.4.1, two natural choices arise, either the functions or their derivatives vanish. As also discussed there, in probability theory, there are two classical boundary behaviors for diffusion processes. Either the process is killed on the boundary, or it is reflected. (There are of course other possibilities, such as jumping inside the interior following a law depending on the exit point, as well as other options which shall not be considered here.)

The so-called *Dirichlet boundary conditions* correspond to the case when the semigroup is stable on functions vanishing at the boundary. The associated Markov process is therefore killed on the boundary. Processes killed on the boundary are rather easy to describe. Indeed, given the solution $(X_t)_{t \geq 0}$ of the stochastic differential equation

$$dX_t = \sqrt{2}\, dB_t + c(X_t) dt$$

inside some open domain $\mathcal{O}$, starting from $x \in \mathcal{O}$, consider the stopping time $T = \inf\{s \geq 0 ; X_s \notin \mathcal{O}\}$. Then define the semigroup killed at the boundary by

$$P_t^D f(x) = \mathbb{E}_x \left(f(X_t) \mathbb{1}_{\{t < T\}} \right), \quad t \geq 0, \ x \in \mathcal{O}. \tag{2.6.3}$$

The reasoning which identified this semigroup with the semigroup with Dirichlet boundary conditions on the half-line (Sect. 2.4.1) may be repeated here with no change.

The *Neumann boundary conditions* correspond to the case when the semigroup $(P_t^N)_{t \geq 0}$ is stable on functions with derivative equal to 0 on the boundary (in higher dimensions, with normal derivative equal to 0 on the boundary), with associated reflected Markov processes. In this case, it is not enough to solve an ordinary stochastic differential equation in order to describe the semigroup and it is necessary to add to the equation a drift term called local time, which is a measure on $\mathbb{R}_+$ supported by the set of times where the process is at the boundary, and which is singular with respect to the Lebesgue measure. We will not go into these considerations here.

It is not easy to connect Neumann or Dirichlet conditions and processes reflected or killed at the boundary. It is not so difficult to see that for the process killed at the boundary, starting from a continuous function f vanishing at the boundary of $\mathcal{O}$, and provided the coefficients of the stochastic differential equation are smooth and bounded, then $P_t^D f$ also vanishes at the boundary of $\mathcal{O}$. However, for the semigroup associated with the reflected process, it is in general much more difficult to make sure that whenever f has a derivative vanishing at the boundary of $\mathcal{O}$, then the same holds for $P_t^N f$. We do not investigate this question here, which would lead to a further study of local times.

In the case of Sturm-Liouville operators on an interval of the real line, we systematically consider below and throughout this work the Neumann boundary conditions, and thus the Neumann semigroup which we denote by $(P_t)_{t \geq 0}$ for simplicity. These conditions will indeed make it possible for the constant function $\mathbb{1}$ to belong to the domain and to satisfy $P_t(\mathbb{1}) = \mathbb{1}$.

The semigroup $(P_t)_{t \geq 0}$ associated with a Sturm-Liouville operator L on $[-1, +1]$ as in (2.6.2) with smooth and bounded coefficients is always Hilbert-Schmidt in $\mathbb{L}^2(\mu)$ for the invariant probability measure μ (which is also reversible in this one-dimensional case). This will be studied later, since the operator actually satisfies a Sobolev-type inequality, and in this case the density kernels $p_t(x, y)$ will actually be bounded for $t > 0$. Its spectrum is therefore discrete. To understand the structure of the eigenvectors, one has to solve the second order differential equation $Lf = -\lambda f$. But, for any λ, and for every choice of $f(-1)$ and $f'(-1)$, there is a unique solution in the open interval $(-1, +1)$. For Neumann conditions, one has to impose $f'(-1) = 0$. Then, there exists an infinite sequence of values of λ for which the solutions of this equation satisfy $f'(1) = 0$ (the value of $f(-1)$ does not matter and can be fixed). This infinite sequence consists of the eigenvalues of $-L$ with Neumann boundary conditions, and the associated eigenfunctions, suitably normalized, form an orthonormal basis of $\mathbb{L}^2(\mu)$. In particular, the constant function (equal to 1) is an eigenfunction with eigenvalue 0.

Denote by $(\lambda_k)_{k \in \mathbb{N}}$ the sequence of eigenvalues of $-L$ as just described, and by $(f_k)_{k \in \mathbb{N}}$ the sequence of the corresponding eigenvectors (normalized in $\mathbb{L}^2(\mu)$). Then the kernel densities of $(P_t)_{t \geq 0}$ with respect to the invariant measure μ are

2.6 Sturm-Liouville Semigroups on an Interval

given, following (1.7.3), p. 32, by

$$p_t(x,y) = \sum_{k\in\mathbb{N}} e^{-\lambda_k t} f_k(x) f_k(y), \quad t>0, \ (x,y) \in [-1,+1]\times[-1,+1]. \quad (2.6.4)$$

However, very few cases yield explicit expressions for λ_k and f_k, and hence of this heat kernel. The later three sections devoted to the Hermite, Laguerre and Jacobi operators are such instances.

2.6.3 Essential Self-adjointness

For many interesting examples, it often happens that the drift coefficient is singular at the boundary, with a boundary repulsion force on the process. There is a nice way to determine whether the associated Markov process $\mathbf{X} = \{X_t^x \, ; t \geq 0, x \in [-1,+1]\}$ reaches the boundary or not in terms of harmonic functions, that is, solutions of $Lh = 0$. If L is of the form (2.6.2), so that $Lh = h'' + c(x)h'$, then the harmonic functions h are solutions of $\frac{h''}{h'} = -c(x)$, namely

$$h(x) = \int_a^x \exp\left(-\int_b^y c(r)dr\right) dy.$$

Since the density w of the invariant measure μ satisfies $\frac{w'}{w} = c$, a harmonic function h thus satisfies $h' = \frac{1}{w}$. In terms of $\mathbf{X}$, assuming that $(X_t)_{t\geq 0}$ starts from an interior point (say 0), then $(h(X_t))_{t\geq 0}$ is a local martingale. If T_u and T_v are the hitting times of $u, v \in (-1,+1)$ respectively, and if $T_{u,v} = T_u \wedge T_v$ is the hitting time of the boundary of $[u,v] \subset (-1,1)$, we get, since h is harmonic,

$$\mathbb{E}\big(h(X_{T_{u,v}})\big) = h(0) = h(u)\,\mathbb{P}(T_u < T_v) + h(v)\,\mathbb{P}(T_v < T_u). \quad (2.6.5)$$

Hence, if $h(u) \to \infty$ as $u \to -1$, the probability of reaching -1 before reaching any other point in the interior is zero, and therefore the process never hits the boundary. From a practical point of view, the critical case (for -1 say) occurs when $c(x) \simeq \frac{\alpha}{x+1}$ as $x \to -1$. The process never hits the boundary, if $\alpha \geq 1$, and it reaches it when $\alpha < 1$.

As already pointed out earlier for Brownian motion on the half-line, not reaching the boundary is not the same as being essentially self-adjoint. Indeed, when the drift is singular, the definition of the generator L on the open interval may be enough to fully describe the Markov semigroup and process. In analytical terms, the $\mathcal{C}^\infty$ compactly supported functions on the open interval $(-1,+1)$ form a core of the domain, and thus the operator is essentially self-adjoint. (In particular, whenever L is self-adjoint, there is no need to care about the boundary conditions when solving the heat equation.) However, this is usually not that easy to prove. The following proposition provides a sufficient criterion. It is similar to Proposition 2.4.1 above.

Proposition 2.6.1 *Let* $Lf = f'' + cf'$ *be a Sturm-Liouville operator on* $(-1,+1)$. *Assume that c is smooth in $(-1,+1)$ and that there exist $C_1, C_2 > 0$ such that for every $x \in (-1,+1)$,*

$$c'(x) + \frac{c^2(x)}{2} \geq C_1 \min\big((1+x)^{-2}, (1-x)^{-2}\big) - C_2.$$

Then L *is essentially self-adjoint.*

Observe that when $c(x) \simeq \frac{\alpha_-}{1+x}$ at $x = -1$ and $c(x) \simeq -\frac{\alpha_+}{1-x}$ at $x = +1$, then the above condition requires that $\min(\alpha_-, \alpha_+) > 2$, whereas the condition for the process not to reach the boundary requires $\min(\alpha_-, \alpha_+) \geq 1$. Such arguments are of course very specific to dimension one, since explicit computations can be performed in this case. In higher dimensions, the idea is often to mimic the one-dimensional case by using comparison arguments.

2.7 Diffusion Semigroups Associated with Orthogonal Polynomials

On the real line $\mathbb{R}$, a probability measure μ with exponential moments (that is, such that $\int_\mathbb{R} e^{\alpha|x|} d\mu(x) < \infty$ for some $\alpha > 0$) has a more or less canonical Hilbertian basis for the space $\mathbb{L}^2(\mu)$ consisting of orthogonal polynomials, since polynomials are then dense in $\mathbb{L}^2(\mu)$. Such a basis is obtained by orthogonalization of the sequence of polynomials x^k, $k \in \mathbb{N}$, in the scalar product of $\mathbb{L}^2(\mu)$. Up to normalization and change of sign, these orthogonal polynomials are unique, and in the following we always choose them so that they are normalized in $\mathbb{L}^2(\mu)$ with strictly positive leading coefficient.

There are only a few cases of orthogonal polynomials which are also eigenvectors of diffusion operators. In dimension one, there are only, up to affine transformations, the Hermite, Laguerre and Jacobi polynomials. This is why we give special attention to these polynomials in the following sub-sections. Moreover, these examples belong to the few cases for which there is a complete description of the sequences of eigenvalues and eigenvectors for a Sturm-Liouville operator. There is a huge literature on orthogonal polynomials, especially on these three main families. The following only outlines what will be useful and relevant for the rest of the book.

As will become clear in the following, the Hermite, Laguerre and Jacobi operators analyzed here may be considered on various classes of functions, $\mathcal{C}^\infty$ functions, $\mathcal{C}^\infty$ rapidly decreasing functions, or even polynomials. Again, Chap. 3 will present the suitable framework and conditions covering such examples.

2.7 Diffusion Semigroups Associated with Orthogonal Polynomials

2.7.1 The Ornstein-Uhlenbeck Semigroup

The Ornstein-Uhlenbeck (or Hermite) semigroup is one of the simplest Sturm-Liouville semigroups on the real line (which can then be extended, by product, to $\mathbb{R}^n$) for which all the characteristic elements are explicit. It will serve as a main reference example in the investigation of many functional inequalities in later chapters.

The *Ornstein-Uhlenbeck operator* on the real line acts on smooth functions f on $\mathbb{R}$ by

$$L_{OU} f = f'' - x f'.$$

According to the preceding section, or as can be immediately checked, its carré du champ operator is $\Gamma(f) = f'^2$ and its invariant and reversible probability measure is the standard Gaussian probability distribution on the real line $d\mu(x) = e^{-x^2/2} \frac{dx}{\sqrt{2\pi}}$.

The preceding definition may be extended to $\mathbb{R}^n$ by the product procedure outlined in Sect. 1.15.3, p. 59. The Ornstein-Uhlenbeck operator on $E = \mathbb{R}^n$ is then given by $L_{OU} = \Delta - D$ where D is the dilation vector field acting on every smooth function f by $Df(x) = \sum_{i=1}^{n} x_i \partial_i f(x)$, $x \in \mathbb{R}^n$. In other words, for every smooth function f on $\mathbb{R}^n$,

$$L_{OU} f = \Delta f - x \cdot \nabla f. \qquad (2.7.1)$$

To lighten the notation, we write $L_{OU} = L$ throughout this and the next sub-section.

The carré du champ operator is the usual gradient operator $\Gamma(f) = |\nabla f|^2$. Since $-D$ is the gradient of $-|x|^2/2$, the invariant and reversible probability measure is given by

$$d\mu(x) = e^{-|x|^2/2} \frac{dx}{(2\pi)^{n/2}}$$

that is μ is the standard Gaussian measure on $\mathbb{R}^n$ (with mean zero and covariance the identity matrix). We may also characterize the Ornstein-Uhlenbeck operator L as the unique operator such that

$$L x_i = -x_i, \qquad \Gamma(x_i, x_j) = \delta_{ij}, \quad 1 \leq i, j \leq n,$$

on the system of coordinates $(x_i)_{1 \leq i \leq n}$.

The associated Markov semigroup and process may be explicitly described. The *Ornstein-Uhlenbeck semigroup* $\mathbf{P} = (P_t)_{t \geq 0}$ with infinitesimal generator $L = L_{OU}$ is conveniently presented by means of the associated Markov process $\mathbf{X} = \{X_t^x ; t \geq 0, x \in \mathbb{R}^n\}$ which admits the nice explicit representation

$$X_t^x = e^{-t} \left(x + \sqrt{2} \int_0^t e^s dB_s \right) \qquad (2.7.2)$$

where $(B_t)_{t \geq 0}$ is a standard Brownian motion (in $\mathbb{R}^n$) starting at the origin. This process is indeed the solution of the stochastic differential equation

$$dX_t = \sqrt{2} dB_t - X_t dt, \quad X_0 = x,$$

with the generator $L = \Delta - x \cdot \nabla$, defining the *Ornstein-Uhlenbeck process*.

The Ornstein-Uhlenbeck semigroup $(P_t)_{t\geq 0}$ admits an explicit integral representation since the distribution of each X_t^x is Gaussian with covariance matrix $(1 - e^{-2t})\,\mathrm{Id}$ centered at $e^{-t}x$ so that, for suitable functions $f : \mathbb{R}^n \to \mathbb{R}$,

$$P_t f(x) = \mathbb{E}\big(f\big(e^{-t}x + \sqrt{1 - e^{-2t}}\,G\big)\big)$$
$$= \int_{\mathbb{R}^n} f\big(e^{-t}x + \sqrt{1 - e^{-2t}}\,y\big)d\mu(y), \quad t \geq 0, \; x \in \mathbb{R}^n, \quad (2.7.3)$$

where G is a standard Gaussian variable on $\mathbb{R}^n$ (with distribution μ). This formula should be compared with (2.1.2) which describes a similar form for the Euclidean Brownian semigroup. There are of course many connections between these two semigroups. In particular, the Ornstein-Uhlenbeck semigroup $(P_t)_{t\geq 0}$ admits kernel densities (Mehler kernel) with respect to the Lebesgue measure given, for every $t > 0$ and $(x, y) \in \mathbb{R}^n \times \mathbb{R}^n$, by

$$\frac{1}{(2\pi(1 - e^{-2t}))^{n/2}} \exp\left(-\frac{|y - e^{-t}x|^2}{2(1 - e^{-2t})}\right).$$

The densities with respect to the Gaussian invariant measure μ are on the other hand

$$p_t(x, y) = \frac{1}{(1 - e^{-2t})^{n/2}} \exp\left(-\frac{|x|^2 - 2e^t x \cdot y + |y|^2}{2(e^{2t} - 1)}\right). \quad (2.7.4)$$

It can be verified that

$$\int_{\mathbb{R}^n} \int_{\mathbb{R}^n} p_t^2(x, y)d\mu(x)d\mu(y) < \infty$$

so that the operators P_t, $t > 0$, of the Ornstein-Uhlenbeck semigroup are Hilbert-Schmidt (cf. Sect. A.6, p. 483).

The Ornstein-Uhlenbeck generator L of (2.7.1) satisfies the curvature condition $CD(1, \infty)$ of (1.16.6), p. 72. Indeed, one can easily check that for every smooth enough function f on $\mathbb{R}^n$,

$$\Gamma_2(f) = |\nabla\nabla f|^2 + |\nabla f|^2 \geq |\nabla f|^2 = \Gamma(f).$$

It is not difficult to show that L cannot satisfy any $CD(\rho, m)$ condition where $\rho \in \mathbb{R}$ and $m < \infty$, expressing that in a sense the operator is intrinsically infinite-dimensional (independently of the dimension of the state space). The curvature condition $CD(1, \infty)$ may also be described by the commutation property on the semigroup

$$\nabla P_t f = e^{-t} P_t(\nabla f) \quad (2.7.5)$$

for every $t \geq 0$ and f sufficiently regular. This property may be verified directly from the representation (2.7.3). Alternatively, starting with the commutation relation

$$[L, \nabla] = L\nabla - \nabla L = \nabla,$$

2.7 Diffusion Semigroups Associated with Orthogonal Polynomials

the vector-valued map $\Lambda(s) = P_s(\nabla P_{t-s} f)$, $0 \leq s \leq t$, satisfies $\Lambda'(s) = \Lambda(s)$. Hence $e^{-t}\Lambda(t) = \Lambda(0)$ which amounts to (2.7.5). As a consequence of (2.7.5),

$$|\nabla P_t f| \leq e^{-t} P_t(|\nabla f|),$$

which is linked to the curvature condition $CD(1, \infty)$ in Theorem 3.3.18, p. 163 (see also Sect. 4.7, p. 206, and Sect. 5.5, p. 257, below).

As announced, of particular interest are the eigenvectors of the Ornstein-Uhlenbeck generator which are described in terms of the Hermite polynomials orthogonal with respect to the Gaussian measure. Here we begin with dimension one, so that μ denotes below the standard Gaussian measure on $\mathbb{R}$ (satisfying in particular $\int_{\mathbb{R}} e^{\alpha|x|} d\mu(x) < \infty$ for some, or even every, $\alpha > 0$). For every integer k, the Ornstein-Uhlenbeck operator L sends the space $\mathcal{P}_k$ of polynomials of degree less than or equal to k into itself. If the finite-dimensional vector space $\mathcal{P}_k$ is equipped with the scalar product of $\mathbb{L}^2(\mu)$, L defines a symmetric operator in a Euclidean space. It may be diagonalized with respect to an orthonormal basis. By recurrence over k, one then constructs a sequence of orthonormal polynomials in $\mathbb{L}^2(\mu)$ which are eigenvectors of L. Comparing the step from k to $k+1$, the polynomial added at $k+1$ is none other than the polynomial of degree $k+1$ orthogonal to $\mathcal{P}_k$. The sequence of orthonormal polynomials $(H_k)_{k \in \mathbb{N}}$ obtained in this way is therefore the orthonormal polynomials for the Gaussian measure μ known as the *Hermite polynomials*. These polynomials are unique provided their signs are prescribed, and by convention will be taken so that the coefficient of the term of highest degree is strictly positive.

For each $k \in \mathbb{N}$, H_k is of degree k and is an eigenvector of $-L$, that is

$$-L H_k = \lambda_k H_k$$

for some positive number λ_k. Inspecting the term of highest degree shows that $\lambda_k = k$ (H_0 is a constant). Alternatively, the sequence $(H_k)_{k \in \mathbb{N}}$ of Hermite polynomials may be defined via the generating series

$$e^{sx - s^2/2} = \sum_{k \in \mathbb{N}} \frac{s^k}{\sqrt{k!}} H_k(x), \quad x \in \mathbb{R}, \; s \in \mathbb{R},$$

and it is then an exercise to check using this representation that $-LH_k = kH_k$ for every k. There are many other useful representations of the Hermite polynomials $H_k, k \in \mathbb{N}$, for instance

$$H_k(x) = \frac{1}{\sqrt{k!}} \mathbb{E}\big((x + iG)^k\big) = \frac{1}{\sqrt{k!}} \int_{\mathbb{R}} (x + iy)^k d\mu(y), \quad x \in \mathbb{R}$$

(observe that $\int_{\mathbb{R}} e^{s(x+iy)} d\mu(y) = e^{sx - s^2/2}$, $s \in \mathbb{R}$). For example $H_0(x) = 1$, $H_1(x) = x$, $H_2(x) = \frac{1}{\sqrt{2}}(x^2 - 1)$, and so forth. One may also deduce in this way the transition densities (2.7.4) from the general expansion (2.6.4).

On the basis of this orthogonal decomposition of $L^2(\mu)$, the functions in the $L^2(\mu)$-domain $\mathcal{D}(L)$ of L are easily described. Any function in $L^2(\mu)$ can therefore be developed as a series of Hermite polynomials

$$f = \sum_{k \in \mathbb{N}} a_k H_k \tag{2.7.6}$$

with $a_k = \int_{\mathbb{R}} f H_k d\mu$, $k \in \mathbb{N}$, and $\|f\|_2^2 = \sum_{k \in \mathbb{N}} a_k^2$. In this case, the semigroup P_t acts, for every $t \geq 0$, as

$$P_t f = \sum_{k \in \mathbb{N}} e^{-kt} a_k H_k.$$

Similarly, it is expected that

$$-Lf = \sum_{k \in \mathbb{N}} k\, a_k H_k,$$

so that the functions in the $L^2(\mu)$-domain of L are the functions f with decomposition (2.7.6) such that the series describing Lf is in $L^2(\mu)$, that is, such that $\sum_{k \in \mathbb{N}} k^2 a_k^2 < \infty$. This condition therefore requires rapidly decreasing coefficients a_k. Translating this condition into the regularity properties of a given function $f \in L^2(\mu)$ requires more care. The latter is actually equivalent to saying that $f \in \mathcal{C}^1$ and that its second derivative in the distributional sense is a function such that $f'' - xf'$ is in $L^2(\mu)$.

For a function f decomposed in $L^2(\mu)$ as (2.7.6), the Dirichlet form of Definition 1.7.1, p. 30, is given by

$$\mathcal{E}(f) = \sum_{k \in \mathbb{N}} k\, a_k^2$$

(use that $\mathcal{E}(f) = \int f(-Lf) d\mu$). Hence functions with $\mathcal{E}(f) < \infty$ are a priori less regular than functions in the domain (if one agrees that the regularity of a function is described in terms of the decay of its $L^2(\mu)$-coefficients).

Finally, for any function $f \in L^2(\mu)$, $P_t f$ is, for every $t > 0$, in the domain $\mathcal{D}(L)$ and

$$\|LP_t f\|_2^2 \leq \frac{C}{t^2} \|f\|_2^2.$$

To prove this assertion, note that the function re^{-r} is bounded on $r \in \mathbb{R}_+$, say by C, so that for every $k \in \mathbb{N}$ and $t > 0$, $k^2 e^{-2kt} \leq \frac{C}{t^2}$. Then compute the norm of $LP_t f$ as above in its spectral decomposition. This property is actually not particular to the Ornstein-Uhlenbeck operator, and the same argument may be applied to any symmetric semigroup provided the spectral representation may be used. Hence, a symmetric semigroup is always smoothing in the L^2-sense (which is however weak). Below, stronger regularizing properties will be investigated, depending on the ellipticity of the generator. This comment emphasizes more the difference with

2.7 Diffusion Semigroups Associated with Orthogonal Polynomials

non-symmetric semigroups. For example, a dynamical system which pushes Dirac masses onto Dirac masses can never be regularizing.

The spectral structure of the Ornstein-Uhlenbeck generator and semigroup in dimension n is similar. Products of Hermite polynomials

$$\prod_{i=1}^{n} H_{k_i}(x_i), \quad (k_1, \ldots, k_n) \in \mathbb{N}^n, \ (x_1, \ldots, x_n) \in \mathbb{R}^n,$$

are eigenfunctions, with eigenvalues $-\sum_{i=1}^{n} k_i$ of L, and define an orthonormal basis of $\mathbb{L}^2(\mu)$, where now μ is the standard Gaussian measure in $\mathbb{R}^n$ (the product of the one-dimensional Gaussian measures on each coordinate). The spectrum is again $\mathbb{N}$, but this time the vector space associated with a given eigenvalue is no longer one-dimensional.

It is worth mentioning that several useful integration by parts formulas for the Gaussian measure μ may actually be established with the help of Hermite polynomials. The simplest one is obtained using the fact that $L(x_i) = -x_i$ ($1 \leq i \leq n$) to get that for $f \in \mathcal{D}(L)$,

$$\int_{\mathbb{R}^n} x_i f \, d\mu = -\int_{\mathbb{R}^n} L(x_i) f \, d\mu = \int_{\mathbb{R}^n} \Gamma(x_i, f) d\mu = \int_{\mathbb{R}^n} \partial_i f \, d\mu. \quad (2.7.7)$$

Of course, such an identity also follows from a standard integration by parts on the Gaussian density. While this standard argument may be iterated in various ways, the Hermite polynomials as eigenvalues of L provide an efficient systematic tool. For example, using similarly that $L(|x|^2 - n) = -2(|x|^2 - n)$ yields that

$$\int_{\mathbb{R}^n} |x|^2 f \, d\mu = n \int_{\mathbb{R}^n} f \, d\mu + \int_{\mathbb{R}^n} \Delta f \, d\mu \quad (2.7.8)$$

and so on.

The Ornstein-Uhlenbeck operator is related to another famous and well-studied operator, the *harmonic oscillator* in $\mathbb{R}^n$, given on smooth functions f by

$$Hf = \Delta f - |x|^2 f. \quad (2.7.9)$$

This is an operator with potential, symmetric with respect to the Lebesgue measure, and corresponding to the simplest model of quantum mechanics. Denote by $(K_t)_{t \geq 0}$ its associated semigroup. Observing that $H(U_0) = -n U_0$ where $U_0 = e^{-|x|^2/2}$, the ground state transformation described in Sect. 1.15.6, p. 63, may be performed. Thus consider the semigroup

$$R_t f = e^{nt} \frac{1}{U_0} K_t(U_0 f), \quad t \geq 0,$$

with generator given on smooth functions f by

$$Lf = nf + \frac{1}{U_0} H(U_0 f).$$

Now

$$H(U_0 f) = -n U_0 f + U_0 \Delta f + 2 \nabla U_0 \cdot \nabla f$$

so that

$$L f = \Delta f + 2 \nabla \log U_0 \cdot \nabla f$$

which is precisely the Ornstein-Uhlenbeck operator $\Delta - x \cdot \nabla$. The transformation $f \mapsto U_0 f$ carries the analysis of H into the analysis of the Ornstein-Uhlenbeck operator in terms of Hermite polynomials.

2.7.2 The Infinite-Dimensional Ornstein-Uhlenbeck Semigroup

Various functional inequalities in this book are concerned with the dependence of constants with respect to dimension. Independence with respect to dimension then usually allows for infinite-dimensional extensions. Actually, in most cases, infinite-dimensional functional inequalities appear as limits of finite-dimensional inequalities with constants controlled in terms of dimension. Since we will not discuss many infinite-dimensional examples, it is of interest to describe at least one genuine infinite-dimensional example, the infinite-dimensional Ornstein-Uhlenbeck operator. Nevertheless, the functional inequalities for Gaussian measures emphasized in the future chapters will be restricted to their (dimension-free) finite-dimensional statements. Infinite-dimensional versions may be reached either by a limiting argument on the finite-dimensional results, or by suitable adaptations of the formalism and of the main principles of proofs as they can be drawn from the material discussed here.

As presented in the finite-dimensional case in the previous Section, the Ornstein-Uhlenbeck operator in $\mathbb{R}^n$ may be described as the diffusion operator L such that $L x_i = -x_i$ and $\Gamma(x_i, x_j) = \delta_{ij}$ where x_i, $1 \leq i \leq n$, are linear forms corresponding to an orthonormal basis. Moreover, these coordinates x_i define independent standard normal variables under the invariant measure. In a somewhat more abstract way, L may be described by its action on linear forms $\ell(x) = \ell \cdot x$, $x \in \mathbb{R}^n$, as

$$L(\ell) = -\ell, \qquad \Gamma(\ell, \ell') = \ell \cdot \ell',$$

thus independent of the chosen orthonormal basis. This discussion provides an opportunity to mention that if μ is a centered Gaussian measure on $\mathbb{R}^n$ with non-singular covariance matrix Q, then it is the invariant measure of the Ornstein-Uhlenbeck generator

$$\sum_{i,j=1}^n M^{ij} \partial_{ij}^2 - \sum_{i=1}^n (M Q^{-1} x)_i \partial_i$$

2.7 Diffusion Semigroups Associated with Orthogonal Polynomials

where $M = (M^{ij})_{1 \leq i,j \leq n}$ is any symmetric positive-definite matrix. For M the identity matrix, the associated Ornstein-Uhlenbeck semigroup takes the form, for $f : \mathbb{R}^n \to \mathbb{R}$ (cf. (2.7.3)),

$$P_t f(x) = \int_{\mathbb{R}^n} f\bigl(e^{-tQ^{-1}}x + \bigl(\text{Id} - e^{-2tQ^{-1}}\bigr)^{1/2} y\bigr) d\mu(y), \quad t \geq 0, \ x \in \mathbb{R}^n.$$

From this description, the extension from $\mathbb{R}^n$ to infinite-dimensional (Hilbert or Banach) spaces requires suitable systems of coordinates. However, some care has to be taken, in particular with respect to the existence itself of an invariant measure. Recall first that a probability measure μ on the Borel sets of a (real separable) Banach space $\mathcal{B}$ is said to be (centered) Gaussian if the linear functions $x \mapsto \ell(x) = \langle \ell, x \rangle$, $\ell \in \mathcal{B}^*$, are, under μ, (centered) real Gaussian variables. Since weak limits of Gaussian variables are also Gaussian, it is enough to have this property for a dense subset of $\mathcal{B}^*$. Now, for example, let $\mathcal{H}$ be a (real) separable Hilbert space with scalar product $\langle \cdot, \cdot \rangle$ equipped with a centered Gaussian measure μ. Given an orthonormal basis $(e_i)_{i \geq 1}$ of $\mathcal{H}$, the corresponding functions $X_i(x) = \langle e_i, x \rangle$, $i \geq 1$, $x \in \mathcal{H}$, should be, by orthogonality, independent standard normal variables under μ. But then, the strong law of large numbers would show that, μ-almost everywhere, $\frac{1}{k} \sum_{i=1}^k X_i^2 \to 1$ as $k \to \infty$, so that $\sum_{i \geq 1} X_i^2 = \infty$, contradicting the hypothesis that μ is supported on $\mathcal{H}$. This suggests that the Gaussian measure μ should be supported on a larger space.

In a Banach space setting, one must first address the issue of defining Hilbertian coordinates. These questions have been investigated in the context of abstract Wiener spaces with the study of Gaussian measures in Banach spaces and their associated reproducing kernel Hilbert spaces (which yield the required orthonormal system of coordinates with the scalar product induced by the covariance structure of the Gaussian measure). We do not cover the general theory here, but describe how to overcome these difficulties in the concrete and basic example of the Wiener measure. One of the most used and classical examples of Gaussian measure in a Banach space is the Wiener measure, the law of Brownian motion on the space $\mathcal{C}_0([0, 1])$ of continuous functions $w : [0, 1] \to \mathbb{R}$ such that $w(0) = 0$ (the so-called canonical space). One crucial feature of the construction (not addressed here) is that the Brownian paths may be shown to be almost surely continuous so that the law of Brownian motion $(B_t)_{t \geq 0}$ (restricted to $t \in [0, 1]$) indeed defines a probability measure μ on $\mathcal{C}_0([0, 1])$ identifying $w(t)$ with $B_t(w)$, called the *Wiener measure*. The dual space of $\mathcal{C}_0([0, 1])$ consists of bounded (signed) measures, via the duality bracket $\langle v, w \rangle = \int_0^1 w(s) dv(s)$, and for linear combinations $v = \sum_{i=1}^p a_i \delta_{t_i}$, $a_i \in \mathbb{R}$, $t_i \in [0, 1]$, $\langle v, w \rangle = \sum_{i=1}^p a_i B_{t_i}$ which defines a real centered Gaussian variable with variance

$$\mathbb{E}\left(\left|\sum_{i=1}^p a_i B_{t_i}\right|^2\right) = \sum_{i,j=1}^p a_i a_j (t_i \wedge t_j) = \int_0^1 \left|\sum_{i=1}^p a_i \mathbb{1}_{[0, t_i]}\right|^2 dt.$$

By a density argument, this basic identity may be used to define a Hilbertian subset of $\mathcal{C}_0([0, 1])$ with scalar product induced by $\mathbb{L}^2(\mu)$ known as the *repro-*

ducing kernel Hilbert space $\mathcal{H}$ associated with the Wiener measure μ. Indeed, if $h(t) = \sum_{i=1}^{p} a_i(t_i \wedge t)$, $t \in [0, 1]$, h defines an absolutely continuous function in $\mathcal{C}_0([0, 1])$ such that

$$|h|_{\mathcal{H}}^2 = \int_0^1 \dot{h}^2(t) dt = \mathbb{E}\left(\left|\sum_{i=1}^{p} a_i B_{t_i}\right|^2\right) = \mathbb{E}\left(\left|\int_0^1 \dot{h}(t) dB_t\right|^2\right).$$

The Hilbert space $\mathcal{H}$ is identified with the set of absolutely continuous elements h of $\mathcal{C}_0([0, 1])$ such that $\int_0^1 \dot{h}^2(t) dt < \infty$, traditionally called the *Cameron-Martin Hilbert space* (of the Wiener measure μ).

For any orthonormal sequence $(e_i)_{i \geq 1}$ in $\mathbb{L}^2([0, 1], dt)$, the Wiener integrals (random variables on the canonical space)

$$X_i(w) = \int_0^1 e_i(s) dB_s(w), \quad i \geq 1,$$

are independent standard normal random variables (see Sect. B.1, p. 487). Note that when the functions e_i have one derivative in $\mathbb{L}^2([0, 1], dt)$, Itô's formula indicates that

$$X_i(w) = e_i(1) w(1) - \int_0^1 e_i'(s) w(s) ds.$$

The sequence $(X_i)_{i \geq 1}$ defines a set of orthonormal coordinates on the canonical space such that, if $v_i(t) = \int_0^t e_i(s) ds$, $t \in [0, 1]$, $i \geq 1$ (elements of $\mathcal{H}$), the series $\sum_{i \geq 1} v_i X_i$ is almost everywhere convergent with distribution μ (although $\sum_{i \geq 1} X_i^2 = \infty$ μ-almost everywhere).

On $\mathbb{L}^2(\mu)$ with state space $E = \mathcal{C}_0([0, 1])$ and probability measure the Wiener measure μ, consider the algebra $\mathcal{A}_0$ of functions $f : \mathcal{C}_0([0, 1]) \to \mathbb{R}$ which are polynomials in a finite number of such X_i's, that is $f(w) = P_k(X_1(w), \ldots, X_n(w))$. The Ornstein-Uhlenbeck operator L may then be defined as acting on such f's by

$$\mathrm{L} f = \mathrm{L} P_k(X_1(w), \ldots, X_n(w))$$

where on the right L is the usual n-dimensional Ornstein-Uhlenbeck operator. It is easily seen that $\mathcal{A}_0$ is dense in $\mathbb{L}^2(\mu)$, stable under L and that L is a diffusion operator such that

$$\mathrm{L} X_i = -X_i, \quad \Gamma(X_i, X_j) = \delta_{ij}, \quad i, j \geq 1.$$

The operator L generates a semigroup $(P_t)_{t \geq 0}$ (infinite-dimensional Ornstein-Uhlenbeck semigroup) described exactly as in (2.7.3)

$$P_t f(x) = \int_E f\left(e^{-t} x + \sqrt{1 - e^{-2t}}\, y\right) d\mu(y), \quad t \geq 0, \ x \in E.$$

This representation may be verified directly for functions $f \in \mathcal{A}_0$ since it then boils down to the corresponding representation in $\mathbb{R}^n$. Observe that $\mathcal{A}_0$ is stable under

2.7 Diffusion Semigroups Associated with Orthogonal Polynomials

the action of $(P_t)_{t \geq 0}$, and therefore is a core for L. The representation is therefore independent of the chosen base $(e_i)_{i \geq 1}$ of $\mathbb{L}^2([0, 1], dt)$.

The $CD(1, \infty)$ curvature property of the finite-dimensional Ornstein-Uhlenbeck operator immediately extends to the infinite-dimensional setting. Furthermore, since the operator is an infinite sum of one-dimensional Ornstein-Uhlenbeck operators, its eigenvectors and eigenvalues are quite easy to describe. The eigenvalues of $-L$ are the positive integers. However, the corresponding eigenspaces are now infinite-dimensional. For example, the eigenspace associated with the eigenvalue $\lambda_1 = 1$ is the space of linear functionals. The infinite-dimensional Ornstein-Uhlenbeck operator is thus an example of an operator with a purely point spectrum but which is not discrete in the terminology of Sect. A.4, p. 478, in Appendix A.

2.7.3 The Laguerre Semigroup

The next example of a Sturm-Liouville operator which may be diagonalized in orthogonal polynomials is the Laguerre operator. This example has many similarities with the Ornstein-Uhlenbeck example (so that not every detail is supplied here).

The *Laguerre operator* on $E = \mathbb{R}_+$ acts on smooth functions f on $\mathbb{R}_+$ (or restrictions to $\mathbb{R}_+$ of smooth functions on a neighborhood of $\mathbb{R}_+$) by

$$L_\alpha f = x f'' + (\alpha - x) f', \qquad (2.7.10)$$

where $\alpha > 0$. Of course, any linear change $x \mapsto ax$ ($a > 0$) would modify the coefficients a little, the chosen normalization being justified below. Its invariant and reversible measure is the gamma distribution on $\mathbb{R}_+$

$$d\mu_\alpha(x) = \gamma_\alpha^{-1} x^{\alpha-1} e^{-x} dx.$$

The carré du champ operator is given on smooth functions by $\Gamma(f) = x f'^2$.

It may be observed that the Laguerre operator for suitable parameters may be constructed from the Ornstein-Uhlenbeck operator of the preceding section. More precisely, $2L_{n/2}$ appears as the Ornstein-Uhlenbeck operator $L_{OU} = \Delta - x \cdot \nabla$ on $\mathbb{R}^n$ acting on radial functions of the form $f(\frac{|x|^2}{2})$ where $f : \mathbb{R}_+ \to \mathbb{R}$. In particular, it is therefore of no surprise that the Laguerre operator L_α satisfies the curvature condition $CD(\frac{1}{2}, \infty)$, at least when 2α is an integer. It is easily checked that this property remains true for any $\alpha \geq \frac{1}{2}$, and that the curvature $\frac{1}{2}$ is the best possible.

For the other values, the nature of the boundary conditions has to be identified. On the form (2.7.10) itself, it is however not immediately clear which conditions should be imposed since $\Gamma(f) = 0$ at the boundary (the metric degenerates). The change of variable $x \mapsto \frac{y^2}{4}$ shows that we are dealing with Neumann boundary conditions. Indeed, as explained in Sect. 2.6, this change of variable transforms L_α into the operator $f'' + \frac{4\alpha - 2 - y^2}{2y} f'$ on $\mathbb{R}_+$. In fact, following Sect. 2.6, the process associated with this operator lives on $(0, \infty)$ as soon as $\alpha \geq 1$, and is essentially self-adjoint on $(0, \infty)$ as soon as $\alpha > \frac{3}{2}$. Comparing with the case of the hyperbolic Laplacian on $\mathbb{R}^{n-1} \times (0, \infty)$ (Sect. 2.3) shows that this issue really depends on the completeness of the space for the natural distance induced by the operator L as described in (3.3.9), p. 166. In the Laguerre case, the distance from the boundary $\{x = 0\}$ to any point in $(0, \infty)$ is finite, while in the case of the hyperbolic Laplacian it is infinite.

As in the Ornstein-Uhlenbeck case, the Laguerre operator L_α may be diagonalized with respect to a basis of orthogonal polynomials, namely the Laguerre orthogonal polynomials. Indeed, as before, L_α preserves the finite-dimensional space of polynomial functions with degree less than k, and is symmetric on it when endowed with the Euclidean structure inherited from $\mathbb{L}^2(\mu_\alpha)$. Therefore there exists an orthonormal basis $(L_k)_{k \in \mathbb{N}}$ of $\mathbb{L}^2(\mu_\alpha)$ consisting of polynomials L_k of degree k which are eigenvectors of L_α. After normalization, this sequence is unique and defines the *Laguerre orthonormal polynomials* associated with μ_α. As usual $L_0 = \mathbb{1}$, and the eigenvalues are computed just by looking at the highest degree term so that

$$-L_\alpha L_k = k L_k$$

for every $k \in \mathbb{N}$. Laguerre polynomials may be described by generating series.

Comparing with the Ornstein-Uhlenbeck operator from which it comes when $\alpha = \frac{n}{2}$, the spectrum of $2 L_\alpha$ is identical to the even part of the Ornstein-Uhlenbeck spectrum. That is, the Laguerre polynomial of degree k corresponds to a $2k$-degree Hermite polynomial in dimension n which is radial. Indeed, the $2k$-degree Hermite polynomial H_{2k} is an even polynomial which may therefore be written as $Q_k(\frac{x^2}{2})$ for some degree k polynomial Q_k. As the Ornstein-Uhlenbeck operator L_{OU} on $\mathbb{R}^n$ is invariant under rotations, for any unit vector $e \in \mathbb{S}^{n-1}$, the map $x \in \mathbb{R}^n \mapsto Q_k(\frac{(e \cdot x)^2}{2})$ is an eigenvector of L_{OU} with eigenvalue $2k$, and hence so is the map sending x to

$$\int_{\mathbb{S}^{n-1}} Q_k \left(\frac{(e \cdot x)^2}{2} \right) d\sigma_{n-1}(e)$$

where σ_{n-1} is the uniform measure on the sphere $\mathbb{S}^{n-1} \subset \mathbb{R}^n$. This function may be written as $R_k(\frac{|x|^2}{2})$ where R_k is a further polynomial of degree k. Then R_k satisfies $2 L_{n/2} R_k = L R_k = -2k R_k$, and up to a multiplicative constant, it is precisely the Laguerre polynomial with degree k and parameter $\alpha = \frac{n}{2}$.

The image of the Laguerre operator as a projection of the n-dimensional Ornstein-Uhlenbeck operator leads to further relationships between the two families. Since $\int_{\mathbb{R}_+} x d\mu_\alpha = \alpha$, recentering and scaling the variable as $x = \alpha + \sqrt{\alpha}\, y$ transforms the Laguerre operator L_α into the operator

$$\left(1 + \frac{y}{\sqrt{\alpha}} \right) f''(y) - y f'(y)$$

on $[-\sqrt{\alpha}, \infty)$. As $\alpha \to \infty$, this operator converges to the one-dimensional Ornstein-Uhlenbeck generator. Indeed, under this change of variable, the measure

μ_α converges to the standard Gaussian measure, the orthogonal Laguerre polynomials themselves converge to the corresponding Hermite polynomials, and the convergence is furthermore valid at the level of the semigroups. For $\alpha = \frac{n}{2}$, the geometric picture of the preceding simply boils down to the central limit theorem for sums $\sum_{i=1}^{n} X_i^2$ where the X_i's are independent standard normal variables. The same phenomenon will be observed for the Jacobi semigroups below (Sect. 2.7.4). In the same way, many of the properties of the Hermite polynomials may be translated to Laguerre polynomials with integer parameter, and very often extended to the general parameter.

The semigroup with generator L_α is called the *Laguerre semigroup* and is denoted by $(P_t^\alpha)_{t \geq 0}$. In this setting, it may actually be defined explicitly from its action on the basis of Laguerre orthogonal polynomials, as was the case for the Ornstein-Uhlenbeck semigroup. It coincides with the semigroup of the unique essentially self-adjoint extension in the range $\alpha > \frac{3}{2}$. The Laguerre semigroup admits kernel densities with respect to the invariant measure.

Like the Ornstein-Uhlenbeck semigroup, the family of Laguerre semigroups satisfies a number of remarkable identities. For example, the identity $\partial_x L_\alpha = (L_{\alpha+1} - \text{Id})\partial_x$ on the generators L_α and $L_{\alpha+1}$ leads to the commutation property on the associated semigroups

$$\partial_x P_t^\alpha = e^{-t} P_t^{\alpha+1} \partial_x$$

which may be compared to the corresponding formula (2.7.5) for the Ornstein-Uhlenbeck semigroup. Similarly, the h-transform

$$x^{\alpha-1} L_\alpha(x^{1-\alpha} f) = (L_{2-\alpha} + \alpha - 1)f$$

translates into

$$x^{1-\alpha} P_t^\alpha(x^{\alpha-1} f) = e^{t(\alpha-1)} P_t^{2-\alpha} f$$

which exhibits a duality between L_α and $L_{2-\alpha}$ ($0 < \alpha < 2$) very similar to the duality observed for Bessel semigroups or generators (see Sect. 2.4.2).

Products of L_α on $\mathbb{R}_+^n$ may be considered similarly.

2.7.4 The Jacobi Semigroup

The last family of Sturm-Liouville operators presented here is the one associated with the so-called Jacobi polynomials. As mentioned earlier, it plays an important role in the analysis of the Laplacian on spheres. It is, after the Ornstein-Uhlenbeck and Laguerre operators, the third family of one-dimensional diffusion operators which may be diagonalized with respect to a family of orthogonal polynomials. It is a remarkable fact that these three operators are the only operators which have this property.

The *Jacobi operator* on $E = [-1, +1]$ acts on smooth functions f on $[-1, +1]$ (or restrictions to $[-1, +1]$ of smooth functions on a neighborhood of $[-1, +1]$) by

$$\mathrm{L}_{\alpha,\beta} f = (1 - x^2) f'' - [(\alpha + \beta)x + \alpha - \beta] f', \qquad (2.7.11)$$

where $\alpha, \beta > 0$. Its invariant and reversible measure is the beta distribution on $[-1, +1]$

$$d\mu_{\alpha,\beta}(x) = C_{\alpha,\beta} (1 - x)^{\alpha - 1} (1 + x)^{\beta - 1} dx$$

where $C_{\alpha,\beta} > 0$ is a suitable normalization constant. The carré du champ operator is given on smooth functions f by $\Gamma(f) = (1 - x^2) f'^2$. When $\alpha = \beta$, we sometimes speak of symmetric Jacobi operators (although all Jacobi operators are symmetric with respect to the invariant measure).

While the Jacobi operators (2.7.11) may be considered for any values of $\alpha, \beta > 0$, the boundary issues are more easily described after a suitable change of variable. Setting $x = \cos \theta$, we obtain in the new variable θ an operator on $[0, \pi]$ which reads

$$\partial_\theta^2 + \frac{(\alpha + \beta - 1) \cos \theta + \alpha - \beta}{\sin \theta} \partial_\theta. \qquad (2.7.12)$$

According to Proposition 2.6.1, this operator is self-adjoint on the interval $(0, \pi)$ (and hence the Jacobi operator $\mathrm{L}_{\alpha,\beta}$ is self-adjoint on the interval $(-1, +1)$) as soon as $\min(\alpha, \beta) > \frac{3}{2}$ while the associated Markov process does not reach the boundary as soon as $\min(\alpha, \beta) \geq 1$.

As in the case of the Ornstein-Uhlenbeck operator, $\mathrm{L}_{\alpha,\beta}$ sends the family of polynomials of degree less than or equal to k into itself, and thus admits a basis of eigenvectors consisting of polynomials. These polynomials are the orthogonal polynomials associated with the measure $\mu_{\alpha,\beta}$, called the *Jacobi polynomials* $(J_k)_{k \in \mathbb{N}}$ (with implicit parameters α, β). Jacobi polynomials may be described by generating series (see (2.7.14) below in the symmetric case).

The Ornstein-Uhlenbeck generator can be seen as a limit of Jacobi operators. Indeed, for $\alpha = \beta = \frac{n}{2}$, and after changing x into $\frac{x}{\sqrt{n}}$, $\frac{1}{n} \mathrm{L}_{\alpha,\beta}$ converges as $n \to \infty$ to the (one-dimensional) Ornstein-Uhlenbeck operator L_{OU}. (In fact, the convergence is much more comprehensive, the measures, the eigenvectors and the eigenvalues converge, and the Markov semigroups themselves converge, together with the finite-dimensional laws of the associated Markov processes.) In the same way, if we translate $[-1, +1]$ to $[0, 2]$, and then change x into $\frac{x}{n}$ and α into n (with β fixed), then $\frac{1}{n} \mathrm{L}_{n,\beta}$ converges to the Laguerre operator $\mathrm{L}_{2\beta}$ with the same strength of convergence as described in the remark above.

The eigenvalues of the Jacobi operators are easy to describe by considering the action of $\mathrm{L}_{\alpha,\beta}$ on the highest degree terms of the orthogonal polynomials. The eigenvalues of $-\mathrm{L}_{\alpha,\beta}$ are given by the sequence

$$\lambda_k = k(k + \alpha + \beta - 1), \quad k \in \mathbb{N}.$$

There is an analogy with the eigenvalues of the spherical Laplacian $\Delta_{\mathbb{S}^n}$, which should come as no surprise. When n is an integer ($n \geq 1$), and $\alpha = \beta = \frac{n}{2}$, the Jacobi

2.7 Diffusion Semigroups Associated with Orthogonal Polynomials 115

operator $L_{\alpha,\beta}$ may indeed be seen as an image of $\Delta_{\mathbb{S}^n}$ via its action on functions depending only on one coordinate in $\mathbb{R}^{n+1}$, that is the functions which are invariant under rotations leaving one point fixed (for example $(1, 0, \ldots, 0)$). These functions are called *zonal*. This may easily be seen via (2.2.2) acting on functions depending only on x^1. Hence, if J_k is a Jacobi polynomial of degree k, that is, an eigenvector for $L_{\alpha,\alpha}$ for $\alpha = \frac{n}{2}$, then the restriction to $\mathbb{S}^n$ of the function $J_k(x^1)$ defined on $\mathbb{R}^{n+1}$ is an eigenvector for $\Delta_{\mathbb{S}^n}$.

This interpretation of the Jacobi operator $L_{\alpha,\alpha}$ as the projection of $\Delta_{\mathbb{S}^n}$ on zonal functions leads to the fact that it satisfies the curvature-dimension condition $CD(2\alpha - 1, 2\alpha)$ for $\alpha = \frac{n}{2}$ (which may of course be checked directly from the representation (2.7.12) and the general formula (1.16.9), p. 73). Observe furthermore that (1.16.9) (which characterizes the curvature-dimension condition) is in fact an equality, as it was for the radial part of the hyperbolic Laplace operator. Therefore the symmetric Jacobi operator is a very good one-dimensional candidate for testing properties related to curvature-dimension.

For non-symmetric Jacobi operators ($\alpha \neq \beta$), there is also a representation from spheres at least when α and β are half-integers $\alpha = \frac{n}{2}$, $\beta = \frac{p}{2}$. The Jacobi operator is then interpreted as a quotient of the sphere of dimension $n + p - 1$ (up to a factor of 4). The eigenvalues of degree k correspond to eigenfunctions on the sphere of dimension $2k$ as may be seen from the relation $4k(k + \alpha + \beta - 1) = 2k(2k + n + p - 2)$. Of course, this observation in itself is not enough to claim that the generator comes from the sphere models, but it is a strong indication. To really understand the construction, let the spherical Laplacian on $\mathbb{S}^{n+p-1} \subset \mathbb{R}^{n+p}$ act on functions of $y = x_1^2 + \cdots + x_p^2$ producing in this way an operator defined on functions of $y \in [0, 1]$. After the change of variable $y = 2x - 1$ towards the interval $[-1, +1]$, one obtains (up to a factor of 4) the Jacobi operator $L_{\alpha,\beta}$ with parameters $\alpha = \frac{n}{2}$, $\beta = \frac{p}{2}$. It is then seen to satisfy the curvature-dimension condition $CD(\frac{n+p-2}{4}, n + p - 1)$, this property being easily verified even when n and p are no longer integers. However, in this dissymmetric situation, there are many other curvature-dimension conditions which are not immediately comparable with one another. For example, for $n = p$, the first interpretation leads to an optimal condition $CD(n - 1, n)$ for $L_{n/2, n/2}$ whereas the second one leads to $CD(\frac{n-1}{2}, 2n - 1)$.

From this geometric interpretation, Ornstein-Uhlenbeck or Laguerre operators may also be viewed as limits of spherical operators when the dimension n goes to infinity on spheres scaled to have radius $\sqrt{n}$. As a significant illustration, and a source of inspiration for many results on the (infinite-dimensional) curvature condition $CD(\rho, \infty)$, (normalized) uniform measures on spheres of dimension n and radius $\sqrt{n}$ converge as $n \to \infty$, when projected on k coordinates, to the standard Gaussian measure on $\mathbb{R}^k$. As mentioned above, a similar convergence holds at the level of the operators and semigroups. This observation confirms the infinite-dimensional character of the Ornstein-Uhlenbeck operator (as well as the Laguerre operator) from the curvature-dimensional condition viewpoint.

The semigroup with generator $L_{\alpha,\beta}$ is called the *Jacobi semigroup* denoted by $(P_t^{\alpha,\beta})_{t \geq 0}$. Like the Ornstein-Uhlenbeck and Laguerre semigroups, the Jacobi semigroup may actually be defined explicitly from its action on the basis of Jacobi orthogonal polynomials (and coincides similarly with the semigroup of the unique essentially self-adjoint extension in the range $\min(\alpha, \beta) > \frac{3}{2}$). The Jacobi semigroup $(P_t^{\alpha,\beta})_{t \geq 0}$ admits kernel densities $p_t^{\alpha,\beta}(x, y)$, $t > 0$, $(x, y) \in [-1, +1] \times [-1, +1]$, with respect to the invariant measure.

The family of Jacobi semigroups also satisfies some remarkable relations. Observe for example that

$$\partial_x \mathrm{L}_{\alpha,\beta} = \bigl(\mathrm{L}_{\alpha+1,\beta+1} - (\alpha + \beta)\,\mathrm{Id}\bigr)\partial_x.$$

This identity translates on the associated semigroups $(P_t^{\alpha,\beta})_{t \geq 0}$ as

$$\partial_x P_t^{\alpha,\beta} = e^{-(\alpha+\beta)t} P_t^{\alpha+1,\beta+1} \partial_x$$

and therefore also leads to a simple relation between $\partial_x J_k^{\alpha,\beta}$ and $J_{k-1}^{\alpha+1,\beta+1}$. In turn, these relations connect the kernels $p_t^{\alpha,\beta}(x, y)$ and $p_t^{\alpha+1,\beta+1}(x, y)$. Indeed, for smooth functions f compactly supported in $(-1, +1)$, writing

$$\int_{[-1,+1]} \partial_x p_t^{\alpha,\beta}(x, y) f(y) d\mu_{\alpha,\beta}(y)$$

$$= e^{-(\alpha+\beta)t} \int_{[-1,+1]} p_t^{\alpha+1,\beta+1}(x, y) \partial_y f(y) d\mu_{\alpha+1,\beta+1}(y)$$

yields after integration by parts

$$e^{(\alpha+\beta)t} \frac{C_{\alpha,\beta}}{C_{\alpha+1,\beta+1}} \partial_x p_t^{\alpha,\beta}(x, y)$$

$$= -\bigl(1 - y^2\bigr)\partial_y p_t^{\alpha+1,\beta+1}(x, y) + \bigl[(\alpha + \beta + 2)y + \alpha - \beta\bigr] p_t^{\alpha+1,\beta+1}(x, y)$$

where $C_{\alpha,\beta} > 0$ is the normalization constant of the beta distribution $\mu_{\alpha,\beta}$. Now, since $\int_{[-1,+1]} \mathrm{L}(x) d\mu_{\alpha,\beta} = \int_{[-1,+1]} \mathrm{L}(x^2) d\mu_{\alpha,\beta} = 0$,

$$\frac{C_{\alpha,\beta}}{C_{\alpha+1,\beta+1}} = \int_{[-1,+1]} (1 - x^2) d\mu_{\alpha,\beta} = \frac{1}{\alpha + \beta + 1}\left(1 - \frac{(\alpha - \beta)^2}{\alpha + \beta}\right),$$

providing an explicit formula linking $p_t^{\alpha+1,\beta+1}$ to $p_t^{\alpha,\beta}$. This formula takes in particular a simpler form for $y = \pm 1$ and $\alpha = \beta$, namely

$$p_t^{\alpha+1,\alpha+1}(x, \pm 1) = \pm \frac{e^{2\alpha t}}{2(\alpha + 1)(2\alpha + 1)} \partial_x p_t^{\alpha,\alpha}(x, \pm 1). \qquad (2.7.13)$$

This formula will be translated below in (2.7.15) into an identity for heat kernels on the sphere.

The representation of Jacobi operators from the spherical Laplacian when the parameters are half-integers leads to nice formulas for the family of Jacobi orthogonal

2.7 Diffusion Semigroups Associated with Orthogonal Polynomials

polynomials via the Poisson formula (2.2.6). Let us for example illustrate the picture when $\alpha = \beta = \frac{n}{2}$, $n \geq 1$. Any function $f(x_1)$ defined on $[-1, +1]$ may be lifted to a zonal function $\hat{f}$ on the sphere by $\hat{f}(x) = f(x \cdot e_1)$ where $e_1 = (1, 0, \ldots, 0) \in \mathbb{S}^n$ (the variable x_1 being with respect to as the first coordinate of a point $x \in \mathbb{S}^n$). Then expand $f \in \mathbb{L}^2(\mu_{\alpha,\alpha})$ with respect to the basis of Jacobi polynomials $(J_k)_{k \in \mathbb{N}}$, say $f(x_1) = \sum_{k \in \mathbb{N}} a_k J_k(x_1)$. Since $\hat{J}_k$ is the restriction to $\mathbb{S}^n$ of a homogeneous harmonic polynomial of degree k, $F_k(X) = r^k \hat{J}_k(x)$ is harmonic in $\mathbb{R}^{n+1}$, where $X = rx$, $x \in \mathbb{S}^n$ (polar coordinates). Then $Q_r f(x_1) = \sum_{k \in \mathbb{N}} a_k r^k J_k(x_1)$, $0 \leq r \leq 1$, is such that $\widehat{Q_r f}$ is the harmonic extension to the unit ball of $\hat{f}$ (read in polar coordinates). This construction is represented by the kernel

$$Q_r f(x_1) = \int_{[-1,+1]} q_r(x_1, y_1) d\mu_{\alpha,\alpha}(y_1)$$

where $q_r(x_1, y_1) = \sum_{k \in \mathbb{N}} r^k J_k(x_1) J_k(y_1)$. On the other hand, by the Poisson formula (2.2.6),

$$\widehat{Q_r f}(x) = \int_{\mathbb{S}^n} \frac{1 - r^2}{|rx - y|^{n+1}} \hat{f}(y) d\sigma_n(y).$$

Choosing $x = e_1$ in the previous formula leads by identification to

$$\sum_{k \in \mathbb{N}} r^k J_k(1) J_k(y_1) = \frac{1 - r^2}{(1 + r^2 - 2ry_1)^{(n+1)/2}} \quad (2.7.14)$$

which is known as the generating function for the symmetric Jacobi polynomials.

The general formula for the Poisson kernel is however a bit more complicated. It requires the representation of a point on $\mathbb{S}^n$ as $(x_1, \sqrt{1 - x_1^2}\,\hat{x})$, where $\hat{x} \in \mathbb{S}^{n-1}$, which then leads to

$$\sum_{k \in \mathbb{N}} r^k J_k(x_1) J_k(y_1)$$

$$= \int_{[-1,+1]} \frac{1 - r^2}{[1 + r^2 - 2rx_1 y_1 + 2rs\sqrt{(1 - x_1^2)(1 - y_1^2)}]^{\frac{n+1}{2}}} d\mu_{\alpha-1, \alpha-1}(s).$$

Extending this formula to the general case where n is no longer an integer is then easy and amounts to observing that both sides are solutions of the differential equation $(\partial_r^2 + \frac{n}{r}\partial_r + \frac{1}{r^2} L_{\alpha,\alpha}) F = 0$, the remaining argument being similar.

The case of non-symmetric Jacobi operators may essentially be treated similarly (at the expense of more complicated expressions), starting from the half-integer case for the geometric picture and then extending to arbitrary parameters.

Conversely, the interpretation of the Jacobi operator as the spherical Laplacian acting on functions which depend only on the first coordinate (that is, functions depending only on the distance from $(1, 0, \ldots, 0)$ on the sphere) leads to useful information about the sphere $\mathbb{S}^n$ itself. First, the image measure of the uniform measure σ_n on $\mathbb{S}^n$ via the map $x = (x_1, \ldots, x_{n+1}) \mapsto x_1$ is the measure $\mu_{n/2, n/2}$, and its image measure via $(x_1, \ldots, x_{n+1}) \mapsto 2 \sum_{i=1}^{p} x_i^2 - 1$ is $\mu_{(n+1-p)/2, p/2}$. Moreover, the spherical heat kernels $p_t(x, y)$, which, as we saw in Sect. 2.2, depend only on the distance $d(x, y)$ between $x, y \in \mathbb{S}^n$, may be expressed in terms of the kernel densities associated with the Jacobi operators. More precisely, if

$$p_t(x, y) = s_n(t, d(x, y)) = \hat{s}_n(t, x \cdot y), \quad t > 0, \ (x, y) \in \mathbb{S}^n \times \mathbb{S}^n,$$

then $\hat{s}_n(t, u) = p_t^{n/2, n/2}(1, u)$ where $p_t^{n/2, n/2}(x, y)$ is the kernel associated with the Jacobi operator $\mathrm{L}_{n/2, n/2}$. Now, the recurrence formula (2.7.13) leads to a recurrence formula for $\hat{s}_n$, namely

$$\hat{s}_{n+2}(t, u) = \frac{e^{nt}}{(n+1)(n+2)} \partial_u \hat{s}_n(t, u), \qquad (2.7.15)$$

very similar to the corresponding formula for the hyperbolic space given in (2.3.2).

2.8 Notes and References

Again, this chapter gathers in an informal way various models and results from different areas and perspectives. The references below only include a few pointers to the literature.

A standard and most valuable introduction to Euclidean and spherical harmonic analysis covering most of Sects. 2.1 and 2.2 is the monograph [389]. The more geometric (and Riemannian) aspects of the sphere and hyperbolic models are discussed in e.g. [159, 194, 240, 260, 346, 354, 409]. The recent [187] is a good source on hyperbolic space and Laplacian from the probabilistic viewpoint. The explicit expression for the hyperbolic heat kernels may be found in [144].

Dirichlet and Neumann boundary conditions are standard conditions in the analysis of partial differential equations on bounded open sets in $\mathbb{R}^n$ (cf. [96, 179, 203]). The probabilistic aspects of Brownian motion killed or reflected at the boundary of a half-line or an interval, briefly described in Sects. 2.4 and 2.5, may be found in the standard references on stochastic calculus, including [255, 263, 350, 358].

A recent reference on Sturm-Liouville operators on the line is [447].

The examples of Sect. 2.7 are discussed in most textbooks on orthogonal polynomials including [131, 402] and the recent [268] where many useful formulas are displayed. The characterization of diffusion operators on the line which are diagonalized with respect to orthogonal polynomials, leading to the three examples of the Hermite, Laguerre and Jacobi semigroups, was achieved by O. Mazet [301]. Recent developments in higher dimensions are investigated in [41]. Infinite-dimensional Ornstein-Uhlenbeck operators (Sect. 2.7.2) are presented in [26, 82, 142, 257, 308, 334, 417], and in the context of the analysis of Brownian paths on a manifold in [251, 297, 391].

Chapter 3
Symmetric Markov Diffusion Operators

The aim of this chapter is to provide a general framework for the investigation of symmetric Markov diffusion semigroups and operators. It is based on the early observations of Chap. 1 and on the investigation of the model examples in Chap. 2. This chapter develops all the fundamental properties which will justify the computations in the following chapters in the most convenient framework. It is not intended to be read linearly, and could be skipped on first reading. The interested reader should come back to this material when specific technical details have to be understood (such as, for example, the use of curvature conditions towards gradient bounds, which lies at the heart of much of the subsequent analysis). A summary of the framework is presented at the end of the chapter.

The various questions addressed here are actually somewhat delicate. While the chapter aims at describing a general abstract setup, it clearly cannot encompass all the instances of interest. Moreover, each specific example often requires an independent analysis with suitable assumptions and hypotheses. Therefore, we do not try to be maximally general at each level, instead emphasizing the tools and methods which may then be suitably adapted to the cases of interest. These issues in particular concern the classes of functions on which the various operators and semigroups are analyzed. Again, we present the main features of the classes of functions required to develop the investigation rather than carefully describing each example. In particular, the exposition freely emphasizes the formal and inspiring arguments of the Γ-calculus and heat flow monotonicity on which most of the following functional inequalities will be based. Nevertheless, on the basis of the analysis of diffusion operators on complete manifolds in Sect. 3.2, Sect. 3.3 develops a complete and self-contained framework fully justifying the various results and conclusions

In Chap. 1, we investigated the analysis of Markov semigroups starting from the knowledge of the semigroup itself (sometimes in a concrete instance), and introduced the infinitesimal generator together with its domain as a tool to describe it. In this chapter, we adopt the reverse point of view, analyzing the semigroup based on the properties of an operator (a second order differential operator), or rather its carré du champ operator Γ, leading to the basic Markov Triple structure (E, μ, Γ) which will form the natural environment for future investigations.

The operator and carré du champ operator are initially defined on some set $\mathcal{A}_0$ of functions (typically the set of smooth compactly supported functions). The analysis of differential diffusion operators on smooth manifolds leads us to consider a larger class $\mathcal{A}$ (playing the role of smooth functions without support conditions), and to precisely analyze the relationships between the two classes $\mathcal{A}_0$ and $\mathcal{A}$, in particular with regard to integration by parts formulas. Throughout, when working with classes of (real-valued) functions (rather than points), it should be kept in mind that equalities and inequalities between functions should always be understood to hold almost everywhere (with respect to the underlying invariant measure).

The first section describes the main setting consisting of a triple (E, μ, Γ) with state space E, σ-finite measure μ, carré du champ operator Γ acting on an algebra $\mathcal{A}_0$ of functions, and the diffusion property in force throughout this work. This allows us to describe what we call a Standard Markov Triple, which is a convenient setting allowing us to develop the formalism of Markov semigroups towards functional inequalities and convergence to equilibrium. Section 3.2 presents, for the concrete example of second order differential operators on smooth complete connected manifolds, some of the basic issues, properties and hypotheses needed to work with diffusion operators and semigroups on natural classes of functions. This analysis anticipates the further introduction of the extended function algebra $\mathcal{A}$, which enables the translation into the general framework of some central features such as connexity, completeness, weak hypo-ellipticity etc. With the tool of essential self-adjointness at the center of the construction, Sect. 3.3 then emphasizes the relevant hypotheses and properties of Full Markov Triples, covering in particular the examples illustrated in Chaps. 1 and 2. Section 3.4 summarizes the various hypotheses and definitions, introducing the Full Markov Triple as the convenient framework in which most results of the book are presented. The reader may easily refer to it while progressing through the subsequent chapters.

3.1 Markov Triples

This section presents the basic setting for the analysis of (symmetric diffusion) Markov generators and semigroups. As already mentioned, when compared with Chap. 1 and the various model examples in Chap. 2, the definitions put forward here appear in the reverse order, emphasizing on the underlying invariant measure and the carré du champ operator. These are actually the central objects of interest in the investigation of functional inequalities and of their links with curvature conditions. The exposition below makes clear how this framework encompasses the semigroup and generator viewpoint.

3.1.1 Initial Structure

The following definition describes the initial setting of investigation. Recall the notion of a good measurable space (p. 7).

3.1 Markov Triples

Definition 3.1.1 A triple (E, μ, Γ) consists of a good measurable space $(E, \mathcal{F})$ equipped with a σ-finite measure μ, a vector space $\mathcal{A}_0$ of (real-valued) bounded functions dense in all $\mathbb{L}^p(\mu)$-spaces, $1 \leq p < \infty$, stable under products (that is $\mathcal{A}_0$ is an algebra), and a symmetric bilinear map $\Gamma : \mathcal{A}_0 \times \mathcal{A}_0 \to \mathcal{A}_0$ (the carré du champ operator) mapping two functions of $\mathcal{A}_0$ to an element of $\mathcal{A}_0$, such that $\Gamma(f, f) \geq 0$ for every $f \in \mathcal{A}_0$. The fundamental identity

$$\int_E \Gamma(g, f^2) d\mu + 2 \int_E g\, \Gamma(f, f) d\mu = 2 \int_E \Gamma(fg, f) d\mu \tag{3.1.1}$$

holds for any $f, g \in \mathcal{A}_0$. Moreover, there exists an increasing sequence of functions in $\mathcal{A}_0$ converging μ-almost everywhere to the constant function $\mathbb{1}$.

Most often $\Gamma(f, f)$ will be abbreviated as $\Gamma(f)$.

The identity (3.1.1) of the previous definition, while curious at first glance, has a very clear interpretation in the context of Markov chains on a finite set E as presented in Sect. 1.9, p. 33. Indeed, in this case a carré du champ operator has the form

$$\Gamma(f, g)(x) = \frac{1}{2} \sum_{y \in E} a(x, y) [f(y) - f(x)] [g(y) - g(x)], \quad x \in E,$$

for some matrix $a(x, y)$, $(x, y) \in E \times E$. Then, if $\mu(x) > 0$ for any $x \in E$, (3.1.1) is translated into the identity, valid for every pair $(x, y) \in E \times E$,

$$\mu(x) a(x, y) = \mu(y) a(y, x),$$

which amounts to the reversibility (or detailed balance) condition (1.9.1), p. 35, in $\mathbb{L}^2(\mu)$ of the associated Markov operator $Lf(x) = \sum_{y \in E} a(x, y) [f(y) - f(x)]$. If $a(x, y)$, $(x, y) \in E \times E$, does not satisfy this balance condition, the symmetrized matrix

$$\widehat{a}(x, y) = a(x, y) + \frac{\mu(y)}{\mu(x)} a(y, x), \quad (x, y) \in E \times E,$$

does, and if $\widehat{\Gamma}$ is then the associated carré du champ operator, for any function $f : E \to \mathbb{R}$, $\int_E \Gamma(f) d\mu = \int_E \widehat{\Gamma}(f) d\mu$ so that everything is coherent.

It should be emphasized that Definition 3.1.1 involves an algebra $\mathcal{A}_0$ on which the carré du champ operator Γ is acting. This class is most often clear from the context, although its properties should be carefully described, as will be developed in this chapter. Typically, on a smooth manifold, $\mathcal{A}_0$ will be the set of smooth compactly supported functions. On the basis of this example, and in order to develop a suitable calculus on the carré du champ operator, it is necessary to deal with expressions such as $\Gamma(fg, h)$ or $\Gamma(\psi(f), g)$ for smooth functions ψ on the real line. When the measure μ is infinite, constant functions are not in $\mathbb{L}^2(\mu)$ and therefore it is necessary to restrict the functions ψ to those satisfying $\psi(0) = 0$. It is therefore natural to assume that $\mathcal{A}_0$ is an algebra stable under the action of smooth functions vanishing at 0. Note that, on the other hand, it is also necessary to assert that $\Gamma(f, \mathbb{1}) = 0$, so that Γ will have to be extended later to a wider class containing constant functions.

Remark 3.1.2 Working with classes of functions rather than points, it should be emphasized that the preceding setting already ensures that the algebra $\mathcal{A}_0$ is rich enough in a measurable sense. For example, if $f \in \mathbb{L}^1(\mu)$ and if $\int_E gf d\mu \geq 0$ for every positive $g \in \mathcal{A}_0$, then $f \geq 0$ (μ-almost everywhere). Indeed, given $A \in \mathcal{F}$ with $\mu(A) < \infty$, there exists a sequence $(f_k)_{k \in \mathbb{N}}$ in $\mathcal{A}_0$ converging to $\mathbb{1}_A$ in $\mathbb{L}^1(\mu)$, and therefore, up to the choice of a subsequence, μ-almost everywhere. By the stability by composition with functions vanishing at 0, $g_k = 2f_k^2(1 + f_k^2)^{-1}$, $k \in \mathbb{N}$, forms a sequence of functions in $\mathcal{A}_0$, uniformly bounded from above and converging μ-almost everywhere to $\mathbb{1}_A$. By dominated convergence, $\int_E g_k f d\mu \to \int_A f d\mu$ as $k \to \infty$ and the conclusion follows. In the same spirit, whenever $\mu(A) > 0$, there exists a positive g in $\mathcal{A}_0$ such that $\int_A g d\mu > 0$.

3.1.2 Diffusion Property

The central diffusion hypothesis will be in force in most of this book. While presented for the generator in Sect. 1.11, p. 42, it is defined here equivalently on the carré du champ operator.

Definition 3.1.3 (Diffusion property) Let (E, μ, Γ) be a triple as in Definition 3.1.1 relative to an algebra $\mathcal{A}_0$ stable under the action of smooth $(\mathcal{C}^\infty)$ functions vanishing at 0. The carré du champ operator Γ is said to have the diffusion property if for any choice $f_1, \ldots, f_k, g$ of functions in $\mathcal{A}_0$, and any smooth $(\mathcal{C}^\infty)$ function $\Psi : \mathbb{R}^k \to \mathbb{R}$ vanishing at 0,

$$\Gamma\big(\Psi(f_1, \ldots, f_k), g\big) = \sum_{i=1}^k \partial_i \Psi(f_1, \ldots, f_k) \Gamma(f_i, g). \qquad (3.1.2)$$

Recall that equalities between functions (in $\mathcal{A}_0$ here) are understood μ-almost everywhere. In other words, $\Gamma(f, g)$ is a derivation on each of its arguments. Restricted to polynomial functions Ψ, the preceding boils down to the *chain rule*

$$\Gamma(fg, h) = f \Gamma(g, h) + g \Gamma(f, h) \qquad (3.1.3)$$

for any choice of f, g, h in $\mathcal{A}_0$. Under the diffusion hypothesis, it is clear that the fundamental identity (3.1.1) is satisfied.

3.1.3 Diffusion Operators

On the basis of some of the material of Chap. 1, we now describe how the preceding framework actually includes the more usual description in terms of semigroups and

3.1 Markov Triples

generators. This discussion will lead us to enrich the underlying algebra $\mathcal{A}_0$ with several properties which, in addition to the diffusion property, will then form the basic (diffusion) Markov Triple structure. The various properties of $\mathcal{A}_0$ presented below will be summarized in Sect. 3.4.

Given a triple (E, μ, Γ) as in Definition 3.1.1 together with the diffusion property from Definition 3.1.2, the associated symmetric operator L may then be defined, on $\mathcal{A}_0$, by the integration by parts formula,

$$\int_E g\, Lf\, d\mu = -\int_E \Gamma(f, g)\, d\mu \qquad (3.1.4)$$

for every $f, g \in \mathcal{A}_0$ (cf. (1.6.3), p. 27). In order to uniquely define the generator L on $\mathcal{A}_0$ from (3.1.4), it is necessary to assume that for any $f \in \mathcal{A}_0$, there exists a finite constant $C(f)$ such that for any $g \in \mathcal{A}_0$,

$$\left| \int_E \Gamma(f, g)\, d\mu \right| \leq C(f)\, \|g\|_2 \qquad (3.1.5)$$

so that $g \mapsto \int_E \Gamma(f, g)\, d\mu$ may be extended to a continuous linear form on $\mathbb{L}^2(\mu)$. (Note that Lf is then, as usual, only defined as a class of functions in $\mathbb{L}^2(\mu)$.)

The next basic requirement on $\mathcal{A}_0$ is that for any $f \in \mathcal{A}_0$, $Lf \in \mathcal{A}_0$. Then, the fundamental identity (3.1.1) of Definition 3.1.1 yields that, for every $f \in \mathcal{A}_0$,

$$L(f^2) = 2f\, Lf + 2\Gamma(f) \qquad (3.1.6)$$

(multiply (3.1.1) by $g \in \mathcal{A}_0$ and integrate against μ). By polarisation,

$$L(fg) = f\, Lg + g\, Lf + 2\Gamma(f, g)$$

for all $f, g \in \mathcal{A}_0$.

It is also necessary to impose that

$$\int_E Lf\, d\mu = 0$$

for any $f \in \mathcal{A}_0$, what was identified as the *invariance* (or *stationarity*) of the measure μ in Chap. 1. Of course, if constant functions belong to $\mathcal{A}_0$, this is automatically true due to integration by parts (since then $\Gamma(f, \mathbb{1}) = 0$ thanks to the chain rule). It is however not wise in general to assume that constant functions belong to $\mathcal{A}_0$. This property will be automatic once the richer class $\mathcal{A}$ has been introduced in Sect. 3.3 below. By symmetry of Γ, it is clear that the invariant measure μ is *reversible for* L in the sense that

$$\int_E g\, Lf\, d\mu = \int_E f\, Lg\, d\mu \qquad (3.1.7)$$

for every pair $(f, g) \in \mathcal{A}_0 \times \mathcal{A}_0$.

By the diffusion property of the carré du champ operator Γ (Definition 3.1.3), the operator L satisfies the analogous diffusion property (1.11.4), p. 43, that is

$$L\big(\Psi(f_1,\ldots,f_k)\big) = \sum_{i=1}^{k} \partial_i \Psi(f_1,\ldots,f_k) L f_i \qquad (3.1.8)$$
$$+ \sum_{i,j=1}^{k} \partial_{ij}^2 \Psi(f_1,\ldots,f_n) \Gamma(f_i,f_j)$$

for $\Psi : \mathbb{R}^k \to \mathbb{R}$ smooth (of class $\mathcal{C}^\infty$), $\Psi(0) = 0$, and $f_1, \ldots, f_k$ in $\mathcal{A}_0$. To verify this assertion, consider for simplicity the case where Ψ depends only on one variable, so that the formula to be satisfied is

$$L\big(\Psi(f)\big) = \Psi'(f) L f + \Psi''(f) \Gamma(f) \qquad (3.1.9)$$

(for every f in $\mathcal{A}_0$). Now, by the integration by parts formula (3.1.4), this amounts to

$$\int_E \Gamma\big(\Psi(f), g\big) d\mu = \int_E \Gamma\big(f, \Psi'(f)g\big) d\mu - \int_E g \Psi''(f) \Gamma(f) d\mu$$

for every $g \in \mathcal{A}_0$, which appears as an immediate consequence of the diffusion property for Γ. Conversely, starting from a generator L, the change of variables formula (3.1.9) for $\Psi(r) = r^2$ yields in return the carré du champ operator Γ in the form

$$2\Gamma(f) = L(f^2) - 2f L f.$$

Everything is therefore coherent.

Remark 3.1.4 It is perhaps worth deciphering the relationship between the change of variables formulas (3.1.2) and (3.1.8) for Γ and L respectively, and the mere chain rule (3.1.3), as given on p. 43.

On a commutative algebra such as $\mathcal{A}_0$, a first order differential operator (with no 0 order term), also often called a derivation or a vector field, is a linear operator $Z : \mathcal{A}_0 \to \mathcal{A}_0$ satisfying the chain rule $Z(fg) = f Zg + g Zf$. An operator $X : \mathcal{A}_0 \to \mathcal{A}_0$ is a second order differential operator if its associated carré du champ operator $\Gamma_X(f,g) = \frac{1}{2}[X(fg) - f Xg - g Xf]$ is a first order operator in each argument. (One could define in this way differential operators of any order.) A first order operator is then simply a second order operator with vanishing carré du champ operator. Now, it is quite easy to see that, for any first order operator Z, the change of variables formula $Z(\Psi(f_1,\ldots,f_k)) = \sum_{i=1}^{k} \partial_i \Psi(f_1,\ldots,f_k) Zf_i$ is valid for any polynomial function Ψ with $\Psi(0) = 0$ (extended without this restriction provided $\mathcal{A}_0$ contains a unit $\mathbb{1}$ for which $Z(\mathbb{1}) = 0$). In the same way, the change of variable formula (3.1.8) is also valid for any second order differential operator X with polynomial functions Ψ. Therefore, the change of variables formula is simply the extension of the chain rule from polynomials to smooth functions. With these definitions, L is a second order differential operator, and Γ a first order operator in each of its arguments. Moreover, the Γ_2 operator (Definition 3.3.12 below) is a second order differential in each of its arguments, and the Hessian operator $H(f)(g,h)$ (Definition 3.3.13 below) is of second order in f and of first order in g and h.

3.1.4 Dirichlet Form and Domains

Given a triple (E, μ, Γ) as in Definition 3.1.1 with algebra $\mathcal{A}_0$ satisfying the diffusion property and the preceding requirements, the next step is to understand on which natural domain the (diffusion) operator L, given on $\mathcal{A}_0$, may be extended. The first task is to construct the Dirichlet form $\mathcal{E}$ and its domain $\mathcal{D}(\mathcal{E})$ by completion from $\mathcal{A}_0$, and then to build the domain $\mathcal{D}(L)$ of the generator L from this Dirichlet domain. This procedure yields a self-adjoint operator L with $\mathbb{L}^2(\mu)$-domain $\mathcal{D}(L)$, from which the semigroup $\mathbf{P} = (P_t)_{t \geq 0}$ with infinitesimal generator L and invariant and reversible measure μ is then constructed following the lines presented in Appendix A.

To start with, recall the Dirichlet form $\mathcal{E}$ introduced in Definition 1.7.1, p. 30, as

$$\mathcal{E}(f,g) = \int_E \Gamma(f,g) d\mu = -\int_E f \, Lg \, d\mu, \quad (f,g) \in \mathcal{A}_0 \times \mathcal{A}_0. \quad (3.1.10)$$

Also write $\mathcal{E}(f)$ for $\mathcal{E}(f,f)$. Note that by the Cauchy-Schwarz inequality (both for the quadratic form Γ and in $\mathbb{L}^2(\mu)$),

$$|\mathcal{E}(f,g)| \leq \mathcal{E}(f)^{1/2} \mathcal{E}(g)^{1/2}.$$

The Dirichlet form $\mathcal{E}$ is a priori defined only on $\mathcal{A}_0 \times \mathcal{A}_0$ but may actually be extended to a wider class, the so-called Dirichlet domain $\mathcal{D}(\mathcal{E})$. In fact, if $\mathcal{A}_0$ is endowed with the Dirichlet norm

$$\|f\|_{\mathcal{E}} = \left[\|f\|_2^2 + \mathcal{E}(f) \right]^{1/2}, \quad (3.1.11)$$

we may take the completion of $\mathcal{A}_0$ with respect to this norm to turn it into a Hilbert space embedded in $\mathbb{L}^2(\mu)$. For this to be possible, the form must be closable, that is, if a sequence $(f_k)_{k \in \mathbb{N}}$ in $\mathcal{A}_0$ is such that $\|f_k\|_2 \to 0$ and if $\mathcal{E}(f_k - f_\ell) \to 0$ when $k, \ell \to \infty$, then $\mathcal{E}(f_k) \to 0$. But this is actually automatic with the help of the operator L. Indeed, whenever $k > \ell$,

$$\mathcal{E}(f_k) = -\int_E f_k \, L f_\ell \, d\mu + \mathcal{E}(f_k, f_k - f_\ell)$$

$$\leq \|f_k\|_2 \|L f_\ell\|_2 + \mathcal{E}(f_k)^{1/2} \mathcal{E}(f_k - f_\ell)^{1/2}.$$

Fixing ℓ and letting k go to infinity yields the conclusion.

Via this construction, the Dirichlet form $\mathcal{E}$ is now defined on a domain $\mathcal{D}(\mathcal{E})$ which contains $\mathcal{A}_0$, and in which $\mathcal{A}_0$ is dense (in the topology $\|\cdot\|_{\mathcal{E}}$ of the domain as described above). Any function f in $\mathcal{D}(\mathcal{E})$ may thus be approximated in this topology by a sequence $(f_k)_{k\in\mathbb{N}}$ in $\mathcal{A}_0$.

The bilinear carré du champ operator Γ may be extended to $\mathcal{D}(\mathcal{E})$. Indeed, for a Cauchy sequence $(f_k)_{k\in\mathbb{N}}$ in $\mathcal{D}(\mathcal{E})$, since Γ is positive, for every $k, \ell \in \mathbb{N}$,

$$\left|\Gamma(f_k)^{1/2} - \Gamma(f_\ell)^{1/2}\right|^2 \leq \Gamma(f_k - f_\ell)$$

so that the sequence $(\Gamma(f_k)^{1/2})_{k\in\mathbb{N}}$ is a Cauchy sequence in $\mathbb{L}^2(\mu)$. This allows us to define in a proper way $\Gamma(f)$, and then $\Gamma(f, g)$ for $f, g \in \mathcal{D}(\mathcal{E})$, as an element of $\mathbb{L}^1(\mu)$ such that $\mathcal{E}(f) = \int_E \Gamma(f) d\mu$.

At this stage, it is certainly worth making the following observation.

Proposition 3.1.5 *If $f \in \mathcal{D}(\mathcal{E})$ and $h \in \mathcal{A}_0$, then $hf \in \mathcal{D}(\mathcal{E})$ (in other words, $\mathcal{D}(\mathcal{E})$ is an $\mathcal{A}_0$-module). More precisely, whenever a sequence $(f_k)_{k\in\mathbb{N}}$ in $\mathcal{A}_0$ converges in $\mathcal{D}(\mathcal{E})$ to f, then the sequence $(hf_k)_{k\in\mathbb{N}}$ converges in $\mathcal{D}(\mathcal{E})$ to hf. Moreover, for any f, g in $\mathcal{D}(\mathcal{E})$ and any $h \in \mathcal{A}_0$, $\Gamma(hf, g) = h\Gamma(f, g) + f\Gamma(h, g)$.*

Proof From the diffusion property, for $f, h \in \mathcal{A}_0$,

$$\mathcal{E}(hf) = \int_E \Gamma(hf) d\mu \leq 2 \int_E \left(h^2 \Gamma(f) + f^2 \Gamma(h)\right) d\mu$$

$$\leq 2 \max\left(\|h\|_\infty^2, \|\Gamma(h)\|_\infty\right) \|f\|_\mathcal{E}.$$

This inequality applied to an approximating sequence yields the result. The second claim of the proposition is achieved in the same way. □

As an illustration, if the underlying state space E is $\mathbb{R}^n$ (or a manifold), and μ has a density with respect to the Lebesgue measure, $\mathcal{D}(\mathcal{E})$ corresponds to the set of functions with one derivative in $\mathbb{L}^2(\mu)$ (or more precisely for which $\Gamma(f) \in \mathbb{L}^1(\mu)$). We shall be making strong use of the Dirichlet domain $\mathcal{D}(\mathcal{E})$ throughout this work. Any formula involving only $\mathcal{E}(f)$, as is the case for many of the functional inequalities discussed in this book, will automatically extend from functions in $\mathcal{A}_0$ to functions in this Dirichlet domain.

Once the Dirichlet form $\mathcal{E}$ and its domain $\mathcal{D}(\mathcal{E})$ have been constructed, the operator L may be extended from $\mathcal{A}_0$ to its domain $\mathcal{D}(L)$ which is defined in the following way. According to (3.1.5), say that a function $f \in \mathcal{D}(\mathcal{E})$ belongs to $\mathcal{D}(L)$ if there exists a finite constant $C(f)$ such that

$$\mathcal{E}(f, g) \leq C(f) \|g\|_2$$

for every $g \in \mathcal{D}(\mathcal{E})$. The extension of L to $\mathcal{D}(L)$ (again denoted by L) is then achieved as above via the integration by parts formula (3.1.4). It turns out that such an operator L is then always self-adjoint and is the infinitesimal generator of a symmetric semigroup $\mathbf{P} = (P_t)_{t\geq 0}$ of contractions in $\mathbb{L}^2(\mu)$ with invariant reversible measure μ. Moreover $\mathcal{D}(L)$ coincides with the $\mathbb{L}^2(\mu)$-domain as the generator of

3.1 Markov Triples

a semigroup of operators on a Banach space (cf. Sect. 1.4, p. 18, and Sect. A.1, p. 473). On $\mathcal{D}(L)$, the semigroup $(P_t)_{t \geq 0}$ with generator L solves the heat equation

$$\partial_t P_t = L P_t = P_t L$$

(and accordingly is often called the heat semigroup or heat flow with respect to L).

It now remains to check the positivity preserving and Markov properties of $(P_t)_{t \geq 0}$, which will be addressed in the next sub-sections.

3.1.5 Positivity Preserving Property

That the associated semigroup $(P_t)_{t \geq 0}$ is positivity preserving (i.e. $P_t f \geq 0$ whenever $f \geq 0$) is not at all clear from the above construction. The proof given via (1.4.5), p. 21, is purely formal and cannot be turned into a real proof without further hypotheses on the triple (E, μ, Γ), thus we require another approach. Indeed, in the general theory of Dirichlet forms, in order for $(P_t)_{t \geq 0}$ (constructed from $\mathcal{E}$) to be positivity preserving it suffices that the Dirichlet domain $\mathcal{D}(\mathcal{E})$ is stable under the maps $f \mapsto \psi(f)$, where $\psi : \mathbb{R} \to \mathbb{R}$ is a smooth function such that $\psi(0) = 0$ and $|\psi'| \leq 1$, and moreover that for such ψ's,

$$\mathcal{E}(\psi(f)) \leq \mathcal{E}(f). \tag{3.1.12}$$

In particular, $\mathcal{E}(f - \psi(f), f) \geq 0$, and whenever $f \in \mathcal{D}(L)$, $\int_E (f - \psi(f)) L f \, d\mu \leq 0$. Indeed, assuming (3.1.12), let us show that whenever $f \in \mathbb{L}^2(\mu)$ is such that $f \leq 1$, then $P_t f \leq 1$. To this end, we introduce, for $\lambda > 0$, the resolvent operator $R_\lambda = \int_0^\infty e^{-\lambda s} P_s \, ds$, which is bounded on $\mathbb{L}^2$ by spectral decomposition (cf. (A.1.2), p. 474). It is known that $R_\lambda = (\lambda \operatorname{Id} - L)^{-1}$, and that for any $f \in \mathbb{L}^2(\mu)$, $R_\lambda(f) \in \mathcal{D}(L)$. Moreover

$$P_t = \lim_{\lambda \to \infty} e^{-\lambda t} \sum_{k \in \mathbb{N}} \frac{t^k}{k!} (\lambda^2 R_\lambda)^k$$

which is clear from spectral decomposition (cf. Appendix A), and therefore, in order to prove that $P_t f \leq 1$, it is enough to show that $g = \lambda R_\lambda(f) \leq 1$. Now, $\lambda g = L g + \lambda f$, and from the preceding, for every smooth ψ with $\psi(0) = 0$ and $|\psi'| \leq 1$,

$$\lambda \int_E g(g - \psi(g)) d\mu = \int_E L g (g - \psi(g)) d\mu + \lambda \int_E f(g - \psi(g)) d\mu$$

$$\leq \lambda \int_E f(g - \psi(g)) d\mu.$$

Thus

$$\int_E (f - g)(g - \psi(g)) d\mu \geq 0.$$

This property, established for any such smooth function ψ, easily extends to the function $\psi(r) = r \wedge 1$, $r \in \mathbb{R}$, from which

$$\int_{\{g>1\}} (f-g)(g-1) d\mu \geq 0.$$

Since $f \leq 1$, it follows that $\mu(g > 1) = 0$. By homogeneity, for any $a > 0$, whenever $f \leq a$, we have $P_t f \leq a$, and therefore if $f \leq 0$, then $P_t f \leq 0$ which is the announced positivity property. The argument ensures in the same way that the associated semigroup is sub-Markov (that is $P_t(\mathbb{1}) \leq \mathbb{1}$, where $\mathbb{1}$ is understood in the sense of Definition 3.1.1).

Now, restricting to the case of diffusion operators, (3.1.12) easily holds. Indeed, it is not hard to show that if ψ is smooth with $\psi(0) = 0$ and ψ' bounded, and if $f \in \mathcal{D}(\mathcal{E})$, then $\psi(f) \in \mathcal{D}(\mathcal{E})$ and $\Gamma(\psi(f)) = \psi'^2(f)\Gamma(f)$. To see why $\psi(f) \in \mathcal{D}(\mathcal{E})$, for a sequence $(f_k)_{k \in \mathbb{N}} \in \mathcal{A}_0$ converging to $f \in \mathcal{D}(\mathcal{E})$ in the Dirichlet norm and μ-almost everywhere, and for $k, \ell \in \mathbb{N}$, we have

$$\mathcal{E}(\psi(f_k) - \psi(f_\ell)) \leq \int_E \big[\psi'(f_k)^2 \Gamma(f_k - f_\ell) + \big(\psi'(f_k)^2 - \psi'(f_\ell)^2\big)\Gamma(f_k) \\ + 2\big|\psi'(f_k)\big|\big|\psi'(f_k) - \psi'(f_\ell)\big|\Gamma(f_k)^{1/2}\Gamma(f_\ell)^{1/2}\big] d\mu.$$

Since the sequence $(\Gamma(f_k))_{k \in \mathbb{N}}$ is uniformly integrable, it follows that $(\psi(f_k))_{k \in \mathbb{N}}$ is a Cauchy sequence in $\mathcal{D}(\mathcal{E})$ and therefore convergent. Hence (3.1.12) is satisfied, as announced.

Recall that the semigroup just constructed is moreover sub-Markov (that is, it is positivity preserving and satisfies $P_t(\mathbb{1}) \leq \mathbb{1}$) rather than Markov ($P_t(\mathbb{1}) = \mathbb{1}$). Criteria for the latter to hold (Definition 3.1.12) will be presented later.

The preceding construction is the so-called Friedrichs self-adjoint extension of the operator L initially given on $\mathcal{A}_0$ as described in Sect. A.3, p. 477, for which

$$\mathcal{A}_0 \subset \mathcal{D}(L) \subset \mathcal{D}(\mathcal{E}).$$

Basically, the domain $\mathcal{D}(\mathcal{E})$ of the Dirichlet form is the $\mathbb{L}^2(\mu)$-domain of $(-L)^{1/2}$.

3.1.6 Semigroup and Spectral Decomposition

As already emphasized in Sect. 1.4, p. 18, and Sect. 1.6, p. 24, and above, the construction of the Dirichlet form $\mathcal{E}$ with domain $\mathcal{D}(\mathcal{E})$ and of the self-adjoint operator L with $\mathbb{L}^2(\mu)$-domain $\mathcal{D}(L)$ yields a Markov or sub-Markov semigroup $\mathbf{P} = (P_t)_{t \geq 0}$ on $(E, \mathcal{F})$ with infinitesimal generator L and invariant reversible measure μ. The semigroup viewpoint has several advantages and actually allows us to identify further properties of interest.

3.1 Markov Triples

Consider, for example, for a given function f in $\mathbb{L}^2(\mu)$ and $t > 0$, the expression

$$\frac{1}{t}\int_E f(f - P_t f)d\mu = \frac{1}{t}\left(\int_E f^2 d\mu - \int_E (P_{t/2}f)^2 d\mu\right). \tag{3.1.13}$$

By (1.7.2), p. 30, this expression is decreasing in t. The domain $\mathcal{D}(\mathcal{E})$ of the Dirichlet form may then be described equivalently as the set of functions $f \in \mathbb{L}^2(\mu)$ for which the limit of this expression is finite when $t \to 0$. If f is in the $\mathbb{L}^2(\mu)$-domain $\mathcal{D}(L)$ of the generator L with semigroup $\mathbf{P} = (P_t)_{t \geq 0}$, it is also in the Dirichlet domain $\mathcal{D}(\mathcal{E})$ by the integration by parts formula $\mathcal{E}(f) = -\int_E f L f \, d\mu$.

The expression (3.1.13) may moreover be rewritten using the kernels $p_t(x, dy)$ of the representation (1.2.4), p. 12, as

$$\frac{1}{2t}\int_E \int_E [f(x) - f(y)]^2 p_t(x, dy) d\mu(x). \tag{3.1.14}$$

In this form, we now observe that the expression is decreasing if f is replaced by $\psi(f)$ where $\psi : \mathbb{R} \to \mathbb{R}$ is a contraction in the sense that $|\psi(r) - \psi(s)| \leq |r - s|$, $r, s \in \mathbb{R}$. Therefore, the Dirichlet domain $\mathcal{D}(\mathcal{E})$ is stable under the action of such contractions ψ and, for any f in $\mathcal{D}(\mathcal{E})$,

$$\mathcal{E}(\psi(f)) \leq \mathcal{E}(f). \tag{3.1.15}$$

With respect to (3.1.12), the latter therefore also covers contractions ψ which are not necessarily smooth. Moreover, using Fatou's lemma, whenever a sequence $(f_k)_{k \in \mathbb{N}}$ in $\mathcal{D}(\mathcal{E})$ converges in $\mathbb{L}^2(\mu)$ to $f \in \mathcal{D}(\mathcal{E})$, then, since the expression (3.1.14) is decreasing in t,

$$\mathcal{E}(f) \leq \liminf_{k \to \infty} \mathcal{E}(f_k). \tag{3.1.16}$$

In particular, for any $a \in \mathbb{R}$, there is a sequence of smooth approximations $\psi_k(r)$, $k \geq 1$, of the function $r \mapsto r$ such that ψ_k' vanishes on $(a - \frac{1}{k}, a + \frac{1}{k})$. Applying (3.1.16) to $f_k = \psi_k(f)$, $k \geq 1$, then yields that, for any $f \in \mathcal{D}(\mathcal{E})$, $\mathcal{E}(f) \leq \int_{\{f \neq a\}} \Gamma(f) d\mu$. Therefore, for every $f \in \mathcal{D}(\mathcal{E})$ and every $a \in \mathbb{R}$,

$$\int_{\{f=a\}} \Gamma(f) d\mu = 0. \tag{3.1.17}$$

This result captures the fact that $\Gamma(f) = 0$ as long as f remains constant. This will be further developed in Chap. 8 as a weak co-area formula.

In this language, the contraction property (3.1.15) along the Dirichlet form may be translated into a contraction property for Γ itself (even without the diffusion assumption). Indeed, for $f \in \mathcal{A}_0$ and $t_k \to 0$, μ-almost everywhere in $x \in E$,

$$\Gamma(f)(x) = \lim_{k \to \infty}\left(\frac{1}{2t_k} P_{t_k}(f^2)(x) - P_{t_k}(f)(x)^2\right).$$

But, if $P_t f(x) = \int_E f(y) p_t(x, dy)$, this last expression may be written as

$$\frac{1}{4t_k} \int_E \int_E \bigl[f(y_1) - f(y_2)\bigr]^2 p_{t_k}(x, dy_1)\, p_{t_k}(x, dy_2),$$

and is therefore clearly decreasing under the action of a contraction ψ. Hence, for any $\psi : \mathbb{R} \to \mathbb{R}$ such that $|\psi(r) - \psi(s)| \leq |r - s|$, $r, s \in \mathbb{R}$, and every $f \in \mathcal{A}_0$,

$$\Gamma\bigl(\psi(f)\bigr) \leq \Gamma(f), \tag{3.1.18}$$

which then extends to functions in $\mathcal{D}(\mathcal{E})$.

We already observed that when ψ is smooth with bounded derivative, and under the diffusion property, the formula $\Gamma(\psi(f)) = \psi'(f)^2 \Gamma(f)$ also immediately extends from $\mathcal{A}_0$ to $\mathcal{D}(\mathcal{E})$. Furthermore, this point of view also yields immediate access to the expression of Γ in terms of the generator as expressed by (3.1.6). Indeed, under the ESA property of Definition 3.1.10, any $f \in \mathcal{D}(L)$ is such that f^2 belongs to the $\mathbb{L}^1(\mu)$-domain and

$$L(f^2) = 2 f L f + 2 \Gamma(f) \tag{3.1.19}$$

as is easily checked using any sequence in $\mathcal{A}_0$ which converges to f in the domain topology. In particular, for any $f \in \mathbb{L}^2(\mu)$ and any $t > 0$, $(P_t f)^2$ belongs to the $\mathbb{L}^1(\mu)$-domain.

The quadratic form $\Gamma(f)$ may be thought of in different ways according as $f \in \mathcal{A}_0$ or only in $\mathcal{D}(\mathcal{E})$, but the properties of Γ which may be checked on $\mathcal{A}_0$ are in general easily extended to $\mathcal{D}(\mathcal{E})$, provided some care is taken. For example, let $\psi : \mathbb{R} \to \mathbb{R}$ be smooth so that ψ'' is bounded. When $\mathcal{A}_0$ is dense in $\mathcal{D}(L)$, if $f \in \mathcal{D}(L)$, then $\psi'(f) \in \mathbb{L}^2(\mu)$ and $\psi(f) \in \mathbb{L}^1(\mu)$. In this case, $\psi(f)$ also belongs to the $\mathbb{L}^1(\mu)$-domain of L (defined in Sect. 1.4.1, p. 18) and

$$L\bigl(\psi(f)\bigr) = \psi'(f) L f + \psi''(f) \Gamma(f).$$

Again, this identity is proved by approximation by functions in $\mathcal{A}_0$ exactly as in (3.1.19) above (corresponding to $\psi(r) = r^2$). We thus recover along these lines the diffusion property for L.

This semigroup language will feature in numerous basic proofs relying on interpolation along $(P_t)_{t \geq 0}$ via a formula sometimes known as *Duhamel's formula*, at the heart of *heat flow monotonicity* or *semigroup interpolation*. A typical example, already used in (1.7.2), p. 30, of particular use and importance throughout this monograph, is the following. Fix $t > 0$, and consider, for $f \in \mathbb{L}^2(\mu)$,

$$\Lambda(s) = P_s\bigl((P_{t-s} f)^2\bigr), \quad s \in [0, t]$$

(at almost any point, omitted as usual). Then, from (3.1.19),

$$\Lambda'(s) = P_s\bigl(L(P_{t-s}f)^2 - 2 P_{t-s} f L P_{t-s} f\bigr) = 2 P_s\bigl(\Gamma(P_{t-s}f)\bigr) \tag{3.1.20}$$

3.1 Markov Triples   131

in $\mathbb{L}^1(\mu)$. Taking integrals between 0 and t yields a useful variance representation along the semigroup,

$$P_t(f^2) - (P_t f)^2 = 2 \int_0^t P_s\big(\Gamma(P_{t-s}f)\big) ds. \qquad (3.1.21)$$

Such representations and arguments will be developed at length in Chaps. 4, 5 and 6 in order to establish heat kernel inequalities.

The spectral decompositions of L and its associated semigroup $\mathbf{P} = (P_t)_{t \geq 0}$ are further useful tools towards the description of the domains $\mathcal{D}(\mathcal{E})$ and $\mathcal{D}(L)$, as well as of the semigroup $(P_t)_{t \geq 0}$ itself. Several elementary properties in this regard are summarized in the next proposition.

Recall the spectral decomposition $-L = \int_0^\infty \lambda \, dE_\lambda$ of the positive self-adjoint operator $-L$ on $\mathbb{L}^2(\mu)$ as developed in Sect. A.4, p. 478. Also recall from there that $f \in \mathbb{L}^2(\mu)$ if and only if $\int_0^\infty d\langle E_\lambda f, f\rangle < \infty$, and that whenever $g = \int_0^\infty h(\lambda) \, dE_\lambda(f)$, we have $\int_E g^2 d\mu = \int_0^\infty h^2(\lambda) \, d\langle E_\lambda f, f\rangle$.

Proposition 3.1.6 *Let $-L = \int_0^\infty \lambda \, dE_\lambda$ be the spectral decomposition of the opposite of the self-adjoint operator L on $\mathbb{L}^2(\mu)$. The following assertions hold.*

(i) *The domain $\mathcal{D}(L)$ of L is the set of functions $f \in \mathbb{L}^2(\mu)$ such that*

$$\int_E (Lf)^2 d\mu = \int_0^\infty \lambda^2 d\langle E_\lambda f, f\rangle < \infty$$

and $-Lf = \int_0^\infty \lambda \, dE_\lambda f$.

(ii) *The Dirichlet domain $\mathcal{D}(\mathcal{E})$ is the set of functions $f \in \mathbb{L}^2(\mu)$ such that*

$$\mathcal{E}(f) = \int_0^\infty \lambda \, d\langle E_\lambda f, f\rangle < \infty.$$

(iii) *For every $t \geq 0$ and every $f \in \mathbb{L}^2(\mu)$,*

$$P_t f = \int_0^\infty e^{-\lambda t} dE_\lambda f.$$

In particular, for any $t > 0$, $P_t f \in \mathcal{D}(L)$ and

$$\int_E (LP_t f)^2 d\mu \leq \frac{1}{e^2 t^2} \int_E f^2 d\mu.$$

(iv) *For every $t \geq 0$ and every $f \in \mathbb{L}^2(\mu)$,*

$$\mathcal{E}(P_t f) = \int_0^\infty \lambda e^{-2\lambda t} d\langle E_\lambda f, f\rangle.$$

Therefore $\mathcal{E}(P_t f) \leq \frac{1}{2et} \int_E f^2 d\mu$ for $t > 0$.

(v) *For every $f \in \mathbb{L}^2(\mu)$, the map $t \mapsto \mathcal{E}(P_t f)$ is decreasing.*

More or less, all items in this proposition are immediate from the definition of the spectral decomposition. (iii) is a simple consequence of the bound $r^2 e^{-r} \leq 4 e^{-2}$ on $\mathbb{R}_+$, (iv) follows from the bound $r e^{-r} \leq e^{-1}$, while (v), which follows from the fact that $t \mapsto \lambda e^{-2\lambda t}$ is decreasing, may be seen more directly using that

$$\partial_t \mathcal{E}(P_t f) = -\partial_t \int_E P_t f \, \mathrm{L} P_t f \, d\mu = -2 \int_E (\mathrm{L} P_t f)^2 d\mu \leq 0.$$

It may finally be pointed out that Proposition 3.1.6 takes a much simpler form in the case of a discrete spectrum. Indeed, in this case there exists a sequence $(E_k)_{k \in \mathbb{N}}$ of orthogonal closed subspaces of $\mathbb{L}^2(\mu)$ with $\bigoplus_{k \in \mathbb{N}} E_k = \mathbb{L}^2(\mu)$ which are eigenspaces of L associated with the eigenvalues $-\lambda_k$, $\lambda_k \geq 0$, $k \in \mathbb{N}$. When a function $f \in \mathbb{L}^2(\mu)$ is decomposed into $f = \sum_{k \in \mathbb{N}} f_k$, then

$$-\mathrm{L} f = \sum_{k \in \mathbb{N}} \lambda_k f_k, \quad P_t f = \sum_{k \in \mathbb{N}} e^{-\lambda_k t} f_k, \quad \mathcal{E}(f) = \sum_{k \in \mathbb{N}} \lambda_k \| f_k \|_2^2$$

and so on.

Remark 3.1.7 Using the same arguments as in Proposition 3.1.6, it may be shown that if $f \in \mathbb{L}^2(\mu)$, then $P_t f \in \mathcal{D}(\mathrm{L}^k)$, $t > 0$, for any $k \in \mathbb{N}$. Therefore, at the $\mathbb{L}^2(\mu)$ level, even in the absence of any hypo-ellipticity condition, the semigroup is smoothing as an effect of symmetry only. Moreover, if $f \in \mathcal{D}(\mathcal{E})$, then $P_t f$ converges to f in $\mathcal{D}(\mathcal{E})$ as $t \to 0$. This is also true if $f \in \mathcal{D}(\mathrm{L})$ as $\bigcap_k \mathcal{D}(\mathrm{L}^k)$ is dense in $\mathcal{D}(\mathrm{L}^p)$ for any $p \in \mathbb{N}$, again by means of the smoothing properties of the semigroup.

3.1.7 Diffusion Markov Triple

At this point of the construction, we introduce the notion of a symmetric *Diffusion Markov Triple* which summarizes the hypotheses on the triple (E, μ, Γ) and the underlying algebra $\mathcal{A}_0$ presented so far. Section 3.4 below will summarize the various hypotheses on the algebra $\mathcal{A}_0$. Although the semigroup $(P_t)_{t \geq 0}$ is not yet Markov (mass conservative), the somewhat abusive terminology at this point is convenient.

Definition 3.1.8 (Diffusion Markov Triple) A Diffusion Markov Triple is a triple (E, μ, Γ) as in Definition 3.1.1 with the diffusion hypothesis and linear operator $\mathrm{L} : \mathcal{A}_0 \to \mathcal{A}_0$ defined by and satisfying the integration by parts formula

$$\int_E g \mathrm{L} f \, d\mu = -\int_E \Gamma(f, g) d\mu$$

3.1 Markov Triples   133

for every $f, g \in \mathcal{A}_0$. The measure μ is invariant and reversible with respect to L. The associated Dirichlet form $\mathcal{E}$ is defined by

$$\mathcal{E}(f, g) = \int_E \Gamma(f, g) d\mu$$

for every $(f, g) \in \mathcal{A}_0 \times \mathcal{A}_0$. The domain $\mathcal{D}(\mathcal{E})$ of the Dirichlet form $\mathcal{E}$ is the completion of $\mathcal{A}_0$ with respect to the norm $\|f\|_\mathcal{E} = [\|f\|_2^2 + \mathcal{E}(f)]^{1/2}$, and $\mathcal{E}$ is extended to $\mathcal{D}(\mathcal{E})$ by continuity together with the carré du champ operator Γ. The symmetric semigroup of contractions $\mathbf{P} = (P_t)_{t \geq 0}$ with infinitesimal generator L defined on its $\mathbb{L}^2(\mu)$-domain $\mathcal{D}(L)$ is positivity preserving but not necessarily Markov (in general it is only sub-Markov, i.e. $P_t(\mathbb{1}) \leq \mathbb{1}$).

3.1.8 Essential Self-adjointness

The following addresses the question of essential self-adjointness which will turn out to be central in the abstract construction developed in Sect. 3.3 below, in particular concerning mass conservation. Besides the operator L constructed so far, it is necessary to introduce an adjoint operator L^*, where the duality refers both to the measure μ and the algebra $\mathcal{A}_0$.

Definition 3.1.9 (Adjoint operator) A function $f \in \mathbb{L}^2(\mu)$ belongs to $\mathcal{D}(L^*)$ if there exists a finite constant $C(f)$ such that for any $g \in \mathcal{A}_0$,

$$\left| \int_E f \, L g \, d\mu \right| \leq C(f) \|g\|_2.$$

In this situation, $L^* f$ is the unique element of $\mathbb{L}^2(\mu)$ such that for any $g \in \mathcal{A}_0$,

$$\int_E f \, L g \, d\mu = \int_E g \, L^* f \, d\mu.$$

By symmetry of L (cf. (3.1.7)), we always have that $\mathcal{D}(L) \subset \mathcal{D}(L^*)$, and $L^* = L$ on $\mathcal{D}(L)$. However, while L is self-adjoint on $\mathcal{D}(L)$, this does not mean that $L = L^*$ (L^* being defined from $\mathcal{A}_0$). The self-adjointness property of L refers to the dual operator constructed in the same way as L^*, but where $\mathcal{D}(L)$ replaces $\mathcal{A}_0$. In other words, it is not true in general that the algebra $\mathcal{A}_0$ is dense in $\mathcal{D}(L)$ with respect to the norm

$$\|f\|_{\mathcal{D}(L)} = \left[\|f\|_2^2 + \|Lf\|_2^2 \right]^{1/2}. \tag{3.1.22}$$

This happens if and only if $\mathcal{D}(L^*) = \mathcal{D}(L)$. In this case, the extension of L from $\mathcal{A}_0$ to a larger domain as a self-adjoint operator is unique. This property will thus play a central role our analysis, and is called *essential self-adjointness*, which therefore depends on the choice of $\mathcal{A}_0$. The density of $\mathcal{A}_0$ in $\mathcal{D}(L)$ is equivalent to the condition that the operator L defined on $\mathcal{A}_0$ has a unique self-adjoint extension, a property also often referred to as uniqueness extension (see Appendix A, p. 476).

Definition 3.1.10 (Essential self-adjointness—ESA) The operator L is said to be essentially self-adjoint if $\mathcal{D}(L) = \mathcal{D}(L^*)$ (recall that $\mathcal{D}(L^*)$ is defined with respect to $\mathcal{A}_0$). Equivalently, $\mathcal{A}_0$ is dense in $\mathcal{D}(L)$ in the $\mathcal{D}(L)$-topology, that is $\mathcal{A}_0$ is a core for L.

In order to ensure the ESA property, extra conditions are needed. The next section will demonstrate how this property holds for second order elliptic differential operators on smooth complete connected manifolds where $\mathcal{A}_0$ is the class of smooth compactly supported functions. Section 3.3 will address this issue in the present abstract framework.

Examples in Chap. 2 show that there may be many different self-adjoint extensions $\widehat{L}$ of a given operator L defined on $\mathcal{A}_0$, and conditions for the ESA property to hold may depend on the behavior of the measure at the boundary of the domain (Sect. 2.4, p. 92, and Sect. 2.6, p. 97). In general, for various self-adjoint extensions $\widehat{L}$, their domains $\mathcal{D}(\widehat{L})$ are not comparable. To each self-adjoint extension $\widehat{L}$ corresponds a Dirichlet form with domain $\mathcal{D}(\widehat{\mathcal{E}})$ and an associated semigroup (not necessarily Markov), using the procedure described by (3.1.13) below. In contrast to the different domains $\mathcal{D}(\widehat{L})$ however, the associated Dirichlet domains $\mathcal{D}(\widehat{\mathcal{E}})$ may be compared. The domain $\mathcal{D}(\mathcal{E})$ of the Friedrichs extension that we constructed is the minimal one, and corresponds in practice to operators with the Dirichlet boundary conditions. There is also a maximal extension corresponding to the Neumann boundary conditions. In the context developed here, namely if the operator is essentially self-adjoint, the minimal and maximal extensions coincide and there is only one Dirichlet domain (referred to as the unique extension).

3.1.9 Ergodicity and Mass Conservation

It is of great importance in many applications to understand the behavior of the semigroup P_t as $t \to \infty$. From the previous spectral decomposition, it is clear that for any function $f \in \mathbb{L}^2(\mu)$, $P_t f$ converges in $\mathbb{L}^2(\mu)$ to the projection of f on the space of functions which satisfies $Lf = 0$ (see (iii) in Proposition 3.1.6). In many natural situations, such a function will be constant, and therefore vanishes when the measure is infinite. This property will be called *ergodicity* (see Definition 3.1.11 below). When the measure μ is finite (and therefore under our convention a probability measure), this projection should be $\int_E f d\mu$. But for this to hold, it would require the constant function $\mathbb{1}$ to be in the domain of the operator, which implies that $P_t(\mathbb{1}) = \mathbb{1}$. (Recall that according to Definition 3.1.1, $\mathbb{1}$ is understood as the increasing limit of a sequence of elements in $\mathcal{A}_0$.) This property is known as *mass conservation* (or, depending on the context, *Markov, stochastic completeness* or *non-explosion*). Even in the presence of the ESA property, it is not always satis-

3.1 Markov Triples

fied. For instance, Sect. 2.6, p. 97, gives examples of operators which are essentially self-adjoint but nevertheless not conservative.

These observations lead to the following two definitions in the setting of a Diffusion Markov Triple (E, μ, Γ) with associated operator L and semigroup $\mathbf{P} = (P_t)_{t \geq 0}$.

Definition 3.1.11 (Ergodicity) The operator L is said to be ergodic if every $f \in \mathcal{D}(L)$ such that $Lf = 0$ is constant (in particular $f = 0$ if μ has infinite mass). Equivalently, any $f \in \mathcal{D}(\mathcal{E})$ such that $\Gamma(f) = 0$ is constant.

To verify the equivalence of the two possible definitions, note that if $f \in \mathcal{D}(L)$ is such that $Lf = 0$, then $\mathcal{E}(f) = 0$ and therefore $\Gamma(f) = 0$. Conversely, if $f \in \mathcal{D}(\mathcal{E})$ is such that $\Gamma(f) = 0$, then for any $h \in \mathcal{D}(\mathcal{E})$, $\mathcal{E}(f, h) = 0$, and therefore f belongs to $\mathcal{D}(L)$ with $Lf = 0$.

Definition 3.1.12 (Mass conservation) The semigroup $\mathbf{P} = (P_t)_{t \geq 0}$ is said to be conservative if for every $t \geq 0$, $P_t(\mathbb{1}) = \mathbb{1}$.

Observe that since P_t is defined on any bounded measurable function, the definition makes sense (and is in fact of great importance) even when $\mu(E) = \infty$. Section 3.2.4 below presents a useful criterion ensuring mass conservation for second order differential operators on a complete Riemannian manifold. Section 3.3 further develops the abstract framework to deduce this property from essential self-adjointness and curvature conditions. On the other hand, when solving a stochastic differential equation, it is sometimes easy to directly check mass conservation. In particular, we presented in the second chapter (Sect. 2.6, p. 97) criteria in dimension one which are almost necessary and sufficient conditions for this to happen.

From the preceding, the following proposition is immediate.

Proposition 3.1.13 *Assume that ergodicity holds.*

(i) *If μ is infinite, for any $f \in \mathbb{L}^2(\mu)$, $\lim_{t \to \infty} P_t f = 0$ in $\mathbb{L}^2(\mu)$.*
(ii) *If μ is finite and the semigroup is conservative, for any $f \in \mathbb{L}^2(\mu)$,*

$$\lim_{t \to \infty} P_t f = \int_E f d\mu \quad \text{in } \mathbb{L}^2(\mu).$$

This proposition will be related later to a connexity property (Sect. 3.2.1 and Proposition 3.3.10).

Remark 3.1.14 It may be observed that under the ergodicity and the ESA properties, the set $\{Lh \,;\, h \in \mathcal{A}_0\}$ is dense in the space $\mathbb{L}_0^2(\mu)$ of functions in $\mathbb{L}^2(\mu)$ which are orthogonal to constants (that is $\mathbb{L}^2(\mu)$ itself when $\mu(E) = \infty$). Indeed, if $f \in \mathbb{L}^2(\mu)$ is such that, for any $h \in \mathcal{A}_0$, $\int_E f \, Lh \, d\mu = 0$, then $f \in \mathcal{D}(L^*)$ with $L^* f = 0$. From the ESA property, $f \in \mathcal{D}(L)$ and $Lf = 0$. Ergodicity allows us to conclude that f is constant. It is then 0 when $\mu(E) = \infty$. When $\mu(E) = 1$, since for any $h \in \mathcal{A}_0$, $Lh \in \mathbb{L}_0^2(\mu)$, the same conclusion follows working in $\mathbb{L}_0^2(\mu)$ instead of $\mathbb{L}^2(\mu)$.

3.1.10 Standard Markov Triple

At this point of the construction, we emphasize the definition of a *Standard Markov Triple* which is essentially a Diffusion Markov Triple with the ergodicity and mass conservation properties. To lighten the terminology, "Diffusion" is not repeated. The complete definition will be summarized in Sect. 3.4 below.

Definition 3.1.15 (Standard Markov Triple) A Standard Markov Triple is a Diffusion Markov Triple (E, μ, Γ) with an algebra $\mathcal{A}_0$ (the properties of $\mathcal{A}_0$ being summarized in Sect. 3.4) such that the associated operator L and semigroup $\mathbf{P} = (P_t)_{t \geq 0}$ are ergodic and Markov (conservative).

The structure of a Markov Triple induces on the probabilistic side corresponding Markov processes. However, some care has to be taken in the description of the laws of these processes and their regularity properties on classes of functions. Indeed, as described in the first chapter, Sect. 1.3, p. 16, once the Markov semigroup $\mathbf{P} = (P_t)_{t \geq 0}$ is constructed, one may consider the associated Markov process $\mathbf{X} = \{X_t^x ; t \geq 0, x \in E\}$, or more precisely the law of the process $(X_t^x)_{t \geq 0}$ starting from any point $x \in E$. In the preceding context, recall that elements of $\mathcal{A}_0$ are not functions, but classes of functions with respect to the relation of μ-almost everywhere equality. Therefore, properties such as continuity or regularity of the processes $(f(X_t^x))_{t \geq 0}$ for functions $f \in \mathcal{A}_0$ have no meaning per se. What is clearly defined however is the law of the process when the initial distribution μ_0 of X_0 is absolutely continuous with respect to μ, for example $d\mu_0 = \frac{\mathbb{1}_A}{\mu(A)} d\mu$ for any A such that $0 < \mu(A) < \infty$. The law of $(X_t)_{t \geq 0}$ with initial condition $X_0 = x$ is then defined up to sets of μ-measure 0 as conditional expectations. In practice, most of the time the state space is actually a topological space, and moreover is often locally compact. The conditional laws of $(X_t^x)_{t \geq 0}$ then have a version which is continuous (in x) for the weak topology. In the abstract context, while it is meaningless to talk about the law of $(X_t^x)_{t \geq 0}$ for any $x \in E$, when the measure decomposition theorem applies, however, it makes sense for μ-almost every x. In particular, an identity such that $P_t f(x) = \mathbb{E}(f(X_t) \mid X_0 = x)$ is understood μ-almost everywhere as already emphasized.

3.1.11 Intermediate Value Theorem and Slicing

We close this section with two further properties which are relevant in this context.
Even though there is no topology on the state space E, functions in $\mathcal{D}(\mathcal{E})$ share some minimal regularity as a consequence of the diffusion property. In particular they satisfy a measurable form of the intermediate value theorem.

Theorem 3.1.16 (Intermediate value) *Let (E, μ, Γ) be an ergodic Diffusion Markov Triple. Let $f \in \mathcal{D}(\mathcal{E})$. If, for some $a < b$, $\mu(f \leq a) > 0$ and $\mu(f \geq b) > 0$, then for any $a < c < d < b$, $\mu(f \in [c, d]) > 0$.*

Proof Assume that $\mu(f \in [c,d]) = 0$ and choose $\psi : \mathbb{R} \to \mathbb{R}$ smooth and constant outside the interval $[c,d]$ with $\psi(a) \neq \psi(b)$, such that $\psi'(f) = 0$. Since by the diffusion property

$$\int_E \Gamma\big(\psi(f)\big) d\mu = \int_E \psi'(f)^2 \Gamma(f) d\mu = 0,$$

$\psi(f)$ is constant μ-almost everywhere, contradicting the hypotheses $\mu(f \leq a) > 0$ and $\mu(f \geq b) > 0$. The theorem is established. □

The slicing procedure described in the next result will prove to be particularly useful.

Proposition 3.1.17 (Slicing) *Let (E, μ, Γ) be an ergodic Diffusion Markov Triple. For any strictly increasing sequence $(a_k)_{k \in \mathbb{N}}$ of strictly positive numbers, given a function $f : E \to \mathbb{R}_+$, set $f_k = (f - a_k)_+ \wedge (a_{k+1} - a_k)$, $k \in \mathbb{N}$. Then, if $f \in \mathcal{D}(\mathcal{E})$,*

$$\sum_{k \in \mathbb{N}} \mathcal{E}(f_k) \leq \mathcal{E}(f). \tag{3.1.23}$$

The proof of this result is an easy consequence of (3.1.16) and the change of variable formula. Indeed, using some smooth approximation of the function $r \mapsto (r - a_k)_+ \wedge (a_{k+1} - a_k)$, and (3.1.16), it is easily seen that

$$\mathcal{E}(f_k) \leq \int_{\{f \in (a_k, a_{k+1})\}} \Gamma(f) d\mu.$$

An important feature used here is that $\Gamma(f) = 0$ on the set $\{f = a\}$, which is a consequence of (3.1.17).

3.2 Second Order Differential Operators on a Manifold

Throughout this book, manifolds mean smooth connected manifolds (not necessarily complete since, although we do not investigate boundary behaviors, specific drifts as observed in Chap. 2 can make the second order differential operator under investigation essentially self-adjoint).

In this section however, to illustrate the general setting presented in the previous section, we concentrate on the example of a (smooth connected) complete Riemannian manifold equipped with the class $\mathcal{A}_0$ of smooth ($\mathcal{C}^\infty$) compactly supported functions in order to identify the precise issues that need to be addressed in the definition of a Markov Triple (E, μ, Γ). (The next section will cover non-complete instances at the expense of a deeper investigation.) The task will then be to translate directly to the Markov generator L and its carré du champ operator Γ most of the required properties on the manifold, in order to ensure in particular properties such that the density of $\mathcal{A}_0$ in the domain $\mathcal{D}(L)$ or the Markov property $P_t(1) = 1$. In what follows, the manifold could of course be $\mathbb{R}^n$, or some open set in $\mathbb{R}^n$. Indeed, as soon as the coefficients $g^{ij}(x)$ in (3.2.1) below are not constant, the relevant quantities do not differ much in $\mathbb{R}^n$ or in a general manifold. This section makes significant use of the material presented in Appendices A and C.

In the manifold context, here and throughout, the algebra $\mathcal{A}_0$ consists of the smooth $\mathcal{C}^\infty$ compactly supported functions. The degree of smoothness may actually depend on the smoothness of the manifold, however it will be at least $\mathcal{C}^2$. For simplicity, we only consider $\mathcal{C}^\infty$ manifolds and function algebras, for which the generic terminology "smooth" will be used. More general instances may be adapted accordingly.

We start with some general observations. As already pointed out in the last section, when dealing with diffusion operators L and semigroups $(P_t)_{t \geq 0}$ on manifolds, classes of functions which are larger than the class $\mathcal{A}_0$, such as the class of smooth functions without special growth conditions, may be used. It is then necessary to be cautious when using integration by parts. Indeed, as discussed in Sect. 1.6, p. 24, for a symmetric diffusion operator L on a smooth manifold, the integration by parts formula holds as soon as f and g are smooth and compactly supported, and it is actually enough that one of the functions be compactly supported. If neither of the smooth functions f and g is compactly supported, the validity of the formula requires some knowledge on the behavior of the functions f and g at infinity depending on the underlying (reversible) measure μ. The statement that a function f is in the domain $\mathcal{D}(L)$ of L hides two different aspects, a regularity assumption, and a growth condition. Moreover, in the presence of boundaries, there are additional requirements at the boundary (for example, normal derivatives at the boundary have to vanish in the case of Neumann boundary conditions).

Similar comments hold for the associated semigroup $\mathbf{P} = (P_t)_{t \geq 0}$. Indeed, when the manifold is not compact, $P_t f$, $t > 0$, would never be compactly supported even if f is. But the smoothness of $P_t f$ for smooth compactly supported functions f is in general a consequence of two types of criteria: when the semigroup is given as the law of the solution of a stochastic differential equation with global properties such as non-explosion (see Sect. B.4, p. 495), or from local considerations as a consequence of ellipticity or hypo-ellipticity (see Sect. 1.12, p. 49).

It is therefore important to determine a setting in which the integration by parts formula (3.1.4) and other properties hold beyond the preceding class $\mathcal{A}_0$. For example, in the core of the analysis, one is often led to consider quantities such as $\Gamma(P_t f)$. If we restrict ourselves to smooth functions f in the $\mathbb{L}^2(\mu)$-domain of L, $\Gamma(P_t f)$ is certainly not in general in $\mathbb{L}^2(\mu)$. Furthermore, unless L may be defined on any smooth function, the meaning of $L\Gamma(P_t f)$ is not entirely clear. What would then guarantee for a function such as $g = \Gamma(P_t f)$ that $\partial_t P_t g = P_t(Lg)$? Indeed, coming back to the basic examples of symmetric diffusions with density kernels as presented in the preceding chapters, so that for $t > 0$ and $x \in E$,

$$P_t f(x) = \int_E f(y) \, p_t(x, y) d\mu(y)$$

3.2 Second Order Differential Operators on a Manifold

and $L_x p_t(x, y) = L_y p_t(x, y)$, then, assuming that derivation and integration commute,

$$\partial_t P_t f(x) = \int_E f(y) L_x p_t(x, y) d\mu(y)$$
$$= \int_E f(y) L_y p_t(x, y) d\mu(y) = \int_E L f(y) p_t(x, y) d\mu(y).$$

But the last step here is in fact integration by parts and requires some a priori knowledge on the behavior with respect to μ of both $p_t(x, y)$ and $f(y)$ at infinity.

As mentioned above, this discussion may also be illustrated in probabilistic terms on the diffusion semigroups given by solutions of stochastic differential equations as in Sect. 1.10, p. 38. From the representation (1.1.1), p. 8, of the semigroup in terms of the expectation $\mathbb{E}(f(X_t^x))$, it is clear that as soon as the associated Markov process $\mathbf{X} = \{X_t^x ; t \geq 0, x \in E\}$ is continuous and f and Lf are bounded and continuous, then $\partial_t P_t f = P_t L f$. Fortunately, with a minimum of assumptions on the state space emphasized below, this property still holds for smooth functions f in $\mathbb{L}^2(\mu)$ as soon as Lf is also in $\mathbb{L}^2(\mu)$.

Motivated by this discussion, the two main questions and properties which should be addressed consistently are therefore the uniqueness of the symmetric extension of L given the algebra $\mathcal{A}_0$ and the mass conservation (Markov) property $P_t(\mathbb{1}) = \mathbb{1}$. The uniqueness or essentially self-adjointness question (Definition 3.1.10), that is whether $\mathcal{A}_0$ is a core for the operator L (cf. Sect. A.5, p. 481), is an important issue since, when satisfied, any formula valid in $\mathcal{A}_0$ and continuous under the topology of the domain $\mathcal{D}(L)$ would still hold on this domain. The Markov property, on the other hand, allows for the correct probabilistic setting and interpretations.

The aim of this section is to establish these two properties in the context of second order diffusion operators on a smooth complete connected manifold M, for the class $\mathcal{A}_0$ of smooth ($\mathcal{C}^\infty$) compactly supported functions. Therefore, throughout this section, we deal with a second order differential operator L acting in a local system of coordinates on functions $f : M \to \mathbb{R}$ in $\mathcal{A}_0$ as

$$Lf = \sum_{i,j=1}^n g^{ij} \partial_{ij}^2 f + \sum_{i=1}^n b^i \partial_i f \tag{3.2.1}$$

where $(g^{ij}(x))_{1 \leq i, j \leq n}$ is a symmetric positive-definite matrix and $(b^i(x))_{1 \leq i \leq n}$ a vector, both depending smoothly on $x \in M$. The (elliptic) operator L is furthermore assumed to be symmetric with respect to a (σ-finite) measure μ on the Borel sets of M. In $\mathbb{R}^n$ or a Riemannian manifold, μ will be assumed to have a smooth strictly positive density with respect to the Lebegue or Riemannian measure. As announced, this analysis deals simultaneously with another algebra of functions, namely the class $\mathcal{A}$ of smooth ($\mathcal{C}^\infty$) functions on M, which is an extension of $\mathcal{A}_0$, and to which there is in general no difficulty in extending L and the associated carré du champ operator Γ (provided the coefficients defining L are smooth enough). Recall that, on such functions, Γ takes the form

$$\Gamma(f,g) = \sum_{i,j=1}^{n} g^{ij} \partial_i f \, \partial_j g, \qquad (3.2.2)$$

and clearly L and Γ satisfy the diffusion property of Definition 3.1.3. The associated semigroup $\mathbf{P} = (P_t)_{t \geq 0}$ with generator L as constructed from the Dirichlet form in Sect. 3.1 may already be assumed to be sub-Markov ($P_t(\mathbb{1}) \leq \mathbb{1}$).

The exposition here in the manifold case points towards the general setting developed in the next section. We try to identify the various properties used on the manifold and the differential operator in order to express them directly in terms of the operator and its carré du champ operator. The hypotheses put forward in Sect. 3.3 fully rely on this main example. The three basic properties required to achieve this program and uniqueness of the symmetric extension are connexity, completeness and (weak-) hypo-ellipticity, which are discussed in the following subsections.

3.2.1 Connexity

Recalling that $\Gamma(f)$ stands for $|\nabla f|^2$ (for Laplacians on manifolds, for example), a minimum requirement is that if $f : M \to \mathbb{R}$ is constant then $\Gamma(f) = 0$. This is an easy consequence of the diffusion property (3.1.8). Conversely, it is also a natural requirement that if a function f satisfies $\Gamma(f) = 0$, then it is constant. As illustrated next, this property actually hides the connexity of the underlying state space together with some intrinsic form of hypo-ellipticity.

In the context of finite Markov chains (see Sect. 1.9, p. 33), connexity is equivalent to the fact that there is only one recurrence class, or in other words that the Markov chain has a strictly positive probability of moving from any point to any other point in a finite number of steps. In some abstract situation, connexity together with the diffusion property implies some minimal continuity for functions in the Dirichlet space. It shows in particular that the diffusion property may never hold on a discrete space.

If a diffusion operator L is elliptic in the sense of (1.12.2), p. 50, a function $f : M \to \mathbb{R}$ satisfying $\Gamma(f) = 0$ is locally constant. As announced, that locally constant functions are constant entails connexity of the state space. For example, if M is just two copies of $\mathbb{R}$, that is $M = \mathbb{R} \times \{-1, +1\}$ and if, for $\epsilon = \pm 1$, $Lf(x, \epsilon) = \partial_x^2 f(x, \epsilon)$, then it is clear that any function f satisfying $\Gamma(f) = 0$ is constant on each copy of $\mathbb{R}$ (each connected component of M) but may not be globally constant.

The picture may be even more complicated. Consider, on $\mathbb{R}^2$, the operator $Lf = \partial_x^2 f$ for which $\Gamma(f) = (\partial_x f)^2$ so that the connexity property does not hold. Note that L is not hypo-elliptic in the sense described in Sect. 1.12, p. 49. However, in this context, the situation is rather simple. The variable y is just a parameter, and the associated Markov process $(X_t, Y_t)_{t \geq 0}$ starting from (x, y) amounts to $(X_t, y)_{t \geq 0}$ where $(X_t)_{t \geq 0}$ is a one-dimensional Brownian motion (with speed 2)

3.2 Second Order Differential Operators on a Manifold

starting from x. The state space may thus be restricted to $\mathbb{R} \times \{y\}$ and L may be considered as an elliptic operator on $\mathbb{R}$ satisfying the connexity condition. Let now $\mathrm{L} f = \partial_x^2 f + x \partial_y f$ for which, still, $\Gamma(f) = (\partial_x f)^2$ so that the operator is still non-connected. But now, the associated Markov process is of the form $(X_t, y + \int_0^t X_s ds)_{t \geq 0}$ and no longer preserves the second component. It is therefore not possible to restrict here the analysis to $\mathbb{R} \times \{y\}$, and the connexity property does not hold although the operator is now hypo-elliptic.

It may be thought that connexity is related to ellipticity. The following example indicates that it is not the case. On $\mathbb{R}^3$, consider the two vector fields

$$Z_1 = \partial_x - \frac{y}{2} \partial_z, \qquad Z_2 = \partial_y + \frac{x}{2} \partial_z, \qquad (3.2.3)$$

and the operator $\mathrm{L} = Z_1^2 + Z_2^2$. This operator is not elliptic, but connected. Indeed, $\Gamma(f) = (Z_1 f)^2 + (Z_2 f)^2$, and a smooth function f which satisfies $\Gamma(f) = 0$ is such that $Z_1 f = Z_2 f = 0$. Now the commutator $[Z_1, Z_2] = Z_3$ is $Z_3 f = \partial_z f$ so that f also satisfies $Z_3 f = 0$ and is therefore constant. This model is classically known as the Heisenberg (hypo-elliptic) operator.

Connexity will play a fundamental role in many functional inequalities below (related to tightness), and will be used to establish convergence to equilibrium (ergodicity). An abstract presentation of the connexity property will be further developed in Sect. 3.3.

3.2.2 Completeness and Weak Hypo-ellipticity

Uniqueness of the self-adjoint extension (the ESA property of Definition 3.1.10) of second order differential operators L of the form (3.2.1) on a manifold M will be achieved by completeness of M and hypo-ellipticity of the operator. As illustrated in Sects. 2.4 and 2.6, p. 92 and p. 97, there are situations where completeness is not satisfied while ESA actually holds.

The standard completeness assumption on a Riemannian manifold $(M, \mathfrak{g})$ refers to the associated Riemannian metric (cf. Sect. C.5, p. 511). To translate completeness in our setting, we use an argument given in Sect. C.4, p. 509. Namely, according to Proposition C.4.1, p. 511, on a complete Riemannian manifold $(M, \mathfrak{g})$, there exists an increasing sequence $(\zeta_k)_{k \in \mathbb{N}}$ of smooth compactly supported functions such that $\lim_{k \to \infty} \zeta_k = 1$ and $|\nabla \zeta_k| \leq \frac{1}{k}$ for every $k \geq 1$. Therefore, *completeness* in our setting will refer to the existence of a sequence $(\zeta_k)_{k \geq 1}$ of positive functions in $\mathcal{A}_0$, increasing to 1 and such that $\Gamma(\zeta_k) \leq \frac{1}{k}$ for every $k \geq 1$.

The second important property used here is *hypo-ellipticity* of the operator L as presented in Sect. 1.12, p. 49. With respect to the general definition, we use this local property here in the sense that any solution of $\mathrm{L}^* f = \lambda f$ is smooth. Recall that L^* is the adjoint of L on $\mathcal{A}_0$ with respect to μ as presented in Definition 3.1.9, that is $L^* f = g$ if and only if for all $h \in \mathcal{A}_0$,

$$\int_M f\, Lh\, d\mu = \int_M g h\, d\mu.$$

This weak form of hypo-ellipticity, again expressed in an intrinsic form, is actually enough for later purposes (see Definition 3.3.8 below for an abstract formulation).

The following proposition links completeness and self-adjointness for hypo-elliptic operators.

Proposition 3.2.1 *Assume that* L *is a hypo-elliptic diffusion operator of the form* (3.2.1), *symmetric with respect to* μ, *on a smooth complete connected manifold* M. *Then, the operator* L *is essentially self-adjoint on the set* $\mathcal{A}_0$ *of smooth compactly supported function. In other words,* $\mathcal{A}_0$ *is a core for the domain* $\mathcal{D}(L)$ *of* L, *that is, the ESA property holds.*

Proof In order to see that the operator L is essentially self-adjoint on $\mathcal{A}_0$, we make use of the criterion described in Proposition A.5.3, p. 482. (This criterion has already been used in the proof of Proposition 2.4.1, p. 95.) Namely, the aim is to prove that for some $\lambda > 0$ any $\mathbb{L}^2(\mu)$-solution f of $L^* f = \lambda f$ is zero for the adjoint operator L^* of L. Assume for simplicity that $\lambda = 1$ (any $\lambda > 0$ would work similarly). By hypo-ellipticity, f is smooth and $L^* f = Lf$. Now, for $\zeta \in \mathcal{A}_0$,

$$\int_M f^2 \zeta^2 d\mu = \int_M (Lf) f \zeta^2 d\mu = -\int_M \Gamma(f, f\zeta^2) d\mu,$$

integration by parts being justified by the fact that ζ is compactly supported. Together with the chain rule formula,

$$0 \leq \int_M f^2 \zeta^2 d\mu = -\int_M \zeta^2 \Gamma(f) d\mu - 2 \int_M f \zeta \, \Gamma(f, \zeta) d\mu.$$

As a consequence,

$$\int_M \zeta^2 \Gamma(f) d\mu \leq 2 \int_M |f\zeta| \Gamma(f)^{1/2} \Gamma(\zeta)^{1/2} d\mu.$$

Now, if $\Gamma(\zeta) \leq \varepsilon$, $\varepsilon > 0$, by the Cauchy-Schwarz inequality,

$$\int_M \zeta^2 \Gamma(f) d\mu \leq 2\sqrt{\varepsilon}\, \|f\|_2 \left(\int_M \zeta^2 \Gamma(f) d\mu \right)^{1/2}.$$

Hence, $\int_M \zeta^2 \Gamma(f) d\mu \leq 4\varepsilon \|f\|_2^2$. Replacing ζ with the functions ζ_k, $k \geq 1$, of the completeness property, it follows that $\int_M \Gamma(f) d\mu = 0$, and therefore that f is constant which in turn yields that $f = 0$ since $Lf = f$. The proposition is established. $\square$

As an immediate consequence, we have

3.2 Second Order Differential Operators on a Manifold

Corollary 3.2.2 *For* $L = \Delta_\mathfrak{g} - \nabla W \cdot \nabla$ *on a smooth complete connected Riemannian manifold* $(M, \mathfrak{g})$ *where* $W : M \to \mathbb{R}$ *is a smooth potential*, L *is essentially self-adjoint on the set of smooth compactly supported functions for the measure* $e^{-W} d\mu_\mathfrak{g}$, *where* $d\mu_\mathfrak{g}$ *is the Riemannian measure on* $(M, \mathfrak{g})$.

Recall that the setting of this corollary consisting of a Riemannian manifold $(M, \mathfrak{g})$ with the measure $d\mu = e^{-W} d\mu_\mathfrak{g}$ is sometimes referred to as a weighted Riemannian manifold. Any elliptic operator L symmetric with respect to μ may be decomposed in the form $L = \Delta_\mathfrak{g} - \nabla W \cdot \nabla$ with $\mathfrak{g} = (g^{ij})$ a Riemannian (co-)metric and $\Delta_\mathfrak{g}$ a Laplace-Beltrami operator (cf. Sect. 1.11.3, p. 46). The important aspect is that the resulting properties only depend on the second order part of L and not of W. In particular, in $\mathbb{R}^n$, any operator of the form $L = \Delta - \nabla W \cdot \nabla$ with smooth W is essentially self-adjoint.

As already mentioned, the standard completeness assumption refers to the distance d on M defined from the Riemannian metric associated with any elliptic operator of the form (3.2.1) as described in Sect. C.5, p. 511. In the dual formulation put forward in Sect. C.4, p. 509,

$$d(x, y) = \sup_{f \in \mathcal{A}, |\nabla f| \leq 1} [f(x) - f(y)], \quad (x, y) \in M \times M,$$

where $\mathcal{A}$ is the set of smooth functions on M. The advantage of this representation is that it will immediately lead below to a general definition replacing $|\nabla f| \leq 1$ by $\Gamma(f) \leq 1$ and functions satisfying $\Gamma(f) \leq 1$ will be called Lipschitz functions. The diameter, when it exists, may be defined in a similar way as a supremum of this distance function $d(x, y)$ on $E \times E$. Saying that the diameter is bounded therefore amounts to placing bounds on Lipschitz functions.

3.2.3 Gradient Bounds

Proposition 3.2.1 above ensures that the hypo-elliptic diffusion operator L from (3.2.1) on a smooth complete connected manifold M is essentially self-adjoint and is therefore the infinitesimal generator in $\mathbb{L}^2(\mu)$ of a symmetric semigroup $\mathbf{P} = (P_t)_{t \geq 0}$ as discussed in Appendix A. While the Dirichlet form construction of Sect. 3.1 already ensures that $(P_t)_{t \geq 0}$ may be assumed sub-Markov, one central question is to determine whether or not $P_t(\mathbb{1}) = \mathbb{1}$ (for every $t \geq 0$). As already pointed out, depending on the context, this property is variously referred to as mass conservation, Markov, stochastic completeness or non-explosion. Indeed, even in the absence of boundary, it may happen that the associated Markov process $(X_t^x)_{t \geq 0}$ goes to infinity in finite time. Denoting by T this explosion time, if the manifold M is complete, there is just one symmetric semigroup $(P_t)_{t \geq 0}$ with generator L, and in this case, $P_t f(x) = \mathbb{E}(f(X_t^x)\mathbb{1}_{\{t < T\}})$ so that the Markov property $P_t(\mathbb{1}) = \mathbb{1}$ does not hold when $T < \infty$. As developed in this section, the validity of the mass conservation or Markov property in a smooth manifold setting is closely related to gradient bounds. The mass conservation property will require more information on the set $\mathcal{A}_0$ than the mere completeness of the manifold and makes use of the extended algebra $\mathcal{A}$ of smooth functions on the manifold M.

A useful criterion towards mass conservation is the use of the curvature condition $CD(\rho, \infty)$ of Definition 1.16.1, p. 72. This investigation will raise in addition further issues concerning the algebras $\mathcal{A}_0$ and $\mathcal{A}$. Recall to start with the Γ_2 operator from (1.16.1), p. 71, defined for $f \in \mathcal{A}_0$ by

$$\Gamma_2(f) = \Gamma_2(f, f) = \frac{1}{2}\big[L\Gamma(f) - 2\Gamma(f, Lf)\big].$$

The Riemannian content of the Γ and Γ_2 operators is emphasized in Sect. C.6, p. 513, as $\Gamma(f) = |\nabla f|^2$ and

$$\Gamma_2(f) = |\nabla\nabla f|^2 + (\mathrm{Ric}_\mathfrak{g} - \nabla_S Z)(\nabla f, \nabla f)$$

where $\mathrm{Ric}_\mathfrak{g}$ is the Ricci tensor and $\nabla_S Z$ the symmetric part of ∇Z. As for L and Γ, the Γ_2 operator may be extended in the manifold case to the class $\mathcal{A}$ of smooth functions. Recall then the curvature condition $CD(\rho, \infty)$, $\rho \in \mathbb{R}$,

$$\Gamma_2(f) \geq \rho\, \Gamma(f)$$

(for all $f \in \mathcal{A}$ thus) which suitably extends the notion of Ricci curvature lower bound on Riemannian manifolds.

One fundamental result with respect to the curvature condition $CD(\rho, \infty)$ is the following gradient bound or commutation between the actions of the semigroup and the carré du champ operator. Simple examples are the Brownian and Ornstein-Uhlenbeck semigroups in $\mathbb{R}^n$ with $\rho = 0$ or $\rho = 1$ respectively (cf. e.g. (2.7.5), p. 104). The operator L (from (3.2.1)) is assumed to be elliptic in the sense of (1.12.2), p. 50, since in general no curvature bounds may hold for non-elliptic operators.

Theorem 3.2.3 (Gradient bound) *Let L be an elliptic diffusion operator with semigroup $\mathbf{P} = (P_t)_{t \geq 0}$, symmetric with respect to μ, on a smooth complete connected manifold M. If L satisfies the curvature condition $CD(\rho, \infty)$ for some $\rho \in \mathbb{R}$, then, for every $f \in \mathcal{A}_0$ and every $t \geq 0$,*

$$\Gamma(P_t f) \leq e^{-2\rho t} P_t\big(\Gamma(f)\big).$$

It is not difficult to see that, conversely, the gradient bound of Theorem 3.2.3 for every $t \geq 0$ implies in return the curvature condition $CD(\rho, \infty)$ (see Corollary 3.3.19 below or Theorem 4.7.2, p. 209, in the context of local Poincaré inequalities). Below, we present a formal and easy proof of Theorem 3.2.3, and then we address the question of how this formal argument can be justified.

The formal proof is very elementary, and relies on the basic interpolation argument along the semigroup extensively developed throughout this work. Consider, for $f \in \mathcal{A}_0$ and fixed $t > 0$, the function defined by

$$\Lambda(s) = P_s\big(\Gamma(P_{t-s} f)\big), \quad s \in [0, t]$$

3.2 Second Order Differential Operators on a Manifold

(at any fixed point). Then,

$$\Lambda'(s) = LP_s\big(\Gamma(P_{t-s}f)\big) - 2P_s\big(\Gamma(P_{t-s}f, LP_{t-s}f)\big).$$

The first term comes from the derivative of P_s and the heat equation, the second one from the derivative of $P_{t-s}f$ applied to the quadratic form Γ. Rewriting this identity with $g = P_{t-s}f$,

$$\Lambda'(s) = P_s\big(L\Gamma(g) - 2\Gamma(g, Lg)\big) = 2P_s\big(\Gamma_2(g)\big).$$

It is worthwhile to observe that this identity is the analogue at the level of the Γ_2 operator of the Duhamel formula (3.1.20) for Γ. Now the $CD(\rho, \infty)$ condition simply reads $\Lambda' \geq 2\rho \Lambda$, from which the conclusion immediately follows.

Now, we have to wonder why and when such a proof could be justified, that is, for which classes of functions f the preceding argument may be developed. The operator P_s acts a priori on bounded functions, or on functions which are in some $\mathbb{L}^p(\mu)$-space, but there is no guarantee that differentiation and commutation of L and P_s are valid for those functions. More precisely, the first identities to be understood are $LP_s(\Gamma(g)) = P_s(L\Gamma(g))$ as well as differentiation of $\Gamma(P_{t-s}f)$ under P_s (which consists, as already mentioned earlier, of an integration by parts together with the derivation of an integral, and hence requires some justification). Of course, working with an operator L with smooth coefficients on a compact smooth manifold, the preceding is fully justified for sufficiently smooth functions. Indeed, in this case, if f is smooth, so is $P_s f$, and so are $g = P_{t-s}f$ and $\Gamma(g)$. They are moreover bounded, and in the domain of L, so that the conclusion follows. However, we would like to assert that this result is true for a generic operator L with smooth coefficients on a (a priori non-compact) manifold. If f is smooth and compactly supported, $g = P_{t-s}f$ and $\Gamma(g)$ are still smooth, but no longer compactly supported (it is not too hard to see that, under the $CD(\rho, \infty)$ hypothesis, the kernel measures $p_t(x, dy)$ have strictly positive densities everywhere for any x). Therefore care has to be taken here when considering $P_s(\Gamma(g))$.

Before moving on to a precise analysis of the arguments which justify the ingredients used in the proof of Theorem 3.2.3, let us mention a reinforced form of it which we shall also use extensively.

Theorem 3.2.4 (Strong gradient bound) *In the setting of Theorem 3.2.3, if L satisfies the curvature condition $CD(\rho, \infty)$ for some $\rho \in \mathbb{R}$, then, for every $f \in \mathcal{A}_0$ and every $t \geq 0$,*

$$\sqrt{\Gamma(P_t f)} \leq e^{-\rho t} P_t\big(\sqrt{\Gamma(f)}\big).$$

This is clearly an improvement upon Theorem 3.2.3 since, by Jensen's inequality, $P_t(g^2) \geq (P_t g)^2$ (P_t is sub-Markov). The formal proof of Theorem 3.2.4 is similar to that of Theorem 3.2.3, except that we now consider

$$\Lambda(s) = P_s\big(\Gamma(P_{t-s}f)^{1/2}\big), \quad s \in [0, t].$$

The formal derivation yields in this case

$$\Lambda'(s) = P_s\big(L\big(\Gamma(P_{t-s}f)^{1/2}\big) - \Gamma(P_{t-s}f)^{-1/2}\Gamma(P_{t-s}f, LP_{t-s}f)\big).$$

Using the change of variables formula (diffusion property) for L, this may be rewritten as

$$P_s\bigg(\Gamma(P_{t-s}f)^{-1/2}\bigg(\Gamma_2(P_{t-s}f) - \frac{\Gamma(\Gamma(P_{t-s}f))}{4\Gamma(P_{t-s}f)}\bigg)\bigg).$$

From (C.6.4), p. 515, the $CD(\rho, \infty)$ curvature hypothesis actually yields the reinforced inequality

$$4\Gamma(g)\big[\Gamma_2(g) - \rho\Gamma(g)\big] \geq \Gamma\big(\Gamma(g)\big) \tag{3.2.4}$$

(for any g in $\mathcal{A}$). Applying this inequality to $g = P_{t-s}f$ then yields the differential inequality $\Lambda' \geq \rho\Lambda$, from which the conclusion follows again.

The following now tries to fully justify the proof of the gradient bounds (Theorems 3.2.3 and 3.2.4) in the smooth context of this section. According to the discussion in Sect. 1.11.3, p. 46, the elliptic operator L from (3.2.1) takes the form $L = \Delta_{\mathfrak{g}} + Z$ where $\Delta_{\mathfrak{g}}$ is the Laplace-Beltrami operator on M associated with the (co-) metric $\mathfrak{g} = (g^{ij})$ and Z is a smooth gradient field $Zf = -\nabla W \cdot \nabla f$. The invariant measure of L is given by $d\mu = e^{-W}d\mu_{\mathfrak{g}}$ where $d\mu_{\mathfrak{g}}$ is the Riemannian measure on $(M, \mathfrak{g})$. Recall the class $\mathcal{A}_0$ of smooth ($\mathcal{C}^\infty$) compactly supported functions on M, and the class $\mathcal{A}$ of smooth ($\mathcal{C}^\infty$) functions. The carré du champ operator is simply given by $\Gamma(f) = |\nabla f|^2$ (on $\mathcal{A}_0$ or $\mathcal{A}$). Recall that, according to Sect. C.6, p. 513, the curvature condition $CD(\rho, \infty)$ expresses, when $Z = 0$, a lower bound on the Ricci curvature of the manifold.

Proof of Theorem 3.2.4 As a consequence of the completeness assumption, by Proposition C.4.1, p. 511, Appendix C, there is a sequence $(\zeta_k)_{k\geq 1}$ of positive functions in $\mathcal{A}_0$, increasing to the constant function $\mathbb{1}$ and satisfying $\sqrt{\Gamma(\zeta_k)} = |\nabla\zeta_k| \leq \frac{1}{k}$ for every $k \geq 1$. Observe that this property only controls $\Gamma(\zeta_k)$, but not for example $L\zeta_k$. This drawback will make the analysis below rather delicate.

The $CD(\rho, \infty)$ curvature assumption will be used first to extend to second order derivatives the integration by parts formula $\int_M fLg\,d\mu = -\int_M \Gamma(f,g)d\mu$. Indeed, from the very definition of the operator Γ_2 and integration by parts,

$$\int_M \Gamma_2(f)d\mu = \int_M (Lf)^2 d\mu,$$

which is valid for all functions $f \in \mathcal{A}_0$. The first step is to extend this property to smooth functions of the $\mathbb{L}^2(\mu)$-domain $\mathcal{D}(L)$ of L. In fact, the following considerations will reveal that in our analysis we then have to replace the Dirichlet form, $\int_M |\nabla f|^2 d\mu$, by quantities such as $\int_M |\nabla\nabla f|^2 d\mu$ involving Hessians. This aspect increases the complexity of the integration by parts formulas. Some of the difficulties are resolved by observing that

$$\Gamma_2(f) - \rho\Gamma(f) \geq |\nabla\nabla f|^2 \tag{3.2.5}$$

3.2 Second Order Differential Operators on a Manifold

which follows from the explicit description of the Γ_2 operator in this Riemannian context (cf. (C.5.3), p. 512).

To start with, note that for $f \in \mathcal{A}_0$, from (3.2.5),

$$\|\nabla\nabla f\|_2^2 = \int_M |\nabla\nabla f|^2 d\mu$$

$$\leq -\rho \int_M \Gamma(f) d\mu + \int_M (\mathrm{L}f)^2 d\mu$$

$$\leq |\rho| \, \|f\|_2 \, \|\mathrm{L}f\|_2 + \|\mathrm{L}f\|_2^2 \tag{3.2.6}$$

$$\leq \frac{|\rho|}{2} \|f\|_2^2 + \left(1 + \frac{|\rho|}{2}\right) \|\mathrm{L}f\|_2^2$$

$$\leq \left(1 + \frac{|\rho|}{2}\right) \|f\|_{\mathcal{D}(\mathrm{L})}^2.$$

Hence, if a sequence $(f_k)_{k \in \mathbb{N}}$ of functions in $\mathcal{A}_0$ converges as $k \to \infty$ to f in the domain norm $\| \cdot \|_{\mathcal{D}(\mathrm{L})}$, the sequence $\nabla\nabla f_k$, $k \in \mathbb{N}$, is a Cauchy sequence in the set of symmetric tensors equipped with the norm

$$\|\nabla\nabla f\|_2 = \left(\int_M |\nabla\nabla f|^2 d\mu\right)^{1/2}.$$

It therefore converges to some symmetric tensor K. The limit does not depend on the approximating sequence, and is thus denoted Kf. If $f \in \mathcal{A}_0$, then $Kf = \nabla\nabla f$. This remains true for any smooth function. To see why this is, it is enough to observe that given $h \in \mathcal{A}_0$,

$$\nabla\nabla(h f_k) = h \nabla\nabla f_k + \nabla h \otimes \nabla f_k + \nabla f_k \otimes \nabla h + f_k \nabla\nabla h$$

for every k. In the limit as $k \to \infty$,

$$\nabla\nabla(hf) = h Kf + \nabla h \otimes \nabla f + \nabla f \otimes \nabla h + f \nabla\nabla h,$$

from which the identification $Kf = \nabla\nabla f$ follows. Then the bound (3.2.6) extends to functions f in the $\mathbb{L}^2(\mu)$-domain of L by density.

Having made these preliminary observations, we can now begin the proof of Theorem 3.2.4. Choose a function $f \in \mathcal{A}_0$ and fix $\varepsilon > 0$. For $t > 0$, consider the function in $s \in [0, t]$ given by

$$\Lambda(s) = \left(e^{-2\rho s} \Gamma(P_{t-s} f) + \varepsilon\right)^{1/2}.$$

By the reinforced inequality (3.2.4), it easily seen that

$$\mathrm{L}\Lambda + \partial_s \Lambda \geq 0. \tag{3.2.7}$$

We may also observe, from basic calculus on smooth functions, that

$$e^{\rho s}|\nabla \Lambda(s)| \leq |\nabla \nabla P_{t-s} f| \qquad (3.2.8)$$

for every $s \in [0, t]$. From (3.2.6), $|\nabla \nabla P_{t-s} f|$ is in $\mathbb{L}^2(\mu)$ with $\mathbb{L}^2(\mu)$-norm bounded from above by $C_1 \|f\|_{\mathcal{D}(L)}$ for some $C_1 > 0$ depending only on ρ.

Choose now two auxiliary positive functions $\xi, \zeta : M \to \mathbb{R}$, smooth and compactly supported. Let, for $s \geq 0$,

$$G(s) = \int_M \xi P_s(\zeta \Lambda(s)) d\mu = \int_M \zeta \Lambda(s) P_s \xi \, d\mu.$$

Since ζ is compactly supported, there is no difficulty in differentiating this expression to get (with $\Lambda = \Lambda(s)$ and $\Lambda' = \Lambda'(s)$ for simplicity)

$$G'(s) = \int_M \zeta \Lambda' P_s \xi \, d\mu + \int_M \zeta \Lambda L P_s \xi \, d\mu.$$

From (3.2.7), for every $s \geq 0$,

$$G'(s) \geq \int_M [L(\zeta \Lambda) - \zeta L \Lambda] P_s \xi \, d\mu = \int_M [\Lambda L \zeta + 2\Gamma(\Lambda, \zeta)] P_s \xi \, d\mu.$$

After integration by parts,

$$G'(s) \geq \int_M \Gamma(\zeta, \Lambda) P_s \xi \, d\mu - \int_M \Lambda \Gamma(\zeta, P_s \xi) d\mu.$$

We therefore obtain the lower bound, for all $s \geq 0$,

$$G'(s) \geq -\|\sqrt{\Gamma(\zeta)}\|_\infty \left(\|\sqrt{\Gamma(\Lambda)}\|_2 \|P_s \xi\|_2 + \|\Lambda\|_2 \|\sqrt{\Gamma(P_s \xi)}\|_2 \right),$$

and together with (3.2.8) and (3.2.6),

$$G'(s) \geq -C_2(t) \|\sqrt{\Gamma(\zeta)}\|_\infty \|\xi\|_{\mathcal{D}(L)} \|f\|_{\mathcal{D}(L)}$$

where $C_2(t)$ only depends on t. Therefore,

$$G(t) - G(0) \geq C_3(t) \|\sqrt{\Gamma(\zeta)}\|_\infty \|\xi\|_{\mathcal{D}(L)} \|f\|_{\mathcal{D}(L)}.$$

Now replace ζ by the terms ζ_k, $k \geq 1$ of the sequence which is used to characterize completeness. In the limit as $k \to \infty$,

$$\int_M \xi [P_t(\Lambda(t)) - \Lambda(0)] d\mu \geq 0.$$

This being true for any positive $\xi \in \mathcal{A}_0$, it finally yields $P_t(\Lambda(t)) \geq \Lambda(0)$. It remains to let ε go to 0 to get the desired inequality. The strong gradient bound of Theorem 3.2.4 is established. □

3.2 Second Order Differential Operators on a Manifold

There is therefore a huge difference between the formal proof of Theorem 3.2.4 given earlier and the actual, more honest proof, that justifies in practice the formal computations. For example, the same result would also be true on a compact manifold with convex boundary (with the Neumann conditions at the boundary), but yet another proof would be required.

In Sect. 3.3, in the abstract framework of a Markov Triple, the preceding gradient bounds will be extended from the complete case to the essentially self adjoint case. However, this extension requires a much more refined analysis of the relations between the classes $\mathcal{A}_0$ and $\mathcal{A}$.

Theorem 3.2.4 may be strengthened by letting ρ vary, i.e. by assuming the existence of a (smooth) function $\rho : M \to \mathbb{R}$ such that $\Gamma_2(f) \geq \rho \Gamma(f)$ at any point x and for every $f \in \mathcal{A}$. In a Riemannian setting with $L = \Delta_g - \nabla W \cdot \nabla$, the best function $\rho = \rho(x)$ for which the preceding is satisfied is the minimal eigenvalue at the point x of the symmetric tensor $\mathrm{Ric} + \nabla \nabla W$. The next proposition is the announced extension, together with its probabilistic interpretation.

Proposition 3.2.5 *In the setting of Theorem 3.2.3 under the curvature condition $CD(\rho, \infty)$ for some $\rho = \rho(x)$ bounded from below by $c \in \mathbb{R}$, for any $f \in \mathcal{A}_0$ and $t \geq 0$,*

$$\sqrt{\Gamma(P_t f)} = |\nabla P_t f| \leq \widehat{P}_t(|\nabla f|)$$

where $(\widehat{P}_t)_{t \geq 0}$ is the semigroup with generator $L - \rho \, \mathrm{Id}$. In particular, if $\{X_t^x ; t \geq 0, x \in M\}$ denotes the Markov process with generator L, for any $t \geq 0$ and $x \in M$,

$$|\nabla P_t f|(x) \leq \mathbb{E}_x \left(|\nabla f|(X_t) \, e^{-\int_0^t \rho(X_s) ds} \right). \tag{3.2.9}$$

The proof of this proposition closely follows the previous proof of Theorem 3.2.4 working now with

$$\Lambda(s) = \widehat{P}_s \big(\Gamma(P_{t-s} f)^{1/2} \big), \quad s \in [0, t].$$

Since $\rho(x) \geq c$ for every $x \in M$, we already have an a priori bound $\sqrt{\Gamma(P_t f)} \leq e^{-ct} P_t(\sqrt{\Gamma(f)})$ which justifies the various derivatives of Λ. Then, it remains to apply the Feynman-Kac formula of Sect. 1.15.6, p. 63.

3.2.4 Mass Conservation

On the basis of Theorems 3.2.3 and 3.2.4 of the last sub-section, we now consider the mass conservation property $P_t(\mathbb{1}) = \mathbb{1}$ under a $CD(\rho, \infty)$ curvature condition for some $\rho \in \mathbb{R}$. Formally, conservation of mass reflects the fact that $L(\mathbb{1}) = 0$. However, when the measure μ is infinite, the constant function $\mathbb{1}$ cannot be in the domain $\mathcal{D}(L)$.

Theorem 3.2.6 (Mass conservation) *Let L be an elliptic diffusion operator with semigroup* $\mathbf{P} = (P_t)_{t \geq 0}$*, symmetric with respect to* μ*, on a smooth complete connected manifold M. If the* $CD(\rho, \infty)$ *curvature condition holds for some* $\rho \in \mathbb{R}$*, then for all* $t \geq 0$*,* $P_t(\mathbb{1}) = \mathbb{1}$*.*

Proof To make sense of the statement, recall that $P_t(\mathbb{1})$ is understood here as the (increasing) limit of $P_t \zeta_k$ as $k \to \infty$ where $(\zeta_k)_{k \geq 1}$ is any sequence of positive functions in $\mathcal{A}_0$ increasing to $\mathbb{1}$. Choose two positive functions ξ and ζ in $\mathcal{A}_0$, and consider, for $t \geq 0$, the quantity $\int_M (P_t \zeta - \zeta) \xi \, d\mu$. By integration by parts,

$$\left| \int_M (P_t \zeta - \zeta) \xi \, d\mu \right| = \left| \int_0^t \int_M \xi L P_s \zeta \, d\mu \, ds \right|$$

$$\leq \int_0^t \int_M |\Gamma(\xi, P_s \zeta)| \, d\mu \, ds \qquad (3.2.10)$$

$$\leq \left\| \sqrt{\Gamma(\xi)} \right\|_1 \left\| \sqrt{\Gamma(\zeta)} \right\|_\infty \int_0^t e^{-\rho s} \, ds$$

where the last step uses the curvature condition via Theorem 3.2.4. Now replace ζ by the terms ζ_k, $k \geq 1$, of the sequence which is used to characterize completeness of the manifold M. Since $\sqrt{\Gamma(\zeta_k)} \leq \frac{1}{k}$, in the limit as $k \to \infty$,

$$\int_M \big(P_t(\mathbb{1}) - \mathbb{1} \big) \xi \, d\mu = 0,$$

which in turn yields $P_t(\mathbb{1}) = \mathbb{1}$ since $\xi \geq 0$ is arbitrary in $\mathcal{A}_0$. The proof is complete. □

Another interesting consequence of the preceding tools is the following statement, which ensures that the invariant measure is necessarily finite in spaces with strictly positive curvature (such as the sphere, for example).

Theorem 3.2.7 *Let L be an elliptic diffusion operator with Markov semigroup* $\mathbf{P} = (P_t)_{t \geq 0}$*, symmetric with respect to* μ*, on a smooth complete connected manifold M. If the* $CD(\rho, \infty)$ *curvature condition holds for some* $\rho > 0$*, then the invariant measure* μ *is finite.*

Proof When f is in $\mathbb{L}^2(\mu)$ and $t \to \infty$, $P_t f$ converges in $\mathbb{L}^2(\mu)$ to $P_\infty f$ which is the projection of f on the space of invariant functions, that is, an element of the domain $\mathcal{D}(L)$ satisfying $L P_\infty f = 0$ (cf. Sect. 1.6, p. 24). Now, by hypo-ellipticity, such a function $g = P_\infty f$ is smooth, and being in $\mathbb{L}^2(\mu)$ together with Lg, one has $\int_M \Gamma(g) d\mu = 0$ and thus g is constant by connexity. If $\mu(E)$ is infinite, then this constant must be 0. Now, as in the proof of mass conservation Theorem 3.2.6, choose two functions ξ and ζ in $\mathcal{A}_0$. By (3.2.10) at $t = \infty$,

$$\left| \int_M (\xi - P_\infty \xi) \zeta \, d\mu \right| \leq \left\| \sqrt{\Gamma(\xi)} \right\|_1 \left\| \sqrt{\Gamma(\zeta)} \right\|_\infty \int_0^\infty e^{-\rho s} \, ds.$$

Replacing as before ζ by the smoothing functions ζ_k, as $k \to \infty$, and whenever $\mu(E) = \infty$, it follows that $\int_M \xi \, d\mu = 0$ for any $\xi \in \mathcal{A}_0$, yielding a contradiction. Theorem 3.2.7 is proved. □

3.3 Heart of Darkness

This section attempts to describe a general setting for the investigation of symmetric Markov diffusion operators and semigroups, which will be convenient for the analysis of various questions related to functional inequalities, convergence to equilibrium, heat kernel bounds and so on. Although we do not want to restrict ourselves to the case of smooth manifolds, in view of several important applications (in particular in infinite dimension), we are now in a position to extract from the previous analysis in manifolds the main features of the spaces of smooth compactly supported ($\mathcal{A}_0$) and smooth ($\mathcal{A}$) functions used there. The abstract formalism put forward here is a self-consistent framework in which the intuition behind the Γ-calculus and semigroup monotonicity may easily be developed, without having to take too much care over the various families of functions involved in the analysis. In particular, this setting is not intended for an investigation in itself. Applications will develop the ideas emphasized through this formalism rather than the technical properties which have to be investigated somewhat case by case.

Recall also that we do not want to impose a topology on the basic measure space $(E, \mathcal{F}, \mu)$ on which the Markov operators act, or on the space in which the Markov processes live. Indeed, every notion should be invariant under measurable bijections of the space (although such generality may in practice be quite useless). In particular, equalities and inequalities between functions have to be understood μ-almost everywhere. The various classes of (real-valued) functions involved in the investigation actually aim at replacing a pointwise analysis by a measurable one. In particular, the classes should be rich enough to develop a relevant calculus, leading below to the necessary hypotheses in this regard (not always transparent). In applications of interest, such as in a smooth manifold context, the various relevant inequalities usually do hold pointwise.

The following thus describes a suitable framework enabling us to smoothly develop the Γ-calculus at the root of this investigation. The next Sect. 3.4 contains a summary of the various definitions and hypotheses. The starting point for the analysis is the underlying algebra $\mathcal{A}_0$ of a symmetric Diffusion Markov Triple (E, μ, Γ) (Definition 3.1.8). As already pointed out, this class $\mathcal{A}_0$ is in practise the class of smooth (C^∞) functions with compact support. On $\mathcal{A}_0$ is given the carré du champ operator Γ. From the carré du champ operator is constructed the generator L together with the Dirichlet form $\mathcal{E}$ defined on its domain $\mathcal{D}(\mathcal{E})$. Many properties of functions defined on $\mathcal{A}_0$ automatically extend to the Dirichlet space $\mathcal{D}(\mathcal{E})$. The semigroup $\mathbf{P} = (P_t)_{t \geq 0}$ with infinitesimal generator L in $\mathbb{L}^2(\mu)$ is symmetric with respect to μ and sub-Markov (positivity preserving and such that $P_t(\mathbb{1}) \leq \mathbb{1}$ for every $t \geq 0$).

A Diffusion Markov Triple together with the additional ergodicity and mass conservation properties yields the notion of a Standard Markov Triple defined in Definition 3.1.15.

But in the initial setting of a Diffusion Markov Triple (E, μ, Γ), many properties (such as essential self-adjointness (ESA), ergodicity, mass conservation or gradient bounds as developed in the smooth case) may or may not hold. Criteria for these properties to hold actually require the introduction of a new larger class $\mathcal{A}$ of functions, viewed as an extension of $\mathcal{A}_0$, on which the operator L is defined as an extension together with the associated Γ operator, regardless of integrability properties. In the smooth manifold case, this class $\mathcal{A}$ is the class of smooth ($\mathcal{C}^\infty$) functions. The following develops the abstract construction of such a class $\mathcal{A}$. The ESA property will appear as the cornerstone of the construction. Indeed, while positivity of the carré du champ operator Γ is the main ingredient of the extension results, at the second order of the Γ_2 operator, curvature conditions together with essential self-adjointness allow for a parallel treatment.

3.3.1 Extended Algebra $\mathcal{A}$

An extension of $\mathcal{A}_0$ is a class $\mathcal{A}$ of measurable functions f on E containing the constants, stable under the action of smooth multivariate functions, and such that $hf \in \mathcal{A}_0$ for all $h \in \mathcal{A}_0$. In other words, $\mathcal{A}_0$ is an ideal in $\mathcal{A}$. The extension $\mathcal{A}$ of $\mathcal{A}_0$ is not unique, and many algebras $\mathcal{A}$ suitably extending $\mathcal{A}_0$ may be considered. The largest possible class $\mathcal{A}$ would be the set of functions f such that $hf \in \mathcal{A}_0$ for any $h \in \mathcal{A}_0$. It is however not always wise or necessary to work with this largest class $\mathcal{A}$. In many examples the ideal property is not suitable either, and may be relaxed in some situations (at the price of some extra and tedious hypotheses on $\mathcal{A}$).

Beyond the properties of $\mathcal{A}_0$ and Γ described in Sect. 3.1 and summarized in Sect. 3.4 below, the requirements on $\mathcal{A}$ are the following.

Definition 3.3.1 (Extended algebra $\mathcal{A}$) $\mathcal{A}$ is an algebra of measurable functions on E containing $\mathcal{A}_0$, containing the constant functions and satisfying the following requirements.

(i) Whenever $f \in \mathcal{A}$ and $h \in \mathcal{A}_0$, $hf \in \mathcal{A}_0$ (ideal property).
(ii) For any $f \in \mathcal{A}$, if $\int_E hf d\mu \geq 0$ for every positive $h \in \mathcal{A}_0$, then $f \geq 0$.
(iii) $\mathcal{A}$ is stable under composition with smooth ($\mathcal{C}^\infty$) functions $\Psi : \mathbb{R}^k \to \mathbb{R}$.
(iv) The operator L : $\mathcal{A} \to \mathcal{A}$ is an extension of L on $\mathcal{A}_0$. The carré du champ operator Γ is also defined on $\mathcal{A} \times \mathcal{A}$ by the formula, for every $(f, g) \in \mathcal{A} \times \mathcal{A}$,

$$\Gamma(f, g) = \frac{1}{2}\big[L(fg) - f L(g) - g L(f)\big] \in \mathcal{A}.$$

(v) For every $f \in \mathcal{A}$, $\Gamma(f) \geq 0$.
(vi) The operators Γ and L satisfy the change of variables formulas (3.1.2) and (3.1.8) for any smooth ($\mathcal{C}^\infty$) function $\Psi : \mathbb{R}^k \to \mathbb{R}$.

3.3 Heart of Darkness

(vii) For every $f \in \mathcal{A}$ and every $g \in \mathcal{A}_0$, the integration by parts formula

$$\int_E \Gamma(f, g) d\mu = -\int_E g \, L f \, d\mu = -\int_E f \, L g \, d\mu \qquad (3.3.1)$$

holds true.

(viii) For every $f \in \mathcal{A}_0$ and every $t \geq 0$, $P_t f \in \mathcal{A}$.

The ideal property (i) ensures in particular that whenever $f \in \mathcal{A}$ and $h \in \mathcal{A}_0$, $\Gamma(f, h) \in \mathcal{A}_0$. In the same way, in many formulas (involving L, Γ and later Γ_2, Hessians etc.) where some function $h \in \mathcal{A}_0$ appears, the result is in $\mathcal{A}_0$. This will allow us to freely use integration by parts arguments.

In view of Remark 3.1.2, (ii) is not a strong requirement, and is easy to check in any practical situation. However, we do not want at this stage to impose on $\mathcal{A}_0$ any kind of locality, and neither do we want to impose on $\mathcal{A}$ any kind of local integrability. The ideal property ensures that (ii) makes sense without further assumptions.

The most important property in Definition 3.3.1 is (viii). In some sense, the algebra $\mathcal{A}$ is the smallest class, according to each specific context, in which computations are easy to make and where functions in $\mathcal{A}_0$ are mapped through the action of P_t. Note indeed that beyond the case of compact spaces, $\mathcal{A}_0$ is in general not stable under the semigroup. The possible stability of $\mathcal{A}$ under $(P_t)_{t \geq 0}$ reflects in general smoothness properties of the coefficients of a given operator. Such operators with smooth coefficients should not be confused with hypo-elliptic operators (cf. Definition 3.3.8 below), hypo-ellipticity being in general a much stronger requirement.

Note finally that the validity of the integration by parts formula (3.3.1) may still hold for some $g \notin \mathcal{A}_0$ but then usually requires some extra analysis (see for example Proposition 3.3.16).

Remark 3.3.2 It might be worthwhile to mention again that the ideal and stability properties of the extended algebra $\mathcal{A}$ provide an alternative understanding of the density of $\mathcal{A}_0$. For example, the non-negativity of Γ on $\mathcal{A}_0$ may be extended to $\mathcal{A}$ in the following way. Letting $f \in \mathcal{A}$ and $g \in \mathcal{A}_0$, $g \geq 0$, by the chain rule, for every integer $k \geq 1$,

$$0 \leq \Gamma(f^k g) = k^2 f^{2k-2} g^2 \Gamma(f) + 2k f^{2k-1} g \Gamma(f, g) + f^{2k} \Gamma(g).$$

Changing f into $f + 1$, dividing by $k^2(f+1)^{2k-2}$ and letting $k \to \infty$ yields $g^2 \Gamma(f) \geq 0$, hence $g \Gamma(f) \geq 0$, from which $\Gamma(f) \geq 0$ by (ii) of Definition 3.3.1. (See below in Remark 3.3.15 the corresponding extension for the curvature-dimension condition with similar arguments.)

Before developing the analysis relating $\mathcal{A}_0$, $\mathcal{A}$, L and Γ, it should be emphasized that the preceding setting is not the only one that may be considered, and some situations could be handled with other environments. For instance, allowing constants to belong to $\mathcal{A}_0$ will lead (through the ideal property) to $\mathcal{A} = \mathcal{A}_0$, and therefore to the stability of $\mathcal{A}_0$ under $(P_t)_{t\geq 0}$. This is of course the most desirable situation (and is valid, for example, on compact manifolds) and many properties (such as the ESA property) are then automatic. It requires the measure μ to be finite. Relying on this hypothesis actually allows us to favor intuition without taking too much care over technical details.

There are also cases where $\mathcal{A}_0$ may be assumed to be stable under $(P_t)_{t\geq 0}$ without containing the constant functions. For instance, in $\mathbb{R}^n$ and for the standard Laplace operator Δ, or the Ornstein-Uhlenbeck operator, the Schwartz space of rapidly decreasing functions together with all their derivatives is such an example. With some extra work (depending on our knowledge of the model), this may also be achieved in a fairly large setting. But it is unlikely that this will hold on a non-compact manifold only under the curvature condition, which is at the heart of the analysis developed in this monograph. For the Ornstein-Uhlenbeck, Laguerre and Jacobi operators of Chap. 2, it is in general better to work with the algebra $\mathcal{A}_0$ of polynomials in order to take advantage of the particular structure of the associated semigroups. The functions in $\mathcal{A}_0$ are then no longer stable under composition with smooth functions. The reader may easily adapt the various hypotheses to those particular cases.

Remark 3.3.3 When the measure μ is finite and the semigroup $(P_t)_{t\geq 0}$ is conservative, constant functions belong to $\mathcal{D}(L)$. Then $\mathcal{A}_0$ may be replaced by the larger class $\mathcal{A}_0^{\text{const}} = \{f + c\,;\, f \in \mathcal{A}_0, c \in \mathbb{R}\}$, extending Γ with $\Gamma(f, \mathbb{1}) = 0$ and $L(\mathbb{1}) = 0$. In the same spirit, the change of variables formulas (3.1.2) and (3.1.8) easily extend to functions f in $\mathcal{A}_0^{\text{const}}$ without the restriction that $\Psi(0) = 0$. However, care has to be taken not to apply the ideal property to $\mathcal{A}_0^{\text{const}}$ (since if it applies we would be back to $\mathcal{A} = \mathcal{A}_0$). Indeed, even the integration by parts property $\int_E g \mathrm{L} f d\mu = -\int_E \Gamma(f, g)\, d\mu$ does not in general make sense for $f \in \mathcal{A}$ when g is constant. It is also worth observing that the properties $\Gamma(f, \mathbb{1}) = 0$ and $L(\mathbb{1}) = 0$ are direct consequences of the change of variables formula applied with $\Psi(r) = 1$, $r \in \mathbb{R}$. In particular, integration by parts (3.3.1) entails the invariance of μ as $\int_E \mathrm{L} f d\mu = 0$ for any $f \in \mathcal{A}_0$. When the underlying invariant measure is finite, it will also be convenient to deal in the later chapters with the family $\mathcal{A}_0^{\text{const}+} = \{f + c\,;\, f \in \mathcal{A}_0, f \geq 0, c > 0\}$.

Remark 3.3.4 The ideal property (i) of Definition 3.3.1 is convenient for the analysis on non-compact Riemannian manifolds, but would not be suitable in some other settings. For example, in many models of statistical mechanics, where an infinite number of variables is considered, $\mathcal{A}_0$ would be the space of (smooth compactly supported) functions depending on a finite number of coordinates. Then, $\mathcal{A}$ would be the space of smooth functions depending on an infinite number of variables and the ideal property does not hold. Fortunately, in many infinite-dimensional instances, this difficulty is easily overcome. For example, for the infinite-dimensional Ornstein-Uhlenbeck semigroup described in Sect. 2.7.2, p. 108, the algebra $\mathcal{A}$ would be the space of smooth functions depending on a finite number of coordinates, and the stability under $(P_t)_{t\geq 0}$ is valid due to the fact that the different components

3.3 Heart of Darkness

do not interact. In general, however, when interaction is in action, a lot of analysis has to be developed on approximating models with a finite number of particles, and the major challenge is then to obtain functional inequalities where the control on the constants does not depend on the dimension (in this context the number of particles). While we do not address any specific infinite-dimensional models in this work, the formalism put forward here, and more importantly some of the basic principles of investigation, may be adapted to some extent to cover numerous examples of interest.

3.3.2 Domains

At this point, there are two definitions of the operators L and Γ, one on $\mathcal{D}(L)$ (or $\mathcal{D}(\mathcal{E})$) and one on the extended algebra $\mathcal{A}$. The first task is to verify that they actually coincide. Recall the adjoint L^* of L (with respect to μ and $\mathcal{A}_0$) as presented in Definition 3.1.9.

Proposition 3.3.5

(i) If $f \in \mathcal{A} \cap \mathcal{D}(L^*)$, then $L^* f = Lf$.
(ii) If $f \in \mathcal{A} \cap \mathcal{D}(\mathcal{E})$, both definitions of Γ, in $\mathcal{D}(\mathcal{E})$ and in $\mathcal{A}$, coincide.
(iii) If $f \in \mathcal{A} \cap \mathcal{D}(\mathcal{E})$ and $Lf \in \mathbb{L}^2(\mu)$, then $f \in \mathcal{D}(L)$.

Proof The first assertion is immediate as a consequence of integration by parts. Indeed, for any $h \in \mathcal{A}_0$, $\int_E h L^* f d\mu = \int_E f Lh d\mu = \int_E h Lf d\mu$. The result follows from the density of $\mathcal{A}_0$ in $\mathbb{L}^2(\mu)$.

The second assertion (ii) is a bit more delicate. For the moment, denote by $\Gamma_\mathcal{A}$ the carré du champ operator on $\mathcal{A}$, keeping the notation Γ for the carré du champ operator on $\mathcal{D}(\mathcal{E})$. We first check that for every $h \in \mathcal{A}_0$, $\Gamma(f, h) = \Gamma_\mathcal{A}(f, h)$. To this end, it is enough to prove that for any $k \in \mathcal{A}_0$, $\int_E k \Gamma(f, h) d\mu = \int_E k \Gamma_\mathcal{A}(f, h) d\mu$. By Proposition 3.1.5, $kf \in \mathcal{D}(\mathcal{E})$ and $\Gamma(kf, h) = k\Gamma(f, h) + f \Gamma(k, h)$, while the same identity holds for $\Gamma_\mathcal{A}$ from the diffusion property. Therefore, we need to check that

$$\int_E \Gamma(kf, h) d\mu - \int_E f \Gamma(k, h) d\mu = \int_E \Gamma_\mathcal{A}(kf, h) d\mu - \int_E f \Gamma_\mathcal{A}(k, h) d\mu.$$

But from the ideal property $kf \in \mathcal{A}_0$, so that $\Gamma(kf, h) = \Gamma_\mathcal{A}(kf, h)$, and similarly $f \Gamma(h, k) = f \Gamma_\mathcal{A}(h, k)$, hence equality holds. The same argument then allows us to pass from $\Gamma(f, h)$ with $f \in \mathcal{A} \cap \mathcal{D}(\mathcal{E})$ and $h \in \mathcal{A}_0$ to $\Gamma(f, g)$ with f and g in $\mathcal{A} \cap \mathcal{D}(\mathcal{E})$.

For the last point (iii), for every $h \in \mathcal{A}_0$,

$$\left| \mathcal{E}(h, f) \right| = \left| \int_E h Lf d\mu \right| \leq \|Lf\|_2 \|h\|_2,$$

which extends to any $h \in \mathcal{D}(\mathcal{E})$ by density of $\mathcal{A}_0$ in $\mathcal{D}(\mathcal{E})$ with respect to the $\mathcal{D}(\mathcal{E})$-topology. This is precisely the definition of $f \in \mathcal{D}(L)$. □

It is in general wrong to assert that $f \in \mathcal{A} \cap \mathbb{L}^2(\mu)$ and $\Gamma(f) \in \mathbb{L}^1(\mu)$ suffice to assert that $f \in \mathcal{D}(\mathcal{E})$ as may be seen by the example of the Laplace operator on $[0, 1]$. This implication actually holds under the essential self-adjointness assumption ESA (Definition 3.1.10) as described by the next statement.

Proposition 3.3.6 *Assume that the ESA property holds. Then*

(i) *If $f \in \mathcal{A} \cap \mathbb{L}^2(\mu)$ and $Lf \in \mathbb{L}^2(\mu)$, then $f \in \mathcal{D}(L)$.*
(ii) *If $f \in \mathcal{A} \cap \mathbb{L}^2(\mu)$ and $\Gamma(f) \in \mathbb{L}^1(\mu)$, then $f \in \mathcal{D}(\mathcal{E})$.*

Proof (i) is immediate since under the hypotheses $f \in \mathcal{D}(L^*)$ (by the very definition of $\mathcal{D}(L^*)$) and by the ESA property $\mathcal{D}(L) = \mathcal{D}(L^*)$.

Turning to (ii), consider the linear map $\ell_f : h \mapsto \int_E h L f d\mu$ defined on $\mathcal{A}_0$. From the integration by parts formula and the Cauchy-Schwarz inequality,

$$|\ell_f(h)| \leq \mathcal{E}(h)^{1/2} \left(\int_E \Gamma(f) d\mu \right)^{1/2}.$$

The linear map ℓ_f thus extends to a continuous linear form on the Hilbert space $\mathcal{D}(\mathcal{E})$, and may be represented as $\mathcal{E}(g, h)$ for some $g \in \mathcal{D}(\mathcal{E})$. Then for any $h \in \mathcal{A}_0$, $\int_E (f - g) L h d\mu = 0$, and therefore $f - g \in \mathcal{D}(L^*)$ (with $L^*(f - g) = 0$). Since $\mathcal{D}(L^*) = \mathcal{D}(L)$ by the ESA assumption, f may be written as sum of an element of $\mathcal{D}(L)$ and an element of $\mathcal{D}(\mathcal{E})$, hence in $\mathcal{D}(\mathcal{E})$. The proof is complete. □

3.3.3 Connexity, Weak Hypo-ellipticity and Completeness

Having presented the extended algebra $\mathcal{A}$ and some of its properties, we next describe the connexity, weak hypo-ellipticity and completeness properties based on the picture provided by the smooth manifold case of the previous section. Thus in the following, the triple (E, μ, Γ) denotes a Diffusion Markov Triple and $\mathcal{A}$ the extended algebra of Definition 3.3.1.

Definition 3.3.7 (Connexity) The Diffusion Markov Triple (E, μ, Γ) is said to be connected if $f \in \mathcal{A}$ and $\Gamma(f) = 0$ imply that f is constant.

It should be mentioned here that with respect to the ergodicity property of Definition 3.1.11, connexity is a local property for functions in $\mathcal{A}$ whereas for ergodicity the property holds for functions in $\mathcal{D}(\mathcal{E})$.

The second definition describes weak hypo-ellipticity in this context.

3.3 Heart of Darkness

Definition 3.3.8 (Weak hypo-ellipticity) The Diffusion Markov Triple (E, μ, Γ) is said to be weakly hypo-elliptic if for every $\lambda \in \mathbb{R}$, any $f \in \mathcal{D}(L^*)$ which satisfies $L^* f = \lambda f$ belongs to $\mathcal{A}$.

Very often (for example in the case of manifolds with hypo-elliptic operators, as in Sect. 1.12, p. 49), it is also true that for any bounded measurable function f and any $t > 0$, $P_t f \in \mathcal{A}$. However, this property will not be used below, and passing from weak hypo-ellipticity to this seemingly stronger form would require extra assumptions on the algebra $\mathcal{A}$.

The third definition is that of completeness in this context, following the characterization of Proposition C.4.1, p. 511, in Appendix C, as extensively used in the previous section on Riemannian manifolds.

Definition 3.3.9 (Completeness) The Diffusion Markov Triple (E, μ, Γ) is said to be complete if there exists an increasing sequence $(\zeta_k)_{k \geq 1}$ of positive functions in $\mathcal{A}_0$ such that $\lim_{k \to \infty} \zeta_k = \mathbb{1}$ (μ-almost everywhere) and $\Gamma(\zeta_k) \leq \frac{1}{k}$, $k \geq 1$.

Connexity and weak hypo-ellipticity yield ergodicity as defined in Definition 3.1.11.

Proposition 3.3.10 *Under the connexity and weak hypo-ellipticity assumptions, ergodicity occurs.*

For a proof, let $f \in \mathcal{D}(L)$ be such that $Lf = 0$. Then $f \in \mathcal{A}$ by the weak hypo-ellipticity property. By integration by parts, $0 = \int_E f L f d\mu = -\int_E \Gamma(f) d\mu$. Hence $\Gamma(f) = 0$ and therefore f is constant from the connexity assumption.

As a further easy observation, the proof of Proposition 3.2.1 in the manifold case immediately extends to a useful criterion for essential self-adjointness (the ESA property).

Proposition 3.3.11 *Under the connexity, weak hypo-ellipticity and completeness assumptions, the ESA property holds.*

3.3.4 Curvature-Dimension Conditions

To establish the gradient bounds described in Sect. 3.2 in the smooth Riemannian manifold setting, we investigate curvature conditions in this abstract context. To this end, we first define the Γ_2 operator, already introduced in (1.16.1), p. 71, on the algebra $\mathcal{A}$, and define in the same way Hessians.

Definition 3.3.12 (Γ_2 operator) The Γ_2 operator is the bilinear map $\Gamma_2 : \mathcal{A} \times \mathcal{A} \to \mathcal{A}$ defined by

$$\Gamma_2(f, g) = \frac{1}{2} \Big[L \Gamma(f, g) - \Gamma(f, Lg) - \Gamma(Lf, g) \Big]$$

for $(f, g) \in \mathcal{A} \times \mathcal{A}$. As for the carré du champ operator Γ, we often write more simply $\Gamma_2(f) = \Gamma_2(f, f)$.

The change of variables formula for Γ_2 ((1.16.3), p. 71) is valid on $\mathcal{A}$ (being a direct consequence of the change of variables formula for L)

$$\Gamma_2(\psi(f)) = \psi'(f)^2 \Gamma_2(f) + \psi'(f)\psi''(f) \Gamma(f, \Gamma(f)) + \psi''(f)^2 \Gamma(f)^2 \quad (3.3.2)$$

(with $f \in \mathcal{A}$ and $\psi : \mathbb{R} \to \mathbb{R}$ smooth). Its multi-dimensional extension (C.6.5), p. 516, is also similar.

The subsequent analysis will require us to consider, besides the Γ_2 operator, analogues of Hessians (on $\mathcal{A}$). The following definition is the suitable abstract formulation of the manifold case as described in Remark C.5.1, p. 512.

Definition 3.3.13 (Hessian) For a function $f \in \mathcal{A}$, the Hessian of f is the bilinear map $H(f) : \mathcal{A} \times \mathcal{A} \to \mathcal{A}$ defined by

$$H(f)(g, h) = \frac{1}{2}\big[\Gamma\big(g, \Gamma(f, h)\big) + \Gamma\big(h, \Gamma(f, g)\big) - \Gamma\big(f, \Gamma(g, h)\big)\big]. \quad (3.3.3)$$

Such Hessians appear in the chain rule formula for Γ_2 ((C.6.6), p. 516) in the form

$$\Gamma_2(hf, g) = h\Gamma_2(f, g) + f\Gamma_2(h, g) + 2H(g)(h, f). \quad (3.3.4)$$

Many formulas involving the Γ operator twice may also be seen as Hessians, such as for example

$$\Gamma\big(h, \Gamma(f)\big) = 2H(f)(f, h), \qquad \Gamma\big(f, \Gamma(f, h)\big) = H(h)(f, f) - H(f)(f, h).$$

It is straightforward to observe that $H(f)(g, h)$ is a first order operator in the sense that, for $k \in \mathcal{A}$, $H(f)(kg, h) = kH(f)(g, h) + gH(f)(k, h)$. It is also a second order differential operator with respect to the variable f, and satisfies the chain rule

$$H(kf)(g, h) = kH(f)(g, h) + fH(k)(g, h) + \Gamma(k, g)\Gamma(f, h) + \Gamma(f, g)\Gamma(k, h).$$

Furthermore, the integration by parts formula, valid as soon as one of the functions f, g, h belongs to $\mathcal{A}_0$, reads

$$\int_E H(f)(g, h) d\mu$$
$$= \frac{1}{2} \int_E f\big[L\Gamma(g, h) + \Gamma(h, Lg) + \Gamma(g, Lh) + 2Lg\,Lh\big] d\mu \quad (3.3.5)$$
$$= \frac{1}{2} \int_E g\big[\Gamma(f, Lh) - \Gamma(h, Lf) - L\Gamma(f, h)\big] d\mu.$$

3.3 Heart of Darkness

These chain rules extend to change of variables formulas for smooth functions as a consequence of the formula for the Γ operator (see Remark 3.1.4).

The curvature-dimension condition is defined using the Γ_2-operator, as in Sect. 1.16, p. 70.

Definition 3.3.14 (Curvature-dimension condition) The curvature-dimension condition $CD(\rho, n)$, for $\rho \in \mathbb{R}$ and $n \in [1, \infty]$, holds if for every $f \in \mathcal{A}$,

$$\Gamma_2(f) \geq \rho\, \Gamma(f) + \frac{1}{n}(\mathrm{L}f)^2.$$

As presented in Sect. C.6, p. 513, this definition suitably extends Ricci curvature lower bounds in dimensional Riemannian manifolds. Furthermore, the technique developed for (C.6.4), p. 515, yields a reinforced version of the $CD(\rho, \infty)$ curvature condition of the form

$$4\,\Gamma(f)\big[\Gamma_2(f) - \rho\,\Gamma(f)\big] \geq \Gamma\big(\Gamma(f)\big) \tag{3.3.6}$$

for every $f \in \mathcal{A}$. Moreover, under the $CD(\rho, \infty)$ condition, it also holds that (cf. (C.6.8), p. 517)

$$H(f)(g, h)^2 \leq \big[\Gamma_2(f) - \rho\,\Gamma(f)\big]\Gamma(g)\Gamma(h) \tag{3.3.7}$$

for $f, g, h \in \mathcal{A}$.

Remark 3.3.15 In the same way as the positivity of Γ extends from $\mathcal{A}_0$ to $\mathcal{A}$ via the ideal property, the change of variables formula and condition (ii) of Definition 3.3.1 (see Remark 3.3.2) may be used to extend the $CD(\rho, n)$ inequality. The argument is slightly more involved due to the second order terms in the change of variables formula for Γ_2. Let us illustrate the principle on the simpler curvature inequality $CD(\rho, \infty)$. For a given function $f \in \mathcal{A}$, it is enough to show that for any positive function $h \in \mathcal{A}_0$, $hK(f) \geq 0$ where $K(f, f) = \Gamma_2(f) - \rho\Gamma(f)$. From the positivity of K on $\mathcal{A}_0$ and the Cauchy-Schwarz inequality (with respect to the positive quadratic form K), $K(hf, hf) - 2fK(hf, h) + f^2 K(h, h) \geq 0$. Let then $\psi : \mathbb{R} \to \mathbb{R}$ be smooth such that $\psi(0) = 0$ and apply the latter to $\psi(h)$. By the change of variables formula, after some simplification,

$$\psi^2(h)K(f) + 4\psi(h)\psi'(h)H(f)(h, f) + 2\psi'^2(h)\big(\Gamma(f)\Gamma(h) + \Gamma(h, f)^2\big) \geq 0.$$

Now fix $\varepsilon > 0$ and choose ψ (smooth) such that $\psi(0) = 0$, and $\psi(r) = 1$ and $\psi'(r) = 0$ when $r \geq \varepsilon$. It follows that $K(f) \geq 0$ on $\{h \geq \varepsilon\}$, and it remains to let $\varepsilon \to 0$ to obtain the result.

3.3.5 Extensions

In order to develop the proof of the strong gradient bound (Theorem 3.2.4) in this abstract setting, it is necessary to extend the Γ_2 and the Hessian operators from

$\mathcal{A}_0$ to $\mathcal{D}(L)$, in the same way as Γ extends from $\mathcal{A}_0$ to $\mathcal{D}(\mathcal{E})$. In a sense, the approach here parallels at the second order what has been developed in Sect. 3.1 at the first order. At this second order level, positivity of Γ has to be replaced by the $CD(\rho, \infty)$ curvature condition together with essential self-adjointness (ESA). This goal is achieved in the following proposition.

Proposition 3.3.16 *Assume the ESA assumption and the $CD(\rho, \infty)$ condition for some $\rho \in \mathbb{R}$.*

(i) *The Γ_2 operator extends to a continuous bilinear operator on $\mathcal{D}(L)$ satisfying*

$$\int_E \Gamma_2(f) d\mu = \int_E (Lf)^2 d\mu \qquad (3.3.8)$$

for all $f \in \mathcal{D}(L)$.

(ii) *The $CD(\rho, \infty)$ condition extends to $\mathcal{D}(L)$.*

(iii) *Whenever $(f_\ell)_{\ell \in \mathbb{N}}$ is a sequence in $\mathcal{A}_0$ converging to $f \in \mathcal{D}(L)$, then $(\Gamma_2(f_\ell))_{\ell \in \mathbb{N}}$ converges in $\mathbb{L}^1(\mu)$ to $\Gamma_2(f)$.*

(iv) *For any pair $(g, h) \in \mathcal{A}_0 \times \mathcal{A}_0$, the linear operator $H(f)(g, h)$ defined for $f \in \mathcal{A}_0$ extends to a bounded bilinear operator on $\mathcal{D}(L) \times \mathcal{A}_0 \times \mathcal{A}_0$, and may be further extended to $\mathcal{D}(L) \times \mathcal{D}(\mathcal{E}) \times \mathcal{A}_0$. The Hessian operator $H(f)(g, h)$ defined for $(f, g, h) \in \mathcal{A}_0 \times \mathcal{A}_0 \times \mathcal{A}_0$ can be extended to $\mathcal{D}(L) \times \mathcal{A}_0 \times \mathcal{A}_0$, and further to $\mathcal{D}(L) \times \mathcal{D}(\mathcal{E}) \times \mathcal{A}_0$. It satisfies the extended inequality (3.3.7).*

(v) *For g and h in $\mathcal{A}_0$, whenever $(f_\ell)_{\ell \in \mathbb{N}}$ is a sequence in $\mathcal{A}_0$ converging to $f \in \mathcal{D}(L)$, then $(H(f_\ell)(g, h))_{\ell \in \mathbb{N}}$ converges to $H(f)(g, h)$ in $\mathbb{L}^2(\mu)$. When $h \in \mathcal{A}_0$ and $g \in \mathcal{D}(\mathcal{E})$, then the convergence takes place in $\mathbb{L}^1(\mu)$. For fixed $f \in \mathcal{D}(L)$ and $h \in \mathcal{A}_0$, if $(g_\ell)_{\ell \in \mathbb{N}}$ converges to g in $\mathcal{D}(\mathcal{E})$, then $(H(f)(g_\ell, h))_{\ell \in \mathbb{N}}$ converges to $H(f)(g, h)$ in $\mathbb{L}^1(\mu)$.*

(vi) *For every $f \in \mathcal{D}(L)$, $h, k \in \mathcal{A}_0$ and $g \in \mathcal{D}(\mathcal{E})$,*

$$H(f)(hg, k) = h H(f)(g, k) + g H(f)(h, k).$$

(vii) *For every $h \in \mathcal{A}_0$ and $f \in \mathcal{D}(L)$, $hf \in \mathcal{D}(L)$ and, for any $g \in \mathcal{D}(L)$,*

$$\Gamma_2(hf, g) = h \Gamma_2(f, g) + f \Gamma_2(h, g) + 2 H(g)(f, h).$$

(viii) *For every $f \in \mathcal{D}(L)$, $g \in \mathcal{D}(\mathcal{E})$ and $h \in \mathcal{A}_0$,*

$$H(f)(g, h)^2 \leq [\Gamma_2(f) - \rho \Gamma(f)] \Gamma(g) \Gamma(h).$$

Proof In Sect. 3.1, the bilinear Γ operator was extended from $\mathcal{A}_0$ to $\mathcal{D}(\mathcal{E})$ using that, on $\mathcal{A}_0$, $\mathcal{E}(f) = \int_E \Gamma(f) d\mu$, $\Gamma(f) \geq 0$ and $\mathcal{A}_0$ is dense in $\mathcal{D}(\mathcal{E})$ with respect to the $\mathcal{D}(\mathcal{E})$-topology. Under the $CD(\rho, \infty)$ and ESA conditions, the same procedure may be performed on the bilinear positive map $\Gamma_2(f) - \rho \Gamma(f)$ using the density of $\mathcal{A}_0$ in $\mathcal{D}(L)$. The basic ingredient here is the formula $\int_E \Gamma_2(f) d\mu = \int_E (Lf)^2 d\mu$ which holds for any $f \in \mathcal{A}_0$. The extension of the $CD(\rho, \infty)$ condition to $\mathcal{D}(L)$ is

3.3 Heart of Darkness

then straightforward. From (3.3.7), the extension of $H(f)(g, h)$ follows the same lines, and the change of variables formula (3.3.4) goes to the limit. □

The next step, as was the case for the Γ operator (Propositions 3.3.5 and 3.3.6), is to identify the two definitions of Γ_2 and the Hessian (on $\mathcal{D}(L)$ and on $\mathcal{A}$), and moreover to extend the integration by parts formula (3.3.8) to $\mathcal{A}$. The following proposition fulfills this task.

Proposition 3.3.17 *Assume the ESA assumption and the $CD(\rho, \infty)$ condition for some $\rho \in \mathbb{R}$. If $f \in \mathcal{A} \cap \mathcal{D}(L)$, the two definitions of $\Gamma_2(f)$ (on $\mathcal{D}(L)$ and on $\mathcal{A}$) coincide. The same is true for the Hessian $H(f)(g, h)$ whenever $f \in \mathcal{A} \cap \mathcal{D}(L)$, $g \in \mathcal{A} \cap \mathcal{D}(\mathcal{E})$ and $h \in \mathcal{A}_0$.*

Moreover, if $f \in \mathcal{A} \cap \mathcal{D}(L)$ then $\Gamma_2(f) \in \mathbb{L}^1(\mu)$ and $\Gamma(f) \in \mathbb{L}^1(\mu)$, in which case

$$\int_E \Gamma_2(f) d\mu = \int_E (Lf)^2 d\mu.$$

Conversely, and under the additional ergodicity property, for $f \in \mathbb{L}^2(\mu) \cap \mathcal{A}$, if both $\Gamma_2(f)$ and $\Gamma(f)$ belong to $\mathbb{L}^1(\mu)$, then $f \in \mathcal{D}(L)$.

Proof We follow the same lines as the identification of Γ on $\mathcal{A}$ and $\mathcal{D}(\mathcal{E})$ (Proposition 3.3.5). Start with the identifications of the Hessians $H(f)(g, h)$ for $f \in \mathcal{D}(L) \cap \mathcal{A}$ and $g, h \in \mathcal{A}_0$. To this end, denote by $H(f)$ the Hessian for $f \in \mathcal{D}(L)$ and by $H_\mathcal{A}(f)$ the Hessian computed for $f \in \mathcal{A}$. We aim to prove that for every $k \in \mathcal{A}_0$,

$$\int_E k H(f)(g, h) d\mu = \int_E k H_\mathcal{A}(f)(g, h) d\mu.$$

By the chain rule,

$$\int_E k H(f)(g, h) d\mu = \int_E H(kf)(g, h) d\mu - \int_E f H(k)(g, h) d\mu$$
$$- \int_E \big(\Gamma(f, g)\Gamma(h, k) + \Gamma(f, h)\Gamma(k, g)\big) d\mu.$$

By the integration by parts formula (3.3.5), this expression is reduced to an integral of the form $\int_E f E(g, h, k) d\mu$ where $E(g, h, k)$ (even if tedious to write down explicitly) belongs to $\mathcal{A}_0$. Now similarly $\int_E k H_\mathcal{A}(f)(g, h) d\mu = \int_E f E(g, h, k) d\mu$ with the same expression $E(g, h, k)$. Then, changing f into a sequence $(f_\ell)_{\ell \in \mathbb{N}}$ of functions in $\mathcal{A}_0$ converging to $f \in \mathcal{D}(L)$, the identity of the two integrals follows in the limit. Fixing now $f \in \mathcal{D}(L)$ and $h \subset \mathcal{A}_0$, and replacing $g \in \mathcal{A}_0$ by $g \in \mathcal{D}(\mathcal{E})$, the same procedure (using the other chain rule for H and the other integration by parts formula (3.3.5)) yields the announced coincidence of H and $H_\mathcal{A}$. The identity of the Γ_2 operators follow along the same lines.

We already know from Sect. 3.1 that if $f \in \mathcal{D}(L)$, then $Lf \in \mathbb{L}^2(\mu)$ (by definition) and $\Gamma(f) \in \mathbb{L}^1(\mu)$. Moreover, in $\mathcal{D}(L)$, $\int_E \Gamma_2(f) d\mu = \int_E (Lf)^2 d\mu$ and

$-\int_E f\,\mathrm{L}f\,d\mu = \int_E \Gamma(f)\,d\mu$. Therefore both $\Gamma(f)$ and $\Gamma_2(f) - \rho\Gamma(f)$ are in $\mathbb{L}^1(\mu)$, and so too is $\Gamma_2(f)$.

It remains to verify the last claim of the proposition, that is, under ergodicity, when $f \in \mathcal{A} \cap \mathbb{L}^2(\mu)$ with $\Gamma_2(f)$ and $\Gamma(f)$ in $\mathbb{L}^1(\mu)$, then $f \in \mathcal{D}(\mathrm{L})$. Towards this goal, following Proposition 3.3.6, it is enough to show that $\mathrm{L}f \in \mathbb{L}^2(\mu)$. But under the ESA and ergodicity properties $\{\mathrm{L}h\,;\,h \in \mathcal{A}_0\}$ is dense in the space $\mathbb{L}^2(\mu)$ of $\mathbb{L}^2(\mu)$-functions orthogonal to constants. Restrict for simplicity to the case $\mu(E) = \infty$, in which case $\{\mathrm{L}h\,;\,h \in \mathcal{A}_0\}$ is dense in the space $\mathbb{L}^2(\mu)$ (the proof is easily adapted to the case $\mu(E) = 1$). Consider the bilinear form $K(f, f) = \Gamma_2(f, f) - \rho\Gamma(f, f)$ and the linear form

$$\ell_f(h) = \int_E (\mathrm{L}f + \rho f)\,\mathrm{L}h\,d\mu = \int_E K(f, h)\,d\mu, \quad h \in \mathcal{D}(\mathrm{L}).$$

Since K is a positive bilinear form, it holds that $K(f, h)^2 \leq K(f, f)K(h, h)$, which implies for any $h \in \mathcal{D}(\mathrm{L})$, $|\ell_f(h)| \leq C(f)\|h\|_{\mathcal{D}(\mathrm{L})}$. Then, there exists some $g \in \mathcal{D}(\mathrm{L})$ such that $\ell_f(h) = \int_E [\mathrm{L}g\,\mathrm{L}h + gh]\,d\mu$. But it is easily seen from the spectral decomposition that in fact $g = \mathrm{L}k$, where $k = (\mathrm{L}^2 + \mathrm{Id})^{-1}(\mathrm{L} + \rho\,\mathrm{Id})f$. Again from the spectral decomposition, the operators

$$\left(\mathrm{L}^2 + \mathrm{Id}\right)^{-1}(\mathrm{L} + \rho\,\mathrm{Id}) \quad \text{and} \quad \left(\mathrm{L}^2 + \mathrm{Id}\right)^{-1}\left(\mathrm{L}^2 + \rho\,\mathrm{L}\right)$$

are bounded in $\mathbb{L}^2(\mu)$, so that, for every $h \in \mathcal{A}_0$,

$$|\ell_f(h)| = \left|\int_E (\mathrm{L}g + k)\,\mathrm{L}h\,d\mu\right| \leq \left(\|\mathrm{L}g\|_2 + \|k\|_2\right)\|\mathrm{L}h\|_2 \leq C'(f)\|\mathrm{L}h\|_2.$$

Thanks to the ergodicity and ESA properties, the linear form ℓ_f can be extended to $\mathbb{L}^2(\mu)$, which shows that $\mathrm{L}f + \rho f \in \mathbb{L}^2(\mu)$, and therefore $\mathrm{L}f \in \mathbb{L}^2(\mu)$. The proof is complete. □

Note from the preceding proof that whenever $\rho \geq 0$, for a function $f \in \mathcal{A} \cap \mathbb{L}^2(\mu)$ such that $\Gamma_2(f) \in \mathbb{L}^1(\mu)$, the condition $\Gamma(f) \in \mathbb{L}^1$ is not needed to ensure that $\mathrm{L}f \in \mathbb{L}^2(\mu)$.

3.3.6 Gradient Bounds and Mass Conservation

On the basis of the previous curvature conditions, we now address the announced gradient bounds (or commutation between P_t and Γ) which were developed in the smooth manifold case in Sect. 3.2.3 under the hypotheses of connexity and completeness. The point here is that the result will be extended in the abstract framework to the case when the underlying operator is essentially self-adjoint. As we have seen in Proposition 3.3.11, essential self-adjointness holds under connexity, weak hypo-ellipticity and completeness, so that the approach here goes further than the manifold case.

3.3 Heart of Darkness

Although the gradient bound $\Gamma(P_t f) \leq e^{-2\rho t} P_t(\Gamma(f))$ seems easier to obtain than the strong bound, the main difficulty is that we have no a priori control on the $\mathbb{L}^2(\mu)$-norm of it, in contrast to that of $\sqrt{\Gamma(P_t f)}$. Unfortunately, this leads to some extra complications. One main result is emphasized in the following theorem and its corollary.

Theorem 3.3.18 (Strong gradient bound) *Let (E, μ, Γ) be a Diffusion Markov Triple satisfying the ESA property, with extended algebra $\mathcal{A}$. Then, under the $CD(\rho, \infty)$ condition, for every $f \in \mathcal{A}_0$ and every $t \geq 0$,*

$$\sqrt{\Gamma(P_t f)} \leq e^{-\rho t} P_t\left(\sqrt{\Gamma(f)}\right).$$

Corollary 3.3.19 (Gradient bound) *Let (E, μ, Γ) be a Diffusion Markov Triple satisfying the ESA property, with extended algebra $\mathcal{A}$. Then, under the $CD(\rho, \infty)$ condition, for every $f \in \mathcal{A}_0$ and every $t \geq 0$,*

$$\Gamma(P_t f) \leq e^{-2\rho t} P_t\left(\Gamma(f)\right).$$

Although the corollary seems a weaker result than the theorem (by Jensen's inequality for P_t, which is sub-Markov), we will see later (cf. Theorem 4.7.2, p. 209) that they are in fact both equivalent to the curvature $CD(\rho, \infty)$ condition. Actually, taking the derivative at $t = 0$ of $\Gamma(P_t f) \leq e^{-2\rho t} P_t(\Gamma(f))$ yields

$$L\Gamma(f) \geq 2\Gamma(f, Lf) + 2\rho \Gamma(f)$$

which amounts to $\Gamma_2(f) \geq \rho \Gamma(f)$.

The proof of Theorem 3.3.18 relies on the following lemmas.

Lemma 3.3.20 *Under the ESA and $CD(\rho, \infty)$ assumptions, for every $f \in \mathcal{A}_0$, and any $t \geq 0$ and $\varepsilon > 0$, $\sqrt{\Gamma(P_t f) + \varepsilon^2} - \varepsilon \in \mathcal{D}(\mathcal{E})$.*

Proof Set $G_\varepsilon = \sqrt{\Gamma(P_t f) + \varepsilon^2} - \varepsilon$. Since $P_t f \in \mathcal{A}$, by composition with a smooth function, $G_\varepsilon \in \mathcal{A}$, and from the ESA property, it is enough to prove that $G_\varepsilon \in \mathbb{L}^2(\mu)$ and $\Gamma(G_\varepsilon) \in \mathbb{L}^1(\mu)$ (Proposition 3.3.6 (ii)). The first claim follows since we have $\sqrt{r + \varepsilon^2} - \varepsilon \leq \sqrt{r}$ for any $r \geq 0$ and $\int_E \Gamma(P_t f) d\mu = \mathcal{E}(P_t f) \leq \mathcal{E}(f) < \infty$. The second claim is a consequence of the bounds

$$\Gamma(G_\varepsilon) = \frac{\Gamma(\Gamma(P_t f))}{4(\Gamma(P_t f) + \varepsilon^2)} \leq \Gamma_2(P_t f) - \rho \Gamma(P_t f),$$

by the extended Γ_2 inequality (3.2.4), and

$$\int_E \left[\Gamma_2(P_t f) - \rho \Gamma(P_t f)\right] d\mu \leq \|Lf\|_2^2 + |\rho| \mathcal{E}(f).$$

The lemma is therefore established. □

Lemma 3.3.21 *Under the ESA and $CD(0, \infty)$ assumptions, for any $f \in \mathcal{A} \cap \mathbb{L}^2(\mu)$ such that $\mathrm{L}f \in \mathcal{D}(\mathcal{E})$ and $\sqrt{\Gamma(f) + \varepsilon^2} - \varepsilon \in \mathcal{D}(\mathcal{E}) \cap \mathcal{A}$ for some $\varepsilon > 0$, and for any h positive and bounded in $\mathcal{D}(\mathcal{E})$,*

$$\mathcal{E}\left(h, \sqrt{\Gamma(f) + \varepsilon^2} - \varepsilon\right) + \int_E h \frac{\Gamma(f, \mathrm{L}f)}{\sqrt{\Gamma(f) + \varepsilon}} d\mu \leq 0.$$

Proof Using the same notation $G_\varepsilon = \sqrt{\Gamma(f) + \varepsilon^2} - \varepsilon$, for $h \in \mathcal{A}_0$

$$\mathcal{E}(h, G_\varepsilon) = -\int_E h \mathrm{L} G_\varepsilon d\mu$$

$$= -\int_E h \left(\frac{4\Gamma_2(f)(\Gamma(f) + \varepsilon^2) - \Gamma(\Gamma(f))}{4(\Gamma(f) + \varepsilon^2)^{3/2}} + \frac{\Gamma(f, \mathrm{L}f)}{\sqrt{\Gamma(f) + \varepsilon^2}} \right) d\mu.$$

Applying once again the extended Γ_2 inequality (3.2.4) the announced inequality follows in this case. It may then be extended to any $h \in \mathcal{D}(\mathcal{E})$ by density. □

Proof of Theorem 3.3.18 We only prove the theorem when $\rho = 0$, the extension to the general case being straightforward. Fix $\varepsilon > 0$, $f \in \mathcal{A}_0$ and $h \geq 0$ in $\mathcal{A}_0$. For $t > 0$ fixed, consider

$$\Lambda(s) = \int_E h P_s G_\varepsilon d\mu = \int_E G_\varepsilon P_s h \, d\mu, \quad s \in [0, t],$$

with the notation $G_\varepsilon = \sqrt{\Gamma(P_{t-s}f) + \varepsilon^2} - \varepsilon$ as above. The task will be to show that Λ is increasing. From Lemma 3.3.20, $G_\varepsilon \in \mathcal{D}(\mathcal{E})$. By spectral analysis, for every $G \in \mathcal{D}(\mathcal{E})$, $\partial_s \int_E G P_s h d\mu = -\mathcal{E}(P_s h, G)$. On the other hand,

$$\partial_s G_\varepsilon = -\frac{\Gamma(P_{t-s}f, P_{t-s}\mathrm{L}f)}{G_\varepsilon + \varepsilon}$$

while

$$\partial_s^2 G_\varepsilon = \frac{\Gamma(P_{t-s}\mathrm{L}f, P_{t-s}\mathrm{L}f) + \Gamma(P_{t-s}f, P_{t-s}\mathrm{L}^2 f)}{G_\varepsilon + \varepsilon} - \frac{\Gamma(P_{t-s}f, P_{t-s}\mathrm{L}f)^2}{(G_\varepsilon + \varepsilon)^3}.$$

For any $f \in \mathcal{A}_0$, $\Gamma(P_t f) \in \mathbb{L}^1(\mu)$ showing that $\|\partial_s G_\varepsilon\|_1 + \|\partial_s^2 G_\varepsilon\|_1 \leq C$ for some constant C depending only on $f \in \mathcal{A}_0$. Since $P_t h$ is bounded, derivation of Λ is then justified and, by the usual change of variables formula,

$$\Lambda(s)' = -\mathcal{E}(P_s h, G_\varepsilon) - \int_E P_s h \frac{\Gamma(P_{t-s}f, P_{t-s}\mathrm{L}f)}{(\Gamma(P_{t-s}f) + \varepsilon)^{1/2}} d\mu.$$

This is positive from Lemma 3.3.21. Hence $\Lambda(t) \geq \Lambda(0)$ for any positive $h \in \mathcal{A}_0$ and any $\varepsilon > 0$, leading to the conclusion. The theorem is established. □

3.3 Heart of Darkness

The gradient bounds of Theorem 3.3.18 immediately extend to functions f in $\mathcal{D}(\mathcal{E})$ via an approximation argument. However, it is not clear that they apply to functions in $\mathcal{A}$ without further hypotheses.

Under the $CD(\rho, \infty)$ condition, further approximation procedures for functions in the $\mathcal{D}(L)$-domain by elements of $\mathcal{A}_0$ may be developed. Namely, from Corollary 3.3.19 together with the Duhamel formula (3.1.21), for any f in $\mathcal{A}_0$,

$$P_t(f^2) - (P_t f)^2 \geq 2 \int_0^t e^{2\rho s} ds\, \Gamma(P_t f) = \frac{e^{2\rho t} - 1}{\rho} \Gamma(P_t f).$$

(This type of local Poincaré inequality will be extensively examined in Sect. 4.7, p. 206.) In particular, whenever $(f_\ell)_{\ell \in \mathbb{N}}$ (in $\mathcal{A}_0$) converges to f in $\mathbb{L}^p(\mu)$ for some $p > 2$, then $\Gamma(P_t f_\ell) \to \Gamma(P_t f)$ in $\mathbb{L}^q(\mu)$ for any $q < \frac{p}{2}$ and any $t > 0$.

Remark 3.3.22 The gradient bounds of Theorem 3.3.18 will play a major role in the forthcoming chapters. In particular cases, similar results may be reached under weaker hypotheses. For example, when solving stochastic differential equations, the route proposed in Proposition 3.2.5 provides such bounds with stochastic calculus tools. Similar results may be obtained for semigroups on bounded domains with Neumann boundary conditions, provided there is enough information about the geometry of the boundary. Such an example would however never be covered by the arguments presented here, due to the lack of the ESA property in this context. The conclusion of Corollary 3.3.19 is still valid in non-diffusion instances (for example for Markov chains on a finite space). In some examples, such as the Heisenberg model (cf. (3.2.3)), the gradient bounds also hold up to a constant $K > 1$ as

$$\sqrt{\Gamma(P_t f)} \leq K e^{-\rho t} P_t\left(\sqrt{\Gamma(f)}\right)$$

(while there is in general no $CD(\rho, \infty)$ condition). In these examples, the weaker gradient bound from Corollary 3.3.19 is usually far easier to obtain than the strong version.

With the preceding gradient bounds, Theorem 3.2.6 of the previous section on the *mass conservation* property extends to the present abstract setting. Recall that the semigroup $\mathbf{P} = (P_t)_{t \geq 0}$ is conservative if $P_t(\mathbb{1}) = \mathbb{1}$ for every $t \geq 0$. As mentioned earlier, the mass conservation property is also known as the *Markov, stochastic completeness* or *non-explosion* property. The following statement also covers Theorem 3.2.7 in this abstract framework and is established similarly.

Theorem 3.3.23 (Mass conservation) *Let (E, μ, Γ) be a Diffusion Markov Triple with extended algebra $\mathcal{A}$. Under the completeness, ESA and $CD(\rho, \infty)$ assumptions, the semigroup $\mathbf{P} = (P_t)_{t \geq 0}$ is conservative. Moreover, if $\rho > 0$, then $\mu(E) < \infty$.*

3.3.7 Intrinsic Distance and Lipschitz Functions

According to the corresponding discussion in smooth manifolds, a distance function associated with a Markov generator L and the underlying extended algebra $\mathcal{A}$ may be defined as

$$d(x, y) = \operatorname{esssup}[f(x) - f(y)], \quad (x, y) \in E \times E, \tag{3.3.9}$$

the esssup running over all bounded functions f in $\mathcal{A}$ such that $\Gamma(f) \leq 1$. (Here, the esssup is defined as the least measurable function on $(E \times E, \mathcal{F} \otimes \mathcal{F})$ which is larger than $f(x) - f(y)$ for $\mu \otimes \mu$-almost every $(x, y) \in E \times E$ over the given class of functions f.) The choice of an esssup (which thus rules out sets of μ-measure 0) instead of a mere supremum is due to the fact that we are considering only classes of functions. This distance, often called the *intrinsic distance* (although depending on the underlying algebra $\mathcal{A}$), coincides with the Riemannian distance in the case of diffusions on a manifold with elliptic coefficients (see Sect. C.4, p. 509), and also in some hypo-elliptic cases with the so-called Carnot-Carathéodory distance. The diameter is defined as the $\mathbb{L}^\infty(E \times E, \mu \otimes \mu)$-norm of the distance function $d(x, y)$, $(x, y) \in E \times E$. We make no claim that (3.3.9) effectively defines a distance (beyond the triangle inequality which is obviously satisfied), and moreover that this distance has anything to do with the completeness hypothesis. This would require many more hypotheses on $\mathcal{A}$ and $\mathcal{A}_0$ (and would certainly be useless). In particular, in some infinite-dimensional settings such as the Ornstein-Uhlenbeck example of Sect. 2.7.2, p. 108, this distance is almost everywhere infinite. However, it will play a key role in many functional inequalities or estimates on heat kernels in particular by means of the associated Lipschitz functions. Distance to a measurable set A as $d(x, A) = \operatorname{esssup} f(x)$, $x \in E$, where the esssup runs over all bounded functions f in $\mathcal{A}$ such that $\Gamma(f) \leq 1$ and $f \mathbb{1}_A = 0$, may be considered similarly. In the above mentioned example of the infinite-dimensional Ornstein-Uhlenbeck semigroup, this distance is μ-almost everywhere finite as soon as $\mu(A) > 0$. It will not be used below but any result on Lipschitz function would also apply to these distances from sets.

Definition 3.3.24 (Lipschitz function) A function f on a Diffusion Markov Triple (E, μ, Γ) with extended algebra $\mathcal{A}$ is called Lipschitz if $f \in \mathcal{A}$ and $\Gamma(f) \in \mathbb{L}^\infty(\mu)$. Denote by $\|f\|_{\operatorname{Lip}} = \|\Gamma(f)\|_\infty^{1/2}$ its Lipschitz semi-norm. f is said to be 1-Lipschitz if $\|f\|_{\operatorname{Lip}} \leq 1$.

In this definition of Lipschitz function, the class $\mathcal{A}$ may be replaced alternatively by $\mathcal{D}(\mathcal{E})$, leading in practice to similar results. Dealing with functions in the Dirichlet domain is somewhat easier with respect to contraction properties (cf. (3.1.15)). In the classical case of $\mathbb{R}^n$, Rademacher's Theorem tells us that Lipschitz functions coincide with functions with almost everywhere bounded gradients.

The notion of the diameter of a Diffusion Markov Triple (E, μ, Γ) is defined in the usual way where the set E is regarded as a metric space with respect to the intrinsic distance d, and is described equivalently as

$$D(E, \mu, \Gamma) = \sup_{\|f\|_{\text{Lip}} \leq 1} \|\tilde{f}\|_\infty \qquad (3.3.10)$$

where $\tilde{f}(x, y) = f(x) - f(y)$, $(x, y) \in E \times E$, the $\mathbb{L}^\infty$-norm being that of $\mathbb{L}^\infty(E \times E, \mu \otimes \mu)$. In particular, if the diameter is finite, every Lipschitz function which is finite μ-almost everywhere is bounded.

3.3.8 Full Markov Triple

To conclude this construction, starting from the (symmetric) Diffusion Markov Triple Definition 3.1.8 given on some algebra $\mathcal{A}_0$, in order to fully develop the tools related to curvature conditions and gradient bounds, and to many other topics such as integration by parts when functions do not belong to $\mathcal{A}_0$, it has thus been necessary to deal with an extended algebra $\mathcal{A}$. The main assumptions on $\mathcal{A}_0$ and $\mathcal{A}$ are summarized in Sect. 3.4. This basis gives rise to the definition of a Full Markov Triple which includes the connexity and ESA properties. Note that a Standard Markov Triple includes the mass conservation property while it was established previously that a Diffusion Markov Triple with the ESA property satisfying a curvature condition $CD(\rho, \infty)$ for some $\rho \in \mathbb{R}$ is actually conservative.

As illustrated in Sect. 3.2, the Full Markov Triple structure covers the important example of second order differential operators on smooth complete connected Riemannian manifolds with $\mathcal{A}_0$ the class of smooth $\mathcal{C}^\infty$ compactly supported functions and $\mathcal{A}$ the class of smooth $\mathcal{C}^\infty$ functions. Throughout this work, such examples will always be considered for these algebras of functions under the generic name of "smooth" functions. More general degrees of smoothness, at least $\mathcal{C}^2$, may be investigated similarly.

Definition 3.3.25 (Full Markov Triple) A Full Markov Triple (E, μ, Γ) is a Standard Markov Triple (thus with algebra $\mathcal{A}_0$ and extended algebra $\mathcal{A}$) satisfying moreover the connexity and ESA properties.

Occasionally we shall also make the further assumptions of completeness or weak hypo-ellipticity, however we shall always explicitly mention when these additional assumptions have been made. Sometimes, to make things easier, it is also convenient to assume that $\mathcal{A} = \mathcal{A}_0$ (in which case the ESA property is automatic and the gradient bounds are much easier to reach), corresponding to compact manifolds (without boundary) in the smooth case.

Definition 3.3.26 (Compact Markov Triple) A Compact Markov Triple (E, μ, Γ) is a Full Markov Triple for which $\mathcal{A} = \mathcal{A}_0$.

For a Compact Markov Triple, $\mathcal{A}_0$ is stable under $(P_t)_{t \geq 0}$ and therefore automatically dense in $\mathcal{D}(L)$ with respect to its topology.

3.4 Summary of Hypotheses (Markov Triple)

In this last section, we summarize the previous investigation and present the typical framework for the analysis of symmetric Markov diffusion operators in which we will be working in most parts of this monograph. Although various results might hold in broader settings, without in particular the symmetry or diffusion assumptions, we will always work with a Standard or Full Markov Triple (E, μ, Γ) as presented in Definitions 3.1.15 and 3.3.25 with underlying algebra $\mathcal{A}_0$ and extended algebra $\mathcal{A}$ respectively (the properties of which are recalled below). This framework includes the Dirichlet form $\mathcal{E}$ with domain $\mathcal{D}(\mathcal{E})$ defined by $\mathcal{E}(f) = \int_E \Gamma(f) d\mu$, the associated diffusion operator L with $\mathbb{L}^2(\mu)$-domain $\mathcal{D}(L)$, infinitesimal generator of the Markov semigroup $\mathbf{P} = (P_t)_{t \geq 0}$ (with invariant and reversible measure μ), and associated Markov process, or family of processes, $\mathbf{X} = \{X_t^x; t \geq 0, x \in E\}$. The semigroup $\mathbf{P} = (P_t)_{t \geq 0}$ may be represented according to (1.2.4), p. 12, by probability kernels as

$$P_t f(x) = \int_E f(y) \, p_t(x, dy), \quad t \geq 0, \; x \in E.$$

The kernels $p_t(x, dy)$ describe the distribution at time t of the Markov process X_t^x starting at x. Functions are understood as classes of functions, and equalities and inequalities between them hold μ-almost everywhere.

Throughout, we adopt the Full Markov Triple structure as a convenient framework which allows us to describe the main results and freely develop the central ideas and principles of the Γ-calculus and heat flow monotonicity.

Standard Markov Triples, or even just Diffusion Markov Triples, suffice for many statements. Similarly, some results only require a number of specific properties of the Full Markov Triple definition, but for convenience we stick to the latter. If there is any need to distinguish between Full and Standard Triples, generally speaking Full Markov Triples are necessary as soon as gradient bounds, the Γ_2 operator and curvature-dimension conditions enter into play, otherwise Standard Markov Triples may be used.

As mentioned earlier, it is sometimes appropriate to work in the Compact Markov Triple setting (Definition 3.3.26), although the conclusions will actually hold in the full case.

Finally, for simplicity of exposition, we mostly use the reduced terminology "Markov Triple", assuming that Full Markov Triple is meant (or only Standard Markov Triple if this is enough for the purpose of the given results according to the preceding remarks).

We present below a short synthesis of the different objects of interest and hypotheses described in this first part, to which the reader may refer while progressing through the following chapters.

3.4 Summary of Hypotheses (Markov Triple)

3.4.1 Diffusion Markov Triple (Definition 3.1.8)

A Diffusion Markov Triple (E, μ, Γ) is composed of a measure space $(E, \mathcal{F}, \mu)$, a class $\mathcal{A}_0$ of real-valued measurable functions on E and a symmetric bilinear operator (the carré du champ operator) $\Gamma : \mathcal{A}_0 \times \mathcal{A}_0 \to \mathcal{A}_0$ satisfying properties **D1** to **D9** below.

D1. The state space $(E, \mathcal{F}, \mu)$ is a good measure space as defined in Sect. 1.14, p. 53. The measure μ is σ-finite and, when finite, assumed to be a probability.

D2. $\mathcal{A}_0$ is a vector space of bounded measurable functions $f : E \to \mathbb{R}$, which is dense in every $\mathbb{L}^p(\mu)$, $1 \leq p < \infty$, stable under products (that is $\mathcal{A}_0$ is an algebra) and stable under the action of smooth (C^∞) functions $\Psi : \mathbb{R}^k \to \mathbb{R}$ such that $\Psi(0) = 0$.

D3. The carré du champ operator Γ is a bilinear symmetric map $\mathcal{A}_0 \times \mathcal{A}_0 \to \mathcal{A}_0$ such that $\Gamma(f, f) \geq 0$ for all f in $\mathcal{A}_0$. $\Gamma(f, f)$ is abbreviated as $\Gamma(f)$. It satisfies the diffusion hypothesis of Definition 3.1.3: for every smooth function $\Psi : \mathbb{R}^k \to \mathbb{R}$ such that $\Psi(0) = 0$ and every $f_1, \ldots, f_k, g \in \mathcal{A}_0$,

$$\Gamma\big(\Psi(f_1, \ldots, f_k), g\big) = \sum_{i=1}^{k} \partial_i \Psi(f_1, \ldots, f_k) \Gamma(f_i, g). \tag{3.4.1}$$

D4. For every f in $\mathcal{A}_0$, there exists a finite constant $C(f)$ such that for every $g \in \mathcal{A}_0$,

$$\left| \int_E \Gamma(f, g) d\mu \right| \leq C(f) \|g\|_2.$$

The Dirichlet form $\mathcal{E}$ is defined for every $(f, g) \in \mathcal{A}_0 \times \mathcal{A}_0$ by

$$\mathcal{E}(f, g) = \int_E \Gamma(f, g) d\mu.$$

$\mathcal{E}(f, f)$ is abbreviated as $\mathcal{E}(f)$. The domain $\mathcal{D}(\mathcal{E})$ of the Dirichlet form $\mathcal{E}$ is the completion of $\mathcal{A}_0$ with respect to the norm $\|f\|_\mathcal{E} = [\|f\|_2^2 + \mathcal{E}(f)]^{1/2}$. The Dirichlet form $\mathcal{E}$ is extended to $\mathcal{D}(\mathcal{E})$ by continuity together with the carré du champ operator Γ.

D5. L is a linear operator on $\mathcal{A}_0$ defined by and satisfying the integration by parts formula

$$\int_E g\, \mathrm{L} f\, d\mu = -\int_E \Gamma(f, g) d\mu$$

for all $f, g \in \mathcal{A}_0$. The change of variables formula (3.1.8) for L is a consequence of the change of variables formula (3.4.1) for Γ.

D6. For the operator L defined in **D5**, $\mathrm{L}(\mathcal{A}_0) \subset \mathcal{A}_0$.

D7. The domain $\mathcal{D}(\mathrm{L})$ of the operator L is defined as the set of $f \in \mathcal{D}(\mathcal{E})$ for which there exists a finite constant $C(f)$ such that for any $g \in \mathcal{D}(\mathcal{E})$

$$|\mathcal{E}(f, g)| \leq C(f) \|g\|_2.$$

On $\mathcal{D}(L)$, L is extended via the integration by parts formula for every $g \in \mathcal{D}(\mathcal{E})$. L defined on $\mathcal{D}(L)$ is always self-adjoint.

D8. For every $f \in \mathcal{A}_0$, $\int_E L f d\mu = 0$.

D9. The semigroup $\mathbf{P} = (P_t)_{t \geq 0}$ is the symmetric semigroup with infinitesimal generator L defined on its domain $\mathcal{D}(L)$. It is positivity preserving but not necessarily Markov (in general it is only sub-Markov, i.e. $P_t(\mathbb{1}) \leq \mathbb{1}$). Moreover, $\mathcal{A}_0$ is not necessarily dense in the domain $\mathcal{D}(L)$ with respect to the domain norm $\|f\|_{\mathcal{D}(L)} = [\|f\|_2^2 + \|Lf\|_2^2]^{1/2}$. One may use the spectral decomposition to study $(P_t)_{t \geq 0}$ and various properties of functions in $\mathbb{L}^2(\mu)$ (Proposition 3.1.6).

3.4.2 Standard Markov Triple (Definition 3.1.15)

A Standard Markov Triple is a Diffusion Markov Triple satisfying in addition the ergodicity and mass conservation properties **S10** and **S11**.

S10. The operator L is ergodic in the sense that any $f \in \mathcal{D}(L)$ such that $Lf = 0$ is constant. Equivalently, any $f \in \mathcal{D}(\mathcal{E})$ such that $\Gamma(f) = 0$ is constant (Definition 3.1.11).

S11. The semigroup $\mathbf{P} = (P_t)_{t \geq 0}$ is conservative in the sense that $P_t(\mathbb{1}) = \mathbb{1}$ for every $t \geq 0$ (Definition 3.1.12).

In this context, further properties, not required in the Standard Markov Triple structure, may be considered.

Adjoint Operator (Definition 3.1.9) The domain $\mathcal{D}(L^*)$ is the set of functions $f \in \mathbb{L}^2(\mu)$ such that there exists a finite constant $C(f)$ for which, for every $g \in \mathcal{A}_0$,

$$\left| \int_E f L g \, d\mu \right| \leq C(f) \|g\|_2.$$

On this domain, the adjoint operator L^* is defined by integration by parts: for any $g \in \mathcal{A}_0$,

$$L^*(f) = \int_E f L g \, d\mu = \int_E g L^* f \, d\mu.$$

It holds that $\mathcal{D}(L) \subset \mathcal{D}(L^*)$ and L^* is an extension of L.

Essential Self-adjointness (ESA) (Definition 3.1.10) $\mathcal{D}(L) = \mathcal{D}(L^*)$, or equivalently $\mathcal{A}_0$ is dense in $\mathcal{D}(L)$ with respect to the $\mathcal{D}(L)$ topology ($\mathcal{A}_0$ is a core for L). It is also equivalent to require that L defined on $\mathcal{A}_0$ as a symmetric operator has a unique self-adjoint extension.

3.4 Summary of Hypotheses (Markov Triple)

3.4.3 Full Markov Triple (Definition 3.3.25)

A Full Markov Triple is a Standard Markov Triple for which there is an extended algebra $\mathcal{A} \supset \mathcal{A}_0$ of functions, with no requirements of integrability for elements of $\mathcal{A}$, satisfying the following requirements (Definition 3.3.1).

F1. Whenever $f \in \mathcal{A}$ and $h \in \mathcal{A}_0$, $hf \in \mathcal{A}_0$ (ideal property).

F2. Whenever $f \in \mathcal{A}$ satisfies $\int_E hf\, d\mu \geq 0$ for any positive $h \in \mathcal{A}_0$, then $f \geq 0$.

F3. $\mathcal{A}$ is stable under composition with smooth ($\mathcal{C}^\infty$) functions $\Psi : \mathbb{R}^k \to \mathbb{R}$.

F4. The operator $\mathrm{L} : \mathcal{A} \to \mathcal{A}$ is an extension of L on $\mathcal{A}_0$. The carré du champ operator Γ is also defined on $\mathcal{A} \times \mathcal{A}$ by the formula

$$\Gamma(f, g) = \frac{1}{2}\big[\mathrm{L}(fg) - f\mathrm{L}(g) - g\mathrm{L}(f)\big].$$

F5. For any $f \in \mathcal{A}$, $\Gamma(f) = \Gamma(f, f) \geq 0$.

F6. The operators Γ and L satisfy the change of variables formulas (3.1.2) and (3.1.8) respectively on $\mathcal{A}$ for any smooth ($\mathcal{C}^\infty$) $\Psi : \mathbb{R}^k \to \mathbb{R}$.

F7. For every $f \in \mathcal{A}$ and $g \in \mathcal{A}_0$, the integration by parts formula

$$\int_E \Gamma(f, g)\, d\mu = -\int_E g\,\mathrm{L} f\, d\mu = -\int_E f\,\mathrm{L} g\, d\mu$$

holds true.

F8. For every $f \in \mathcal{A}_0$, and every $t \geq 0$, $P_t f \in \mathcal{A}$.

F9. The Markov Triple is connected in the sense that if $f \in \mathcal{A}$, $\Gamma(f) = 0$ implies that f is constant (Definition 3.3.7).

F10. The ESA property holds (Definition 3.1.10).

3.4.4 Compact Markov Triple (Definition 3.3.26)

A Compact Markov Triple is a Full Markov Triple such that $\mathcal{A} = \mathcal{A}_0$. The ESA property is then automatic.

3.4.5 Miscellaneous

Algebra $\mathcal{A}_0^{\mathrm{const}}$ (Remark 3.3.3) When μ is a probability measure, $\mathcal{A}_0$ may be replaced by $\mathcal{A}_0^{\mathrm{const}} = \{f + c\,;\, f \in \mathcal{A}_0, c \in \mathbb{R}\}$, extending Γ by defining $\Gamma(f, \mathbb{1}) = 0$ and $\mathrm{L}(\mathbb{1}) = 0$. The change of variables formulas (3.1.2) and (3.1.8) extend to functions $f \in \mathcal{A}_0^{\mathrm{const}}$ without the restriction that $\Psi(0) = 0$. The ideal property does not apply to $\mathcal{A}_0^{\mathrm{const}}$ (in which case $\mathcal{A} = \mathcal{A}_0$). We also sometimes work with $\mathcal{A}_0^{\mathrm{const}+} = \{f + c\,;\, f \in \mathcal{A}_0, f \geq 0, c > 0\}$.

Γ₂ Operator (Definition 3.3.12) The Γ_2 operator is the bilinear map $\mathcal{A} \times \mathcal{A} \to \mathcal{A}$ defined by

$$\Gamma_2(f, g) = \frac{1}{2}\big[\mathrm{L}\Gamma(f, g) - \Gamma(f, \mathrm{L}g) - \Gamma(\mathrm{L}f, g)\big].$$

$\Gamma_2(f)$ denotes $\Gamma_2(f, f)$.

Hessian (Definition 3.3.13) The Hessian is a trilinear operator $\mathcal{A} \times \mathcal{A} \times \mathcal{A} \to \mathcal{A}$ defined by

$$H(f)(g, h) = \frac{1}{2}\big[\Gamma\big(g, \Gamma(f, h)\big) + \Gamma\big(h, \Gamma(f, g)\big) - \Gamma(f, \Gamma(g, h))\big].$$

$H(f)(g, h)$ is a first order operator in g and h and second order in f satisfying the chain rule, for $f, g, h, k \in \mathcal{A}$,

$$H(kf)(g, h) = k H(f)(g, h) + f H(k)(g, h) + \Gamma(f, g)\,\Gamma(k, h) + \Gamma(f, h)\,\Gamma(k, g).$$

The Γ_2 operator satisfies the chain rule formula, for $f, g, h \in \mathcal{A}$,

$$\Gamma_2(hf, g) = f\,\Gamma_2(h, g) + h\,\Gamma_2(f, g) + 2 H(g)(h, f).$$

Curvature-Dimension Condition (Definition 3.3.14) A Standard Markov Triple (E, μ, Γ) satisfies the curvature-dimension condition $CD(\rho, n)$ for $\rho \in \mathbb{R}$ and $n \in [1, \infty]$ if for any $f \in \mathcal{A}_0$

$$\Gamma_2(f) \geq \rho\,\Gamma(f) + \frac{1}{n}\,(\mathrm{L}f)^2.$$

The curvature condition $CD(\rho, \infty)$ amounts to $\Gamma_2(f) \geq \rho\,\Gamma(f)$ for every $f \in \mathcal{A}_0$. The $CD(\rho, n)$ condition automatically extend to $\mathcal{A}$ in the setting of a Full Markov Triple (see Remark 3.3.15).

By the diffusion hypothesis, the $CD(\rho, \infty)$ condition extends to Hessian inequalities in the form

$$H(f)(g, h)^2 \leq \big[\Gamma_2(f) - \rho\,\Gamma(f)\big]\Gamma(g)\Gamma(h)$$

for f, g, h in $\mathcal{A}_0$ (or $\mathcal{A}$). As a consequence, the reinforced curvature condition holds

$$\Gamma\big(\Gamma(f)\big) \leq 4\big[\Gamma_2(f) - \rho\,\Gamma(f)\big]\Gamma(f).$$

Weak Hypo-ellipticity (Definition 3.3.8) The weak hypo-ellipticity property is the fact that for every $\lambda \in \mathbb{R}$, any $f \in \mathcal{D}(\mathrm{L}^*)$ which is solution of $\mathrm{L}^* f = \lambda f$ belongs to $\mathcal{A}$.

Completeness (Definition 3.3.9) The completeness assumption is the existence of a sequence $(\zeta_k)_{k \geq 1}$ of positive functions in $\mathcal{A}_0$, increasing (μ-almost everywhere) to the constant function $\mathbb{1}$ and such that $\Gamma(\zeta_k) \leq \frac{1}{k}$, $k \geq 1$.

Intrinsic Distance (cf. (3.3.9)) The distance function on (E, μ, Γ) is defined by

$$d(x, y) = \operatorname{esssup}\bigl[f(x) - f(y)\bigr], \quad (x, y) \in E \times E,$$

where the essential supremum (with respect to the measure μ) is computed on bounded functions f in $\mathcal{A}$ such that $\Gamma(f) \leq 1$.

Lipschitz Function (Definition 3.3.24) A Lipschitz function is an element f of $\mathcal{A}$ (or $\mathcal{D}(\mathcal{E})$) such that $\|\Gamma(f)\|_\infty < \infty$. The Lipschitz constant $\|f\|_{\text{Lip}}$ is $\|\Gamma(f)\|_\infty^{1/2}$.

Diameter (cf. (3.3.10)) The diameter $D = D(E, \mu, \Gamma)$ of the Markov Triple (E, μ, Γ) is defined as

$$D = \operatorname{esssup}\bigl[f(x) - f(y)\bigr],$$

where the essential supremum (with respect to the measure μ) runs over all Lipschitz functions f with $\|f\|_{\text{Lip}} \leq 1$.

3.5 Notes and References

This chapter collects and formalizes a selection of definitions and properties extracted from Chaps. 1 and 2. In particular, several references relevant to this chapter may already be found there.

The abstract framework of a Markov Triple (E, μ, Γ) emphasized in Sects. 3.1 and 3.3 is inspired by the Dirichlet form theory as developed in [91, 189, 190, 294]. It essentially summarizes and extends the early description put forward in the lecture notes [26] (see also [27, 28]). The intermediate value Theorem 3.1.16 is recorded in [39]. The slicing argument of Proposition 3.1.17 is part of the folklore on Dirichlet forms and capacities (cf. [91, 303] etc.).

Section 3.2 emphasizes some fundamental regularity results in the manifold setting. The uniqueness Proposition 3.2.1 and its Corollary 3.2.2 are suitable extractions from [247, 248]. In this manifold setting, completion and self-adjointness have been deeply investigated by R. Strichartz [390]. Gradient bounds and commutation properties between gradient and semigroup (Theorems 3.2.3 and 3.2.4), of fundamental importance throughout this work, appeared in the probabilistic context via the Bismut representation formula of Proposition 3.2.5 [69, 171, 172] (see [251, 391]). The Γ_2 approach to the gradient bounds originates in [36] and has been promoted in [22, 24] (see also [25]) in connection with the analysis of the boundedness of Riesz transforms. The mass conservation Theorem 3.2.6 under a lower bound on the Ricci curvature essentially goes back to S.-T. Yau [443] (see earlier [191]). It has been widely studied and extended under Ricci or volume growths by numerous authors, and a detailed history is presented for example in [215, 218]. The monograph [217] by A. Grigor'yan is a comprehensive investigation of heat kernel bounds on Riemannian manifolds and metric spaces in which the reader will find complete references and historical developments.

The general setting put forward in Sect. 3.3 is largely inspired by the early contribution [36] and already outlined in the lecture notes [26] (see also [27, 28]). The complete and self-contained exposition developed here is new, and emphasizes in particular essential self-adjointness as a critical tool in this context. The intrinsic distance is also a classical feature of the theory of Dirichlet forms [91, 189, 190]. See also [243, 395, 396]. On gradient bounds for the hypo-elliptic Heisenberg model (cf. Remark 3.3.22), see [30, 51, 282, 283, 425, 446].

Part II
Three Model Functional Inequalities

Chapter 4
Poincaré Inequalities

This chapter investigates the first important family of functional inequalities for Markov semigroups, the Poincaré or spectral gap inequalities. These will provide the first results towards convergence to equilibrium, and illustrate, at a mild and accessible level, some of the basic ideas and techniques on Markov semigroups and functional inequalities developed throughout this monograph, at the interplay between analysis, probability theory and geometry.

Following the conclusions of Chap. 3 and the summary in Sect. 3.4, p. 168, it is most convenient to present the results of this chapter, as in most parts of the book, in the Full Markov Triple framework. Furthermore, according to the convention set forth there, we use the terminology "Markov Triple" for Full Markov Triple. Results with specific hypotheses will be clearly indicated.

More general settings may be considered for the analysis of spectral gap inequalities. In order to state the Poincaré inequalities, it is actually enough to deal with the minimal structure (E, μ, Γ) with finite invariant measure μ, even without the diffusion property. Standard Markov Triples suffice for many of the results described below, Full Markov Triples being necessary when dealing with inequalities involving gradient bounds and curvature conditions.

Recall that the Markov Triple setting includes a triple (E, μ, Γ) with state space E, invariant reversible measure μ and carré du champ operator Γ acting on an algebra of bounded measurable functions $\mathcal{A}_0$. The setting involves the Dirichlet form

$$\mathcal{E}(f) = \int_E \Gamma(f) d\mu$$

with domain $\mathcal{D}(\mathcal{E})$ and the associated diffusion operator L with domain $\mathcal{D}(L)$, generator of the Markov semigroup $\mathbf{P} = (P_t)_{t \geq 0}$. Most inequalities are usually stated and established for functions in $\mathcal{D}(\mathcal{E})$. In general, they are established in fact only for the functions of the algebra $\mathcal{A}_0$, which by construction is dense in the Dirichlet domain $\mathcal{D}(\mathcal{E})$. The extended algebra $\mathcal{A} \supset \mathcal{A}_0$ allows for the Γ-calculus and gradient bounds. In concrete examples, $\mathcal{A}_0$ plays the role of the set of smooth compactly supported functions while $\mathcal{A}$ represents smooth functions with no restriction on integrability and support. The associated Markov semigroup $\mathbf{P} = (P_t)_{t \geq 0}$ with infinitesimal generator L and invariant reversible measure μ admits a kernel representation

$$P_t f(x) = \int_E f(y) \, p_t(x, dy), \quad t \geq 0, \ x \in E.$$

The *Poincaré*, or *spectral gap, inequality* is the simplest inequality which quantifies ergodicity and controls convergence to equilibrium of the semigroup $\mathbf{P} = (P_t)_{t \geq 0}$ towards the invariant measure μ (in other words, the convergence of the kernels $p_t(x, dy)$, $x \in E$, as $t \to \infty$, towards $d\mu(y)$). Since the kernels $p_t(x, dy)$ describe the distribution at time t of the associated Markov process X_t^x starting at x, this convergence is translated equivalently as a convergence of X_t^x towards the equilibrium.

On the basis of the example of the Ornstein-Uhlenbeck semigroup (Sect. 4.1), Sect. 4.2 introduces the formal definition of Poincaré or spectral gap inequalities in the context of a Markov Triple (E, μ, Γ) as above. Exponential decay in $\mathbb{L}^2(\mu)$ along the semigroup is part of the first main equivalent descriptions of Poincaré inequalities. Tensorization properties are presented next. The example of the exponential measure on the line is discussed in Sect. 4.4, together with exponential integrability properties of Lipschitz functions under a Poincaré inequality. The next section studies Poincaré inequalities for measures on the real line or on an interval of the real line. Section 4.5.1 first presents a characterization of measures on the line satisfying a Poincaré inequality. Then, in Sect. 4.5.2, Poincaré inequalities on an interval with respect to Neumann, Dirichlet and periodic boundary conditions are discussed. Section 4.6 presents the Lyapunov function method of obtaining Poincaré inequalities. Local Poincaré inequalities for heat kernel measures under curvature conditions are investigated next in Sect. 4.7 via the basic semigroup interpolation scheme. Global Poincaré inequalities for the invariant measure under curvature-dimension conditions are developed in Sect. 4.8. Further inequalities of Brascamp-Lieb-type are presented in Sect. 4.9. Finally, Sect. 4.10 is a somewhat in depth investigation of the description of the bottom of the spectrum and essential spectrum of Markov generators together with criteria of general interest which establish the discreteness of spectra via the notion of a Persson operator.

4.1 The Example of the Ornstein-Uhlenbeck Semigroup

The example of the Ornstein-Uhlenbeck semigroup will give the flavour of some of the spectral properties and results that will be defined and investigated throughout the chapter in greater generality. This elementary model will moreover provide an opportunity to detail, step by step, some of the reasoning and tools which will be developed later in more general cases.

The Ornstein-Uhlenbeck semigroup $\mathbf{P} = (P_t)_{t \geq 0}$ on the real line (for simplicity), described in Sect. 2.7.1, p. 103, admits the standard Gaussian probability measure $d\mu(x) = (2\pi)^{-1/2} e^{-x^2/2} dx$ as invariant and reversible measure, $Lf = f'' - xf'$ as infinitesimal generator and $\Gamma(f) = f'^2$ as carré du champ operator (acting on smooth functions $f : \mathbb{R} \to \mathbb{R}$). The Dirichlet form is given accordingly by

4.1 The Example of the Ornstein-Uhlenbeck Semigroup

$\mathcal{E}(f) = \int_{\mathbb{R}} f'^2 d\mu$. (While the set of smooth compactly supported functions is a natural choice for the algebra $\mathcal{A}_0$, it could be convenient here to work alternatively, after some adaptations, with the algebra of polynomials.)

Convergence to equilibrium is best presented first on the spectral decomposition of **P** and L which has been described in terms of Hermite polynomials. Recall (see Sect. 2.7.1, p. 103) that a function f in $\mathbb{L}^2(\mu)$ can be decomposed with respect to the basis of Hermite orthonormal polynomials $(H_k)_{k \in \mathbb{N}}$ as

$$f = \sum_{k \in \mathbb{N}} a_k H_k,$$

where, for every integer k, $a_k = \int_{\mathbb{R}} f H_k d\mu$ and

$$\|f\|_2^2 = \int_{\mathbb{R}} f^2 d\mu = \sum_{k \in \mathbb{N}} a_k^2 < \infty.$$

In this case,

$$P_t f = \sum_{k \in \mathbb{N}} e^{-kt} a_k H_k, \quad t \geq 0,$$

so that $P_t f$ converges in $\mathrm{L}^2(\mu)$ towards $a_0 = \int_{\mathbb{R}} f H_0 d\mu = \int_{\mathbb{R}} f d\mu$ as $t \to \infty$ (recall that $H_0 = 1$). This property expresses equivalently that the distribution of the associated Ornstein-Uhlenbeck process X_t^x converges as $t \to \infty$ towards the invariant measure μ (from any initial point x).

The task now is to control and quantify this convergence. On the basis of the preceding decomposition, observe that $f - \int_{\mathbb{R}} f d\mu = \sum_{k \geq 1} a_k H_k$ so that, for every $t \geq 0$,

$$\left\| P_t \left(f - \int_{\mathbb{R}} f d\mu \right) \right\|_2^2 = \sum_{k \geq 1} e^{-2kt} a_k^2 \leq e^{-2t} \sum_{k \geq 1} a_k^2 = e^{-2t} \left\| f - \int_{\mathbb{R}} f d\mu \right\|_2^2.$$

This inequality describes an exponential convergence of $P_t f$ towards the mean of $f \in \mathbb{L}^2(\mu)$ as $t \to \infty$.

The latter family of inequalities may be analyzed infinitesimally. Assume for simplicity that the mean of f is equal to 0 so that

$$\int_{\mathbb{R}} (P_t f)^2 d\mu \leq e^{-2t} \int_{\mathbb{R}} f^2 d\mu$$

for every $t \geq 0$. Since both terms of this inequality are equal at $t = 0$, we can perform a first order Taylor expansion at $t = 0$ of both sides, for any function f in the domain $\mathcal{D}(L)$. More precisely, since $P_t f = f + t L f + o(t)$, we get that

$$\int_{\mathbb{R}} (P_t f)^2 d\mu = \int_{\mathbb{R}} f^2 d\mu + 2t \int_{\mathbb{R}} f L f d\mu + o(t)$$

$$= \int_{\mathbb{R}} f^2 d\mu - 2t \, \mathcal{E}(f) + o(t)$$

and, clearly,

$$e^{-2t} \int_{\mathbb{R}} f^2 d\mu = \int_{\mathbb{R}} f^2 d\mu - 2t \int_{\mathbb{R}} f^2 d\mu + o(t).$$

Hence $\int_{\mathbb{R}} f^2 d\mu \leq \mathcal{E}(f)$. Since $\mathcal{E}(f+c) = \mathcal{E}(f)$ for any real c, it follows, after centering, that for any function $f \in \mathcal{D}(L)$, and thus by extension for every f in the Dirichlet domain $\mathcal{D}(\mathcal{E})$ of the Ornstein-Uhlenbeck operator,

$$\int_{\mathbb{R}} f^2 d\mu - \left(\int_{\mathbb{R}} f d\mu \right)^2 \leq \mathcal{E}(f). \qquad (4.1.1)$$

This inequality (4.1.1) (or rather family of inequalities) may be read in two ways. First, for smooth functions f, by construction of the carré du champ operator Γ,

$$\int_{\mathbb{R}} f^2 d\mu - \left(\int_{\mathbb{R}} f d\mu \right)^2 \leq \mathcal{E}(f) = \int_{\mathbb{R}} \Gamma(f) d\mu = \int_{\mathbb{R}} f'^2 d\mu.$$

On the other hand, (4.1.1) expresses equivalently that the first non-trivial eigenvalue of (the opposite generator) $-L$ is greater than or equal to 1. Indeed, if f is non-constant such that $-Lf = \lambda f$ for some λ (≥ 0), then $\int_{\mathbb{R}} f d\mu = 0$ and $\int_{\mathbb{R}} \Gamma(f) d\mu = \int_{\mathbb{R}} f(-Lf) d\mu = \lambda \int_{\mathbb{R}} f^2 d\mu$ so that $\lambda \geq 1$. Inequality (4.1.1) will be called the Poincaré, or spectral gap, inequality for the Gaussian measure μ (with respect to the carré du champ operator Γ or the Dirichlet form $\mathcal{E}$).

Note that (4.1.1) may be established directly on the spectral decomposition in Hermite polynomials since

$$\mathcal{E}(f) = \sum_{k \geq 1} k a_k^2 \geq \sum_{k \geq 1} a_k^2 = \int_{\mathbb{R}} f^2 d\mu - \left(\int_{\mathbb{R}} f d\mu \right)^2,$$

where, again, the last equality is a consequence of the fact that $a_0 = \int_{\mathbb{R}} f d\mu$. Note also that (4.1.1) is optimal in the sense that functions of the form $f = a H_0 + b H_1 = a + bx$ for $a, b \in \mathbb{R}$ achieve equality in (4.1.1), thus describing the class of extremal functions for this inequality.

Observe that similar arguments on multiple Hermite polynomials (or tensorization tools as will be developed in Sect. 4.3 below) yield a Poincaré or spectral gap inequality for the standard Gaussian probability measure $d\mu(x) = (2\pi)^{-n/2} e^{-|x|^2/2} dx$ on $\mathbb{R}^n$ with carré du champ operator $\Gamma(f) = |\nabla f|^2$ and Dirichlet form $\mathcal{E}(f) = \int_{\mathbb{R}^n} |\nabla f|^2 d\mu$ as described in the next proposition.

Proposition 4.1.1 (Poincaré inequality for the Gaussian measure) *Let μ be the standard Gaussian measure on the Borel sets of $\mathbb{R}^n$. For every function $f : \mathbb{R}^n \to \mathbb{R}$ in the Dirichlet domain $\mathcal{D}(\mathcal{E})$,*

$$\int_{\mathbb{R}^n} f^2 d\mu - \left(\int_{\mathbb{R}^n} f d\mu \right)^2 \leq \mathcal{E}(f) = \int_{\mathbb{R}^n} |\nabla f|^2 d\mu.$$

If μ is a centered Gaussian measure on $\mathbb{R}^n$ with covariance matrix Q, by a simple change of variables,

$$\int_{\mathbb{R}^n} f^2 d\mu - \left(\int_{\mathbb{R}^n} f\, d\mu \right)^2 \leq \int_{\mathbb{R}^n} Q\nabla f \cdot \nabla f\, d\mu$$

for every smooth function $f : \mathbb{R}^n \to \mathbb{R}$.

4.2 Poincaré Inequalities

This section formally introduces the notion of Poincaré or spectral gap inequalities for Markov operators and their invariant measures, and some of their first properties, in the context of Markov Triples (E, μ, Γ). While spectral properties may be analyzed for general measures, the formal notion of a Poincaré inequality deals with invariant probability measures.

4.2.1 Poincaré and Spectral Gap Inequalities

For a probability measure ν on a measurable space $(E, \mathcal{F})$, we define the variance of a function f in $\mathbb{L}^2(\nu)$ as

$$\text{Var}_\nu(f) = \int_E f^2 d\nu - \left(\int_E f\, d\nu \right)^2. \quad (4.2.1)$$

Definition 4.2.1 (Poincaré inequality) A Markov Triple (E, μ, Γ) (with μ a probability measure) is said to satisfy a Poincaré, or spectral gap, inequality $P(C)$ with constant $C > 0$, if for all functions $f : E \to \mathbb{R}$ in the Dirichlet domain $\mathcal{D}(\mathcal{E})$,

$$\text{Var}_\mu(f) \leq C\, \mathcal{E}(f).$$

The best constant $C > 0$ for which such an inequality holds is sometimes referred to as the *Poincaré constant* (of the Markov Triple).

It is enough to state such a Poincaré inequality for a family of functions f which is dense in the domain $\mathcal{D}(\mathcal{E})$ of the Dirichlet form $\mathcal{E}$. This principle will be used almost automatically when establishing Poincaré (and more general functional) inequalities, and typically functions from the underlying algebra $\mathcal{A}_0$ will be used to this task.

The terminology between Poincaré and spectral gap inequality is somewhat fluctuant. We often say that the probability measure μ satisfies a Poincaré inequality $P(C)$ with constant C (with respect to the carré du champ operator Γ or the Dirichlet form $\mathcal{E}$). For example, by Proposition 4.1.1, the standard Gaussian measure μ on $\mathbb{R}^n$ satisfies a Poincaré inequality $P(1)$ with respect to the standard Dirichlet form

$\mathcal{E}(f) = \int_{\mathbb{R}^n} |\nabla f|^2 d\mu$. If f is an eigenfunction of $-L$ with eigenvalue λ, applying the spectral gap inequality $P(C)$ to f shows that, if $\lambda \neq 0$, then $C\lambda \geq 1$. Indeed, $\int_E f d\mu = 0$ by invariance and

$$\int_E f^2 d\mu = \mathrm{Var}_\mu(f) \leq C\mathcal{E}(f) = C \int_E f(-Lf) d\mu = C\lambda \int_E f^2 d\mu.$$

Hence, under $P(C)$, every non-zero eigenvalue of $-L$ is greater than or equal to $\frac{1}{C}$. Indeed, even when it is not discrete, the spectrum of the symmetric positive operator $-L$, as defined in Sect. A.4, p. 478, is included in $\{0\} \cup [\frac{1}{C}, \infty)$, and $P(C)$ thus describes a *gap* in its spectrum.

We shall return to the following remark in several different contexts later.

Remark 4.2.2 A function with zero variance is constant. Therefore, one important consequence of a spectral gap inequality for the invariant measure μ is that if $\mathcal{E}(f) = 0$, then f is (μ-almost everywhere) constant. In particular, such a conclusion reinforces the ergodicity property of Definition 3.1.11, p. 135. It is also a stronger form of connexity (Definition 3.3.7, p. 156). Functional inequalities with this property will be called tight. A general principle, to be developed throughout this book, is that tight inequalities have something to say about convergence to equilibrium (cf. Sect. 1.8, p. 32).

Actually, as will be developed later for other functional inequalities, one can consider non-tight Poincaré inequalities such as

$$\int_E f^2 d\mu \leq a \left(\int_E f d\mu \right)^2 + C\mathcal{E}(f), \quad f \in \mathcal{D}(\mathcal{E}).$$

This would not be of much help here, at least for probability measures, since this inequality applied for f to the constant function $f = 1$ shows that $a \geq 1$. Then, applied to $g = f - \int_E f d\mu$, we are back to the case $a = 1$. On the other hand, the Nash-type inequalities investigated later in Sect. 7.4.1, p. 364,

$$\int_E f^2 d\mu \leq a \left(\int_E |f| d\mu \right)^2 + C\mathcal{E}(f), \quad f \in \mathcal{D}(\mathcal{E}),$$

will be of much greater interest.

In this spirit, the following elementary observation will prove useful in several instances. We refer to Sect. 4.10.1 for probabilistic interpretations of these kind of inequalities.

Lemma 4.2.3 *Under the Poincaré inequality $P(C)$, for a function $f \in \mathcal{D}(\mathcal{E})$ such that $f = 0$ outside a set $A \in \mathcal{F}$ with $\mu(A) < 1$, it holds that*

$$\int_A f^2 d\mu \leq \frac{C}{1 - \mu(A)} \mathcal{E}(f).$$

4.2 Poincaré Inequalities

The proof is a simple application of the Cauchy-Schwarz inequality. Indeed,

$$\left(\int_E f\,d\mu\right)^2 = \left(\int_A f\,d\mu\right)^2 \leq \mu(A) \int_E f^2 d\mu,$$

so that $(1 - \mu(A)) \int_E f^2 d\mu \leq \text{Var}_\mu(f)$ from which the claim follows by definition of the Poincaré inequality $P(C)$.

Remark 4.2.4 Inequalities of the form

$$\int_A f^2 d\mu \leq \Phi\big(\mu(A)\big)\,\mathcal{E}(f)$$

for functions $f \in \mathcal{D}(\mathcal{E})$ with support in A as in the previous lemma may hold in various contexts. For example, the so-called Faber-Krahn inequalities, in the context of Nash inequalities, belong to this family (see Remark 6.2.4, p. 284, and Remark 8.2.2, p. 399).

4.2.2 Variance Decay

As illustrated earlier for on the Ornstein-Uhlenbeck semigroup, a Poincaré inequality for a probability measure μ describes (and in fact is equivalent to) exponential convergence of the semigroup $(P_t)_{t \geq 0}$ to equilibrium in $\mathbb{L}^2(\mu)$. The same proof may be used to derive the following main statement which formalizes this equivalence in the general framework.

Theorem 4.2.5 (Exponential decay in variance) *Given a Markov Triple (E, μ, Γ) with associated Markov semigroup $\mathbf{P} = (P_t)_{t \geq 0}$, the following assertions are equivalent:*

(i) *(E, μ, Γ) satisfies a Poincaré inequality $P(C)$.*
(ii) *For every function $f : E \to \mathbb{R}$ in $\mathbb{L}^2(\mu)$, and every $t \geq 0$,*

$$\text{Var}_\mu(P_t f) \leq e^{-2t/C}\,\text{Var}_\mu(f).$$

(iii) *For every function $f \in \mathbb{L}^2(\mu)$, there exists a constant $c(f) > 0$ (possibly depending on f) such that, for every $t \geq 0$,*

$$\text{Var}_\mu(P_t f) \leq c(f)\, e^{-2t/C}.$$

Proof Exactly as for the Ornstein-Uhlenbeck semigroup in Sect. 4.1, assertion (i) follows from assertion (ii) by a first order Taylor expansion at $t = 0$. The converse implication relies on the formula

$$\frac{d}{dt}\text{Var}(P_t f) = -2\,\mathcal{E}(P_t f) \tag{4.2.2}$$

for every $f \in \mathbb{L}^2(\mu)$ and $t > 0$. Recall that according to Proposition 3.1.6 and Remark 3.1.7, p. 131 and p. 132, P_t, $t > 0$, maps $\mathbb{L}^2(\mu)$ into $\mathcal{D}(L^k)$ for every integer k. Assume then for simplicity that f has mean zero. Since by invariance, $\int_E P_t f \, d\mu = \int_E f \, d\mu = 0$,

$$\Lambda(t) = \mathrm{Var}_\mu(P_t f) = \int_E (P_t f)^2 d\mu, \quad t \geq 0.$$

By the heat equation and integration by parts,

$$\Lambda'(t) = 2 \int_E P_t f \, L P_t f \, d\mu = -2\mathcal{E}(P_t f)$$

which yields (4.2.2). The Poincaré inequality $P(C)$ therefore translates into the differential inequality $\Lambda(t) \leq -\frac{C}{2} \Lambda'(t)$, $t > 0$, which amounts to saying that the function $e^{2t/C} \Lambda(t)$ is decreasing in $t \geq 0$. The conclusion follows by comparing the values of this function at t and at $t = 0$.

To prove the equivalence of the third statement (iii), it is convenient to record the following lemma of independent interest.

Lemma 4.2.6 *For any non-zero function $f \in \mathbb{L}^2(\mu)$, the map $t \mapsto \log(\|P_t f\|_2)$ is convex on $\mathbb{R}_+$.*

The proof is immediate. Indeed, if $\Lambda(t) = \int_E (P_t f)^2 d\mu$, $t > 0$, as before, $\Lambda'(t) = 2 \int_E P_t f \, L P_t f \, d\mu$ and

$$\Lambda''(t) = 4 \int_E (L P_t f)^2 d\mu.$$

Hence $\Lambda'(t)^2 \leq \Lambda(t) \Lambda''(t)$ by the Cauchy-Schwarz inequality which amounts to the convexity of $\log \Lambda(t)$.

Returning to the proof of Theorem 4.2.5, in order to deduce (ii) from (iii), it suffices from the preceding lemma to observe that if a convex function ϕ (on the positive half-line) is bounded from above by $\beta - \alpha t$, then it is bounded by $\phi(0) - \alpha t$. The proof of the theorem is therefore complete. □

For later purposes, it is useful to draw from the preceding arguments a general inequality on the $\mathbb{L}^2(\mu)$-norm of $f - P_t f$ (which sheds some further light on the preceding equivalences). Namely, for every $f \in \mathcal{D}(\mathcal{E})$ and every $t \geq 0$,

$$\int_E (f - P_t f)^2 d\mu \leq 2t \, \mathcal{E}(f). \tag{4.2.3}$$

The elementary proof follows the same pattern. For a function $f \in \mathcal{D}(\mathcal{E})$, the derivative of $\int_E (f - P_t f)^2 d\mu$ in t for $t > 0$ is equal, by symmetry and integration by parts, to

$$-2 \int_E (f - P_t f) L P_t f \, d\mu = 2\mathcal{E}(P_{t/2} f) - 2\mathcal{E}(P_t f) \leq 2\mathcal{E}(f),$$

since $t \mapsto \mathcal{E}(P_t f)$ is decreasing (see Proposition 3.1.6, p. 131).

It is, in general, not so easy to establish a Poincaré inequality for a probability measure μ with respect to a given carré du champ operator Γ. One first approach is to compare it to a simpler, or known, inequality. A first trivial observation in this direction is that whenever μ satisfies a Poincaré inequality $P(C)$ with respect to Γ, and if Γ_1 is another carré du champ operator such that $\Gamma \leq a\Gamma_1$ for some $a > 0$, then (μ, Γ_1) satisfies $P(aC)$.

The following proposition is a further simple, but important, observation concerning stability by perturbation of Poincaré inequalities.

Proposition 4.2.7 (Bounded perturbation) *Assume that the Markov Triple (E, μ, Γ) satisfies a Poincaré inequality $P(C)$. Let μ_1 be a probability measure with density h with respect to μ such that $\frac{1}{b} \leq h \leq b$ for some constant $b > 0$. Then μ_1 satisfies $P(b^3 C)$ (with respect to Γ).*

The proof is elementary, since by duplication, for every probability measure ν,

$$\mathrm{Var}_\nu(f) = \frac{1}{2} \int_E \int_E [f(x) - f(y)]^2 d\nu(x) d\nu(y). \tag{4.2.4}$$

With the somewhat more refined Lemma 5.1.7, p. 240, below, b^3 may be replaced by b^2 in the preceding. An even more precise statement is that if $h = e^k$, then b^2 may be replaced by $e^{\mathrm{osc}(k)}$ where $\mathrm{osc}(k) = \sup k - \inf k$ is the *oscillation* of the function k. (As usual, sup and inf stand here respectively for esssup and essinf with respect to the measure μ.)

These first elementary observations already yield Poincaré inequalities for measures μ on $\mathbb{R}$ or $\mathbb{R}^n$ with a bounded density with respect to the standard Gaussian measure, and for Dirichlet forms, on the line say, of the form $\int_\mathbb{R} a(x) f'^2(x) d\mu(x)$ where the function $a(x)$ is bounded from below by a strictly positive constant.

4.3 Tensorization of Poincaré Inequalities

This section develops tensorization properties of Poincaré inequalities. One important feature of Poincaré inequalities is their stability under products.

Proposition 4.3.1 (Stability under products) *If (E_1, μ_1, Γ_1) and (E_2, μ_2, Γ_2) satisfy Poincaré inequalities with respective constants C_1 and C_2, then the product Markov Triple $(E_1 \times E_2, \mu_1 \otimes \mu_2, \Gamma_1 \oplus \Gamma_2)$ satisfies a Poincaré inequality with constant $C = \max(C_1, C_2)$.*

The proof will give a meaning to $\Gamma_1 \oplus \Gamma_2$, more precisely to the underlying product Dirichlet form.

Proof As a first indication of the result, consider the elementary case when the corresponding (opposite) generators $-L_1$ and $-L_2$ have discrete spectra, $(\lambda_k^1)_{k \in \mathbb{N}}$ and $(\lambda_\ell^2)_{\ell \in \mathbb{N}}$. The (opposite) generator $-(L_1 \oplus L_2)$ on the product space $E_1 \times E_2$ has spectrum $(\lambda_k^1 + \lambda_\ell^2)_{k,\ell \in \mathbb{N}}$. Its smallest strictly positive eigenvalue is thus the minimum of the smallest strictly positive eigenvalues of $(\lambda_k^1)_{k \in \mathbb{N}}$ and $(\lambda_\ell^2)_{\ell \in \mathbb{N}}$. The conclusion follows in this case with $C_1 = \frac{1}{\lambda_1^1}$ and $C_2 = \frac{1}{\lambda_1^2}$.

The general case is not much harder. The main point is to understand that the Dirichlet form $\mathcal{E}$ on the product $(E_1 \times E_2, \mu_1 \otimes \mu_2, \Gamma_1 \oplus \Gamma_2)$ is given, on suitable functions $f : E_1 \times E_2 \to \mathbb{R}$, by

$$\mathcal{E}(f) = \int_{E_2} \mathcal{E}_1(f) d\mu_2 + \int_{E_1} \mathcal{E}_2(f) d\mu_1.$$

Here $\mathcal{E}_1$ is the Dirichlet form on E_1 applied to the first variable (with the second one being fixed), and similarly $\mathcal{E}_2$ is the Dirichlet form on E_2 applied to the second variable (with the first one being fixed). First we need the following lemma.

Lemma 4.3.2 *If $f = f(x_1, x_2)$ is a function of two variables $(x_1, x_2) \in E_1 \times E_2$, measurable in x_2 and in the Dirichlet space $\mathcal{D}(\mathcal{E}_1)$ for the first variable (that is, for every $x_2 \in E_2$, $\mathcal{E}_1(f(\cdot, x_2))$ is finite), then the function*

$$g(x_1) = \int_{E_2} f(x_1, x_2) d\mu_2(x_2), \quad x_1 \in E_1,$$

is in the Dirichlet space $\mathcal{D}(\mathcal{E}_1)$ and moreover

$$\mathcal{E}_1(g) \le \int_{E_2} \mathcal{E}_1(f(\cdot, x_2)) d\mu_2(x_2).$$

Proof Rather than giving a formal proof, let us sketch the main idea. Dirichlet forms are bilinear and positive, sending functions from the Dirichlet spaces into $\mathbb{R}_+$. Such quadratic forms are convex, and if $\phi : \mathbb{R} \to \mathbb{R}$ is convex, then by Jensen's inequality (under suitable integrability properties),

$$\phi\left(\int_{E_2} F d\mu_2\right) \le \int_{E_2} \phi(F) d\mu_2.$$

This inequality applied to $F(x_2) = f(\cdot, x_2)$ then yields the announced claim. □

We now conclude the proof of Proposition 4.3.1. Let $f : E_1 \times E_2 \to \mathbb{R}$ be in the Dirichlet space of the product space, and let $g(x_1)$, $x_1 \in E_1$, be as in Lemma 4.3.2. The variance of f is easily decomposed as

$$\mathrm{Var}_{\mu_1 \otimes \mu_2}(f) = \mathrm{Var}_{\mu_1}(g) + \int_{E_1} \left(\int_{E_2} [f(x_1, x_2) - g(x_1)]^2 d\mu_2(x_2)\right) d\mu_1(x_1).$$

4.4 The Example of the Exponential Measure, and Exponential Integrability

By the Poincaré inequality for μ_1, the first term on the right-hand side of this identity is bounded from above by $C_1 \mathcal{E}_1(g)$, and by Lemma 4.3.2, $\mathcal{E}_1(g) \leq \int_{E_2} \mathcal{E}_1(f) d\mu_2$. By the Poincaré inequality for μ_2 for every fixed $x_1 \in E_1$, the second term on the right-hand side is bounded from above by $C_2 \int_{E_1} \mathcal{E}_2(f) d\mu_1$. Therefore

$$\mathrm{Var}_{\mu_1 \otimes \mu_2}(f) \leq C_1 \int_{E_2} \mathcal{E}_1(f) d\mu_2 + C_2 \int_{E_1} \mathcal{E}_2(f) d\mu_1 \leq \max(C_1, C_2) \mathcal{E}(f).$$

Proposition 4.3.1 is established. □

Remark 4.3.3 A somewhat different route may actually be used to establish Proposition 4.3.1. On the product space $E = E_1 \times E_2$ equipped with the product probability measure $\mu = \mu_1 \otimes \mu_2$, consider, for every $t \geq 0$ and $x \in E$, the probability measure $\nu_t(dx, dy) = p_t(x, dy) \mu(dx)$ where $p_t(x, dy)$ denotes the kernel of the semigroup $(P_t)_{t \geq 0}$ (on E). As for the variance, it is easily checked that

$$\int_E \int_E [f(x) - f(y)]^2 \nu_t(dx, dy) = 2 \left(\int_E f^2 d\mu - \int_E f P_t f d\mu \right).$$

When $t = \infty$, $\nu_\infty(dx, dy) = \mu(dx)\mu(dy)$, which simply corresponds to the idea that $P_\infty(x, dy) = \mu(dy)$. Now, since by symmetry $\int_E f P_t f d\mu = \int_E (P_{t/2} f)^2 d\mu$, the Poincaré inequality $P(C)$ for μ is equivalent by Theorem 4.2.5 to

$$\int_E \int_E [f(x) - f(y)]^2 \nu_t(dx, dy)$$
$$\geq (1 - e^{-t/C}) \int_E \int_E [f(x) - f(y)]^2 \nu_\infty(dx, dy)$$

(for every $t \geq 0$). Since it is clear that $\nu_t = \nu_t^1 \otimes \nu_t^2$ corresponds to the tensor product of the respective semigroups on E_1 and E_2, the preceding inequality is immediately stable under products.

The tensorization Proposition 4.3.1 may for example be used immediately to recover the Poincaré inequality of Proposition 4.1.1 for the standard Gaussian measure on $\mathbb{R}^n$ from its one-dimensional counterpart (4.1.1). Similarly, the Poincaré inequality for the exponential measure investigated in the next section may be tensorized to multiple coordinates.

4.4 The Example of the Exponential Measure, and Exponential Integrability

The preceding elementary tools are usually not sufficient to cover all the instances of Poincaré inequalities. Another approach is to try, in a given context, to find some particular functions with pleasant properties (called Lyapunov functions in a Markov theory setting). We illustrate this strategy here with the example of the exponential measure on the real line (and positive half-line), generalized later in Sect. 4.6. On the basis of this example, we shall derive some integrability properties of Lipschitz functions and measure concentration under a Poincaré inequality.

4.4.1 The Exponential Measure

The exponential measure $d\mu(x) = e^{-x}dx$ on the positive half-line $\mathbb{R}_+$, as a particular example of a gamma distribution with parameter 1, has been discussed in Sect. 2.7.3, p. 111, on the Laguerre operator, in connection with spectral decompositions along Laguerre orthogonal polynomials. In particular, following the discussion of Sect. 4.1 on the Gaussian measure, similar spectral decomposition arguments may be used to show that for every function $f : \mathbb{R}_+ \to \mathbb{R}$ in the Dirichlet domain of the Laguerre operator,

$$\mathrm{Var}_\mu(f) \leq \int_{\mathbb{R}_+} x f'^2 d\mu \qquad (4.4.1)$$

(the right-hand side describing the carré du champ operator and Dirichlet form associated to the Laguerre operator). The spectrum here is discrete (equal to $\mathbb{N}$) and the Poincaré inequality (4.4.1) expresses the gap between 0 and the first non-zero eigenvalue 1.

However, one may also investigate the Poincaré inequality for the exponential measure with respect to the standard carré du champ operator $\Gamma(f) = f'^2$ and its associated Dirichlet form $\mathcal{E}$ which is the content of the next proposition.

Proposition 4.4.1 *Let $d\mu(x) = e^{-x}dx$ on the positive half-line $\mathbb{R}_+$. For every function $f : \mathbb{R}_+ \to \mathbb{R}$ in the Dirichlet domain $\mathcal{D}(\mathcal{E})$ associated to the carré du champ operator $\Gamma(f) = f'^2$,*

$$\mathrm{Var}_\mu(f) \leq 4 \int_{\mathbb{R}_+} f'^2 d\mu. \qquad (4.4.2)$$

It should be pointed out that the Dirichlet domain in this case generates a sub-Markov semigroup so that Proposition 4.4.1 is not really in the realm of Poincaré inequalities. Moreover the essential self-adjointness property (Definition 3.1.10, p. 134) is certainly not satisfied here (cf. Sect. 2.4, p. 92).

Proof It is enough to consider smooth compactly supported functions $f : \mathbb{R}_+ \to \mathbb{R}$. Note first that

$$\mathrm{Var}_\mu(f) \leq \int_{\mathbb{R}_+} \big[f - f(0)\big]^2 d\mu$$

since the variance minimizes the distance to constants in $\mathbb{L}^2(\mu)$. Since $\int_{\mathbb{R}_+} f'^2 d\mu$ is not modified by adding a constant to f, it will be enough to establish the desired

4.4 The Example of the Exponential Measure, and Exponential Integrability

inequality for functions f such that $f(0) = 0$. Then, by Fubini's Theorem,

$$\int_{\mathbb{R}_+} f^2 d\mu = \int_0^\infty f(x)^2 e^{-x} dx$$

$$= \int_0^\infty \left(2 \int_0^x f'(t) f(t) dt \right) e^{-x} dx$$

$$= 2 \int_0^\infty f'(t) f(t) e^{-t} dt = 2 \int_{\mathbb{R}_+} ff' d\mu.$$

The conclusion follows by applying the Cauchy-Schwarz inequality to the last integral. □

We next somewhat broaden the spirit of the preceding proof. For this discussion, let us consider the symmetric exponential measure on the whole real line, $d\mu(x) = \frac{1}{2} e^{-|x|} dx$. The density of μ with respect to the Lebesgue measure is not quite smooth and, if necessary, the potential $x \mapsto |x|$ may be replaced by a smooth function with the same behavior at infinity. However, we work with the given density since the argument which is developed is simpler in this case. The associated Markov generator is $Lf = f'' - \text{sign}(x) f'$, where $\text{sign}(x)$ denotes the sign of x, with carré du champ operator $\Gamma(f) = f'^2$. Its spectrum is not discrete. This is actually not due to the non-smooth coefficient, but rather to the behavior of $|x|$ at infinity so that the decay of μ is not strong enough. For example, it is easily seen that as soon as $\lambda \geq \frac{1}{4}$, there are no solutions to $f'' - f' = -\lambda f$ which are square integrable on $(0, \infty)$.

Nevertheless, the operator L admits a Poincaré inequality (and thus a gap in the spectrum) as can be seen from the following argument directly inspired by the proof of Proposition 4.4.1. What follows is a bit formal, but may be easily justified. Consider the function $h(x) = |x|$, $x \in \mathbb{R}$. Then $Lh = 2\delta_0 - 1$ where δ_0 is the Dirac mass at 0, so that integration against it with a (smooth) function f such that $f(0) = 0$ gives 0. Let f be such a function, and write

$$\int_\mathbb{R} f^2 d\mu = -\int_\mathbb{R} f^2 Lh \, d\mu = \int_\mathbb{R} \Gamma(f^2, h) d\mu = 2 \int_\mathbb{R} f \Gamma(f, h) d\mu.$$

Now $\Gamma(h) = 1$ so that $|\Gamma(f, h)| \leq \Gamma(f)^{1/2}$. Hence, after an application of the Cauchy-Schwarz inequality as in the proof of Proposition 4.4.1,

$$\int_\mathbb{R} f^2 d\mu \leq 2 \left(\int_\mathbb{R} f^2 d\mu \right)^{1/2} \left(\int_\mathbb{R} \Gamma(f) d\mu \right)^{1/2},$$

and therefore

$$\text{Var}_\mu(f) \leq 4 \int_\mathbb{R} \Gamma(f) d\mu = 4 \int_\mathbb{R} f'^2 d\mu \quad (4.4.3)$$

which is the announced Poincaré inequality for the symmetric exponential measure μ on $\mathbb{R}$. It is worth pointing out that the significant component of the argument is that $-Lh$ is bounded from below by a strictly positive constant away from 0 (outside a compact set containing 0 would have been enough) and that $\Gamma(h)$ is bounded from above. This principle will be amplified in Sect. 4.6 below, which deals with the so-called Lyapunov functions.

4.4.2 Exponential Integrability

One miracle of the previous argument is that the constant 4 in Proposition 4.4.1 and (4.4.3) is optimal. To see why, we examine another consequence of Poincaré inequalities, namely the integrability properties of Lipschitz functions. Recall the notion of a Lipschitz function f from Definition 3.3.24, p. 166, with Lipschitz coefficient $\|f\|_{\text{Lip}} = \|\Gamma(f)\|_\infty^{1/2}$. A function f is said to be 1-Lipschitz if $\|f\|_{\text{Lip}} \leq 1$.

Under a Poincaré inequality for a measure μ, Lipschitz functions are exponentially integrable. This is the content of the following statement.

Proposition 4.4.2 (Exponential integrability) *If μ satisfies a Poincaré inequality $P(C)$ for some $C > 0$, then for every 1-Lipschitz function f and every $s < \sqrt{\frac{4}{C}}$,*

$$\int_E e^{sf} \, d\mu < \infty.$$

Remark 4.4.3 Lipschitz functions belong either to $\mathcal{A}$ or to $\mathcal{D}(\mathcal{E})$. It is part of the result here that Lipschitz functions are indeed in $\mathbb{L}^2(\mu)$, and therefore also belong to $\mathcal{D}(\mathcal{E})$, so that the statement may not be restricted to Lipschitz functions $f \in \mathcal{D}(\mathcal{E})$. While the result is stated for $f \in \mathcal{A}$, it applies with the same argument to Lipschitz functions in $\mathcal{D}(\mathcal{E})$. In fact, in the core of the proof below, it is convenient to work with functions in $\mathcal{D}(\mathcal{E})$ allowing for a simple truncation procedure, which however does not preserve functions in $\mathcal{A}$. The proof for functions in $\mathcal{A}$ is easily adapted using a smooth truncation procedure.

Proof We first show that, under the Poincaré inequality $P(C)$, for every integrable Lipschitz function f with $\|f\|_{\text{Lip}} \leq 1$, and every $s < \sqrt{\frac{4}{C}}$,

$$\int_E e^{sf} \, d\mu \leq e^{s \int_E f d\mu} \prod_{\ell=0}^{\infty} \left(1 - \frac{Cs^2}{4^{\ell+1}}\right)^{-2^\ell}. \tag{4.4.4}$$

It is enough to establish this inequality for a bounded function f (indeed one can work with $f_k = (f \wedge k) \vee (-k)$, $k \in \mathbb{N}$, for which $\Gamma(f_k) \leq 1$ since $\Gamma(\psi(f)) \leq \Gamma(f)$ for any contraction $\psi : \mathbb{R} \to \mathbb{R}$, and let then $k \to \infty$ by Fatou's lemma). Apply the Poincaré inequality $P(C)$ to $g = e^{sf/2}$, $s \in \mathbb{R}$. Since

$$\mathcal{E}(g) = \frac{s^2}{4} \int_E e^{sf} \Gamma(f) d\mu \leq \frac{s^2}{4} \int_E e^{sf} d\mu,$$

4.4 The Example of the Exponential Measure, and Exponential Integrability

we get, setting $Z(s) = \int_E e^{sf} d\mu$,

$$\left(1 - \frac{Cs^2}{4}\right) Z(s) \leq Z\left(\frac{s}{2}\right)^2.$$

If $1 - \frac{Cs^2}{4} > 0$, then

$$Z(s) \leq \left(1 - \frac{Cs^2}{4}\right)^{-1} Z\left(\frac{s}{2}\right)^2.$$

Iterating the procedure by replacing s by $\frac{s}{2}$ yields the claim since $\lim_{\ell \to \infty} Z(\frac{s}{2^\ell})^{2^\ell} = e^{s \int_E f d\mu}$.

To reach the full conclusion of the statement, it remains to show that every 1-Lipschitz function f is indeed integrable (provided it is finite μ-almost everywhere). To this end, apply the Poincaré inequality to $f_k = |f| \wedge k$ for every $k \in \mathbb{N}$. In particular, since $\Gamma(f_k) \leq \Gamma(|f|) \leq \Gamma(f) \leq 1$,

$$\int_E \left(f_k - \int_E f_k d\mu\right)^2 d\mu \leq c_1$$

for some constant $c_1 > 0$ independent of k, and therefore

$$\mu\left(\left|f_k - \int_E f_k d\mu\right| \geq \sqrt{2c_1}\right) \leq \frac{1}{2}.$$

Let $c_2 > 0$ be large enough so that $\mu(|f| \geq c_2) < \frac{1}{2}$ (since f is μ-almost everywhere finite). Hence, $\mu(|f_k| \geq c_2) < \frac{1}{2}$ for every k. Now, if two measurable sets S_1 and S_2 are such that $\mu(S_1) \geq \frac{1}{2}$ and $\mu(S_2) > \frac{1}{2}$, then, $\mu(S_1 \cap S_2) > 0$, and in particular $S_1 \cap S_2 \neq \emptyset$. Therefore, for every $k \in \mathbb{N}$, there exists an $x \in E$ such that $|f_k(x) - \int_E f_k d\mu| < \sqrt{2c_1}$ and $|f_k(x)| < c_2$, and hence

$$\int_E f_k d\mu < \sqrt{2c_1} + c_2.$$

Since c_1 and c_2 are independent of k, it follows from Fatou's lemma that $\int_E |f| d\mu < \infty$. Hence (4.4.4) applies, and since the infinite product is finite under the condition $s < \sqrt{\frac{4}{C}}$, the conclusion follows. The proof of Proposition 4.4.2 is complete. $\square$

A careful examination of the infinite product of (4.4.4) shows that (in the setting described in Proposition 4.4.2), for every $s < \sqrt{\frac{4}{C}}$,

$$\int_E e^{s(f - \int_E f d\mu)} d\mu \leq \frac{2 + s\sqrt{C}}{2 - s\sqrt{C}}. \tag{4.4.5}$$

Returning to the example of the exponential measure $d\mu(x) = \frac{1}{2}e^{-|x|}dx$ on $\mathbb{R}$, it satisfies $P(C)$ with $C = 4$ by (4.4.3). This corresponds to the critical exponent $s = 1$ in Proposition 4.4.2 achieved for the Lipschitz function $f(x) = |x|$. This indicates that both the spectral gap constant in (4.4.3) as well as the exponent in Proposition 4.4.2 are optimal.

To conclude this sub-section, we briefly observe that new Poincaré inequalities may be obtained from a given one by a contraction principle via Lipschitz functions.

Proposition 4.4.4 (Contraction principle) *If (μ, Γ) satisfies a Poincaré inequality $P(C)$ on a state space E, and if $F : E \to \mathbb{R}$ is 1-Lipschitz with respect to Γ, then the image measure μ_F of μ by F satisfies a Poincaré inequality $P(C)$ with respect to the usual carré du champ operator on the real line.*

Proof The proof is immediate. For every suitable function g on $\mathbb{R}$,
$$\operatorname{Var}_{\mu_F}(g) = \operatorname{Var}_\mu(g \circ F).$$
By the Poincaré inequality $P(C)$ for μ and the change of variables formula,
$$\operatorname{Var}_\mu(g \circ F) \leq C \int_E \Gamma(g \circ F) d\mu$$
$$= C \int_E g'^2 \circ F \, \Gamma(F) d\mu$$
$$\leq C \int_E g'^2 \circ F \, d\mu = C \int_{\mathbb{R}} g'^2 d\mu_F,$$
which is the announced claim. $\square$

4.4.3 Exponential Measure Concentration

Proposition 4.4.2 may be reformulated in the form of an exponential *measure concentration* property for the measure μ satisfying a Poincaré inequality $P(C)$ for some $C > 0$. Namely, for every Lipschitz function f and every $r \geq 0$,
$$\mu\left(f \geq \int_E f d\mu + r\right) \leq 3 e^{-r/\sqrt{C} \|f\|_{\text{Lip}}}. \tag{4.4.6}$$

This is an immediate application of (4.4.5) with, for simplicity, $s = \frac{1}{\sqrt{C}}$ together with Markov's exponential inequality and homogeneity. The numerical constants are not sharp. The inequality (4.4.6) is actually more a deviation inequality of f from its mean. A concentration inequality around the mean is easily obtained together with the same inequality for $-f$ yielding
$$\mu\left(\left|f - \int_E f d\mu\right| \geq r\right) \leq 6 e^{-r/\sqrt{C} \|f\|_{\text{Lip}}}, \quad r \geq 0.$$

As a consequence of the measure concentration inequality (4.4.6), the image under a Lipschitz map of a measure satisfying a spectral gap inequality is a measure with tails at most exponential. This is in particular the case for the exponential measure itself. Together with the tensorization principle of Sect. 4.3, various applications of interest may be produced. For example, if $f : \mathbb{R} \to \mathbb{R}$ is Lipschitz with respect to Γ, then

$$F(x) = \frac{1}{\sqrt{n}}\big[f(x_1) + \cdots + f(x_n)\big], \quad x = (x_1, \ldots, x_n) \in \mathbb{R}^n,$$

is Lipschitz on $\mathbb{R}^n$ with respect to the tensored Γ on $\mathbb{R}^n$ with the same Lipschitz parameter. Now, if μ is the probability distribution of some real-valued random variable Z_1, $\mu^{\otimes n}$ is the distribution on $\mathbb{R}^n$ of the sample $Z = (Z_1, \ldots, Z_n)$ of independent variables each with the same law as Z_1. Therefore, if μ satisfies a Poincaré inequality $P(C)$ with respect to the standard carré du champ operator on the line, and if $\|f\|_{\mathrm{Lip}} \leq 1$ for example, then

$$\mathbb{P}\big(F(Z) \geq \mathbb{E}\big(F(Z)\big) + r\big) \leq 3\,e^{-r/\sqrt{C}}$$

for every $r \geq 0$. This is a consequence of the concentration inequality (4.4.6) applied to the Lipschitz function F since $\mu^{\otimes n}$ satisfies $P(C)$ on $\mathbb{R}^n$ by Proposition 4.3.1. The preceding bound is independent of the dimension n. The same principle applies to other Lipschitz functions of the sample Z, such as for example $\max_{1 \leq i \leq n} Z_i$. Non-identically distributed samples may be considered similarly.

Remark 4.4.5 One may wonder how far the exponential integrability of Proposition 4.4.2, or the tail estimate (4.4.6), are from a Poincaré inequality. The Muckenhoupt characterization presented in the next section clearly indicates that they are not sufficient in general to entail a Poincaré inequality. However, under a curvature condition $CD(0, \infty)$, the concentration bound (4.4.6) surprisingly turns out to imply in return a Poincaré inequality $P(C')$, which moreover has proportional constants. This will be analyzed in Sect. 8.7, p. 425, on the basis of the local heat kernel inequalities of Sect. 4.7 below (in particular their reverse forms).

4.5 Poincaré Inequalities on the Real Line

This section investigates Poincaré inequalities on the real line or on an interval of the real line. We first discuss a useful criterion, known as Muckenhoupt's criterion, which characterizes Poincaré inequalities. Following that, we describe optimal Poincaré inequalities on a bounded interval in terms of Neumann, Dirichlet and periodic boundary conditions.

4.5.1 The Muckenhoupt Criterion

This section describes a criterion for a measure μ on the real line $\mathbb{R}$ to satisfy a Poincaré inequality with respect to the usual carré du champ operator $\Gamma(f) = f'^2$ (recall that it is always possible to reduce to this case by a suitable change of variables, cf. Sect. 2.6, p. 97). This criterion, going back to the work of B. Muckenhoupt, yields a useful necessary and sufficient condition for the Poincaré inequality to hold (although it is not always completely straightforward to check). We only state it for measures absolutely continuous with respect to the Lebesgue measure (see however below). The $\mathbb{R}^n$ case will be treated in Chap. 8 in terms of more general measure-capacity inequalities (which, however, are less useful in practice).

Theorem 4.5.1 (Muckenhoupt's criterion) *Let μ be a probability measure on the Borel sets of $\mathbb{R}$ with density p with respect to the Lebesgue measure, and let m be a median of μ (i.e. $\mu([m, +\infty)) \geq \frac{1}{2}$ and $\mu((-\infty, m]) \geq \frac{1}{2}$). Set*

$$B_+ = \sup_{x > m} \mu([x, +\infty)) \int_m^x \frac{1}{p(t)} dt$$

and

$$B_- = \sup_{x < m} \mu((-\infty, x]) \int_x^m \frac{1}{p(t)} dt.$$

Then, a necessary and sufficient condition in order that μ satisfies a Poincaré inequality $P(C)$ is that $B = \max(B_+, B_-) < \infty$. Moreover, $B \leq 2C \leq 8B$ where C is the Poincaré constant of μ.

It should be mentioned that whenever $B < \infty$, the median is unique, since the measure of any interval included in the support is strictly positive. A careful look at the criterion indicates that in order for it to be satisfied, $\frac{1}{p}$ has to be locally integrable on the support of μ with respect to the Lebesgue measure. Hence the measure μ should not have "holes" in its support. For example, it is immediately checked that whenever $p = 0$ on some interval $[x_0, x_1]$ with $m \leq x_0 < x_1$ such that $\mu([x_1, +\infty)) > 0$, or when at some point x_0, $p \simeq |x - x_0|$, then the condition is not satisfied. It is a good exercise to check that the measure on $\mathbb{R}$ with density $c_\alpha e^{-|x|^\alpha}$ with respect to the Lebesgue measure satisfies the Muckenhoupt criterion if and only if $\alpha \geq 1$.

It is not necessary for a probability measure μ to have a density with respect to the Lebesgue measure in order to satisfy a Poincaré inequality (use for example the contraction Proposition 4.4.4 with a Lipschitz map which is constant on some interval of positive measure). In fact, Theorem 4.5.1 may be extended and stated without any change by taking p to be the density of the absolutely continuous part of μ. For simplicity, and to make the exposition of the proof a little easier, we only deal with measures with densities, and we assume that the suprema in B_+ and B_- are taken over points x in the smallest closed interval which carries μ. The general case is left to the reader.

4.5 Poincaré Inequalities on the Real Line

We turn to the proof of Theorem 4.5.1 under these assumptions. We start with the following simpler result. Set $h_+(x) = \int_m^x \frac{1}{p(t)} dt$, defined for $x \geq m$, and similarly $h_-(x) = \int_x^m \frac{1}{p(t)} dt$, defined for $x \leq m$.

Proposition 4.5.2 *If μ with density p is such that*

$$\int_m^\infty h_+(x) p(x) dx < \infty \quad \text{and} \quad \int_{-\infty}^m h_-(x) p(x) dx < \infty$$

where m is a median of μ, then it satisfies a Poincaré inequality.

Proof As in the proof of the Poincaré inequality for the exponential measure, it is enough to bound from above

$$\int_m^\infty [f(x) - f(m)]^2 d\mu \quad \text{and} \quad \int_{-\infty}^m [f(x) - f(m)]^2 d\mu$$

by $C\mathcal{E}(f) = C \int_{\mathbb{R}} f'^2 d\mu$. By the Cauchy-Schwarz inequality, for every $x > m$,

$$[f(x) - f(m)]^2 = \left(\int_m^x f'(t) dt\right)^2$$
$$\leq \int_{\mathbb{R}} f'^2(t) p(t) dt \int_m^x \frac{1}{p(t)} dt = \mathcal{E}(f) h_+(x).$$

Therefore,

$$\int_m^\infty [f(x) - f(m)]^2 d\mu \leq \mathcal{E}(f) \int_m^\infty h_+(x) p(x) dx.$$

Together with a similar treatment for the second integral, we immediately obtain a Poincaré inequality with constant

$$C = \max\left(\int_m^\infty h_+(x) p(x) dx, \int_{-\infty}^m h_-(x) p(x) dx\right). \qquad \square$$

Next, we turn to the proof of Theorem 4.5.1 itself.

Proof of Theorem 4.5.1 First, the condition $B < \infty$ is sufficient for a Poincaré inequality $P(C)$ to hold. To establish this claim, we proceed as in the proof of Proposition 4.5.2 to bound $[f(x) - f(m)]^2$ for every $x > m$. However, we need a somewhat sharper bound. Set $g(x) = \sqrt{h_+(x)}$ and $k(x) = g(x) p(x)$, $x > m$. Then, by the Cauchy-Schwarz inequality,

$$[f(x) - f(m)]^2 = \left(\int_m^x f'(t) dt\right)^2 \leq \int_m^x f'^2(t) k(t) dt \int_m^x \frac{1}{k(t)} dt.$$

Therefore, by Fubini's Theorem,

$$\int_m^\infty [f(x) - f(m)]^2 p(x) dx \leq \int_m^\infty f'^2(t) k(t) \left(\int_t^\infty \left(\int_m^x \frac{1}{k(u)} du \right) p(x) dx \right) dt.$$

The choice of the function k is justified by the fact that $g' = \frac{1}{2k}$, and thus that $\int_m^x \frac{1}{k(u)} du = 2g(x)$. The definition of B_+ indicates that, for every $x > m$,

$$g(x) \leq \sqrt{\frac{B_+}{\mu([x, +\infty))}}$$

so that, for $t > m$,

$$\int_t^\infty \left(\int_m^x \frac{1}{k(u)} du \right) p(x) dx \leq 2\sqrt{B_+} \int_t^\infty \frac{p(x)}{\sqrt{\mu([x, +\infty))}} dx.$$

Setting $r(x) = \sqrt{\mu([x, +\infty))}$,

$$\frac{p(x)}{\sqrt{\mu([x, +\infty))}} = -2 r'(x).$$

As a consequence of the preceding,

$$\int_m^\infty [f(x) - f(m)]^2 p(x) dx \leq 4\sqrt{B_+} \int_m^\infty f'^2(t) k(t) r(t) dt.$$

But

$$k(t) r(t) = p(t) \sqrt{\mu([t, +\infty))} h_+(t) \leq \sqrt{B_+} \, p(t),$$

so that finally

$$\int_m^\infty [f(x) - f(m)]^2 p(x) dx \leq 4 B_+ \int_m^\infty f'^2(t) \, p(t) dt.$$

The term $\int_{-\infty}^m p(x) [f(x) - f(m)]^2 dx$ is treated similarly with B_- so that the measure μ indeed satisfies a Poincaré inequality with constant $C = 4B$.

Turning to the converse, the idea is to apply the Poincaré inequality $P(C)$ to suitable test functions f. We first replace the Poincaré inequality with the easiest consequence of it given by Lemma 4.2.3 with $A = [m, \infty)$ where m is the median of μ (hence $\mu(A) \leq \frac{1}{2}$ and $\frac{1}{1-\mu(A)} \leq 2$). Therefore, under $P(C)$, for any function $f \in \mathcal{D}(\mathcal{E})$ supported on $[m, \infty)$,

$$\int_\mathbb{R} f^2 d\mu \leq 2C \int_\mathbb{R} f'^2 d\mu.$$

4.5 Poincaré Inequalities on the Real Line

Then take a function $g \in \mathbb{L}^2(\mu)$, and apply the latter to $f(x) = \int_m^x g(t)dt$, $x > m$. It follows that

$$\int_m^\infty \left(\int_m^x g(t)dt \right)^2 d\mu(x) \leq 2C \int_m^\infty g^2(x) d\mu(x).$$

Fix $\varepsilon > 0$ and $r > m$, and apply this inequality to the function

$$g(x) = \frac{1}{p(x) + \varepsilon} \mathbb{1}_{[m,r]}(x), \quad x \in \mathbb{R}.$$

The left-hand side of the previous inequality is therefore bounded from below by

$$\int_r^\infty \left(\int_m^x g(t)dt \right)^2 p(x)dx = \left(\int_m^r \frac{1}{p(t)+\varepsilon} dt \right)^2 \mu\bigl([r,\infty)\bigr),$$

while the right-hand side is equal to

$$2C \int_m^r \frac{p(x)}{(p(x)+\varepsilon)^2} dx \leq 2C \int_m^r \frac{1}{p(x)+\varepsilon} dx.$$

Comparing the two sides,

$$\mu\bigl([r,\infty)\bigr) \int_m^r \frac{1}{p(t)+\varepsilon} dt \leq 2C.$$

As $\varepsilon \to 0$, the first part of the conclusion, namely $B_+ \leq 2C$, follows. The second part $B_- \leq 2C$ is achieved similarly, yielding the announced bound $B \leq 2C$. Theorem 4.5.1 is established. □

4.5.2 Poincaré Inequalities on an Interval

The Muckenhoupt criterion indicates that many probability measures on a bounded interval of the real line satisfy a Poincaré inequality. This section is concerned with optimal constants in certain instances. Dealing with intervals with boundaries, the Full Markov Triple assumption (which requires essential self-adjointness) is not assumed to hold in this sub-section, and instead we work with Diffusion Markov Triples.

Before discussing the Poincaré inequalities themselves, it may be observed first that, starting for example from the Poincaré inequality for the Gaussian measure on the real line, the simple Lemma 4.2.3 yields inequalities of the form

$$\int_I f^2 d\mu \leq C\,\mathcal{E}(f)$$

on any bounded interval I, for any smooth function f compactly supported in I and for the usual carré du champ operator $\Gamma(f) = f'^2$, and for any measure μ with strictly positive and bounded density. To understand the meaning of such inequalities (which will be further explored in Sect. 4.10), consider for simplicity a bounded interval $I \subset \mathbb{R}$, with the usual carré du champ operator $\Gamma(f) = f'^2$ and a reference probability measure μ with a smooth density w bounded from below by some constant $a > 0$. As described in Sect. 2.6, p. 97, this situation is related to the analysis of the Sturm-Liouville operator $\mathrm{L}f = f'' + \frac{w'}{w} f'$ acting on smooth functions f on an interval $I \subset \mathbb{R}$. The following statement holds.

Proposition 4.5.3 *For the Sturm-Liouville operator $\mathrm{L}f = f'' + \frac{w'}{w} f'$ acting on smooth functions on a bounded interval $I \subset \mathbb{R}$ where w is smooth and bounded from below by a constant $a > 0$, with invariant probability measure $d\mu = w dx$, there is a constant $C > 0$ such that, for any smooth function f compactly supported in the interval I,*

$$\int_I f^2 d\mu \leq C \int_I f'^2 d\mu. \tag{4.5.1}$$

The best constant C in this inequality is $\frac{1}{\lambda_0}$, where $-\lambda_0$ is the lowest eigenvalue of the Sturm-Liouville operator L with Dirichlet boundary conditions (functions which vanish at the boundary).

Proof From the analysis performed in Sect. 2.6, p. 97, the operator L defined on the class $\mathcal{A}_0$ of smooth and compactly supported functions in the interval I extends to a self-adjoint operator L (the Friedrichs extension) which is such that $\mathcal{A}_0$ is dense in the Dirichlet domain $\mathcal{D}(\mathcal{E})$ of this extension, corresponding to the Dirichlet boundary conditions. Moreover, the spectrum of (the opposite generator) $-\mathrm{L}$ is discrete, consisting of an infinite increasing sequence of positive eigenvalues $(\lambda_k)_{k \in \mathbb{N}}$ associated to eigenvectors $(f_k)_{k \in \mathbb{N}}$ which form an orthonormal basis of $\mathbb{L}^2(\mu)$. It requires a bit of further analysis (not developed here) to see that in fact λ_0 is strictly positive, simple, and corresponds to a positive eigenvector f_0, strictly positive in the interior of I (λ_0 and f_0 are the analogues of the Perron-Frobenius eigenvalue and eigenvector in the context of finite Markov chains, see Sect. 1.9, p. 33).

Now, the standard spectral decomposition picture applies. Namely, if $f \in \mathbb{L}^2(\mu)$ is developed in the basis $(f_k)_{k \in \mathbb{N}}$ in the form $f = \sum_{k \in \mathbb{N}} a_k f_k$, then

$$-\mathrm{L}f = \sum_{k \in \mathbb{N}} a_k \lambda_k f_k.$$

Therefore,

$$\mathcal{E}(f) = -\int_I f \mathrm{L}f \, d\mu = \sum_{k \in \mathbb{N}} a_k^2 \lambda_k \geq \lambda_0 \sum_{k \in \mathbb{N}} a_k^2 = \lambda_0 \int_I f^2 d\mu.$$

Inequality (4.5.1) of Proposition 4.5.3 follows with $C = \frac{1}{\lambda_0}$, where $\frac{1}{\lambda_0}$ appears to be the best possible (since f_0 saturates the inequality). On the other hand, if (4.5.1) holds for all $f \in \mathcal{A}_0$, then it extends to every function in the Dirichlet domain $\mathcal{D}(\mathcal{E})$ by density. The proof of the proposition is therefore complete. $\square$

4.5 Poincaré Inequalities on the Real Line

Of course, what is described in the previous proof extends similarly to bounded domains on a manifold with any smooth elliptic operator and smooth density measure.

We next give an interpretation of the Poincaré inequality of Definition 4.2.1 for the invariant measure of a Sturm-Liouville operator on an interval of the real line in terms of Neumann boundary conditions. We again refer to Sect. 2.6, p. 97, for the required background.

Proposition 4.5.4 *For the Sturm-Liouville operator* $Lf = f'' + \frac{w'}{w} f'$ *acting on smooth functions on an bounded interval* $I \subset \mathbb{R}$ *where* w *is smooth and bounded from below by a constant* $a > 0$, *with invariant probability measure* $d\mu = w dx$, *there is a constant* $C > 0$ *such that, for every smooth function* f *on* I *(or restriction to* I *of a smooth function defined on a neighborhood of* I*),*

$$\operatorname{Var}_\mu(f) \leq C \int_I f'^2 d\mu. \tag{4.5.2}$$

The best constant C *in this inequality is* $\frac{1}{\lambda_1}$, *where* $-\lambda_1$ *is the first non-zero eigenvalue of the Sturm-Liouville operator* L *with Neumann boundary conditions (functions whose derivatives vanish at the boundary).*

Proof This relies on the same analysis as that given in Proposition 4.5.3. The only point to observe is that one may approximate a smooth function f defined in a neighborhood of I by a sequence of smooth functions with derivatives which vanish in a neighborhood of the boundaries, in such a way that the corresponding quantities in the inequality (4.5.2) converge to the quantities for f. At a technical level, the idea is to first approximate the derivatives. The details are left to the reader. □

It is worth mentioning that there are cases when (4.5.2) is valid only for functions which are compactly supported in I. As is illustrated by the next statement, this situation corresponds to a spectral gap of the operator with periodic boundary conditions. The next statement actually investigates more precisely the different values of the constants in the preceding two propositions with the example of the Lebesgue measure on the interval $I = [0, 1]$.

Proposition 4.5.5 *Let* $Lf = f''$ *be the Sturm-Liouville operator acting on smooth functions* f *on* $[0, 1]$ *with invariant measure the Lebesgue measure* dx.

(i) *For any smooth function* $f : [0, 1] \to \mathbb{R}$ *such that* $f(0) = f(1) = 0$,

$$\int_{[0,1]} f^2 dx \leq \frac{1}{\pi^2} \int_{[0,1]} f'^2 dx. \tag{4.5.3}$$

The inequality is optimal and the sharp constant is attained by the function $f(x) = \sin(\pi x)$.

(ii) For any smooth function $f : [0, 1] \to \mathbb{R}$,

$$\operatorname{Var}_{dx}(f) \leq \frac{1}{\pi^2} \int_{[0,1]} f'^2 dx. \tag{4.5.4}$$

The inequality is optimal and the sharp constant is attained by the function $f(x) = \cos(\pi x)$.

(iii) For any smooth function $f : [0, 1] \to \mathbb{R}$ such that $f(0) = f(1)$,

$$\operatorname{Var}_{dx}(f) \leq \frac{1}{4\pi^2} \int_{[0,1]} f'^2 dx. \tag{4.5.5}$$

The inequality is optimal and the sharp constant is attained by functions of the form $f(x) = a \cos(2\pi x) + b \sin(2\pi x) + c$ where a, b, c are real constants.

Item (i) corresponds to the bottom of the spectrum with Dirichlet boundary conditions (Proposition 4.5.3) while (ii) corresponds to the spectral gap for Neumann boundary conditions (Proposition 4.5.4). Property (iii) corresponds to an estimate on the spectral gap for periodic functions and is usually referred to as *Wirtinger's inequality* in the literature (see the Notes and References).

Proof The assertion (iii) is proved directly using a Fourier decomposition of a periodic function similar to the Hermite expansion in Sect. 4.1. Namely if

$$f(x) = a_0 + \sum_{k \geq 1} \left[a_k \cos(2\pi kx) + b_k \sin(2\pi kx) \right]$$

(take if necessary a finite sum to start with), then assuming that $a_0 = \int_{[0,1]} f \, dx = 0$,

$$\int_{[0,1]} f^2 dx = \frac{1}{2} \sum_{k \geq 1} (a_k^2 + b_k^2) \quad \text{and} \quad \int_{[0,1]} f'^2 dx = 2\pi^2 \sum_{k \geq 1} k^2 (a_k^2 + b_k^2).$$

The conclusion immediately follows. Note that any smooth function f on $[0, 1]$ such that $f(0) = f(1)$ may be extended to a continuous 1-periodic function. While it may happen that this extension is not C^1, it may be approximated by a sequence of smooth periodic functions such that each term in the inequality converges to the corresponding one for f. Therefore, the optimal (Poincaré) constant in (4.5.5) is also the optimal constant in the Poincaré inequality on the interval $[0, 1]$ over the class of smooth 1-periodic functions. The second set (ii) of inequalities, without any boundary condition, appears as a consequence of (i) by symmetrization and periodization (for $f : [0, 1] \to \mathbb{R}$ arbitrary, define $g : [-1, +1] \to \mathbb{R}$ by $g(x) = f(x)$ for $x \in [0, +1]$, $g(x) = f(-x)$ for $x \in [-1, 0]$, and apply (i) to g on the interval $[-1, +1]$ after re-scaling). Finally (i) is a consequence of (iii) by anti-symmetrization and periodization (for $f : [0, 1] \to \mathbb{R}$ such that $f(0) = f(1) = 0$, define $g : [-1, +1] \to \mathbb{R}$ by $g(x) = f(x)$ for $x \in [0, +1]$, $g(x) = -f(-x)$ for $x \in [-1, 0]$). □

4.6 The Lyapunov Function Method

This section investigates a method of establishing Poincaré inequalities close to the method used for the exponential measure in Sect. 4.4.1. The form is however a bit different.

4.6.1 The Lyapunov Function Method

The main principle is the following. In general, on $\mathbb{R}^n$ for example, it is not that difficult to reach Poincaré inequalities on compact subsets, balls for example, by comparison with known measures or gradients. Accordingly, we say that a Markov Triple (E, μ, Γ), where μ is not necessarily a probability measure, with underlying algebra $\mathcal{A}_0$, satisfies a local Poincaré inequality on a measurable subset K with $0 < \mu(K) < \infty$ of the state space E if for some constant $C_K > 0$ and every function f in $\mathcal{A}_0$,

$$\int_K (f - m_K)^2 d\mu \leq C_K \int_K \Gamma(f) d\mu \tag{4.6.1}$$

where $m_K = \frac{1}{\mu(K)} \int_K f d\mu$. In other words, this is the Poincaré inequality with respect to the carré du champ operator Γ for functions restricted to K and with respect to the measure μ restricted to K and normalized to be a probability measure. Actually, only a much weaker form of (4.6.1) will be used below, namely that for some (measurable) set $L \supset K$ and any $f \in \mathcal{A}_0$,

$$\int_K (f - m_K)^2 d\mu \leq C_{K,L} \int_L \Gamma(f) d\mu \tag{4.6.2}$$

for some $C_{K,L} > 0$.

Remark 4.6.1 On $\mathbb{R}^n$ equipped with the Lebesgue measure, or a manifold equipped with a measure equivalent to the Lebesgue measure in any system of local coordinates, whenever the set K is compact with smooth boundary, the local Poincaré inequality (4.6.1) on K is nothing else but the Poincaré inequality on K with Neumann boundary conditions, as it is in dimension one (see Proposition 4.5.4). This may be shown along the same lines, although with more technicalities. On the one hand, it is always possible to approximate a smooth function f defined in a neighborhood of K by a sequence $(f_\ell)_{\ell \in \mathbb{N}}$ of smooth functions satisfying the Neumann boundary conditions in such a way that $\int_K (f_\ell^2 - \frac{1}{\mu(K)} \int_K f_\ell d\mu)^2 d\mu$ and $\int_K \Gamma(f_\ell) d\mu$ both converge to the corresponding quantities for f (this step is not so easy and requires the use of normal coordinates in a neighborhood of the boundary). This shows that the Poincaré inequality with Neumann boundary condition extends to the inequality (4.6.1) valid for the restriction to K of smooth functions defined in a neighborhood of K. On the other hand, it is quite easy to extend any smooth function defined on K and satisfying the Neumann boundary conditions into a smooth function defined in a neighborhood of K, showing that the two inequalities are in fact equivalent. If one is not concerned with optimal constants, it is easier to use the simpler spectral inequality (4.10.1) in Sect. 4.10 below in a larger domain $L \supset K$, for which the constant corresponds to the Dirichlet boundary conditions. This immediately provides the weaker inequality (4.6.2).

We say that a function $J : E \to [1, \infty)$ (in the extended algebra $\mathcal{A}$) is a *Lyapunov function* (for the underlying generator L) if there are constants $\lambda, b > 0$ and a measurable set $K \subset E$ on which there is a local Poincaré inequality (4.6.2) (for some $L \supset K$ with constant $C_{K,L}$) such that J satisfies

$$1 \le -\frac{LJ}{\lambda J} + b \mathbb{1}_K.$$

Theorem 4.6.2 (Lyapunov's criterion) *Let (E, μ, Γ) be a Markov Triple with μ a probability measure satisfying the completeness property of Definition 3.3.9, p. 157. If there exists a Lyapunov function J as above, then μ satisfies a Poincaré inequality $P(C)$ on the whole space E with constant $C = \frac{1}{\lambda} + b C_{K,L}$.*

Proof By density, it suffices to prove that for every $f \in \mathcal{A}_0$,

$$\int_E (f - m)^2 d\mu \le C \int_E \Gamma(f) d\mu$$

for some $m = m(f)$ to be chosen later. Using of the Lyapunov function J, write

$$\int_E (f - m)^2 d\mu \le -\frac{1}{\lambda} \int_E \frac{LJ}{J} (f - m)^2 d\mu + b \int_K (f - m)^2 d\mu. \quad (4.6.3)$$

Choose then $m = \frac{1}{\mu(K)} \int_K f d\mu$. The second term on the right-hand side of (4.6.3) is then bounded from above thanks to (4.6.2) by $b C_{K,L} \int_L \Gamma(f) d\mu$. In order to deal with the first term, we show that for every $g \in \mathcal{A}_0$,

$$-\int_E \frac{LJ}{J} g^2 d\mu \le \int_E \Gamma(g) d\mu. \quad (4.6.4)$$

To this end, integrate by parts to obtain

$$-\int_E \frac{LJ}{J} g^2 d\mu = \int_E \Gamma\left(\frac{g^2}{J}, J\right) d\mu. \quad (4.6.5)$$

Observe then that

$$\Gamma\left(\frac{g^2}{J}, J\right) = \frac{2g}{J} \Gamma(g, J) - \frac{g^2}{J^2} \Gamma(J) \le \Gamma(g).$$

The first equality here follows from the diffusion property of the Γ operator, while the second is a direct consequence of the Cauchy-Schwarz inequality for Γ, (1.4.3), p. 21.

The proof is then completed provided (4.6.4) and (4.6.5) may be applied to $g = f - m$, for which $\Gamma(g) = \Gamma(f)$. However, g is not in general in $\mathcal{A}_0$ even if f is (recall that $\mathcal{A}_0$ stands in general for the set of smooth compactly supported functions). To overcome this difficulty, use the completeness property and replace g by $g \zeta_k$, $k \in \mathbb{N}$, where ζ_k are the functions used to define completeness and then pass to the limit as $k \to \infty$. In this way, Theorem 4.6.2 is established. □

One may wonder whether the condition $J \geq 1$ has indeed been used in the preceding proof. First, since the hypotheses in Theorem 4.6.2 are unchanged when J is replaced by cJ, the preceding condition essentially amounts to the fact that J is bounded from below by a strictly positive constant. This condition then ensures that if we divide by J the resulting function $\frac{g^2}{J}$ will be in $\mathcal{A}$. Moreover, the completeness assumption is used only to justify the integration by parts formula (4.6.5). This formula may hold without it, for example as soon as $\mathrm{L}(\log J)$ and $\Gamma(\log J)$ are in $\mathbb{L}^1(\mu)$ and $\int_E \mathrm{L}(\log J) d\mu = 0$. Following Proposition 3.3.6, p. 156, it would be enough for this purpose that $\mathrm{L}(\log J) \in \mathbb{L}^2(\mu)$.

The preceding Lyapunov criterion (Theorem 4.6.2) admits a number of variations, under somewhat different hypotheses and conclusions. Note for example that it may be applied to the standard Gaussian measure on $\mathbb{R}^n$ with the choice of $J = 1 + |x|^2$ and K some Euclidean ball. The resulting Poincaré constant is however far from optimal (in particular it depends on the dimension). The criterion also applies to the exponential measure with for example $J = e^{c|x|}$ for some sufficiently small $c > 0$. Comparing with the previous (and optimal) proof of the Poincaré inequality for the exponential measure given in Proposition 4.4.1, setting $J = e^{ch}$, the hypothesis on J may be deduced from a uniform bound on $\Gamma(h)$, an upper bound on $-\mathrm{L}h$ outside a compact set and a proper choice of the parameter c. The Lyapunov hypothesis on J is then more general, although the proof given earlier, following the same lines, was more precise.

4.6.2 Log-Concave Measures

A nice and powerful illustration of the Lyapunov function method is the following statement for log-concave measures. A probability measure $d\mu = e^{-W} dx$ on the Borel sets of $\mathbb{R}^n$ such that W is a smooth convex function is called *log-concave*.

Theorem 4.6.3 (Kannan-Lovász-Simonovits-Bobkov Theorem) *Let μ be a log-concave probability measure on $\mathbb{R}^n$. Then μ satisfies a Poincaré inequality with respect the usual carré du champ operator $\Gamma(f) = |\nabla f|^2$.*

Proof According to Theorem 4.6.2, it is enough to exhibit a suitable Lyapunov function. The idea is to choose $c, R > 0$ and $J : \mathbb{R}^n \to \mathbb{R}$ smooth with $J(x) = e^{c|x|}$ for $|x| \geq R$ and $J(x) \geq 1$ for $|x| \leq R$. To verify that, for some $c > 0$, J is a Lyapunov function in the preceding sense amounts to showing that for a convex function W such that $\int_{\mathbb{R}^n} e^{-W} dx < \infty$,

$$\liminf_{|x| \to \infty} \frac{x \cdot \nabla W}{|x|} > 0,$$

which is left here as an exercise. □

Note for further purposes that the dependence of the Poincaré constant on the dimension n provided by the preceding proof is quite poor.

Theorem 4.6.3 may be applied to the uniform measure on a convex body in $\mathbb{R}^n$. Indeed let K be convex and compact with non-empty interior in $\mathbb{R}^n$ and consider the convex function W (or rather a smooth approximation of it) defined by $W(x) = 1$ if $x \in K$ and $W(x) = \infty$ otherwise. The uniform normalized Lebesgue measure μ_K on K thus satisfies a Poincaré inequality

$$\text{Var}_{\mu_K}(f) \le C \int_K |\nabla f|^2 d\mu_K \qquad (4.6.6)$$

for some $C > 0$ and all smooth functions $f : K \to \mathbb{R}$. The classical Payne-Weinberger Theorem, relying on a refined geometric analysis of the spectral decomposition of the Laplace operator on K, describes the optimal estimate

$$C \le \frac{D^2}{\pi^2} \qquad (4.6.7)$$

in terms (only) of the diameter D of K. This inequality is optimal since it becomes an equality when K is a ball, and includes in particular (4.5.4) in dimension one. Even if (4.6.7) is optimal, the constant depends on the diameter and this estimate cannot be used to prove a Poincaré inequality for general log-concave measures (Theorem 4.6.3). Observe furthermore that (4.6.7) is false in general for non-convex domains. A compact connected domain K in $\mathbb{R}^n$ still satisfies a Poincaré inequality (4.6.6) but the constant may depend on more than only the diameter (think for example of a dumbbell).

4.6.3 A Glueing Property

According to the previous comments, the local Poincaré inequalities (4.6.1) on a measurable subset $K \subset E$ with $0 \le \mu(K) < \infty$ thus strongly depend on the shape of the set K. In particular, the Poincaré constants C_K do not have any kind of monotonicity: a set K with small C_K may contain a subset L with large C_L. However, they satisfy a useful glueing property illustrated by the following statement with which we close this sub-section. Recall that μ is not necessarily a probability measure here.

4.6 The Lyapunov Function Method

Proposition 4.6.4 *Let K and L be two measurable subsets of E such that $\mu(K \cap L) > 0$ and $\mu(K \cup L) < \infty$. Then*

$$C_{K \cup L} \leq \frac{\mu(K \cup L)}{\mu(K \cap L)} \max(C_K, C_L).$$

Here, the Poincaré inequalities (4.6.1) are assumed to hold on some suitable class $\mathcal{A}_1$ of functions on E, not necessarily $\mathcal{A}_0$. The proof relies on the following lemma, which is of independent interest.

Lemma 4.6.5 *Let K and L be two measurable subsets in a measure space $(E, \mathcal{F}, \mu)$ with $\mu(K \cap L) > 0$ and $\mu(K \cup L) < \infty$. Then, for any measurable function $f : E \to \mathbb{R}$ such that $\int_{K \cup L} f^2 d\mu < \infty$ and $\int_{K \cup L} f d\mu = 0$,*

$$\frac{1}{\mu(K)} \left(\int_K f d\mu \right)^2 + \frac{1}{\mu(L)} \left(\int_L f d\mu \right)^2 \leq \left(1 - \frac{\mu(K \cap L)}{\mu(K \cup L)} \right) \int_{K \cup L} f^2 d\mu.$$

Proof Using homogeneity, we reduce to the case where $\mu(K \cup L) = 1$ and $\int_{K \cup L} f^2 d\mu = 1$. Setting $a = \mu(K \setminus L)$, $b = \mu(L \setminus K)$, $x = \int_{K \setminus L} f d\mu$, $y = \int_{L \setminus K} f d\mu$ and $z = \int_{K \cap L} f d\mu$, we have $x + y + z = \int_{K \cup L} f d\mu = 0$ and, by Cauchy-Schwarz,

$$x^2 \leq a \int_{K \setminus L} f^2 d\mu, \quad y^2 \leq b \int_{L \setminus K} f^2 d\mu, \quad z^2 \leq (1 - a - b) \int_{K \cap L} f^2 d\mu.$$

Therefore, provided that $a > 0$ and $b > 0$,

$$\frac{x^2}{a} + \frac{(x+y)^2}{1-a-b} + \frac{y^2}{b} = \frac{x^2}{a} + \frac{z^2}{1-a-b} + \frac{y^2}{b} \leq \int_{K \cup L} f^2 d\mu = 1.$$

The left-hand side of the inequality to be established may be rewritten as

$$\frac{1}{1-b}(x+z)^2 + \frac{1}{1-a}(y+z)^2 = \frac{y^2}{1-b} + \frac{x^2}{1-a}.$$

After some tedious details, it appears that the supremum of $\frac{y^2}{1-b} + \frac{x^2}{1-a}$ on the set $\frac{x^2}{a} + \frac{(x+y)^2}{1-a-b} + \frac{y^2}{b} \leq 1$ is bounded from above by $a + b = 1 - \mu(K \cap L)$ which yields the claim. This bound is actually not optimal, but the optimal bound (given a and b) has a less pleasant form. The argument is similar if either $a = 0$ or $b = 0$. □

To complete the proof of Proposition 4.6.4, it is necessary to control the constant $C_{K \cup L}$ in the inequality

$$\int_{K \cup L} f^2 d\mu \leq \frac{1}{\mu(K \cup L)} \left(\int_{K \cup L} f d\mu \right)^2 + C_{K \cup L} \int_{K \cup L} \Gamma(f) d\mu$$

for all functions f (in some class $\mathcal{A}_1$). To this end, assuming $\int_{K \cup L} f d\mu = 0$, simply write

$$\int_{K \cup L} f^2 d\mu \leq \int_K f^2 d\mu + \int_L f^2 d\mu$$

and use the definition of C_K and C_L together with the lemma.

4.7 Local Poincaré Inequalities

In the framework of this chapter, this section deals with Poincaré inequalities not for the invariant measure μ but for the heat kernel measures $p_t(x, dy)$, $t \geq 0$, $x \in E$, of the Markov semigroup $\mathbf{P} = (P_t)_{t \geq 0}$ (from the representation

$$P_t f(x) = \int_E f(y) \, p_t(x, dy)$$

of (1.2.4), p. 12) with respect to the carré du champ operator Γ. These inequalities along the semigroup will be called *local* with reference to the (initial) point x and the time variable t. (In particular, there is no need here to speak of an invariant or reversible measure.) These are important illustrations of semigroup monotonicity based on the Duhamel interpolation principle.

The inequalities investigated here are close in spirit to the local Poincaré inequalities (4.6.1) where the measure restricted to a set K is replaced by the heat kernel measure $p_t(x, dy)$ around a point x (highly concentrated around this point, at least for small t's), but play a different and fundamental role. Indeed, as will be shown here, these inequalities are in general easier to reach, and provide a lot of important information on the behavior of the semigroup. They are moreover a powerful tool in the study of partial differential equations.

The local heat kernel inequalities will be presented and established for functions in the algebra $\mathcal{A}_0$ of the underlying Markov Triple (E, μ, Γ). Depending on the context, they may then be extended to any function in $\mathcal{D}(\mathcal{E})$ or $\mathcal{D}(L)$, or even to any bounded measurable function. These extensions shall not be discussed below and are left to the reader.

4.7.1 Poincaré Inequalities for Gaussian Kernels

To better introduce these ideas, we first start with the example of the Ornstein-Uhlenbeck semigroup $\mathbf{P} = (P_t)_{t \geq 0}$ presented in Sect. 2.7.1, p. 103, whose generator, on $\mathbb{R}^n$, is given by $Lf = \Delta f - x \cdot \nabla f$ with carré du champ operator $\Gamma(f) = |\nabla f|^2$ (on an algebra $\mathcal{A}_0$ of smooth functions). In this simple and concrete example, both the semigroup $(P_t)_{t \geq 0}$ and the kernels $p_t(x, dy)$ are explicit. In particular, the representation (2.7.3), p. 104, yields the commutation property $\nabla P_t f = e^{-t} P_t(\nabla f)$

4.7 Local Poincaré Inequalities

from which, for every $t \geq 0$ and every f in $\mathcal{A}_0$,

$$\Gamma(P_t f) = |\nabla P_t f|^2 \leq e^{-2t} P_t(|\nabla f|^2) = e^{-2t} P_t(\Gamma(f)) \qquad (4.7.1)$$

(where we used the fact that, by convexity, $(P_t h)^2 \leq P_t(h^2)$).

Hence the gradient of P_t decays exponentially in t. This property has interesting consequences via the Duhamel-type formula (3.1.21), p. 131. Let us recall the principle. For fixed $t > 0$, and for a function f in $\mathcal{A}_0$, consider

$$\Lambda(s) = P_s\big((P_{t-s} f)^2\big), \quad s \in [0, t]$$

(as usual at almost any point $x \in E$, omitted everywhere). Taking the derivative of Λ yields

$$\Lambda'(s) = P_s\big(L((P_{t-s} f)^2) - 2 P_{t-s} f \, L P_{t-s} f\big) = 2 P_s\big(\Gamma(P_{t-s} f)\big)$$

as an immediate consequence of the basic chain rule

$$2 \Gamma(g) = L(g^2) - 2 g \, L g$$

applied to $g = P_{t-s} f$. Therefore, integrating the preceding in $s \in [0, t]$,

$$P_t(f^2) - (P_t f)^2 = \Lambda(t) - \Lambda(0) = \int_0^t \Lambda'(s) ds = 2 \int_0^t P_s\big(\Gamma(P_{t-s} f)\big) ds.$$

On the basis of this interpolation formula, recall the commutation relation (4.7.1)

$$\Gamma(P_{t-s} f) \leq e^{-2(t-s)} P_{t-s}(\Gamma(f))$$

for every $s \in [0, t]$. Furthermore, by the semigroup property,

$$P_s\big(\Gamma(P_{t-s} f)\big) \leq e^{-2(t-s)} P_t(\Gamma(f)).$$

As a consequence,

$$P_t(f^2) - (P_t f)^2 \leq 2 P_t(\Gamma(f)) \int_0^t e^{-2(t-s)} ds = \big(1 - e^{-2t}\big) P_t(\Gamma(f)). \qquad (4.7.2)$$

This inequality is a Poincaré inequality $P(C)$, with constant $C = 1 - e^{-2t}$, not for the invariant (Gaussian) measure μ, but for the heat kernel measure $p_t(x, dy)$ of the representation $P_t f(x) = \int_{\mathbb{R}^n} f(y) \, p_t(x, dy)$, for any $x \in E$. However, as $t \to \infty$, we recover the Poincaré inequality of Proposition 4.1.1 since $p_t(x, dy)$ converges to $\mu(dy)$. Recall that the initial point x is implicit in (4.7.2) (as is the case for all of the local inequalities considered here).

The preceding argument via (4.7.1) may also be used to establish a converse inequality. Indeed, by (4.7.1) again, for every $s \geq 0$ and every $g \in \mathcal{A}_0$,

$$P_s(\Gamma(g)) \geq e^{2s} \Gamma(P_s g).$$

Therefore, with $g = P_{t-s}f$,

$$P_s\big(\Gamma(P_{t-s}f)\big) \geq e^{2s}\,\Gamma(P_t f).$$

As a consequence this time,

$$P_t(f^2) - (P_t f)^2 \geq 2\,\Gamma(P_t f)\int_0^t e^{2s}\,ds = \big(e^{2t} - 1\big)\Gamma(P_t f). \qquad (4.7.3)$$

The latter inequality, which may be considered as a *reverse Poincaré inequality*, is of some particular interest. Indeed if for example f is bounded, then $P_t f$ is Lipschitz for every $t > 0$, with a Lipschitz norm improving as $t \to \infty$, and of the order of $t^{-1/2}$ as $t \to 0$. We know that if $f \in \mathbb{L}^2(\mu)$, then $P_t f$ belongs, for every t, to the Dirichlet space $\mathcal{D}(\mathcal{E})$, but here more is available: whenever f is bounded, $P_t f$ for $t > 0$ has a bounded gradient with an explicit bound. The semigroup $(P_t)_{t \geq 0}$ is thus in some sense regularizing.

The preceding semigroup interpolation argument describes a heat flow monotonicity proof of the Poincaré inequality for the Gaussian measure relying on the commutation or gradient bound (4.7.1). This commutation has a curvature flavor which is involved in geometric interpretations of the preceding heat kernel inequalities. More precisely, the gradient bound (4.7.1) was already described in Chap. 3 as being equivalent to the curvature condition $CD(1, \infty)$ of Definition 3.3.14, p. 159 (cf. Corollary 3.3.19, p. 163). It is worth recalling the various steps in this concrete context of the Ornstein-Uhlenbeck operator. Namely, at $t = 0$, the two sides of (4.7.1) are actually equal, and thus the asymptotics as $t \to 0$ yields new information of the form

$$\mathrm{L}\Gamma(f) \geq 2\,\Gamma(f, \mathrm{L}f) + 2\,\Gamma(f).$$

Recall now the Γ_2 operator defined in (1.16.1), p. 71, by

$$\Gamma_2(f) = \Gamma_2(f, f) = \frac{1}{2}\big[\mathrm{L}\Gamma(f) - 2\,\Gamma(f, \mathrm{L}f)\big].$$

The preceding inequality expresses that $\Gamma_2(f) \geq \Gamma(f)$ for every f in $\mathcal{A}_0$. In other words, the generator L satisfies the curvature condition $CD(1, \infty)$. In this Ornstein-Uhlenbeck example, and as already presented in Sect. 2.7.1, p. 103, this curvature condition may be checked directly, since for every sufficiently smooth function $f : \mathbb{R}^n \to \mathbb{R}$,

$$\Gamma_2(f) = |\nabla\nabla f|^2 + |\nabla f|^2 \geq |\nabla f|^2 = \Gamma(f).$$

In particular, the curvature condition $CD(1, \infty)$ is independent of the dimension n of the underlying state space $\mathbb{R}^n$. The point here is that the commutation relation (4.7.1) is formally equivalent to the curvature condition $CD(1, \infty)$.

The fact that the invariant measure μ is reversible does not play any particular role in the preceding developments, neither does the fact that it is a probability measure. Integration by parts has not been used either. The calculus is only developed for the heat kernel measures $p_t(x, dy)$. In particular, the preceding applies

4.7 Local Poincaré Inequalities

in a completely similar way to the heat (Brownian) semigroup on $\mathbb{R}^n$ with the classical Laplacian $\mathrm{L} = \Delta$ as infinitesimal generator (cf. Sect. 2.1, p. 78). In this case $P_t \nabla = \nabla P_t$, and hence, for every function f in the corresponding algebra $\mathcal{A}_0$ (smooth compactly supported functions for example), and every $t \geq 0$,

$$\Gamma(P_t f) = |\nabla P_t f|^2 \leq P_t(|\nabla f|^2) = P_t(\Gamma(f)).$$

As for (4.7.2), and its reverse form (4.7.3), we deduce that, for every $t \geq 0$,

$$2t\,\Gamma(P_t f) \leq P_t(f^2) - (P_t f)^2 \leq 2t\,P_t(\Gamma(f)). \tag{4.7.4}$$

Differentiating at $t = 0$ yields here that $\Gamma_2(f) \geq 0$ (which is of course immediate since $\Gamma_2(f) = |\nabla\nabla f|^2$), expressing the curvature condition $CD(0, \infty)$ of the Brownian semigroup.

Remark 4.7.1 Since the kernels $p_t(x, dy)$ are Gaussian measures, the preceding Poincaré inequalities (4.7.4) and (4.7.2) are just variations (with different means and variances) of the Poincaré inequality for the standard Gaussian measure (Proposition 4.1.1) applied to functions $y \mapsto f(x + ty)$. The approach is however completely different, emphasizing interpolation along the semigroup, the commutation relations and the connection with the Γ_2 operator and curvature conditions.

4.7.2 Local Poincaré Inequalities

On the basis of the Brownian and Ornstein-Uhlenbeck semigroup examples, we next present a general formulation of the preceding heat kernel inequalities following the discussion on gradient bounds and curvature conditions in Corollary 3.3.19, p. 163 (in particular, the equivalence between (i) and (ii) below is part of the latter). Recall the curvature condition $CD(\rho, \infty)$ expressing that $\Gamma_2(f) \geq \rho\,\Gamma(f)$ for every $f \in \mathcal{A}_0$, or equivalently $f \in \mathcal{A}$. All the local inequalities are stated for functions in the algebra $\mathcal{A}_0$, but actually extend to functions in $\mathcal{A}$ according to the extension procedures described in Sect. 3.3, p. 151. Moreover, the reverse inequality (4.7.6) (for $t > 0$) extends to bounded measurable functions.

Theorem 4.7.2 (Local Poincaré inequalities) *Let (E, μ, Γ) be a Markov Triple with semigroup $\mathbf{P} = (P_t)_{t \geq 0}$. The following assertions are equivalent.*

(i) *The curvature condition $CD(\rho, \infty)$ holds for some $\rho \in \mathbb{R}$.*
(ii) *For every function f in $\mathcal{A}_0$, and every $t \geq 0$,*

$$\Gamma(P_t f) \leq e^{-2\rho t} P_t(\Gamma(f)).$$

(iii) *For every function f in $\mathcal{A}_0$, and every $t \geq 0$,*

$$P_t(f^2) - (P_t f)^2 \leq \frac{1 - e^{-2\rho t}}{\rho} P_t(\Gamma(f)). \tag{4.7.5}$$

(iv) *For every function f in $\mathcal{A}_0$, and every $t \geq 0$,*

$$P_t(f^2) - (P_t f)^2 \geq \frac{e^{2\rho t} - 1}{\rho} \Gamma(P_t f). \tag{4.7.6}$$

Further equivalent descriptions of the curvature condition $CD(\rho, \infty)$ will be discussed throughout this monograph, in particular in the next chapter on logarithmic Sobolev inequalities (cf. Theorem 5.5.2, p. 259). Both $\frac{1-e^{-2\rho t}}{\rho}$ and $\frac{e^{2\rho t}-1}{\rho}$ have to be understood as $2t$ when $\rho = 0$.

Proof As mentioned before, the implication from (i) to (ii) is the content of the gradient bound of Corollary 3.3.19, p. 163. The proof from the gradient bound (ii) to the local Poincaré inequalities (iii) and (iv) has been discussed above for the Brownian and Ornstein-Uhlenbeck semigroups and is similar here for any value of ρ. In the same way, the local Poincaré inequality (iii) and its reverse form (iv) yield in the limit as $t \to 0$ the curvature condition $CD(\rho, \infty)$. Indeed, using a second order Taylor expansion

$$P_t h = h + t \operatorname{L} h + \frac{t^2}{2} \operatorname{L}^2 h + o(t^2)$$

on (iii) for example, the left-hand side is given by

$$t \operatorname{L}(f^2) + \frac{t^2}{2} \operatorname{L}^2(f^2) - t^2 (\operatorname{L} f)^2 - 2tf \operatorname{L} f - t^2 f \operatorname{L}^2 f + o(t^2)$$

while the right-hand side reads

$$2t \, \Gamma(f) - 2\rho t^2 \Gamma(f) + 2t^2 \operatorname{L} \Gamma(f) + o(t^2).$$

By definition of $\Gamma(f)$, the first order terms vanish, yielding

$$\frac{1}{2} \operatorname{L}^2(f^2) - (\operatorname{L} f)^2 - f \operatorname{L}^2 f \leq -2\rho \, \Gamma(f) + 2 \operatorname{L} \Gamma(f).$$

Developing further the chain rule, we obtain

$$2 \Gamma(f, \operatorname{L} f) + \operatorname{L} \Gamma(f) \leq -2\rho \, \Gamma(f) + 2 \operatorname{L} \Gamma(f)$$

which amounts to $\Gamma_2(f) \geq \rho \Gamma(f)$ by definition of $\Gamma_2(f)$. The proof of the theorem is complete. □

Note that in general little is known about the heat kernel measures $p_t(x, dy)$, $t \geq 0$, $x \in E$. However, under a curvature condition, Theorem 4.7.2 ensures that they satisfy Poincaré-type inequalities, and thus for example, by Proposition 4.4.2, have exponential tails. On the other hand, as mentioned in the Ornstein-Uhlenbeck example, the reverse Poincaré inequalities (4.7.6) of Theorem 4.7.2 provide a useful quantitative regularization property in the form of

$$\Gamma(P_t f) \leq \frac{\rho}{e^{2\rho t} - 1} \qquad (4.7.7)$$

for any bounded measurable function f such that $|f| \leq 1$ and any $t > 0$ (cf. Proposition 8.6.1, p. 422, for a somewhat sharper bound). In particular, $\Gamma(P_t f) = O(t^{-1/2})$ as $t \to 0$.

Remark 4.7.3 It is worth mentioning that the implications from (ii) to (iii) and (iv) of Theorem 4.7.2 work similarly if (ii) only holds up to some constant $K \geq 1$, that is, for every $t \geq 0$ and all functions f in $\mathcal{A}_0$,

$$\Gamma(P_t f) \leq K e^{-2\rho t} P_t(\Gamma(f)).$$

The proof is entirely similar, and (iii), for example, then simply reads

$$P_t(f^2) - (P_t f)^2 \leq \frac{K(1 - e^{-2\rho t})}{\rho} P_t(\Gamma(f)).$$

This observation is of interest for some hypo-coercive examples as described in Remark 3.3.22, p. 165. The point is that only $K = 1$ transfers back to the curvature condition $CD(\rho, \infty)$.

Remark 4.7.4 The preceding proof shows that each of the assertions of Theorem 4.7.2 are equivalent to the statement that there exist a $t_0 > 0$ and a function $c(t) = 2t - 2\rho t^2 + o(t^2)$ such that, for any $t \in (0, t_0)$ and any f in $\mathcal{A}_0$,

$$P_t(f^2) - (P_t f)^2 \leq c(t) P_t(\Gamma(f))$$

(respectively

$$P_t(f^2) - (P_t f)^2 \geq c(t) \Gamma(P_t f).$$

4.8 Poincaré Inequalities Under a Curvature-Dimension Condition

The preceding section describes Poincaré inequalities under the curvature condition $CD(\rho, \infty)$ for the heat kernel measures $p_t(x, dy)$ of a Markov semigroup $(P_t)_{t \geq 0}$. In the well behaved cases, typically when $\rho > 0$ (as for the Ornstein-Uhlenbeck semigroup discussed above), these inequalities extend as $t \to \infty$ to inequalities for the invariant measure μ as a basic instance of ergodicity as emphasized earlier in Sect. 1.8, p. 32. In particular, as discussed there, only the minimal connexity condition (if $\Gamma(f) = 0$ then f is μ-almost everywhere constant) is required for this to be the case. Moreover, it has been shown in Theorem 3.3.23, p. 165, that whenever $\rho > 0$, the invariant measure μ is always finite (and thus a probability measure by normalization) and ergodicity then ensures that $P_\infty f = \int_E f d\mu$ for every f in $\mathbb{L}^2(\mu)$. Under this assumption, the local Poincaré inequalities (4.7.5) for $p_t(x, dy)$ of Theorem 4.7.2 extend as $t \to \infty$ to a Poincaré inequality $P(\frac{1}{\rho})$ for the invariant (probability) measure μ. The following statement summarizes this consequence. Although it is formally contained in the stronger Theorem 4.8.4 below (corresponding to $n = \infty$), it is worth stating independently.

Proposition 4.8.1 (Poincaré inequality under $CD(\rho, \infty)$) *Under the curvature condition $CD(\rho, \infty)$ with $\rho > 0$, the Markov Triple (E, μ, Γ) satisfies a Poincaré inequality $P(C)$ with constant $C = \frac{1}{\rho}$. That is, for every function $f \in \mathcal{D}(\mathcal{E})$,*

$$\mathrm{Var}_\mu(f) \leq \frac{1}{\rho} \mathcal{E}(f).$$

This proposition covers the example of the standard Gaussian measure on $\mathbb{R}^n$ for which $\rho = 1$ thus yielding the optimal Poincaré inequality of Proposition 4.1.1. It actually covers more generally probability measures $d\mu = e^{-W} dx$ on $\mathbb{R}^n$ with respect to the standard carré du champ operator where W is a smooth potential such that $\nabla \nabla W \geq \rho \,\mathrm{Id}$ (as symmetric matrices) for some $\rho > 0$ (equivalently, $W(x) - \frac{\rho |x|^2}{2}$ is convex). Indeed, according to Sect. 1.11.3, p. 46, the Markov diffusion operator with invariant and symmetric measure μ is of the form $Lf = \Delta f - \nabla W \cdot \nabla f$ with carré du champ operator $\Gamma(f) = |\nabla f|^2$ on smooth functions f. As a consequence of (1.16.5), p. 72, or an immediate verification in this simple case, on every smooth function $f : \mathbb{R}^n \to \mathbb{R}$,

$$\Gamma_2(f) = |\nabla \nabla f|^2 + \nabla \nabla W(\nabla f, \nabla f)$$

where $\nabla \nabla W$ is the Hessian of W. By the convexity assumption,

$$\nabla \nabla W(\nabla f, \nabla f) \geq \rho |\nabla f|^2.$$

In other words, the Markov Triple consisting of $\mathbb{R}^n$ with the measure $d\mu = e^{-W} dx$ for which $\nabla \nabla W \geq \rho \,\mathrm{Id}$ and the standard carré du champ operator therefore satisfies the curvature condition $CD(\rho, \infty)$. We present this conclusion as an independent statement. It will also appear as a consequence of the stronger Brascamp-Lieb inequality below (Theorem 4.9.1).

Corollary 4.8.2 *Let $d\mu = e^{-W} dx$ be a probability measure on the Borel sets of $\mathbb{R}^n$ where $W : \mathbb{R}^n \to R$ is a smooth potential such that $\nabla \nabla W \geq \rho \,\mathrm{Id}$ for some $\rho > 0$. Then μ satisfies the Poincaré inequality $P(\frac{1}{\rho})$.*

A similar statement, with the same proof, holds on a weighted Riemannian manifold $(M, \mathfrak{g})$ for the operator $L = \Delta_\mathfrak{g} - \nabla W \cdot \nabla$ with the reversible (probability)

4.8 Poincaré Inequalities Under a Curvature-Dimension Condition

measure μ having density e^{-W} with respect to the Riemannian measure under the curvature condition $\mathrm{Ric}(L) = \mathrm{Ric}_\mathfrak{g} + \nabla\nabla W \geq \rho\,\mathfrak{g}$ with $\rho > 0$ (cf. Sect. 1.16, p. 70, and Sect. C.6, p. 513).

The main purpose of this section is to investigate Poincaré inequalities for a Markov Triple (E, μ, Γ) under the stronger curvature-dimension condition $CD(\rho, n)$ of Definition 3.3.14, p. 159, for some $\rho > 0$ and some finite dimension $n \geq 1$. Recall that this condition expresses that

$$\Gamma_2(f) \geq \rho\,\Gamma(f) + \frac{1}{n}(Lf)^2$$

for all $f \in \mathcal{A}_0$ (or $\mathcal{A}$). In particular, it will be observed how the (finite) dimension will improve upon the preceding Poincaré constant in Proposition 4.8.1.

To this end, we start from the $CD(\rho, n)$ condition for $\rho > 0$ and $n > 1$ integrated with respect to μ to get that, for every f in $\mathcal{A}_0$,

$$\int_E \Gamma_2(f)d\mu \geq \rho \int_E \Gamma(f)d\mu + \frac{1}{n}\int_E (Lf)^2 d\mu.$$

Now, by integration by parts and the construction of the operators Γ and Γ_2 (cf. Sect. 3.3, p. 151),

$$\int_E \Gamma(f)d\mu = \int_E f(-Lf)d\mu \quad \text{and} \quad \int_E \Gamma_2(f)d\mu = \int_E (Lf)^2 d\mu,$$

so that the preceding inequality reads

$$\int_E (Lf)^2 d\mu \geq \frac{\rho n}{n-1}\int_E f(-Lf)d\mu.$$

As a first intuitive approach, apply this inequality to a (non-zero) eigenfunction f of $-L$ with (positive) eigenvalue λ to see that

$$\lambda\left(\lambda - \frac{\rho n}{n-1}\right) \geq 0.$$

Hence, if $\lambda \neq 0$, $\lambda \geq \frac{\rho n}{n-1}$. By the correspondence between spectral gap and Poincaré inequality, this bound improves upon Proposition 4.8.1 (note that the two Poincaré constants agree when $n = \infty$).

As a consequence of the preceding, the first non-trivial eigenvalue λ_1 of the Laplace operator on an n-dimensional (compact) Riemannian manifold with Ricci curvature bounded from below by $\rho > 0$ satisfies

$$\lambda_1 \geq \frac{\rho n}{n-1}, \tag{4.8.1}$$

which is the essence of *Lichnerowicz' eigenvalue comparison Theorem*. For the Laplacian on the sphere $\mathbb{S}^n$ in $\mathbb{R}^{n+1}$ (cf. Sect. 2.2, p. 81), or for the symmetric Jacobi operator with dimension n (cf. Sect. 2.7.4, p. 113), the preceding lower bound

$\frac{\rho n}{n-1}$ of the spectral gap is optimal since these two models both satisfy a curvature-dimension condition $CD(n-1, n)$, yielding a Poincaré constant equal to $\frac{1}{n}$ which corresponds to the inverse of the first non-trivial eigenvalue. In particular, Lichnerowicz's Theorem (4.8.1) compares the first non-trivial eigenvalue of the Laplacian on a (compact) n-Riemannian manifold having a strictly positive lower bound on the Ricci curvature by the one of the sphere with the same dimension and curvature. Obata's Theorem expresses conversely that when there is equality in (4.8.1), the manifold is isometric to this sphere. Note that the result also holds on $\mathbb{S}^1$ ($n = 1$) by Proposition 4.5.5 via a simple dilation from $[0, 1]$ to $[0, 2\pi]$. But this may appear as a simple coincidence, since the condition $CD(0, m)$, $m \geq 1$, does not imply a spectral gap inequality in general.

The preceding spectral argument implicitly assumes that the spectrum of L is discrete. (We will actually establish later in Corollary 6.8.2, p. 306, that this is indeed the case under these hypotheses.) We next provide a less formal approach relying on semigroup tools and heat flow monotonicity. Along these lines, we start with a useful dual description of the Poincaré inequality (also implicit from the previous argument) which develops the heat flow argument via the second derivative of the variance along the semigroup, or equivalently the first derivative of the Dirichlet form, giving rise to the Γ_2 operator.

Proposition 4.8.3 *A Markov Triple (E, μ, Γ) satisfies a Poincaré inequality $P(C)$ for some $C > 0$ if and only if*

$$\mathcal{E}(f) = \int_E \Gamma(f) d\mu \leq C \int_E (Lf)^2 d\mu \qquad (4.8.2)$$

for every function f in the domain $\mathcal{D}(L)$.

Proof As already used above, for a given $f \in \mathcal{A}_0$, consider the function $\Lambda(t) = \int_E (P_t f)^2 d\mu$, $t \geq 0$, for which

$$\Lambda'(t) = -2 \int_E \Gamma(P_t f) d\mu \quad \text{and} \quad \Lambda''(t) = 4 \int_E (LP_t f)^2 d\mu.$$

Under (4.8.2), $\Lambda''(t) \geq -\frac{2}{C} \Lambda'(t)$, $t \geq 0$, which integrates into the Poincaré inequality $P(C)$ since

$$\text{Var}_\mu(f) = -\int_0^\infty \Lambda'(t) dt \leq \frac{C}{2} \int_0^\infty \Lambda''(t) dt = -\frac{C}{2} \Lambda'(0) = C \int_E \Gamma(f) d\mu.$$

Conversely, for every $f \in \mathcal{D}(L)$ with mean zero as it can be assumed,

$$\int_E \Gamma(f) d\mu = \int_E f(-Lf) d\mu \leq \text{Var}_\mu(f)^{1/2} \left(\int_E (Lf)^2 d\mu \right)^{1/2}$$

by the Cauchy-Schwarz inequality, and the conclusion follows from the application of $P(C)$. □

Proposition 4.8.3 is actually a purely spectral statement only relying on ergodicity. Under the ESA property of Definition 3.1.10, p. 134, (4.8.2) for every $f \in \mathcal{A}_0$ extends immediately to any $f \in \mathcal{D}(L)$. In this case, it is equivalent to

$$\mathcal{E}(f) \leq C \int_E \Gamma_2(f) d\mu \qquad (4.8.3)$$

which is valid for any $f \in \mathcal{A}_0$. Moreover, under a curvature condition $CD(\rho, \infty)$ for some $\rho \in \mathbb{R}$, the Γ_2 operator and (4.8.3) extend to $\mathcal{D}(L)$.

The following main statement under the curvature-dimension condition $CD(\rho, n)$, which includes in particular the Lichnerowicz bound (4.8.1), follows.

Theorem 4.8.4 (Poincaré inequality under $CD(\rho, n)$) *Under the curvature-dimension condition $CD(\rho, n)$, $\rho > 0$, $n > 1$, the Markov Triple (E, μ, Γ) satisfies a Poincaré inequality $P(C)$ with constant $C = \frac{n-1}{\rho n}$.*

Proof Integration of the $CD(\rho, n)$ condition yields as before

$$\int_E \Gamma_2(f) d\mu \geq \rho \int_E \Gamma(f) d\mu + \frac{1}{n} \int_E (Lf)^2 d\mu$$

for any f in $\mathcal{A}_0$. Since $\int_E (Lf)^2 d\mu = \int_E \Gamma_2(f) d\mu$, (4.8.2) of Proposition 4.8.3 holds with $C = \frac{n-1}{\rho n}$. The theorem is established. □

4.9 Brascamp-Lieb Inequalities

In this section, we deal with a stronger variant of Poincaré inequalities, known as Brascamp-Lieb-type inequalities.

4.9.1 The Matrix Brascamp-Lieb Inequality

We first state and prove the classical version in $\mathbb{R}^n$.

Theorem 4.9.1 (Brascamp-Lieb inequality) *For a probability measure $d\mu = e^{-W} dx$ on $\mathbb{R}^n$ for which the smooth potential $W : \mathbb{R}^n \to \mathbb{R}$ is strictly convex,*

$$\mathrm{Var}_\mu(f) \leq \int_{\mathbb{R}^n} (\nabla\nabla W)^{-1}(\nabla f, \nabla f) d\mu$$

for every smooth compactly supported function f on $\mathbb{R}^n$, where $(\nabla\nabla W)^{-1}$ is the inverse of the Hessian of W.

In this Euclidean context, for the operator $L = \Delta - \nabla W \cdot \nabla$ with $\Gamma(f) = |\nabla f|^2$, the curvature condition $CD(\rho, \infty)$ expresses that $\nabla \nabla W \geq \rho \operatorname{Id}$ in the sense of symmetric matrices. Hence, if $\rho > 0$,

$$(\nabla \nabla W)^{-1}(\nabla f, \nabla f) \leq \frac{1}{\rho} |\nabla f|^2$$

so that the Brascamp-Lieb inequality improves upon Corollary 4.8.2 in this context.

To address the proof of Theorem 4.9.1, the following general lemma of independent interest will be useful.

Lemma 4.9.2 *Let A_1 and A_2 be symmetric operators defined on a common dense domain $\mathcal{D}_0$ in a Hilbert space $\mathcal{H}$. Assume that for some $\varepsilon > 0$ and any $u \in \mathcal{D}_0$,*

$$\varepsilon \langle u, u \rangle \leq \langle A_1 u, u \rangle \leq \langle A_2 u, u \rangle.$$

Consider then the self-adjoint Friedrichs extensions of A_1 and A_2 (denoted in the same way) with respective domains $\mathcal{D}(A_1)$ and $\mathcal{D}(A_2)$. The inverse operators A_1^{-1} and A_2^{-1}, defined by spectral decomposition, are bounded with norms bounded from above by $\frac{1}{\varepsilon}$. Then, for any $u \in \mathcal{H}$,

$$\langle A_2^{-1} u, u \rangle \leq \langle A_1^{-1} u, u \rangle. \tag{4.9.1}$$

Proof It is enough to prove (4.9.1) for $u = A_2 v$ with $v \in \mathcal{D}(A_2)$. Then the inequality boils down to $\langle v, A_2 v \rangle \leq \langle A_1^{-1} A_2 v, A_2 v \rangle$. Now, it is easily checked that $\mathcal{D}(A_2) \subset \mathcal{D}(A_1)$, and setting $B = A_2 - A_1$, the latter then reads $\langle A_1^{-1} Bv, A_2 v \rangle \geq 0$. In other words,

$$\langle A_1^{-1} Bv, Bv \rangle + \langle Bv, v \rangle \geq 0$$

which is immediate since A_1 and B are positive. □

Proof of Theorem 4.9.1 The proof is only sketched. It uses in particular the extension of the operator $L = \Delta - \nabla W \cdot \nabla$ and its associated semigroup $(P_t)_{t \geq 0}$ to vector-valued functions, and everything will ultimately rely on the commutation property, easily checked on smooth functions f,

$$\nabla L f = (L - \nabla \nabla W) \nabla f \tag{4.9.2}$$

(where $\nabla f = (\partial_i f)_{1 \leq i \leq n}$).

To give a better sense to the latter, consider the Hilbert space $\mathcal{H}$ of vectors $u = (f_1, \ldots, f_n)$ of functions in $\mathbb{L}^2(\mu)$ with norm

$$\|u\|^2 = \int_{\mathbb{R}^n} \sum_{i=1}^n f_i^2 \, d\mu.$$

4.9 Brascamp-Lieb Inequalities

Denote by $\langle u, v \rangle$ the scalar product in $\mathcal{H}$. For functions f_i, $i = 1, \ldots, n$, smooth and compactly supported on $\mathbb{R}^n$, introduce the operator

$$\widetilde{L}(f_1, \ldots, f_n) = (Lf_1, \ldots, Lf_n).$$

The operator $\widetilde{L}$ is symmetric in $\mathcal{H}$, and for $u = (u_1, \ldots, u_n)$ and $v = (v_1, \ldots, v_n)$ in $\mathcal{H}$,

$$\langle -\widetilde{L}u, v \rangle = \int_{\mathbb{R}^n} \nabla u \cdot \nabla v \, d\mu = \int_{\mathbb{R}^n} \sum_{i=1}^n \nabla u_i \cdot \nabla v_i \, d\mu$$

so that $-\widetilde{L}$ is positive.

Denote by $\mathcal{M} = \mathcal{M}(x)$ multiplication by the matrix $\nabla\nabla W = \nabla\nabla W(x)$ on elements of $\mathcal{H}$. The aim is to apply Lemma 4.9.2 with $A_1 = \mathcal{M}$ and $A_2 = \mathcal{M} - \widetilde{L}$, or rather $A_1 = \varepsilon + \mathcal{M}$ and $A_2 = \varepsilon + \mathcal{M} - \widetilde{L}$ for some $\varepsilon > 0$. The operators $\mathcal{M}$ and $\mathcal{M} - \widetilde{L}$ are symmetric on $\mathcal{H}$. Since $\nabla\nabla W$ is strictly positive-definite, $\mathcal{M}$ is a positive operator. In particular, $\mathcal{M} - \widetilde{L}$ is also positive. Therefore, by Lemma 4.9.2,

$$\langle (\varepsilon + \mathcal{M} - \widetilde{L})^{-1} u, u \rangle \leq \langle (\varepsilon + \mathcal{M})^{-1} u, u \rangle \quad (4.9.3)$$

for every $u \in \mathcal{H}$.

Now let $f \in \mathcal{A}_0$, where $\mathcal{A}_0$ is the set of smooth compactly supported functions on $\mathbb{R}^n$. Set

$$g_\varepsilon = (\varepsilon - L)^{-1/2} f = c \int_0^\infty e^{-\varepsilon s} s^{-1/2} P_s f \, ds$$

for which

$$\int_{\mathbb{R}^n} f^2 d\mu = \varepsilon \int_{\mathbb{R}^n} g_\varepsilon^2 d\mu + \int_{\mathbb{R}^n} |\nabla g_\varepsilon|^2 d\mu.$$

(Here, the operator ∇ has to be properly extended from $\mathcal{A}_0$ to $\mathcal{D}(\mathcal{E})$ as an operator with values in $\mathcal{H}$, but this is immediate.) Moreover, for $f \in \mathcal{A}_0$, the identity $\nabla L f = (\widetilde{L} - \mathcal{M})\nabla$ from (4.9.2) extends to

$$\nabla g_\varepsilon = \nabla\big((\varepsilon - L)^{-1/2} f\big) = (\varepsilon + \mathcal{M} - \widetilde{L})^{1/2} \nabla f.$$

Indeed, if $(\widehat{P}_t)_{t \geq 0}$ is the contraction semigroup on $\mathcal{H}$ with generator $\widetilde{L} - \mathcal{M}$, the latter relies on the commutation $\widehat{P}_t \nabla f = \nabla P_t f$, which is valid for $f \in \mathcal{A}_0$ and from the fact that $\nabla P_t f \in \mathcal{D}(\widetilde{L} - \mathcal{M})$, which is easily checked from the definition of $\mathcal{D}(\mathcal{M} - \widetilde{L})$ and the techniques developed in Chap. 3. As a consequence, together with (4.9.3),

$$\int_{\mathbb{R}^n} f^2 d\mu = \varepsilon \int_{\mathbb{R}^n} g_\varepsilon^2 d\mu + \langle (\varepsilon + \mathcal{M} - \widetilde{L})^{-1} \nabla f, \nabla f \rangle$$

$$\leq \varepsilon \int_{\mathbb{R}^n} g_\varepsilon^2 d\mu + \langle (\varepsilon + \mathcal{M})^{-1} \nabla f, \nabla f \rangle.$$

It remains to let ε go to 0, with the further observation that whenever $\int_{\mathbb{R}^n} f \, d\mu = 0$, then $\lim_{\varepsilon \to 0} \varepsilon \|g_\varepsilon\|_2^2 = 0$. Indeed, $\lim_{s \to \infty} \|P_s f\|_2 = 0$ and

$$\|\varepsilon^{1/2} g_\varepsilon\|_2 = c \left\| \int_0^\infty e^{-s} s^{-1/2} P_{s/\varepsilon} f \, ds \right\|_2 \leq c \int_0^\infty e^{-s} s^{-1/2} \|P_{s/\varepsilon} f\|_2 \, ds$$

which tends to 0 with ε by dominated convergence. The proof is complete. □

The Brascamp-Lieb inequality (Theorem 4.9.1) admits a simple extension to elliptic operators on a manifold $(M, \mathfrak{g})$ which we briefly describe now. Recall from Sect. 1.11.3, p. 46, that in the elliptic case, $L = \Delta_\mathfrak{g} - \nabla W \cdot \nabla$, where the reversible (probability) measure μ has density e^{-W} with respect to the Riemannian measure. Recall also (Sect. 1.16, p. 70) that in this case

$$\Gamma_2(f) = |\nabla \nabla f|^2 + \mathrm{Ric}(L)(\nabla f, \nabla f),$$

with $\mathrm{Ric}(L) = \mathrm{Ric}_\mathfrak{g} + \nabla \nabla W$. The next statement is then the analogue of the Brascamp-Lieb inequality in this context.

Theorem 4.9.3 *In this Riemannian setting, if the tensor $\mathrm{Ric}(L)$ is strictly positive everywhere, denoting by $\mathrm{Ric}(L)^{-1}$ its inverse tensor, then for every smooth compactly supported function f on M,*

$$\mathrm{Var}_\mu(f) \leq \int_M \mathrm{Ric}(L)^{-1}(\nabla f, \nabla f) \, d\mu.$$

The proof is rather similar to the Euclidean case of Theorem 4.9.1, replacing all vectors u by 1-forms w and the operator $\widetilde{L}$ by the operator $\widetilde{L}$ given in a local system of coordinates (and with Einstein's summation notation) by

$$\widetilde{L} w_j = \nabla^i \nabla_i w_j - \nabla^i(W) \cdot \nabla_i w_j.$$

The same commutation formula holds

$$d L f = \bigl(L - \mathrm{Ric}(L)\bigr) df,$$

and the argument is then completed along the same lines.

4.9.2 Probabilistic Representations

A consequence of the Brascamp-Lieb inequality of Theorem 4.9.1, and of Theorem 4.9.3 in the manifold setting M, is that if $\rho(x)$ is the least eigenvalue, at the point x, of the matrix $\nabla \nabla W(x)$ (in case of Theorem 4.9.1), or $\mathrm{Ric}(L)$ (in case of Theorem 4.9.3), then, for every sufficiently smooth function $f : M \to \mathbb{R}$,

$$\mathrm{Var}_\mu(f) \leq \int_M \frac{1}{\rho(x)} |\nabla f|^2 d\mu. \tag{4.9.4}$$

4.9 Brascamp-Lieb Inequalities

This inequality may then be related to Proposition 3.2.5, p. 149, and the semigroup $(\widehat{P}_t)_{t \geq 0}$ with generator $\widehat{L} = L - \rho\,\mathrm{Id}$. Indeed, by the development $f = -\int_0^\infty LP_t f\,dt$ of a smooth mean-zero function f and integration by parts,

$$\mathrm{Var}_\mu(f) = \int_0^\infty \int_M \nabla f \cdot \nabla P_t f\, d\mu\, dt.$$

Now, by means of the gradient bound

$$|\nabla P_t f| \leq \widehat{P}_t(|\nabla f|) \tag{4.9.5}$$

of Proposition 3.2.5, p. 149, it holds that

$$\mathrm{Var}_\mu(f) \leq \int_M \int_0^\infty |\nabla f| \widehat{P}_t(|\nabla f|) dt\, d\mu = \int_M |\nabla f|(-\widehat{L})^{-1}(|\nabla f|)\, d\mu$$

from which (4.9.4) follows by an analysis similar to the one developed for Theorem 4.9.1.

The gradient commutation bound (4.9.5) actually has an even more precise form of a probabilistic nature, which may prove useful in a wide variety of contexts, although it is not so easy to express. Namely, starting from the commutation formula (4.9.2), the semigroup $(\widetilde{Q}_t)_{t \geq 0}$ with generator $\widetilde{L} - \mathcal{M}$ acting on vector-valued functions (or on 1-forms in the manifold case) can be rewritten as the (exact) formula

$$\nabla P_t f = Q_t \nabla f, \quad t \geq 0$$

(on suitable functions $f : M \to \mathbb{R}$). Therefore, (4.9.5) amounts to $|Q_t u| \leq \widehat{P}_t(|u|)$ for vector-valued functions u. A direct approach to this proceeds via the Feynman-Kac-type probabilistic representations as described in Sect. 1.15.6, p. 63. Let us briefly illustrate the principle in the flat Euclidean case $M = \mathbb{R}^n$ with $L = \Delta - \nabla W \cdot \nabla$, corresponding to Theorem 4.9.1, borrowing the notation from its statement and proof.

Indeed, the semigroup $(Q_t)_{t \geq 0}$ admits a nice probabilistic interpretation in terms of the Markov process $\{X_t^x\,;\, t \geq 0,\, x \in \mathbb{R}^n\}$ with semigroup $(P_t)_{t \geq 0}$ and generator L. Consider its vector-valued extension (with generator $\widetilde{L}$) given on a vector-valued function $u = (f_1, \ldots, f_n)$ by

$$\widetilde{P}_t u(x) = \mathbb{E}_x\big(u(X_t)\big) = \big(\mathbb{E}_x\big(f_1(X_t)\big), \ldots, \mathbb{E}_x\big(f_n(X_t)\big)\big), \quad t \geq 0,\ x \in \mathbb{R}^n.$$

The vector-valued version of the Feynman-Kac formula of Sect. 1.15.6, p. 63, then takes the form

$$Q_t u(x) = \mathbb{E}_x\big(A_t\big(u(X_t)\big)\big) \tag{4.9.6}$$

where now A_t, $t \geq 0$, is a (random) matrix, being a solution of the ordinary differential equation $\partial_t A_t = -A_t \mathcal{M}(X_t)$, with $A_0 = \mathrm{Id}$. (Note that this is not in general $\exp(-\int_0^t \mathcal{M}(X_s)\,ds)$ since the matrices $\mathcal{M}(X_s)$ do not commute, and the latter expression is not even symmetric even if the matrices $\mathcal{M}(X_s)$ are.) By a standard lemma from ordinary differential equation theory, if for any $s \geq 0$, $\langle \mathcal{M}(X_s)u, u \rangle \geq \rho(X_s)|u|^2$, then the matrix A_t satisfies

$$|A_t u| \leq e^{-\int_0^t \rho(X_s)ds} |u|.$$

As a consequence, the announced inequality $|Q_t u| \leq \widehat{P}_t(|u|)$ holds.

In the manifold setting (that is, as soon as the second order terms in L are no longer constant), the preceding construction requires additional tools such as stochastic parallel transport in order to give a reasonable meaning to the various expressions. Indeed, the matrix operator $\text{Ric}(L)(X_s)$ acts naturally on the tangent space above X_s, and one has to carry back these operations on the same vector space (in general the tangent space above X_0), causing extra trouble (and serious technicalities) which will not be developed here.

4.10 Further Spectral Inequalities

In the last section of this chapter, we investigate some further spectral properties related to symmetric Markov semigroups. Essentially, Poincaré inequalities concern Markov semigroups which are symmetric with respect to a given probability measure μ. In this context, the constant function $\mathbb{1}$ is always an eigenvector associated with the 0 eigenvalue. Hence, Poincaré inequalities explore the gap between 0 and the rest of the spectrum, and more precisely describe when the spectrum of $-L$ is included in $\{0\} \cup [\lambda_1, \infty)$ for some $\lambda_1 > 0$. But it may happen that the spectrum itself is included in $[\lambda_0, \infty)$ for some $\lambda_0 > 0$. This is in general the case for sub-Markov semigroups, such as those related to Dirichlet boundary conditions on compact sets, but it may also happen when the measure μ is infinite. These questions are addressed in Sect. 4.10.1.

Furthermore, it is often of great interest to know when the spectrum is discrete, that is when the essential spectrum is empty. Recall (cf. Appendix A) that this property conceals several different facts. First that the spectrum is purely punctual, and that to every point in the spectrum corresponds at least one non-zero eigenfunction (which is in $\mathbb{L}^2(\mu)$ by definition). Second, that every point λ in the spectrum is isolated, that is, there exists an $\varepsilon > 0$ such that no point of the spectrum lies in $(\lambda - \varepsilon, \lambda + \varepsilon)$. Moreover, for any point in the spectrum, the corresponding eigenspace is finite-dimensional. Section 4.10.2 investigates in this regard the bottom of the essential spectrum, providing simple criteria related to Nash-type inequalities (further developed in Chap. 7).

In many situations, the essential spectrum depends only on "what happens at infinity". To give a precise meaning to this statement, we describe in Sect. 4.10.3 the notion of a Persson operator, which leads once again to useful criteria for emptiness of the essential spectrum.

We refer to Appendix A, and more precisely to Sect. A.4, p. 478, for the various notions and definitions of the spectrum of a positive self-adjoint operator. The

4.10 Further Spectral Inequalities   221

first two sub-sections below are concerned with the minimal Markov triple structure (E, μ, Γ) as introduced in the first part of Sect. 3.1, p. 120, and focus on the self-adjoint operator L associated with the triple.

4.10.1 Bottom of the Spectrum

As already mentioned, Poincaré inequalities make sense only when the reference measure μ is finite (in fact a probability measure), and when the semigroup $\mathbf{P} = (P_t)_{t \geq 0}$ is Markov. It could nevertheless be the case, either for infinite measures (or for sub-Markov semigroups), that the following inequality

$$\int_E f^2 d\mu \leq C \mathcal{E}(f) \qquad (4.10.1)$$

holds for some $C > 0$ and some suitable class functions f on E. One such example has already appeared in Lemma 4.2.3 above as a consequence of a Poincaré inequality on a larger set. When such an inequality holds for a class of functions f which is dense in the Dirichlet domain $\mathcal{D}(\mathcal{E})$ of the given Dirichlet form $\mathcal{E}$ (and therefore holds for every function f in it), it amounts to saying that the spectrum of the underlying (opposite) generator $-L$ lies in the interval $[\frac{1}{C}, \infty)$. Here, L is understood as the minimal extension of the operator defined from the Dirichlet form $\mathcal{E}$, that is, its Friedrichs extension as described in Sect. 3.1, p. 120.

In the case of a bounded domain (in $\mathbb{R}^n$ or a manifold), this extension corresponds to the Dirichlet boundary conditions. The connection between the spectrum and (4.10.1) has been illustrated for Sturm-Liouville operators on an interval of the real line in Sect. 4.5.2, giving rise to a useful connection between eigenvalues with Dirichlet or Neumann boundary conditions. This section provides further material on the spectrum of a diffusion operator.

From a spectral analysis point of view, the best constant C in (4.10.1) is $\frac{1}{\lambda_0}$ where λ_0 is the bottom of the spectrum of $-L$ (which is different from the spectral gap related to Poincaré inequalities). This observation applies only when $\mathbb{1}$ is no longer an eigenvector for $-L$ which may indeed happen in a wide variety of situations: either the measure μ is infinite, or the boundary conditions imposed on the domain $\mathcal{D}(L)$ prevent $\mathbb{1}$ from belonging to it, in particular in the case of Dirichlet boundary conditions.

Remark 4.10.1 For an elliptic operator L on an open domain $K \subset \mathbb{R}^n$ with regular boundary, the natural probabilistic interpretation of the constant λ_0 for the operator L with Dirichlet boundary conditions in K may be described in terms of the expectation of the hitting time of the boundary. To fix the ideas, assume that L is the usual Laplace operator Δ in some open relatively compact set $\mathcal{O} \subset \mathbb{R}^n$, and consider the associated semigroup $(P_t^D)_{t \geq 0}$ with Dirichlet boundary conditions on $\mathcal{O}$. If $(X_t^x)_{t \geq 0}$ is Brownian motion starting from $x \in \mathcal{O}$, and if

$$T = \inf\{t \geq 0 \,;\, X_t \notin \mathcal{O}\},$$

then for all $\lambda < \lambda_0$, $\mathbb{E}_x(e^{\lambda T}) < \infty$ while for all $\lambda \geq \lambda_0$, $\mathbb{E}_x(e^{\lambda T}) = \infty$. To understand this result, the basic observation is that the bottom λ_0 of the spectrum of $-\Delta$ corresponds to some strictly positive eigenvector f_0 which vanishes at the boundary (this strictly positive eigenvector corresponds to the ground state for Schrödinger operators and to the Perron-Frobenius eigenvector in the case of finite Markov chains). Then $e^{\lambda_0 t} f_0(X_t)$, $t \geq 0$, defines a martingale, and since f_0 vanishes at the boundary, the finiteness of $\mathbb{E}_x(e^{\lambda_0 T})$ would yield a contradiction. Hence $\mathbb{E}_x(e^{\lambda T}) = \infty$ for every $\lambda \geq \lambda_0$. On the other hand, for any $\lambda < \lambda_0$, there exists a strictly positive eigenvector solution of $-Lf = \lambda f$ (not vanishing identically at the boundary), which may be thought of as the bottom eigenvector on some enlarged set. The same reasoning with the martingale $e^{\lambda t} f(X_t)$, $t \geq 0$, leads to the fact that $\mathbb{E}_x(e^{\lambda T})$ must be finite. This type of argument may be extended to more general settings, replacing the Brownian motion by the Markov process with generator L.

One of the most important geometric examples of an inequality such as (4.10.1) is given by the hyperbolic Laplace operator discussed in Sect. 2.3, p. 88. That is, the state space E is the hyperbolic space $\mathbb{H}^n$ ($n \geq 2$) which may be represented as the upper half-space $\mathbb{R}^{n-1} \times (0, \infty)$ with reference measure $d\mu(x) = x_n^{-n} dx$ and carré du champ operator $\Gamma(f) = x_n^2 |\nabla f|^2$. Here x_n stands for the last coordinate (the one which is strictly positive), dx for the standard Lebesgue measure and $|\nabla f|^2$ for the standard Euclidean carré du champ operator of a smooth function f. Accordingly, the Dirichlet form is given by $\mathcal{E}(f) = \int_{\mathbb{H}^n} \Gamma(f) d\mu$.

Proposition 4.10.2 *For the hyperbolic space $\mathbb{H}^n$ of dimension n, and for any smooth compactly supported function f on $\mathbb{H}^n$,*

$$\int_{\mathbb{H}^n} f^2 d\mu \leq \frac{4}{(n-1)^2} \mathcal{E}(f).$$

Proof The argument is similar to the proof of the Poincaré inequality for the exponential measure (Sect. 4.4.1). In the chosen upper half-space representation, the hyperbolic Laplacian may be explicitly represented as

$$Lf = x_n^2 \Delta f - (n-2) x_n \partial_n f.$$

Now, the coordinate function x_n satisfies $L(x_n) = -(n-2)x_n$ and $\Gamma(x_n) = x_n^2$. Therefore, if f is a smooth and compactly supported function on $\mathbb{H}^n$,

$$\int_{\mathbb{H}^n} f^2 d\mu = \frac{1}{n-2} \int_{\mathbb{H}^n} (-Lx_n) x_n^{-1} f^2 d\mu = \frac{1}{n-2} \int_{\mathbb{H}^n} \Gamma(x_n, x_n^{-1} f^2) d\mu.$$

Now,

$$\Gamma(x_n, x_n^{-1} f^2) = -f^2 x_n^{-2} \Gamma(x_n) + 2f x_n^{-1} \Gamma(x_n, f)$$
$$= -f^2 + 2f x_n^{-1} \Gamma(x_n, f)$$
$$\leq -f^2 + 2|f| \Gamma(f)^{1/2}$$

4.10 Further Spectral Inequalities

since $\Gamma(x_n) = x_n^2$. It follows that

$$\left(1 + \frac{1}{n-2}\right)\int_{\mathbb{H}^n} f^2 d\mu \leq \frac{2}{n-2}\int_{\mathbb{H}^n} |f|\Gamma(f)^{1/2} d\mu$$

$$\leq \frac{2}{n-2}\left(\int_{\mathbb{H}^n} f^2 d\mu\right)^{1/2} \mathcal{E}(f)^{1/2}$$

where we used the Cauchy-Schwarz inequality in the last step. The conclusion immediately follows and the proposition is therefore established. One may check that the constant $\frac{4}{(n-1)^2}$ is indeed optimal in this inequality. □

4.10.2 Bottom of the Essential Spectrum

Here we return to the spectral information contained in a Poincaré inequality. As pointed out earlier, the Poincaré inequality $P(C)$ with constant $C > 0$ for a Markov operator L indicates that the spectrum of (the opposite generator) $-L$ lies in $\{0\} \cup [\frac{1}{C}, \infty)$. Now, recall (Sect. A.4, p. 478) that the spectrum of a positive self-adjoint operator such as $-L$ lies in $\mathbb{R}_+$ and may be divided into two parts, the discrete spectrum, which corresponds to isolated eigenvalues with finite-dimensional eigenspaces, and the rest which is called the essential spectrum and is denoted by $\sigma_{\text{ess}}(-L)$. The spectrum is said to be discrete if it essential spectrum is empty (Definition A.4.3, p. 480).

The following statement is a first description of some useful information on the essential spectrum $\sigma_{\text{ess}}(-L)$ in terms of related families of spectral inequalities.

Proposition 4.10.3 *Assume that $\sigma_{\text{ess}}(-L) \subset [\sigma_0, \infty)$ for some $\sigma_0 > 0$. Then, for every $r > \frac{1}{\sigma_0}$, there exists a finite-dimensional vector subspace $F \subset \mathbb{L}^2(\mu)$ such that, for any $f \in \mathcal{D}(\mathcal{E})$,*

$$\int_E f^2 d\mu \leq r \mathcal{E}(f) + \int_E (\Pi_F f)^2 d\mu \quad (4.10.2)$$

where Π_F denotes the orthogonal projection onto F. Conversely, if there exists a finite-dimensional vector subspace $F \subset \mathbb{L}^2(\mu)$ such that (4.10.2) holds for some $r > 0$ and all $f \in \mathcal{D}(\mathcal{E})$, then the essential spectrum $\sigma_{\text{ess}}(-L)$ lies in $[\frac{1}{r}, \infty)$.

Proof The first part is straightforward. Indeed, given $r > \frac{1}{\sigma_0}$, let F be the direct sum of all the eigenspaces corresponding to eigenvalues $\lambda < \frac{1}{r}$. By definition of $\sigma_{\text{ess}}(-L)$, there is only a finite number of such eigenvalues and the subspace F is finite-dimensional. Let $f \in \mathcal{D}(\mathcal{E})$, and set for simplicity $g = \Pi_F(f)$, $h = f - \Pi_F(f)$. Clearly $\mathcal{E}(g,h) = 0$ and $\mathcal{E}(f) = \mathcal{E}(g) + \mathcal{E}(h) \geq \mathcal{E}(h)$. On the other hand, by construction, $\mathcal{E}(h) \geq \frac{1}{r}\int_E h^2 d\mu$. It follows that

$$\int_E f^2 d\mu = \int_E g^2 d\mu + \int_E h^2 d\mu \leq \int_E g^2 d\mu + r\mathcal{E}(h) \leq \int_E g^2 d\mu + r\mathcal{E}(f)$$

which is (4.10.2). Conversely, under (4.10.2) for some $r > 0$ and some subspace F, assume that there exists a $\lambda \in \sigma_{\text{ess}}(-L)$ such that $\lambda < \frac{1}{r}$. Recall from Weyl's criterion (Theorem A.4.4, p. 481) that for any $\varepsilon > 0$, there exists an orthonormal sequence $(f_k)_{k \in \mathbb{N}}$ such that $\|\lambda f_k + L f_k\|_2 \leq \varepsilon$ for every k. Since f_k has norm 1 in $\mathbb{L}^2(\mu)$, by the Cauchy-Schwarz inequality $-\int_E (\lambda f_k + L f_k) f_k d\mu \leq \varepsilon$ which amounts to $\mathcal{E}(f_k) \leq \lambda + \varepsilon$, $k \in \mathbb{N}$. Choose then ε such that $r(\lambda + \varepsilon) < 1$. Applying (4.10.2) to f_k for every k yields

$$1 \leq r(\lambda + \varepsilon) + \|\Pi_E f_k\|_2^2.$$

But an orthonormal sequence in a Hilbert space is always weakly converging to 0, so that its projection on any finite-dimensional subspace converges (strongly) to 0. Therefore $\|\Pi_E f_k\|_2 \to 0$ which produces a contradiction. The proof of Proposition 4.10.3 is complete. □

A more precise look at the preceding result yields an interesting consequence of the fact that $\sigma_{\text{ess}}(-L) = \emptyset$.

Proposition 4.10.4 *If the spectrum of $-L$ is discrete (i.e. $\sigma_{\text{ess}}(-L) = \emptyset$), there exist a positive function $w \in \mathbb{L}^2(\mu)$ and a map $\beta : (0, \infty) \to (0, \infty)$ such that, for any $r \in (0, \infty)$ and any $f \in \mathcal{D}(\mathcal{E})$,*

$$\int_E f^2 d\mu \leq r \mathcal{E}(f) + \beta(r) \left(\int_E |f| w \, d\mu \right)^2. \tag{4.10.3}$$

Equivalently, there exists a positive function $w \in \mathbb{L}^2(\mu)$ and an increasing concave function $\Phi : (0, \infty) \to (0, \infty)$ satisfying $\lim_{r \to \infty} \frac{\Phi(r)}{r} = 0$ such that for any function f in $\mathcal{D}(\mathcal{E})$ with $\int_E |f| w d\mu = 1$,

$$\int_E f^2 d\mu \leq \Phi(\mathcal{E}(f)). \tag{4.10.4}$$

Converses of this proposition will be studied below when the semigroup with Markov generator L has a density kernel (Theorem 4.10.5) or for Persson operators (Theorem 4.10.8).

Proof For any $r > 0$, denote by F_r the vector space spanned by all the eigenvectors with eigenvalues less than $\frac{1}{r}$ and by $n(r)$ its dimension. Let $(f_k)_{k \in \mathbb{N}}$ be a sequence of eigenvectors such that, for any r, $(f_0, \ldots, f_{n(r)})$ defines an orthonormal basis of F_r. Let $(\alpha_k)_{k \in \mathbb{N}}$ be a decreasing sequence of strictly positive real numbers such that $\sum_{k \in \mathbb{N}} \alpha_k < \infty$. Set $w = \sum_{k \in \mathbb{N}} \alpha_k |f_k|$. The function w belongs to $\mathbb{L}^2(\mu)$ and, for any $0 \leq k \leq n(r)$ and any $f \in \mathbb{L}^2(\mu)$, setting $c_k = \int_E f f_k d\mu$, $k \in \mathbb{N}$,

$$|c_k| \leq \frac{1}{\alpha_{n(r)}} \int_E |f| w \, d\mu.$$

Since $\|\Pi_{F_r}(f)\|_2^2 = \sum_{k=0}^{n(r)} c_k^2$, it follows that $\|\Pi_{E_r}(f)\|_2^2 \leq n(r)\alpha_{n(r)}^{-2}$. It then suffices to choose $\beta(r) = n(r)\alpha_{n(r)}^{-2}$, $r > 0$, and to follow the proof of Proposition 4.10.3 above. Inequality (4.10.4) is obtained with $\Phi(r) = \inf_{s>0}[rs + \beta(s)]$, which is clearly increasing and concave and satisfies $\lim_{r\to\infty} \frac{\Phi(r)}{r} = 0$. Conversely, if (4.10.4) holds for some Φ, it may be assumed to be $\mathcal{C}^1$, therefore satisfying $\lim_{r\to\infty} \Phi'(r) = 0$ (namely, replace Φ by $\widetilde{\Phi}(r) = \frac{2}{r}\int_0^r \Phi(s)ds$ which is still increasing and concave and satisfies $\frac{1}{2}\widetilde{\Phi}(r) \leq \Phi(r) \leq \widetilde{\Phi}(r)$). The version (4.10.3) then follows from the family of inequalities (concavity)

$$\Phi(s) \leq \Phi'(s_0)s + \Phi(s_0) - s_0\Phi'(s_0), \quad s > 0,$$

setting $r = \Phi'(s_0)$. The proposition is established. □

The family of inequalities described in Proposition 4.10.4 will be further investigated in Chap. 7 (Sect. 7.4.3, p. 370) under the name of generalized weighted Nash inequalities associated to the growth functions β or Φ.

As announced, we next examine some converses to Proposition 4.10.4. It is not true that the spectrum of $-L$ is always discrete under a generalized weighted Nash inequality (4.10.3) or (4.10.4). For example, any infinite product of Gaussian measures provides a case where a logarithmic Sobolev inequality holds (see Chap. 5), and thus a generalized weighted Nash inequality, whereas the eigenspace associated with the first non-zero eigenvalue is infinite-dimensional. However, imposing an extra assumption on the semigroup will be enough. Recall (see Definition 1.2.4, p. 14) that a symmetric operator P has an $\mathbb{L}^2$-density kernel with respect to μ if there exists a (symmetric, positive) measurable function $p(x, y)$ on the product space $E \times E$ such that, for μ-almost every $x \in E$, $\int_E p(x, y)^2 d\mu(y) < \infty$ and

$$Pf(x) = \int_E f(y)\, p(x, y) d\mu(y), \quad x \in E.$$

Note that it is not required here that $\int_E \int_E p^2(x, y) d\mu(x) d\mu(y) < \infty$, which would imply that P is Hilbert-Schmidt and therefore compact (cf. Sect. A.6, p. 483).

Theorem 4.10.5 *Assume that for some $r > 0$ there exists a positive function $w \in \mathbb{L}^2(\mu)$ such that for any $f \in \mathcal{D}(\mathcal{E})$,*

$$\int_E f^2 d\mu \leq r\mathcal{E}(f) + \left(\int_E |f| w\, d\mu\right)^2. \tag{4.10.5}$$

Assume moreover that there exists a $t > 0$ for which P_t has an $\mathbb{L}^2$-density kernel with respect to μ. Then the essential spectrum $\sigma_{ess}(-L)$ of $-L$ lies in $[\frac{1}{r}, \infty)$.

In particular, if there exists a positive function $w \in \mathbb{L}^2(\mu)$ such that for every $r > 0$, there exists a $\beta(r) > 0$ such that for all $f \in \mathcal{D}(\mathcal{E})$,

$$\int_E f^2 d\mu \leq r\mathcal{E}(f) + \beta(r)\left(\int_E |f| w\, d\mu\right)^2,$$

and if P_t has an $\mathbb{L}^2$-density kernel with respect to μ for some $t > 0$, then the spectrum of $-\mathrm{L}$ is discrete.

Proof The second part of the statement follows from the first part by applying it to the weight $\beta(r)^{1/2}w$ (for any $r > 0$). We thus concentrate on the first assertion. Let $G_w = \{g \in \mathbb{L}^2(\mu)\,;\,|g| \leq w\}$. First observe that for an operator P with a positive density kernel in $\mathbb{L}^2(\mu)$, the closure $\overline{P(G_w)}$ of $P(G_w)$ is compact. Indeed, given any sequence $(g_k)_{k \in \mathbb{N}}$ in G_w, thus bounded in $\mathbb{L}^2(\mu)$, there exists a subsequence $(g_{k_\ell})_{\ell \in \mathbb{N}}$ which converges weakly, say to some function $g \in \mathbb{L}^2(\mu)$. Then, P having a density kernel, Pg_{k_ℓ} converges μ-almost everywhere to Pg, and moreover $|Pg_{k_\ell}| \leq Pw \in \mathbb{L}^2(\mu)$. From the dominated convergence Theorem, Pg_{k_ℓ} then converges to Pg in $\mathbb{L}^2(\mu)$ as $\ell \to \infty$.

Now (4.10.5) applied to $P_t f$ yields, with $K = \overline{P_t(G_w)}$,

$$\|P_t f\|_2^2 \leq r\,\mathcal{E}(P_t f) + \sup_{g \in K}\left(\int_E fg\,d\mu\right)^2. \tag{4.10.6}$$

Indeed, with $h = \mathrm{sign}(P_t f)w$, $\int_E |P_t f|w\,d\mu = \int_E P_t f h\,d\mu = \int_E f P_t h\,d\mu$, so that

$$\int_E |P_t f|w\,d\mu \leq \sup_{g \in K}\left|\int_E fg\,d\mu\right|.$$

From this point, we may proceed as in the proof of Proposition 4.10.3 and choose by contradiction $\lambda < \frac{1}{r}$ such that $\lambda \in \sigma_{ess}(-\mathrm{L})$. For any $\varepsilon > 0$, there exists an orthonormal sequence $(f_k)_{k \in \mathbb{N}} \subset \mathcal{D}(L)$ such that $\|\lambda f_k + \mathrm{L}f_k\|_2 \leq \varepsilon$ for every $k \in \mathbb{N}$. Looking at the derivative in s of $e^{-2\lambda(t-s)}\|P_s f_k\|_2^2$, and recalling that $\int_E f_k^2 d\mu = 1$, it follows that for every $k \in \mathbb{N}$,

$$\left|\|P_t f_k\|_2^2 - e^{-2\lambda t}\right| \leq 2t\sqrt{\varepsilon}.$$

Combined with $|\mathcal{E}(P_t f_k) - \lambda\|P_t f_k\|_2^2| \leq \sqrt{\varepsilon}$, it follows that

$$\left|\mathcal{E}(P_t f_k) - \lambda e^{-2\lambda t}\right| \leq (1 + 2\lambda t)\sqrt{\varepsilon}.$$

Now, since $(f_k)_{k \in \mathbb{N}}$ converges weakly to 0 as $k \to \infty$, for any compact set $K \subset \mathbb{L}^2(\mu)$,

$$\limsup_{k \to \infty}\sup_{g \in K}\left|\int_E f_k g\,d\mu\right| = 0.$$

Applying (4.10.6) to f_k, letting $k \to \infty$, and choosing sufficiently small ε, we obtain a contradiction. Theorem 4.10.5 is established. □

In Theorem 4.10.5, the hypothesis that P_t has a density kernel may be replaced by the fact that there exists a sequence $(P^\ell)_{\ell \in \mathbb{N}}$ of operators with density kernels (not necessarily positive) such that, in operator norm, $\lim_{\ell \to \infty}\|P_t - P^\ell\| = 0$. Indeed, this assumption still ensures that $P_t(G_w)$ is relatively compact.

4.10.3 Persson Operators and the Essential Spectrum

In practice, Theorem 4.10.5 is not that useful since to establish that P_t has an $\mathbb{L}^2(\mu)$ density kernel, it is often necessary to prove first that it is Hilbert-Schmidt, and therefore compact, which already implies that $-L$ has a discrete spectrum. In the following we develop some further techniques which will enable us to reach the same conclusion. Actually, to complete the picture, the reader should be aware of the huge difference that may exist, in the analysis of Markov semigroups, between finite and infinite dimension. In particular, in reasonable situations, the essential spectrum only depends on what happens at infinity. To allow such an analysis, we introduce the notion of a *Persson operator*.

Definition 4.10.6 (Persson operator) Let (E, μ, Γ) be a Markov Triple. Let $(A_k)_{k \in \mathbb{N}}$ be an increasing sequence of measurable subsets of E with $E = \bigcup_{k \in \mathbb{N}} A_k$ and $\mu(A_k) < \infty$ for every $k \in \mathbb{N}$. Let λ_{ess} be the infimum of the essential spectrum $\sigma_{\text{ess}}(-L)$ of $-L$. We say that L is Persson with respect to $(A_k)_{k \in \mathbb{N}}$ if $\lambda_{\text{ess}} = \sup_{k \in \mathbb{N}} \lambda_k$ where $\lambda_k = \inf \mathcal{E}(f)$, the infimum being taken over all functions $f \in \mathcal{D}(\mathcal{E})$ which are 0 on A_k and such that $\int_E f^2 d\mu = 1$.

In concrete instances, $(A_k)_{k \in \mathbb{N}}$ will typically be any increasing sequence of compact subsets of E (in finite dimension), or even just balls. Observe that since the sequence $(A_k)_{k \in \mathbb{N}}$ is increasing, the sequence $(\lambda_k)_{k \in \mathbb{N}}$ is increasing too. As a consequence, for a Persson operator with respect to a sequence $(A_k)_{k \in \mathbb{N}}$, to assert that the spectrum is discrete it is enough to prove that the corresponding increasing sequence $(\lambda_k)_{k \in \mathbb{N}}$ converges to ∞.

Given a sequence $(A_k)_{k \in \mathbb{N}}$ as in Definition 4.10.6, introduce for each $k \in \mathbb{N}$ the space $\mathbb{L}^2(A_k, \mu)$ of all measurable functions $f : E \to \mathbb{R}$ such that $\int_{A_k} f^2 d\mu < \infty$ with associated norm $\|f\|_{2, A_k}$. Define furthermore

$$\mathcal{H}^1(A_k) = \left\{ f \in \mathcal{D}(\mathcal{E}) \cap \mathbb{L}^2(A_k, \mu) \,;\, \int_{A_k} \Gamma(f) d\mu < \infty \right\}.$$

One main result concerning Persson operators is the following.

Proposition 4.10.7 *For a sequence $(A_k)_{k \in \mathbb{N}}$ as above, assume that for any $k \in \mathbb{N}$, there exists a function $\psi_k \in \mathcal{D}(\mathcal{E})$ with values in $[0, 1]$ such that $\psi_k = 0$ on A_k, $\psi_k = 1$ on A_{k+1}^c, and such that $\Gamma(\psi_k)$ is bounded. Assume furthermore that the embedding from $\mathcal{H}^1(A_k)$ into $\mathbb{L}^2(A_k, \mu)$ is compact for each k. Then the operator L is Persson with respect to the sequence $(A_k)_{k \in \mathbb{N}}$.*

Proof Let $\sigma = \sup_{k \in \mathbb{N}} \lambda_k$ in Definition 4.10.6. The inequality $\lambda_{\text{ess}} \leq \sigma$ is universal and requires nothing more than the fact that $E = \bigcup_{k \in \mathbb{N}} A_k$. Indeed, choose any $\lambda < \lambda_{\text{ess}}$. By definition of the essential spectrum (Definition A.4.3, p. 480), the spectral space $\mathcal{H}_\lambda$ (that is the image of $\mathbb{L}^2(\mu)$ under the spectral projection E_λ) is finite-dimensional, thus its unit ball is compact. Now, if K is any compact subset of $\mathbb{L}^2(\mu)$, $\lim_{k\to\infty} \sup_{g\in K} \int_{A_k^c} g^2 d\mu = 0$. Therefore, given any $\varepsilon > 0$, there exists a k large enough so that, for any $g \in \mathcal{H}_\lambda$ with $\int_E g^2 d\mu = 1$, $\int_{A_k^c} g^2 d\mu \leq \varepsilon$. Let then $f \in \mathcal{D}(\mathcal{E})$ be supported in A_k^c with $\int_E f^2 d\mu = 1$. Denote by g the projection of f on $\mathcal{H}_\lambda$, and set $h = f - g$. The functions h and g are orthogonal in $\mathbb{L}^2(\mu)$ with norm less than or equal to 1. Moreover, since $g + h = 0$ on $\mathcal{A}_k$,

$$\int_{A_k} g^2 d\mu = -\int_{A_k} gh\, d\mu = \int_{A_k^c} gh\, d\mu \leq \varepsilon.$$

Therefore $\int_E g^2 d\mu \leq 2\varepsilon$, so that $\int_E h^2 d\mu \geq 1 - 2\varepsilon$. Now,

$$\mathcal{E}(f) = \mathcal{E}(g) + \mathcal{E}(h) \geq \mathcal{E}(h) \geq \lambda \int_E h^2 d\mu \geq \lambda(1 - 2\varepsilon).$$

This being valid for any $f \in \mathcal{D}(\mathcal{E})$ with support in A_k^c, we get $\lambda(1 - 2\varepsilon) \leq \lambda_k$, and the announced inequality $\lambda_{\text{ess}} \leq \sigma$ follows.

Turning to the converse inequality $\lambda_{\text{ess}} \geq \sigma$, let λ belong to the essential spectrum σ_{ess}. Applying again Weyl's criterion (Theorem A.4.4, p. 481), for any $\varepsilon > 0$, there exists an orthonormal sequence $(f_\ell)_{\ell \in \mathbb{N}}$ in $\mathcal{D}(L)$ such that $\|\lambda f_\ell + L f_\ell\|_2 \leq \varepsilon$ for every ℓ. As in previous similar proofs, $|\mathcal{E}(f_\ell) - \lambda| \leq \varepsilon$, and therefore the sequence $(\mathcal{E}(f_\ell))_{\ell \in \mathbb{N}}$ is bounded together with the sequence $(\|f_\ell\|_2)_{\ell \in \mathbb{N}}$. Hence, for any $k \in \mathbb{N}$, the sequence $(f_\ell)_{\ell \in \mathbb{N}}$ is also bounded in $\mathcal{H}^1(A_k)$, so that by the compactness assumption, there exists a subsequence which converges in $\mathbb{L}^2(A_k, \mu)$. As the sequence is orthonormal in $\mathbb{L}^2(\mu)$, it converges weakly to 0 and therefore the limit of this subsequence is 0. This may be done for any subsequence of $(f_\ell)_{\ell \in \mathbb{N}}$, which shows that $(f_\ell)_{\ell \in \mathbb{N}}$ converges to 0 in $\mathbb{L}^2(A_k, \mu)$ for any k.

Now fix $k \in \mathbb{N}$ and consider, according to the hypotheses, ψ_k with values in $[0, 1]$ vanishing on A_k, equal to 1 on A_{k+1}^c, and with bounded gradient. For every ℓ, let $g_\ell = \psi_k f_\ell$. Then

$$\mathcal{E}(g_\ell) = \int_E \psi_k^2 \Gamma(f_\ell) d\mu + 2\int_E \psi_k f_\ell \Gamma(f_\ell, \psi_k) d\mu + \int_E f_\ell^2 \Gamma(\psi_k) d\mu.$$

Since $\Gamma(\psi_k) = 0$ on A_{k+1}^c, $\mathcal{E}(g_\ell) \leq \mathcal{E}(f_\ell) + \varepsilon_\ell$ with $\lim_{\ell \to \infty} \varepsilon_\ell = 0$. Moreover $\mathcal{E}(f_\ell) \leq \lambda \int_E f_\ell^2 d\mu + \varepsilon = \lambda + \varepsilon$. It follows that

$$\mathcal{E}(g_\ell) \leq \lambda + \varepsilon + \varepsilon_\ell.$$

On the other hand, $\int_E (1 - \psi_k)^2 f_\ell^2 d\mu \leq \int_{A_{k+1}} f_\ell^2 d\mu$ and thus $\lim_{\ell \to \infty} \int_E g_\ell^2 d\mu = 1$. Therefore, as $\ell \to \infty$, $\lambda_k \leq \lambda + \varepsilon$ and the announced inequality follows. The proof of Proposition 4.10.7 is complete. □

The use of the Persson hypothesis may be combined with the weighted Nash inequalities of Theorem 4.10.5 to reach the following main characterization.

4.10 Further Spectral Inequalities

Theorem 4.10.8 *Assume that* L *is Persson. Then, the spectrum of* $-L$ *is discrete if and only if there exist a positive function* $w \in \mathbb{L}^2(\mu)$ *such that for every* $r > 0$, *there exists a* $\beta(r) > 0$ *such that for all* $f \in \mathcal{D}(\mathcal{E})$,

$$\int_E f^2 d\mu \leq r \mathcal{E}(f) + \beta(r) \left(\int_E |f| w \, d\mu \right)^2.$$

Proof Denote by $(A_k)_{k \in \mathbb{N}}$ a sequence for which L is Persson. The task is to bound from below the corresponding values λ_k, $k \in \mathbb{N}$, defined in Definition 4.10.6. To this end, let $r > 0$ be fixed, and choose $k = k(r)$ such that $\beta(r) \int_{A_k^c} w^2 d\mu \leq \frac{1}{2}$. Then, by the hypothesis applied to a function f with support in A_k^c, and the Cauchy-Schwarz inequality,

$$\int_E f^2 d\mu \leq r \mathcal{E}(f) + \beta(r) \left(\int_{A_k^c} |f| w \, d\mu \right)^2 \leq r \mathcal{E}(f) + \frac{1}{2} \int_E f^2 d\mu,$$

whence $\lambda_{k(r)} \geq \frac{1}{2r}$. Since $r > 0$ is arbitrary, the increasing sequence $(\lambda_k)_{k \in \mathbb{N}}$ converges to ∞. The converse is achieved via Proposition 4.10.4. The proof is therefore complete. □

The compactness hypothesis in Proposition 4.10.7 ensures that, for fixed $k \in \mathbb{N}$, given any sequence $(f_\ell)_{\ell \in \mathbb{N}}$ in $\mathcal{D}(\mathcal{E})$ such that $\int_{A_k} [\Gamma(f_\ell) + f_\ell^2] d\mu$ is uniformly bounded, there exists a subsequence which converges in $\mathbb{L}^2(A_k, \mu)$. In practice when dealing with elliptic operators on $\mathbb{R}^n$ or on open subsets $\mathcal{O} \subset \mathbb{R}^n$ with smooth coefficients, $(A_k)_{k \in \mathbb{N}}$ will be any increasing sequence of "nice" sets such as balls. Then the embedding is nothing else than the classical Rellich-Kondrachov Theorem (Theorem 6.4.3, p. 291). When the space $\mathcal{C}_c^\infty(\mathcal{O})$ of $\mathcal{C}^\infty$ compactly supported functions in $\mathcal{O}$ is dense in the domain $\mathcal{D}(\mathcal{E})$, it is easily seen, playing with the functions ψ_k between A_k and A_{k+1}, that the infimum in the definition of λ_k may be restricted to functions $f \in \mathcal{C}_c^\infty(A_k^c)$ (with $\int_E f^2 d\mu = 1$). As a consequence, Proposition 4.10.7 yields the following useful corollary.

Corollary 4.10.9 *Let* $L = \Delta - \nabla W \cdot \nabla$ *in* $\mathbb{R}^n$ *be such that*

$$\lim_{x \to \infty} \left[-\Delta W(x) + \frac{1}{2} |\nabla W|^2(x) \right] = \infty.$$

Then, the spectrum of $-L$ *is discrete. The same is true if* $\mathbb{R}^n$ *is replaced by an open subset* $\mathcal{O} \subset \mathbb{R}^n$ *provided that* L *is essentially self-adjoint for* $\mathcal{C}_c^\infty(\mathcal{O})$, *under the preceding limit at the boundary of* $\mathcal{O}$.

Proof We first deal with the case of $\mathbb{R}^n$, for which we already know that L is essentially self-adjoint thanks to Corollary 3.2.2, p. 143. Recall that here $d\mu = e^{-W} dx$. Choose as the sequence $(A_k)_{k \in \mathbb{N}}$ the sequence of balls centered at the origin with radius k which clearly satisfies the hypotheses of Proposition 4.10.7. It is therefore enough to show that for any smooth function f compactly supported in A_k^c,

$$\int_{\mathbb{R}^n} |\nabla f|^2 d\mu \geq c_k \int_{\mathbb{R}^n} f^2 d\mu$$

for some $c_k \to \infty$. Changing f into $ge^{W/2}$, and after integration by parts (justified since g is compactly supported), this amounts to proving that for any g smooth and compactly supported in A_k^c,

$$\int_{\mathbb{R}^n} [|\nabla g|^2 + Rg^2] dx \geq c_k \int_{\mathbb{R}^n} g^2 dx$$

where $R = -\frac{1}{2}\Delta W + \frac{1}{4}|\nabla W|^2$. This is however a direct consequence of the hypothesis.

For an open set $\mathcal{O} \subset \mathbb{R}^n$, let $(A_k)_{k \in \mathbb{N}}$ consisting of the sets of $A_k = \{J \leq k\}$ satisfying the hypotheses of Proposition 4.10.7. The argument then reduces to examining the ratio $\int_{\mathcal{O}} |\nabla f|^2 d\mu / \int_{\mathcal{O}} f^2 d\mu$ for functions which are compactly supported in A_k^c and the same transform of $f \mapsto fe^{W/2}$ applies. The proof is complete. $\square$

Similar proofs may be adapted to more general settings, in particular to manifolds. As an example, for every $\alpha > 1$, the probability measure in $\mathbb{R}^n$, $d\mu_\alpha(x) = c_\alpha e^{-|x|^\alpha}$, has a discrete spectrum. For $\alpha = 1$, the discrete spectrum is empty as already observed on p. 189.

4.11 Notes and References

Due to their interplay with spectral properties, Poincaré inequalities form a vast subject. They have been investigated and studied in a wide variety of settings, in particular in the analysis of partial differential equations, and under various names in the literature. This chapter only focuses on the aspects of Poincaré inequalities emphasized in this book, namely functional inequalities, convergence to equilibrium and heat kernel bounds. In particular, we do not address Poincaré inequalities on balls (local Poincaré inequalities), doubling properties and the related analysis on metric spaces. For these topics, we refer, for example, to [9, 13, 125, 217, 230, 236, 376, 397, 422] and to the references therein.

Poincaré inequalities also sometimes refer to the comparison of the $\mathbb{L}^2$-norm of a function with its gradient (in the form (4.10.1)) for functions vanishing at the boundary of some domain. This form is related to the Dirichlet eigenvalue problem in partial differential equations.

The Poincaré inequalities investigated in this chapter, and throughout this book, mainly deal with Neumann boundary conditions comparing the $\mathbb{L}^2$-norm of a function with mean zero with the one of its gradient. In this form, they may probably

4.11 Notes and References

be traced back to the contributions [348, 349] where H. Poincaré used duplication ((4.2.4)) and a clever change of variables to establish the inequality on bounded convex domains in $\mathbb{R}^n$. They are sometimes called Poincaré-Wirtinger inequalities in reference to W. Wirtinger who seemingly analyzed functions on the circle (corresponding to (4.5.5)) using a spectral decomposition along the trigonometric system. W. Wirtinger, who is mentioned in the book [70], also considered the extension on the sphere using spherical harmonics. The term "Poincaré inequality" first appears in the book by R. Courant and D. Hilbert [140]. See [8, 300] for a short historical introduction to Poincaré inequalities in the context of partial differential equations. We refer in addition to [317] for a comprehensive survey of inequalities that involve a relationship between a function and its derivatives or integrals, including in particular Poincaré inequalities.

The name of Poincaré inequality has been used for numerous inequalities of the same form or which are obtained similarly. In particular, the Poincaré inequality for the Gaussian measure (Proposition 4.1.1) may be traced back to the early 30s in the physics literature along with the Hermite eigenfunction expansion. It appears later in papers of J. Nash [323], H. Chernoff [130], L. Chen [127] and in many places today.

The main classical exponential decay Theorem 4.2.5 in Sect. 4.2.2 is a standard result in Fourier-type analysis, and a fundamental part of the theory of Poincaré inequalities. Lemma 4.2.6 may be found in [139].

The stability by bounded perturbations and tensorization (Sect. 4.3) properties of Poincaré inequalities form part of the folklore of the subject.

The Poincaré inequality for the exponential measure of Proposition 4.4.1 is pointed out in [403]. Its interpretation as a Lyapunov criterion has been emphasized in, for example, [29] (see below). Exponential integrability under a Poincaré inequality goes back to the works [89] and [222]. It was revived and improved in the corresponding investigation under logarithmic Sobolev inequalities (see the next chapter) by S. Aida and D. Stroock [7], L. Gross and O. Rothaus [227] and others (cf. [276]). The proof presented here is extracted from [79]. More on applications to measure concentration may be found in [90, 276, 278].

The Muckenhoupt criterion presented in Sect. 4.5.1 also has a long history starting with early papers by G. Hardy [231, 232], later developed in [321, 405, 410]. In particular, B. Muckenhoupt presented in [321] a characterization with respect to two measures on $\mathbb{R}_+$ (see also Chap. 8 in this regard). More details on those results can be found in [269]. In the form of Theorem 4.5.1, the criterion is recalled in [77] by comparison with the criterion for logarithmic Sobolev inequalities (see the next chapter). See also [14] for further details. More precise comparisons between the Muckenhoupt functional and the optimal Poincaré constant are developed in [310]. The Poincaré inequalities on a bounded interval of the real line presented in Sect. 4.5.2 are more or less classical and may be found in various places in the literature, including [203].

The Lyapunov method discussed in Sect. 4.6 has its origin in the Markov chain context (cf. e.g. [309]) where it has been used as a tool towards convergence to equilibrium. It has also been developed by L. Hörmander in partial differential equations, already with the aim of establishing functional inequalities. The method has been quite successful in the context of functional inequalities in recent years, see e.g. [29, 33, 119] and the forthcoming monograph [118] for further developments. Theorem 4.6.3 on Poincaré inequalities for log-concave measures is due to R. Kannan, L. Lovász and M. Simonovits [262] in the case of the uniform measure on a convex body and to S. Bobkov [74] in the general case. The Poincaré constant from these works, based on the variance of the measure, improves upon the mere diameter bound (as in (4.6.6)) but is still far from the famous KLS conjecture from [262] asserting that the sharp constant should be achieved on affine functions (see [228] for a recent short survey). In dimension one, an another estimate in terms of the mean value of the distance to the median has been pointed out in [186]. The Lyapunov proof of Theorem 4.6.3 is given in [29]. The optimal dependence on the diameter (4.6.6) for a convex domain is due to L. Payne and H. Weinberger [53, 343]. The glueing Proposition 4.6.4 is taken from [187].

The local Poincaré inequalities for heat kernel measures of Sect. 4.7 were put forward in [27] in parallel with their counterparts for logarithmic Sobolev inequalities discussed in the next chapter (see also [28, 277, 394]). The basic semigroup interpolation argument goes back to the seminal work [36] in the context of logarithmic Sobolev inequalities (see next chapter). Illustrations of the underlying Duhamel principle from the 19th century may be found, for example, in [258]. Local inequalities (here and in the next chapters) associated to operators of the form $L = \Delta_\mathfrak{g} + Z$ for a general vector field Z may be investigated similarly, although the non-compact case raises several delicate technical issues (cf. [431, 434, 435]). The local Poincaré inequalities for heat kernel measures may be seen as counterparts of local Poincaré inequalities on balls. The latter have been investigated deeply in the analysis of spaces endowed with measures with doubling properties (without curvature assumptions). In particular, the works of A. Grigor'yan [214] and L. Saloff-Coste [374] connect them with Sobolev-type and Harnack inequalities (cf. [217, 376]).

Poincaré inequalities under a curvature-dimension condition in Sect. 4.8 have their origin in the Lichnerowicz eigenvalue comparison Theorem (4.8.1) [286] for manifolds with a strictly positive lower bound on the Ricci curvature (see, for example [121, 123, 126, 194, 346]). See also [246, 249]. Obata's Theorem, stating that whenever there is equality in the Lichnerowiz lower bound (4.8.1) then the manifold is isometric to a sphere, goes back to [335]. In the context of the Γ-calculus, Poincaré inequalities under a curvature-dimension condition first appeared in [36] (cf. [26] for a first synthesis). In particular, Theorem 4.8.4 includes the Wirtinger inequality for the standard n-sphere. Sharper spectral comparison theorems in this context are developed in [42] (see the references therein).

The Brascamp-Lieb inequality (Theorem 4.9.1) goes back to [93] (see [288]). The proof here is inspired by [385] and was quite successful under the name of the Witten Laplacian method in the study of models in statistical physics (cf. [237–239]). The probabilistic representation (4.9.6) of the gradient of P_t is known as Bismut's formula [69] and is detailed in a manifold setting in, for example, [251, 391, 394].

4.11 Notes and References

Section 4.10 collects a variety of results and observations on spectral properties scattered all over the literature. Details on the positivity of eigenvectors may be found in [355]. In the regular case, results and history on the existence of strictly positive harmonic functions are discussed in [347]. The Poincaré inequality on the hyperbolic space (Proposition 4.10.2) is proved in a general context in [306] (see also [121]). Theorem 4.10.5 is due to F.-Y. Wang [429] and expands on earlier work of A. Persson [345] on Schrödinger operators (see also [163] and [219]).

Chapter 5
Logarithmic Sobolev Inequalities

After Poincaré inequalities, logarithmic Sobolev inequalities are amongst the most studied functional inequalities for semigroups. Indeed, they contain much more information than Poincaré inequalities, and are at the same time sufficiently general to be available in numerous cases of interest, in particular in infinite dimension (as limits of Sobolev inequalities on finite-dimensional spaces). Moreover, they entail remarkable smoothing properties of the semigroup in the form of hypercontractivity.

The structure of this chapter is quite similar to the preceding one on Poincaré inequalities. In particular, the setting is that of a Full Markov Triple, abbreviated as "Markov Triple", (E, μ, Γ) with associated Dirichlet form $\mathcal{E}$, infinitesimal generator L, Markov semigroup $\mathbf{P} = (P_t)_{t \geq 0}$ and underlying function algebras $\mathcal{A}_0$ and $\mathcal{A}$ (cf. Sect. 3.4, p. 168). Logarithmic Sobolev inequalities under the invariant measure only concern finite (normalized) measures μ. It should be mentioned that logarithmic Sobolev inequalities involve entropy and Fisher information, which deal with strictly positive functions. It will therefore be convenient to deal with the class $\mathcal{A}_0^{\text{const}+}$ of Remark 3.3.3, p. 154, consisting of functions which are sums of a positive function in $\mathcal{A}_0$ and a strictly positive constant.

Once again, several definitions and properties make sense for more general triples (E, μ, Γ), and the Standard Markov Triple assumption suffices for a number of results. Contrary to the preceding chapter on Poincaré inequalities, the diffusion property can however not be discarded.

The first section introduces the basic definition of a logarithmic Sobolev inequality together with its first properties. Section 5.2 presents the exponential decay in entropy and the fundamental equivalence between logarithmic Sobolev inequality and hypercontractivity. The next sections discuss integrability properties of eigenvectors and of Lipschitz functions under a logarithmic Sobolev inequality and present, as in the case of Poincaré inequalities, a criterion for measures on the real line to satisfy a logarithmic Sobolev inequality (for the usual gradient). Sections 5.5 and 5.7 deal with curvature conditions, first for the local logarithmic Sobolev inequalities for heat kernel measures, then for the invariant measure with additional dimensional information. Local hypercontractivity and some applications of the local logarithmic Sobolev inequalities towards heat kernel bounds are further presented. Section 5.6

develops Harnack-type inequalities under the infinite-dimensional curvature conditions $CD(\rho, \infty)$, and describes links with reverse local logarithmic Sobolev inequalities.

5.1 Logarithmic Sobolev Inequalities

This section formally introduces the notion of a logarithmic Sobolev inequality for a Markov Triple (E, μ, Γ) and presents some of its first properties and equivalent formulations.

5.1.1 Logarithmic Sobolev Inequalities

For a measure ν (not necessarily finite) on a measurable space $(E, \mathcal{F})$, define for all positive integrable functions f such that $\int_E f |\log f| d\nu < \infty$,

$$\mathrm{Ent}_\nu(f) = \int_E f \log f \, d\nu - \int_E f \, d\nu \log\left(\int_E f \, d\nu\right). \tag{5.1.1}$$

(Recall that $0 \log 0$ is interpreted as 0 in this definition.) When ν is a probability measure, since the function $\phi(r) = r \log r$ is strictly convex on $\mathbb{R}_+$, by Jensen's inequality, $\mathrm{Ent}_\nu(f) \geq 0$ and is equal to 0 only if f is constant. Note furthermore that $\mathrm{Ent}_\nu(cf) = c \, \mathrm{Ent}_\nu(f)$ for every $c \geq 0$. For future purposes, let us also mention the classical *entropic inequality* for a probability measure ν,

$$\int_E fg \, d\nu \leq \mathrm{Ent}_\nu(f) + \int_E f \, d\nu \log\left(\int_E e^g d\nu\right) \tag{5.1.2}$$

whenever $f \geq 0$ and g are suitably integrable. For a sketch of the proof, assume that $\int_E f d\nu = 1$ and set $d\tilde{\nu} = f d\nu$ so that $\int_E e^g d\nu = \int_E e^g f^{-1} d\tilde{\nu}$. Then apply Jensen's inequality with respect to the probability measure $\tilde{\nu}$ and the concave function $\log$. The entropic inequality (5.1.2) implies the useful variational formula, for ν a probability measure, and every positive measurable function $f : E \to \mathbb{R}$,

$$\mathrm{Ent}_\nu(f) = \sup \int_E fg \, d\nu, \tag{5.1.3}$$

where the supremum runs over all functions g such that $\int_E e^g d\nu \leq 1$.

The following definition introduces the notion of a *logarithmic Sobolev inequality*.

Definition 5.1.1 (Logarithmic Sobolev inequality) A Markov Triple (E, μ, Γ) (with μ a probability measure) is said to satisfy a logarithmic Sobolev inequality $LS(C, D)$ with constants $C > 0$, $D \geq 0$, if for all functions $f : E \to \mathbb{R}$ in the

5.1 Logarithmic Sobolev Inequalities

Dirichlet domain $\mathcal{D}(\mathcal{E})$,

$$\operatorname{Ent}_\mu(f^2) \leq 2C\,\mathcal{E}(f) + D\int_E f^2 d\mu.$$

When $D = 0$, the logarithmic Sobolev inequality will be called tight, and will be denoted by $LS(C)$. When $D > 0$, the logarithmic Sobolev inequality $LS(C, D)$ is called defective.

The best constant $C > 0$ for which such an inequality $LS(C)$ holds is sometimes referred to as the *logarithmic Sobolev constant* (of the Markov Triple). Note that it is part of the information contained in the logarithmic Sobolev inequality that provided $f \in \mathcal{D}(\mathcal{E})$, then $\int_E f^2 \log(1 + f^2) d\mu < \infty$, that is f belongs to the Orlicz space $\mathbb{L}^2 \log \mathbb{L}(\mu)$ (observe further that $\mathbb{L}^2 \log \mathbb{L}(\mu) \supset \mathbb{L}^p(\mu)$ for every $p > 2$ for the comparison with Sobolev inequalities in the next chapter). As usual, it is enough to state (and prove) such inequalities for a family of functions f which is dense in the Dirichlet domain $\mathcal{D}(\mathcal{E})$ (typically the algebra $\mathcal{A}_0$).

We often simply say that the probability measure μ satisfies a logarithmic Sobolev inequality $LS(C, D)$ or $LS(C)$ (with respect to the underlying carré du champ operator Γ or Dirichlet form $\mathcal{E}$). The normalization constant 2 is chosen for further comparisons. By the homogeneity property of entropy, the logarithmic Sobolev inequality of Definition 5.1.1 is homogeneous (invariant under a change of f to cf). The tight logarithmic Sobolev inequality $LS(C)$ implies that if $\mathcal{E}(f) = 0$, then f is constant. This observation is in agreement with the corresponding property for Poincaré inequalities (cf. Remark 4.2.2, p. 182, and the link with connexity in Sect. 3.2.1, p. 140).

The logarithmic Sobolev inequality $LS(C, D)$ is often presented equivalently for $\sqrt{f}$, $f \geq 0$ (see Remark 5.1.2 below), and then takes the form, by the chain rule formula,

$$\operatorname{Ent}_\mu(f) \leq \frac{C}{2} \int_E \frac{\Gamma(f)}{f} d\mu + D \int_E f\, d\mu. \tag{5.1.4}$$

Since f may be changed by homogeneity into cf, it can be assumed moreover that $\int_E f d\mu = 1$. Setting $dv = f d\mu$ then defines a probability measure on $(E, \mathcal{F})$. The quantity

$$\mathrm{H}(v \mid \mu) = \operatorname{Ent}_\mu(f) = \int_E \log f\, dv \tag{5.1.5}$$

is then the *relative entropy* of the probability measure v with respect to μ whereas

$$\mathrm{I}(v \mid \mu) = \mathrm{I}_\mu(f) = \int_E \frac{\Gamma(f)}{f} d\mu = 4 \int_E \Gamma(\sqrt{f}) d\mu \tag{5.1.6}$$

is called the *Fisher information* of v with respect to μ. A logarithmic Sobolev inequality $LS(C)$ thus compares the entropy to the information

$$\mathrm{H}(v \mid \mu) \leq \frac{C}{2} \mathcal{I}(v \mid \mu) \tag{5.1.7}$$

for every probability measure ν absolutely continuous with respect to μ, a language often used in the context of information theory (where logarithmic Sobolev inequalities were first considered).

Remark 5.1.2 The meaning of $I_\mu(f)$ where the function f may vanish should be clarified. It actually makes perfect sense when $f = g^2$ for any $g \in \mathcal{A}_0$, or even $g \in \mathcal{D}(\mathcal{E})$, since then $I_\mu(f) = 4 \int_E \Gamma(g) \, d\mu$. In a concrete setting, it also makes sense for any smooth strictly positive function f as the integral $\int_E \frac{\Gamma(f)}{f} \, d\mu$, where the latter integral may or may not be convergent. When f is only positive, observe that $I_\mu(f) = \lim_{\varepsilon \searrow 0} I_\mu(f + \varepsilon)$. Therefore when dealing with such quantities, it is often enough to consider $I_\mu(f)$ for f in $\mathcal{A}_0^{\text{const}+}$. Throughout this section, inequalities involving a Fisher-type expression will usually be established first for functions in this class and then extended to more general functions taking limits. This procedure will often be implicit.

The prototypical logarithmic Sobolev inequality is that for the standard Gaussian measure on $\mathbb{R}$ or $\mathbb{R}^n$ for which $LS(C)$ holds with $C = 1$, independently of the dimension n of the underlying state space. (The value $C = 1$ justifies the normalization chosen in Definition 5.1.1. This normalization will also be convenient for the comparison with Poincaré inequalities below.) However, while the Poincaré inequality for the Gaussian measure may be established by a simple spectral expansion along the Hermite polynomials, the proof of the logarithmic Sobolev inequality is more delicate, and is thus postponed until later in the chapter (Proposition 5.5.1).

The name logarithmic Sobolev inequality describes a weak form of the Sobolev inequalities studied in Chap. 6. However, while Sobolev inequalities will typically hold only in finite-dimensional spaces, logarithmic Sobolev inequalities may be investigated in infinite-dimensional contexts (in probability theory, statistical mechanics, statistics, and in numerous settings dealing with an infinite number of coordinates), where they provide useful bounds and controls. It will be shown later how they really appear as limits of Sobolev inequalities as the dimension tends to infinity (see Remark 6.2.6, p. 285, and Sect. 6.5, p. 291). Logarithmic Sobolev inequalities are also of interest in finite dimension as will be discussed in the chapter and throughout the book.

5.1.2 Tightening and Perturbation

The tight logarithmic Sobolev inequality $LS(C)$ is stronger than the Poincaré inequality $P(C)$ (Definition 4.2.1, p. 181). This is the content of the following proposition, which furthermore shows that a logarithmic Sobolev inequality $LS(C, D)$ together with a Poincaré inequality actually yield a tight logarithmic Sobolev inequality.

Proposition 5.1.3 (Tightening with a Poincaré inequality) *A tight logarithmic Sobolev inequality $LS(C)$ implies a Poincaré inequality $P(C)$. Furthermore, a log-*

5.1 Logarithmic Sobolev Inequalities

arithmic Sobolev inequality $LS(C, D)$ together with a Poincaré inequality $P(C')$ implies a tight logarithmic Sobolev inequality $LS(C + C'(\frac{D}{2} + 1))$.

Proof For the first assertion, apply $LS(C)$ to $f = 1 + \varepsilon g$ where $g \in \mathcal{D}(\mathcal{E})$ with $\int_E g \, d\mu = 0$. As $\varepsilon \to 0$, it is not difficult to check by a Taylor expansion that

$$\mathrm{Ent}_\mu(f^2) = 2\varepsilon^2 \int_E g^2 d\mu + o(\varepsilon^2)$$

while, clearly, $\mathcal{E}(f) = \varepsilon^2 \mathcal{E}(g)$. It follows that

$$2 \int_E g^2 d\mu \leq 2C \, \mathcal{E}(g)$$

which amounts to the Poincaré inequality $P(C)$.

Turning to the second claim, the following lemma will be useful.

Lemma 5.1.4 (Rothaus' Lemma) *Let $f : E \to \mathbb{R}$ be measurable and such that $\int_E f^2 \log(1 + f^2) \, d\mu < \infty$. For every $a \in \mathbb{R}$,*

$$\mathrm{Ent}_\mu\big((f+a)^2\big) \leq \mathrm{Ent}_\mu(f^2) + 2 \int_E f^2 d\mu.$$

Proof With f and a fixed, the aim is to verify that the function

$$\psi(r) = \mathrm{Ent}_\mu\big((rf+a)^2\big) - \mathrm{Ent}_\mu\big((rf)^2\big) - 2 \int_E (rf)^2 d\mu, \quad r \in \mathbb{R},$$

is negative. To this end, observe that ψ is concave. Namely,

$$\frac{\psi''(r)}{2} = \int_E f^2 \log v^2 d\mu - \int_E f^2 d\mu \log\left(\int_E v^2 d\mu\right)$$
$$- \mathrm{Ent}_\mu(f^2) - 2 \left(\int_E v^2 d\mu\right)^{-1} \left(\int_E fv \, d\mu\right)^2$$

where $v = rf + a$. Now, by concavity and Jensen's inequality,

$$\mathrm{Ent}_\mu(f^2) \geq \int_E f^2 \log v^2 d\mu - \int_E f^2 d\mu \log\left(\int_E v^2 d\mu\right).$$

Hence ψ is concave, and since, as is easily checked, $\psi(0) = \psi'(0) = 0$, the announced claim follows. Note that whenever $\int_E v^2 d\mu = 0$ in this argument, then $rf + a = 0$ μ-almost everywhere, and the conclusion is immediate in this case. The lemma is proved. □

Applied to $\hat{f} = f - \int_E f d\mu$ with $a = \int_E f d\mu$, the previous lemma yields

$$\text{Ent}_\mu(f^2) \leq \text{Ent}_\mu(\hat{f}^2) + 2\text{Var}_\mu(f). \tag{5.1.8}$$

On this basis, the proof of the second assertion of Proposition 5.1.3 is immediate. Indeed, by the logarithmic Sobolev inequality $LS(C, D)$ applied to $\hat{f}$, and (5.1.8),

$$\text{Ent}_\mu(f^2) \leq 2C\, \mathcal{E}(f) + D \int_E \hat{f}^2 d\mu + 2\,\text{Var}_\mu(f),$$

and since $\int_E \hat{f}^2 d\mu = \text{Var}_\mu(f)$, it remains to use the Poincaré inequality $P(C')$. Proposition 5.1.3 is therefore established. □

Remark 5.1.5 Examples will be provided below for which the logarithmic Sobolev constant is strictly larger than the (finite) Poincaré constant (or is even infinite). In this respect, observe that whenever the smallest (non-trivial) eigenvalue λ_1 of $-L$ (provided it exists) admits an eigenfunction h such that $\int_E h^3 d\mu \neq 0$, then the Taylor expansion to the next term around h used in the proof of Proposition 5.1.3 yields such an instance for which the logarithmic Sobolev constant of μ is necessarily strictly larger than the spectral gap constant (equal to $\frac{1}{\lambda_1}$).

As for Poincaré inequalities, logarithmic Sobolev inequalities for a probability measure μ with respect to a given carré du champ operator Γ are often established by comparison with known ones. For example, if μ satisfies a logarithmic Sobolev inequality $LS(C, D)$ with respect to Γ, and if Γ_1 is another carré du champ operator such that $\Gamma \leq a\Gamma_1$ for some $a > 0$, then (μ, Γ_1) satisfies $LS(aC, D)$.

The following perturbation result is an analogue of Proposition 4.2.7, p. 185.

Proposition 5.1.6 (Bounded perturbation) *Assume that the Markov Triple (E, μ, Γ) satisfies a logarithmic Sobolev inequality $LS(C, D)$. Let μ_1 be a probability measure with density h with respect to μ such that $\frac{1}{b} \leq h \leq b$ for some constant $b > 0$. Then μ_1 satisfies $LS(b^2 C, b^2 D)$ (with respect to Γ).*

The proposition is a simple consequence of the following variational formula. As for the Poincaré inequality, if $h = e^k$, b^2 may be replaced by $e^{\text{osc}(k)}$ where again $\text{osc}(k) = \sup k - \inf k$.

Lemma 5.1.7 *Let $\phi : I \to \mathbb{R}$ on some open interval $I \subset \mathbb{R}$ be convex of class $\mathcal{C}^2$. For every (bounded or suitably integrable) measurable function $f : E \to \mathbb{R}$ with values in I,*

$$\int_E \phi(f) d\mu - \phi\left(\int_E f d\mu\right) = \inf_{r \in I} \int_E \big[\phi(f) - \phi(r) - \phi'(r)(f - r)\big] d\mu.$$

The important issue here, for both the proof of the lemma and its application, is that $\phi(s) - \phi(r) - \phi'(r)(s - r) \geq 0$, $r, s \in I$, by convexity. With $s = \int_E f d\mu$, the

5.1 Logarithmic Sobolev Inequalities

left-hand side of the equality of the lemma is less than or equal to its right-hand side, while equality is achieved precisely for $r = \int_E f d\mu$. Now for the function $\phi(r) = r \log r$ on $I = (0, \infty)$, Proposition 5.1.6 is easily established since

$$\mathrm{Ent}_{\mu_1}(f) = \inf_{r \in I} \int_E \big[\phi(f) - \phi(r) - \phi'(r)(f - r)\big] d\mu_1 \leq b\, \mathrm{Ent}_\mu(f)$$

(while obviously $\int_E \Gamma(f) d\mu \leq b \int_E \Gamma(f) d\mu_1$).

5.1.3 Slicing of Logarithmic Sobolev Inequalities

To conclude this section, we present some alternative forms of logarithmic Sobolev inequalities. Like many of the functional inequalities encountered throughout this book, logarithmic Sobolev inequalities may appear in a number of different forms, and it is not always easy to recognize them in disguise. Below we present one of many examples, included in the so-called Nash family which we shall meet later in Chaps. 6 and 7.

Proposition 5.1.8 *If (E, μ, Γ) satisfies a logarithmic Sobolev inequality $LS(C, D)$, then, for any $p \in [1, 2)$ and any $f \in \mathcal{D}(\mathcal{E})$,*

$$\|f\|_2^2 \log\left(\frac{\|f\|_2^2}{\|f\|_p^2}\right) \leq C_1 \mathcal{E}(f) + D_1 \|f\|_2^2, \tag{5.1.9}$$

with $C_1 = \frac{(2-p)C}{p}$ and $D_1 = \frac{(2-p)D}{p}$. Conversely, if (5.1.9) holds for some $p \in [1, 2)$ and every $f \in \mathcal{D}(\mathcal{E})$, then, for constants $c(p) > 0$ and $d(p) > 0$ depending only on p, $LS(C, D)$ holds with $C = c(p)C_1$ and $D = c(p)D_1 + d(p)$.

Proof We first show that $LS(C, D)$ implies (5.1.9). The argument only relies on the rephrasing of Hölder's inequality as the convexity of the map $r \mapsto \phi(r) = \log(\|f\|_{1/r})$, $r \in (0, 1]$ (for $f \neq 0$). Therefore, for $1 \leq p < 2$,

$$\frac{\phi(\frac{1}{p}) - \phi(\frac{1}{2})}{\frac{1}{p} - \frac{1}{2}} \geq \phi'\left(\frac{1}{2}\right).$$

Now, derivatives of $\mathbb{L}^p(\mu)$-norms give rise to entropy as

$$\phi'\left(\frac{1}{2}\right) = -\frac{\mathrm{Ent}_\mu(f^2)}{\|f\|_2^2},$$

from which (5.1.9) follows with $C_1 = \frac{(2-p)C}{p}$ and $D_1 = \frac{(2-p)D}{p}$. Note that $D_1 = 0$ for a tight logarithmic Sobolev inequality $LS(C)$.

To get the reverse inequality, we introduce a slicing technique, already mentioned in Proposition 3.1.17, p. 137, and which will be used later in the study of Sobolev inequalities and formalized in a more efficient way in Chaps. 6 and 8. Let $f \in \mathcal{D}(\mathcal{E})$, which may be assumed positive as $\mathcal{E}(|f|) \leq \mathcal{E}(f)$. By homogeneity, it can also be assumed that $\|f\|_2 = 1$. The idea is to apply (5.1.9) to the family of functions $f_k = (f - 2^k)_+ \wedge 2^k$, $k \in \mathbb{Z}$ (which still belong to the Dirichlet domain $\mathcal{D}(\mathcal{E})$). To this end, let $\mathcal{N}_k = \{f > 2^k\}$, $k \in \mathbb{Z}$. Since $f = \sum_{k \in \mathbb{Z}} \mathbb{1}_{\mathcal{N}_k \setminus \mathcal{N}_{k+1}} f$, observe first that

$$\frac{3}{4} \sum_{k \in \mathbb{Z}} 2^{2k} \mu(\mathcal{N}_k) \leq \int_E f^2 d\mu \leq 3 \sum_{k \in \mathbb{Z}} 2^{2k} \mu(\mathcal{N}_k).$$

In other words, since $\int_E f^2 d\mu = 1$, setting $\alpha_k = 2^{2k} \mu(\mathcal{N}_k)$, $k \in \mathbb{Z}$,

$$\frac{1}{3} \leq \sum_{k \in \mathbb{Z}} \alpha_k \leq \frac{4}{3}.$$

In particular $\alpha_k \leq \frac{4}{3}$ for every k. Now, for every $k \in \mathbb{Z}$,

$$2^k \mathbb{1}_{\mathcal{N}_{k+1}} \leq f_k \leq 2^k \mathbb{1}_{\mathcal{N}_k},$$

so that, letting $\beta_k = \|f_k\|_2^2$, $\frac{\alpha_{k+1}}{4} \leq \beta_k \leq \alpha_k$ and, for $1 \leq p < 2$,

$$\|f_k\|_p^2 \leq 2^{2k} \mu(\mathcal{N}_k)^{2/p} = 2^{-2k((2/p)-1)} \alpha_k^{2/p}$$

for every $k \in \mathbb{Z}$. Therefore,

$$\frac{\|f_k\|_2^2}{\|f_k\|_p^2} \geq 2^{2k((2/p)-1)} \left(\frac{3}{4}\right)^{(2/p)-1} \frac{\beta_k}{\alpha_k}$$

where we used that $\alpha_k^{2/p} \leq (\frac{3}{4})^{1-(2/p)} \alpha_k$. Now apply (5.1.9) to f_k for every $k \in \mathbb{Z}$ to get that

$$k \beta_k \leq c(p) \left(\beta_k \log\left(\frac{\alpha_k}{\beta_k}\right) + C_1 \mathcal{E}(f_k) + \left(D_1 + d(p)\right) \beta_k \right)$$

for constants $c(p), d(p) > 0$ only depending on p. Decomposing f again as $f = \sum_{k \in \mathbb{Z}} \mathbb{1}_{\mathcal{N}_k \setminus \mathcal{N}_{k+1}} f$ shows similarly (after some details) that

$$\int_E f^2 \log f^2 d\mu \leq 3 \log 4 \left(\sum_{k \in \mathbb{Z}} k \alpha_k + 8 \right).$$

(The numerical constant 8 is far from optimal in this last inequality.)

It thus follows from the preceding that, for possibly new values of $c(p)$ and $d(p)$,

$$\int_E f^2 \log f^2 d\mu \leq c(p) \left(\sum_{k \in \mathbb{Z}} \beta_k \log\left(\frac{\alpha_k}{b_k}\right) + C_1 \sum_{k \in \mathbb{Z}} \mathcal{E}(f_k) + D_1 + d(p) \right).$$

5.2 Entropy Decay and Hypercontractivity

By Jensen's inequality for the concave function log and the fact that $\frac{\alpha_{k+1}}{4} \leq \beta_k \leq \alpha_k$ for every k,

$$\sum_{k \in \mathbb{Z}} \beta_k \log\left(\frac{\alpha_k}{\beta_k}\right) \leq \left(\sum_{k \in \mathbb{Z}} \beta_k\right) \log\left(\frac{\sum_{k \in \mathbb{Z}} \alpha_k}{\sum_{k \in \mathbb{Z}} \beta_k}\right) \leq \frac{4}{3} \log 4$$

from which it follows that

$$\int_E f^2 \log f^2 d\mu \leq c(p)\left(C_1 \sum_{k \in \mathbb{Z}} \mathcal{E}(f_k) + d(p) + D_1\right).$$

With the help of Proposition (3.1.17), p. 137, $\sum_{k \in \mathbb{Z}} \mathcal{E}(f_k) \leq \mathcal{E}(f)$, so that

$$\int_E f^2 \log f^2 d\mu \leq c(p) C_1 \mathcal{E}(f) + c(p) D_1 + d(p),$$

which amounts by homogeneity to the announced logarithmic Sobolev inequality. Proposition 5.1.8 is established. □

In the preceding proof, going from one form of the inequality to the other and back does not preserve the constants. While the proof shows that the tight logarithmic Sobolev inequality $LS(C)$ implies $D_1 = 0$, conversely (5.1.9) with $D_1 = 0$ in the preceding proof does not yield back $LS(C)$ for some $C > 0$. However, a Taylor expansion $f = 1 + \varepsilon g$ in (5.1.9) with $D_1 = 0$ yields a Poincaré inequality $P(\frac{C_1}{2-p})$, so that, together with Proposition 5.1.3, we actually reach a tight logarithmic Sobolev inequality in this case.

5.2 Entropy Decay and Hypercontractivity

Like Poincaré inequalities (cf. Sect. 4.2.2, p. 183), logarithmic Sobolev inequalities have something to tell about convergence to equilibrium of the semigroup $\mathbf{P} = (P_t)_{t \geq 0}$ towards the invariant (probability) measure μ. In addition, logarithmic Sobolev inequalities ensure smoothing properties of the semigroup in the form of hypercontractivity.

5.2.1 Exponential Decay in Entropy

The following statement is the analogue for entropy of the exponential decay in $\mathbb{L}^2(\mu)$ produced by a Poincaré inequality (Theorem 4.2.5, p. 183). Furthermore, the proof of this result exhibits the fundamental relation between decay of the entropy along the semigroup and the Fisher information (5.1.6).

Theorem 5.2.1 (Exponential decay in entropy) *The logarithmic Sobolev inequality $LS(C)$ for the probability measure μ is equivalent to saying that for every positive function f in $\mathbb{L}^1(\mu)$ (with finite entropy),*

$$\mathrm{Ent}_\mu(P_t f) \leq e^{-2t/C} \mathrm{Ent}_\mu(f)$$

for every $t \geq 0$.

It is not obvious a priori that the convergence in this statement is stronger than the convergence in variance as produced by a Poincaré inequality. One point is that the convergence holds for a larger class of functions. For example, if μ is the standard Gaussian distribution on the real line (for which $LS(C)$ holds with $C = 1$) and if $f(x) = e^{cx^2/2}$, $x \in \mathbb{R}$, $c \in (0, 1)$, then $\mathrm{Var}_\mu(f) = \infty$ if $c \geq \frac{1}{2}$ whereas $\mathrm{Ent}_\mu(f) = \frac{1}{2}(c(1-c)^{-3/2} + (1-c)^{-1/2} \log(1-c))$ for every $c \in (0, 1)$.

Furthermore, convergence in entropy yields convergence in total variation according to the following scheme. For probability measures μ, ν on $(E, \mathcal{F})$, introduce, according to (5.1.5), the relative entropy $\mathrm{H}(\nu \,|\, \mu)$ of ν with respect to μ as

$$\mathrm{H}(\nu \,|\, \mu) = \int_E \log \frac{d\nu}{d\mu} \, d\nu \qquad (5.2.1)$$

whenever ν is absolutely continuous with respect to μ with Radon-Nikodym derivative $f = \frac{d\nu}{d\mu}$, and $\mathrm{H}(\nu \,|\, \mu) = \infty$ if not. In other words, if $\nu \ll \mu$ with Radon-Nikodym derivative f, $\mathrm{H}(\nu \,|\, \mu) = \mathrm{Ent}_\mu(f)$. Then Theorem 5.2.1 may be reinterpreted as

$$\mathrm{H}(\nu_t \,|\, \mu) \leq e^{-2t/C} \mathrm{H}(\nu_0 \,|\, \mu)$$

where $d\nu_t = P_t f d\mu$, $t \geq 0$.

Now, the celebrated *Pinsker-Csizsár-Kullback inequality* indicates that for μ, ν probability measures on $(E, \mathcal{F})$,

$$\|\mu - \nu\|_{\mathrm{TV}}^2 \leq \frac{1}{2} \mathrm{H}(\nu \,|\, \mu) \qquad (5.2.2)$$

where $\|\mu - \nu\|_{\mathrm{TV}}$ is the total variation distance

$$\|\mu - \nu\|_{\mathrm{TV}} = \sup_{A \in \mathcal{F}} \big|\mu(A) - \nu(A)\big|.$$

Therefore, under a control of $\mathrm{H}(\nu_0 \,|\, \mu)$, Theorem 5.2.1 implies the stronger convergence of ν_t towards μ in total variation.

For the sake of completeness, here is a brief proof of (5.2.2) (which shares some features with the proof of Lemma 5.1.4 above). If $f \geq 0$ denotes the Radon-Nikodym derivative $\frac{d\nu}{d\mu}$, then, thanks to the identity $\int_E |1 - \frac{d\nu}{d\mu}| d\mu = 2\|\mu - \nu\|_{\mathrm{TV}}$, (5.2.2) translates into

$$\left(\int_E |1 - f| d\mu \right)^2 \leq 2 \mathrm{Ent}_\mu(f).$$

5.2 Entropy Decay and Hypercontractivity

Set $f_s = sf + 1 - s$, $s \in [0, 1]$, and consider

$$\Lambda(s) = 2\,\text{Ent}_\mu(f_s) - \left(\int_E |1 - f_s| d\mu\right)^2 = 2\,\text{Ent}_\mu(f_s) - s^2\left(\int_E |1 - f| d\mu\right)^2.$$

Then, $\Lambda(0) = \Lambda'(0) = 0$ and

$$\Lambda''(s) = 2\int_E \frac{(1-f)^2}{f_s} d\mu - 2\left(\int_E |1 - f| d\mu\right)^2, \quad s \in [0, 1).$$

By the Cauchy-Schwarz inequality, $\Lambda'' \geq 0$ so that Λ is positive, which completes the proof. $\blacksquare$

We next address the proof of Theorem 5.2.1. The principle of proof is the same as that for Theorem 4.2.5, p. 183, this time differentiating entropy $\Lambda(t) = \text{Ent}_\mu(P_t f)$, $t \geq 0$, rather than variance. The argument relies on the following main identity which is an analogue of (4.2.2), p. 183, for entropy. Recall the Fisher information I_μ from (5.1.6).

Proposition 5.2.2 (de Bruijn's identity) *For every positive f in $\mathcal{D}(\mathcal{E})$,*

$$\frac{d}{dt}\text{Ent}_\mu(P_t f) = -I_\mu(P_t f). \tag{5.2.3}$$

The proof is immediate. Indeed, by the heat equation and integration by parts, for any $f \in \mathcal{A}_0^{\text{const}+}$ to start with, and then extending to $f \in \mathcal{D}(\mathcal{E})$,

$$\frac{d}{dt}\text{Ent}_\mu(P_t f) = \int_E (1 + \log P_t f) L P_t f \, d\mu$$

$$= -\int_E \Gamma(P_t f, 1 + \log P_t f) d\mu$$

$$= -\int_E \frac{\Gamma(P_t f)}{P_t f} d\mu = -I_\mu(P_t f).$$

This basic relation between the time derivative of entropy and the Fisher information in particular tells us that entropy is decreasing along the flow of the semigroup, a property often referred to as the *Boltzmann (H-) Theorem*.

On the basis of de Bruijn's identity, the logarithmic Sobolev inequality $LS(C)$ in its Fisher information formulation (5.1.4)

$$\text{Ent}_\mu(g) \leq \frac{C}{2} I_\mu(g)$$

translates into the differential inequality $\Lambda(t) \leq -\frac{C}{2}\Lambda'(t)$, $t > 0$, from which the inequality of Theorem 5.2.1 easily follows. Conversely, differentiating this inequality at $t = 0$ yields the logarithmic Sobolev inequality by the same argument.

5.2.2 Hypercontractivity

The evolution of the semigroup $\mathbf{P} = (P_t)_{t \geq 0}$ under a logarithmic Sobolev inequality may be described in another, new, formulation, at the origin of the notion of logarithmic Sobolev inequality, known as the hypercontractivity property. This is a smoothing property that exactly translates the logarithmic Sobolev inequality. The next theorem presents this fundamental equivalence.

Theorem 5.2.3 (Gross' hypercontractivity Theorem) *For parameters $C > 0$, $D \geq 0$, $1 < p < \infty$, let $q(t)$, $1 < q(t) < \infty$, and $M(t) \geq 0$ be defined for every $t \geq 0$ by*

$$\frac{q(t) - 1}{p - 1} = e^{2t/C}, \qquad M(t) = D\left(\frac{1}{p} - \frac{1}{q(t)}\right).$$

For a Markov Triple (E, μ, Γ) with semigroup $\mathbf{P} = (P_t)_{t \geq 0}$, the following assertions are equivalent.

(i) *The logarithmic Sobolev inequality $LS(C, D)$ holds.*
(ii) *For some (every) $1 < p < \infty$, and for every $t \geq 0$ and every $f \in \mathbb{L}^p(\mu)$,*

$$\|P_t f\|_{q(t)} \leq e^{M(t)} \|f\|_p.$$

The second assertion of the theorem is thus called *hypercontractivity* of the semigroup $(P_t)_{t \geq 0}$. Note that $M(t) = 0$ under a tight logarithmic Sobolev inequality ($D = 0$) so that $LS(C)$ holds if and only if for every $t > 0$ and $f \in \mathbb{L}^p(\mu)$,

$$\|P_t f\|_q \leq \|f\|_p$$

for (some or any) $1 < p < q < \infty$ such that $e^{2t/C} \geq \frac{q-1}{p-1}$. What Theorem 5.2.3 tells us is that, under a logarithmic Sobolev inequality, not only are the operators P_t contractions in all $\mathbb{L}^p(\mu)$-spaces, but they improve integrability for large t (in particular they are contractions from $\mathbb{L}^p(\mu)$ to $\mathbb{L}^q(\mu)$ when $D = 0$). This may be shown to be optimal for some examples, such as the Ornstein-Uhlenbeck semigroup (see below). Note that the theorem does not tell us anything in the case where $p = 1$.

Proof Once the correct starting point has been identified, the proof is rather simple. To show that $LS(C, D)$ implies hypercontractivity, we use the traditional strategy of differentiating a suitable functional. Below, f is a function in $\mathcal{A}_0^{\text{const}+}$ (the result being extended to functions in $\mathbb{L}^p(\mu)$ by standard density arguments). A first observation, already used earlier in this chapter, is that the derivative of the norm $\|\cdot\|_q$ along its parameter q gives rise to the entropy. Namely,

$$\partial_q \|f\|_q^q = \partial_q \int_E f^q d\mu$$

$$= \int_E f^q \log f \, d\mu = \frac{1}{q}\left(\text{Ent}_\mu(f^q) + \int_E f^q d\mu \log \int_E f^q d\mu\right).$$

5.2 Entropy Decay and Hypercontractivity

On the other hand, for every fixed $q > 1$, by integration by parts and the diffusion property,

$$\partial_t \int_E (P_t f)^q d\mu = q \int_E (P_t f)^{q-1} L P_t f \, d\mu$$

$$= -q(q-1) \int_E (P_t f)^{q-2} \Gamma(P_t f) d\mu.$$

Finally, the logarithmic Sobolev inequality $LS(C, D)$ applied to $f^{q/2}$ ($q > 0$) yields, again by the change of variables formula,

$$\mathrm{Ent}_\mu(f^q) \leq \frac{Cq^2}{2} \int_E f^{q-2} \Gamma(f) d\mu + D \int_E f^q d\mu. \tag{5.2.4}$$

On the basis of these three observations, the proof may easily be completed. For f in $\mathcal{A}_0^{\mathrm{const}+}$ as before, consider

$$\Lambda = \Lambda(t, q) = \int_E (P_t f)^q d\mu, \quad t \geq 0, \, q > 1.$$

By the preceding, the $LS(C, D)$ inequality tells us that

$$\partial_q \Lambda \leq -\frac{C}{2(q-1)} \partial_t \Lambda + \frac{D}{q} \Lambda + \frac{1}{q} \Lambda \log \Lambda.$$

For the given expressions of $q(t)$ and $M(t)$, it follows immediately from this inequality that $H(t) = \frac{1}{q(t)} \log(\Lambda(t, q(t))) - M(t)$ is decreasing in $t \geq 0$. Hypercontractivity is then simply the inequality $H(t) \leq H(0)$.

This shows that the logarithmic Sobolev inequality $LS(C, D)$ implies hypercontractivity for any $p > 1$. Conversely, start from the hypercontractivity inequality

$$\|P_t f\|_{q(t)} \leq e^{M(t)} \|f\|_p$$

for some $p > 1$. Taking the derivative at $t = 0$ gives back (5.2.4) with $q = p$ which amounts to the logarithmic Sobolev inequality $LS(C, D)$ after changing of f into $f^{2/p}$. The proof is complete. □

Remark 5.2.4 The above proof may easily be modified to yield a reverse hypercontractivity property. Indeed, choosing a tight logarithmic Sobolev inequality $LS(C)$ for simplicity, $LS(C)$ holds if and only if for every $t > 0$ and every strictly positive measurable function $f : E \to \mathbb{R}$,

$$\|P_t f\|_q \geq \|f\|_p$$

for (some or any) $-\infty < q < p < 1$ such that $e^{2t/C} = \frac{q-1}{p-1}$.

In Theorem 5.2.3, in order for the logarithmic Sobolev inequality to hold it is enough that hypercontractivity is satisfied for some $t_0 > 0$ and not for any $t > 0$. This is the content of the following statement. Of course, the exact correspondence between constants is then lost.

Theorem 5.2.5 *Given a Markov Triple (E, μ, Γ) with Markov semigroup $\mathbf{P} = (P_t)_{t \geq 0}$, if for some $t_0 > 0$ and some $q > 2$, there exists a constant $M > 0$ such that*

$$\|P_{t_0} f\|_q \leq M \|f\|_2$$

for all $f \in \mathbb{L}^2(\mu)$, then a logarithmic Sobolev inequality $LS(C, D)$ holds (with $C = \frac{2q\, t_0}{q-2}$ and $D = \frac{2q \log M}{q-2}$).

This theorem may be shown to follow from a general relationship between entropy and energy along the semigroup, which is an analogue of (4.2.3), p. 184.

Proposition 5.2.6 (Cattiaux's inequality) *For every function f in $\mathcal{D}(\mathcal{E})$ and every $t \geq 0$,*

$$\int_E f^2 \log f^2 d\mu \leq 2t\, \mathcal{E}(f) + \int_E f^2 \log(P_t f)^2 d\mu.$$

More generally, for all functions f, g in $\mathcal{D}(\mathcal{E})$ and every $t \geq 0$,

$$\int_E f^2 \log g^2 d\mu \leq 2t\, \mathcal{E}(f) + \int_E f^2 \log(P_t g)^2 d\mu.$$

Proof We only establish the first assertion, the second one being proved in the same way. As usual, it is convenient to work with $f \in \mathcal{A}_0^{\text{const}+}$ and to extend to $\mathcal{D}(\mathcal{E})$ by density. As in many of the earlier proofs, consider a functional along the semigroup given here by $\Lambda(t) = \int_E f^2 \log P_t f \, d\mu$, $t \geq 0$. Then, by the integration by parts and diffusion properties,

$$\Lambda'(t) = \int_E f^2 \frac{L P_t f}{P_t f} d\mu = \int_E \left(-\frac{2f}{P_t f} \Gamma(f, P_t f) + \frac{f^2}{(P_t f)^2} \Gamma(P_t f) \right) d\mu.$$

Now, by the Cauchy-Schwarz inequality for the quadratic form Γ,

$$-\frac{2f}{P_t f} \Gamma(f, P_t f) + \frac{f^2}{(P_t f)^2} \Gamma(P_t f) \geq -\Gamma(f)$$

so that $\Lambda'(t) \geq -\int_E \Gamma(f) d\mu = -\mathcal{E}(f)$ for every t. Integrating from 0 to t yields the result. □

5.2 Entropy Decay and Hypercontractivity

Proof of Theorem 5.2.5 Let $q > 2$ and $f \in \mathcal{D}(\mathcal{E})$ with $\|f\|_2 = 1$. By Proposition 5.2.6, for every $t \geq 0$,

$$\int_E f^2 \log f^2 d\mu \leq 2t\,\mathcal{E}(f) + \int_E f^2 \log(P_t f)^2 d\mu$$

$$= 2t\,\mathcal{E}(f) + \frac{2}{q} \int_E f^2 \log |P_t f|^q d\mu.$$

By Jensen's inequality with respect to the probability measure $f^2 d\mu$,

$$\int_E f^2 \log |P_t f|^q d\mu \leq \int_E f^2 \log f^2 d\mu + \log \int_E |P_t f|^q d\mu.$$

Therefore

$$\left(1 - \frac{2}{q}\right) \int_E f^2 \log f^2 d\mu \leq 2t\,\mathcal{E}(f) + \log\bigl(\|P_t f\|_q^2\bigr),$$

from which Theorem 5.2.5 immediately follows. □

Note that Theorem 5.2.5 (with possibly different bounds) may be shown to follow alternatively from Lemma 4.2.6, p. 184, and Proposition 5.1.8 (use by duality that $\|P_t f\|_2 \leq M \|f\|_{q^*}$ where q^* is the conjugate exponent of q).

5.2.3 Tensorization of Logarithmic Sobolev Inequalities

To conclude this section, we observe that, like Poincaré inequalities (Sect. 4.3, p. 185), logarithmic Sobolev inequalities are stable under products. This dimension-free feature is one fundamental aspect justifying the importance of logarithmic Sobolev inequalities in infinite-dimensional contexts. It technically allows for the extension of logarithmic Sobolev bounds on finite-dimensional space to infinite dimension (such as, for example, the extension from finite-dimensional to infinite dimensional Gaussian measures).

Proposition 5.2.7 (Stability under product) *If $(\mathbb{E}_1, \mu_1, \Gamma_1)$ and $(\mathbb{E}_2, \mu_2, \Gamma_2)$ satisfy logarithmic Sobolev inequalities $LS(C_1, D_1)$ and $LS(C_2, D_2)$ respectively, then the product $(\mathbb{E}_1 \times \mathbb{E}_2, \mu_1 \otimes \mu_2, \Gamma_1 \oplus \Gamma_2)$ satisfies a logarithmic Sobolev inequality $LS(\max(C_1, C_2), D_1 + D_2)$.*

Proof A quick proof may be provided via the equivalence with hypercontractivity (Theorem 5.2.3). Indeed, if $(P_t^1)_{t \geq 0}$ and $(P_t^2)_{t \geq 0}$ denote the respective Markov semigroups, and if P_t^i is bounded from $\mathbb{L}^p(\mu)$ into $\mathbb{L}^q(\mu)$ with norm e^{M_i} for t such that $e^{2t/C_i} \geq \frac{q-1}{p-1}$, $i = 1, 2$, it is immediate that $P_t^1 \otimes P_t^2$ is bounded from

$\mathbb{L}^p(\mu_1 \otimes \mu_2)$ into $\mathbb{L}^q(\mu_1 \otimes \mu_2)$ with norm bounded from above by $e^{M_1+M_2}$ for $e^{2t/\max(C_1,C_2)} \geq \frac{q-1}{p-1}$.

The proposition may also be proved in the same way as Proposition 4.3.1, p. 185, with a traditional scheme which is often used when dealing with correlations in statistical mechanics models for example. Indeed, using the notation therein,

$$\mathrm{Ent}_{\mu_1 \otimes \mu_2}(f^2) = \mathrm{Ent}_{\mu_1}(g^2)$$
$$+ \int_{E_1} \left(\int_{E_2} f^2(x_1, x_2) \log f^2(x_1, x_2) d\mu_2(x_2) \right.$$
$$\left. - \int_{E_2} f^2(x_1, x_2) d\mu_2(x_2) \log \int_{E_2} f^2(x_1, x_2) d\mu_2(x_2) \right) d\mu_1(x_1)$$

where now $g \geq 0$ is defined by $g^2(x_1) = \int_{E_2} f^2(x_1, x_2) d\mu_2(x_2)$, $x_1 \in E_1$. Using that $\mathcal{E}_1(g) \leq \int_{E_2} \mathcal{E}_1(f) d\mu_2$, the conclusion easily follows. However, the latter requires more here in the form of the change of variables formula for carré du champ operators and the Cauchy-Schwarz inequality yielding that

$$\Gamma_1(g)(x_1) \leq \int_{E_2} \Gamma_1(f(x_1, x_2)) d\mu_2(x_2). \qquad \square$$

5.3 Integrability of Eigenvectors

While the spectral content of Poincaré inequalities is quite clear, this is not so obvious for logarithmic Sobolev inequalities. In this short section, we draw integrability properties of the eigenvectors of a diffusion operator L with Markov semigroup $(P_t)_{t \geq 0}$ from the hypercontractive Theorem 5.2.3. Provided that the spectrum is discrete and that the eigenvalues go to infinity fast enough, these integrability properties of eigenvectors are actually equivalent to hypercontractive bounds, and therefore to logarithmic Sobolev inequalities.

We illustrate the picture in the Gaussian case first. As mentioned above, the standard Gaussian measure $d\mu(x) = (2\pi)^{-n/2} e^{-|x|^2/2} dx$ on the Borel sets of $\mathbb{R}^n$ will be shown below (Proposition 5.5.1) to satisfy $LS(1)$ (with respect to the standard carré du champ operator). Therefore, by Theorem 5.2.3, the associated Ornstein-Uhlenbeck semigroup $\mathbf{P} = (P_t)_{t \geq 0}$ (as presented in Sect. 2.7.1, p. 103) is hypercontractive in the sense that for every $1 < p < q < \infty$ and $t > 0$ such that $e^{2t} \geq \frac{q-1}{p-1}$, and every $f \in \mathbb{L}^p(\mu)$,

$$\|P_t f\|_q \leq \|f\|_p. \qquad (5.3.1)$$

Eigenfunctions of the Ornstein-Uhlenbeck operator are described by Hermite polynomials. Recall from Sect. 2.7.1, p. 103, that the Hermite polynomials $(H_k)_{k \in \mathbb{N}}$ form an orthogonal basis of $\mathbb{L}^2(\mu)$ on the real line. Similarly, the polynomials

$$H(x) = H_{k_1}(x_1) \cdots H_{k_n}(x_n), \quad x = (x_1, \ldots, x_n) \in \mathbb{R}^n, \ k_1, \ldots, k_n \in \mathbb{N},$$

5.3 Integrability of Eigenvectors

form an orthogonal basis of $\mathbb{L}^2(\mu)$ on $\mathbb{R}^n$. Now let $(H_\ell)_{\ell \in L_k}$ be a given family of such polynomials with the property that $k_1 + \cdots + k_n = k$ for some fixed $k \in \mathbb{N}$, and set

$$Q = \sum_{\ell \in L_k} a_\ell H_\ell.$$

Such a polynomial Q is an eigenvector of the Ornstein-Uhlenbeck operator with eigenvalue k, that is $-LQ = kQ$. Hence, the action of the Ornstein-Uhlenbeck semigroup $(P_t)_{t \geq 0}$ on Q is given by $P_t Q = e^{-kt} Q$, $t \geq 0$. In particular, by (5.3.1), for every $t > 0$,

$$e^{-kt} \|Q\|_q = \|P_t Q\|_q \leq \|Q\|_2$$

where $q = 1 + e^{2t}$. Hence, for every $q > 2$,

$$\|Q\|_q \leq (q-1)^k \|Q\|_2. \tag{5.3.2}$$

This moment inequality furthermore provides sharp exponential tails on the distribution of Q ($\mu(|Q| \geq r)$ is bounded from above by $e^{-cr^{2/k}}$ as $r \to \infty$). When considering infinitely many coordinates x_i, this inequality extends to convergent (in $\mathbb{L}^2(\mu)$) sums Q, and the resulting tail inequalities are not so easy to obtain by other means.

These integrability properties of eigenvectors are not particularly tied to the example of the Ornstein-Uhlenbeck semigroup and the Hermite polynomials. They hold for any eigenvector of the generator under the hypercontractive estimate of Theorem 5.2.3. In fact, in contrast to the family of Sobolev inequalities studied in the next chapter, logarithmic Sobolev inequalities do not in general say anything about the existence of eigenvectors (or the existence of any point in the spectrum), unless extra properties such as that of Persson are imposed (cf. Sect. 4.10.3, p. 227). There are indeed examples, in infinite dimension, where hypercontractivity holds but the spectrum of L is not discrete.

However, when there are eigenvalues with associated eigenvectors, logarithmic Sobolev inequalities do produce $\mathbb{L}^p$-bounds. Indeed, as in the Gaussian example, if h is an eigenvector of $-L$ associated with eigenvalue λ, under the logarithmic Sobolev inequality $LS(C, D)$, and thus hypercontractivity (Theorem 5.2.3), for any $q > 2$

$$\|h\|_q \leq e^{D(\frac{1}{2} - \frac{1}{q})} (q-1)^{\lambda C/2} \|h\|_2.$$

We simply apply Theorem 5.2.3 with $p = 2$ to h, for which $P_t h = e^{-\lambda t} h$, choosing t according to the rule $q - 1 = e^{2t/C}$. This shows in particular that any eigenvector is in every $\mathbb{L}^q(\mu)$-space.

The preceding admits a somewhat surprising converse. Assume that the operator L admits a point spectral decomposition in the sense that any $f \in \mathbb{L}^2(\mu)$ may be decomposed into an orthogonal sequence

$$f = \sum_{k \in \mathbb{N}} \Pi_k(f)$$

where $-L\Pi_k(f) = \lambda_k \Pi_k(f)$ for every $k \in \mathbb{N}$. (Comparing with the discrete spectrum hypothesis of Definition A.4.3, p. 480, it should be mentioned that this point spectral decomposition does not assume that the corresponding eigenspaces are finite-dimensional.) Assume moreover that the spectral values λ_k, $k \in \mathbb{N}$, increase fast enough to infinity such that, for some $\varepsilon > 0$, $\sum_{k \in \mathbb{N}} e^{-\varepsilon \lambda_k} < \infty$. In this context, as soon as, for some $q > 2$ and some $a, C > 0$,

$$\|\Pi_k(f)\|_q \leq C e^{a\lambda_k} \|f\|_2 \qquad (5.3.3)$$

for every $k \in \mathbb{N}$, then, by the triangle inequality,

$$\|P_t f\|_q = \left\|\sum_{k \in \mathbb{N}} e^{-\lambda_k t} \Pi_k(f)\right\|_q \leq C \|f\|_2 \sum_{k \in \mathbb{N}} e^{(a-t)\lambda_k} \leq C_1 \|f\|_2$$

for every $t \geq a + \varepsilon$. Therefore, for such a large t, the operator P_t is bounded from $\mathbb{L}^2(\mu)$ into $\mathbb{L}^q(\mu)$, and, by Theorem 5.2.5, a logarithmic Sobolev inequality $LS(C, D)$ holds, and hence we have full hypercontractivity. Notice that (5.3.3) is nothing else than the similar upper bound on the eigenvectors in the associated eigenspaces. In this setting, hypercontractivity appears somewhat surprisingly as a consequence of spectral analytic bounds on the generator (via information on the size of the eigenvectors).

5.4 Logarithmic Sobolev Inequalities and Exponential Integrability

As shown in Sect. 4.4.2, p. 190, Lipschitz functions are exponentially integrable whenever μ satisfies a Poincaré inequality $P(C)$. We investigate in this section the corresponding property under a logarithmic Sobolev inequality and the associated Gaussian measure concentration.

5.4.1 Exponential Integrability

Exponential integrability under a logarithmic Sobolev inequality is achieved through a method known as the Herbst argument. For a given Lipschitz function $f \in \mathcal{D}(\mathcal{E}) \cup \mathcal{A}$, recall its Lipschitz semi-norm $\|f\|_{\text{Lip}} = \|\Gamma(f)\|_\infty^{1/2}$. A function $f \in \mathcal{D}(\mathcal{E}) \cup \mathcal{A}$ is said to be 1-Lipschitz if $\|f\|_{\text{Lip}} \leq 1$.

Proposition 5.4.1 (Herbst's argument) *If μ satisfies a logarithmic Sobolev inequality $LS(C)$ for some $C > 0$, then for every 1-Lipschitz function f and every $\sigma^2 < \frac{1}{C}$,*

$$\int_E e^{\sigma^2 f^2 / 2} d\mu < \infty.$$

5.4 Logarithmic Sobolev Inequalities and Exponential Integrability

More precisely, any 1-*Lipschitz function* f *is integrable and for every* $s \in \mathbb{R}$,

$$\int_E e^{sf} d\mu \leq e^{s \int_E f d\mu + Cs^2/2}. \tag{5.4.1}$$

At the expense of further technical arguments not developed here, the exponential integrability result $\int_E e^{\sigma^2 f^2/2} d\mu < \infty$ for every $\sigma^2 < \frac{1}{C}$ still holds under a defective logarithmic Sobolev inequality $LS(C, D)$.

With respect to Poincaré inequalities (Proposition 4.4.2, p. 190), squares of Lipschitz functions are here exponentially integrable (and not just the Lipschitz functions themselves). In particular, the exponential measure, which satisfies a Poincaré inequality, cannot satisfy a logarithmic Sobolev inequality (for the standard carré du champ operator). Note that the integrability level in Proposition 5.4.1 is optimal since the standard Gaussian measure satisfies $LS(1)$ (cf. Proposition 5.5.1 below), and (5.4.1) is sharp in this example with f linear.

Proof The proof is perhaps even simpler than the corresponding one for Poincaré inequalities. We start with (5.4.1), and similarly restrict ourselves to bounded Lipschitz functions in $\mathcal{D}(\mathcal{E})$ (or $\mathcal{A}$) the general case being reached through truncation and approximation (cf. Remark 4.4.3, p. 190). Therefore, given a bounded 1-Lipschitz function $f \in \mathcal{D}(\mathcal{E})$, set for $s \in \mathbb{R}$, $Z(s) = \int_E e^{sf} d\mu$. The aim is to apply $LS(C)$ to $e^{sf/2}$ for every s. Towards this goal, observe that

$$Z'(s) = \int_E f e^{sf} d\mu$$

while

$$\mathrm{Ent}_\mu(e^{sf}) = s \int_E f e^{sf} d\mu - Z(s) \log Z(s) = s Z'(s) - Z(s) \log Z(s).$$

On the other hand, since $\Gamma(f) \leq 1$,

$$\mathcal{E}(e^{sf/2}) = \frac{s^2}{4} \int_E e^{sf} \Gamma(f) d\mu \leq \frac{s^2}{4} Z(s).$$

The logarithmic Sobolev inequality $LS(C)$ then expresses the differential inequality in $s \in \mathbb{R}$,

$$s Z' \leq Z \log Z + \frac{Cs^2}{2} Z.$$

It remains to integrate this inequality. To this end, set $F(s) = \frac{1}{s} \log Z(s)$ (with $F(0) = \int_E f d\mu$) so that $F'(s) \leq \frac{C}{2}$ for every s. It follows that

$$F(s) \leq \int_E f d\mu + \frac{Cs}{2}$$

which amounts to the control (5.4.1) of the Laplace transform.

Integrating (5.4.1) along the centered Gaussian measure with variance σ^2 in the s variable yields, by Fubini's Theorem, that

$$\int_E e^{\sigma^2 f^2/2} d\mu \leq \frac{1}{\sqrt{1 - C\sigma^2}} \exp\left[\frac{\sigma^2}{2(1 - C\sigma^2)} \left(\int_E f\, d\mu\right)^2\right]$$

for every $\sigma^2 < \frac{1}{C}$. Hence the first claim of the proposition holds and the proof is complete. $\square$

Proposition 5.4.1 admits a simple and useful variation in the form of moment estimates. Namely, if $R(p) = \|f\|_p^2$, $p \geq 2$, for a function $f \in \mathbb{L}^p(\mu)$ with $\int_E \Gamma(f)^{p/2} d\mu < \infty$, the derivative $R'(p)$ of R in p is given by

$$\frac{2}{p^2} R(p)^{1-(p/2)} \operatorname{Ent}_\mu(f^p).$$

By the logarithmic Sobolev inequality $LS(C)$ and Hölder's inequality,

$$R'(p) \leq C \left(\int_E \Gamma(f)^{p/2} d\mu\right)^{2/p}.$$

By integration and monotonicity, the following statement holds.

Proposition 5.4.2 (Moment bounds) *If μ satisfies a logarithmic Sobolev inequality $LS(C)$ for some $C > 0$, then for every $p \geq 2$ and every $f \in \mathbb{L}^p(\mu)$,*

$$\|f\|_p^2 \leq \|f\|_2^2 + C(p - 2)\left(\int_E \Gamma(f)^{p/2} d\mu\right)^{2/p}.$$

If f is Lipschitz, then

$$\|f\|_p^2 \leq \|f\|_2^2 + C(p - 2)\|f\|_{\operatorname{Lip}}^2$$

for every $p \geq 2$. By a series expansion, the growth in p amounts to the same level of integrability as that in Proposition 5.4.1.

The next statement is an analogue of Proposition 4.4.4, p. 192, in the setting of logarithmic Sobolev inequalities. The proof is entirely similar.

Proposition 5.4.3 (Contraction principle) *If (μ, Γ) satisfies a logarithmic Sobolev inequality $LS(C, D)$ on a state space E, and if $F : E \to \mathbb{R}$ is Lipschitz with respect to Γ with $\|F\|_{\operatorname{Lip}} \leq 1$, then the image measure μ_F of μ under F satisfies a logarithmic Sobolev inequality $LS(C, D)$ with respect to the usual carré du champ operator on the real line.*

5.4.2 Gaussian Measure Concentration

As in Sect. 4.4.3, p. 192, for Poincaré inequalities, Proposition 5.4.1 may be reformulated in the form of a Gaussian *measure concentration* property for the measure μ satisfying a logarithmic Sobolev inequality $LS(C)$ for some $C > 0$. Indeed, for every Lipschitz function f and every $r \geq 0$,

$$\mu\left(f \geq \int_E f\,d\mu + r\right) \leq e^{-r^2/2C\|f\|_{\text{Lip}}^2}. \tag{5.4.2}$$

This is an immediate application of (5.4.1) together with Markov's exponential inequality and homogeneity. The numerical constant 2 is sharp. A concentration inequality around the mean is easily obtained together with the same inequality for $-f$ yielding

$$\mu\left(\left|f - \int_E f\,d\mu\right| \geq r\right) \leq 2\,e^{-r^2/2C\|f\|_{\text{Lip}}^2}, \quad r \geq 0.$$

These Gaussian concentration inequalities have of course to be compared to the corresponding exponential concentration inequalities (4.4.6), p. 192, under Poincaré inequalities.

Again as in Sect. 4.4.3, p. 192, as a consequence of the tensorization Proposition 5.2.7, if μ on E satisfies a logarithmic Sobolev inequality $LS(C)$ with respect to Γ, then the product measure $\mu^{\otimes n}$ on E^n satisfies $LS(C)$ with respect to the product carré du champ operator, independently of the dimension n of the state space E^n. Tail inequalities for Lipschitz functions of a sample $Z = (Z_1, \ldots, Z_n)$ of independent random variables with common distribution μ on $\mathbb{R}$ satisfying $LS(C)$ may then be deduced from Proposition 5.4.1. For example, if $f : \mathbb{R} \to \mathbb{R}$, $\|f\|_{\text{Lip}} \leq 1$, and

$$F(x) = \frac{1}{\sqrt{n}}\big[f(x_1) + \cdots + f(x_n)\big], \quad x = (x_1, \ldots, x_n) \in \mathbb{R}^n,$$

then for every $r \geq 0$,

$$\mathbb{P}\big(F(Z) \geq \mathbb{E}(F(Z)) + r\big) \leq e^{-r^2/2C}.$$

In particular, these dimension-free Gaussian tail inequalities suitably describe the central limit theorem. More general Lipschitz functions as well as non-identically distributed samples may be considered similarly.

Remark 5.4.4 In analogy with Remark 4.4.5, p. 193, in the context of Poincaré inequalities, one may wonder how far the concentration bounds of the Herbst argument (Proposition 5.4.1 and (5.4.2)) are from the logarithmic Sobolev inequality $LS(C)$ (or $LS(C, D)$). Again, the Muckenhoupt characterization below (Theorem 5.4.5) indicates that they are not sufficient in general to entail a logarithmic

Sobolev inequality. However, under a curvature condition $CD(0, \infty)$, they do imply a logarithmic Sobolev inequality with moreover proportional constants. This result will be presented in Sect. 8.7, p. 425, together with the corresponding result for Poincaré inequalities on the basis of the local heat kernel inequalities of Sect. 5.5 below (in particular the reverse forms).

5.4.3 Muckenhoupt's Criterion

We conclude this section with an analogue of the Muckenhoupt criterion for logarithmic Sobolev inequalities. As for Poincaré inequalities (Sect. 4.5.1, p. 194), it is possible to characterize measures μ on the real line $\mathbb{R}$ satisfying a logarithmic Sobolev inequality $LS(C)$ with respect to the usual carré du champ operator $\Gamma(f) = f'^2$. The next statement is an analogue of Theorem 4.5.1, p. 194, with a similar although more involved proof not presented here. As for the corresponding equivalence for Poincaré inequalities, this result appears as the one-dimensional version of more general relationships between capacities and measures discussed in Chap. 8.

Theorem 5.4.5 (Muckenhoupt's criterion) *Let μ be a probability measure on the Borel sets of $\mathbb{R}$ with density p with respect to the Lebesgue measure, and let m be a median for μ. Set*

$$B_+ = \sup_{x>m} \mu\big([x, +\infty)\big) \log\left(\frac{1}{\mu([x, +\infty))}\right) \int_m^x \frac{1}{p(t)}\, dt$$

and

$$B_- = \sup_{x<m} \mu\big((-\infty, x]\big) \log\left(\frac{1}{\mu((-\infty, x])}\right) \int_x^m \frac{1}{p(t)}\, dt.$$

Then, a necessary and sufficient condition in order that μ satisfies a logarithmic Sobolev inequality $LS(C)$ is that $B = \max(B_+, B_-) < \infty$. Moreover, $cB \leq C \leq c'B$ for numerical $c, c' > 0$ where C is the logarithmic Sobolev constant of μ.

It is a good exercise to verify that the probability measure on $\mathbb{R}$ with density $c_\alpha e^{-|x|^\alpha}$ with respect to the Lebesgue measure satisfies the Muckenhoupt criterion for the logarithmic Sobolev inequality if and only if $\alpha \geq 2$. Recall that the corresponding criterion for the Poincaré inequality yields in this case $\alpha \geq 1$, thus distinguishing clearly the two families of inequalities. As a further example of interest, the spectral decomposition of the Laguerre operator in Sect. 2.7.3, p. 111, shows that the exponential measure μ on $\mathbb{R}_+$ has spectral gap 1 for the carré du champ operator $\Gamma(f) = xf'^2$. Since the Laguerre operator is of curvature $CD(\frac{1}{2}, \infty)$, according to Proposition 5.7.1 below, μ satisfies a logarithmic Sobolev inequality with constant $C = 2$. This constant is optimal since the 1-Lipschitz map $f(x) = 2\sqrt{x}$ (with respect to Γ) is such that $\int_{\mathbb{R}_+} e^{\sigma^2 f^2/2} d\mu < \infty$ for every $\sigma^2 < \frac{1}{2}$ and nothing better (Proposition 5.4.1).

5.5 Local Logarithmic Sobolev Inequalities

Following the treatment of Poincaré inequalities in Sect. 4.7, p. 206, of the previous chapter, this section investigates logarithmic Sobolev inequalities under the semigroup $\mathbf{P} = (P_t)_{t \geq 0}$, that is with respect to the heat kernel measures of the representation (1.2.4), p. 12,

$$P_t f(x) = \int_E f(y) \, p_t(x, dy), \quad t \geq 0, \ x \in E.$$

The approach further develops the main principle of heat flow monotonicity using the Duhamel interpolation formula, here in the context of entropy.

Before addressing the general framework, we first consider the Ornstein-Uhlenbeck example leading to the logarithmic Sobolev inequalities for Gaussian measure. The section closes with a version of hypercontactivity for heat kernel measures and some heat kernel bounds. The local heat kernel inequalities will be presented and established for functions in the algebra $\mathcal{A}_0$ of the Markov Triple (E, μ, Γ), and we technically work with the class $\mathcal{A}_0^{\text{const}+}$.

5.5.1 Logarithmic Sobolev Inequalities for Gaussian Kernels

We commence our analysis of local logarithmic Sobolev inequalities, again by starting with the example of the Ornstein-Uhlenbeck semigroup $\mathbf{P} = (P_t)_{t \geq 0}$ on $\mathbb{R}^n$ with generator $Lf = \Delta f - x \cdot \nabla f$ and carré du champ operator $\Gamma(f) = |\nabla f|^2$ on an algebra $\mathcal{A}_0$ of smooth functions (cf. Sect. 2.7.1, p. 103). Recall from (2.7.5), p. 104, that $\nabla P_t f = e^{-t} P_t(\nabla f)$ so that, for every $f \in \mathcal{A}_0$, and every $t \geq 0$,

$$|\nabla P_t f| \leq e^{-t} P_t(|\nabla f|). \tag{5.5.1}$$

As for the local Poincaré inequalities in Sect. 4.7, p. 206, consider now, for $t > 0$ and $f \in \mathcal{A}_0^{\text{const}+}$,

$$\Lambda(s) = P_s\big(\psi(P_{t-s}f)\big), \quad s \in [0, t],$$

where $\psi(r) = r \log r$, $r \in \mathbb{R}_+$. Then, setting $g = P_{t-s}f$, by the heat equation and the change of variables formula,

$$\Lambda'(s) = P_s\big(L\psi(g) - \psi'(g) Lg\big) = P_s\big(\psi''(g) \Gamma(g)\big) = P_s\left(\frac{\Gamma(g)}{g}\right).$$

Now, by (5.5.1),

$$\Gamma(g) = |\nabla P_{t-s}f|^2 \leq e^{-2(t-s)} \big(P_{t-s}(|\nabla f|)\big)^2.$$

Since $p_{t-s}(x,dy)$ is a probability measure, by the Cauchy-Schwarz inequality, for every h,
$$(P_{t-s}h)^2 \leq P_{t-s}f \, P_{t-s}\left(\frac{h^2}{f}\right).$$

With $h = |\nabla f|$, this amounts to
$$\frac{(P_{t-s}(|\nabla f|))^2}{P_{t-s}f} \leq P_{t-s}\left(\frac{|\nabla f|^2}{f}\right) = P_{t-s}\left(\frac{\Gamma(f)}{f}\right).$$

As a consequence,
$$\Lambda'(s) \leq e^{-2(t-s)} P_s\left(P_{t-s}\left(\frac{\Gamma(f)}{f}\right)\right) = e^{-2(t-s)} P_t\left(\frac{\Gamma(f)}{f}\right).$$

By definition of $\Lambda(s)$, $s \in [0,t]$, it follows that, for every $t \geq 0$,
$$P_t(f \log f) - P_t f \log P_t f \leq \frac{1 - e^{-2t}}{2} P_t\left(\frac{\Gamma(f)}{f}\right). \tag{5.5.2}$$

It should be emphasized that with respect to the corresponding Poincaré inequality (4.7.2), p. 207, the logarithmic Sobolev inequality (5.5.2) makes use of the commutation (5.5.1) rather than only (4.7.1), p. 207 (strong gradient bound versus gradient bound in the terminology of Chap. 3). This aspect will be further developed in Theorem 5.5.2 below.

The preceding inequality is thus a (tight) logarithmic Sobolev inequality $LS(1 - e^{-2t})$ for the heat kernel measures $p_t(x, dy)$, with a constant independent of the initial point x (implicit throughout the argument), which converges to 1 as $t \to \infty$. Since the kernels $p_t(x, dy)$ converge to the standard Gaussian measure μ, as seen from (2.7.3), p. 104, we therefore conclude that the latter satisfies $LS(1)$.

Proposition 5.5.1 (Logarithmic Sobolev inequality for the Gaussian measure) *The standard Gaussian measure μ on the Borel sets of $\mathbb{R}^n$ satisfies $LS(1)$. In other words, for every function $f : \mathbb{R}^n \to \mathbb{R}$ in the Dirichlet space of the standard carré du champ operator $\Gamma(f) = |\nabla f|^2$,*
$$\mathrm{Ent}_\mu(f^2) \leq 2\mathcal{E}(f) = 2\int_{\mathbb{R}^n} |\nabla f|^2 d\mu.$$

This proposition states the logarithmic Sobolev inequality for the centered Gaussian measure μ on $\mathbb{R}^n$ with identity as covariance matrix (sometimes called the *Gaussian logarithmic Sobolev inequality*). If μ is a centered Gaussian measure on $\mathbb{R}^n$ with covariance matrix Q, a simple change of variables shows that for every smooth function f on $\mathbb{R}^n$,
$$\mathrm{Ent}_\mu(f^2) \leq 2\int_{\mathbb{R}^n} Q\nabla f \cdot \nabla f \, d\mu.$$

5.5 Local Logarithmic Sobolev Inequalities

The constant in the Gaussian logarithmic Sobolev inequality of Proposition 5.5.1 is sharp. This may be shown in various ways. For example, by comparison with the Poincaré inequality (Proposition 5.1.3) and the first eigenvector (with eigenvalue 1). It may also be shown that functions $f(x) = e^{ax}$ (on the line) satisfy equality in $LS(1)$ (and are actually the only extremal functions). Alternatively, Proposition 5.4.1 on the exponential integrability of Lipschitz functions implies that $C = 1$ cannot be improved by the simple example of $f(x) = x$.

Another approach to optimality would be via the equivalence with hypercontractivity of Theorem 5.2.3, namely that for every $t > 0$, P_t is a contraction from $\mathbb{L}^2(\mu)$ into $\mathbb{L}^{q(t)}(\mu)$ for $q(t) = 1 + e^{2t}$. Since the Ornstein-Uhlenbeck semigroup is one of the rare examples for which the semigroup and its kernel are explicitly known (cf. Sect. 2.7.1, p. 103), the optimality of this embedding may be established directly. It is however a bit delicate. Much easier is to observe that if $q > q(t)$, then P_t is unbounded from $\mathbb{L}^2(\mu)$ into $\mathbb{L}^q(\mu)$, so that the norm of P_t as an operator from $\mathbb{L}^2(\mu)$ into $\mathbb{L}^q(\mu)$ is equal to 1 for $q \in [2, q(t)]$ and to $+\infty$ for $q > q(t)$.

As for local Poincaré inequalities, the preceding logarithmic Sobolev inequality (5.5.2) for P_t may be reversed in the form of

$$P_t(f \log f) - P_t f \log P_t f \geq \frac{e^{2t} - 1}{2} \cdot \frac{\Gamma(P_t f)}{P_t f} \tag{5.5.3}$$

for every f positive in $\mathcal{A}_0$ (where the right-hand side is understood, according to Remark 5.1.2, as $\lim_{\varepsilon \searrow 0} \frac{\Gamma(P_t f)}{P_t f + \varepsilon}$).

In a further similarity with the chapter on Poincaré inequalities, the same families of inequalities may be established for the Euclidean (Brownian) heat semigroup $\mathbf{P} = (P_t)_{t \geq 0}$ on $\mathbb{R}^n$. Recall that in this case $|\nabla P_t f| \leq P_t(|\nabla f|)$. The corresponding local inequalities for the heat kernel measures then take the form

$$t \frac{\Gamma(P_t f)}{P_t f} \leq P_t(f \log f) - P_t \log P_t f \leq t P_t\left(\frac{\Gamma(f)}{f}\right).$$

Since at time $t = \frac{1}{2}$, the distribution of the heat semigroup P_t is the standard Gaussian measure, the first inequality recovers Proposition 5.5.1.

5.5.2 Local Logarithmic Sobolev Inequalities

The preceding analysis may be performed for any diffusion Markov semigroup $\mathbf{P} = (P_t)_{t \geq 0}$ under the curvature condition $CD(\rho, \infty)$ of Definition 3.3.14, p. 159. With respect to Theorem 4.7.2, p. 209, dealing with local Poincaré inequalities, the condition $CD(\rho, \infty)$ is enriched by several new (stronger) equivalences, in particular the reinforced curvature condition (ii) and the strong gradient bound (iii).

Theorem 5.5.2 (Local logarithmic Sobolev inequalities) *Let (E, μ, Γ) be a Markov Triple with semigroup $\mathbf{P} = (P_t)_{t \geq 0}$. The following assertions are equivalent.*

(i) *The curvature condition $CD(\rho, \infty)$ holds for some $\rho \in \mathbb{R}$.*
(ii) *For every function f in $\mathcal{A}_0$,*

$$4\,\Gamma(f)\bigl[\Gamma_2(f) - \rho\,\Gamma(f)\bigr] \geq \Gamma\bigl(\Gamma(f)\bigr).$$

(iii) *For every function f in $\mathcal{A}_0$, and every $t \geq 0$,*

$$\sqrt{\Gamma(P_t f)} \leq e^{-\rho t}\, P_t\bigl(\sqrt{\Gamma(f)}\bigr). \tag{5.5.4}$$

(iv) *For every positive function f in $\mathcal{A}_0$, and every $t \geq 0$,*

$$P_t(f \log f) - P_t f \log P_t f \leq \frac{1 - e^{-2\rho t}}{2\rho}\, P_t\!\left(\frac{\Gamma(f)}{f}\right). \tag{5.5.5}$$

(v) *For every positive function f in $\mathcal{A}_0$, and every $t \geq 0$,*

$$P_t(f \log f) - P_t f \log P_t f \geq \frac{e^{2\rho t} - 1}{2\rho}\, \frac{\Gamma(P_t f)}{P_t f}. \tag{5.5.6}$$

When $\rho = 0$, the quantities $\frac{1-e^{-2\rho t}}{2\rho}$ and $\frac{e^{2\rho t}-1}{2\rho}$ have to be replaced by t (as is the case for the heat semigroup on $\mathbb{R}^n$ of curvature $CD(0, \infty)$). Also recall from Remark 5.1.2 that the Fisher-type expressions such as $P_t(\frac{\Gamma(f)}{f})$ are understood as $\lim_{\varepsilon \searrow 0} P_t(\frac{\Gamma(f)}{f+\varepsilon})$. The inequalities in Theorem 5.5.2 extend to positive functions f in $\mathcal{A}$ following the various extension procedures described in Sect. 3.3, p. 151. Moreover, as for the corresponding Poincaré inequality (4.7.6), p. 210, the reverse logarithmic Sobolev inequality (5.5.6) (for $t > 0$) extends to measurable positive functions.

Proof We refer to Sect. C.6, p. 513, in Appendix C for the equivalence between the $CD(\rho, \infty)$ condition and (ii) already used in Theorem 3.3.18, p. 163, and at the heart of the (strong) gradient bound (iii). As explained there, this apparent reinforcement of the curvature condition only relies on the chain rule formulas for L and Γ, that is, the diffusion property. The proof from the gradient bound (iii) to the local logarithmic Sobolev inequality (iv) and its reverse form (v) is exactly the same as the one presented above for the Ornstein-Uhlenbeck and Brownian semigroups (with the only modification being the constant ρ). To complete the circle of equivalences, proceed as usual to the asymptotics as $t = 0$ of the local logarithmic Sobolev inequality or the reverse inequality yielding the curvature condition $CD(\rho, \infty)$. More simply, use that these (local) logarithmic Sobolev inequalities imply the corresponding (local) Poincaré inequalities (by Theorem 5.1.3) and the corresponding result for Poincaré inequalities (Theorem 4.7.2, p. 209). The proof is therefore complete. □

As in the corresponding Remark 4.7.3, p. 211, on local Poincaré inequalities, the implications from (iii) to (iv) and (v) of Theorem 5.5.2 work similarly if (iii) only holds up to some constant $K \geq 1$, that is, for every $t \geq 0$, and all functions f in $\mathcal{A}_0$,

$$\sqrt{\Gamma(P_t f)} \leq \sqrt{K}\, e^{-\rho t}\, P_t\bigl(\sqrt{\Gamma(f)}\bigr).$$

5.5 Local Logarithmic Sobolev Inequalities

The proof is entirely similar, and (5.5.5) for example, then simply reads

$$P_t(f \log f) - P_t f \log P_t f \leq \frac{K(1 - e^{-2\rho t})}{2\rho} P_t\left(\frac{\Gamma(f)}{f}\right).$$

Similarly, the analogue of Remark 4.7.4, p. 211, indicates that all the assertions of Theorem 5.5.2 are also equivalent to saying that there exist $t_0 > 0$ and a function $c(t) = t - \rho t^2 + o(t^2)$ such that, for any $t \in (0, t_0)$ and any positive f in $\mathcal{A}_0$,

$$P_t(f \log f) - P_t f \log P_t f \leq c(t) P_t\left(\frac{\Gamma(f)}{f}\right),$$

respectively

$$P_t(f \log f) - P_t f \log P_t f \geq c(t) \frac{\Gamma(P_t f)}{P_t f}.$$

While the proof of Theorem 5.5.2 is based on the reinforced curvature condition in terms of the commutation (5.5.4) of P_t and $\sqrt{\Gamma}$, the standard definition of $CD(\rho, \infty)$ is actually enough to reach the local logarithmic Sobolev inequalities of Theorem 5.5.2 (as for Poincaré inequalities). It is worth emphasizing the argument here since the principle will be used again in subsequent developments. The following proposition is the key technical step towards this goal, illustrating the role of the Γ_2 operator in the heat flow monotonicity principle. It will allow us later to reach stronger conclusions such as Sobolev-type inequalities as well as Harnack-type inequalities.

Proposition 5.5.3 *Let f be positive in $\mathcal{A}_0$, and let for $t > 0$,*

$$\Lambda(s) = P_s(P_{t-s} f \log P_{t-s} f), \quad s \in [0, t].$$

Then,

$$\Lambda'(s) = P_s\big(P_{t-s} f \, \Gamma(\log P_{t-s} f)\big)$$

and

$$\Lambda''(s) = 2 P_s\big(P_{t-s} f \, \Gamma_2(\log P_{t-s} f)\big).$$

Now, on the basis of these identities, under the curvature condition $CD(\rho, \infty)$, that is $\Gamma_2(f) \geq \rho \Gamma(f)$ for every $f \in \mathcal{A}_0$ (or $\mathcal{A}$), it follows that $\Lambda'' \geq 2\rho \Lambda'$. The local inequalities of Theorem 5.5.2 are then easily derived in this alternative way.

Proof As usual, we work with $f \in \mathcal{A}_0^{\text{const}+}$. The first derivative is a fairly general property, already used in the proof of Theorem 5.5.2. Indeed, whenever $\psi : \mathbb{R} \to \mathbb{R}$ is smooth enough, setting $\Lambda(s) = P_s(\psi(P_{t-s} f))$ and $g = P_{t-s} f$, $s \in [0, t]$, then

$$\Lambda'(s) = P_s\big(L\psi(g) - \psi'(g) Lg\big) = P_s\big(\psi''(g) \Gamma(g)\big)$$

by the diffusion property. For $\psi(r) = r \log r$, $r > 0$, $\psi'' = \frac{1}{r}$ and $\frac{1}{g}\Gamma(g) = g\,\Gamma(\log g)$, which yields the claim.

The second derivative is more specific to the function $\psi(r) = r \log r$. We have, with $\Psi = \psi''$, again by the heat equation,

$$\Lambda''(s) = P_s\big(\mathrm{L}\big(\Psi(g)\,\Gamma(g)\big) - \Psi'(g)\,\mathrm{L}g\,\Gamma(g) - 2\Psi(g)\,\Gamma(g, \mathrm{L}g)\big).$$

It remains to consider the quantity

$$E(g) = \mathrm{L}\big(\Psi(g)\,\Gamma(g)\big) - \Psi'(g)\,\mathrm{L}g\,\Gamma(g) - 2\Psi(g)\,\Gamma(g, \mathrm{L}g)$$

which by the change of variables formula (3.1.9), p. 124, for L may be written as

$$E(g) = 2\Psi(g)\,\Gamma_2(g) + 2\Psi'(g)\,\Gamma\big(g, \Gamma(g)\big) + \Psi''(g)\,\Gamma(g)^2.$$

For the choice of $\psi(r) = r \log r$, and thus $\Psi(r) = \frac{1}{r}$, this expression may be directly compared to the change of variables formula (3.3.2), p. 158, for the Γ_2 operator (consequence of the diffusion property of L), that is

$$\Gamma_2\big(\psi_1(g)\big) = \psi_1'^{\,2}(g)\,\Gamma_2(g) + \psi_1'(g)\,\psi_1''(g)\,\Gamma\big(g, \Gamma(g)\big) + \psi_1''^{\,2}(g)\,\Gamma(g)^2.$$

With $\psi_1(r) = \log r$, $E(g) = 2g\,\Gamma_2(\log g)$ from which the conclusion follows. □

Remark 5.5.4 The functions $\psi(r) = r^2$ ($r \in \mathbb{R}$) and $\psi(r) = r \log r$ ($r \in \mathbb{R}_+$) are the only ones for which the second derivative of $\Lambda(s) = P_s(\psi(P_{t-s}f))$, $s \in [0, t]$, takes such a simple form. For $\psi(r) = r^2$, even the subsequent derivatives have nice expressions yielding the iterated Gamma operators $\Gamma_3, \Gamma_4 \ldots$ (constructed by the same rule as Γ_2). This is no longer the case for the entropy $\psi(r) = r \log r$ where the third derivative already has a complicated form. Nevertheless, analogues of the preceding local inequalities may be obtained by replacing r^2 or $r \log r$ by r^α, $1 < \alpha < 2$ (or even more general functions as in Proposition 7.6.1, p. 383, below). Such inequalities (modified a little) are of interest for heat kernel decays as well as tail bounds between exponential and Gaussian behaviour (cf. Chap. 7).

5.5.3 Local Hypercontractivity

Under a logarithmic Sobolev inequality, the Markov semigroup $\mathbf{P} = (P_t)_{t \geq 0}$ is hypercontractive (Theorem 5.2.3). The same argument may be developed under a local logarithmic Sobolev inequality yielding hypercontractive characterizations of the curvature bound $CD(\rho, \infty)$. One advantage of this description is that it extends to positive measurable functions, and is therefore stable under various kinds of semigroup convergences. Furthermore, as shown by its proof, Theorem 5.5.5 provides a semigroup characterization of the curvature condition $CD(\rho, \infty)$ without any derivation argument.

Theorem 5.5.5 (Local hypercontractivity) *Let (E, μ, Γ) be a Markov Triple with semigroup $\mathbf{P} = (P_t)_{t \geq 0}$. The following assertions are equivalent.*

5.5 Local Logarithmic Sobolev Inequalities

(i) *The curvature condition $CD(\rho, \infty)$ holds for some $\rho \in \mathbb{R}$.*
(ii) *For all $t > 0$, $0 < s \leq t$ and $1 < p < q < \infty$ such that*

$$\frac{q-1}{p-1} = \frac{e^{2\rho t} - 1}{e^{2\rho s} - 1}, \qquad (5.5.7)$$

$$\left[P_s\big((P_{t-s}f)^q\big)\right]^{1/q} \leq \left[P_t(f^p)\right]^{1/p} \qquad (5.5.8)$$

for all positive measurable functions f.

As usual, when $\rho = 0$, then $\frac{1-e^{-2\rho t}}{2\rho}$ has to be replaced by t and $\frac{e^{2\rho t}-1}{e^{2\rho s}-1}$ by $\frac{t}{s}$.

Proof Theorem 5.5.2 indicates that (i) is equivalent to the local logarithmic Sobolev inequality (5.5.5). Assume then that (5.5.5) holds. Let f be in $\mathcal{A}_0^{\text{const}+}$ and consider

$$\Lambda(s) = P_s\big((P_{t-s}f)^{q(s)}\big)^{1/q(s)}, \quad s \in [0, t],$$

where $q : [0, t] \to (1, \infty)$. If $g = P_{t-s}f$,

$$\Lambda^{q-1}(s) = \operatorname{Ent}_{P_s}(g^q) + \frac{q^2(q-1)}{q'} P_s\big(g^{q-2}\Gamma(g)\big)$$

$$\leq q^2 \left(\frac{1 - e^{-2\rho s}}{2\rho} + \frac{q-1}{q'}\right) P_s\big(g^{q-2}\Gamma(g)\big)$$

by the local logarithmic Sobolev inequality (5.5.5). Now, if $q > 1$ satisfies $\frac{1-e^{-2\rho s}}{2\rho} + \frac{q-1}{q'} = 0$, that is,

$$\frac{q(s) - 1}{q(t) - 1} = \frac{e^{2\rho t} - 1}{e^{2\rho s} - 1},$$

then Λ is increasing which amounts to (5.5.8) (with $q(s) = q$ and $q(t) = p$). Conversely, assuming (ii), a first order Taylor expansion (with t fixed) as $p = 2$, $q = 2(1+\varepsilon)$ and $s = t(1-\alpha\varepsilon) + o(\varepsilon)$ with $\alpha = \frac{(1-e^{-2\rho t})}{\rho t}$ as $\varepsilon \to 0$ (for which (5.5.7) is satisfied in the limit), yields the local logarithmic Sobolev inequality (5.5.5) which is equivalent to $CD(\rho, \infty)$. Inequality (5.5.8) for functions in $\mathcal{A}_0^{\text{const}+}$ extends by density to positive measurable functions. The proof of Theorem 5.5.5 is complete. □

The same proof shows that

$$\left[P_t(f^p)\right]^{1/p} \leq \left[P_s\big((P_{t-s}f)^q\big)\right]^{1/q},$$

$t > 0$, $0 < s \leq t$, $-\infty < q < p < 1$ satisfying (5.5.7) and f strictly positive, may be added to the equivalences of Theorem 5.5.5.

It should be observed that Theorem 5.5.5 is not a direct consequence of the hypercontractivity Theorem 5.2.3 applied to the local logarithmic Sobolev inequalities, which would correspond to the semigroup with same carré du champ operator Γ and reversible measure $p_t(x, dy)$. On the other hand, Theorem 5.5.5 applied with $s = t - u$ yields as $t \to \infty$ with u fixed the hypercontractive bound $\|P_u f\|_q \leq \|f\|_p$ of Theorem 5.2.3.

This comment implies that (5.5.8) of Theorem 5.5.5 is optimal for the Ornstein-Uhlenbeck semigroup in the sense that $\rho = 1$ is the optimal parameter such that (5.5.8) holds. For the Brownian semigroup ($\rho = 0$), Theorem 6.7.7, p. 304, will indicate how this inequality becomes optimal with Gaussian extremal functions and parameters depending upon the dimension.

5.5.4 Some Heat Kernel Bounds

The application discussed here concerns some cheap but rough heat kernel bounds which may be produced from the exponential integrability of heat kernel measures. Assume that the Markov semigroup $\mathbf{P} = (P_t)_{t \geq 0}$ is such that P_t, $t > 0$, admits a density kernel $p_t(x, y)$ with respect to the invariant measure μ as

$$P_t f(x) = \int_E f(y)\, p_t(x, y)\, d\mu(y), \quad t > 0, \ x \in E,$$

where $p_t(x, y)$ is symmetric since the measure μ is reversible (Definition 1.2.4, p. 14).

By the Chapman-Kolmogorov equation (1.3.2), p. 17, for every $t > 0$ and $(x, y) \in E \times E$,

$$p_{2t}(x, y) = \int_E p_t(x, z)\, p_t(z, y)\, d\mu(z).$$

Then, for $c(t) > 0$ to be specified below, and with d the intrinsic distance on (E, μ, Γ) (Sect. 3.3.7, p. 166),

$$p_{2t}(x, y) \leq e^{-d(x,y)^2/c(t)} \int_E p_t(x, z)\, e^{2d(x,z)^2/c(t)}\, p_t(z, y)\, e^{2d(z,y)^2/c(t)}\, d\mu(z).$$

By the Cauchy-Schwarz inequality, and symmetry,

$$p_{2t}(x, y) \leq e^{-d(x,y)^2/c(t)} \int_E p_t(x, z)^2\, e^{4d(x,z)^2/c(t)}\, d\mu(z).$$

Assume now a uniform bound on the kernel $p_t(x, z) \leq K(t)$ (such bounds will be discussed later in Chaps. 6 and 7). Furthermore, assume that we are in a setting in which the local logarithmic Sobolev inequalities (5.5.5) of Theorem 5.5.2 hold for the measures $p_t(\cdot, z) d\mu(z)$. Therefore, together with the exponential integrability result for the Lipschitz distance function (Proposition 5.4.1), there exists a $c(t) > 0$ such that

$$\int_E p_t(x, z)\, e^{4d(x,z)^2/c(t)}\, d\mu(z) < \infty$$

and the bound is uniform in x. As a consequence of this discussion,

$$p_{2t}(x, y) \leq K'(t) e^{-d(x,y)^2/c(t)} \qquad (5.5.9)$$

where $K'(t) > 0$ only depends on t (and the choice of $c(t)$ in accordance with geometric features). This bound emphasizes, at a mild level, the standard Gaussian bounds which may be expected on heat kernels. Such bounds will be addressed later in Chap. 7. It should be added that the only features used in the previous argument are the triangle inequality for the distance function d and the fact that $z \mapsto d(x, z)$ is 1-Lipschitz.

5.6 Infinite-Dimensional Harnack Inequalities

This section is concerned with Harnack-type inequalities under the curvature condition $CD(\rho, \infty)$ and with various applications of the heat kernel logarithmic Sobolev inequalities of the previous section.

The main result is a Harnack-type inequality, a consequence of the gradient bounds under curvature conditions (cf. Theorem 3.3.18, p. 163, or (iii) of Theorem 5.5.2). We state it for convenience in the setting of smooth complete connected Riemannian manifolds (in order to freely speak of geodesics joining two points) although the Markov Triple version may also be reached by different means (see below). The algebra $\mathcal{A}_0$ then stands for the family of smooth compactly supported functions.

Theorem 5.6.1 (Wang's Harnack inequality) *Let $\mathbf{P} = (P_t)_{t \geq 0}$ be a Markov semigroup with infinitesimal generator $\mathrm{L} = \Delta_{\mathfrak{g}} - \nabla W \cdot \nabla$ on a complete connected Riemannian manifold $(M, \mathfrak{g})$. The following are equivalent.*

(i) *The curvature condition $CD(\rho, \infty)$ holds for some $\rho \in \mathbb{R}$.*
(ii) *For every positive measurable function f on M, every $t > 0$, every $x, y \in M$ and every $\alpha > 1$,*

$$(P_t f)^\alpha (x) \leq P_t(f^\alpha)(y) \exp\left(\frac{\alpha \rho\, d(x, y)^2}{2(\alpha - 1)(e^{2\rho t} - 1)}\right) \qquad (5.6.1)$$

where $d(x, y)$ is the Riemannian distance from x to y.

Inequality (5.6.1) is a *Harnack-type inequality* for infinite-dimensional diffusion operators (compare with the dimensional Harnack inequality (6.7.6), p. 302).

Proof We prove the implication from (i) to (ii) for $\alpha = 2$ and $\rho = 0$. Hence, the inequality to establish is, for a positive smooth function f,

$$(P_t f)^2(x) \leq P_t(f^2)(y)\, e^{d^2/2t}$$

where $d = d(x, y)$. The idea of the proof relies on a variation of the usual interpolation scheme by joining x and y by a constant speed geodesic $(x_s)_{s \in [0,t]}$ and then analyzing the function

$$\Lambda(s) = P_s\bigl((P_{t-s} f)^2\bigr)(x_s), \quad s \in [0, t].$$

Setting $g = P_{t-s} f$ as usual,

$$\Lambda'(s) = 2 P_s\bigl(|\nabla g|^2\bigr)(x_s) + x'_s \cdot \nabla P_s\bigl(g^2\bigr)(x_s)$$

$$\geq 2 P_s\bigl(|\nabla g|^2\bigr) - \frac{d}{t} \bigl|\nabla P_s\bigl(g^2\bigr)\bigr|.$$

Under the curvature condition $CD(0, \infty)$, by the gradient bound (iii) of Theorem 5.5.2,

$$\bigl|\nabla P_s\bigl(g^2\bigr)\bigr| \leq P_s\bigl(\bigl|\nabla\bigl(g^2\bigr)\bigr|\bigr) = 2 P_s\bigl(g |\nabla g|\bigr).$$

Therefore, for every $s \in [0, t]$,

$$\Lambda'(s) \geq 2 P_s\biggl(|\nabla g|^2 - \frac{d}{t} g|\nabla g|\biggr) \geq -\frac{d^2}{2t^2} P_s\bigl(g^2\bigr) = -\frac{d^2}{2t^2} \Lambda(s).$$

Integrating this differential inequality immediately yields the conclusion. The general case, $\alpha > 1$ and $\rho \neq 0$, is similar, the only difference is that we have to choose a geodesic $(x_s)_{s \in [0,1]}$ with a non-constant speed.

For the converse implication from (ii) to (i), assume again for simplicity that $\rho = 0$. Let $\alpha = 1 + \varepsilon$ and let y_ε be the exponential map starting from x with initial tangent vector v. A Taylor expansion (as ε goes to 0) of (5.6.1) then yields at the point $x \in M$,

$$P_t(f \log f) - P_t f \log P_t f \geq -v \cdot \nabla P_t(f) - P_t f \frac{|v|^2}{4t}.$$

Choosing $v = -2t \frac{\nabla P_t f}{P_t f}$ yields the logarithmic Sobolev inequality in its reverse form (5.5.6) which is equivalent to the curvature condition $CD(\rho, \infty)$. The proof is therefore complete. $\square$

Remark 5.6.2 (Log-Harnack inequality) The same method applied to $\Lambda(s) = P_s(\exp(P_{t-s} \log f))(x_s)$ where $(x_s)_{s \in [0,t]}$ is a geodesic between x and y, provides a log-Harnack inequality under the condition $CD(\rho, \infty)$,

$$P_t(\log f)(x) \leq \log P_t f(y) + \frac{\rho \, d(x, y)^2}{2(e^{2\rho t} - 1)} \tag{5.6.2}$$

for any strictly positive bounded measurable function f on M, any $t > 0$ and any $x, y \in M$. This inequality may be proved alternatively by letting $\alpha \to \infty$ in (5.6.1) (after the changing f into $f^{1/\alpha}$).

5.6 Infinite-Dimensional Harnack Inequalities

The latter log-Harnack inequality has a simple interesting consequence for lower bounds on the kernel density $p_t(x, y)$ of P_t with respect to μ, which we assume to be a probability density. Namely, applying it to $f(z) = p_t(x, z) + \varepsilon$, $z \in M$, and letting $\varepsilon \to 0$, it implies that for any $x, y \in M$,

$$\int_M p_t(x, z) \log p_t(x, z) d\mu(z) \leq \log p_{2t}(x, y) + \frac{\rho d(x, y)^2}{2(e^{2\rho t} - 1)}.$$

By Jensen's inequality for the convex function $r \mapsto r \log r$, $r \in \mathbb{R}_+$, with respect to μ, since $p_t(x, \cdot)$ is a probability density, the left-hand side of this inequality is positive. Replacing t by $\frac{t}{2}$, it follows that, for all $t > 0$ and $(x, y) \in M \times M$,

$$p_t(x, y) \geq \exp\left(-\frac{\rho d(x, y)^2}{2(e^{\rho t} - 1)}\right). \tag{5.6.3}$$

We conclude this section with a useful observation concerning the local logarithmic Sobolev inequalities of Theorem 5.5.2 and their relationships with the previous Harnack inequalities. It appears that the reverse logarithmic Sobolev inequalities (5.5.6) of Theorem 5.5.2 provide rather precise information on the kernels as $t \to 0$, which in turn get close to the Harnack inequality (5.6.1) of Theorem 5.6.1.

Consider a function $f \in \mathcal{A}_0^{\text{const}+}$ such that $\varepsilon \leq f \leq 1$ for some $\varepsilon > 0$. Then $P_t(f \log f) \leq 0$ and thus (5.5.6) ensures that

$$-P_t f \log P_t f \geq \frac{e^{2\rho t} - 1}{2\rho} \frac{\Gamma(P_t f)}{P_t f}.$$

It then follows by elementary means that

$$\Gamma\left(\sqrt{\log \frac{1}{P_t f}}\right) \leq \frac{\rho}{2(e^{2\rho t} - 1)}. \tag{5.6.4}$$

This gradient bound is an analogue of (4.7.7), p. 211, in the context of local Poincaré inequalities.

With the aim of applying (5.6.4) to Harnack inequalities, we assume for simplicity that $\rho = 0$. Then, by definition of the intrinsic distance (3.3.9), p. 166, for every $t > 0$ and $x, y \in E$,

$$\sqrt{\log \frac{1}{P_t f(x)}} \leq \sqrt{\log \frac{1}{P_t f(y)}} + \frac{d(x, y)}{2t}.$$

After some work, it may then be shown that for each $\varepsilon > 0$, there exists a $C(\varepsilon) > 0$ such that

$$(P_t f)^2(x) \leq C(\varepsilon) P_t(f^2)(y) e^{d(x,y)^2/(2+\varepsilon)t},$$

which is as close as possible to (5.6.1) (for $\alpha = 2$). Employing more refined arguments of an isoperimetric nature, the optimal bound will be achieved using this principle in Sect. 8.6, p. 421.

5.7 Logarithmic Sobolev Inequalities Under a Curvature-Dimension Condition

In this section we present an analysis corresponding to that of the Poincaré inequality of Sect. 4.8, p. 211, and investigate logarithmic Sobolev inequalities for the invariant probability measure μ under a curvature-dimension condition $CD(\rho, n)$ (Definition 3.3.14, p. 159).

To start with however, observe that under the condition $CD(\rho, \infty)$ with $\rho > 0$, one may let $t \to \infty$ in the local inequalities of Theorem 5.5.2 to get in the limit that for every positive function f in $\mathcal{A}_0$,

$$\mathrm{Ent}_\mu(f) \leq \frac{1}{2\rho} \int_E \frac{\Gamma(f)}{f} \, d\mu. \tag{5.7.1}$$

In other words (changing f into f^2 and extending to the domain $\mathcal{D}(\mathcal{E})$), μ satisfies $LS(\frac{1}{\rho})$. This limiting procedure requires the ergodicity properties described in Sect. 1.8, p. 32, which hold in particular under connexity (cf. e.g. Sect. 3.2.1, p. 140).

Although this logarithmic Sobolev inequality is contained in Theorem 5.7.4 below (corresponding to $n = \infty$), it is worth stating it independently.

Proposition 5.7.1 (Logarithmic Sobolev inequality under $CD(\rho, \infty)$) *Under the curvature condition $CD(\rho, \infty)$, $\rho > 0$, the Markov Triple (E, μ, Γ) satisfies a logarithmic Sobolev inequality $LS(C)$ with constant $C = \frac{1}{\rho}$. That is, for every function $f \in \mathcal{D}(\mathcal{E})$,*

$$\mathrm{Ent}_\mu(f^2) \leq \frac{2}{\rho} \mathcal{E}(f).$$

The constant is optimal for the example of the Ornstein-Uhlenbeck semigroup with invariant measure the standard Gaussian measure corresponding to the curvature condition $CD(1, \infty)$. The proposition covers probability measures $d\mu = e^{-W} dx$ with respect to the standard carré du champ operator on $\mathbb{R}^n$, where W satisfies $\nabla \nabla W \geq \rho \, \mathrm{Id}$ (as symmetric matrices) for some $\rho > 0$ (equivalently, $W(x) - \frac{\rho |x|^2}{2}$ is convex).

Corollary 5.7.2 *Let $d\mu = e^{-W} dx$ be a probability measure on the Borel sets of $\mathbb{R}^n$ where $W : \mathbb{R}^n \to R$ is a smooth potential such that $\nabla \nabla W \geq \rho \, \mathrm{Id}$ for some $\rho > 0$. Then μ satisfies the logarithmic Sobolev inequality $LS(\frac{1}{\rho})$.*

As for the corresponding Poincaré inequality (Proposition 4.8.2, p. 212), with respect to which it is actually a strengthening, a similar statement, with the same proof, holds on a weighted Riemannian manifold $(M, \mathfrak{g})$ for the operator $L = \Delta_\mathfrak{g} - \nabla W \cdot \nabla$ with the reversible (probability) measure μ having density e^{-W}

5.7 Logarithmic Sobolev Inequalities Under a Curvature-Dimension Condition

with respect to the Riemannian measure under the curvature condition

$$\text{Ric}(L) = \text{Ric}_\mathfrak{g} + \nabla\nabla W \geq \rho\, \mathfrak{g}$$

(cf. Sect. 1.16, p. 70, and Sect. C.6, p. 513).

Towards further developments under a curvature-dimension condition, it is worthwhile to formulate an analogue of Proposition 4.8.3, p. 214, which amounts to the second derivative of entropy along the semigroup, that is according to Bruijn's identity (Proposition 5.2.2), to the derivative of the Fisher information. This further emphasizes the semigroup interpolation principle and its connection with the Γ_2 operator, now with respect to the invariant measure. Indeed, for $f \in \mathcal{A}_0^{\text{const}+}$ say, consider $\Lambda(t) = \int_E P_t f \log P_t f \, d\mu, \, t \geq 0$. As in Proposition 5.5.3,

$$\Lambda'(t) = -\int_E P_t f\, \Gamma(\log P_t f)\, d\mu = -\mathrm{I}_\mu(P_t f)$$

and

$$\Lambda''(t) = 2\int_E P_t f\, \Gamma_2(\log P_t f)\, d\mu$$

where we recall the Fisher information I_μ from (5.1.6). Now, if $\Lambda''(t) \geq -\frac{2}{C}\Lambda'(t)$ for some $C > 0$, then $e^{2t/C}\Lambda'(t)$ is increasing in $t \geq 0$, and therefore

$$\mathrm{I}_\mu(P_t f) = -\Lambda'(t) \leq -e^{-2t/C}\Lambda'(0) = e^{-2t/C}\mathrm{I}_\mu(f) \tag{5.7.2}$$

for every $t \geq 0$. Writing then from de Bruijn's identity, for f positive in $\mathcal{A}_0$,

$$\text{Ent}_\mu(f) = \int_0^\infty \mathrm{I}_\mu(P_t f)\, dt, \tag{5.7.3}$$

the logarithmic Sobolev inequality $LS(C)$ follows.

Note that integrating the gradient bound (iii) of Theorem 5.5.2 with respect to the invariant measure μ together with the Cauchy-Schwarz inequality yields that, under the curvature condition $CD(\rho, \infty)$ for some $\rho \in \mathbb{R}$,

$$\mathrm{I}_\mu(P_t f) \leq e^{-2\rho t}\mathrm{I}_\mu(f) \tag{5.7.4}$$

for every positive function f in $\mathcal{A}_0$ and every $t \geq 0$. We therefore recover in this way Proposition 5.7.1 whenever $\rho > 0$.

As announced, from the expressions of Λ' and Λ'' above, an analogue of Proposition 4.8.3, p. 214, may be given at this stage.

Proposition 5.7.3 *A Markov Triple (E, μ, Γ) satisfies a logarithmic Sobolev inequality $LS(C)$ for some $C > 0$ if*

$$\int_E f\, \Gamma(\log f)\, d\mu \leq C \int_E f\, \Gamma_2(\log f)\, d\mu \tag{5.7.5}$$

for every positive function f in $\mathcal{A}$ which is bounded together with $\Gamma(f)$ and Lf.

At a technical level, (5.7.5) should actually be used for $P_t f$ whenever $f \in \mathcal{A}_0^{\text{const}+}$. Now, under a curvature condition $CD(\rho, \infty)$ for some $\rho \in \mathbb{R}$, the analysis developed in Sect. 3.3, p. 151, indicates that for $t > 0$, $P_t f$ is indeed in $\mathcal{A}$ and moreover bounded together with $\Gamma(P_t f)$ and $LP_t f$. Proposition 5.7.5 will only be used below under curvature bounds.

Inequality (5.7.5) is very similar indeed to the dual inequality

$$\int_E \Gamma(f) d\mu \leq C \int_E \Gamma_2(f) d\mu$$

of Proposition 4.8.3, p. 214, used for Poincaré inequalities. But whereas the latter expresses a dual equivalent formulation of the Poincaré inequality, (5.7.5) turns out to be strictly stronger in general than the logarithmic Sobolev inequality $LS(C)$.

On the basis of the previous semigroup tools, we next investigate logarithmic Sobolev inequalities under curvature-dimension conditions. As for Poincaré inequalities (cf. Theorem 4.8.4, p. 215), the constant $C = \frac{1}{\rho}$ in Proposition 5.7.1 may be improved under a curvature-dimension hypothesis $CD(\rho, n)$ with $n < \infty$. Again by Proposition 5.1.3, the next statement strengthens the Poincaré inequality from Theorem 4.8.4, p. 215. The proof is a prototypical example of the power of the change of variables formula for the Γ_2 operator (and several integral identities involving Γ_2 in the proof below will turn out to be useful later in this monograph).

Theorem 5.7.4 (Logarithmic Sobolev inequality under $CD(\rho, n)$) *Under the curvature-dimension condition $CD(\rho, n)$, $\rho > 0$, $n > 1$, the Markov Triple (E, μ, Γ) satisfies a logarithmic Sobolev inequality $LS(C)$ with constant $C = \frac{n-1}{\rho n}$.*

Proof The objective is to reach the criterion of Proposition 5.7.3. The main argument relies on the chain rule formula (3.3.2), p. 158, for the Γ_2 operator used here with $\psi(r) = e^{ar}$ for a real parameter a so to get, for $g \in \mathcal{A}_0$,

$$\Gamma_2(e^{ag}) = a^2 e^{2ag} \big[\Gamma_2(g) + a\,\Gamma\big(g, \Gamma(g)\big) + a^2\,\Gamma(g)^2\big]. \tag{5.7.6}$$

Note that $\Gamma_2(e^{ag})$ actually takes place in the extended algebra $\mathcal{A}$. Now on the other hand, by Proposition 3.3.17, p. 161, which applies to e^{ag} with $g \in \mathcal{A}_0$, and the change of variables for L,

$$\int_E \Gamma_2(e^{ag}) d\mu = \int_E \big(\mathrm{L}(e^{ag})\big)^2 d\mu$$

$$= a^2 \int_E e^{2ag} \big[\mathrm{L}g + a\,\Gamma(g)\big]^2 d\mu$$

$$= a^2 \int_E e^{2ag} \big[(\mathrm{L}g)^2 + 2a\,\mathrm{L}g\,\Gamma(g) + a^2 \Gamma(g)^2\big] d\mu.$$

5.7 Logarithmic Sobolev Inequalities Under a Curvature-Dimension Condition

Observe that since $g \in \mathcal{A}_0$, all the terms in the latter integral are bounded and therefore integrable with respect to the finite measure μ. Integrating by parts,

$$\int_E e^{2ag} \mathrm{L}g\, \Gamma(g) d\mu = -\int_E e^{2ag}\, \Gamma\bigl(g,\Gamma(g)\bigr) d\mu - 2a \int_E e^{2ag}\, \Gamma(g)^2 d\mu. \quad (5.7.7)$$

After some elementary algebra, it follows from the preceding identities that

$$\int_E e^{2ag} (\mathrm{L}g)^2 d\mu = \int_E e^{2ag} \bigl[\Gamma_2(g) + 3a\, \Gamma\bigl(g,\Gamma(g)\bigr) + 4a^2\, \Gamma(g)^2\bigr] d\mu. \quad (5.7.8)$$

This identity turns out to be most useful, providing a way to classify many kinds of quantities which are related to each other after integration by parts.

Now, by (5.7.6) again, the curvature-dimension condition $CD(\rho, n)$ applied to e^{ag} yields that

$$\Gamma_2(g) + a\, \Gamma\bigl(g, \Gamma(g)\bigr) + a^2\, \Gamma(g)^2 \geq \rho\, \Gamma(g) + \frac{1}{n} \bigl[\mathrm{L}g + a\, \Gamma(g)\bigr]^2.$$

After multiplication by e^g and integration with respect to μ,

$$\int_E e^g \bigl[\Gamma_2(g) + a\, \Gamma\bigl(g,\Gamma(g)\bigr) + a^2\, \Gamma(g)^2\bigr] d\mu$$

$$\geq \rho \int_E e^g\, \Gamma(g) d\mu$$

$$+ \frac{1}{n} \int_E e^g \bigl[(\mathrm{L}g)^2 + 2a\, \mathrm{L}g\, \Gamma(g) + a^2 \Gamma(g)^2\bigr] d\mu \quad (5.7.9)$$

$$= \rho \int_E e^g\, \Gamma(g) d\mu$$

$$+ \frac{1}{n} \int_E e^g \bigl[(\mathrm{L}g)^2 - 2a\, \Gamma\bigl(g,\Gamma(g)\bigr) + (a^2 - 2a) \Gamma(g)^2\bigr] d\mu$$

where in the last step, (5.7.7) (with $a = \frac{1}{2}$) was used again. In the latter, replace now the term $\int_E e^g (\mathrm{L}g)^2 d\mu$ by (5.7.8) (with $a = \frac{1}{2}$) to get that

$$\int_E e^g\, \Gamma_2(g) d\mu \geq \frac{\rho n}{n-1} \int_E e^g\, \Gamma(g) d\mu$$

$$+ \frac{3 - 2(n+2)a}{2(n-1)} \int_E e^g\, \Gamma\bigl(g, \Gamma(g)\bigr) d\mu \quad (5.7.10)$$

$$+ \frac{(a-1)^2 - na^2}{n-1} \int_E e^g\, \Gamma(g)^2 d\mu.$$

Choosing $a = \frac{3}{2(n+2)}$ (for which the term $\int_E e^g \Gamma(g, \Gamma(g)) d\mu$ disappears) yields that

$$\int_E e^g \Gamma_2(g) d\mu \geq \frac{\rho n}{n-1} \int_E e^g \Gamma(g) d\mu + \frac{4n-1}{4(n+2)^2} \int_E e^g \Gamma(g)^2 d\mu \quad (5.7.11)$$

and in particular

$$\int_E e^g \Gamma_2(g) d\mu \geq \frac{\rho n}{n-1} \int_E e^g \Gamma(g) d\mu. \quad (5.7.12)$$

This is almost the required (5.7.5) although for $f = e^g$ where $g \in \mathcal{A}_0$. Now, by Proposition 3.3.6, p. 156, and Lemma 3.3.16, p. 160, under the curvature condition (5.7.12) extends to functions $f \in \mathcal{A}$ which are bounded, together with $\Gamma(f)$ and Lf (such functions belong to $\mathcal{D}(L)$). The proof of Theorem 5.7.4 is complete. □

Theorem 5.7.4 applies to the sphere $\mathbb{S}^n$ in $\mathbb{R}^{n+1}$ (Sect. 2.2, p. 81), and for the symmetric Jacobi operator with dimension n (Sect. 2.7.4, p. 113), the lower bound $\frac{\rho n}{n-1}$ of the logarithmic Sobolev constant being optimal since these two models both satisfy a curvature-dimension condition $CD(n-1, n)$ and their Poincaré constant (equal to $\frac{1}{n}$) is already optimal. The case $n = 1$ corresponding to the one-dimensional torus, for which $\rho = 0$, may easily be integrated into the picture. Indeed, by Proposition 4.5.5, p. 199, the Poincaré inequality $P(1)$ holds in this case, and by the dual formulation of Proposition 4.8.3, p. 214, for every f in $\mathcal{D}(L)$,

$$\int_E \Gamma_2(f) d\mu \geq \int_E \Gamma(f) d\mu.$$

This inequality applied to $f = e^{g/2}$ (for $g \in \mathcal{A}_0$) yields by (5.7.6) that

$$\int_E e^g \left[\Gamma_2(g) + \frac{1}{2} \Gamma(g, \Gamma(g)) + \frac{1}{4} \Gamma(g)^2 \right] d\mu \geq \int_E e^g \Gamma(g) d\mu. \quad (5.7.13)$$

Now, under $CD(0, 1)$, (5.7.9) with $a = \frac{1}{3}$ shows that

$$\int_E e^g \Gamma_2(g) d\mu \geq \int_E e^g \left[(Lg)^2 - \Gamma(g, \Gamma(g)) - \frac{2}{3} \Gamma(g)^2 \right] d\mu.$$

Comparing with (5.7.8) with $a = \frac{1}{2}$ indicates that

$$\int_E e^g \Gamma(g, \Gamma(g)) d\mu + \frac{2}{3} \int_E e^g \Gamma(g)^2 d\mu \leq 0.$$

Therefore, by (5.7.13),

$$\int_E e^g \Gamma_2(g) d\mu \geq \int_E e^g \Gamma(g) d\mu,$$

that is (5.7.12) with the convention that $\frac{\rho n}{n-1} = 1$. The logarithmic Sobolev inequality with constant 1 (equal to the Poincaré constant) follows similarly. Of course, simple integrations by parts on the torus may be used to provide a direct proof. The constant is optimal since it is already optimal for the Poincaré inequality.

By analogy with the Poincaré inequalities of Proposition 4.5.5, p. 199, we may emphasize the preceding conclusion for $n = 1$ on the unit interval (after scaling).

Proposition 5.7.5 *Let* $Lf = f''$ *be the Sturm-Liouville operator acting on smooth functions* f *on* $[0, 1]$ *with invariant measure the Lebesgue measure* dx.

(i) *For any smooth function* $f : [0, 1] \to \mathbb{R}$,

$$\mathrm{Ent}_{dx}(f^2) \leq \frac{2}{\pi^2} \int_{[0,1]} f'^2 dx$$

and the constant is optimal.

(ii) *For any smooth function* $f : [0, 1] \to \mathbb{R}$ *such that* $f(0) = f(1)$,

$$\mathrm{Ent}_{dx}(f^2) \leq \frac{1}{2\pi^2} \int_{[0,1]} f'^2 dx$$

and the constant is optimal.

These two results correspond respectively to Neumann and periodic boundary conditions on the unit interval. The first assertion follows from the second one by symmetrization and periodization. Sharpness of the constants is a consequence of the corresponding result for Poincaré inequalities.

5.8 Notes and References

The notion of entropy has a long run in mathematics and information theory. Landmark contributions involving entropy are due to L. Boltzmann in mathematical physics back in the 19th century, and to C. Shannon in information theory [383].

Logarithmic Sobolev inequalities have emerged under different forms and names in the second half of the 20th century. Although not identified as such, one of the first occurrences of logarithmic Sobolev inequalities (in their Euclidean version discussed in the next chapter) may be found in the work [387] by A. Stam in the context of information theory (see [106, 141, 157, 424, 426]). A further early observation is due to P. Federbush [183] in mathematical physics. However, it is really with the seminal paper of L. Gross [224] that logarithmic Sobolev inequalities were clearly identified and emphasized as a central object of interest. One main result of [224] is the equivalence (Theorem 5.2.3) with hypercontractivity (in a rather general context). Hypercontractivity for the Ornstein-Uhlenbeck semigroup had been put forward earlier by E. Nelson [325, 326] (see also later [327, 328]), while a first hypercontractive result goes back to J. Glimm [204] and to B. Simon and R. Høegh-

Krohn [384] (in the form of Theorem 5.2.5). Hyperboundedness in the sense of the hypothesis of Theorem 5.2.5 implies a spectral gap, as conjectured in [384]. This was recently established in [311] (see [432] for a prior weaker result), and considerably improves upon Proposition 5.1.3 (although not quantitatively). The existence and uniqueness of ground states for Schrödinger operators [223] was a further important step in the theory. The interplay between logarithmic Sobolev inequalities and hypercontractivity has actually been quite rich and fruitful. The reference [145] gives an account of some of the main aspects of this connection, while the reviews [225] and [226] survey the developments of logarithmic Sobolev inequalities over the last decades and collect references. Further general references on logarithmic Sobolev inequalities are [372] and [14] as an introduction, and [229, 238] with an emphasis on statistical physics models (for bounded and unbounded spins).

The tightening Proposition 5.1.3 has been observed in [158] (on the basis of an independent proof of (5.1.8)). Lemma 5.1.4 is due to O. Rothaus [370] as is Remark 5.1.5 [367–369]. The stability by perturbation property, Proposition 5.1.6, is due to R. Holley and D. Stroock [245]. The general proof presented here on the basis of Lemma 5.1.7 is taken from [14] (see also [372]). Proposition 5.1.8 is part of the slicing technique further developed in the context of Sobolev-type inequalities in Chap. 6.

Theorem 5.2.1 goes back to the origin of entropy and to the Boltzmann H-Theorem (see [86, 424]). The Pinsker-Csiszár-Kullback inequality (5.2.2) can be found in numerous references including e.g. [14, 141, 426]. de Bruijn's identity of Proposition 5.2.2 is recorded in [387] (see also [157]) for the Euclidean heat semigroup and extensively developed in a more general framework in [36]. Exponential decays in uniform norms for models from statistical mechanics may be developed with the same principle (see e.g. [229]). The main equivalence Theorem 5.2.3 with hypercontractivity is due to L. Gross [224]. Reverse hypercontractivity (Remark 5.2.4) was first observed by C. Borell and S. Janson [88]. Theorem 5.2.5 goes back to [384]. Proposition 5.2.6 is due to P. Cattiaux [115], who gave a different proof. A main feature of logarithmic Sobolev inequalities, stability under products, had already been emphasized and used in a critical way in [224]. The dimension-free property is in particular a powerful argument in statistical mechanics (cf. [229, 372]).

Integrability of Wiener chaos via hypercontractivity was emphasized by C. Borell [88]. The discussion at the end of Sect. 5.3 on spectral features of hypercontractive bounds is taken from [81].

The so-called Herbst argument from Proposition 5.4.1 in Sect. 5.4 goes back to an unpublished letter (1975) of I. Herbst to L. Gross. The argument is presented in [146], and emphasized later in the paper [6] which revived interest in the question and gave rise to several subsequent contributions, in particular related to measure concentration, synthesized in the notes [276] (to which we refer for more details). See also [90] for applications to concentration inequalities. Proposition 5.4.2 is taken from [7]. The Muckenhoupt criterion of Theorem 5.4.5 is due to S. Bobkov and F. Götze [77] and is based on the corresponding result for Poincaré inequalities (see the preceding chapter).

5.8 Notes and References

There are at least fifteen different proofs of the Gaussian logarithmic Sobolev inequality (Proposition 5.5.1), some of them mentioned in this book, including the proof of L. Gross [224] relying on the two-point space inequality, tensorization and the central limit Theorem. The proof given here using heat flow monotonicity along the Ornstein-Uhlenbeck semigroup goes back to the seminal paper [36] and is partially inspired by the stochastic calculus proof of Nelson's hypercontractivity by J. Neveu [329]. A careful and detailed exposition of the argument in the language of partial differential equations is given in [412] in the Gaussian case, and later in [16] for strictly convex potentials in $\mathbb{R}^n$. The paper [412] actually emphasized the importance of logarithmic Sobolev inequalities towards convergence to equilibrium in this context. The logarithmic Sobolev inequality for Gaussian measures (as well as the corresponding Poincaré inequality) has been extended to the infinite-dimensional setting of the Wiener measure (cf. Sect. 2.7.2, p. 108), and further to Brownian paths on a Riemannian manifold. With this task in mind, finite-dimensional approximations may be used. Alternatively, the semigroup scheme may also be developed directly in the infinite-dimensional context, in particular by means of the Clark-Ocone probabilistic interpolation formula for functionals of the Brownian paths (cf. [5, 105, 182, 250, 417] and [251] for a synthesis). The local heat kernel inequalities of Sect. 5.5 have been analyzed in [27] (see also [28, 277, 394]) on the basis of the ideas in [36]. The latter paper, which introduces the basic interpolation scheme along the semigroup and the importance of the Γ_2 operator, also describes the logarithmic Sobolev inequalities under the curvature-dimension condition of Sect. 5.7. See [26] for a first synthesis. Remark 5.5.4 is developed in [274] for entropy. Local hypercontractivity (Theorem 5.5.5) is taken from [31].

The Harnack inequalities for an infinite-dimensional diffusion operator from Theorem 5.6.1 are due to F.-Y. Wang [428]. The log-Harnack inequality (5.6.2) can be found in [76] as well as in [434] as a limit case of the previous Harnack inequalities. The lower bound (5.6.3) is pointed out in [435]. Discussions and extensions are developed in a series of papers by this author [431, 433, 435]. The useful observation (5.6.4) is due to M. Hino [242] and will be employed in a critical way in Sect. 8.7, p. 425. Heat kernel bounds of the type (5.5.9) are surveyed in [217].

Logarithmic Sobolev inequalities under curvature-dimension condition (Sect. 5.7) go back to [26, 36] (see also [371]). That the integral criterion from Proposition 5.7.3 is not equivalent to the corresponding logarithmic Sobolev inequality has been pointed out by B. Helffer ([14, 238]). Proposition 5.7.5 on logarithmic Sobolev inequalities on an interval may be found in [367, 368, 439] and [173]. Hypercontractivity on the n-sphere via ultraspherical polynomials is studied in [322].

Chapter 6
Sobolev Inequalities

Following our study of Poincaré and logarithmic Sobolev inequalities, this chapter is devoted to the investigation of Sobolev inequalities. Sobolev inequalities play a central role in analysis, providing in particular compact embeddings and tight connections with heat kernels bounds. They are also deeply linked with the geometric structure of the underlying state space through conformal invariance. Here our study only covers a small fraction of the vast subject of Sobolev inequalities, with a specific focus on the main theme of Markov diffusion operators and semigroups. As in the preceding chapters, we work here in the context of a Full Markov Triple ("Markov Triple") (E, μ, Γ) with Dirichlet form $\mathcal{E}$, infinitesimal generator L, Markov semigroup $\mathbf{P} = (P_t)_{t \geq 0}$ and underlying function algebras $\mathcal{A}_0$ and $\mathcal{A}$ as summarized in Sect. 3.4, p. 168. Once again, several properties and results described in this chapter remain valid in more general settings (in particular for Standard Markov Triples), and the reader may adapt the statements and proofs if necessary.

The chapter starts with a brief exposition of the classical Sobolev inequalities on the model spaces, namely the Euclidean, spherical and hyperbolic spaces. These examples will both give hints and provide a better understanding of further developments. The next section investigates the various definitions of Sobolev-type inequalities in the Markov Triple context, emphasizing in particular (logarithmic) entropy-energy and Nash-type inequalities. These inequalities will be further investigated in Chap. 7. Section 6.3 presents the basic equivalence between Sobolev inequalities and (uniform) heat kernel bounds (ultracontractivity), and Sect. 6.4 highlights applications to compact embeddings. Sections 6.5 and 6.6 address issues on tensorization properties of Sobolev-type inequalities and diameter, Lipschitz functions and volume estimates. In particular, with respect to Poincaré and logarithmic Sobolev inequalities, tensorization of Sobolev inequalities have to take into account a dimensional parameter. Local inequalities under the semigroup $(P_t)_{t \geq 0}$ are investigated next, providing in particular a heat flow approach to the celebrated Li-Yau parabolic inequality. Section 6.8 establishes the sharp Sobolev inequality under the curvature-dimension condition $CD(\rho, n)$ ($\rho > 0, n < \infty$), covering the example of the standard sphere. The subsequent section describes the conformal invariance properties of Sobolev inequalities, and, as a consequence, the sharp Sobolev inequalities in the Euclidean and hyperbolic spaces on the basis of the inequality on the sphere. Gagliardo-Nirenberg inequalities form a further family of equivalent Sobolev-type inequalities, well-suited to non-linear porous medium and fast diffusion equations. These topics, and their geometric counterparts, are presented and detailed in Sects. 6.10 and 6.11, providing in particular a fast diffusion approach to several Sobolev inequalities of interest.

6.1 Sobolev Inequalities on the Model Spaces

The fundamental example of a Sobolev inequality takes place in the Euclidean space $\mathbb{R}^n$ equipped with the Lebesgue measure dx and the standard carré du champ operator $\Gamma(f) = |\nabla f|^2$. The basic Sobolev inequality expresses here that for every smooth compactly supported function f on $\mathbb{R}^n$, $n > 2$ (and thus every f in the Dirichlet domain),

$$\|f\|_p^2 \leq C_n \|\,|\nabla f|\,\|_2^2 = C_n \int_{\mathbb{R}^n} |\nabla f|^2 dx \qquad (6.1.1)$$

where $p = \frac{2n}{n-2}$ and $C_n > 0$ is an explicit constant whose optimal value is known (and will be explicitly calculated below in Theorem 6.9.4). The norms are of course understood here with respect to the Lebesgue measure. For simplicity $\|\,|\nabla f|\,\|_2$ is denoted $\|\nabla f\|_2$ below.

The exponent p takes the value $\frac{2n}{n-2}$ ($n > 2$) for a very good reason. It is only for this value that the Sobolev inequality (6.1.1) is invariant under dilation, that is by the change of $f(x)$ into $f_s(x) = f(sx)$, $s > 0$. Indeed, $\|f_s\|_p = s^{-n/p}\|f\|_p$ while $\|\nabla f_s\|_2^2 = s^{2-n}\|\nabla f\|_2^2$, and it is only for the exponent $p = \frac{2n}{n-2}$ that the inequality can hold, any other value leading to a contradiction as $s \to 0$ or ∞. Dilations actually play a central role in the analysis of Sobolev inequalities in Euclidean space as will be illustrated throughout this chapter.

On the unit sphere $\mathbb{S}^n$ in $\mathbb{R}^{n+1}$, with the normalized uniform measure μ and the spherical carré du champ operator Γ as defined in Sect. 2.2, p. 81, there is also a Sobolev inequality which takes a somewhat different form, but with the same exponent $p = \frac{2n}{n-2}$. Namely, for every smooth function f on $\mathbb{S}^n$ (in the Dirichlet domain)

$$\|f\|_p^2 \leq \|f\|_2^2 + \frac{4}{n(n-2)} \int_{\mathbb{S}^n} \Gamma(f) d\mu. \qquad (6.1.2)$$

The norms denote here the norms in $\mathbb{L}^r(\mu)$, $r \geq 1$. In this form, the Sobolev inequality is closer to the Poincaré and logarithmic Sobolev inequalities discussed in the preceding chapters, and may be established for large families of Markov Triples. In the absence of dilations, the value of $p = \frac{2n}{n-2}$ is rather mysterious. By monotonicity (Jensen's inequality), different values of $2 \leq p \leq \frac{2n}{n-2}$ may actually be considered, however not exceeding the critical value $\frac{2n}{n-2}$. Note also that (6.1.2) is tight in the sense of Remark 4.2.2, p. 182, implying that the only functions f such that $\Gamma(f) = 0$ are constant.

Finally, there is also a Sobolev inequality on the hyperbolic space $\mathbb{H}^n$ (cf. Sect. 2.3, p. 88) which takes the form

$$\|f\|_p^2 \leq -A\|f\|_2^2 + C_n \int_{\mathbb{H}^n} \Gamma(f) d\mu \qquad (6.1.3)$$

for some (explicit) $A > 0$, $C_n > 0$, and every smooth function $f : \mathbb{H}^n \to \mathbb{R}$.

These inequalities will be established as the chapter unfolds (cf. Theorem 6.9.4 below, where the various constants will be made explicit). Actually, the (deep) connections between the three Sobolev inequalities, on the three model spaces of geometry, Euclidean, spherical and hyperbolic, as well as the explicit values of the constants and of the extremizers, will be discussed later in Sect. 6.9 by means of conformal invariance, and form the basis of the geometric analysis of Sobolev inequalities.

6.2 Sobolev and Related Inequalities

Following the preceding introductory section, we present families of Sobolev inequalities, and some of their equivalent forms, in the setting of a Markov Triple (E, μ, Γ). Recall the norm $\|\cdot\|_p$ in $\mathbb{L}^p(\mu)$, $1 \leq p \leq \infty$.

6.2.1 Sobolev Inequalities

The next definition introduces the notion of a *Sobolev inequality* in a general context.

Definition 6.2.1 (Sobolev inequality) A Markov Triple (E, μ, Γ) is said to satisfy a Sobolev inequality $S(p; A, C)$ with exponent $p > 2$ and constants $A \in \mathbb{R}$, $C > 0$, if for all functions f in the Dirichlet domain $\mathcal{D}(\mathcal{E})$,

$$\|f\|_p^2 \leq A\|f\|_2^2 + C\,\mathcal{E}(f).$$

On the basis of the model examples, the exponent p will often take the form $p = \frac{2n}{n-2}$ for some $n > 2$ (not always an integer), and we then speak of a Sobolev inequality $S_n(A, C)$ of *dimension* n (> 2) and constants $A \in \mathbb{R}$, $C > 0$. Under a Sobolev inequality $S(p; A, C)$, if $f \in \mathcal{D}(\mathcal{E})$ then $f \in \mathbb{L}^p(\mu)$. As usual, it is enough to state (and prove) such inequalities for a family of functions f which is dense in the Dirichlet domain $\mathcal{D}(\mathcal{E})$ (typically the algebra $\mathcal{A}_0$).

When the measure μ is finite and normalized into a probability measure, applying the Sobolev inequality from Definition 6.2.1 to $f = \mathbb{1}$ shows that $A \geq 1$. (In particular, a Sobolev inequality $S(p; A, C)$ with $A = 0$ (or $A \leq 0$) can only hold if the measure μ has infinite mass.) The inequality is called *tight* if $A = 1$, and is then denoted by $S(p;C)$ (respectively $S_n(C)$) for simplicity. The best constant $C > 0$ for which such an inequality $S(p;C)$ or $S_n(C)$ holds is sometimes referred to as the *Sobolev constant* (of the Markov Triple).

As for logarithmic Sobolev inequalities, tightness only holds if there is a Poincaré inequality (Definition 4.2.1, p. 181).

Proposition 6.2.2 (Tightening with a Poincaré inequality) *Let (E, μ, Γ) be a Markov Triple with μ a probability measure, and let $p > 2$. A tight Sobolev inequality $S(p;C)$ implies a Poincaré inequality $P(\frac{C}{p-2})$. Furthermore, a Sobolev inequality $S(p;A,C)$ together with a Poincaré inequality $P(C')$ imply a tight Sobolev inequality $S(p;(p-1)(AC'+C))$.*

It should be noted that the application of the first assertion to the Sobolev inequality (6.1.2) on the sphere $\mathbb{S}^n$ yields the Poincaré inequality $P(\frac{1}{n})$ with its sharp constant (cf. Theorem 4.8.4, p. 215). In particular, both constants (1 and $\frac{4}{n(n-2)}$) in the Sobolev inequality (6.1.2) are optimal.

Proof The proof is similar to Proposition 5.1.3, p. 238. For the first assertion, apply $S(p;C)$ to $f = 1 + \varepsilon g$ where $g \in \mathcal{D}(\mathcal{E})$ with $\int_E g\,d\mu = 0$. As $\varepsilon \to 0$, it is not difficult to verify by a Taylor expansion that

$$\|f\|_p^2 - \|f\|_2^2 = (p-2)\varepsilon^2 \int_E g^2 d\mu + o(\varepsilon^2)$$

while, clearly, $\mathcal{E}(f) = \varepsilon^2 \mathcal{E}(g)$. It follows that

$$(p-2)\int_E g^2 d\mu \leq C\,\mathcal{E}(g)$$

which amounts to the Poincaré inequality $P(\frac{C}{p-2})$.

The second assertion relies on the analogue of (5.1.8), p. 240, stating that, for $p > 2$ and every $f \in \mathbb{L}^p(\mu)$,

$$\|f\|_p^2 \leq \left(\int_E f\,d\mu\right)^2 + (p-1)\|\hat{f}\|_p^2 \qquad (6.2.1)$$

where $\hat{f} = f - \int_E f\,d\mu$. (This inequality actually covers (5.1.8) as $p \to 2$.) For a proof, consider a bounded function f such that $\int_E f\,d\mu = 1$ represented as $f = 1 + rg$, $r \in \mathbb{R}$, with $\int_E g\,d\mu = 0$ and $\int_E g^2 d\mu = 1$. The inequality to be established is then

$$\psi(r) = \|1+rg\|_p^2 \leq 1 + (p-1)r^2 \|g\|_p^2$$

which may be shown by elementary calculus, proving that $\psi''(r) \leq 2(p-1)\|g\|_p^2$.

6.2 Sobolev and Related Inequalities

Using (6.2.1), apply $S(p\,;A,C)$ to $\hat{f}$ (in $\mathcal{D}(\mathcal{E})$) and then the Poincaré inequality $P(C')$ to get

$$\|f\|_p^2 \leq \left(\int_E f\,d\mu\right)^2 + (p-1)\bigl[A\,\|\hat{f}\|_2^2 + C\,\mathcal{E}(f)\bigr]$$

$$\leq \left(\int_E f\,d\mu\right)^2 + (p-1)\bigl(A\,C' + C\bigr)\mathcal{E}(f).$$

By Jensen's inequality, this inequality is even better than the expected Sobolev inequality $S(p\,;(p-1)(AC'+C))$. The proof of the proposition is complete. □

6.2.2 Logarithmic Entropy-Energy and Nash Inequalities

We next turn to an important feature of Sobolev-type inequalities, namely that they may be presented equivalently (up to constants) in various forms, such as families of logarithmic Sobolev inequalities or other functional inequalities. Further equivalent formulations will appear throughout this chapter as well as in Chaps. 7 and 8. While these various formulations mostly originate from inequalities in Euclidean space, they may be addressed in an abstract framework. For the matter of comparison with the most familiar inequalities in Euclidean space, we use the notation $S_n(A,C)$ where $n > 2$ reflects a dimension. Recall the entropy notation Ent_μ introduced in (5.1.1), p. 236.

Proposition 6.2.3 *For a Markov Triple* (E, μ, Γ), *the following implications hold.*

(i) *Under* $S_n(A,C)$, $A \geq 0$, $C > 0$, *for every function* f *in* $\mathcal{D}(\mathcal{E})$ *with* $\int_E f^2\,d\mu = 1$,

$$\text{Ent}_\mu(f^2) \leq \frac{n}{2}\log\bigl(A + C\,\mathcal{E}(f)\bigr). \qquad (6.2.2)$$

(ii) *Under* (6.2.2), *for every function* f *in* $\mathcal{D}(\mathcal{E})$,

$$\|f\|_2^{n+2} \leq \bigl[A\,\|f\|_2^2 + C\,\mathcal{E}(f)\bigr]^{n/2}\,\|f\|_1^2. \qquad (6.2.3)$$

(iii) *Conversely, under* (6.2.3), *a Sobolev inequality* $S_n(A_1, C_1)$ *holds with constants* $A_1 \geq 0$ *and* $C_1 > 0$ *only depending on* n, A *and* C. *Moreover* $A_1 = 0$ *whenever* $A = 0$.

This proposition introduces in particular the *logarithmic entropy-energy inequality* (6.2.2) and the *Nash inequality* (6.2.3) on the basis of their classical counterparts in $\mathbb{R}^n$. The classical Nash inequality in $\mathbb{R}^n$ says that for every smooth compactly supported function $f : \mathbb{R}^n \to \mathbb{R}$,

$$\|f\|_2^{n+2} \leq C_n\,\|\nabla f\|_2^n\,\|f\|_1^2 \qquad (6.2.4)$$

where the sharp constant C_n is known (see the Notes and References). While Nash inequalities with $A < 0$ may be considered similarly, we stick for simplicity to the case $A \geq 0$. By concavity, the logarithmic entropy-energy inequality (6.2.2) is equivalent to the family of (defective) logarithmic Sobolev inequalities

$$\mathrm{Ent}_\mu(f^2) \leq \Phi'(r)\mathcal{E}(f) + \Psi(r), \quad r \in (0, \infty), \tag{6.2.5}$$

for every f in $\mathcal{D}(\mathcal{E})$ with $\int_E f^2 d\mu = 1$, where $\Phi(r) = \frac{n}{2}\log(A + Cr)$ and $\Psi(r) = \Phi(r) - r\Phi'(r)$, $r \in (0, \infty)$. The classical logarithmic entropy-energy inequality in $\mathbb{R}^n$ is discussed below in Proposition 6.2.5 as an equivalent form of the logarithmic Sobolev inequality for Gaussian measures. Generalized entropy-energy and Nash-type inequalities will be thoroughly studied in Chap. 7.

When μ is a probability, the logarithmic entropy-energy and Nash inequalities (6.2.2) and (6.2.3) are called tight if $A = 1$. A tight logarithmic entropy-energy inequality implies a logarithmic Sobolev inequality $LS(\frac{Cn}{4})$. Applied to $f = 1 + \varepsilon g$ with $\varepsilon \to 0$, the tight Nash inequality implies a Poincaré inequality $P(\frac{Cn}{2})$.

Proposition 6.2.3 thus indicates that, up to constants (but we will however see that optimal constants play a crucial role in a variety of problems of interest), the preceding three inequalities are equivalent. Many further formulations may be considered here, as is clear from the proof, such as for example the Gagliardo-Nirenberg inequalities presented in Sect. 6.10, the preceding being however amongst the most used ones. Note that both the logarithmic entropy-energy and Nash inequalities make sense for $n \geq 1$, or even $n > 0$, while the Sobolev inequality requires $n > 2$.

Proof of Proposition 6.2.3 We begin with the first implication, proceeding as in Proposition 5.1.8, p. 241, with the convex function $\phi : r \mapsto \log(\|f\|_{1/r})$, $r \in (0, 1]$, for which $\phi'(\frac{1}{2}) = -\mathrm{Ent}_\mu(f^2)$ (under $\|f\|_2 = 1$). Thus, by convexity with $p = \frac{2n}{n-2} > 2$,

$$n\left[\phi\left(\frac{1}{2}\right) - \phi\left(\frac{1}{p}\right)\right] \leq \phi'\left(\frac{1}{2}\right),$$

which therefore amounts to

$$\mathrm{Ent}_\mu(f^2) \leq \frac{n}{2}\log(\|f\|_p^2). \tag{6.2.6}$$

It remains to apply the Sobolev inequality $S_n(A, C)$ to obtain the conclusion.

We next turn to the second claim, using the same method. Write here

$$2\left[\phi(1) - \phi\left(\frac{1}{2}\right)\right] \leq \phi'\left(\frac{1}{2}\right),$$

that is, again for a function f such that $\|f\|_2 = 1$,

$$\log\left(\frac{1}{\|f\|_1^2}\right) \leq \mathrm{Ent}_\mu(f^2).$$

An application of (6.2.2) immediately yields (6.2.3).

6.2 Sobolev and Related Inequalities

We are left with the last implication, which is more involved and which relies on the slicing technique already developed in Proposition 5.1.8, p. 241. It is enough to establish the Sobolev inequality for a positive bounded function f in $\mathcal{D}(\mathcal{E})$. The idea is to apply the Nash inequality (6.2.3) to the sequence of functions $f_k = (f - 2^k)_+ \wedge 2^k$, $k \in \mathbb{Z}$. Let $\mathcal{N}_k = \{f > 2^k\}$, $k \in \mathbb{Z}$. Since $2^k \mathbb{1}_{\mathcal{N}_{k+1}} \leq f_k \leq 2^k \mathbb{1}_{\mathcal{N}_k}$,

$$2^{2k} \mu(\mathcal{N}_{k+1}) \leq \int_E f_k^2 \, d\mu \leq 2^{2k} \mu(\mathcal{N}_k) \quad \text{and} \quad \int_E f_k \, d\mu \leq 2^k \mu(\mathcal{N}_k)$$

for every $k \in \mathbb{Z}$. Therefore, (6.2.3) applied to each f_k yields that

$$\left(2^{2k} \mu(\mathcal{N}_{k+1}) \right)^{1+(n/2)} \leq \beta_k^{n/2} \, 2^{2k} \mu(\mathcal{N}_k)^2 \tag{6.2.7}$$

where $\beta_k = A \, 2^{2k} \mu(\mathcal{N}_k) + C \, \mathcal{E}(f_k)$, $k \in \mathbb{Z}$. In other words, with $\alpha_k = 2^{pk} \mu(\mathcal{N}_k)$ ($p = \frac{2n}{n-2}$), for every $k \in \mathbb{Z}$,

$$\alpha_{k+1} \leq 2^p \, \beta_k^{n/(n+2)} \, \alpha_k^{4/(n+2)}.$$

By Hölder's inequality with exponents $\frac{n+2}{n}$ and $\frac{n+2}{2}$, it follows that

$$\sum_{k \in \mathbb{Z}} \alpha_{k+1} \leq 2^p \left(\sum_{k \in \mathbb{Z}} \beta_k \right)^{n/(n+2)} \left(\sum_{k \in \mathbb{Z}} \alpha_k^2 \right)^{2/(n+2)}.$$

By the upper bound $\sum_{k \in \mathbb{Z}} \alpha_k^2 \leq (\sum_{k \in \mathbb{Z}} \alpha_k)^2$, it follows that

$$\left(\sum_{k \in \mathbb{Z}} \alpha_k \right)^{(n-2)/n} \leq 2^{2(n+2)/(n-2)} \sum_{k \in \mathbb{Z}} \beta_k.$$

Now, according to Proposition 3.1.17, p. 137,

$$\sum_{k \in \mathbb{Z}} \mathcal{E}(f_k) \leq \mathcal{E}(f),$$

while on the other hand, since $f = \sum_{k \in \mathbb{Z}} \mathbb{1}_{\mathcal{N}_k \setminus \mathcal{N}_{k+1}} f$,

$$\int_E f^2 \, d\mu \geq \sum_{k \in \mathbb{Z}} 2^{2k} \left(\mu(\mathcal{N}_k) - \mu(\mathcal{N}_{k+1}) \right) = \frac{3}{4} \sum_{k \in \mathbb{Z}} 2^{2k} \mu(\mathcal{N}_k).$$

Hence, by definition of the β_k's,

$$\sum_{k \in \mathbb{Z}} \beta_k \leq \frac{4A}{3} \int_E f^2 \, d\mu + C \, \mathcal{E}(f).$$

It remains to observe that $\int_E f^p \, d\mu \leq 2^p \sum_{k \in \mathbb{Z}} \alpha_k$ to conclude the Sobolev inequality $S_n(A_1, C_1)$ for some A_1 and C_1 only depending on n and the Nash constants

A, C. Clearly $A_1 = 0$ whenever $A = 0$. The proof of Proposition 6.2.3 is thus complete. □

Remark 6.2.4 As for Poincaré inequalities (Lemma 4.2.3, p. 182), the Cauchy-Schwarz inequality may be used in the Nash inequality (6.2.3) to reach spectral inequalities on functions in $\mathcal{D}(\mathcal{E})$ with support in some (measurable) subset B of E. For example, whenever $\mu(B) \leq (2A)^{-n/2}$, for any function $f \in \mathcal{D}(\mathcal{E})$ with support in B,

$$\int_B f^2 d\mu \leq 2C\mu(B)^{2/n} \mathcal{E}(f).$$

In the classical Euclidean case, corresponding to $A = 0$ (and $\mu(E)$ infinite), this inequality is known as a Faber-Krahn inequality. It describes a lower bound on the spectrum of the restriction of the operator L to B with Dirichlet boundary conditions. (See also Remark 8.2.2, p. 399.) Using similar slicing techniques, this inequality, valid for any $B \subset E$, implies in turn the corresponding Sobolev inequality with the same dimension (but different constants).

As mentioned previously, in $\mathbb{R}^n$, the logarithmic entropy-energy inequality (6.2.2) for the Lebesgue measure is actually equivalent to the logarithmic Sobolev inequality for Gaussian measures (Proposition 5.5.1, p. 258).

Proposition 6.2.5 (Euclidean logarithmic Sobolev inequality) *In $\mathbb{R}^n$, for the Lebesgue measure dx and the usual carré du champ operator and Dirichlet form $\mathcal{E}(f) = \int_{\mathbb{R}^n} |\nabla f|^2 dx$,*

$$\mathrm{Ent}_{dx}(f^2) \leq \frac{n}{2} \log\left(\frac{2}{n\pi e} \mathcal{E}(f)\right) \tag{6.2.8}$$

for every function $f : \mathbb{R}^n \to \mathbb{R}$ in the Dirichlet domain $\mathcal{D}(\mathcal{E})$ with $\int_{\mathbb{R}^n} f^2 dx = 1$. The constant $\frac{2}{n\pi e}$ is optimal.

Proof Start from the logarithmic Sobolev inequality for the standard Gaussian measure $d\mu(x) = (2\pi)^{-n/2} e^{-|x|^2/2} dx$ in $\mathbb{R}^n$ (Proposition 5.5.1, p. 258),

$$\mathrm{Ent}_\mu(g^2) \leq 2 \int_{\mathbb{R}^n} |\nabla g|^2 d\mu. \tag{6.2.9}$$

Apply this inequality to $g(x) = (2\pi)^{n/4} e^{|x|^2/4} f(x)$, $x \in \mathbb{R}^n$, where f is smooth with compact support and such that $\int_{\mathbb{R}^n} f^2 dx = 1 \,(= \int_{\mathbb{R}^n} g^2 d\mu)$, to get that

$$\mathrm{Ent}_\mu(g^2) = \mathrm{Ent}_{dx}(f^2) + \frac{1}{2} \int_{\mathbb{R}^n} f^2 |x|^2 dx + \frac{n}{2} \log(2\pi).$$

On the other hand,

$$\int_{\mathbb{R}^n} |\nabla g|^2 d\mu = \int_{\mathbb{R}^n} \left|\nabla f + \frac{x}{2} f\right|^2 dx = \int_{\mathbb{R}^n} \left[|\nabla f|^2 + x \cdot f \nabla f + \frac{|x|^2}{4} f^2\right] dx.$$

6.2 Sobolev and Related Inequalities

Integrate by parts the middle term of the integral on the right-hand side of the preceding equality to get

$$\int_{\mathbb{R}^n} x \cdot f \nabla f \, dx = \frac{1}{4} \int_{\mathbb{R}^n} \nabla(|x|^2) \cdot \nabla(f^2) dx = -\frac{1}{4} \int_{\mathbb{R}^n} f^2 \Delta(|x|^2) dx.$$

Since $\Delta(|x|^2) = 2n$, after (a somewhat miraculous) simplification, the logarithmic Sobolev inequality (6.2.9) for μ implies that

$$\mathrm{Ent}_{dx}(f^2) \leq 2 \mathcal{E}(f) - \frac{n}{2} \log(2\pi e^2).$$

Now change $f(x)$ into $f_s(x) = s^{n/2} f(sx)$, $x \in \mathbb{R}^n$, $s > 0$. Since $\int_{\mathbb{R}^n} f_s^2 dx = \int_{\mathbb{R}^n} f^2 dx = 1$, $\mathcal{E}(f_s) = s^2 \mathcal{E}(f)$ and

$$\mathrm{Ent}_{dx}(f_s^2) = \mathrm{Ent}_{dx}(f^2) + n \log s,$$

applying the preceding inequality to f_s for every $s > 0$, and optimizing ($s^2 = \frac{n}{4\mathcal{E}(f)}$) yields the conclusion. The constant is optimal since this proof clearly indicates that (6.2.8) for the Lebesgue measure is actually equivalent to the logarithmic Sobolev inequality (6.2.9) for the Gaussian measure μ, which is sharp. Since the exponential functions $e^{a \cdot x}$, $a \in \mathbb{R}^n$, $x \in \mathbb{R}^n$, are the extremal functions of the Gaussian logarithmic Sobolev inequality, extremal functions of the Euclidean logarithmic Sobolev inequality are given by Gaussian kernels $e^{a \cdot x - |x|^2/2\sigma^2}$, $a \in \mathbb{R}^n$, $\sigma > 0$, $x \in \mathbb{R}^n$. Proposition 6.2.5 is established. □

Remark 6.2.6 The Euclidean logarithmic Sobolev inequality of Proposition 6.2.5 may actually be shown to follow from the Sobolev inequality in $\mathbb{R}^n$ with its sharp constant. Indeed, apply (6.1.1) to $f^{\otimes k}$ on $\mathbb{R}^{nk}$ for a given (smooth compactly supported) function $f : \mathbb{R}^n \to \mathbb{R}$. With the help of the sharp constant of the Sobolev inequality in Euclidean space (cf. Theorem 6.9.4), (6.2.8) follows in the limit as $k \to \infty$.

The fact that, after the action of dilations, an entropy-energy or logarithmic Sobolev inequality is equivalent to a Sobolev inequality is actually a common phenomenon. It is an illustration of the fact that, for the optimal exponent $p = \frac{2n}{n-2}$, the Sobolev inequality is invariant under dilations in the Euclidean space. The same operation may be performed for any functional inequality on the Euclidean space. After optimization under dilations, it produces in general some new functional inequality which is equivalent to a Sobolev inequality. Further illustrations of this phenomenon will be presented in Sect. 6.10 which deals with Gagliardo-Nirenberg inequalities. Unfortunately, there are very few such functional inequalities for which this operation is effectively tractable and produces useful inequalities.

6.3 Ultracontractivity and Heat Kernel Bounds

Ultracontractivity is for Sobolev inequalities what hypercontractivity is for logarithmic Sobolev inequalities. However, historically ultracontractivity was investigated earlier and expresses a stronger property, namely that under a Sobolev inequality for a Markov Triple (E, μ, Γ), the associated semigroup $(P_t)_{t \geq 0}$ maps $\mathbb{L}^1(\mu)$ into $\mathbb{L}^\infty(\mu)$. By interpolation it then also maps $\mathbb{L}^p(\mu)$ into $\mathbb{L}^q(\mu)$, $1 \leq p < q \leq \infty$. In the following, it will sometimes be convenient to state a number of embedding results with the operator norm

$$\|P_t\|_{p,q} = \sup_{\|f\|_p \leq 1} \|P_t f\|_q \qquad (6.3.1)$$

of P_t from $\mathbb{L}^p(\mu)$ into $\mathbb{L}^q(\mu)$, $1 \leq p, q \leq \infty$.

Under a logarithmic Sobolev inequality, hypercontractivity thus indicates that $\|P_t\|_{p,q} \leq 1$ for a suitable relation between $1 < p < q < \infty$ and $t > 0$ (Theorem 5.2.3, p. 246). *Ultracontractivity* formally expresses that $\|P_t\|_{1,\infty} < \infty$, and more precisely investigates the decay of $\|P_t\|_{1,\infty}$ as a function of $t > 0$. Theorem 6.3.1 below describes the equivalence between this decay and Sobolev inequalities. To get a better feeling for the quantity $\|P_t\|_{1,\infty}$, recall from Proposition 1.2.5, p. 14, that whenever $\|P_t\|_{1,\infty}$ is finite, the operators P_t, $t > 0$, are represented by a density kernel in $\mathbb{L}^\infty(\mu \otimes \mu)$ (thus with respect to the invariant measure μ) in the sense that there exists, for every $t > 0$, a (symmetric) measurable function $p_t(x, y)$ on the product space $E \times E$ such that, for μ-almost every $x \in E$, $\int_E p_t(x,y)^2 d\mu(y) < \infty$,

$$P_t f(x) = \int_E f(y)\, p_t(x, y) d\mu(y)$$

and $\|P_t\|_{1,\infty} = \|p_t(\cdot, \cdot)\|_\infty$ in $\mathbb{L}^\infty(E \times E, \mu \otimes \mu)$.

The next statement presents the equivalence between Sobolev inequalities and uniform heat kernel bounds. With respect to the corresponding hypercontractivity result, exact constants are not really preserved in the equivalence, and thus C below varies from line to line (see however Remark 6.3.2 after the proof).

Theorem 6.3.1 (Ultracontractivity) *Let (E, μ, Γ) be a Markov Triple with semigroup $\mathbf{P} = (P_t)_{t \geq 0}$. Let further $n > 2$ be a (Sobolev) dimension. The following are equivalent.*

(i) *The Sobolev inequality $S_n(A, C)$ holds for some constants $A \geq 0$ and $C > 0$.*
(ii) *There is a constant $C > 0$ such that for every $0 < t \leq 1$,*

$$\|P_t\|_{1,2} \leq \frac{C}{t^{n/4}}.$$

(iii) *There is a constant $C > 0$ such that for every $0 < t \leq 1$,*

$$\|P_t\|_{1,\infty} \leq \frac{C}{t^{n/2}}.$$

6.3 Ultracontractivity and Heat Kernel Bounds

If $A = 0$, $S_n(C)$ is equivalent to (ii) and (iii) holds for every $t > 0$.

Note that the requirement $n > 2$ is only necessary for the Sobolev inequality $S_n(A, C)$ in (i) and not for the ultracontractive bounds (ii) and (iii). Actually, the proof below will transit through Nash inequalities (6.2.3) justifying the extension to any $n > 0$.

Proof Start with the equivalence between (ii) and (iii). Note first that if $\|P_t\|_{1,2} \leq K(t)$, then, by duality, $\|P_t\|_{2,\infty} \leq K(t)$. Since $P_t = P_{t/2} \circ P_{t/2}$, it is therefore bounded from $\mathbb{L}^1(\mu)$ into $\mathbb{L}^\infty(\mu)$ with norm $K^2(\frac{t}{2})$. This shows that (ii) $\Rightarrow$ (iii). The converse implication is obtained by the classical Riesz-Thorin interpolation Theorem which asserts that, for an operator P, whenever $\|P\|_{p_1,q_1} \leq K_1$ and $\|P\|_{p_2,q_2} \leq K_2$, then for every $\theta \in [0, 1]$,

$$\|P\|_{p_\theta,q_\theta} \leq K_1^\theta K_2^{1-\theta}$$

where

$$\frac{1}{p_\theta} = \frac{\theta}{p_1} + \frac{1-\theta}{p_2}, \quad \frac{1}{q_\theta} = \frac{\theta}{q_1} + \frac{1-\theta}{q_2}.$$

Applying this result with $p_1 = p_2 = 1$, $q_1 = \infty$, $q_2 = 1$, $K_1 = Ct^{-n/2}$, $K_2 = 1$ (since P_t is a contraction) and $\theta = \frac{1}{2}$ then yields the claim since $p_\theta = 1$, $q_\theta = 2$ and $K_\theta = C^{1/2} t^{-n/4}$. The same argument shows that $\|P_t\|_{p,q} \leq (Ct^{-\frac{n}{2}})^{(\frac{1}{p}-\frac{1}{q})}$, $t > 0$, $1 \leq p < q \leq \infty$, described in Corollary 6.3.3 below.

Consider now the equivalence between the first two assertions of Theorem 6.3.1. There are numerous ways to deduce ultracontractive bounds from a Sobolev inequality. The simplest is perhaps to use the Nash inequality (6.2.3) (which is a consequence of the Sobolev inequality $S_n(A, C)$ by Proposition 6.2.3). To this end, we make use of the method developed in Theorem 4.2.5, p. 183, to prove the decay of the semigroup in $\mathbb{L}^2(\mu)$ under a Poincaré inequality. That is, given a positive function f in $\mathcal{D}(\mathcal{E})$ with integral 1, set $\Lambda(t) = \int_E (P_t f)^2 d\mu$, $t \geq 0$. Then $\Lambda'(t) = -2\mathcal{E}(P_t f)$, and by invariance of μ, $\int_E P_t f d\mu = 1$ for every $t \geq 0$. The Nash inequality (6.2.3) applied to $P_t f$ then reads

$$\Lambda(t) \leq \left[A\,\Lambda(t) - \frac{C}{2} \Lambda'(t) \right]^{1-\theta}, \quad t \geq 0,$$

where we set $\theta = \frac{2}{n+2}$. It follows that the function $e^{\lambda t}(1 - A\,\Lambda^{-r}(t))$, $t \geq 0$, is decreasing, where

$$\lambda = \frac{2Ar}{C} \quad \text{and} \quad r = \frac{\theta}{1-\theta} = \frac{2}{n}.$$

Therefore, for every $t > 0$,

$$\Lambda(t) \leq \left(\frac{A}{1 - e^{-\lambda t} + \frac{A e^{-\lambda t}}{\Lambda(0)^r}} \right)^{1/r} \leq \left(\frac{A}{1 - e^{-\lambda t}} \right)^{n/2}. \quad (6.3.2)$$

If $A > 0$, the latter is bounded from above by $C' t^{-n/2}$ for some constant $C' > 0$ for every $0 < t \leq 1$. If $A = 0$, $\frac{A}{1-e^{-\lambda t}}$ is replaced by its limit $\frac{C}{2rt} = \frac{Cn}{4t}$ (for all $t > 0$).

Next turn to the converse implication, that is from the decay (ii) to the Nash inequality (6.2.3) equivalent to the Sobolev inequality (i). Using the same notation, we only give the argument for $A > 0$ (the case $A = 0$ being similar). It may be assumed, again for $f \geq 0$ such that $\int_E f d\mu = 1$, that for every $t > 0$, $\Lambda(t) \leq C(1 + t^{-n/2})$. By Lemma 4.2.6, p. 184, the function $\log \Lambda$ is convex. Therefore, for every $\theta \in [0, 1]$, and every $t > 0$,

$$\Lambda(t) \leq \Lambda(0)^{1-\theta} \left[C\left(1 + \left(\frac{t}{\theta}\right)^{-n/2} \right) \right]^{\theta}.$$

Choose then $\theta = \alpha t$ for some fixed α to be specified below, which can be achieved as soon as t is small enough. At $t = 0$, the two sides of the preceding inequality are equal. A Taylor expansion at $t = 0$ then yields

$$-2\mathcal{E}(f) \leq \Lambda(0)\left[-\alpha \log \Lambda(0) + \alpha \log\left(C(1 + \alpha^{n/2}) \right) \right].$$

Choose $\alpha = \frac{\mathcal{E}(f)}{\Lambda(0)}$ so that

$$\log \Lambda(0) \leq 2 + \log\left(C\left[1 + \left(\frac{\mathcal{E}(f)}{\Lambda(0)}\right)^{n/2} \right] \right)$$

and hence

$$\Lambda(0)^{1+(n/2)} \leq K\left[\Lambda(0)^{n/2} + \mathcal{E}(f)^{n/2} \right] \leq K_1 \left[\Lambda(0) + \mathcal{E}(f) \right]^{n/2},$$

for some $K, K_1 > 0$. Since $\Lambda(0) = \int_E f^2 d\mu$ and $\int_E f d\mu = 1$, the announced Nash inequality follows by homogeneity.

The proof of Theorem 6.3.1 is complete. □

Remark 6.3.2 The preceding proof does not attempt to make the constants sharp, and of course the arguments may be tightened at some point. It is nevertheless possible to draw some (non-optimal) information concerning the relationships between the various constants C in Theorem 6.3.1. For example, under the Sobolev inequality $S_n(A, C)$ for some $A \geq 0$, $C > 0$,

$$\|P_t\|_{1,\infty} \leq \frac{C'}{t^{n/2}}$$

for every $0 < t \leq 1$ (every $t > 0$ if $A = 0$) where $C' = (\frac{Cn}{2}(1 + \frac{4A}{Cn}))^{n/2}$.

In the last part of this section, we make a few observations and describe some consequences of Theorem 6.3.1. In particular, we assume in the following that $A \geq 0$ in a given Sobolev inequality $S_n(A, C)$. As a consequence of the Sobolev inequality (6.1.1), one recovers the fact that the standard heat kernels on $\mathbb{R}^n$ are uniformly

6.3 Ultracontractivity and Heat Kernel Bounds

bounded by $Ct^{-n/2}$, $t > 0$, which is of course obvious from the explicit representation (2.1.1), p. 78. A more careful analysis will show in Corollary 7.1.6, p. 354, that the optimal bound $(4\pi t)^{-n/2}$ may actually be deduced from the Euclidean logarithmic Sobolev inequality (6.2.8) with its sharp constant. In Chap. 7, further methods producing ultracontractive bounds from Sobolev inequalities will be developed.

The first corollary describes other forms of ultracontractive bounds.

Corollary 6.3.3 *Under $S_n(A, C)$, for any $1 \leq p < q \leq \infty$,*

$$\|P_t\|_{p,q} \leq \frac{C'}{t^{\frac{n}{2}(\frac{1}{p} - \frac{1}{q})}}$$

for every $0 < t \leq 1$ or $t > 0$ depending on whether $A > 0$ or $A = 0$, where $C' > 0$ depends on the Sobolev constants A, C and n, p, q. Moreover, for any $\lambda > 0$, the resolvent operator $R_\lambda = (\lambda\,\mathrm{Id} - L)^{-1}$ is bounded from $\mathbb{L}^p(\mu)$ into $\mathbb{L}^\infty(\mu)$ as soon as $p > \frac{n}{2}$. When $p \leq \frac{n}{2}$, it is bounded from $\mathbb{L}^p(\mu)$ into $\mathbb{L}^q(\mu)$ for $q < \frac{pn}{n-2p}$.

Proof The first assertion was mentioned in the proof of Theorem 6.3.1 (as a consequence of the Riesz-Thorin Theorem and the fact that P_t is a contraction on all $\mathbb{L}^p(\mu)$-spaces). For the resolvent R_λ, use the integral representation $R_\lambda = \int_0^\infty P_t e^{-\lambda t} dt$ from (A.1.2), p. 474. By the preceding, $\|P_t\|_{p,\infty} \leq Ct^{-n/2p}$, $0 < t \leq 1$. In addition, since the operators P_t, $t \geq 0$, are contractions in all $\mathbb{L}^p(\mu)$-spaces, the norm $\|P_t\|_{p,\infty}$ is decreasing in t, and is in particular bounded for $t \in (1, \infty)$. Consequently, for some constant $C > 0$ only depending on p and the Sobolev constants,

$$\|R_\lambda\|_{p,\infty} \leq \int_0^\infty \|P_t\|_{p,\infty} e^{-\lambda t} dt \leq C \int_0^1 t^{-n/2p} dt + C \int_1^\infty e^{-\lambda t} dt.$$

The claim follows for $p > \frac{n}{2}$. Together with a further interpolation, the same argument leads to the corresponding assertion for $p < \frac{n}{2}$. Corollary 6.3.3 is proved. $\square$

When the measure μ is a probability measure and when the Sobolev inequality is tight (that is $S_n(1, C)$ holds), more precise bounds as $t \to \infty$ may be obtained. Recall that in this setting tightness is equivalent to a Poincaré inequality (Proposition 6.2.2).

Proposition 6.3.4 *Assume that μ is a probability measure satisfying a Sobolev inequality $S_n(A, C_1)$ and a Poincaré inequality $P(C_2)$. Then there are constants $C > 0$ and $T > 0$, depending only on A, C_1, C_2 and n, such that the density kernels $p_t(x, y), t > 0, (x, y) \in E \times E$, of P_t with respect to the measure μ satisfy*

$$|p_t(x, y) - 1| \leq C e^{-t/C_2}$$

for every $t \geq T$ and ($\mu \otimes \mu$-almost) every $(x, y) \in E \times E$.

Proof We prove the result with $T = 2$. For $t > 0$, consider the operator $P_t^0 f = P_t f - \int_E f d\mu$ whose kernel is $p_t(x, y) - 1$. From the hypotheses, $\|P_1^0\|_{1,\infty} \leq C'$ so that $\|P_1^0\|_{1,2} \leq \sqrt{C'}$ and $\|P_1^0\|_{2,\infty} \leq \sqrt{C'}$. On the other hand, the Poincaré inequality indicates by Theorem 4.2.5, p. 183, that

$$\|P_t^0 f\|_2 \leq e^{-t/C_2} \|f\|_2.$$

By composition $P_{2+t}^0 = P_1^0 \circ P_t^0 \circ P_1^0$ is then bounded from $\mathbb{L}^1(\mu)$ into $\mathbb{L}^\infty(\mu)$ with norm $Ce^{-t/2C_2}$, from which the upper bound on $|p_t(x,y) - 1|$ follows. □

It is worth mentioning that the general Proposition 1.2.6, p. 15, applied to P_t combined with an upper bound of the form $p_t(x,y) \leq 1 + Ce^{-t/C_2}$ yields that $p_t(x,y) \geq 1 - 2Ce^{-t/4C_2}$, which is therefore somewhat worse than the bound of the previous proposition.

6.4 Ultracontractivity and Compact Embeddings

Theorem 6.3.1 indicates that as soon as a Sobolev inequality holds, the semigroup $(P_t)_{t \geq 0}$ is ultracontractive, and may therefore be represented by a bounded density kernel $p_t(x, y), t > 0, (x, y) \in E \times E$, as

$$P_t f(x) = \int_E f(y) p_t(x, y) d\mu(y) \quad t > 0, \ x \in E,$$

with respect to the invariant measure μ. When the measure μ is finite, the fact that P_t may be represented by a bounded kernel $p_t(x, y)$ implies that it is Hilbert-Schmidt, since then

$$\int_E \int_E p_t^2(x, y) d\mu(x) d\mu(y) < \infty.$$

Therefore, the operator $P_t, t > 0$, is compact (cf. Sect. A.6, p. 483). We summarize these conclusions in a statement.

Corollary 6.4.1 *Under a Sobolev inequality $S_n(A, C)$ for a finite measure μ, the operators $P_t, t > 0$, are Hilbert-Schmidt and the spectrum of the Markov generator L is discrete.*

It is a specific consequence of this result that the embedding from the Dirichlet space $\mathcal{D}(\mathcal{E})$ into $\mathbb{L}^2(\mu)$ is compact. Recall that $\mathcal{D}(\mathcal{E})$ is endowed with the norm $\|f\|_\mathcal{E} = [\|f\|_2^2 + \mathcal{E}(f)]^{1/2}$.

Theorem 6.4.2 *Under a Sobolev inequality $S_n(A, C)$ for a finite measure μ, for any sequence $(f_k)_{k \in \mathbb{N}}$ in $\mathcal{D}(\mathcal{E})$ bounded with respect to the norm $\|\cdot\|_\mathcal{E}$, there is a subsequence which converges in $\mathbb{L}^2(\mu)$.*

Proof If necessary, normalize μ into a probability measure. Since $(f_k)_{k\in\mathbb{N}}$ is bounded in $\mathbb{L}^2(\mu)$, extract first a subsequence $(f_{k_\ell})_{\ell\in\mathbb{N}}$ which converges weakly to f in $\mathbb{L}^2(\mu)$. Since P_t ($t>0$) is compact, $P_t f_{k_\ell} \to P_t f$ (strongly) in $\mathbb{L}^2(\mu)$. Then, for every ℓ, write

$$\|f_{k_\ell} - f\|_2 \leq \|f_{k_\ell} - P_t f_{k_\ell}\|_2 + \|P_t f_{k_\ell} - P_t f\|_2 + \|P_t f - f\|_2.$$

From (4.2.3), p. 184,

$$\|f_{k_\ell} - P_t f_{k_\ell}\|_2^2 \leq 2t\,\mathcal{E}(f_{k_\ell}).$$

Since $\sup_{\ell\in\mathbb{N}}\mathcal{E}(f_{k_\ell}) < \infty$, first letting $\ell\to\infty$ and then $t\to 0$, it follows that $f_{k_\ell}\to f$ in $\mathbb{L}^2(\mu)$, proving the claim. □

Note that under the Sobolev inequality $S_n(A,C)$, the convergence of the subsequence $(f_{k_\ell})_{\ell\in\mathbb{N}}$ in the preceding proof also takes place in $\mathbb{L}^q(\mu)$, $2\leq q < \frac{2n}{n-2}$.

While Theorem 6.4.2 applies to Sobolev inequalities on spheres $\mathbb{S}^n$ or compact manifolds, it does not apply directly on $\mathbb{R}^n$ since the invariant Lebesgue measure dx is not finite. The same arguments do however apply on bounded sets in $\mathbb{R}^n$, leading to the classical Rellich-Kondrachov Theorem. Given a bounded open set $\mathcal{O}$ in $\mathbb{R}^n$, set, for a smooth and compactly supported function f on $\mathcal{O}$,

$$\|f\|_{\mathcal{H}^1(\mathcal{O})}^2 = \int_{\mathcal{O}} f^2\, dx + \int_{\mathcal{O}} |\nabla f|^2 dx.$$

Let $\mathcal{H}^1(\mathcal{O})$ be the Hilbert completion of the space of smooth compactly supported functions in $\mathcal{O}$ with respect to this norm, considered as a subspace of $\mathbb{L}^2(\mathcal{O},dx)$.

Theorem 6.4.3 (Rellich-Kondrachov Theorem) *In the preceding setting, the embedding from $\mathcal{H}^1(\mathcal{O})$ into $\mathbb{L}^2(\mathcal{O},dx)$ is compact.*

Proof Via the stereographic projection from the sphere $\mathbb{S}^n$ onto $\mathbb{R}^n$ described in Sect. 2.2.2, p. 83, compactly supported functions in $\mathcal{O}$ may be considered as functions defined on $\mathbb{S}^n$. Moreover, $\mathcal{O}$ being bounded, the norm $\|f\|_{\mathcal{H}^1(\mathcal{O})}$ is equivalent to the Dirichlet norm $\|f\|_{\mathcal{E}}$ on the sphere since on a bounded set $\mathcal{O}$, both the measures and the carré du champ operators are comparable for the usual Laplacian and the Laplacian on the sphere. It then remains to apply Theorem 6.4.2 on the sphere to extract, from any bounded sequence in $\mathcal{H}^1(\mathcal{O})$, a subsequence convergent in $\mathbb{L}^2(\mathcal{O},dx)$. □

6.5 Tensorization of Sobolev Inequalities

In contrast to Poincaré or logarithmic Sobolev inequalities, Sobolev inequalities do not tensorize so well. Under Sobolev inequalities $S_{n_1}(A_1,C_1)$ and $S_{n_2}(A_2,C_2)$ on respective spaces E_1 and E_2, the product space $E_1\times E_2$ carries a Sobolev inequality $S_{n_1+n_2}(A,C)$ with constants $A\geq 0$, $C>0$ (depending on $A_i\geq 0$, $C_i>0$, $n_i>2$,

$i = 1, 2$) which are not so easy to describe if one is interested in sharp bounds. However, the parameter n behaves as an expected dimensional parameter (which is an important issue in this product procedure).

There are several ways to reach the preceding tensorization property. The simplest is perhaps to transit through ultracontractivity and Theorem 6.3.1. Indeed, if two operators P_i on E_i are bounded from $L^1(\mu_i)$ into $L^\infty(\mu_i)$ with norms K_i, $i = 1, 2$, then their product $P_1 \otimes P_2$ is bounded from $L^1(\mu_1 \otimes \mu_2)$ into $L^\infty(\mu_1 \otimes \mu_2)$ with norm $K_1 K_2$. The claim then easily follows with $K_i = C_i \, t^{-n_i/2}$.

Another approach may be developed on the basis of the logarithmic entropy-energy inequality (6.2.2), equivalent by Proposition 6.2.3 to the Sobolev inequality $S_n(A, C)$ with the same dimension n. A particularly meaningful statement is obtained when dealing with entropy-energy inequalities with the same constant $A \geq 0$ (for example $A = 0$). Recall from Proposition 4.3.1, p. 185, the Dirichlet form $\mathcal{E}$ with domain $\mathcal{D}(\mathcal{E})$ on a product space $E_1 \times E_2$.

Proposition 6.5.1 *On Markov Triples (E_i, μ_i, Γ_i), $i = 1, 2$, assume that the logarithmic entropy-energy inequalities (6.2.2) hold with respective dimension $n_i > 0$ and constants A_i, C_i with $A_i = A$ and $C_i = \frac{C}{n_i}$ for some $A \geq 0$ and $C > 0$. Then, on the product space $(E_1 \times E_2, \mu_1 \otimes \mu_2, \Gamma_1 \otimes \Gamma_2)$, the logarithmic entropy-energy inequality*

$$\mathrm{Ent}_\mu(f^2) \leq \frac{n}{2} \log\left(A + \frac{C}{n} \mathcal{E}(f)\right)$$

holds for every function f such that $\int_E f^2 d\mu = 1$ in the Dirichlet space $\mathcal{D}(\mathcal{E})$ of the product space $E = E_1 \times E_2$, where $\mu = \mu_1 \otimes \mu_2$ and $n = n_1 + n_2$.

Proof According to (6.2.5), the respective inequalities (6.2.2) are equivalent to the families of (defective) logarithmic Sobolev inequalities

$$\mathrm{Ent}_{\mu_i}(f^2) \leq \Phi_i'(r_i) \mathcal{E}_i(f) + \Psi(r_i), \quad r_i \in (0, \infty),$$

for every f in $\mathcal{D}(\mathcal{E}_i)$ with $\int_{E_i} f^2 d\mu_i = 1$, $i = 1, 2$, where

$$\Phi_i'(r) = \frac{C}{2(A + \frac{Cr}{n_i})}$$

and

$$\Psi_i(r) = \Phi_i(r) - r\, \Phi_i'(r) = \frac{n_i}{2}\left[\log\left(A + \frac{Cr}{n_i}\right) - \frac{\frac{Cr}{n_i}}{A + \frac{Cr}{n_i}}\right].$$

To add dimensions, choose $\frac{r_i}{n_i} = \frac{r}{n}$ for $r > 0$ so that

$$\Phi_1'(r_1) = \Phi_2'(r_2) = \frac{C}{2(A + \frac{Cr}{n})}$$

and
$$\Psi_1(r_1) + \Psi_2(r_2) = \frac{n}{2}\left[\log\left(A + \frac{Cr}{n}\right) - \frac{\frac{Cr}{n}}{A + \frac{Cr}{n}}\right].$$

The conclusion then follows from the tensorization of logarithmic Sobolev inequalities in Proposition 5.2.7, p. 249, and retrieves the logarithmic entropy-energy inequality on the product space. Proposition 6.5.1 is established. $\square$

The previous proposition is of particular interest when $A = 0$, which shows that the Euclidean logarithmic Sobolev inequality (6.2.8) is stable under products. Proposition 6.5.1 may also be applied to tensorize the same logarithmic entropy-energy inequality

$$\mathrm{Ent}_\mu(f^2) \leq \frac{n}{2}\log\big(A + C\,\mathcal{E}(f)\big)$$

k-times, $k \geq 1$, yielding

$$\mathrm{Ent}_{\mu^{\otimes k}}(f^2) \leq \frac{kn}{2}\log\left(A + \frac{C}{k}\,\mathcal{E}(f)\right)$$

for every function f on the k-fold product space E^k such that $\int_{E^k} f^2 d\mu^{\otimes k} = 1$. This is the procedure used in Remark 6.2.6 to reach the optimal Euclidean logarithmic Sobolev inequality from the Sobolev inequality with sharp constants. If $A = 1$, as $k \to \infty$, the latter implies (formally) a logarithmic Sobolev inequality $LS(\frac{Cn}{4})$ on the infinite-dimensional product space. This is another illustration of the importance of logarithmic Sobolev inequalities in infinite dimension (as a form of limiting or dimension-free Sobolev inequalities).

6.6 Sobolev Inequalities and Lipschitz Functions

In the corresponding sections in Chaps. 4 and 5, integrability properties of Lipschitz functions with respect to a measure satisfying either a Poincaré or a logarithmic Sobolev inequality were described. The picture here, under a Sobolev inequality in the finite measure case, is that Lipschitz functions are actually (uniformly) bounded. Recall from Sect. 3.3.7, p. 166, that the Γ operator induces a natural distance $d(x, y)$ on points $(x, y) \in E \times E$ (with respect to the extended algebra $\mathcal{A}$). (For example, for the classical carré du champ operator on $\mathbb{R}^n$, d gives rise to the Euclidean distance.) According to (3.3.10), p. 167, to control the diameter $D = D(E, \mu, \Gamma)$ of the Markov Triple (E, μ, Γ), it is convenient to bound Lipschitz functions as

$$D = D(E, \mu, \Gamma) = \sup_{\|f\|_{\mathrm{Lip}}\leq 1} \|\tilde{f}\|_\infty$$

where $\tilde{f}(x, y) = f(x) - f(y)$, $(x, y) \in E \times E$, and the norm is taken in $\mathbb{L}^\infty(E \times E, \mu \otimes \mu)$.

That the diameter D is finite under a Sobolev inequality with respect to a probability measure μ may, for example, be achieved via logarithmic entropy-energy inequalities (equivalent to Sobolev inequalities by Proposition 6.2.3), much in the spirit of the Herbst argument in Sect. 5.4.1, p. 252.

Proposition 6.6.1 *Let (E, μ, Γ) be a Markov Triple with μ a probability measure. Under the tight logarithmic entropy-energy inequality*

$$\mathrm{Ent}_\mu(f^2) \leq \frac{n}{2} \log(1 + C\mathcal{E}(f)), \quad f \in \mathcal{D}(\mathcal{E}), \ \int_E f^2 d\mu = 1,$$

the diameter $D = D(E, \mu, \Gamma)$ is finite.

Proof Set $\Phi(r) = \frac{n}{2}\log(1 + Cr)$, $r \geq 0$. We proceed as in the proof of Proposition 5.4.1, p. 252, using the same notation. For a (bounded) 1-Lipschitz function f, setting $Z(s) = \int_E e^{sf} d\mu$, $s \in \mathbb{R}$, we reach in the same way the differential inequality

$$F'(s) \leq \frac{1}{s^2} \Phi\left(\frac{s^2}{4}\right), \quad s \neq 0$$

on $F(s) = \frac{1}{s} \log Z(s)$, with $F(0) = \int_E f d\mu$. Integrating, it follows that for every $s \in \mathbb{R}$,

$$\int_E e^{s(f - \int_E f d\mu)} d\mu \leq \exp\left(s \int_0^s \frac{1}{u^2} \Phi\left(\frac{u^2}{4}\right) du\right).$$

Setting $C = \int_0^\infty \frac{1}{u^2} \Phi(\frac{u^2}{4}) du$, which is finite by definition of Φ, it easily follows as $s \to \pm\infty$ that $\|f - \int_E f d\mu\|_\infty \leq C$. The conclusion follows by the very definition of the diameter D. □

The bound produced by Proposition 6.6.1 is not sharp. For example, starting from the Sobolev inequality (6.1.2) on the sphere $\mathbb{S}^n$ (with diameter π), and then the associated logarithmic entropy-energy inequality from Proposition 6.2.3, it may be shown that $D \leq C$ for some (large) $C > \pi$. More refined methods not discussed in this book (see the Notes and References) allow us to reach directly from the Sobolev inequality (6.1.2) the optimal bound $D \leq \pi$. This (more refined) argument actually relies on the application of the Sobolev inequality to some suitable transformation (related to extremal functions of the Sobolev inequality on the sphere) of a Lipschitz function (and not just the exponential function as for entropy-energy inequalities).

Next, we concentrate on further aspects concerning Lipschitz functions under Sobolev-type inequalities. One such aspect deals with volumes of balls.

Proposition 6.6.2 *Given a positive 1-Lipschitz function f on (E, μ, Γ), set $V(r) = \mu(f \leq r)$, $r \geq 0$. Assume that the Sobolev inequality $S_n(A, C)$ with $A \geq 0$*

6.6 Sobolev Inequalities and Lipschitz Functions

holds and that $V(r_0) < \infty$ for some $r_0 > 0$. Then, provided that

$$\limsup_{r \to 0} \frac{\log V(r)}{\log r} < \infty, \tag{6.6.1}$$

there is a constant $c > 0$ such that $V(r) \geq cr^n$ for every $0 \leq r \leq 1$ (every $r \geq 0$ if $A = 0$).

Choosing for f the distance function to a given point, $V(r)$ corresponds to the volume of a ball in the metric space (E, d). The proposition thus indicates that, provided the volume of the balls does not decay too rapidly to 0 as $r \to 0$, the volume of small balls is bounded from below by cr^n where n is the dimension of the Sobolev inequality. This result is already coherent on $\mathbb{R}^n$. In particular, the exponent n is minimal and determined by the volume of (small) balls. This property explains why, for a diffusion operator with smooth coefficients on a Riemannian manifold with dimension n, the Sobolev exponent is at least n.

Proof Assuming f is in $\mathcal{A}$, apply the Sobolev inequality $S_n(A, C)$ to smooth approximations of the functions $g = (1 - \frac{f}{r})_+$, $r > 0$ (which belong, for r small enough, to $\mathcal{D}(\mathcal{E})$ under the condition $V(r_0) < \infty$). Since

$$\frac{1}{2} 1_{\{f \leq \frac{r}{2}\}} \leq g \leq 1_{\{f \leq r\}} \quad \text{and} \quad \Gamma(g) \leq \frac{1}{r^2} 1_{\{f \leq r\}},$$

the inequality $S_n(A, C)$ yields, for some $C_1 > 0$,

$$V\left(\frac{r}{2}\right)^{2/p} \leq C_1 \left(1 + \frac{1}{r^2}\right) V(r) \leq \frac{2C_1}{r^2} V(r)$$

provided $r \leq 1$ (where, as usual, $p = \frac{2n}{n-2}$). After some rewriting, the latter amounts to

$$C_2 \frac{V(\frac{r}{2})}{(\frac{r}{2})^n} \leq \left(C_2 \frac{V(r)}{r^n}\right)^{p/2} \tag{6.6.2}$$

for some suitable constant $C_2 > 0$. Assume now that there exists an $r_1 \leq 1$ such that $C_2 \frac{V(r_1)}{r_1^n} = a < 1$. Then, iterating (6.6.2), for every $k \geq 1$,

$$V\left(\frac{r_1}{2^k}\right) \leq \frac{a^{(p/2)^k} r_1^n}{C_2 2^{kn}}$$

which yields a contradiction with the hypothesis (6.6.1) as $k \to \infty$ since $p > 2$. Therefore $V(r) \geq cr^n$ for some $c > 0$ and every $0 \leq r \leq 1$. The proof is easily modified to see that the preceding holds for every $r \geq 0$ whenever $A = 0$ in the Sobolev inequality. Proposition 6.6.2 is established. □

The proof of the previous proposition actually shows more. Namely, if the volume of small balls is not at least of the order r^n, then it has to decay at least exponentially as $e^{-cr^{-b}}$ as $r \to 0$ for some $b, c > 0$.

6.7 Local Sobolev Inequalities

The local inequalities for heat kernel measures which will be described here are the analogues of the local Poincaré and logarithmic Sobolev inequalities of Sect. 4.7, p. 206, and Sect. 5.5, p. 257, but at the same time are more subtle since they involve a dimensional parameter, as is the case for Sobolev inequalities. While the curvature-dimension condition $CD(\rho, n)$ is closely related to Sobolev-type inequalities as will be illustrated in the following sections, there is however no hope of characterizing it via local Sobolev inequalities on the associated (linear) heat semigroup. This difficulty is due to the fact that the heat semigroup essentially behaves as a Gaussian kernel (and is exactly this in the Euclidean space), and such kernels cannot satisfy Sobolev-type inequalities (the best functional inequalities for them being logarithmic Sobolev inequalities). Furthermore, the methods developed below to reach Sobolev inequalities are non-linear. Entropy-energy inequalities will be established later in Sect. 6.8 by the heat flow method under curvature-dimension conditions, but by means of integration by parts formulas which are not available for local inequalities. The alternative geometric methods of Sects. 6.9 and 6.11 (conformal transformations, porous medium or fast diffusion equations, etc.) do not yield any information on the heat semigroup.

The project pursued here will actually deal with sharpened forms of the local logarithmic Sobolev inequalities of Sect. 5.5, p. 257, introducing the dimensional parameter of the curvature-dimension hypothesis. In particular, the conclusions will be shown to cover sharp Harnack inequalities. Hypercontractivity for heat kernel measures may be investigated similarly.

6.7.1 Dimensional Logarithmic Sobolev Inequality for the Gaussian Measure

As a first step, we are looking for inequalities which hold for Gaussian measures in finite dimension, precisely capturing the dimensional feature. To this end, we begin with the Euclidean logarithmic Sobolev inequality of Proposition 6.2.5. Recall also the standard Gaussian measure on the Borel sets of $\mathbb{R}^n$, $d\mu(x) = (2\pi)^{-n/2} e^{-|x|^2/2} dx$.

Proposition 6.7.1 *Let μ be the standard Gaussian measure on $\mathbb{R}^n$. Then*

$$\mathrm{Ent}_\mu(g) \leq \frac{1}{2} \int_{\mathbb{R}^n} \Delta g \, d\mu + \frac{n}{2} \log\left(1 - \frac{1}{n} \int_{\mathbb{R}^n} g \, \Delta(\log g) d\mu \right) \qquad (6.7.1)$$

6.7 Local Sobolev Inequalities

for every smooth positive function g on $\mathbb{R}^n$ with $\int_{\mathbb{R}^n} g\,d\mu = 1$ and such that Δg and $g\,\Delta(\log g)$ are well-defined and integrable with respect to μ.

Comparing with the general setting, if $\mathcal{A}_0$ denotes the set of smooth compactly supported functions on $\mathbb{R}^n$, Proposition 6.7.1 is established for functions in $\mathcal{A}_0^{\mathrm{const}+}$ (cf. Remark 3.3.3, p. 154), possibly extended then to larger classes of functions.

Note that (6.7.1) improves upon the standard logarithmic Sobolev inequality (6.2.9) for the Gaussian measure μ since $\log(1+r) \leq r$, $r \geq 0$, and

$$g\,\Delta(\log g) = \Delta g - g\bigl|\nabla(\log g)\bigr|^2 = \Delta g - \frac{|\nabla g|^2}{g}.$$

Turning to the proof, which is elementary, set $g = \hat{g}^2$ where $\hat{g} \in \mathcal{A}_0^{\mathrm{const}+}$ and $\int_{\mathbb{R}^n} \hat{g}^2\,d\mu = 1$. (Actually, in this concrete setting, the smooth approximations can be made easily.) Then use the Euclidean logarithmic Sobolev inequality (6.2.8) in which we change f back to $(2\pi)^{-n/4}\hat{g}\,e^{-|x|^2/4}$. After integration by parts for the term $\int_{\mathbb{R}^n} |\nabla f|^2 dx$ (as in the proof of Proposition 6.2.5), we get

$$\mathrm{Ent}_\mu(\hat{g}^2) \leq \frac{1}{2}\int_{\mathbb{R}^n} \hat{g}^2\bigl(|x|^2 + n\log(2\pi)\bigr)d\mu$$

$$+ \frac{n}{2}\log\left(\frac{2}{n\pi e}\int_{\mathbb{R}^n}\left[|\nabla\hat{g}|^2 + \hat{g}^2\left(\frac{n}{2} - \frac{|x|^2}{4}\right)\right]d\mu\right).$$

By a further integration by parts of the form (2.7.8), p. 107, the conclusion easily follows.

This proof does indeed retrieve the logarithmic Sobolev inequality for μ from the Euclidean logarithmic Sobolev inequality. This self-improving property might appear as surprising unless the use of dilations to optimize the Euclidean logarithmic Sobolev inequality is recalled (cf. the proof of Proposition 6.2.5). Extremal functions in (6.7.1) are Gaussian kernels. Note furthermore that (6.7.1) is a second order inequality and that the second order operator which appears is the standard Laplacian Δ on $\mathbb{R}^n$ and not the Ornstein-Uhlenbeck operator $L = \Delta - x \cdot \nabla$ (which is naturally associated with the Gaussian measure μ). Finally, (6.7.1) implicitly contains the fact that for any $g \in \mathcal{A}_0^{\mathrm{const}+}$ (say),

$$\int_{\mathbb{R}^n} g\,\Delta(\log g)\,d\mu < n\int_{\mathbb{R}^n} g\,d\mu = n,$$

in other words

$$\int_{\mathbb{R}^n} \Delta g\,d\mu - \int_{\mathbb{R}^n} g\bigl|\nabla(\log g)\bigr|^2 d\mu < n\int_{\mathbb{R}^n} g\,d\mu.$$

But, from the integration by parts formula (2.7.8), p. 107, which has already been used earlier,

$$\int_{\mathbb{R}^n} \Delta g\,d\mu = \int_{\mathbb{R}^n} g\bigl[|x|^2 - n\bigr]d\mu,$$

and the requested inequality therefore amounts to the obvious

$$\int_{\mathbb{R}^n} g |\nabla(\log g)|^2 d\mu + \int_{\mathbb{R}^n} g |x|^2 d\mu > 0.$$

Remark 6.7.2 Applying (6.7.1) of Proposition 6.7.1 to $1 + \varepsilon g$ and letting $\varepsilon \to 0$ yields

$$\text{Var}_\mu(g)^2 \leq \mathcal{E}(g) - \frac{1}{2n}\left(\int_{\mathbb{R}^n} \Delta g \, d\mu\right)^2. \quad (6.7.2)$$

One may wonder at the nature of this inequality, and ask for a direct proof. It was already mentioned above that $\int_{\mathbb{R}^n} \Delta g \, d\mu = \int_{\mathbb{R}^n} g [|x|^2 - n] d\mu$. Since $u_2 = |x|^2 - n$ is an eigenfunction of $-L$ with eigenvalue 2,

$$\int_{\mathbb{R}^n} \Delta g \, d\mu = -\frac{1}{2}\int_{\mathbb{R}^n} g \, Lu_2 \, d\mu = \frac{1}{2}\int_{\mathbb{R}^n} \nabla g \cdot \nabla u_2 \, d\mu.$$

Furthermore, it is immediately checked that $\int_{\mathbb{R}^n} u_2^2 d\mu = 2n$ (using for example that $|\nabla u_2|^2 = 4|x|^2$). Write then a given function g as $g = \alpha + \beta u_2 + \hat{g}$ where $\hat{g}$ is orthogonal both to the constants and to u_2. Since $\int_{\mathbb{R}^n} \nabla \hat{g} \cdot \nabla u_2 d\mu = 0$, the inequality (6.7.2) reduces to $\int_{\mathbb{R}^n} \hat{g}^2 d\mu \leq \int_{\mathbb{R}^n} |\nabla \hat{g}|^2 d\mu$ which is nothing else but the Poincaré inequality for μ applied to $\hat{g}$.

6.7.2 Local Sobolev Inequalities

The dimensional logarithmic Sobolev inequality of Proposition 6.7.1 may be analyzed further after translation and dilation, leading to abstract versions of heat kernel inequalities. Indeed, applying (6.7.1) to $g(y) = f(x + ty)$, $x, y \in \mathbb{R}^n$, $t \geq 0$, yields, for the heat semigroup $(P_t)_{t \geq 0}$ in the Euclidean space, for any $t > 0$, and all functions $f \in \mathcal{A}_0^{\text{const}+}$,

$$P_t(f \log f) - P_t f \log P_t f \leq t \Delta P_t f + \frac{n}{2} \log\left(1 - \frac{2t}{n} \cdot \frac{P_t(f \Delta(\log f))}{P_t f}\right).$$

Now let $t \to 0$ using the Taylor expansion $P_t f = f + t \Delta f + \frac{t^2}{2}\Delta^2 f$. After a somewhat tedious calculation, one is left with the inequality

$$\Gamma_2(\log f) \geq \frac{1}{n}\left(\Delta(\log f)\right)^2$$

which is precisely the curvature-dimension condition $CD(0, n)$ for the Laplace operator Δ on $\mathbb{R}^n$ (applied to $\log f$). In other words, while the standard Gaussian logarithmic Sobolev inequality (6.2.9) only captures by dilation the $CD(0, \infty)$ curvature condition, the improved inequality (6.7.1) allows us to reach the dimension n. There

6.7 Local Sobolev Inequalities

are few inequalities in Euclidean space with such a property. However, as soon as one is found, it may usually be extended to a general framework.

The following theorem is one such illustration, providing a further example of the semigroup interpolation scheme.

Theorem 6.7.3 (Local Dimensional Logarithmic Sobolev Inequalities) *Let (E, μ, Γ) be a Markov Triple with semigroup $\mathbf{P} = (P_t)_{t \geq 0}$ and generator L. The following assertions are equivalent.*

(i) *The curvature-dimension condition $CD(0, n)$ holds.*
(ii) *For every function f in $\mathcal{A}_0^{\text{const}+}$, and every $t \geq 0$,*

$$P_t f \, \mathrm{L}(\log P_t f) \geq P_t\big(f \, \mathrm{L}(\log f)\big)\left(1 + \frac{2t}{n} \mathrm{L}(\log P_t f)\right). \tag{6.7.3}$$

(iii) *For every function f in $\mathcal{A}_0^{\text{const}+}$, and every $t \geq 0$,*

$$P_t(f \log f) - P_t f \log P_t f$$
$$\leq t \, \mathrm{L} P_t f + \frac{n}{2} P_t f \log\left(1 - \frac{2t}{n} \frac{P_t(f \mathrm{L}(\log f))}{P_t f}\right). \tag{6.7.4}$$

(iv) *For every function f in $\mathcal{A}_0^{\text{const}+}$, and every $t \geq 0$,*

$$P_t(f \log f) - P_t f \log P_t f$$
$$\geq t \, \mathrm{L} P_t f - \frac{n}{2} P_t f \log\left(1 + \frac{2t}{n} \mathrm{L}(\log P_t f)\right). \tag{6.7.5}$$

As usual, restricting the inequalities to functions in $\mathcal{A}_0^{\text{const}+}$ is merely a matter of comfort. In reasonable settings, the various inequalities in Theorem 6.7.3 immediately extend to any positive bounded function f in $\mathcal{A}$ for which every term makes sense, following the various extension procedures described in Sect. 3.3, p. 151. Note that as $n \to \infty$, the theorem covers both the logarithmic Sobolev inequality and the reverse form described in Theorem 5.5.2, p. 259, under the curvature condition $CD(0, \infty)$.

Proof Start with the most important step, namely the proof of (ii) under the curvature-dimension $CD(0, n)$ hypothesis (which will then imply (iii) and (iv)). As usual, set $\Lambda(s) = P_s(\phi(P_{t-s} f))$, $s \in [0, t]$, $t > 0$, where $\phi(r) = r \log r$, $r \in \mathbb{R}_+$, and hence $f \in \mathcal{A}_0^{\text{const}+}$. As already observed earlier in Proposition 5.5.3, p. 261,

$$\Lambda'(s) = P_s\left(\frac{\Gamma(P_{t-s} f)}{P_{t-s} f}\right) = P_s\big(P_{t-s} f \, \Gamma(\log P_{t-s} f)\big)$$

and

$$\Lambda''(s) = 2 P_s\big(P_{t-s} f \, \Gamma_2(\log P_{t-s} f)\big).$$

By the $CD(0,n)$ hypothesis applied to $\log P_{t-s} f$,

$$n \Lambda''(s) \geq 2 P_s \left(P_{t-s} f \left(\mathrm{L}(\log P_{t-s} f) \right)^2 \right) \geq \frac{2 [P_s (P_{t-s} f \, \mathrm{L}(\log P_{t-s} f))]^2}{P_s P_{t-s} f}$$

where the last inequality follows from the Cauchy-Schwarz inequality for the measure P_s with the functions $(P_{t-s} f)^{1/2} \mathrm{L}(\log P_{t-s} f)$ and $(P_{t-s} f)^{1/2}$. The denominator of the expression on the right-hand side of the previous inequality is actually $P_t f$ by the semigroup property (and is thus independent of s). Furthermore, by the change of variables formula,

$$P_{t-s} f \, \mathrm{L}(\log P_{t-s} f) = \mathrm{L} P_{t-s} f - P_{t-s} f \, \Gamma(\log P_{t-s} f),$$

so that the numerator of the same expression takes the form $[\mathrm{L} P_t f - \Lambda'(s)]^2$. This results in the differential inequality

$$\Lambda''(s) \geq \frac{2 [\mathrm{L} P_t f - \Lambda'(s)]^2}{n P_t f}, \quad s \in [0, t]. \tag{6.7.6}$$

In this differential inequality, the terms $\mathrm{L} P_t f$ and $P_t f$ behave as constants, so that we are actually dealing with an inequality of the form $u' \geq (a + bu)^2$, with $u = \Lambda'$. More precisely, setting $\alpha = \frac{2}{n P_t f}$, the differential inequality (6.7.6) leads to the inequality, valid for any $0 \leq u \leq v \leq t$,

$$\left(\Lambda'(v) - \mathrm{L} P_t f \right) - \left(\Lambda'(u) - \mathrm{L} P_t f \right) \\ \geq \alpha(v - u)\left(\Lambda'(0) - \mathrm{L} P_t f \right)\left(\Lambda'(t) - \mathrm{L} P_t f \right), \tag{6.7.7}$$

which holds regardless of the signs of $\Lambda'(v) - \mathrm{L} P_t f$ and $\Lambda'(u) - \mathrm{L} P_t f$. The inequality (6.7.3) of (ii) is then a direct application of (6.7.7) with $u = 0$ and $v = t$. As already obtained in the proof of Theorem 5.5.2, p. 259, under the $CD(0, \infty)$ condition,

$$\frac{\Gamma(P_t f)}{P_t f} \leq P_t \left(\frac{\Gamma(f)}{f} \right),$$

that is $P_t f \, \mathrm{L}(\log P_t f) \geq P_t(f \, \mathrm{L}(\log f))$. Combined with (6.7.3), this immediately yields that

$$1 + \frac{2t}{n} \mathrm{L}(\log P_t f) > 0 \tag{6.7.8}$$

(if $r \geq s(1 + \alpha t r)$ and $r \geq s$, then necessarily $1 + \alpha t r > 0$).

Furthermore, we may deduce from (6.7.7) that

$$\frac{1}{\alpha s - (\Lambda'(0) - \mathrm{L} P_t f)^{-1}} \leq \Lambda'(s) - \mathrm{L} P_t f \leq \frac{1}{\alpha(t - s) - (\Lambda'(t) - \mathrm{L} P_t f)^{-1}},$$

from which the bounds (6.7.4) and (6.7.5) immediately follow by integration between 0 and t.

6.7 Local Sobolev Inequalities

Finally, any of (ii), (iii), (iv) yields $CD(0, n)$ by a Taylor expansion at $t = 0$. The proof is complete. □

As for Proposition 5.5.3, p. 261, it is worth observing that (6.7.4) and (6.7.5) are obtained directly from the curvature-dimension condition without the explicit use of the gradient bound (3.3.18), p. 163, contrary to their weaker forms in Theorem 5.5.2, p. 259. As such, this strategy of proof may be used in settings where gradient bounds are lacking, for example in hypoelliptic settings where on the other hand extensions of the curvature-dimension condition may be available.

Remark 6.7.4 If the condition $CD(0, n)$ is replaced by the more general $CD(\rho, n)$, $\rho \in \mathbb{R}$, the differential inequality on Λ becomes

$$\Lambda''(s) \geq \frac{2[LP_t f - \Lambda'(s)]^2}{nP_t f} + \rho \Lambda'(s), \quad s \in [0, t].$$

This is not harder to solve, except that the corresponding result leads to expressions involving trigonometric and logarithmic functions (depending on the sign of ρ and the values of $LP_t f$ and $P_t f$), and the final result is far less pleasant to express. Fundamentally, such a differential equation (especially when $\rho > 0$) may have no bounded solutions on $[0, t]$ whenever the coefficients (here $LP_t f$ and $P_t f$) are not controlled in some proper way with respect to the initial or final values $\Lambda'(0)$ and $\Lambda'(t)$. When $\rho = 0$, the Li-Yau inequality expresses precisely this non-explosion property, and the equivalent form under $\rho \neq 0$ may similarly be established.

The proof of Theorem 6.7.3, in particular, (6.7.8), includes the celebrated *Li-Yau parabolic inequality*. According to Remark 6.7.4 above, versions under the curvature-dimension $CD(\rho, n)$ condition are also available.

Corollary 6.7.5 (Li-Yau inequality) *Under the curvature-dimension condition $CD(0, n)$, for every function f in $\mathcal{A}_0^{\text{const}+}$, and every $t > 0$,*

$$L(\log P_t f) > -\frac{n}{2t}.$$

Alternatively,

$$\frac{\Gamma(P_t f)}{(P_t f)^2} - \frac{LP_t f}{P_t f} < \frac{n}{2t}. \tag{6.7.9}$$

As above, in reasonable situations, the Li-Yau inequality extends to any strictly positive bounded measurable function f. The Li-Yau inequality is actually an identity for the heat semigroup on $\mathbb{R}^n$ when f is a Dirac mass at x. The inequality is thus optimal and expresses a comparison with the Euclidean model.

6.7.3 Harnack Inequalities

The Li-Yau inequality is one of the main tools in establishing Harnack-type inequalities for solutions of heat equations and Gaussian heat kernel bounds with the correct dependence on dimension (on say Riemannian manifolds). As part of the family of logarithmic Sobolev inequalities for heat kernel measures, it may be qualitatively compared to Wang's (infinite-dimensional) Harnack inequality (Theorem 5.6.1, p. 265). For simplicity, we state the Harnack inequality in the setting of a diffusion operator $L = \Delta_\mathfrak{g} - \nabla W \cdot \nabla$ on a complete connected Riemannian manifold $(M, \mathfrak{g})$.

Corollary 6.7.6 (Harnack inequality under $CD(0, n)$) *Under the curvature-dimension condition $CD(0, n)$, for every positive measurable function f on M, every $x, y \in M$ and every $0 < t < t + s$,*

$$P_t f(x) \leq P_{t+s} f(y) \left(\frac{t+s}{t}\right)^{n/2} e^{d(x,y)^2/4s}, \qquad (6.7.10)$$

where $d(x, y)$ is the Riemannian distance from x to y. Conversely, the Harnack inequality (6.7.10), which holds for every $x, y \in M$ and every $0 < t < t + s$, implies in return the Li-Yau inequality (6.7.9).

For the proof, as in Theorem 5.6.1, p. 265, let $(x_u)_{u \in [t, t+s]}$ be a geodesic with constant speed joining x to y and consider

$$\Lambda(u) = u^{n/2} e^{ud(x,y)^2/4s^2} P_u f(x_u), \quad u \in [t, t+s].$$

Then the Li-Yau inequality tells us precisely that the derivative of $\log \Lambda$ is positive on the interval $[t, t+s]$ from which the claim follows. The converse statement is established as the corresponding assertion in Theorem 5.6.1, p. 265.

Harnack inequalities as in Corollary 6.7.6 are traditionally used to produce sharp off-diagonal Gaussian bounds on heat kernels of the form

$$p_t(x, y) \leq C(n, t, \varepsilon) \, e^{-d(x,y)^2/(4+\varepsilon)t}$$

for $t > 0$, $\varepsilon > 0$, $(x, y) \in M \times M$, where $C(n, t, \varepsilon)$ is of polynomial decay in time depending on geometric features such as volumes of balls. We briefly mention here some results of this type for the Laplacian heat kernels on a Riemannian manifold $(M, \mathfrak{g})$ with dimension n and positive Ricci curvature (referring to the Notes and References for suitable background). A first step in the investigation is that by integration of (6.7.10) over a ball,

$$p_t(x, x) \leq \frac{C_n}{V(x, \sqrt{t})}$$

for all $t > 0$ and $x \in M$, where $V(x, r)$ is the volume of the ball with center $x \in M$ and radius $r > 0$, and $C_n > 0$ only depends upon n. Together with geometric volume

6.7 Local Sobolev Inequalities

comparison theorems in manifolds with positive Ricci curvature, the latter may then be shown to imply uniform heat kernel bounds in the form of ultracontractivity, and thus functional inequalities of Sobolev-type (cf. Sect. 6.3). Off-diagonal upper bounds are achieved using the techniques presented in Sect. 7.2, p. 355, in the next chapter. A typical result in this context is that

$$p_t(x, y) \leq \frac{C_{n,\varepsilon}}{V(x, \sqrt{t})^{1/2} V(y, \sqrt{t})^{1/2}} e^{-d(x,y)^2/(4+\varepsilon)t} \qquad (6.7.11)$$

for all $t > 0$, $\varepsilon > 0$, $(x, y) \in M \times M$.

Lower bounds may be investigated similarly, and improve with a polynomial factor the lower bound (5.6.3), p. 267. For example, Corollary 6.7.6 (for f approaching Dirac mass at x) yields that

$$p_s(x, x) \leq p_{t+s}(x, y) \left(\frac{t+s}{t}\right)^{n/2} e^{d(x,y)^2/4s}$$

for all $s > 0$ and $t > 0$. Provided the local asymptotics $\lim_{s \to 0} (4\pi s)^{n/2} p_s(x, x) = 1$ holds, it follows that

$$p_t(x, y) \geq \frac{1}{(4\pi t)^{n/2}} e^{-d(x,y)^2/4t} \qquad (6.7.12)$$

for all $t > 0$ and $(x, y) \in M \times M$.

These precise upper and lower bounds on heat kernels actually significantly improve milder small time asymptotics at a logarithmic scale, which on the other hand may be considered in greater generality. For example, the classical Laplace-Varadhan small time asymptotics

$$\lim_{t \to 0} 4t \log p_t(x, y) = -d^2(x, y) \qquad (6.7.13)$$

holds for solutions of large classes of second-order differential operators on a Riemannian manifold.

We refer to Sect. 6.12 for more on geometric and non-geometric heat kernel bounds. In the context of this book, such heat kernel bounds will be investigated under functional inequalities in Chap. 7.

6.7.4 Local Dimensional Hypercontractivity

The curvature-dimension condition $CD(0, n)$ may also be used to reinforce, with the dimensional factor n, Theorem 5.5.5, p. 262, on local hypercontractivity by means of the dimensional logarithmic Sobolev inequality (6.7.4). Again, one advantage of local hypercontractivity is that it produces a description of the curvature-dimension $CD(0, n)$ condition which applies to any measurable function. An equivalent form of $CD(\rho, n)$, $\rho \neq 0$, could be established along the same lines, although the result (as for Theorem 6.7.3) is far less simple to state.

Theorem 6.7.7 (Local dimensional hypercontractivity) *Let (E, μ, Γ) be a Markov Triple with semigroup $\mathbf{P} = (P_t)_{t \geq 0}$. The following assertions are equivalent.*

(i) *The curvature-dimension condition $CD(0, n)$ holds.*
(ii) *For all $1 < q_1 < q_2 < \infty$, $u_1, u_2 > 0$ and $\sigma = q_1 u_2 - q_2 u_1 \geq 0$, and for all positive measurable functions f on E,*

$$\left[P_{u_1}\left((P_\sigma f)^{q_2} \right) \right]^{1/q_2} \leq M^{n/2} \left[P_{u_2}\left(f^{q_1} \right) \right]^{1/q_1} \quad (6.7.14)$$

where

$$M = \left(\frac{q_1 - 1}{u_1} \right)^{1 - (1/q_1)} \left(\frac{q_2 - 1}{u_2} \right)^{(1/q_2) - 1} \left(\frac{q_1 u_2 - q_2 u_1}{q_2 - q_1} \right)^{(1/q_2) - (1/q_1)}.$$

Proof We proceed as in the proof of Theorem 5.5.5, p. 262, and only outline the main steps. Starting from the curvature condition (i), let $f \in \mathcal{A}_0^{\text{const}+}$ as usual, and consider

$$\Lambda(s) = \left(P_u (P_{t-s} f)^q \right)^{1/q}, \quad s \in [0, t], \ t > 0,$$

where now $q : [0, t] \to (1, \infty)$ and $u : [0, t] \to [0, \infty)$ are functions of s. Then, with $g = P_{t-s} f$,

$$\frac{q^2}{q'} \Lambda' \Lambda^{q-1} = \text{Ent}_{P_u}(g^q) + \frac{q^2}{q'} (u' - 1) P_u(g^{q-1} L g) \quad (6.7.15)$$
$$+ \frac{q^2}{q'} u'(q - 1) P_u(g^{q-2} \Gamma(g)).$$

Theorem 6.7.3 indicates that (i) is equivalent to the local logarithmic Sobolev inequality (6.7.4). By linearization, the latter amounts to the family of inequalities with parameter $\kappa > 0$,

$$\text{Ent}_{P_u}(f) \leq u \, L P_u f + \frac{n}{2} (\kappa - 1 - \log \kappa) P_u f - u \kappa P_u\big(f \, L(\log f) \big).$$

Now apply these inequalities to g^q and compare with (6.7.15). Choosing $q = q(s), u = u(s)$ and $\kappa = \kappa(s)$ solving the system

$$\begin{cases} \dfrac{q}{q'} (u' - 1) + u(1 - \kappa) = 0 \\ u(q - 1 + \kappa) + \dfrac{q}{q'} u'(q - 1) = 0 \end{cases} \quad (6.7.16)$$

we get that $\frac{q^2}{q'} \Lambda' \Lambda^{-1} \leq \frac{n}{2} A(\kappa)$ with $A(\kappa) = \kappa - 1 - \log \kappa$, $\kappa > 0$. Assuming that q is decreasing, it follows that $\Lambda(s) \leq e^{n M_{s,t}/2} \Lambda(t)$ where

$$M_{s,t} = -\int_s^t \frac{q'(r)}{q(r)^2} A\big(\kappa(r) \big) dr.$$

In other words,

$$P_{u(s)}\big((P_{t-s}f)^{q(s)}\big)^{1/q(s)} \leq e^{nM_{s,t}/2} P_{u(t)}\big(f^{q(t)}\big)^{1/q(t)}.$$

In order to optimize the value of $M_{s,t}$, we choose $q(r)$, $u(r)$ and $\kappa(r)$ on $[s,t]$ of the form

$$q(r) = \frac{r+\gamma}{\alpha r + \beta}, \qquad u(r) = \alpha r + \beta, \qquad \kappa = \frac{1-q}{1-\alpha q},$$

where α, β, γ are constants that we adjust to fit the values of $q(0) = q_2$ and $q(t) = q_1$ (and for which $u(0) = u_1$ and $u(t) = u_2$) (and therefore q is decreasing since it is monotonic). It is easily checked that those functions solve the system (6.7.16). The conclusion then follows after tedious computations. As for Theorem 6.7.3, (ii) yields $CD(0,n)$ by a Taylor expansion as u_1 goes to u_2 and q_1 to q_2. □

The inequalities (6.7.14) in Theorem 6.7.7 are optimal in the sense that if L is the Laplacian in $\mathbb{R}^n$ (therefore satisfying $CD(0,n)$), these inequalities are equalities for $f_\alpha(x) = e^{\alpha|x|^2}$, $x \in \mathbb{R}^n$, $\alpha \in \mathbb{R}$. There is also a version of Theorem 6.7.7 in the range $-\infty < q_2 < q_1 < 0$ or $0 < q_2 < q_1 < 1$.

6.8 Sobolev Inequalities Under a Curvature-Dimension Condition

This section is devoted to Sobolev inequalities for the invariant measure under curvature-dimension hypotheses. As a curvature condition $CD(\rho, \infty)$ with $\rho > 0$ implies a logarithmic Sobolev inequality, a stronger curvature-dimension condition $CD(\rho, n)$ with $\rho > 0$ and $n < \infty$ is expected to ensure the validity of a Sobolev inequality with exponent n.

6.8.1 Logarithmic Entropy-Energy Inequality Under $CD(\rho, n)$

It is not so easy to prove the Sobolev inequality directly. Again, a convenient tool is to establish instead a logarithmic entropy-energy inequality (6.2.2). The following theorem is a first main result in this direction. As usual, we consider a Markov Triple (E, μ, Γ) under the curvature-dimension condition $CD(\rho, n)$, $\rho \in \mathbb{R}$, $n \geq 1$,

$$\Gamma_2(f) \geq \rho\, \Gamma(f) + \frac{1}{n}(Lf)^2$$

for every $f \in \mathcal{A}_0$ (or $\mathcal{A}$) from Definition 3.3.14, p. 159. Recall that since $\rho > 0$ in the statement below, the invariant measure μ is assumed to be finite and normalized into a probability measure (cf. Theorem 3.3.23, p. 165).

Theorem 6.8.1 *Under the curvature-dimension condition $CD(\rho, n)$, $\rho > 0$, $1 \leq n < \infty$, the Markov Triple (E, μ, Γ), with μ a probability measure, satisfies the tight logarithmic entropy-energy inequality*

$$\mathrm{Ent}_\mu(f^2) \leq \frac{n}{2} \log\left(1 + \frac{4}{\rho n} \mathcal{E}(f)\right)$$

for every $f \in \mathcal{D}(\mathcal{E})$ with $\int_E f^2 d\mu = 1$.

As announced, by Proposition 6.2.3, a Sobolev inequality $S_n(A, C)$ of dimension n holds under the curvature-dimension condition $CD(\rho, n)$ with $\rho > 0$ and $n < \infty$. Moreover, together with Proposition 6.6.1, it follows from Theorem 6.8.1 that under the $CD(\rho, n)$ condition with $\rho > 0$ and $n < \infty$, the diameter of (E, μ, Γ) is finite. Finally, thanks to Corollary 6.4.1,

Corollary 6.8.2 *Under the curvature-dimension condition $CD(\rho, n)$ with $\rho > 0$ and $n < \infty$, the spectrum of the generator L is discrete and the Markov operators P_t, $t > 0$, are Hilbert-Schmidt.*

We turn to the proof of Theorem 6.8.1 which is similar to the proof of the logarithmic Sobolev inequality under the $CD(\rho, \infty)$ condition (Theorem 5.7.4, p. 270).

Proof of Theorem 6.8.1 Working with $f = \sqrt{f_0}$, $f_0 \in \mathcal{A}_0^{\mathrm{const+}}$, the inequality to be proved is

$$\mathrm{Ent}_\mu(f) \leq \frac{n}{2} \log\left(1 + \frac{1}{\rho n} \int_E f \Gamma(\log f) d\mu\right) \quad (6.8.1)$$

provided $\int_E f d\mu = 1$. Set then, as usual, $\Lambda(t) = \mathrm{Ent}_\mu(P_t f)$, $t \geq 0$. We already know (cf. the proof of Proposition 5.7.3, p. 269) that for every $t \geq 0$,

$$\Lambda'(t) = -\int_E P_t f \, \Gamma(\log P_t f) d\mu \quad \text{and} \quad \Lambda''(t) = 2 \int_E P_t f \, \Gamma_2(\log P_t f) d\mu.$$

By the curvature-dimension condition $CD(\rho, n)$ applied to $\log P_t f$, for every $t \geq 0$,

$$\Lambda''(t) \geq -2\rho \, \Lambda'(t) + \frac{2}{n} \int_E P_t f \left(\mathrm{L}(\log P_t f)\right)^2 d\mu.$$

Now, since $\int_E P_t f d\mu = 1$, by the Cauchy-Schwarz inequality, with $g = P_t f$ to ease the notation,

$$\int_E g \left(\mathrm{L}(\log g)\right)^2 d\mu \geq \left(\int_E g \mathrm{L}(\log g) d\mu\right)^2 = \left(\int_E g \Gamma(\log g) d\mu\right)^2.$$

The preceding leads to the differential inequality

$$\Lambda'' \geq -2\rho \, \Lambda' + \frac{2}{n} {\Lambda'}^2.$$

6.8 Sobolev Inequalities Under a Curvature-Dimension Condition

In other words, the function

$$e^{-2\rho t}\left[\frac{1}{n} - \frac{\rho}{\Lambda'(t)}\right]$$

is increasing in $t \geq 0$. Recalling that $\Lambda' \leq 0$, it follows that for every $t \geq 0$,

$$-\Lambda'(t) \leq \rho\left[e^{2\rho t}\left(\frac{1}{n} - \frac{\rho}{\Lambda'(0)}\right) - \frac{1}{n}\right]^{-1}.$$

Finally,

$$\Lambda(0) - \Lambda(t) = -\int_0^t \Lambda'(s)ds \leq \frac{n}{2}\log\left(1 - \left(1 - e^{-2\rho t}\right)\frac{\Lambda'(0)}{\rho n}\right).$$

As $t \to \infty$, the announced tight logarithmic entropy-energy inequality (6.8.1) follows. Theorem 6.8.1 is established. □

We next discuss some features of the logarithmic entropy-energy inequality of Theorem 6.8.1. First, using the inequality $\log(1+r) \leq r$, $r \geq 0$, it yields the logarithmic Sobolev inequality $LS(\frac{2}{\rho})$, or alternatively let $n \to \infty$ (corresponding to the curvature condition $CD(\rho, \infty)$). This is however not the optimal logarithmic Sobolev inequality as demonstrated by Theorem 5.7.4, p. 270. On the other hand, the inequality put forward in Theorem 6.8.1 is not the only one of that kind which may be produced by a similar proof. For example, recall (5.7.11), p. 272,

$$\int_E e^g \Gamma_2(g)\,d\mu \geq \frac{\rho n}{n-1}\int_E e^g \Gamma(g)\,d\mu + \frac{4n-1}{4(n+2)^2}\int_E e^g \Gamma(g)^2 d\mu.$$

If $\int_E e^g d\mu = 1$, by Jensen's inequality,

$$\int_E e^g \Gamma(g)^2 d\mu \geq \left(\int_E e^g \Gamma(g)d\mu\right)^2,$$

and the method of proof of Theorem 6.8.1 can then be used in the same way to get this time the differential inequality

$$\Lambda'' \geq -\frac{2\rho n}{n-1}\Lambda' + \frac{(4n-1)}{2(n+2)^2}\Lambda'^2.$$

In this way, we end up with the logarithmic entropy-energy inequality

$$\mathrm{Ent}_\mu(f^2) \leq \frac{p}{2}\log\left(1 + \frac{4(n-1)}{p\rho n}\mathcal{E}(f)\right)$$

for every $f \in \mathcal{D}(\mathcal{E})$ such that $\int_E f^2 d\mu = 1$, where $p = \frac{4(n+2)^2}{4n-1}$. This inequality is of weaker Sobolev dimension but yields this time the sharp logarithmic Sobolev inequality.

More generally, every inequality of the form

$$\int_E e^g \Gamma_2(g) d\mu \geq \Psi\left(\int_E e^g \Gamma(g) d\mu\right)$$

for every $g \in \mathcal{A}_0$ with $\int_E e^g d\mu = 1$ would give, by the same principle, an entropy-energy inequality

$$\mathrm{Ent}_\mu(f^2) \leq \Phi(\mathcal{E}(f)), \quad f \in \mathcal{D}(\mathcal{E}), \quad \int_E f^2 d\mu = 1,$$

for some growth function $\Phi : \mathbb{R}_+ \to \mathbb{R}_+$ of the type studied in Chap. 7. However, the best inequality of this type under the $CD(\rho, n)$ condition does not seem to be known.

6.8.2 Sobolev Inequality Under $CD(\rho, n)$

Via Proposition 6.2.3, Theorem 6.8.1 thus shows how the curvature-dimension condition $CD(\rho, n)$ with $\rho > 0$ and $n < \infty$ produces a Sobolev inequality $S_n(A, C)$ of dimension n. This Sobolev inequality may furthermore be tightened. However, the constants given by this process are not sharp. Actually, the $CD(\rho, n)$ condition, $\rho > 0, n < \infty$, may indeed be shown to reach optimal Sobolev inequalities but using a completely different technique of proof.

Theorem 6.8.3 (Sobolev inequality under $CD(\rho, n)$) *Under the curvature-dimension condition $CD(\rho, n)$, $\rho > 0$, $2 < n < \infty$, the Markov Triple (E, μ, Γ), with μ a probability measure, satisfies a Sobolev inequality $S_n(C)$ with constant $C = \frac{4(n-1)}{\rho n(n-2)}$. That is,*

$$\|f\|_p^2 \leq \|f\|_2^2 + \frac{4}{n(n-2)} \cdot \frac{n-1}{\rho} \mathcal{E}(f) \quad (6.8.2)$$

for every $f \in \mathcal{D}(\mathcal{E})$ where $p = \frac{2n}{n-2}$.

It is not known how to prove this theorem using (linear) heat flow monotonicity. In fact, the semigroup proof of Theorem 6.8.1 (for example) may be adapted to Sobolev inequalities up to some (rather mysterious) exponent $2 < p < \frac{2n}{n-2}$. In Sect. 6.11, alternative evolution equations (fast diffusion equations) will be developed in order to obtain Sobolev inequalities in the form of Gagliardo-Nirenberg inequalities. Here, we use instead non-linear methods on extremal functions. We first sketch the formal argument, and then provide the necessary technical details needed to complete the proof.

Proof We begin with the inequality

$$\|f\|_p^2 - \|f\|_2^2 \leq C \mathcal{E}(f)$$

6.8 Sobolev Inequalities Under a Curvature-Dimension Condition

for all $f \in \mathcal{D}(\mathcal{E})$ with $C > 0$ the optimal constant. If it exists, let f be a positive non-constant extremal function for this inequality, normalized so that $\int_E f^p d\mu = 1$, that is so that

$$\|f\|_p^2 - \|f\|_2^2 = C\mathcal{E}(f)$$

and for every $u \in \mathcal{D}(\mathcal{E})$,

$$\|f+u\|_p^2 - \|f+u\|_2^2 \le C\mathcal{E}(f+u).$$

Changing u into εu and letting $\varepsilon \to 0$ shows that

$$\int_E f^{p-1} u \, d\mu - \int_E fu \, d\mu = 0$$

provided $\mathcal{E}(f, u) = \int_E (-Lf) u \, d\mu = 0$. Therefore, $f \in \mathcal{D}(L)$ and must satisfy the extremal function equation

$$f^{p-1} - f = -CLf. \tag{6.8.3}$$

Later, in the true proof, we shall show that such an extremal function is bounded from above and below, justifying all the integrations below. Setting $f = e^g$, the equation becomes

$$e^{(p-2)g} = 1 - C\big[Lg + \Gamma(g)\big]. \tag{6.8.4}$$

Then multiply both terms of the latter identity by $e^{bg} Lg$, $b \in \mathbb{R}$, and integrate with respect to μ. By integration by parts, it follows that

$$(p-2+b) \int_E e^{(p-2)g} e^{bg} \Gamma(g) d\mu$$

$$= b \int_E e^{bg} \Gamma(g) d\mu + C \int_E e^{bg} (Lg)^2 d\mu$$

$$- C \int_E e^{bg} \Gamma\big(g, \Gamma(g)\big) d\mu - Cb \int_E e^{bg} \Gamma(g)^2 d\mu.$$

The following makes use of various integral formulas on the Γ_2 operator developed in the proof of Theorem 5.7.4, p. 270. For example, by (5.7.8), p. 271, with $2a = b$,

$$\int_E e^{bg} (Lg)^2 d\mu = \int_E e^{bg} \Gamma_2(g) d\mu + \frac{3b}{2} \int_E e^{bg} \Gamma\big(g, \Gamma(g)\big) d\mu$$

$$+ b^2 \int_E e^{bg} \Gamma(g)^2 d\mu.$$

Therefore the preceding equation becomes

$$(p-2+b)\int_E e^{(p-2)g} e^{bg}\Gamma(g)d\mu = b\int_E e^{bg}\Gamma(g)d\mu + C\int_E e^{bg}\Gamma_2(g)d\mu$$
$$+ C\left(\frac{3b}{2}-1\right)\int_E e^{bg}\Gamma\big(g,\Gamma(g)\big)d\mu$$
$$+ Cb(b-1)\int_E e^{bg}\Gamma(g)^2 d\mu.$$

Now replace $e^{(p-2)g}$ by its expression from (6.8.4) and integrate by parts again to get, after some algebra,

$$\int_E e^{bg}\Gamma_2(g)d\mu = \frac{p-2}{C}\int_E e^{bg}\Gamma(g)d\mu$$
$$+ \left(p-1-\frac{b}{2}\right)\int_E e^{bg}\Gamma\big(g,\Gamma(g)\big)d\mu \qquad (6.8.5)$$
$$+ (p-2)(b-1)\int_E e^{bg}\Gamma(g)^2 d\mu.$$

On the other hand, arguing as for (5.7.10), p. 271, the $CD(\rho, n)$ condition yields, for every $a \in \mathbb{R}$,

$$\int_E e^{bg}\Gamma_2(g)d\mu \geq \frac{\rho n}{n-1}\int_E e^{bg}\Gamma(g)d\mu$$
$$+ \frac{3b-2(n+2)a}{2(n-1)}\int_E e^{bg}\Gamma\big(g,\Gamma(g)\big)d\mu \qquad (6.8.6)$$
$$+ \frac{b^2-2ab-(n-1)a^2}{n-1}\int_E e^{bg}\Gamma(g)^2 d\mu.$$

It remains to compare (6.8.5) and (6.8.6), and to adjust the parameters $a, b \in \mathbb{R}$ in such a way that the coefficients in front of the terms $\int_E e^{bg}\Gamma(g,\Gamma(g))d\mu$ and $\int_E e^{bg}\Gamma(g)^2 d\mu$ coincide. Recalling that $p = \frac{2n}{n-2}$, this is achieved by choosing

$$a = \frac{b}{2} - \frac{n-1}{n-2} \quad \text{and} \quad b = \frac{2(n-3)}{n-2}.$$

As a consequence, it follows that

$$\frac{p-2}{C}\int_E e^{bg}\Gamma(g)d\mu \geq \frac{\rho n}{n-1}\int_E e^{bg}\Gamma(g)d\mu.$$

Since $\int_E e^{bg}\Gamma(g)d\mu > 0$ (because f is assumed to be non-constant), $\frac{p-2}{C} \geq \frac{\rho n}{n-1}$, or in other words,

$$C \leq \frac{4(n-1)}{\rho n(n-2)}.$$

6.8 Sobolev Inequalities Under a Curvature-Dimension Condition

This bound concludes the announced Sobolev inequality.

This scheme of proof is rather simple, but unfortunately non-constant extremal functions of Sobolev inequalities do not always exist. In order for the preceding formal proof to work, it is necessary to deal with almost extremal functions at the almost optimal exponent. To justify the procedure, we make use of the logarithmic entropy-energy inequality of Theorem 6.8.1 (and thus of the Sobolev inequality of dimension n but with worse constants).

Consider $2 < q < \frac{2n}{n-2}$ (which will later converge to $p = \frac{2n}{n-2}$) and $\varepsilon > 0$, and denote by $C(q, \varepsilon) > 0$ the best constant in the Sobolev inequality

$$\|f\|_q^2 \leq (1+\varepsilon)\|f\|_2^2 + C(q,\varepsilon)\mathcal{E}(f)$$

for every $f \in \mathcal{D}(\mathcal{E})$. As explained below, choosing $q < \frac{2n}{n-2}$ will ensure the existence of a function saturating this inequality, while $\varepsilon > 0$ will ensure that it is not constant. Moreover, it will be shown that such an extremal function is bounded from above and below, justifying all the integration by parts arguments of the formal proof. The extremal function f, provided it exists, will then satisfy $f^{q-1} = (1+\varepsilon)f - C(q,\varepsilon)\mathrm{L}f$, and with the same arguments as above, one ends up with the upper bound

$$C(q,\varepsilon) \leq \frac{(n-1)(q-2)}{n\rho}.$$

The conclusion follows as $q \to p = \frac{2n}{n-2}$ and $\varepsilon \to 0$.

We are thus left with the question of the existence of such an extremal function. First, recall from Sect. 6.4 that the embedding of $\mathcal{D}(\mathcal{E})$ into $\mathbb{L}^q(\mu)$ is compact for every $q \in [2, p)$. Let then $(f_k)_{k \in \mathbb{N}}$ be a bounded sequence of functions in $\mathcal{D}(\mathcal{E})$ which are almost optimal for the Sobolev inequality, that is such that, for every $k \geq 1$,

$$\|f_k\|_q^2 \geq (1+\varepsilon)\|f_k\|_2^2 + C(q,\varepsilon)\mathcal{E}(f_k) - \frac{1}{k}$$

(and $\|f_k\|_2^2 + \mathcal{E}(f_k) = 1$). Since $\mathcal{E}(|f_k|) \leq \mathcal{E}(f_k)$, it may be assumed that $f_k \geq 0$ for every k. By the compactness property, there is a subsequence, still denoted by $(f_k)_{k \in \mathbb{N}}$, converging almost surely and in $\mathbb{L}^q(\mu)$, and weakly in $\mathcal{D}(\mathcal{E})$, towards a positive function $f \in \mathcal{D}(\mathcal{E})$. For this function, for any $u \in \mathcal{D}(\mathcal{E})$,

$$\int (f^{q-1} - (1+\varepsilon)f) u \, d\mu = C(q,\varepsilon)\mathcal{E}(f, u).$$

This function f is therefore in the domain $\mathcal{D}(\mathrm{L})$ and satisfies the extremal function equation

$$f^{q-1} = (1+\varepsilon)f - C(q,\varepsilon)\mathrm{L}f.$$

Next, we make sure that f is bounded from above and below (by a strictly positive constant). To this end, observe that the equation for extremal functions may be rewritten as

$$f = C(q,\varepsilon)^{-1}\left(\frac{1+\varepsilon}{C(q,\varepsilon)}\operatorname{Id}-L\right)^{-1}(f^{q-1}) = C(q,\varepsilon)^{-1}R_\lambda(f^{q-1}),$$

where $\lambda = \frac{1+\varepsilon}{C(q,\varepsilon)}$ and $R_\lambda = (\lambda\operatorname{Id}-L)^{-1} = \int_0^\infty e^{-\lambda t} P_t\, dt$ is the resolvent operator. For the upper bound, Corollary 6.3.3 already indicates that $R_\lambda(f^{q-1})$ is bounded as soon as $f \in \mathbb{L}^r(\mu)$ for some $r > (q-1)\frac{n}{2}$. But from the same argument, as soon as $f \in \mathbb{L}^r(\mu)$, then $f \in \mathbb{L}^{r'}(\mu)$ for any $r' < \frac{r}{a}$ where $a = (q-1) - \frac{2r}{n}$. Since $a < 1$ as soon as $r > (q-2)\frac{n}{2}$, by a simple iterative procedure, it follows that f is bounded as soon as it belongs to $\mathbb{L}^{q_0}(\mu)$ for some $q_0 > (q-2)\frac{n}{2}$. Since in our case $f \in \mathbb{L}^q(\mu)$ and $q > (q-2)\frac{n}{2}$, it is indeed bounded. For the (strictly positive) lower bound, Proposition 6.3.4 provides a lower bound on the kernel of P_t, and thus, say for $t \geq 1$, $P_t g \geq c \int_E g\, d\mu$ for some $c > 0$ and every positive function g. This in turn implies that

$$R_\lambda(f^{q-1}) \geq \int_1^\infty P_t(f^{q-1}) e^{-\lambda t} dt \geq \frac{Ce^{-\lambda}}{\lambda}\int_E f^{q-1} d\mu$$

from which the claim easily follows. Putting together the various pieces, the proof of Theorem 6.8.3 is complete. □

Remark 6.8.4 The preceding proof similarly shows that, under the curvature-dimension condition $CD(\rho,n)$ with $\rho > 0$ and $2 < n < \infty$, for every $f \in \mathcal{D}(\mathcal{E})$,

$$\frac{\|f\|_q^2 - \|f\|_2^2}{q-2} \leq \frac{(n-1)}{\rho n}\mathcal{E}(f) \tag{6.8.7}$$

for every $2 \leq q \leq \frac{2n}{n-2}$. Such an inequality will be called later in Sect. 7.6, p. 382, a Beckner-type inequality. In this form, it actually holds for every $1 \leq q \leq \frac{2n}{n-2}$, the value $q = 1$ corresponding to the Poincaré inequality, the value $q = 2$ to the logarithmic Sobolev inequality (in the limit), and $q = \frac{2n}{n-2}$ to the Sobolev inequality. By Proposition 6.2.2, any inequality in this family implies the Poincaré inequality. Since on the sphere $\mathbb{S}^n$, $\rho = n-1$, the constant in (6.8.7) is optimal for any q, and in particular Theorem 6.8.3 leads to the optimal Sobolev inequality (6.1.2) in this case. When $1 \leq n \leq 2$, inequality (6.8.7) may actually be considered for any $q \geq 1$ and implies, as $q \to \infty$, Moser-Trudinger-type inequalities of the form

$$\log\left(\int_E e^f d\mu\right) - \int_E f\, d\mu \leq \frac{(n-1)}{2\rho n}\mathcal{E}(f), \quad f \in \mathcal{D}(\mathcal{E}). \tag{6.8.8}$$

Remark 6.8.5 Under the condition $CD(\rho,n)$, for $\rho > 0$, $2 < n < \infty$, the family of inequalities (6.8.7) in the previous remark describes optimal Poincaré, logarithmic Sobolev and Sobolev inequalities. The constant in (6.8.7) may actually be slightly improved together with the spectral gap whenever $q < \frac{2n}{n-2}$. Indeed denoting by λ_1

6.9 Conformal Invariance of Sobolev Inequalities

the first non-trivial eigenvalue of $-L$ (the spectrum of $-L$ is discrete in this case), $\frac{(n-1)}{\rho n}$ in (6.8.7) may be replaced by $\frac{1}{\kappa}$ where

$$\kappa = \theta \, \frac{\rho n}{n-1} + (1-\theta)\lambda_1$$

and

$$\theta = \theta(q) = \frac{(q-1)(n-1)^2}{(q-2)+(n+1)^2} \in [0,1].$$

Note that $\theta(p) = 1$ for the critical exponent $p = \frac{2n}{n-2}$ while $\theta(1) = 0$. In particular, when $q = 2$ this corresponds to the logarithmic Sobolev inequality,

$$\kappa = \frac{(n-1)^2}{(n+1)^2} \cdot \frac{\rho n}{n-1} + \frac{4n}{(n+1)^2} \lambda_1.$$

Since $\lambda_1 \geq \frac{\rho n}{n-1}$ by (4.8.1), p. 213, these bounds improve upon (6.8.7). They are obtained by a simple modification at the end of the (formal) proof of Theorem 6.8.3, using that $\lambda_1 \int_E \Gamma(f) d\mu \leq \int_E (Lf)^2 d\mu$ according to Proposition 4.8.3, p. 214. As a geometric consequence, together with Obata's Theorem, an n-dimensional Riemannian manifold with Ricci curvature bounded from below by $n-1$ for which $\kappa = n$ for some $q < \frac{2n}{n-2}$ (in particular if the logarithmic Sobolev constant is equal to $\frac{1}{n}$) is isometric to the n-sphere $\mathbb{S}^n$ (Obata's Theorem tells us this is the case when $\lambda_1 = n$).

6.9 Conformal Invariance of Sobolev Inequalities

In this section, we will learn how the preceding optimal Sobolev inequality on the sphere (cf. (6.1.2) and Remark 6.8.4), established by means of the curvature-dimension criterion in the abstract framework of a Markov diffusion operator in Theorem 6.8.3, actually yields, and is equivalent to, the corresponding optimal Sobolev inequalities on the Euclidean and hyperbolic spaces. The crucial property used to reach this equivalence is conformal invariance, which is linked to the existence of extremal functions. As already mentioned in the previous section, extremal (non-constant) functions for Sobolev inequalities at the critical exponent do not always exist (actually even for non-critical exponents). However, in the Euclidean space or on the sphere, such extremal functions at the critical exponent do exist. As discussed in this section, this property relies on particular transformations of the underlying state space which preserve the metric (up to dilation). These are the conformal maps, and (suitably modified) Sobolev inequalities are invariant under such conformal maps.

6.9.1 Conformal Invariant

Conformal invariance may be presented in various ways. On the one hand, it may be seen simply as the change on a Riemannian manifold $(M, \mathfrak{g})$ with dimension n

of the co-metric $\mathfrak{g}$ into $c^2\mathfrak{g}$, or of the operator Γ into $c^2\Gamma$, where c is some given (strictly positive) function. At the same time, the measure μ associated with the model has to be changed into $c^{-n}\mu$ (which is the case when dealing with Laplace-Beltrami operators in dimension n and the Riemannian measure defined from the metric). There is actually a way (using the scalar curvature) to present the classical Sobolev inequalities on the three model spaces as invariant under such a family of transformations. On the other hand, such transformations may sometimes be produced by maps $\psi : M \to M$ such that $\Gamma(f \circ \psi) = (c^2 \Gamma(f)) \circ \psi$. In this case, the new metric is just the former one viewed through a change of coordinates. Such a map ψ is called a *conformal map*, and usually induces extra information on the conformal transformation itself.

This section presents the basic material on conformal transformations and Sobolev inequalities, from which extremal functions (and further properties) will be extracted. This will be achieved in the context of a smooth (at least C^2) Riemannian manifold $(M, \mathfrak{g})$ (or some open set in M) of dimension n. The reader may however keep in mind the three model spaces $\mathbb{R}^n$, $\mathbb{S}^n$ and $\mathbb{H}^n$. Sobolev inequalities will always be understood to hold for the class $\mathcal{A}_0$ of smooth (C^∞) compactly supported functions on M. Recall in addition the class $\mathcal{A}$ of smooth (C^∞) functions on M.

In this setting, we begin by recalling the notion of scalar curvature $sc_\mathfrak{g}$ on $(M, \mathfrak{g})$ (see (C.3.2), p. 508).

Definition 6.9.1 (Scalar curvature) The scalar curvature $sc_\mathfrak{g}(x)$ at a point $x \in M$ is the trace of the Ricci tensor. In a local coordinate system,

$$sc_\mathfrak{g}(x) = \sum_{i,j=1}^{n} g^{ij}(x) \, \text{Ric}_{ij}(x)$$

where $(\text{Ric}_{ij}(x))_{1 \leq i,j \leq n}$ is the Ricci tensor at the point $x \in M$.

On the sphere $\mathbb{S}^n \subset \mathbb{R}^{n+1}$ with dimension n and constant Ricci curvature $\text{Ric}_{ij} = (n-1)g_{ij}$, $sc_\mathfrak{g}(x) = n(n-1)$ at any point. Similarly $sc_\mathfrak{g}(x) = 0$ on $\mathbb{R}^n$ and $sc_\mathfrak{g}(x) = -n(n-1)$ on the hyperbolic space $\mathbb{H}^n$. In particular, the (optimal) Sobolev inequality on the sphere (6.1.2) at the critical exponent $p = \frac{2n}{n-2}$ with respect to the uniform probability measure μ and the associated Dirichlet form $\mathcal{E}$ may therefore be rewritten as

$$\|f\|_p^2 \leq \frac{4}{n(n-2)} \left[\frac{(n-2)}{4(n-1)} \int_{\mathbb{S}^n} sc_\mathfrak{g} \, f^2 d\mu + \mathcal{E}(f) \right]$$

for every $f \in \mathcal{D}(\mathcal{E})$. Such a formulation of the Sobolev inequality is actually generic, and emphasizes the scalar curvature as a so-called conformal invariant.

This conformal transformation may be presented in a general context, of independent interest. Consider the framework of a Markov Triple (E, μ, Γ) with generator L and Dirichlet form $\mathcal{E}(f) = \int_E \Gamma(f) d\mu$. The n-conformal class, $n \geq 2$ (it is possible to restrict to $n > 2$, but $n = 2$ is a special important case, even if many formulas

6.9 Conformal Invariance of Sobolev Inequalities

have to be adjusted), of (E, μ, Γ) is the set of all Markov Triples $(E, c^{-n}\mu, c^2\Gamma)$ for all strictly positive functions $c \in \mathcal{A}$. An *n-conformal invariant* of the n-conformal class of (E, μ, Γ) is a map $S = S(\mu, \Gamma) : E \to \mathbb{R}$ depending only on μ and Γ such that, for any function $c = e^\tau \in \mathcal{A}$,

$$S(c^{-n}\mu, c^2\Gamma) = c^2 \left[S(\mu, \Gamma) + \frac{n-2}{2}\left(\mathrm{L}\tau - \frac{n-2}{2}\Gamma(\tau)\right)\right]. \tag{6.9.1}$$

With this definition, the following statement holds.

Proposition 6.9.2 *Setting $p = \frac{2n}{n-2}$ as usual, $n > 2$, let $S(\mu, \Gamma)$ be an n-conformal invariant map. Then, the Sobolev inequality*

$$\|f\|_p^2 \leq C \left[\int_E S(\mu, \Gamma) f^2 d\mu + \mathcal{E}(f) \right], \quad f \in \mathcal{D}(\mathcal{E}),$$

for some $C > 0$, is invariant in the n-conformal class of (E, μ, Γ). That is, if it holds for the pair (μ, Γ), then it holds with the same constant C for the pair $(c^{-n}\mu, c^2\Gamma)$ for every strictly positive function $c \in \mathcal{A}$.

Proof The proof is immediate. It suffices to change f into $c^{(2-n)/2} f$ in the Sobolev inequality and, after the use of (6.9.1), to integrate by parts the mixed term in the change of variables formula for the Γ operator

$$\Gamma(c^{(2-n)/2} f) = c^{2-n} \left[\Gamma(f) - \frac{n-2}{2} \Gamma(f^2, \tau) + \frac{(n-2)^2}{4} f^2 \Gamma(\tau) \right].$$

□

Note that by the change $f \mapsto c^{(2-n)/2} f$ used in Proposition 6.9.2, the generator L associated with (E, μ, Γ) is transformed for $(E, c^{-n}\mu, c^2\Gamma)$ with $c = e^\tau$ into

$$\widehat{\mathrm{L}} = c^2 \big[\mathrm{L}f - (n-2)\Gamma(\tau, f) \big]. \tag{6.9.2}$$

The formula assumes a particularly nice form when $n = 2$.

6.9.2 Geometric Conformal Invariant

According to Proposition 6.9.2 above, the aim here will be to show that, in a Riemannian setting, the scalar curvature is such an n-conformal invariant (thus satisfying (6.9.1)), where n is the dimension of the manifold. We refer to Appendix C for the necessary Riemannian background. In Riemannian geometry, when $\mathfrak{g}$ is the (co-) metric and $d\mu_\mathfrak{g}$ the Riemannian volume element, the change of Γ into $c^2\Gamma$ and $d\mu_\mathfrak{g}$ into $c^{-n} d\mu_\mathfrak{g}$ corresponds to a conformal change of the metric $\mathfrak{g}$ into $c^2 \mathfrak{g}$.

Recall that, for a Laplacian associated to the (co-) metric $\mathfrak{g} = (g^{ij})$, that is when the carré du champ operator takes the form $\Gamma(f) = \sum_{i,j=1}^{n} g^{ij} \partial_i f \partial_j f$, the Riemannian measure is given in local coordinates by $d\mu_{\mathfrak{g}} = \det(\mathfrak{g})^{-1/2} dx$, where dx is the Lebesgue measure in the coordinate system. (Observe that, contrary to our usual convention, it is not normalized whenever it is finite.) The aim is to examine how the scalar curvature is modified by such a conformal transformation. To this end we first study the Ricci curvature. Two approaches are available. One via the Γ_2-calculus for the new operator, and another in local coordinates. Both are heavy and somewhat cumbersome, but it might be easier to work with the Γ_2 operator, which we do below.

While the case of a Laplacian will be our main illustrative example, the Γ_2 formulas which need to be developed are just as easy to describe in the general case of a generator of the form $L = \Delta_{\mathfrak{g}} - \nabla W \cdot \nabla$ for some smooth potential W on a Riemannian manifold $(M, \mathfrak{g})$ with symmetric invariant measure $d\mu = e^{-W} d\mu_{\mathfrak{g}}$. This setting will furthermore be suited to our later investigation of non-geometric examples. We thus discuss the Γ_2 operations below in this more general setting. In this instance, recall from (1.16.4), p. 71 (cf. also Sect. C.6, p. 513), that the Γ_2 operator is given on smooth functions by

$$\Gamma_2(f) = |\nabla \nabla f|^2 + \mathrm{Ric}(L)(\nabla f, \nabla f)$$

where $\mathrm{Ric}(L)$ is a symmetric tensor defined from the Ricci tensor $\mathrm{Ric}_{\mathfrak{g}}$ of the Riemannian manifold $(M, \mathfrak{g})$ by $\mathrm{Ric}(L) = \mathrm{Ric}_{\mathfrak{g}} + \nabla\nabla W$.

In this setting, the first step will be to compute, for any n and any smooth function c, the Ricci tensor associated with $(c^{-n}\mu, c^2 \Gamma)$ from the Ricci tensor associated to (μ, Γ). Although most interesting formulas will be achieved when n is the dimension of M, assume for the moment that it is distinct from this dimension, denoted here by n_0. The computations below are performed as usual on the class $\mathcal{A}_0$ of smooth compactly supported functions on M.

Given c written in the form $c = e^\tau$, the Γ_2 operator associated with $c^2 L$ may be expressed in terms of the Γ_2 operator of L as

$$c^4 \Big[\Gamma_2(f) + 2\Gamma\big(\tau, \Gamma(f)\big) + \big(L\tau + 2\Gamma(\tau)\big) \Gamma(f) - 2Lf\, \Gamma(f,\tau) \Big].$$

According to Proposition 6.9.2, and for a generic n, the aim is to modify this formula in terms of the operator $\widehat{L}$ of (6.9.2), which thus requires us to replace $\Gamma_2(f)$ by $\Gamma_2(f) + (n-2)\nabla\nabla\tau(\nabla f, \nabla f)$ and L by $L - (n-2)\Gamma(\tau, \cdot)$ in the preceding. Hence, denoting by $\widehat{\Gamma}_2$ the Γ_2 operator of $\widehat{L}$,

$$\widehat{\Gamma}_2(f) = c^4 \Big[\Gamma_2(f) + (n-2)\nabla\nabla\tau(\nabla f, \nabla f) + 2\Gamma\big(\tau, \Gamma(f)\big)$$
$$+ \big(L\tau - (n-4)\Gamma(\tau)\big)\Gamma(f) - 2Lf\,\Gamma(f,\tau) + 2(n-2)\Gamma(f,\tau)^2 \Big].$$

The second order terms are gathered in the expression

$$\widehat{\nabla\nabla} f = \nabla\nabla f + 2\nabla f \odot \nabla\tau - \Gamma(f,\tau) \mathfrak{g}$$

6.9 Conformal Invariance of Sobolev Inequalities

where $\nabla f \odot \nabla \tau$ denotes the symmetric tensor product, that is

$$(\nabla f \odot \nabla \tau)(\nabla g, \nabla h) = \frac{1}{2}\big[\Gamma(f,g)\Gamma(\tau,h) + \Gamma(f,h)\Gamma(\tau,g)\big].$$

Hence

$$|\widehat{\nabla\nabla} f|^2 = c^4\Big[|\nabla\nabla f|^2 + 2\Gamma\big(\tau, \Gamma(f)\big) + 2\Gamma(f)\Gamma(\tau)$$
$$+ (n_0 - 2)\,\Gamma(f,\tau)^2 - 2\Delta_{\mathfrak{g}} f\,\Gamma(f,\tau)\Big].$$

In this formula, n_0 (the effective dimension of the manifold) appears as the norm, in the space of symmetric matrices, of the identity matrix in the form of the term $\Gamma(f,\tau)\,\mathfrak{g}$. The term $\Delta_{\mathfrak{g}} f$ appears as the scalar product, still in the space of symmetric matrices, of $\nabla\nabla f$ and $\mathfrak{g}$, that is the trace of $\nabla\nabla f$. Lastly, comparing the two formulas,

$$\mathrm{Ric}(\widehat{L})(\nabla f, \nabla f) = c^4\Big[\mathrm{Ric}(L)(\nabla f, \nabla f) + (n-2)\nabla\nabla\tau(\nabla f, \nabla f)$$
$$+ \big(L\tau - (n-2)\Gamma(\tau)\big)\Gamma(f) + 2\,\nabla W(f)\,\Gamma(f,\tau)$$
$$+ (2n - n_0 - 2)\,\Gamma(f,\tau)^2\Big].$$

In other words, in terms of Ricci tensors with lower indices, that is as symmetric operators acting on the tangent space,

$$\mathrm{Ric}(\widehat{L}) = \mathrm{Ric}(L) + (n-2)\,\nabla\nabla\tau + \big(L\tau - (n-2)\Gamma(\tau)\big)\mathfrak{g}$$
$$+ 2\,\nabla W \odot \nabla\tau + (2n - n_0 - 2)\nabla\tau \odot \nabla\tau.$$

This formula for the Ricci curvature of $\widehat{L}$ is rather heavy, but considerably simplifies when $n = n_0$ is the dimension of the manifold and $L = \Delta_{\mathfrak{g}}$ is a Laplacian ($W = 0$), and it is in this setting that we shall mainly apply the formula. Indeed, after identification,

$$\mathrm{Ric}(\widehat{\Delta}_{\mathfrak{g}}) = \mathrm{Ric}_{\mathfrak{g}} + \mathfrak{g}\,\Delta_{\mathfrak{g}}\tau + (n-2)\big(\nabla\nabla\tau + \nabla\tau \odot \nabla\tau - \Gamma(\tau)\mathfrak{g}\big).$$

Taking the trace to reach scalar curvature, we get the following formula for the scalar curvature $\widehat{\mathrm{sc}}_{\mathfrak{g}}$ of $\widehat{L}$ in terms of $\mathrm{sc}_{\mathfrak{g}}$,

$$\widehat{\mathrm{sc}}_{\mathfrak{g}} = c^2\big[\mathrm{sc}_{\mathfrak{g}} + (n-1)\big(2\Delta_{\mathfrak{g}}\tau - (n-2)\,\Gamma(\tau)\big)\big].$$

The latter may also be rewritten, recalling $c = e^\tau$, as

$$\widehat{\mathrm{sc}}_{\mathfrak{g}} = e^{2\tau}\Big[\mathrm{sc}_{\mathfrak{g}} - \frac{4(n-1)}{n-2}\,e^{(n-2)\tau/2}\Delta_{\mathfrak{g}}\big(e^{-(n-2)\tau/2}\big)\Big]. \tag{6.9.3}$$

This equation (6.9.3) shows that the function $\frac{n-2}{4(n-1)}\,\mathrm{sc}_{\mathfrak{g}}$ satisfies the defining condition (6.9.1) of an n-conformal invariant, which is the announced claim.

Observe that when n is the dimension of the manifold (which is the case here), the n-conformal class of a given Laplacian only contains Laplacians. Note furthermore that, setting $f = e^{-(n-2)\tau/2}$ and as usual $p = \frac{2n}{n-2}$, $n > 2$, (6.9.3) can be rewritten as

$$f^{p-1} \widehat{\mathrm{sc}}_\mathfrak{g} = f\, \mathrm{sc}_\mathfrak{g} - \frac{4(n-1)}{n-2} \Delta_\mathfrak{g} f. \tag{6.9.4}$$

As a consequence of these developments and of Proposition 6.9.2, we may now state the following main conclusion.

Theorem 6.9.3 (Conformal invariant Sobolev inequality) *Let $\Delta_\mathfrak{g}$ be the Laplace-Beltrami operator on a Riemannian manifold $(M, \mathfrak{g})$ with dimension n (> 2) equipped with the Riemannian measure $\mu_\mathfrak{g}$, with associated carré du champ operator $\Gamma(f) = |\nabla f|^2$ (on smooth functions $f : M \to \mathbb{R}$) and Dirichlet form $\mathcal{E}(f) = \int_M \Gamma(f) d\mu_\mathfrak{g}$. Then the scalar curvature $\mathrm{sc}_\mathfrak{g}$ is an n-conformal invariant of the Markov Triple $(M, \mu_\mathfrak{g}, \Gamma)$. That is, if the Sobolev inequality*

$$\|f\|_p^2 \leq C \left[\frac{n-2}{4(n-1)} \int_M \mathrm{sc}_\mathfrak{g} f^2 d\mu + \mathcal{E}_g(f) \right], \quad f \in \mathcal{D}(\mathcal{E}),$$

holds, where $p = \frac{2n}{n-2}$, then the same inequality holds when $\mu_\mathfrak{g}$ is changed into $c^{-n} \mu_\mathfrak{g}$ and Γ into $c^2 \Gamma$ where c is a smooth strictly positive function.

6.9.3 Conformal Invariance and Sharp Sobolev Inequalities

The preceding Theorem 6.9.3 has important consequences, in particular, as announced, it leads to optimal constants and extremal functions in the Sobolev inequalities on the three model spaces. Denote below by ω_n the volume (for the uniform measure) of the standard sphere $\mathbb{S}^n$ in $\mathbb{R}^{n+1}$.

Theorem 6.9.4 (Optimal Sobolev inequalities on the model spaces) *For the Riemannian measures μ on the three model spaces $\mathbb{R}^n$, $\mathbb{S}^n$, $\mathbb{H}^n$ (not normalized on $\mathbb{S}^n$), $n > 2$, the following optimal Sobolev inequalities with $p = \frac{2n}{n-2}$ hold (for every function f in the respective Dirichlet domains $\mathcal{D}(\mathcal{E})$).*

(i) *On the Euclidean space $\mathbb{R}^n$,*

$$\|f\|_p^2 \leq \frac{4}{n(n-2)\omega_n^{2/n}} \mathcal{E}(f).$$

(ii) *On the sphere $\mathbb{S}^n$,*

$$\|f\|_p^2 \leq \frac{1}{\omega_n^{2/n}} \int_{\mathbb{S}^n} f^2 d\mu + \frac{4}{n(n-2)\omega_n^{2/n}} \mathcal{E}(f).$$

6.9 Conformal Invariance of Sobolev Inequalities

(iii) *On the hyperbolic space* $\mathbb{H}^n$,

$$\|f\|_p^2 \leq -\frac{1}{\omega_n^{2/n}} \int_{\mathbb{H}^n} f^2 d\mu + \frac{4}{n(n-2)\,\omega_n^{2/n}} \mathcal{E}(f).$$

Remark 6.9.5 For the Sobolev inequality in $\mathbb{R}^n$ (case (i)), the $\mathbb{L}^2(dx)$ norm of the function f does not appear in the statement. However, the fact that $f \in \mathcal{D}(\mathcal{E})$ (and therefore that $f \in \mathbb{L}^2(dx)$) may not be removed without care since, for example, the inequality cannot hold for a non-zero constant function. However, through standard localization and symmetrization procedures in Euclidean space, the Sobolev inequality (i) may be extended to any smooth function f such that $|\nabla f| \in \mathbb{L}^2(dx)$ and such that $\mathrm{vol}_n(|f| \geq \varepsilon) < \infty$ for any $\varepsilon > 0$.

Proof We start from the optimal Sobolev inequality on the sphere $\mathbb{S}^n$ established in Theorem 6.8.3, rewritten here under the un-normalized uniform measure. Now, it was observed in Sect. 2.2.2, p. 83, that in the stereographic projection, the metric of the sphere is a conformal transformation of the metric of the Euclidean space $\mathbb{R}^n$. The scalar curvature of $\mathbb{R}^n$ is 0, and the resulting optimal Sobolev inequality on $\mathbb{R}^n$ is thus obtained from Theorem 6.9.3. By the conformal invariance principle, these two Sobolev inequalities are therefore equivalent. One can add to the equivalence the Sobolev inequality on the hyperbolic space $\mathbb{H}^n$ since $\mathbb{H}^n$ may be identified with $\mathbb{R}^{n-1} \times (0, \infty)$ with a metric given as a conformal transformation of the Euclidean metric with constant scalar curvature $-n(n-1)$. The proof is complete. □

A further important property and application drawn from Theorem 6.9.3 concerns extremal functions. Assume that the two metrics given by Γ and $c^2 \Gamma$ have constant scalar curvature equal to $n(n-1)$. Then, the equation (6.9.4) of conformal transformation of scalar curvature indicates that the function $f = c^{-(n-2)/2}$ satisfies the equation

$$f^{p-2} = 1 - \frac{4}{n(n-2)} \frac{\mathrm{L}f}{f},$$

which is precisely Eq. (6.8.3) of extremal functions for the Sobolev inequality on the sphere with constant $C = \frac{4}{n(n-2)}$. Now, there exist conformal maps on the sphere, that is diffeomorphisms from $\mathbb{S}^n$ onto itself, such that the image carré du champ operators are of the form $c^2 \Gamma$. But under such a conformal map, the metric $c^2 \Gamma$ is nothing else but the metric Γ viewed under a change of coordinates. The new metric has constant scalar curvature (since curvature is independent of the coordinates and invariant under diffeomorphisms). These observations thus explain the existence of non-constant extremal functions for the Sobolev inequality at the critical exponent on the sphere. (It may be shown that there are no non-constant extremal functions for the Sobolev inequality (6.8.7) below the critical exponent.)

It might be useful to briefly describe the conformal maps from $\mathbb{S}^n$ into itself. Let Q be a point in $\mathbb{R}^{n+1}$ exterior to the sphere $\mathbb{S}^n$. To each $I \in \mathbb{S}^n$ associate the other intersection point J between $\mathbb{S}^n$ and the line QI. The map $I \mapsto J$ is then such a conformal map as a consequence of the following three observations:

(i) The inversions of $\mathbb{R}^{n+1}$ are conformal transformations of $\mathbb{R}^n$ into itself.
(ii) The map $I \mapsto J$ is the restriction to $\mathbb{S}^n$ of the inversion (with center Q and inversion sphere orthogonal to $\mathbb{S}^n$).
(iii) The metric of $\mathbb{S}^n$ is that of $\mathbb{R}^{n+1}$ as an embedded manifold.

These maps may also be interpreted as dilations under stereographic projection.

The same reasoning may be applied to the passage from $\mathbb{R}^n$ to $\mathbb{S}^n$. By stereographic projection (cf. Sect. 2.2.2, p. 83), the spherical carré du champ operator $\widehat{\Gamma}$ is given in terms of the (standard) Euclidean carré du champ operator Γ by

$$\widehat{\Gamma} = \left(\frac{1+|x|^2}{2}\right)^2 \Gamma.$$

Comparing the scalar curvatures, setting $c(x) = \frac{1+|x|^2}{2}$ and $f = c^{-(n-2)/2}$, it follows that

$$f^{p-1} = -\frac{4}{n(n-2)} \Delta f$$

which is the equation of the extremal functions for the optimal Sobolev inequality on $\mathbb{R}^n$. By translation, dilation, and homogeneity, all extremal functions of the Sobolev inequality in $\mathbb{R}^n$ are of the form

$$f_{\sigma,b,x_0}(x) = \left(\sigma^2 + b|x - x_0|^2\right)^{-(n-2)/2}, \quad \sigma > 0, \ b > 0, \ x_0 \in \mathbb{R}^n. \quad (6.9.5)$$

It should be observed that such functions are not in $\mathbb{L}^2(dx)$ when $n = 3, 4$. However, they satisfy the equation for extremal functions and the Euclidean Sobolev inequality (i) applies to them thanks to Remark 6.9.5. These functions will appear naturally when solving fast diffusion equations in $\mathbb{R}^n$ (see Sect. 6.11 below) justifying the role played by those equations in connection with Sobolev inequalities.

Similarly, on the sphere $\mathbb{S}^n$, extremal functions for the optimal Sobolev inequality ((ii) in Theorem 6.9.4) may be written as

$$f_{b,e}(x) = \left(\sigma + b(e \cdot x)\right)^{-(n-2)/2}, \quad (6.9.6)$$

where $x, e \in \mathbb{S}^n$, $\sigma > 1$, $b \in \mathbb{R}$ and $\sigma^2 = 1 + b^2$.

In the same way, extremal functions for the symmetric Jacobi operator on $[-1, +1]$ with invariant reversible measure $d\mu_n(x) = C_n(1 - x^2)^{(n/2)-1} dx$ (Sect. 2.7.4, p. 113) for the Sobolev inequality with exponent $p = \frac{2n}{n-2}$ are $f_b(x) = (\sigma + bx)^{-(n-2)/2}$, where $\sigma^2 = 1 + b^2$. Although it is quite easy in this latter case to check that the extremal function equation (6.8.3) is satisfied with constant $C = \frac{4}{n(n-2)}$, it is not completely obvious that $\int_{[-1,+1]} f_b^p d\mu_n = 1$ (to ensure that it is indeed an extremal function). But this is easily obtained using the change of variable $y = \frac{\alpha+x}{1+\alpha x}$, $|\alpha| < 1$, which leads to

$$\int_{-1}^{+1} (1-x^2)^{(n/2)-1} dx = (1-\alpha^2)^{n/2} \int_{-1}^{+1} (1-y^2)^{(n/2)-2} \frac{dy}{(1-\alpha y)^n}$$

6.9 Conformal Invariance of Sobolev Inequalities

from which the conclusion follows with $\alpha = -\frac{b}{\sigma}$. Observe that, for n integer, $n > 2$, this change of variable from x to y is the trace on zonal functions of the conformal transformation of the sphere.

6.9.4 Non-geometric Conformal Invariant

The conformal invariance result of Theorem 6.9.3 is not restricted to the Laplacian case and to the case where n is the dimension of the manifold. In fact, other general (and sometimes useful) forms of conformal invariance may be considered once a good candidate to replace the scalar curvature is determined. In the last (and somewhat technical) part of this section, we discuss a few relevant instances of interest on the basis of the operations on Γ_2 already developed for Theorem 6.9.3.

In the following examples, we restrict ourselves to the case of elliptic operators on some n_0-dimensional manifold M. Set $\Gamma(f) = \sum_{i,j=1}^{n_0} g^{ij} \partial_i f \partial_j f$ where $\mathfrak{g} = (g^{ij})$ is elliptic, and as usual denote respectively by $\Delta_\mathfrak{g}, \mu_\mathfrak{g}$ and $sc_\mathfrak{g}$ the Laplace operator, Riemannian measure and scalar curvature associated with the (co-) metric $\mathfrak{g}$. Using this notation we look at measures μ which have a density e^{-W} with respect to $\mu_\mathfrak{g}$. The following non-geometric conformal invariance result illustrates a variety of examples.

Proposition 6.9.6 *In the preceding Riemannian context, for $n > n_0$, $d\mu = e^{-W} d\mu_\mathfrak{g}$ and any $\alpha \in \mathbb{R}$, the map*

$$S_\alpha(\mu, \Gamma) = \gamma_n(\alpha) \big[sc_\mathfrak{g} - \alpha \Delta W + \beta_n(\alpha) \Gamma(W) \big] \quad (6.9.7)$$

is n-conformal invariant where

$$\beta_n(\alpha) = \frac{\alpha(n - 2n_0 + 2) - 2(n_0 - 1)}{2(n - n_0)}$$

and

$$\gamma_n(\alpha) = \frac{n - 2}{4(n_0 - 1) - \alpha(n - n_0)}.$$

Proof It is enough to check that $S_\alpha(\mu, \Gamma)$ satisfies (6.9.1). The transformation rules for $sc_\mathfrak{g}$ under the change $(\mu_\mathfrak{g}, \Gamma) \mapsto (c^{-n_0} \mu_\mathfrak{g}, c^2 \Gamma)$ was already considered previously. It remains to observe that if the density of μ is e^{-W} for $(\mu_\mathfrak{g}, \Gamma)$, the density of $c^{-n} \mu$ with respect to $c^{-n_0} \mu_\mathfrak{g}$ is equal to $e^{-W} c^{n_0 - n} = e^{-W - (n - n_0)\tau}$. The computations are then straightforward (although somewhat tedious). □

In fact Proposition 6.9.6 applies for any $n \neq n_0$, but in view of Sobolev inequalities, it is only interesting for $n > n_0$. As alluded to above, restricting to Laplace operators leads to the restriction of conformal transformations to $n = n_0$. Indeed, the change of μ into $c^{-n} \mu$ changes a Riemannian measure into a measure which is not the Riemannian measure associated to $c^2 \Gamma$ as soon as $n \neq n_0$. When $d\mu = e^{-W} d\mu_\mathfrak{g}$, for a conformal change c where c is a function of W (which is often the case), then there are wider classes of conformal invariant maps. For non-elliptic operators, neither the Laplace operator nor the scalar curvature are defined. There is no known general conformal invariant map in this setting.

The non-geometric conformal invariance result of Proposition 6.9.6 may also be used to obtain optimal Sobolev inequalities in situations where the n-dimensional parameter of the Sobolev inequality does not correspond to the dimension of the underlying manifold. The method may actually be used to extend the correspondence between spheres and Euclidean spaces for non-integer values of the dimensional parameter.

Consider, for example, the Sobolev inequality in the Euclidean space $\mathbb{R}^n$. Writing $\mathbb{R}^n = \mathbb{R}^{n_0-1} \times \mathbb{R}^{n-n_0+1}$, this inequality may be considered on functions which are radial in the variable $y \in \mathbb{R}^{n-n_0+1}$. We then get a Sobolev inequality on $\mathbb{R}^{n_0-1} \times \mathbb{R}_+$ for the Euclidean carré du champ operator in $\mathbb{R}^{n_0-1} \times \mathbb{R}_+$, but for the weighted measure $\omega_{n-n_0+1} r^{n-n_0} dx dr$. This Sobolev inequality is still valid, with optimal constants, when n is no longer an integer provided the definition of ω_n is suitably adapted.

To reach this claim, recall that the Sobolev inequality in $\mathbb{R}^n$ may be deduced from the Sobolev inequality on the sphere via stereographic projection, and that the Sobolev inequality on the sphere may be established from the abstract curvature-dimension condition $CD(n-1,n)$. We may therefore try to follow the same path to get this new inequality when $n \notin \mathbb{N}$. To start with, as was mentioned in Sect. 1.16.2, p. 72, on the unit n_0-dimensional sphere $x_1^2 + \cdots + x_{n_0+1}^2 = 1$, the coordinate function $U(x) = x_1$ satisfies $\nabla\nabla U = -U\,\mathrm{Id}$, and hence the operator $\Delta_{\mathbb{S}^{n_0}} + (n-n_0)\nabla \log U$ satisfies the $CD(n-1,n)$ condition on the half-sphere $\{x_1 > 0\}$. Therefore, a tight Sobolev inequality holds on this half-sphere, with dimension n and constant $\frac{4}{n(n-2)}$. The measure on this half-sphere has density $c_{n,n_0} x_1^{n-n_0}$ with respect to the Riemannian measure of the sphere (c_{n,n_0} being the normalizing constant). Next, consider the n-conformal map $S_\alpha(\mu, \Gamma)$ of Proposition 6.9.6 with the choice of $\alpha = \frac{2(n_0-1)}{n-2n_0}$ (corresponding to $(n-n_0)\beta(\alpha) = \alpha$). An easy computation shows that $S_\alpha(\mu, \Gamma) = \frac{n(n-2)}{4}$ (in other words, for this value of α, the corresponding scalar curvature is constant for this weighted manifold). Therefore, with this weighted measure, we are in a situation similar to that for the Riemannian measure, except that n is no longer the dimension of the sphere, but the dimensional parameter of the Sobolev inequality.

To transfer this conclusion from the spherical model to the Euclidean model, apply the stereographic projection of the sphere on $\mathbb{R}^{n_0}$ from the north pole $(0, \ldots, 0, 1)$. The image of the previous half-sphere becomes the half space $\{x_1 > 0\}$. Under the stereographic projection, the function $U = x_1$ takes the form $U(x) = \frac{2x_1}{1+|x|^2}$. Hence, the sharp Sobolev inequality on the half-sphere is transformed into a sharp n-dimensional Sobolev inequality on $\mathbb{R}^{n_0-1} \times \mathbb{R}_+$, with carré

6.10 Gagliardo-Nirenberg Inequalities

du champ operator

$$\Gamma(f) = \left(\frac{1+|x|^2}{2}\right)^2 |\nabla f|^2$$

and measure

$$dv(x) = \frac{2^n}{\omega_{n,n_0}} \cdot \frac{x_1^{n-n_0}}{(1+|x|^2)^n} dx$$

where ω_{n,n_0} is the normalizing constant (so that ν is a probability measure). The resulting inequality takes the form, with $p = \frac{2n}{n-2}$,

$$\left(\int_{\mathbb{R}^{n_0-1}\times\mathbb{R}_+} |f|^p dv\right)^{2/p}$$

$$\leq \frac{4}{n(n-2)} \left(\int_{\mathbb{R}^{n_0-1}\times\mathbb{R}_+} S_\alpha(\mu,\Gamma) f^2 dv + \int_{\mathbb{R}^{n_0-1}\times\mathbb{R}_+} \Gamma(f) dv\right).$$

Apply the n-conformal invariance result to the function $c(x) = \frac{1+|x|^2}{2}$ to obtain a new n-dimensional inequality on $\mathbb{R}^{n_0-1} \times \mathbb{R}_+$, but with the usual carré du champ operator $|\nabla f|^2$ and the measure $x_1^{n-n_0} dx$. For this pair, and for the same α, $S_\alpha = 0$ so that, finally

$$\left(\int_{\mathbb{R}^{n_0-1}\times\mathbb{R}_+} |f|^p \frac{x_1^{n-n_0} dx}{\omega_{n,n_0}}\right)^{2/p}$$

$$\leq \frac{4}{n(n-2)} \int_{\mathbb{R}^{n_0-1}\times\mathbb{R}_+} |\nabla f|^2 \frac{x_1^{n-n_0} dx}{\omega_{n,n_0}}. \tag{6.9.8}$$

This Sobolev inequality is precisely the announced one. In this non-geometric picture, the quantity S_α behaves in the same way as the scalar curvature for Laplace operators as it is constant strictly positive on the sphere and vanishes on $\mathbb{R}^n$.

Further connections between Sobolev inequalities, extremal functions and metrics with constant scalar curvature may be developed along these lines. For example, as already mentioned, there is a strong connection, through (6.9.4), between conformal transformations of a space into itself which preserve constant scalar curvature and extremal solutions of Sobolev inequalities. Furthermore, Sobolev inequalities may be used to control functions c for which the metric associated with $c^2 \Gamma$ has constant scalar curvature.

6.10 Gagliardo-Nirenberg Inequalities

Up to now, Sobolev inequalities on spheres under the curvature-dimension condition $CD(\rho, n)$ were established by means of the heat semigroup $(P_t)_{t\geq 0}$, that is, by solving the differential equation $\partial_t u = \mathrm{L} u$. The sharp constants are however obtained by some variant, nevertheless still dealing with the infinitesimal Markov generator L of $(P_t)_{t \geq 0}$. As discussed in the preceding Sect. 6.9, together with conformal invariance, the sharp Sobolev inequality on the sphere then yields the optimal Sobolev inequalities in Euclidean and hyperbolic spaces. In this picture, the spherical case, with its sharp constants, and the curvature-dimension condition, appear as critical.

However, this methodology has not yet produced any type of Sobolev inequality comparable to the Euclidean inequality directly from the curvature-dimension condition $CD(0, n)$, and similarly for the hyperbolic space under $CD(\rho, n)$ with $\rho < 0$. One explanation for this is the following. When $\rho > 0$, the invariant measure is finite (cf. Theorem 3.3.23, p. 165) and may thus naturally be normalized into a probability measure. It is actually from this normalization that optimal constants are produced. By conformal invariance, this normalization is also reflected in the sharp Sobolev inequalities on $\mathbb{R}^n$ and $\mathbb{H}^n$. However, when the invariant measure has infinite mass, there is no natural normalization, and hence when the measure is multiplied by a (strictly positive) constant, the constants in the Sobolev inequality are modified. Therefore, there is no hope of directly obtaining Sobolev inequalities under curvature-dimension conditions without any further information on the measure.

In this section, we examine how to suitably modify Sobolev inequalities in Euclidean space in order to obtain new families of inequalities which will be shown in the next section to be accessible under the curvature-dimension condition $CD(\rho, n)$ for $\rho \in \mathbb{R}$, via non-linear evolution equations. With this task in mind, we first extend the Sobolev and Nash inequalities to the more general family of *Gagliardo-Nirenberg* inequalities, and then consider entropic formulations of the latter.

6.10.1 Gagliardo-Nirenberg Inequalities

Classical Gagliardo-Nirenberg inequalities in Euclidean space may be defined in the abstract context of Markov Triples (E, μ, Γ). The definition below may be considered for any $p > 2$ but as for Sobolev inequalities we emphasize a dimensional parameter n (> 2) as $p = \frac{2n}{n-2}$.

Definition 6.10.1 (Gagliardo-Nirenberg Inequality) Let $n > 2$, $p = \frac{2n}{n-2}$, and q, s be such that $1 \leq s \leq q \leq p$. The Markov Triple (E, μ, Γ) is said to satisfy a Gagliardo-Nirenberg inequality $GN_n(q, s; A, C)$ with dimension n, parameters q, s, and constants $A \geq 0$, $C > 0$, if for all functions f in $\mathcal{D}(\mathcal{E}) \cap \mathbb{L}^s(\mu)$,

$$\|f\|_q \leq \left[A \|f\|_2^2 + C \mathcal{E}(f) \right]^{\theta/2} \|f\|_s^{1-\theta}$$

where $\theta \in [0, 1]$ is such that

$$\frac{1}{q} = \frac{\theta}{p} + \frac{1-\theta}{s}.$$

6.10 Gagliardo-Nirenberg Inequalities

For simplicity, only $A \geq 0$ is considered here although the formal definition includes $A \in \mathbb{R}$. The Sobolev inequality $S_n(A, C)$ of Definition 6.2.1 belongs to the Gagliardo-Nirenberg family for $\theta = 1$ or $q = s$, while the Nash inequality (6.2.3) is achieved for $q = 2$, $s = 1$. As for the Nash inequality (cf. Proposition 6.2.3), any inequality $GN_n(q, s; A, C)$ is equivalent, up to constants, to the Sobolev inequality $S_n(A, C)$ with the same dimension n.

Proposition 6.10.2 *In the preceding setting, the Sobolev inequality $S_n(A, C)$ implies the Gagliardo-Nirenberg inequality $GN_n(q, s; A, C)$. Conversely, the Gagliardo-Nirenberg inequality $GN_n(q, s; A, C)$ implies a Sobolev inequality $S_n(A', C')$.*

Proof The proof is very similar to the proof of Proposition 6.2.3. For the first implication, use again the concavity of the function $r \mapsto \log(\|f\|_{1/r})$, but this time between the points $s \leq q \leq p$. Conversely, use the slicing procedure, applying $GN_n(q, s; A, C)$ to $(f - 2^k)_+ \wedge 2^k$, $k \in \mathbb{Z}$. The steps of the proof are similar, although a bit heavier. □

By concavity of the logarithm, a Gagliardo-Nirenberg inequality $GN_n(q, s; A, C)$ implies the inequality

$$\|f\|_q^2 \leq \theta \big[A \|f\|_2^2 + C \mathcal{E}(f) \big] + (1 - \theta) \|f\|_s^2, \quad f \in \mathcal{D}(\mathcal{E}) \cap \mathbb{L}^s(\mu). \quad (6.10.1)$$

In $\mathbb{R}^n$, there exist Gagliardo-Nirenberg inequalities $GN_n(q, s; 0, C)$ (due for example to Proposition 6.10.2), and we will be interested in sharp constants for some values of the parameters q, s. But observe first that in this Euclidean (homogeneous) case, the weak form (6.10.1) is as good as the original one. Indeed, changing $f(x)$ into $f_r(x) = f(rx)$, $x \in \mathbb{R}^n$, $r > 0$, in (6.10.1) with $A = 0$, and after a change of variables, we get

$$\|f\|_q^2 \leq C \theta r^{2n(\frac{1}{q} - \frac{1}{p})} \mathcal{E}(f) + (1 - \theta) \|f\|_s^2 r^{2n(\frac{1}{q} - \frac{1}{s})}.$$

Optimizing in $r > 0$ returns the inequality $GN_n(q, s; 0, C)$ with the same constant C.

There is another, more subtle, way to linearize Gagliardo-Nirenberg inequalities which will turn out to be most useful below.

Proposition 6.10.3 *For any $\alpha > 0$, $\alpha \neq 1$, such that $\theta \alpha < 1$, the Gagliardo-Nirenberg inequality $GN_n(q, s; A, C)$ is equivalent to the inequality*

$$\|f\|_q^{2\alpha} \leq \alpha \theta \big[A \|f\|_2^2 + C \mathcal{E}(f) \big] + (1 - \alpha \theta) \|f\|_s^{2\beta}, \quad f \in \mathcal{D}(\mathcal{E}) \cap \mathbb{L}^s(\mu), \quad (6.10.2)$$

where $\beta = \frac{\alpha(1-\theta)}{1-\theta\alpha}$.

Proof Setting $Q = A \|f\|_2^2 + C \mathcal{E}(f)$, the Gagliardo-Nirenberg inequality $GN_n(q, s; A, C)$ takes the form

$$\left(\frac{\|f\|_q^2}{\|f\|_s^2} \right)^\alpha \leq \left(\frac{Q}{\|f\|_s^2} \right)^{\alpha \theta}.$$

As $\theta\alpha < 1$, the function $r^{\alpha\theta}$, $r \geq 0$, is concave and thus, for any $r_0 > 0$,

$$r^{\alpha\theta} \leq \alpha\theta r_0^{\alpha\theta-1} r + (1-\alpha\theta) r_0^{\alpha\theta}.$$

Choose then $r_0 = \|f\|_s^{2\gamma}$ with $\gamma = \frac{\alpha-1}{1-\alpha\theta}$. Conversely, replace f by cf and optimize in $c > 0$. □

It should be observed that constants are preserved in the equivalences of Proposition 6.10.3, and thus so are extremal functions, when they exist.

It is quite remarkable that there is a sub-family of Gagliardo-Nirenberg inequalities in $\mathbb{R}^n$ (for the Lebesgue measure and the standard carré du champ operator Γ and Dirichlet form $\mathcal{E}$) with sharp constants. This is the content of the next statement. Extremal functions will be determined at the same time.

Theorem 6.10.4 (Del Pino-Dolbeault Theorem) *Let $\nu > n > 2$, and set*

$$q = \frac{2\nu}{\nu-2}, \qquad s = \frac{2(\nu-1)}{\nu-2}$$

so that $2 < s < q < p = \frac{2n}{n-2}$ and $q = 2(s-1)$. Then, on $\mathbb{R}^n$, the optimal Gagliardo-Nirenberg inequality $GN_n(q, s; 0, C)$ holds. That is, for every $f \in \mathcal{D}(\mathcal{E}) \cap \mathbb{L}^s(\mu)$,

$$\|f\|_q^2 \leq C \mathcal{E}(f)^\theta \|f\|_s^{2(1-\theta)}$$

where $\frac{1}{q} = \frac{\theta}{p} + \frac{1-\theta}{s}$ and where the (optimal) constant $C > 0$ is the one for which there is equality for the function

$$f(x) = (1 + |x|^2)^{-(\nu-2)/2}, \quad x \in \mathbb{R}^n.$$

The limiting case $\nu \to n$, $\theta \to 1$, corresponds to the standard Sobolev inequality in $\mathbb{R}^n$ with its sharp constant and extremal functions (6.9.5) of the same form. As for the case of the Sobolev inequality, these extremal functions are not in $\mathbb{L}^2(dx)$ when $2\nu \leq n+4$, but the same kind of extension to various classes of functions as the one described in Remark 6.9.5 may be developed similarly.

Proof Although the result is true as stated for $\nu \geq n > 2$, the method of proof developed here only works for $\nu \geq n + \frac{1}{2}$, $n > 2$, to which we restrict below. It is however a quite general strategy to deduce sharp inequalities from other optimal inequalities provided there exist extremal functions. We first deal with the case $\nu = n + \frac{m}{2}$ for some integer $m \geq 1$ and start from the optimal Sobolev inequality on $\mathbb{R}^{n+m}$,

$$\|g\|_p^2 \leq C_{n+m} \mathcal{E}(g)$$

where $p = \frac{2(n+m)}{n+m-2}$. Given a smooth compactly supported function $f : \mathbb{R}^n \to \mathbb{R}_+$, the idea is to apply this Sobolev inequality to

$$g(x, y) = (f(x) + |y|^2)^{-(n+m-2)/2}, \quad (x, y) \in \mathbb{R}^n \times \mathbb{R}^m,$$

6.10 Gagliardo-Nirenberg Inequalities

and to integrate it with respect to the Lebesgue measure in $y \in \mathbb{R}^m$. To this end, observe that

$$|\nabla g|^2 = \frac{(n+m-2)^2}{4}(f+|y|^2)^{-(n+m)}\big[|\nabla f|^2 + 4|y|^2\big].$$

Now, for $a > 0$,

$$\int_{\mathbb{R}^m} (a+|y|^2)^{-(n+m)} dy = c_{n,m}\, a^{-n-(m/2)}$$

and

$$\int_{\mathbb{R}^m} |y|^2 (a+|y|^2)^{-(n+m)} dy = d_{n,m}\, a^{-n-(m/2)+1}$$

where $c_{n,m}, d_{n,m} > 0$ only depend on n, m. In this picture, the measure $(1+|x|^2)^{-(n+m)} dx$ is simply the image of the spherical measure of $\mathbb{R}^{n+m}$ (viewed in stereographic projection from $\mathbb{S}^{n+m}$ onto $\mathbb{R}^{n+m}$, cf. Sect. 2.2.2, p. 83) on $\mathbb{R}^n$ by orthogonal projection.

The integrated Sobolev inequality therefore takes the form

$$\left(\int_{\mathbb{R}^n} f^{-n-(m/2)} dx\right)^{2/p}$$
$$\leq C_1 \int_{\mathbb{R}^n} f^{-n-(m/2)} |\nabla f|^2 dx + C_2 \int_{\mathbb{R}^n} f^{-n-(m/2)+1} dx$$

for constants $C_1, C_2 > 0$ only depending on n, m. The construction ensures that this inequality is an equality for $f = 1 + |x|^2$ since in this case the function g is an extremal function of the Sobolev inequality in $\mathbb{R}^{n+m}$. Changing f into $f^{-4/(2n+m-4)}$ yields the inequality, with further constants $C_3, C_4 > 0$,

$$\left(\int_{\mathbb{R}^n} f^q dx\right)^{2/p} \leq C_3 \int_{\mathbb{R}^n} |\nabla f|^2 dx + C_4 \int_{\mathbb{R}^n} f^s dx$$

for the values q, s specified in the statement (recall that $\nu = n + \frac{m}{2}$), and for which there is now equality for $f = (1+|x|^2)^{-(\nu-2)/2}$. Changing f to cf and optimizing in $c > 0$ yields the announced Gagliardo-Nirenberg inequality. The inequality is optimal since it admits non-constant extremal functions.

The preceding argument however only works when the parameter m is an integer. To cover the general case, it should be observed that we only used functions of $r = |y|$ in $\mathbb{R}^m$ and the Sobolev inequality of dimension $n + m$ in $\mathbb{R}^n \times \mathbb{R}_+$ for the standard carré du champ operator and the measure $d\mu(x, r) = r^{m-1} dx dr$. As presented in Sect. 6.9, this Sobolev inequality may be deduced from the spherical case for $m \geq 1$ via conformal transformations. The proof of Theorem 6.10.4 may therefore be completed in this way. □

Remark 6.10.5 There is a counterpart to Theorem 6.10.4 in the range $q, s < 2$ with $s = 2(q - 1)$ for which the extremal functions are of the form $f(x) = [(\sigma - |x|^2)_+]^{1/(2-q)}$, thus being compactly supported. The proof is very similar. Interestingly enough, letting $q, s \to 2$ (from both sides) gives rise to the Euclidean logarithmic Sobolev inequality (Proposition 6.2.5) with the Gaussian kernels as extremals.

6.10.2 Entropic Form of Gagliardo-Nirenberg Inequalities

The family of Gagliardo-Nirenberg inequalities of Theorem 6.10.4, including the classical Sobolev inequality, shares extremal functions of the same form $(\sigma^2 + b|x|^2)^\alpha$ (from which optimal constants are determined). The existence of explicit extremal functions for this sub-family of Gagliardo-Nirenberg inequalities actually leads to a new, equivalent, entropic formulation which we present next. This formulation will in turn allow for the control of non-linear evolution equations (such as porous medium or fast diffusion equations) as developed in the following section. Moreover, it is suitable for an extension to more general settings under curvature-dimension conditions.

Proposition 6.10.6 *For $\nu \geq n > 2$, set*

$$H(r) = -r^{1-1/\nu} \quad \text{and} \quad \Psi(r) = H'(r) = -\frac{\nu - 1}{\nu} r^{-1/\nu}, \quad r \in (0, \infty).$$

Fix $b > 0$ and let $v_\sigma = h_{\sigma,b}^{-\nu}$, $\sigma > 0$, where $h_{\sigma,b} = h_{\sigma,b}(x) = \sigma^2 + b|x|^2$, $x \in \mathbb{R}^n$. Let $f : \mathbb{R}^n \to \mathbb{R}_+$ be smooth and compactly supported and choose $\sigma > 0$ such that $\int_{\mathbb{R}^n} f\, dx = \int_{\mathbb{R}^n} v_\sigma\, dx$. Then, the Gagliardo-Nirenberg inequality of Theorem 6.10.4 for the parameter ν is equivalent to

$$\int_{\mathbb{R}^n} \big[H(f) - H(v_\sigma) - (f - v_\sigma)\Psi(v_\sigma) \big] dx \leq \frac{b^2 \nu}{4(\nu - 1)} \int_{\mathbb{R}^n} f \big| \nabla(\Psi(f) - \Psi(v_\sigma)) \big|^2 dx \qquad (6.10.3)$$

(holding for every such function f). When $\nu = n$, the latter amounts to the optimal Sobolev inequality on $\mathbb{R}^n$.

Since the function H is (strictly) convex, the left-hand side of (6.10.3) is always positive. It is zero only for $f = v_\sigma$, and the inequality measures, in a certain entropic sense, the distance from f to v_σ.

Proof Fix $b = 1$ (the general case being similar), and to simplify the notation, set $h = h_{\sigma, 1}$. Using integration by parts on the right-hand side, (6.10.3) may be rewritten

6.10 Gagliardo-Nirenberg Inequalities

as

$$\int_{\mathbb{R}^n} f\left(h - \frac{1}{4}|\nabla h|^2\right) dx + \frac{1}{\nu - 1}\int_{\mathbb{R}^n} h^{1-\nu} dx$$
$$\leq \frac{1}{\nu - 1}\int_{\mathbb{R}^n} f^{1-1/\nu}\left(\nu - \frac{1}{2}\Delta h\right) dx + \frac{1}{(\nu - 2)^2}\int_{\mathbb{R}^n} \left|\nabla\left(f^{(\nu-2)/2\nu}\right)\right|^2 dx.$$

The function $h = \sigma^2 + |x|^2$ satisfies $h - \frac{1}{4}|\nabla h|^2 = \sigma^2$ and $\Delta h = 2n$ so that the preceding inequality is equivalent to

$$\sigma^2 \int_{\mathbb{R}^n} f\, dx + \frac{1}{\nu - 1}\int_{\mathbb{R}^n} h^{1-\nu} dx$$
$$\leq \frac{\nu - n}{\nu - 1}\int_{\mathbb{R}^n} f^{1-1/\nu} dx + \frac{1}{(\nu - 2)^2}\int_{\mathbb{R}^n} \left|\nabla\left(f^{(\nu-2)/2\nu}\right)\right|^2 dx.$$

Choosing $f = v_\sigma = h^{-\nu}$ yields equality (while not compactly supported, it is easily checked that the previous computation is still valid for it). But for this particular choice of f,

$$\left|\nabla\left(f^{(\nu-2)/2\nu}\right)\right|^2 = (\nu - 2)^2 h^{-\nu}(h - \sigma^2).$$

Hence $\int_{\mathbb{R}^n} h^{-\nu} dx$ and $\int_{\mathbb{R}^n} h^{1-\nu} dx$ are related as

$$\int_{\mathbb{R}^n} h^{1-\nu} dx = \frac{2\sigma^2(\nu - 1)}{2\nu - n - 2}\int_{\mathbb{R}^n} h^{-\nu} dx.$$

On the other hand, $\int_{\mathbb{R}^n} f\, dx = \int_{\mathbb{R}^n} h^{-\nu} dx = C_{n,\nu}\, \sigma^{n-2\nu}$ where $C_{n,\nu} > 0$ only depends on n and ν. Replacing the values of $\int_{\mathbb{R}^n} h^{1-\nu} dx$ and σ by their values in terms of $\int_{\mathbb{R}^n} f\, dx$, we are left with

$$C'_{n,\nu}\left(\int_{\mathbb{R}^n} f\, dx\right)^{(2\nu-n-2)/(2\nu-n)}$$
$$\leq \frac{\nu - n}{\nu - 1}\int_{\mathbb{R}^n} f^{1-1/\nu} dx + \frac{1}{(\nu - 2)^2}\int_{\mathbb{R}^n} \left|\nabla\left(f^{(\nu-2)/2\nu}\right)\right|^2 dx,$$

where $C'_{n,\nu} > 0$ is a another constant. Finally, changing f into $f^{2\nu/(\nu-2)}$ yields the non-homogeneous Gagliardo-Nirenberg inequality of Proposition 6.10.3 with $\beta = \frac{s}{2}$ and q and s as in Theorem 6.10.4. The resulting inequality is sharp since by construction it admits non-constant extremal functions. For the converse implication, it suffices to translate the Gagliardo-Nirenberg inequality of Theorem 6.10.4 in its non-homogeneous form and to follow the path back. The proof of Proposition 6.10.6 is therefore complete. □

6.11 Fast Diffusion Equations and Sobolev Inequalities

This section will be a bit formal (and at the same time somewhat technical). While we observed that linear heat equations do not allow us to reach sharp Sobolev inequalities, the aim here is to show that possibly non-linear evolution equations may be used to establish optimal Sobolev inequalities under curvature-dimension conditions $CD(0, n)$ for example.

The method may be applied to more general situations and equations, but we only concentrate here on the porous medium and fast diffusion equations, with a particular emphasis on the latter which is well-suited to both curvature-dimension conditions and Sobolev-type inequalities. In addition, this fast diffusion equation plays the same role with respect to Sobolev and curvature-dimension $CD(\rho, n)$ inequalities as the usual heat equation with respect to logarithmic Sobolev and curvature $CD(\rho, \infty)$ inequalities. Deeper relationships are actually underlying the picture. The Ornstein-Uhlenbeck semigroup is a model case to test logarithmic Sobolev inequalities and curvature conditions. Indeed, under $CD(\rho, \infty)$, the heat flow monotonicity method produces in this case logarithmic Sobolev inequalities, both for the heat semigroup $(P_t)_{t \geq 0}$ and for the invariant measure μ (when $\rho > 0$) which are optimal in the example of the Ornstein-Uhlenbeck model on $\mathbb{R}^n$. As already observed in Sect. 5.5, p. 257, this is fully coherent since for this model, on the one side the heat kernels (that is the solutions of the heat equation starting from Dirac masses) are Gaussian measures, for which the logarithmic Sobolev inequality is optimal, and on the other, the logarithmic Sobolev inequality for any Gaussian measure after dilations and translations yields the logarithmic Sobolev inequality for the Ornstein-Uhlenbeck semigroup, which in turn gives in the limit the curvature condition $CD(1, \infty)$. The same picture holds with the Euclidean heat semigroup in place of the Ornstein-Uhlenbeck semigroup with respect to the curvature condition $CD(0, \infty)$. The question now is whether this full set of equivalences and models may be extended to Sobolev and curvature-dimension $CD(\rho, n)$ inequalities (for some finite dimension n). The models for these Sobolev inequalities are now spheres, which may be seen via stereographic projections on the Euclidean space, and for which the references measures are Cauchy measures (cf. Sect. 2.2, p. 81). Although the game with dilations and translations may no longer be played in this context (and should be replaced by conformal transformations on the sphere), the heat equation may still be replaced by the fast diffusion equation described below, which, although non-linear, produces in $\mathbb{R}^n$ the required Cauchy measures when the initial data are Dirac masses. While this method will not be entirely satisfactory, it certainly shows that non-linear fast diffusion equations could play a similar role for Sobolev inequalities as linear heat equations for logarithmic Sobolev inequalities with respect to curvature-dimension $CD(\rho, n)$ inequalities.

The main purpose of this section will be to establish the entropic version of the Gagliardo-Nirenberg inequality of Proposition 6.10.6 (which requires some additional stabilizing function v) under a curvature-dimension condition $CD(0, n)$. As discussed in the previous section, this Gagliardo-Nirenberg inequality is equivalent to the sharp Sobolev inequality in $\mathbb{R}^n$. Recall that stating a Sobolev inequality for an infinite measure requires a given normalization. Here, the choice of the function v in the general context will play the role of this normalization. In some sense, the fast diffusion equation described below is a model case illustrating the interplay between entropy methods for evolution equations, functional inequalities and curvature-dimension conditions. The case of $CD(\rho, n)$ with $\rho > 0$ is briefly addressed next.

6.11.1 Porous Medium and Fast Diffusion Equations

The *porous medium* and *fast diffusion* equations are evolution equations of the form, say on $\mathbb{R}^n$,

$$\partial_t u = \Delta(u^m)$$

for $u = u(t, x) = u_t(x)$, $t \geq 0$, $x \in \mathbb{R}^n$. Porous medium corresponds to $m > 1$ and fast diffusion to $m < 1$. It is not obvious that there exist solutions to such equations at any time, in particular when $m < 1$. Actually, in $\mathbb{R}^n$ and for small m, the solution starting from an initial condition $u_0 > 0$ vanishes in finite time. We do not discuss here existence (or uniqueness) issues for these equations. Our aim is rather to explain how they may be used to reach Sobolev or Gagliardo-Nirenberg inequalities under curvature-dimension $CD(\rho, n)$ conditions, exactly as the heat equation is used to reach Poincaré or logarithmic Sobolev inequalities under curvature conditions $CD(\rho, \infty)$. Namely, via Proposition 6.10.6, functional inequalities of Sobolev-type are related to evolution equations through the control of entropic quantities. As was extensively developed in the preceding chapters, on the one hand, functional inequalities may be used to prove exponential decay of entropy related to the evolution, where the functional inequality is translated into a differential inequality linking the entropy and its time derivative along the evolution equation. On the other hand, functional inequalities may be reached by deriving entropy twice, and using curvature-dimension conditions to directly reach this exponential decay.

In what follows, we present a general framework to control entropy of general evolution equations with stabilizing term v. The general form of the second differential of the entropy is quite heavy, and the comparison with the curvature-dimension $CD(\rho, n)$ condition leads to the rather complicated form of the general Proposition 6.11.6 below. However, a much simplified and surprising form occurs for the fast diffusion equation with parameter $m = 1 - \frac{1}{n}$, leading to the optimal form of the Gagliardo-Nirenberg inequality under the $CD(0, n)$ hypothesis. A slightly modified version also leads directly to the optimal Sobolev inequality under the $CD(\rho, n)$ condition with $\rho > 0$, where no stabilizing term v is required.

We work below with a state space E that will either be $\mathbb{R}^n$ or a smooth manifold, although most of what follows may be extended to more general settings at the expense of several notational complications. Moreover, the various integration by parts performed below, typically for a solution at time $t > 0$ of the associated fast diffusion equation, should be carefully justified to turn this scheme into an actual proof of functional inequalities in the context of general Full Markov Triples. Once again, we will keep the exposition at a formal level in order to better emphasize the main principle of proof.

Thus given a Markov Triple (E, μ, Γ), where $E = \mathbb{R}^n$ or a smooth (weighted) Riemannian manifold, with associated diffusion operator L, and a function Φ at least $\mathcal{C}^2$ on $\mathbb{R}_+$, a first set of conclusions may be developed for the extended porous medium and fast diffusion equations in the form of the non-linear equation

$$\partial_t u = \mathrm{L}\Phi(u), \quad u_0 = f, \qquad (6.11.1)$$

where $u = u(t, x) = u_t(x)$, $t \geq 0$, $x \in E$, with initial condition f.

In $\mathbb{R}^n$ or a Riemannian manifold, the generator L can be written as $-\nabla^*\nabla$ where ∇^* is the adjoint of ∇ on $\mathbb{L}^2(\mu)$ (see (1.11.11), p. 48). Choosing a function Ψ such that $r \Psi'(r) = \Phi'(r)$, the extended equation (6.11.1) may then be rewritten as

$$\partial_t u = -\nabla^*\big(u \nabla \Psi(u)\big).$$

In this form, the heat equation corresponds to $\Psi(r) = \log r$, $r \in (0, \infty)$. A priori, the behavior of u as $t \to \infty$ is unknown, but if the interchange between differentiation and integration is justified, we should at least have

$$\partial_t \left(\int_E u \, d\mu \right) = \int_E \partial_t u \, d\mu = 0.$$

On a compact Riemannian manifold, it may thus be expected that u_t converges at infinity towards $\int_E u \, d\mu$. On the other hand, in $\mathbb{R}^n$, the behavior of u will be forced to a given asymptotic behavior by a suitable choice of function v (to be made explicit below) by modifying the equation as

$$\partial_t u = -\nabla^*\big[u \nabla \big(\Psi(u) - \Psi(v)\big)\big]. \qquad (6.11.2)$$

This function v may be considered as a stabilizing term and plays an important role in the associated functional inequality. For example, for the standard heat equation on $\mathbb{R}^n$, the choice for $\Psi(r) = \log r$ would be $v = e^{-|x|^2/2}$ and the resulting equation is the Fokker-Planck equation (cf. Sect. 1.5, p. 23),

$$\partial_t u = \Delta u + x \cdot \nabla u + nu = \Delta u + \nabla \cdot (ux),$$

describing the evolution equation of the density of the Ornstein-Uhlenbeck semi-group with respect to the Lebesgue measure, which produces a different form than the usual evolution with respect to the reversible (here Gaussian) measure. Setting $u = \hat{u} \, e^{-|x|^2/2}$, it boils down to $\partial_t \hat{u} = \Delta \hat{u} - x \cdot \nabla \hat{u}$, that is, the heat equation associated with the Ornstein-Uhlenbeck operator.

After this transformation, it may be expected that, as $t \to \infty$, u_t will converge towards $\hat{v}$ where $\hat{v}$ is such that $\Psi(v) = \Psi(\hat{v}) + c$ for a certain constant c, since these functions are obviously stable solutions of (6.11.2). Now, since $\int_E \hat{v} \, d\mu = \int_E u \, d\mu$,

6.11 Fast Diffusion Equations and Sobolev Inequalities

these two conditions will in general determine $\hat{v}$ uniquely. Replacing in (6.11.2) v by $\hat{v}$ so that in addition $\int_E \hat{v} d\mu = \int_E f d\mu$ where $f = u_0$ is the initial condition, u_t is expected to converge towards $\hat{v}$, a stationary solution of the equation with conserved mass $\int_E f d\mu$.

In the following, $u = u(t, x) = u_t(x)$ denotes the solution of (6.11.1) and v the stabilizer as constructed above. Moreover, these functions and the functions constructed with them are implicitly assumed to be sufficiently smooth in order for the various expressions and arguments to make sense, and in particular for the validity of the various integrations by parts.

6.11.2 Entropy Decay

In order to quantify the previous convergence along the same lines as the corresponding development for logarithmic Sobolev inequalities in Chap. 5, an analogue of the entropy functional needs to be considered. The quantity that will replace entropy in this case is given by

$$F(u) = \int_E \big[H(u) - u\,\Psi(v)\big]d\mu$$

where H on $(0, \infty)$ is such that $H' = \Psi$. Note that for the heat equation, $H(r)$ is indeed $r \log r - r$, close to the standard entropy. When H is convex (which will be the case below for the fast diffusion equation with $m = 1 - \frac{1}{n}$ where $H(r) = -nr^{1-1/n}$), then $F(u) - F(v) \geq 0$, since

$$F(u) - F(v) = \int_E \big[H(u) - H(v) - (u-v)H'(v)\big]d\mu.$$

Moreover, $F(u) - F(v)$ vanishes only when $u = v$ provided H is strictly convex. The quantity $F(u) - F(v)$ may thus be considered as a distance from u to v.

The first operation to perform in this context is to compute the time derivative of the functional F as an analogue of de Bruijn's identity (Proposition 5.2.2, p. 245) in case of the standard entropy with the suitable (modified) Fisher information denoted by $I_{\mu, F}$.

Proposition 6.11.1 *In the preceding notation, setting $\xi = \Psi(u) - \Psi(v)$,*

$$\frac{d}{dt} F(u) = -I_{\mu, F}(u) = -\int_E u\,\Gamma(\xi)d\mu.$$

Proof We make use of the following simple integration by parts formula, which will be used repeatedly below. Namely, for each smooth function h on E (for example in the algebra $\mathcal{A}_0$),

$$\int_E \partial_t u\, h\, d\mu = -\int_E u\,\Gamma(\xi, h)d\mu. \tag{6.11.3}$$

To prove this formula, use the identity $\partial_t u = \nabla^* u \cdot \nabla \xi$ and, assuming everything is sufficiently smooth, integrate by parts to get

$$\int_E (\partial_t u) h \, d\mu = -\int_E u \nabla \xi \cdot \nabla h \, d\mu = -\int_E u \Gamma(\xi, h) d\mu$$

which is the claim. On the basis of (6.11.3), it then suffices to observe that

$$\frac{d}{dt} F(u) = \int_E \partial_t u \big[\Psi(u) - \Psi(v)\big] d\mu = \int_E (\partial_t u) \, \xi \, d\mu.$$

The proposition is established. □

By comparison with the entropic version of the Gagliardo-Nirenberg inequalities (Proposition 6.10.6), the following statement describes exponential convergence in the entropy F along the fast diffusion equation towards the steady states whose form is closely related to the extremals of the Gagliardo-Nirenberg inequalities.

Proposition 6.11.2 *In $\mathbb{R}^n$, if $H(r) = -n r^{1-1/n}$, $r \in (0, \infty)$, that is, for the equation*

$$\partial_t u = -(n-1) \nabla^* \big[u \nabla \big(u^{-1/n} - v_\sigma^{-1/n} \big) \big]$$

where $v_\sigma = (\sigma^2 + b|x|^2)^{-n}$, provided $\sigma > 0$ is chosen so that $\int_{\mathbb{R}^n} v_\sigma \, dx = \int_{\mathbb{R}^n} u_0 \, dx$, then for every $t \geq 0$,

$$0 \leq F(u_t) - F(v_\sigma) \leq e^{-ct} \big(F(u_0) - F(v_\sigma) \big)$$

with $c = \frac{4(n-1)}{b^2}$. In other words, u_t converges exponentially towards v_σ in the entropic sense as $t \to \infty$.

For the proof, it suffices to observe that Proposition 6.10.6 actually yields that

$$\frac{d}{dt} \big(F(u_t) - F(v_\sigma) \big) \leq -\frac{4(n-1)}{b^2} \big(F(u_t) - F(v_\sigma) \big).$$

As for logarithmic Sobolev inequalities (cf. Theorem 5.2.1, p. 244), the entropy decay of Proposition 6.11.2 implies in return the Gagliardo-Nirenberg inequalities of Proposition 6.10.6 with $\nu = n$ in the limit $t \to 0$.

We next turn to the second step of the analysis, namely computing, as for classical entropy, the second order derivative of the functional F (that is, the derivative of the generalized Fisher information $I_{\mu, F}$) along the fast diffusion evolution with the help of the Γ_2 operator. The curvature-dimension condition will then enter into play when establishing functional inequalities along the lines developed for Poincaré and logarithmic Sobolev inequalities in Sect. 4.8, p. 211, and Sect. 5.7, p. 268, respectively.

This second order differentiation will be performed for some general function Ψ along the evolution (6.11.2) assuming however that the state space E is a smooth manifold M and that the operator L is elliptic (allowing for the use of the differential

6.11 Fast Diffusion Equations and Sobolev Inequalities

calculus described in Sect. C.5, p. 511). Questions of existence, regularity and domains for the solutions of the non-linear equations (6.11.2) are not discussed here. In practice, for any such evolution equation related to Ψ and v, with initial data $u_0 = f$, it may be quite hard to justify the formal computations described below. In particular, for fast diffusion equations, which is the central point of interest below, $\Psi(r)$ is singular at $r = 0$ which causes serious difficulties. Since the point in what follows is to use those non-linear evolution equations to obtain functional inequalities from curvature-dimension conditions $CD(\rho, n)$ together with geometric properties of the function v involved in the equation, it is in general necessary to first approximate the singular map Ψ by some smooth approximation Ψ_ε. For this approximation, all the computations and integrations by parts may be justified by some previous a priori analysis so to yield an approximated functional inequality which converges in the limit $\varepsilon \to 0$ to the expected result. We however do not enter into these quite technical and tedious considerations here, assuming as mentioned above the regularity of u and v and of related expressions justifying the various integration by parts formulas. The aim of this study is rather to show that the modified fast diffusion equation leads, through these formal computations, precisely to the entropic form of the Sobolev or Gagliardo-Nirenberg inequalities described in Proposition 6.10.6.

The next lemma describes the announced second derivative operation on the functional F.

Lemma 6.11.3 *Let R be the function $R(r) = \frac{r\Psi''(r)}{\Psi'(r)}$, $r \in (0, \infty)$. Set $S = -\Psi(u)$, $\xi = \Psi(u) - \Psi(v)$, which are functions (depending on t) from E to $\mathbb{R}$. Then, with the generalized Fisher information $I_{\mu,F}$ of Proposition 6.11.1,*

$$\frac{d}{dt} I_{\mu,F}(u) = -\int_E u \, K \, d\mu$$

where

$$K = 2u\Psi'(u)\,\Gamma_2(\xi) + \Gamma\big(\xi, \Gamma(\xi)\big) - \big(R(u) + 2\big)\Gamma\big(S, \Gamma(\xi)\big)$$
$$- 2R(u)\,\Gamma\big(\xi, \Gamma(\xi, S)\big) + 2\,\frac{R(u) + 1 + uR'(u)}{u\Psi'(u)}\,\Gamma(\xi, S)^2.$$

In the examples of interest, $\Psi \leq 0$ so that $S = -\Psi(u) \geq 0$.

Proof Again, the exchanges between differentiation and integration will not always be fully justified (although care may be developed to this end). Since $\partial_t \xi = \Psi'(u)\partial_t u$,

$$\frac{d}{dt} I_{\mu,F}(u) = \int_E \big[\partial_t u\,\Gamma(\xi) + 2u\,\Gamma\big(\Psi'(u)\partial_t u, \xi\big)\big]d\mu.$$

By integration by parts with respect to the carré du champ operator Γ, for suitable functions f, g (say in the algebra $\mathcal{A}_0$),

$$\int_E u\,\Gamma(f, g)d\mu = -\int_E f\big[u\,\mathrm{L}g + \Gamma(u, g)\big]d\mu.$$

Therefore, the second term on the right-hand side of the preceding identity may be rewritten as

$$-2\int_E \partial_t u\, \Psi'(u)\bigl[u\,L\xi + \Gamma(u,\xi)\bigr]d\mu.$$

From the evolution of u, we have with (6.11.3) that

$$\frac{d}{dt} I_{\mu,F}(u) = -\int_E u\,R_1 d\mu$$

where

$$R_1 = \Gamma\bigl(\xi,\Gamma(\xi)\bigr) + 2\Gamma\bigl(\xi,\Gamma(S,\xi)\bigr) - 2\Gamma\bigl(\xi, u\,\Psi'(u)L\xi\bigr).$$

By the change of variables formula and the definition of the Γ_2 operator,

$$\Gamma\bigl(\xi, u\,\Psi'(u)L\xi\bigr) = u\,\Psi'(u)\,\Gamma(\xi,L\xi) + \bigl(u\Psi''(u) + \Psi'(u)\bigr)L\xi\,\Gamma(\xi,u)$$

$$= u\,\Psi'(u)\,\Gamma(\xi,L\xi) - \bigl(R(u)+1\bigr)L\xi\,\Gamma(\xi,S)$$

$$= u\,\Psi'(u)\left(-\Gamma_2(\xi) + \frac{1}{2}L\Gamma(\xi)\right) - \bigl(R(u)+1\bigr)L\xi\,\Gamma(\xi,S).$$

Furthermore,

$$\int_E u^2\,\Psi'(u)L\Gamma(\xi)d\mu = -\int_E \Gamma\bigl(u^2\Psi'(u),\Gamma(\xi)\bigr)d\mu$$

$$= \int_E u\bigl(R(u)+2\bigr)\Gamma\bigl(S,\Gamma(\xi)\bigr)d\mu.$$

Moreover, via integration by parts,

$$-\int_E u\bigl(R(u)+1\bigr)L\xi\,\Gamma(\xi,S)$$

$$= \int_E \Gamma\bigl(\xi, u\bigl(R(u)+1\bigr)\Gamma(\xi,S)\bigr)d\mu$$

$$= \int_E u\bigl(R(u)+1\bigr)\Gamma\bigl(\xi,\Gamma(\xi,S)\bigr)d\mu - \int_E \frac{R+1+uR'(u)}{\Psi'(u)}\Gamma(\xi,S)^2 d\mu.$$

Putting the various identities together yields the conclusion. Lemma 6.11.3 is established. □

6.11.3 Fast Diffusion and Curvature-Dimension Condition $CD(\rho,n)$

On the basis of Lemma 6.11.3, the aim is now to compare the second derivative of the entropy functional F to quantities arising from the $CD(\rho,n)$ condition. The

6.11 Fast Diffusion Equations and Sobolev Inequalities

situation is much simpler here in the fast diffusion case since then the various quantities involved in the expression K are easier to handle. Indeed, the fast diffusion equation corresponds to $\Phi(r) = r^m$, in which case $\Psi(r) = \frac{m}{m-1} r^{m-1}$, $H(r) = \frac{r^m}{m-1}$ and $R = m - 2$. Therefore $K = K_m$ in Lemma 6.11.3 is equal to

$$K_m = 2(1-m)\, S\, \Gamma_2(\xi) + \Gamma\big(\xi, \Gamma(\xi)\big) - m\, \Gamma\big(S, \Gamma(\xi)\big)$$

$$+ 2(2-m)\, \Gamma\big(\xi, \Gamma(\xi, S)\big) - 2\, \frac{\Gamma(\xi, S)^2}{S}.$$

This quantity K_m may actually be described in more geometrical and tractable terms. Set $\zeta = \Psi(v)$ (and thus $-S = \xi + \zeta$), and $X = \nabla \xi$, $Y = \nabla \zeta$, $M = \nabla \nabla \xi$, $P = \nabla \nabla \zeta$. Using the notation $M \cdot N$ for the scalar product in the space of symmetric tensors, that is in coordinates (and with Einstein's summation notation),

$$M \cdot N = M^{ij} N^{k\ell} g_{ik} g_{j\ell},$$

and the notation $X \odot Y$ for the symmetric tensor product of two vectors $(X \odot Y)^{ij} = \frac{1}{2}(X^i Y^j + X^j Y^i)$, we get that

$$\Gamma_2(\xi) = |M|^2 + \mathrm{Ric}(L) \cdot X \odot X,$$
$$\Gamma\big(S, \Gamma(\xi)\big) = -2M \cdot (X + Y) \odot X,$$
$$\Gamma\big(\xi, \Gamma(\xi)\big) = 2M \cdot X \odot X,$$
$$\Gamma\big(\xi, \Gamma(\xi, S)\big) = -M \cdot X \odot (X + Y) - (M + P) \cdot X \odot X,$$
$$\Gamma(S, \xi) = -X \cdot (X + Y).$$

Then

$$\frac{1}{2} K_m = (1-m)\, S\, \Gamma_2(\xi) + (m-1)\, M \cdot \big[X \odot (3X + 2Y)\big]$$

$$+ (m-2)\, P \cdot (X \odot X) - \frac{(X \cdot (X+Y))^2}{S}.$$

We are now ready to use the curvature-dimension condition $CD(\rho, n)$. We start with an elementary computation which is a further application of the integration by parts formula. Details are omitted.

Lemma 6.11.4 *For every smooth function w,*

$$\int_E w\, (L\xi)^2\, d\mu = \int_E \left[w\, \Gamma_2(\xi) + \frac{1}{2} \Gamma\big(w, \Gamma(\xi)\big) + \Gamma\big(\xi, \Gamma(\xi, w)\big) \right] d\mu.$$

As a consequence,

Proposition 6.11.5 *Under the curvature-dimension condition $CD(\rho, n)$, for every smooth function $w \geq 0$,*

$$\int_E w\, \Gamma_2(\xi)\, d\mu \geq \frac{\rho n}{n-1} \int_E w\, \Gamma(\xi)\, d\mu + \frac{1}{2(n-1)} \int_E \Gamma(w, \Gamma(\xi))\, d\mu$$
$$+ \frac{1}{n-1} \int_E \Gamma(\xi, \Gamma(\xi, w))\, d\mu.$$

In the case of interest, and towards the comparison with K_m, we have $w = (1-m)uS = mu^m$ and $S = \frac{m}{1-m} u^{m-1}$. In terms of X, Y, S, M, P,

$$\nabla w = mu(X+Y), \qquad \nabla \nabla w = mu(P+M) + u^{2-m}(X+Y) \odot (X+Y).$$

Therefore, the inequality of Proposition 6.11.5 reads

$$\int_E (1-m)u\, S\, \Gamma_2(\xi)\, d\mu$$
$$\geq \frac{\rho n}{n-1} \int_E (1-m)u\, S\, \Gamma(\xi) + \frac{m}{n-1} \int_E u M \cdot \bigl[X \odot (3X + 2Y)\bigr]\, d\mu$$
$$+ \frac{m}{n-1} \int_E uP \cdot [X \odot X]\, d\mu + \frac{1}{n-1} \int_E u^{2-m} \bigl[X \cdot (X+Y)\bigr]^2 d\mu.$$

The next proposition summarizes the final relevant inequality on the derivative of the generalized Fisher information.

Proposition 6.11.6 *Under the curvature-dimension condition $CD(\rho, n)$, and for $\Psi(r) = \frac{m}{m-1} r^{m-1}$,*

$$-\frac{1}{2} \frac{d}{dt} I_{\mu, F}(u) \geq \frac{\rho n}{n-1} \int_E mu^m\, \Gamma(\xi)\, d\mu$$
$$+ \left(\frac{m}{n-1} + m - 2 \right) \int_E u P \cdot (X \odot X)\, d\mu$$
$$+ \left(\frac{m}{n-1} + m - 1 \right) \int_E u M \cdot \bigl[X \odot (3X + 2Y)\bigr]\, d\mu$$
$$+ \left(\frac{1}{n-1} + \frac{m-1}{m} \right) \int_E u^{2-m} \bigl[X \cdot (X+Y)\bigr]^2 d\mu,$$

where we recall that $\xi = \Psi(u) - \Psi(v)$, $\zeta = \Psi(v)$, $X = \nabla \xi$, $Y = \nabla \zeta$, $M = \nabla \nabla \xi$ and $P = \nabla \nabla \zeta$.

In the particular case $m = 1 - \frac{1}{n}$, and if $-P = -\nabla \nabla \Psi(v) \geq a \geq 0$, then $-P \cdot (X \odot X) \geq a\, \Gamma(\xi)$, and everything simplifies miraculously to

$$-\frac{1}{2} \frac{d}{dt} I_{\mu, F}(u) \geq \rho \int_E u^m\, \Gamma(\xi)\, d\mu + a\, I(u).$$

6.11 Fast Diffusion Equations and Sobolev Inequalities

On the basis of the preceding proposition, we may now state the resulting functional inequality under a curvature-dimension condition. For simplicity, we only deal with the $CD(0, n)$ hypothesis.

Theorem 6.11.7 *In the preceding setting, under the $CD(0, n)$ condition, for a positive smooth function h such that $\nabla\nabla h \geq a > 0$, then, for $H(r) = -n\, r^{1-1/n}$, $H'(v) = -h$, $f > 0$, and provided $\int_E v\, d\mu = \int_E f\, d\mu$, we have*

$$\int_E \bigl[H(f) - H(v) - (f-v)H'(v)\bigr]d\mu \leq \frac{1}{2a}\int_E f\bigl|\nabla(H'(f) - H'(v))\bigr|^2 d\mu.$$

Proof Note that by the sign choice, the function H is convex, so that the left-hand side in the inequality of the theorem is positive. We stay at a somewhat informal level since justification of each step is rather tedious. Nevertheless, the principle is the following. Under the $CD(0, n)$ condition, Proposition 6.11.6 shows that solving the equation

$$\partial_t u = -\nabla^* u\, \nabla\bigl(\Psi(u) - \Psi(v)\bigr), \quad u_0 = f,$$

leads to the differential inequality

$$\frac{d}{dt} I_{\mu,F}(u_t) \leq -2a\, I_{\mu,F}(u_t).$$

It follows from the latter that

$$I_{\mu,F}(u_t) \leq e^{-2at}\, I_{\mu,F}(f)$$

for every $t \geq 0$. Since $I_{\mu,F}(u_t) = -\frac{d}{dt} F(u_t)$, it follows further that

$$F(u_0) - F(u_\infty) \leq I_{\mu,F}(f) \int_0^\infty e^{-2a(n-1)t} dt = \frac{1}{2a} I_{\mu,F}(f).$$

We are left to show that indeed $u_t \to v$ as $t \to \infty$. A necessary condition is that $\int_E v\, d\mu = \int_E f\, d\mu$. A more precise investigation of the entropy decay (and further technical bounds) indeed ensures that this is the case, so that the preceding amounts to the entropic inequality of the theorem. □

Remark 6.11.8 It is somewhat strange that for the parameters of Theorem 6.10.4, the Gagliardo-Nirenberg inequalities may be deduced either from the Sobolev inequality in $\mathbb{R}^{n+m}$ with $\nu = n + \frac{m}{2}$ relying on the $CD(0, n+m)$ condition, or via the fast diffusion equation from the $CD(0, \nu)$ condition of $\mathbb{R}^n$ (with the impression therefore of a loss of information in the $CD(0, n)$ condition of $\mathbb{R}^n$). But, as already observed throughout this chapter, this loss of information is actually recaptured by the action of the dilation group (which enters in a somewhat subtle way in the equivalence between the entropic inequalities of Proposition 6.10.6 and the

Gagliardo-Nirenberg inequalities). Actually, as already mentioned, most of the functional inequalities on $\mathbb{R}^n$, after optimization under the action of dilations, yield inequalities equivalent to the Sobolev inequality. One illustrative example is the proof of the sharp Euclidean logarithmic Sobolev inequality on $\mathbb{R}^n$ from the Gaussian logarithmic Sobolev inequality, itself a consequence of the $CD(0, \infty)$ condition in Euclidean space.

Remark 6.11.9 Starting from a curvature-dimension condition $CD(\rho, n)$ with $\rho > 0$, there is no need to make use of the stabilizing term v, that is the function h satisfying $\nabla\nabla h \geq a > 0$, to produce a Sobolev or a Gagliardo-Nirenberg inequality. Indeed, using the same evolution equation with v constant, and along the same lines, one may get directly the Sobolev inequality with optimal constants (i.e. with the constants one would get on spheres under the same hypothesis). Namely, for $m = 1 - \frac{1}{n}$,

$$-\frac{1}{2}\frac{d}{dt}I_{\mu,F}(u) \geq \rho \int_E u^m \Gamma(\xi) d\mu.$$

Observe then that

$$\frac{d}{dt}\int_E u^{1-2/n} d\mu = \frac{2(n-2)}{n(n-1)}\int_E u^m \Gamma(\xi) d\mu$$

which leads to

$$\frac{d}{dt}\left(I_{\mu,F}(u) + \frac{\rho n(n-1)}{n-2}\int_E u^{1-2/n} d\mu\right) \leq 0.$$

Therefore,

$$\int_E u_t^{1-2/n} d\mu \leq \int_E u_0^{1-2/n} d\mu + \frac{n-2}{\rho n(n-1)} I_{\mu,F}(u_0).$$

If we accept finally that $\lim_{t\to\infty} u_t = \int_E u_0 d\mu$ (which has to be proved using a more precise analysis than the one provided here), in the limit (after a suitable rewriting of $I_{\mu,F}(u_0)$),

$$\left(\int_E u_0 d\mu\right)^{1-2/n} \leq \int_E u_0^{1-2/n} d\mu + \frac{4}{n(n-2)}\frac{n-1}{\rho}\int_E \Gamma\left(u_0^{(n-2)/(2n)}\right) d\mu.$$

Changing u_0 into $f^{2n/(n-2)}$ leads to the tight Sobolev inequality with the sharp constant of Theorem 6.8.3, thus providing an alternative approach.

6.12 Notes and References

Sobolev inequalities, starting with the founding paper [386] by S. Sobolev, are a central theme in analysis, covering an incredibly large spectrum of both theoretical

6.12 Notes and References

and more applied topics. It is not within the scope of this short notice to present exhaustive comments and historical background on Sobolev inequalities. One major reference is the comprehensive monograph of V. Maz'ya [303] (a recent expansion of the famous [302]) to which we refer for a complete account of the subject. The contributions [1, 2, 96, 288, 413–415] are further relevant references. Related to the geometric aspects emphasized in this work, we mention in addition the monographs by T. Aubin [20, 21], E. Hebey [233, 234], and those of by L. Saloff-Coste [376], A. Grigor'yan [217] and P. Li [284] as well as the references therein.

The optimal constants in the Sobolev inequalities on $\mathbb{R}^n$ are due independently to T. Aubin [18, 19] and G. Talenti [406] (see also [364] for dimension 3). The case of the sphere is examined in [17] (see also [20]), while the hyperbolic case is proved in [235] by similar means as described here, relying on [382]. The Nash inequality in $\mathbb{R}^n$ was introduced by J. Nash [323], the optimal constant having been computed by E. Carlen and M. Loss [110]. The history of the Euclidean logarithmic Sobolev inequality (Proposition 6.2.5) is closely related to that of the Gaussian logarithmic Sobolev inequality (see Chap. 5), which in particular goes back to [387]. Proposition 6.2.5 linking the Euclidean and Gaussian forms has been explicitly or implicitly used by many authors [54, 106, 437, 438]. Forms of the Euclidean logarithmic Sobolev inequality were used by G. Perelman in his solution of the Poincaré conjecture.

The equivalences of Proposition 6.2.3 in Sect. 6.2 between the various forms of Sobolev inequalities classically appeared as consequences of ultracontractive bounds equivalent to the corresponding functional inequalities (Sect. 6.3). The direct approach presented here via the slicing decomposition was introduced in [35] and [149]. See e.g. [376, 377] for more and for bibliographical accounts. The Euclidean logarithmic Sobolev inequality regarded as a limit of the sharp Sobolev inequality (Remark 6.2.6) was pointed out in [58] (see also [57]).

The main ultracontractive Theorem 6.3.1 of Sect. 6.3 is the end result of a series of steps and results by numerous authors including, among others, [26, 108, 138, 144, 181, 319, 323, 421]. The proof presented here is due to Nash, using the corresponding Nash inequality, following [108]. The other authors dealt with either Sobolev or logarithmic Sobolev inequalities, illustrated here in Proposition 6.2.3, or Gagliardo-Nirenberg inequalities. See [26, 144, 217, 376, 377, 422] for further details and historical background.

The Rellich-Kondrachov Theorem 6.4.3 may be found in most classical references such as [96, 179, 203].

Proposition 6.6.1 in Sect. 6.6 is due to L. Saloff-Coste [375] (the proof here being taken from [275]). The sharp version of the diameter bound under a Sobolev inequality (namely that $D \leq \pi$ whenever the Sobolev inequality with the constants of the model space $\mathbb{S}^n$ holds) was obtained in [39], leading to the classical Bonnet-Myers Theorem in this context. Proposition 6.6.2 on the volume growth is part of the folklore and may be found, for example, in [114] (see also [217, 377]).

The celebrated Li-Yau parabolic inequality (Corollary 6.7.5) was established in [285] by means of the maximum principle. The heat flow monotonicity proof presented here via dimensional logarithmic Sobolev inequalities for heat kernels is taken from [40]. More on Harnack inequalities for elliptic operators on Riemannian manifolds and precise heat kernel bounds (of the form (6.7.11) and (6.7.12)) may be found in [144, 217, 284, 378]. The small time asymptotics (6.7.13) go back to [420] (see [332] for the final word). Local hypercontractivity (Theorem 6.7.7) first appeared in [31]. Versions of the Li-Yau inequality in sub-Riemannian geometries by means of heat flow monotonicity have been obtained in [51].

The sharp Sobolev inequality on the sphere (6.1.2) of T. Aubin [17, 20] was extended to Riemannian manifolds with a uniform strictly positive lower bound on the Ricci curvature in [253] via the Lévy-Gromov isoperimetric comparison Theorem (see the Notes and References of Chap. 8). The non-linear proof of Theorem 6.8.3 developed in Sect. 6.8 is adapted from the corresponding analysis by O. Rothaus [371] for logarithmic Sobolev inequalities. The method actually originates in [202] in a partial differential equation context (see later [66]). The final step of the argument is reminiscent of the Nash-Moser iteration principle. The proof is presented in [26], and further developed in [185] where the subsequent Remarks 6.8.4 and 6.8.5 are emphasized, the second one again following [371]. On the sphere, these remarks go back to W. Beckner [56] who used sharp Hardy-Littlewood-Sobolev inequalities for this purpose. See also [66, 160, 161]. Moser-Trudinger inequalities appeared in [320, 416] as the limiting case $n = 2$ (actually any n in the suitable version in higher dimension). See also [35] in the context of the slicing technique of Proposition 6.2.3. On the sphere $\mathbb{S}^2$, the sharp inequality (6.8.8) is sometimes referred to as Onofri's inequality [337] (see also [56]).

Section 6.9 on conformal invariance of Sobolev inequalities describes some of the main ideas involved in the Yamabé program developed in particular by T. Aubin [17, 20] and R. Schoen [380, 381] (see [21, 233, 234]). Its abstract formulation in the context of this book is partly inspired by the paper [382] of R. Schoen and S.-T. Yau. The formal equivalence between the classical Sobolev inequalities in Euclidean, spherical and hyperbolic spaces may also be traced back to [382] and is emphasized in [235]. (Remark 6.9.5 on the (minimal) class of functions in the Sobolev inequality in Euclidean space is part of the classical theory [288, 303, 449].) The references [165, 234] present the (A, B) program which is devoted to the respective sharpness of the constants C and A, in the notation used here, for Sobolev inequalities in manifolds.

The Gagliardo-Nirenberg inequalities recalled in Sect. 6.10 were first put forward in [192, 331]. See for instance [96, 303] for general recent references and historical background. Optimality and extremal functions of the sub-family of Gagliardo-Nirenberg inequalities of Theorem 6.10.4 have been obtained by M. Del Pino and J. Dolbeault [147] by an analytic study of the corresponding minimization problem. Theorem 6.10.4 has been generalized in [330] to a larger class of parameters with arguments based on the mass transportation method of [137]. The reformulation in [147] as an entropy decay along non-linear equations in Proposition 6.10.6 has been an important step in the understanding of the connections between porous medium and fast diffusion equations, functional inequalities and convergence to equilibrium. See also [111], and [424, 426] for links with mass transportation and the references therein.

A comprehensive account of porous medium and fast diffusion equations is the recent monograph [423] to which we refer for a complete background. On the basis

6.12 Notes and References

of the Dolbeault-Del Pino developments, Sect. 6.11, and Theorem 6.11.7, are essentially due to J. Demange [155], who solved in particular all the regularity issues necessary for the proof (see also [28, 156] and [161]). The last Remark 6.11.9 is also due to him (unpublished). Fast diffusion has also been used to obtain Hardy-Littlewood-Sobolev inequalities in [107], relying on the heat flow monotonicity approach to geometric Brascamp-Lieb inequalities (different from the approach described in Sect. 4.9, p. 215) developed in [59, 109].

Part III
Related Functional, Isoperimetric and Transportation Inequalities

Chapter 7
Generalized Functional Inequalities

Part II, devoted to Poincaré, logarithmic Sobolev and Sobolev inequalities, describes how each of these families capture different features of the associated semigroup or the invariant measure, in terms of convergence to equilibrium, estimates on the heat kernels or tail behaviors of the invariant measure. There are many ways to describe intermediate families of functional inequalities which are suited to a wide variety of regimes as well as to more precise, or different, features. This section investigates such families, restricting to three main examples, entropy-energy, generalized Nash and weak Poincaré inequalities. Many other families have been developed in different directions, each of them having its own interest. Some will be mentioned at the end of the section.

In this chapter, we deal as usual with a Full Markov Triple (E, μ, Γ) as presented in Sect. 3.4, p. 168, with Dirichlet form $\mathcal{E}$, infinitesimal generator L, Markov semigroup $\mathbf{P} = (P_t)_{t \geq 0}$ and underlying function algebras $\mathcal{A}_0$ and $\mathcal{A}$. Actually, the natural framework for the investigation here is that of the Standard Markov Triple, and besides the section on off-diagonal heat kernel bounds for which the extended algebra $\mathcal{A}$ is required, all the results may be stated in this framework. We nevertheless simply use the terminology "Markov Triple" to cover the various instances.

The first section describes the family of functional entropy-energy inequalities governed by a growth function Φ, the example of the logarithmic function giving rise to the logarithmic entropy-energy inequality of the previous chapter as an equivalent form of the standard Sobolev inequality. The family of entropy-energy inequalities is in particular well-suited to heat kernel bounds by means of the method developed for hypercontractivity under logarithmic Sobolev inequalities in Chap. 5. Off-diagonal heat kernel estimates may be achieved in the same way (Sect. 7.2). Several examples of both entropy-energy inequalities and their associated heat kernel bounds are presented in Sect. 7.3. The next section investigates generalized Nash inequalities on the basis of the example of the classical Nash inequality in Euclidean space, as well as weighted Nash inequalities which form a further family of interest. Again, heat kernel bounds and tail inequalities of various types may be obtained. Weak Poincaré inequalities are studied in Sect. 7.5 as the main minimal tool to tighten families of standard functional inequalities. Weak Poincaré inequalities may be further studied in their own right from the viewpoint of heat kernel bounds and tail estimates. Section 7.6 briefly describes related families of functional inequalities of interest and their relationships with the previous ones. Various applications of the tools investigated in this chapter are illustrated, for comparison, on the model family of (probability) measures $c_\alpha e^{-|x|^\alpha} dx$, $\alpha > 0$, on $\mathbb{R}^n$ with respect to the standard carré du champ operator $\Gamma(f) = |\nabla f|^2$. Actually, to avoid technical regularity issues of the potential $W(x) = |x|^\alpha$ (at the origin), it will be more convenient to deal with the family

$$d\mu_\alpha(x) = c_\alpha \exp\bigl(-(1+|x|^2)^{\alpha/2}\bigr) dx \qquad (7.0.1)$$

(with $c_\alpha > 0$ the normalization constant) which behave similarly at infinity. The main results for this family in the one-dimensional case are summarized in Sect. 7.7 at the end of the chapter. The multi-dimensional picture is essentially the same with constants depending on the dimension.

Finally, it is worth emphasizing that most inequalities investigated in this chapter compare two functionals via a *growth function* Φ which will always be $\mathcal{C}^1$ increasing and concave from $(0, \infty)$ to $\mathbb{R}$, $\mathbb{R}_+$ or $(0, \infty)$. Although the $\mathcal{C}^1$ hypothesis is not strictly necessary (continuous should be enough), it will be convenient for a number of technical steps. As already (briefly) mentioned in the proof of Proposition 4.10.4, p. 224, a standard procedure in this regard, at least for positive growth functions, is to replace Φ (concave increasing) by $\widetilde{\Phi}(r) = \frac{2}{r}\int_0^r \Phi(s) ds$ which is still increasing and concave and satisfies $\frac{1}{2}\widetilde{\Phi}(r) \leq \Phi(r) \leq \widetilde{\Phi}(r)$). In several parts of this chapter, sharp constants are indeed not an issue and focus is instead placed on growth orders.

7.1 Inequalities Between Entropy and Energy

This section deals with generalized functional inequalities comparing the entropy and energy of a given function in the Dirichlet space. These generalized functional inequalities are then shown to efficiently reach uniform heat kernel bounds.

7.1.1 Entropy-Energy Inequalities

This family compares the entropy and the energy of a function (in the Dirichlet domain $\mathcal{D}(\mathcal{E})$) via a $\mathcal{C}^1$ increasing concave growth function $\Phi : (0, \infty) \to \mathbb{R}$. A first example of such an inequality is the logarithmic entropy-energy inequality (6.2.2) of Proposition 6.2.3, p. 281, with growth function the logarithmic function. The function $\Phi(r)$, $r > 0$, may have a finite limit at $r = 0$, denoted $\Phi(0)$, or converge to $-\infty$ as in the latter example. Recall that the entropy of a measure ν (not necessarily finite) on a measurable space $(E, \mathcal{F})$ has been defined at (5.1.1), p. 236, as

$$\mathrm{Ent}_\nu(f) = \int_E f \log f \, d\nu - \int_E f \, d\nu \log\left(\int_E f \, d\nu\right)$$

for all positive integrable functions $f : E \to \mathbb{R}$ such that $\int_E f |\log f| d\nu < \infty$.

7.1 Inequalities Between Entropy and Energy

Definition 7.1.1 (Entropy-energy inequality) A Markov Triple (E, μ, Γ) is said to satisfy an entropy-energy inequality $EE(\Phi)$ with respect to a growth function $\Phi : (0, \infty) \to \mathbb{R}$ if for every $f \in \mathcal{D}(\mathcal{E})$ such that $\int_E f^2 d\mu = 1$ and $\mathcal{E}(f) > 0$,

$$\mathrm{Ent}_\mu(f^2) \leq \Phi(\mathcal{E}(f)). \tag{7.1.1}$$

As usual, it is enough to state and establish such an inequality for a family of functions f which is dense in the domain $\mathcal{D}(\mathcal{E})$ of the Dirichlet form $\mathcal{E}$ (typically the algebra $\mathcal{A}_0$).

As mentioned previously, the logarithmic entropy-energy inequality with $\Phi(r) = \frac{n}{2} \log(A + Cr)$, $r \in (0, \infty)$, described in Proposition 6.2.3, p. 281, as an equivalent form of a Sobolev inequality, is an instance of Definition 7.1.1. In particular, the Euclidean logarithmic Sobolev inequality with respect to the Lebesgue measure on $\mathbb{R}^n$ of Proposition 6.2.5, p. 284, holds for the optimal function $\Phi(r) = \frac{n}{2} \log(\frac{2r}{n\pi e})$, $r \in (0, \infty)$. Further examples of entropy-energy inequalities have occurred earlier for specific choices of the growth function Φ. For instance, for μ a probability measure, the (defective) logarithmic Sobolev inequality $LS(C, D)$ of Definition 5.1.1, p. 236, is an example of an $EE(\Phi)$ inequality with $\Phi(r) = 2Cr + D$, $r \in (0, \infty)$. Note also that if $EE(\Phi_1)$ and $EE(\Phi_2)$ hold for two concave functions Φ_1 and Φ_2, then $EE(\Phi_1 \wedge \Phi_2)$ holds.

The entropy-energy inequalities from Definition 7.1.1 may be described in equivalent linearized forms. Indeed, since Φ in $EE(\Phi)$ is assumed to be $\mathcal{C}^1$, by concavity $\Phi(s) - \Phi(r) \leq \Phi'(r)(s - r)$ for all $r, s > 0$. The entropy-energy inequality $EE(\Phi)$ of Definition 7.1.1 may then be linearized equivalently as a family of defective logarithmic Sobolev inequalities, for all $f \in \mathcal{D}(\mathcal{E})$ with $\int_E f^2 d\mu = 1$,

$$\mathrm{Ent}_\mu(f^2) \leq \Phi'(r)\mathcal{E}(f) + \Psi(r), \quad r \in (0, \infty), \tag{7.1.2}$$

where $\Psi(r) = \Phi(r) - r\Phi'(r)$.

Whenever Φ' is strictly decreasing, and takes values in some interval $(a, b) \subset \mathbb{R}$, one may choose $s = \Phi'(r)$ and turn $EE(\Phi)$ into the family of inequalities

$$\mathrm{Ent}_\mu(f^2) \leq s\mathcal{E}(f) + \beta(s), \quad s \in (a, b), \tag{7.1.3}$$

$f \in \mathcal{D}(\mathcal{E})$, $\int_E f^2 d\mu = 1$, where $\beta(s) = \Phi(\Phi'^{-1}(s)) - s\,\Phi'^{-1}(s)$. Conversely, (7.1.3) gives rise to the $EE(\Phi)$ inequality with

$$\Phi(r) = \inf_{s \in (a,b)} [sr + \beta(s)], \quad r \in (0, \infty).$$

Observe that for smooth functions, $\beta' = -\Phi'^{-1}$, and so there is a complete equivalence between the two formulations (7.1.2) and (7.1.3). Whenever Φ' is not bijective, Φ'^{-1} has to be replaced by its generalized inverse, where the *generalized inverse* of an increasing function T on an interval $I \subset \mathbb{R}$ is defined by

$$T^{-1}(s) = \inf\{r \in I \,;\, T(r) \geq s\} \qquad (7.1.4)$$

for every s in the range of T.

Inequalities (7.1.2) or (7.1.3) will be considered as the *linearized forms* of the entropy-energy inequality $EE(\phi)$.

When μ is a probability measure, the entropy-energy inequality $EE(\Phi)$ is thus equivalent to a family of defective logarithmic Sobolev inequalities. According to Proposition 5.1.3, p. 238, a tight logarithmic Sobolev inequality $LS(C)$ implies a Poincaré inequality $P(C)$ while conversely, under a Poincaré inequality, a defective logarithmic Sobolev inequality $LS(C, D)$ may be tightened. This principle applied to an entropy-energy inequality $EE(\Phi)$ through the equivalent descriptions (7.1.2) or (7.1.3), shows that, under $EE(\Phi)$, it is equivalent to ask for a Poincaré inequality or for the function Φ to be bounded from above by Cr, $r > 0$, for some constant $C > 0$.

7.1.2 Uniform Heat Kernel Bounds

The main interest in the family of entropy-energy inequalities $EE(\Phi)$ is the flexibility in the choice of the growth function Φ. In particular, this family interpolates between logarithmic Sobolev and Sobolev-type inequalities, and actually allows for intermediate regimes, in particular of heat kernel bounds. The following first and main result is an illustration of this principle. It is a direct extension of the hypercontractivity Theorem 5.2.3, p. 246, and is proved using the same Gross method. If $1 \leq p \leq \infty$, p^* denotes its conjugate exponent, so that $\frac{1}{p} + \frac{1}{p^*} = 1$. The norms are understood as usual in $\mathbb{L}^p(\mu)$, $1 \leq p \leq \infty$. Recall the operator norm notation $\|\cdot\|_{p,q}$ from $\mathbb{L}^p(\mu)$ into $\mathbb{L}^q(\mu)$ from (6.3.1), p. 286.

Theorem 7.1.2 *Let* (E, μ, Γ) *be a Markov Triple satisfying an entropy-energy inequality* $EE(\Phi)$ *for some growth function* Φ. *Recall* $\Psi(r) = \Phi(r) - r\Phi'(r)$, $r \in (0, \infty)$, *from* (7.1.2). *Then, for every* $2 \leq p \leq q \leq \infty$ *and every* $\delta > 0$,

$$\|P_{t(\delta)}\|_{p,q} \leq e^{m(\delta)}$$

where

$$\begin{cases} t(\delta) = \displaystyle\int_{pp^*}^{qq^*} \Phi'(\delta r) \, \frac{dr}{4\sqrt{r(r-4)}}, \\ m(\delta) = \displaystyle\int_{pp^*}^{qq^*} \Psi(\delta r) \, \frac{dr}{r\sqrt{r(r-4)}}. \end{cases} \qquad (7.1.5)$$

When $1 \leq p \leq q \leq 2$, *the integrals* $\int_{pp^*}^{qq^*}$ *in* (7.1.5) *have to be replaced by* $\int_{qq^*}^{pp^*}$ *while when* $1 \leq p \leq 2 \leq q \leq \infty$, *they have to be replaced by* $\int_4^{pp^*} + \int_4^{qq^*}$.

7.1 Inequalities Between Entropy and Energy

Remark 7.1.3 In Theorem 7.1.2, it is sometimes simpler to change r into $s = \sqrt{1 - \frac{4}{r}}$ and δ into 4δ so that (7.1.5) becomes

$$\begin{cases} t(\delta) = \dfrac{1}{2} \displaystyle\int_{1-\frac{2}{p}}^{1-\frac{2}{q}} \Phi'\left(\dfrac{\delta}{1-s^2}\right) \dfrac{ds}{1-s^2}, \\ m(\delta) = \dfrac{1}{2} \displaystyle\int_{1-\frac{2}{p}}^{1-\frac{2}{q}} \Psi\left(\dfrac{\delta}{1-s^2}\right) ds. \end{cases} \quad (7.1.6)$$

In this form, the formulas are valid whatever the values of the parameters $1 \leq p \leq q \leq \infty$, as long as the integrals converge.

We make a few comments before turning to the proof of Theorem 7.1.2. First, the result is compatible with the semigroup property. Namely, since $\|P_t \circ P_s\|_{p,q} \leq \|P_t\|_{p,r} \|P_s\|_{r,q}$ when $p \leq r \leq q$, applying the conclusion separately on $[0, t]$ and on $[0, s]$ yields the same bound as applying it directly on $[0, t+s]$. In particular, the case $p \leq 2 \leq q$ is a direct consequence of the separate results for the pairs $(p, 2)$ and $(2, q)$. Moreover, thanks to the symmetry of the formulas under the change of p and q into their conjugate exponents p^* and q^*, it is also compatible with the symmetry property $\|P_t\|_{p,q} = \|P_t\|_{q^*, p^*}$. Therefore, everything reduces to the case $2 \leq p \leq q \leq \infty$.

The second set of observations actually illustrates the range of applications of Theorem 7.1.2. Note first that whenever $\liminf_{r \to \infty} \Phi'(r) > 0$, as is the case for a logarithmic Sobolev inequality and hypercontractivity (cf. Theorem 5.2.3, p. 246), one may not reach $q = \infty$ in Theorem 7.1.2. There is actually a minimal value of t for which hypercontractivity applies. On the other hand, the conclusion of Theorem 7.1.2 may be applied to $p = 2$ and $q = \infty$, or $p = 1$ and $q = \infty$, as soon as $r^{-1} \Phi'(r)$ is integrable at infinity (which implies that $r^{-2} \Psi(r)$ is also integrable at infinity). It applies in particular to the logarithmic entropy-energy inequality (6.2.2), p. 281, for which $\Phi(r) = \frac{n}{2} \log(A + Cr)$, $r \in (0, \infty)$, and yields in this case ultracontractive bounds of the form $\|P_t\|_{1,\infty} \leq C t^{-n/2}$ as described in Sect. 6.3, p. 286. It also shows that $\Phi(r) = A + Cr^\alpha$, $r \in (0, \infty)$, gives ultracontractive bounds as soon as $\alpha < 1$ from which it appears that the standard logarithmic Sobolev case is a limiting case for ultracontractivity. It may furthermore happen that $\lim_{r \to \infty} \Phi'(r) = 0$ while $\int^\infty \Phi'(r) \frac{dr}{r} = \infty$. In this case, the semigroup is said to be *immediately hypercontractive* in the sense that for any $t > 0$ and any $1 < p < q < \infty$, P_t is a bounded operator from $\mathbb{L}^p(\mu)$ into $\mathbb{L}^q(\mu)$.

Proof of Theorem 7.1.2 According to the preceding comments, it is enough to deal with $2 \leq p \leq q < \infty$ (the case $q = \infty$ being handled by letting $q \to \infty$). The proof closely follows that of Theorem 5.2.3, p. 246, concerning hypercontractivity. For notational convenience, set $q_0 = p \leq q = q_1$. For some positive bounded function f in $\mathcal{D}(\mathcal{E})$, let $q = q(t)$, $t \geq 0$, be smooth enough with $q(0) = q_0 (= p)$. Setting

$\Lambda(t) = \int_E (P_t f)^{q(t)} d\mu$, $t \geq 0$, and arguing as in the proof of Theorem 5.2.3,

$$\Lambda(t)' = \frac{q'(t)}{q(t)} \big(\mathrm{Ent}_\mu\big((P_t f)^{q(t)}\big) + \Lambda(t) \log \Lambda(t)\big) - q(t)\big(q(t) - 1\big) \mathcal{E}_{q(t)}(P_t f)$$

where we set $\mathcal{E}_q(f) = \int_E f^{q-2} \Gamma(f) d\mu$ to ease the notation. Now, (7.1.2) applied to $(P_t f)^{q/2}$ yields that for all $r > 0$,

$$\Lambda' \leq q\bigg(\frac{q'}{4} \Phi(r) - (q-1)\bigg) \mathcal{E}_q(P_t f) + \frac{q'}{q} \Lambda\big(\Psi(r) + \log \Lambda\big).$$

Next choose $q \mapsto r(q) \in (0, \infty)$ and take $q = q(t)$ which solves the differential equation $q' \Phi(r(q)) = 4(q - 1)$. The preceding inequality then reads

$$(\log \Lambda)' \leq \frac{q'}{q} \big(\Psi(r(q)) + \log \Lambda\big),$$

or equivalently

$$\bigg(\frac{1}{q} \log \Lambda\bigg)' \leq \frac{q'}{q^2} \Psi(r(q)).$$

Integrating this differential inequality between 0 and t yields

$$\log \Lambda(t)^{1/q(t)} \leq \log \Lambda(0)^{1/q(0)} + \int_0^t \frac{q'}{q^2} \Psi(r(q)) ds.$$

In other words, for any choice of the function $q \mapsto r(q)$, and using q as a variable instead of t, if

$$t = \int_{q_0}^{q_1} \Phi'(r(q)) \frac{dq}{4(q-1)}$$

with $q_0 < q_1$, then

$$\|P_t\|_{q_0, q_1} \leq \exp\bigg(\int_{q_0}^{q_1} \Psi(r(q)) \frac{dq}{q^2}\bigg).$$

It remains to choose the map $q \mapsto r(q)$ in order to optimize this bound when $q_0 = q(0)$, $q_1 = q(t)$ and t are given. The computation shows that the optimal choice indeed does not depend on Φ, and is given by $q \mapsto r(q) = \frac{8q^2}{q-1}$. The final result is then obtained after a change of variable. Theorem 7.1.2 is established. □

As mentioned previously, Theorem 7.1.2 is of most interest when $r^{-1} \Phi'(r)$ is integrable at infinity in which case $q = \infty$ may be reached in the conclusion. This main instance deserves a separate statement. As for the ultracontractive bounds of Sect. 6.3, p. 286, recall that whenever $\|P_t\|_{1, \infty}$ is finite, the operators P_t, $t > 0$, are represented by (symmetric) density kernels $p_t(x, y)$, $t > 0$, $(x, y) \in E \times E$, in $\mathbb{L}^2(\mu)$ such that $\|P_t\|_{1, \infty} = \|p_t(\cdot, \cdot)\|_{\mathbb{L}^\infty(\mu \otimes \mu)}$.

7.1 Inequalities Between Entropy and Energy

Corollary 7.1.4 *In the setting of Theorem 7.1.2, if the entropy-energy inequality $EE(\Phi)$ holds for a growth function Φ such that $r^{-1}\Phi'(r)$ is integrable at infinity, then, for any $\delta > 0$, $\|P_{t(\delta)}\|_{1,\infty} \leq e^{m(\delta)}$ where $t(\delta)$ and $m(\delta)$ are represented through the parameter $\delta > 0$ as*

$$\begin{cases} t(\delta) = \dfrac{1}{2}\displaystyle\int_1^\infty \Phi'(\delta r)\,\dfrac{dr}{\sqrt{r(r-1)}}, \\ m(\delta) = \dfrac{1}{2}\displaystyle\int_1^\infty \Psi(\delta r)\,\dfrac{dr}{r\sqrt{r(r-1)}}. \end{cases} \quad (7.1.7)$$

Conversely, if for some $t \in (a, \infty)$, $a \geq 0$, $\|P_t\|_{1,\infty} \leq K(t)$, then $EE(\Phi)$ holds for Φ defined by

$$\Phi(r) = \inf_{t>a}\bigl[2tr + \log K(t)\bigr], \quad r \in (0, \infty).$$

Proof The first assertion follows from Theorem 7.1.2 with $p = 1$ and $q = \infty$, changing r into $\frac{r}{4}$ and δ in 4δ. Conversely, as consequence of the Riesz-Thorin Theorem, $\|P_t\|_{1,\infty} \leq K$ implies that $\|P_t\|_{2,\infty} \leq \sqrt{K}$. By Proposition 5.2.6, p. 248, it follows that whenever $\int_E f^2 d\mu = 1$,

$$\mathrm{Ent}_\mu(f^2) \leq 2t\,\mathcal{E}(f) + \log K(t).$$

Optimizing in $t > a$ yields the announced $EE(\Phi)$ inequality. □

The example of $K(t) = B t^{-n/2}$, $t > 0$, in Corollary 7.1.4 (as is the case under a Sobolev inequality, cf. Theorem 6.3.1, p. 286), yields a growth function Φ of the form $\Phi(r) = \frac{n}{2}\log(A + Cr)$, $r \in (0, \infty)$, in the entropy-energy inequality $EE(\Phi)$. In general, the uniform bound on $\|P_t\|_{1,\infty}$ reflects the behavior of Φ at infinity as $t \to 0$, and the behavior of Φ at 0 as $t \to \infty$. In this way, the next statement recovers from Corollary 7.1.4 the ultracontractive bounds under a Sobolev inequality of Theorem 6.3.1, p. 286, through entropy-energy inequalities.

Corollary 7.1.5 *In the setting of Theorem 7.1.2, if the entropy-energy inequality $EE(\Phi)$ holds for a growth function Φ such that $\Phi'(r) - \frac{n}{2r}$ is integrable at infinity, then*

$$\|P_t\|_{1,\infty} \leq \frac{C}{t^{n/2}}, \quad 0 < t \leq 1.$$

In the same way, if $\Phi(r) \sim \frac{n}{2}\log r$ as $r \to 0$, and if $r^{-1}\Phi'(r)$ is integrable at infinity, then

$$\|P_t\|_{1,\infty} \leq \frac{C}{t^{n/2}}, \quad t \geq 1.$$

Proof When $\Phi'(r) - \frac{n}{2r}$ is integrable at infinity, $\Phi(r) = \frac{n}{2}\log r + O(r)$ as $r \to \infty$. It follows that when $\delta \to \infty$, using the form (7.1.7),

$$t(\delta) = \frac{n}{2\delta} + o\left(\frac{1}{\delta}\right), \qquad m(\delta) = \frac{n}{2}\log(\delta) + O(\delta),$$

from which the bound on $\|P_t\|_{1,\infty}$ is easily deduced. The behavior at $t \to \infty$ is treated in the same way. □

A somewhat striking application of the preceding developments concerns the Euclidean logarithmic Sobolev inequality of Proposition 6.2.5, p. 284, which actually yields optimal heat kernel bounds. The proof simply follows from the expression of the optimal function $\Phi(r) = \frac{n}{2}\log(\frac{2r}{n\pi e})$, $r \in (0, \infty)$, and the explicit values of $t(\delta)$ and $m(\delta)$ of (7.1.7) in this case.

Corollary 7.1.6 *Under the Euclidean logarithmic Sobolev inequality $EE(\Phi)$ with $\Phi(r) = \frac{n}{2}\log(\frac{2r}{n\pi e})$, $r \in (0, \infty)$,*

$$\|P_t\|_{1,\infty} \leq \frac{1}{(4\pi t)^{n/2}}, \qquad t > 0.$$

Therefore, in particular, $p_t(x,y) \leq \frac{1}{(4\pi t)^{n/2}}$ uniformly in $t > 0$ and $(\mu \otimes \mu$-almost everywhere) in $(x, y) \in E \times E$.

The function $\Phi(r) = \frac{n}{2}\log(\frac{2r}{n\pi e})$, $r \in (0, \infty)$, thus appears as the best growth function Φ which bounds entropy from energy in the Euclidean case since from the standard representation (2.1.1), p. 78, the heat kernel bound is obviously the best possible. After simple (but tedious) computations, it also follows that, for every $1 \leq p \leq q \leq \infty$ and every $t > 0$,

$$\|P_t\|_{p,q} \leq \left[\frac{p^{\frac{1}{p}}(1-\frac{1}{p})^{1-\frac{1}{p}}}{q^{\frac{1}{q}}(1-\frac{1}{q})^{1-\frac{1}{q}}}\right]^{\frac{n}{2}} \left[\frac{1}{4\pi t}\left(\frac{1}{p}-\frac{1}{q}\right)\right]^{\frac{n}{2}(\frac{1}{p}-\frac{1}{q})}. \qquad (7.1.8)$$

These intermediate bounds are again optimal on the standard heat semigroup on $\mathbb{R}^n$. Indeed, since this bound is optimal when $p = 1$ and $q = \infty$, thanks to the compatibility relations between the (p, q) norms and the semigroup property, any intermediate bound is also optimal.

When the measure μ is finite (and normalized into a probability measure), the growth function Φ in the definition of an entropy-energy inequality $EE(\Phi)$ takes positive values (since $\text{Ent}_\mu(f^2) \geq 0$). Furthermore, as mentioned earlier, the existence of a Poincaré inequality is equivalent to the fact that Φ may be chosen so that $\Phi(r) \leq Cr$ near 0. As it is known from Corollary 6.4.1, p. 290, as soon as $(P_t)_{t\geq 0}$ is ultracontractive and the measure is finite, the operators P_t, $t > 0$, are Hilbert-Schmidt and therefore the generator L has a discrete spectrum. In particular, a Poincaré inequality holds, and it may therefore be assumed that $\Phi(r) \leq Cr$ for some constant C, or equivalently that $\Phi'(0) < \infty$. Note also that, from the general Definition 7.1.1, tightness in $EE(\Phi)$ is achieved as soon as $\Phi(0) = 0$. As a consequence of the discussion in Sect. 7.5 below (Corollary 7.5.7), when $\Phi(0) = 0$ it is possible to choose (some other) growth function Φ such that $\Phi(r) \leq Cr$ for some $C > 0$.

When $\Phi'(0) < \infty$ and $r^{-1}\Phi(r)$ is integrable at infinity, Corollary 7.1.4 actually yields bounds on the convergence to equilibrium. The following statement again follows from a precise analysis of $t(\delta)$ and $m(\delta)$ as $\delta \to 0$ in (7.1.7). The details are left to the reader.

Corollary 7.1.7 *In the setting of Theorem 7.1.2, and whenever μ is a probability measure, if the entropy-energy inequality $EE(\Phi)$ holds for a growth function Φ such that $\Phi'(0) < \infty$ and $r^{-1}\Phi(r)$ is integrable at infinity, then as $t \to \infty$,*

$$\log \|P_t\|_{1,\infty} \leq A\, e^{-\Phi'(0)/2t}\left(1 + o(t)\right)$$

where A and C only depend on Φ and are explicitly given by

$$A = 2e^C \int_0^\infty \Psi(r)\frac{dr}{r^2} \quad \text{and} \quad C = \int_0^1 \left(\frac{\Phi'(r)}{\Phi'(0)} - 1\right)\frac{dr}{r} + \int_1^\infty \frac{\Phi'(r)}{\Phi'(0)}\frac{dr}{r}.$$

7.2 Off-diagonal Heat Kernel Bounds

This section addresses the issue of off-diagonal estimates on the density kernels $p_t(x, y)$, $t > 0$, $(x, y) \in E \times E$, of a Markov semigroup $(P_t)_{t \geq 0}$ (with respect to the invariant measure) under an entropy-energy inequality $EE(\Phi)$. Actually, as soon as uniform bounds are available, that is a control of $p_t(x,x)$ or $p_t(x,y)$ uniformly over $(x, y) \in E \times E$ (almost everywhere), there are in general also off-diagonal estimates taking into account the (intrinsic) distance $d(x, y)$ between x and y associated with the Markov Triple (E, μ, Γ) as considered in (3.3.9), p. 166. Recall that the distance function (Sect. 3.3.7, p. 166) refers to the extended algebra $\mathcal{A}$, justifying the Full Markov Triple assumption (cf. Sect. 3.4, p. 168).

Gaussian bounds of the type

$$p_t(x, y) \leq C(t)\, e^{-d^2(x,y)/ct}, \quad t > 0, \ (x, y) \in E \times E, \qquad (7.2.1)$$

(for constants $C(t) > 0$, $c > 0$) are of interest. Such heat kernel bounds were already considered in Sect. 6.7, p. 296, under geometric features by means of Harnack-type inequalities. In this section, a general method of providing such bounds is developed under functional inequalities in entropy-energy form.

7.2.1 Diameter Bounds

In the finite measure case, whenever $C(t) < 2$ in (7.2.1) (with this in mind, recall Proposition 6.3.4, p. 289), Proposition 1.2.6, p. 15, provides a uniform lower bound on p_{2t}, and therefore such heat kernel bounds can only hold if the diameter $D = D(E, \mu, \Gamma)$ of E with respect to the distance d (cf. (3.3.10), p. 167) is finite. In this respect, we first state a bound on the diameter under an entropy-energy inequality $EE(\Phi)$ for a suitable growth function Φ. It is obtained exactly as Proposition 6.6.1, p. 294, under logarithmic entropy-energy inequalities by means of the Herbst argument (Proposition 5.4.1, p. 252).

Proposition 7.2.1 *Let (E, μ, Γ) be a Markov Triple, with μ a probability measure, satisfying an entropy-energy inequality $EE(\Phi)$ where Φ is such that $\int_0^\infty \Phi(r) \frac{dr}{r^{3/2}} < \infty$. Then, for every integrable 1-Lipschitz function f and every $s > 0$,*

$$\int_E e^{sf} d\mu \leq \exp\left(s \int_E f \, d\mu + \frac{s}{4} \int_0^{s^2/4} \Phi(r) \frac{dr}{r^{3/2}}\right).$$

In particular,

$$D \leq \frac{1}{2} \int_0^\infty \Phi(r) \frac{dr}{r^{3/2}} = \int_0^\infty \Phi'(r) \frac{dr}{r^{1/2}} = \int_0^\infty \Psi(r) \frac{dr}{r^{3/2}}.$$

Recall that a standard logarithmic Sobolev inequality $LS(C)$ with $\Phi(r) = 2Cr$, $r \in (0, \infty)$, does not yield any finite diameter (for example the standard Gaussian measure in $\mathbb{R}^n$). As for Poincaré and logarithmic Sobolev inequalities, the Laplace transform bounds of Proposition 7.2.1 may be used towards tails estimates on $\mu(|f - \int_E f d\mu| \geq r)$, $r > 0$, at rates reflected by the growth of Φ. Further, and stronger, estimates will be illustrated later in the context of generalized Nash inequalities.

7.2.2 Off-diagonal Heat Kernel Bounds

The next theorem presents the announced off-diagonal estimates under an entropy-energy inequality $EE(\Phi)$. According to the previous discussion and Proposition 7.2.1, in the finite measure case the statement is restricted to the case when $EE(\Phi)$ implies a bounded diameter. The technique developed in the proof by a suitable transformation of the semigroup under a Lipschitz function turns out to be most efficient in many situations.

Theorem 7.2.2 *Let (E, μ, Γ) be a Markov Triple satisfying an entropy-energy inequality $EE(\Phi)$ where Φ is such that $\int^\infty \Phi'(r) \frac{dr}{\sqrt{r}} < \infty$. Setting*

7.2 Off-diagonal Heat Kernel Bounds

$U(s) = \int_s^\infty \Phi'(r) \frac{dr}{\sqrt{r}}$, $s > 0$, define $\tau = U^{-1}(d)$, $d = d(x, y)$, $(x, y) \in E \times E$, and

$$T(d) = \frac{1}{2} \int_\tau^\infty \frac{\Phi'(r)}{\sqrt{r(r-\tau)}} dr.$$

Furthermore, for $0 \leq t \leq T(d)$, let

$$H(t, d) = t(\delta - \tau) - \sqrt{\delta} d + \frac{1}{2} \int_\tau^\infty \Phi(r) \left(\sqrt{\delta} - \frac{\delta - \tau}{\sqrt{r + \delta - \tau}} \right) \frac{dr}{r^{3/2}}$$

where $\delta > 0$ is defined from t and d so that

$$t = \frac{1}{2} \int_\tau^\infty \frac{\Phi'(r)}{\sqrt{r(r+\delta-\tau)}} dr.$$

Then, for $0 < t \leq T(d(x, y))$, $(x, y) \in E \times E$, the density kernels $p_t(x, y)$, $t > 0$, of the semigroup $(P_t)_{t \geq 0}$ satisfy

$$\log p_t(x, y) \leq H(t, d(x, y))$$

for $(\mu \otimes \mu$-almost$)$ all $(x, y) \in E \times E$.

Observe that in the finite measure case, the diameter D is bounded from above by $U(0)$ according to Proposition 7.2.1 so that $T(d)$ is well-defined on $[0, D]$. Although the explicit bound put forward in Theorem 7.2.2 appears rather involved, Corollary 7.2.3 below will actually show that it leads to the expected precise off-diagonal bounds under Sobolev-type inequalities.

Proof For a function $h \in \mathcal{A}_0$ such that $\Gamma(h) \leq \delta$ where $\delta > 0$ will be specified later, consider the new semigroup $P_t^h f = e^{-h} P_t(e^h f)$, $t \geq 0$. This semigroup is no longer Markov, but is still positivity preserving. It is symmetric with respect to the measure $e^{2h} d\mu$, but this symmetry will actually not be used since we mostly work with the initial measure μ. The first task is to reach, under the $EE(\Phi)$ inequality, the bound

$$\| P_t^h \|_{1,\infty} \leq e^{m(t, \delta)} \qquad (7.2.2)$$

where $m(t, \delta)$ only depends on t and δ (and of course on the function Φ itself). Hence the density kernel $p_t^h(x, y)$ with respect to μ is uniformly bounded from above by $e^{m(t, \delta)}$. But, by construction, $p_t^h(x, y) = p_t(x, y) e^{h(y) - h(x)}$ so that, for all $t > 0$ and $(x, y) \in E \times E$,

$$p_t(x, y) \leq e^{m(t, \delta) + h(x) - h(y)}.$$

Now, by definition of the distance $d(x, y)$ ((3.3.9), p. 166),

$$\inf_{\Gamma(h) \leq \delta} \left[h(x) - h(y) \right] = -\sqrt{\delta} d(x, y).$$

Minimize then the previous bound on $p_t(x, y)$ on all functions h such that $\Gamma(h) \leq \delta$ to get that

$$p_t(x, y) \leq e^{m(t,\delta) - \sqrt{\delta} d(x,y)}. \tag{7.2.3}$$

The last step will then amount to optimizing δ.

The challenge is therefore to reach the upper bound (7.2.2) on $\|P_t^h\|_{1,\infty}$. To this end, we may proceed as in the proof of Theorem 7.1.2, retaining the same notation. For some positive function $f \in \mathcal{A}_0$, consider $\Lambda(t, q) = \int_E (P_t^h f)^q d\mu$, $t \geq 0$, where $q \in [1, \infty)$. From standard computations and integration by parts, setting $g = g(t) = P_t^h f$,

$$\partial_t \Lambda = -\frac{4(q-1)}{q} \int_E \Gamma(g^{q/2}) d\mu + q \int_E g^q \Gamma(h) d\mu$$

$$- q(q-2) \int_E g^{q-1} \Gamma(g, h) d\mu.$$

By $|\Gamma(g, h)| \leq \Gamma(g)^{1/2} \Gamma(h)^{1/2}$ and the Cauchy-Schwarz inequality in $\mathbb{L}^2(\mu)$,

$$\frac{q}{2} \left| \int_E g^{q-1} \Gamma(g, h) d\mu \right| = \left| \int_E g^{q/2} \Gamma(g^{q/2}, h) d\mu \right|$$

$$\leq \left(\int g^q \Gamma(h) d\mu \right)^{1/2} \left(\int \Gamma(g^{q/2}) d\mu \right)^{1/2}.$$

Linearizing this last inequality, for any $\alpha > 0$,

$$\partial_t \Lambda \leq \left[-\frac{4(q-1)}{q} + \frac{|q-2|}{\alpha} \right] \int_E \Gamma(g^{q/2}) d\mu + (q + \alpha|q-2|) \int_E g^q \Gamma(h) d\mu$$

$$\leq \left[-\frac{4(q-1)}{q} + \frac{|q-2|}{\alpha} \right] \int_E \Gamma(g^{q/2}) d\mu + \delta(q + \alpha|q-2|) \Lambda.$$

For $T > 0$, choose then an increasing function $t \mapsto q(t)$ with $q(0) = 1$ and $q(T) = \infty$ to be specified later and consider on $[0, T]$ the quantity $\tilde{\Lambda}(t) = \Lambda(t, q(t))$, $t \geq 0$. The inequality $EE(\Phi)$ in the form (7.1.2) leads to the differential inequality on $\tilde{\Lambda}$,

$$\tilde{\Lambda}' \leq \left[\frac{q'}{q} \Phi'(r) + \frac{|q-2|}{\alpha} - \frac{4(q-1)}{q} \right] \int_E \Gamma(g^{q/2}) d\mu$$

$$+ \left[\frac{q'}{q} \Psi(r) + \delta(q + \alpha|q-2|) \right] \tilde{\Lambda} + \frac{q'}{q} \tilde{\Lambda} \log \tilde{\Lambda}$$

depending on the two parameters $\alpha > 0$ and $r > 0$. These parameters may be chosen to depend upon q, and by further choosing $q = q(t)$ so that the coefficient of

7.2 Off-diagonal Heat Kernel Bounds

$\int_E \Gamma(g^{q/2})d\mu$ in the previous inequality vanishes, we end up with

$$\left(\frac{1}{q}\log\widetilde{\Lambda}\right)' \leq \frac{q'}{q^2}\Psi(r(q)) + \frac{\delta}{q}(q+\alpha(q)|q-2|)$$

where

$$\alpha(q) = \frac{q|q-2|}{4(q-1) - q'\Phi'(r(q))}. \qquad (7.2.4)$$

Setting

$$\alpha = \frac{q|q-2|s}{4(q-1)(s-1)}$$

where $s = s(q) > 1$ is a new function, (7.2.4) turns into

$$dt = s(q)\frac{\Phi'(r(q))}{4(q-1)}dq.$$

As a consequence, $\|P_T^h f\|_\infty \leq e^{m(T)}\int_E f d\mu$ where

$$T = \int_1^\infty s(q)\Phi'(r(q))\frac{dq}{4(q-1)}$$

and

$$m(T) = \int_1^\infty \left[\Psi(r(q))\frac{4(q-1)}{q^2} + \delta s(p)\Phi'(r(q))\left(1 + \frac{(q-2)^2 s(q)}{4(q-1)(s(q)-1)}\right)\right]\frac{dq}{4(q-1)}.$$

The final step is to choose optimal functions $r(q)$ and $s(q)$ which minimize the quantity $m(T)$ when $T > 0$ is fixed. The optimal choice once again does not depend on Φ, and is given, for some parameter $0 < \kappa < 1$, by

$$s = s(q) = 1 + \sqrt{\frac{qq^* - 4}{qq^* - 4\kappa}}, \qquad r = r(q) = \delta(1-\kappa)\frac{qq^*s}{4(2-s)}$$

where as usual q^* is the conjugate of q, that is $q^* = \frac{q}{q-1}$. After a change of variable, taking $\kappa \in (0,1)$ as a parameter,

$$\begin{cases} t = \frac{1}{2}\int_{1-\kappa}^\infty \Phi'(\delta w)\frac{dw}{\sqrt{w(w+\kappa)}}, \\ m = \delta\kappa t + \frac{1}{2}\int_{1-\kappa}^\infty \Phi(\delta w)\left[1 - \frac{\kappa}{\sqrt{w+\kappa}}\right]\frac{dw}{w^{3/2}}, \end{cases}$$

or, with $\kappa = 1 - \frac{\tau}{\delta}$ and $v = \delta w$,

$$\begin{cases} t = t(\delta, \tau) = \dfrac{1}{2} \displaystyle\int_\tau^\infty \Phi'(v) \dfrac{dv}{\sqrt{v(v+\delta-\tau)}}, \\ m = m(\delta, \tau) = \dfrac{1}{2} \displaystyle\int_\tau^\infty \Phi(v) \left[\sqrt{\delta} - \dfrac{\delta-\tau}{\sqrt{v+\delta-\tau}}\right] \dfrac{dv}{v^{3/2}} + (\delta-\tau)t. \end{cases}$$

Turning back to (7.2.3), the minimum in δ of $m(\delta, \tau) - \sqrt{\delta} d$ when $t(\delta, \tau)$ is fixed is obtained when

$$d = \frac{1}{2} \int_\tau^\infty \frac{\Phi(v)}{v^{3/2}} dv - \frac{\Phi(\tau)}{\sqrt{\tau}} = \int_\tau^\infty \frac{\Phi'(v)}{v^{1/2}} dv$$

from which the theorem is a direct consequence. The proof is complete. □

The following consequence of Theorem 7.2.2 applies to logarithmic entropy-energy inequalities equivalent to classical Sobolev-type inequalities (cf. Proposition 6.2.3, p. 281).

Corollary 7.2.3 *Under an entropy-energy inequality $EE(\Phi)$ with $\Phi(r) = \frac{n}{2}\log(A+Cr), r \in (0, \infty)$, there is a constant $B > 0$ such that for all $0 < t \leq 1$ and $(x, y) \in E \times E$,*

$$p_t(x, y) \leq B t^{-n} e^{-d^2(x,y)/4t}.$$

The details, a direct consequence of Theorem 7.2.2, are left to the reader. It is worth mentioning that the preceding corollary achieves the optimal $4t$ in the exponential factor, although the polynomial term is only of order t^{-n} and not $t^{-n/2}$ as in the uniform estimate (Corollary 7.1.5). This is not a failure of the method. Indeed, on a Riemannian manifold, outside the diagonal, $p_t(x, y)$ behaves when $t \to 0$ in a different way at the cut-locus of x. The correct exponent is $t^{-(n+p)/2}$, where p is the dimension of the manifold of geodesics which go from x to y. For example, $p = n - 1$ on two opposite points on an n-dimensional sphere.

7.2.3 Optimal Bounds for the Harmonic Distance

The distance function d used to obtain the off-diagonal estimates in Theorem 7.2.2 is built from the generator L of the Markov semigroup $(P_t)_{t \geq 0}$. Other kinds of pseudo-distances may be considered in the same way, such as for example, the *harmonic distance* $d^H(x, y)$ defined as

$$d^H(x, y) = \operatorname{esssup}\big[h(x) - h(y)\big], \quad (x, y) \in E \times E, \tag{7.2.5}$$

the esssup running over all functions h in $\mathcal{A}$ such that $Lh = 0$ and $\Gamma(h) \leq 1$. Of course $d^H = 0$ on a compact manifold with respect to the Laplace-Beltrami operator

7.2 Off-diagonal Heat Kernel Bounds

since every harmonic function is then constant. But in $\mathbb{R}^n$ with the usual Laplace operator, $d^H(x, y) = |x - y|$ since the linear functions are harmonic and attain the distance between x and y. With this definition, optimal heat kernel bounds may be achieved.

Proposition 7.2.4 *Let (E, μ, Γ) be a Markov Triple satisfying an entropy-energy inequality $EE(\Phi)$ for some growth function Φ such that $r^{-1}\Phi'(r)$ is integrable at infinity. Then, for every $\delta > 0$ with $m(\delta)$ and $t(\delta)$ given by (7.1.7), and provided both quantities are finite,*

$$p_{t(\delta)}(x, y) \leq e^{m(\delta) - d^H(x,y)^2/4t(\delta)}$$

for all $(x, y) \in E \times E$. In particular, for $\Phi(r) = \frac{n}{2} \log(\frac{2r}{n\pi e})$, $r \in (0, \infty)$, corresponding to the Euclidean logarithmic Sobolev inequality (6.2.8), p. 284, for all $t > 0$

$$p_t(x, y) \leq \frac{1}{(4\pi t)^{n/2}} e^{-d^H(x,y)^2/4t}$$

(which is the exact value of the Euclidean heat kernel).

Proof We only sketch the proof, staying at a somewhat informal level. Following the proof of Theorem 7.2.2, change $P_t f$ into $P_t^\gamma f = e^{-\gamma h} P_t(e^{\gamma h} f)$ where now $\Gamma(h) \leq 1$ and $Lh = 0$, for some real parameter γ. Denoting by L^γ the Markov generator of the semigroup P_t^γ, $t \geq 0$, for every $f > 0$ in the domain $\mathcal{D}(L)$ and in $\mathbb{L}^1(\mu) \cap \mathbb{L}^\infty(\mu)$, and every $q \in (1, \infty)$,

$$\int_E f^{q-1} L^\gamma f \, d\mu = \int_E f^{q-1} L f \, d\mu + \gamma^2 \int_E \Gamma(h) f^q \, d\mu.$$

Therefore, if, for every $r > 0$,

$$\mathrm{Ent}_\mu(f^q) \leq -\Phi'(r) \frac{q^2}{4(q-1)} \int_E f^{q-1} L f \, d\mu + \Psi(r) \int_E f^q \, d\mu,$$

then

$$\mathrm{Ent}_\mu(f^q) \leq -\Phi'(r) \frac{q^2}{4(q-1)} \int_E f^{q-1} L^\gamma f \, d\mu + \left(\Psi(r) + \gamma^2\right) \int_E f^q \, d\mu.$$

Hence, following the proof of Theorem 7.1.2, and with the same values of m and t,

$$\|P_t^\gamma\|_{1,\infty} \leq e^{m+\gamma^2 t},$$

or equivalently for the density kernels

$$p_t(x, y) \leq e^{m+\gamma^2 t - \gamma[h(x) - h(y)]}$$

for every $t > 0$ and $(x, y) \in E \times E$. It then remains to optimize first in h and then in γ to get the desired result. □

7.3 Examples

This section illustrates some of the results of the preceding sections, both at the level of the entropy-energy inequalities themselves and of their consequences for heat kernel bounds. We deal for simplicity with Markov Triples (E, μ, Γ) on the Euclidean space $E = \mathbb{R}^n$ with $\Gamma(f) = |\nabla f|^2$ and $d\mu = e^{-W} dx$ where $W : \mathbb{R}^n \to \mathbb{R}$ is a smooth potential.

The next proposition describes conditions on the growth of the potential W and its derivatives in order for an entropy-energy inequality $EE(\Phi)$ to hold for some function Φ.

Proposition 7.3.1 *In the preceding setting with $d\mu = e^{-W} dx$, let*

$$c(s) = \sup_{x \in \mathbb{R}^n} \left(\frac{1}{4\pi e s} \left[2\Delta W(x) - |\nabla W|^2(x) \right] + W(x) \right), \quad s > 0.$$

Then, an entropy-energy inequality $EE(\Phi)$ holds for any growth function $\Phi : (0, \infty) \to \mathbb{R}$ such that for any $r > 0$, there exists an $s > 0$ with

$$\Phi(r) \geq \frac{n}{2} \left(\log s - 1 + \frac{2r}{n\pi e s} \right) + c(s).$$

In the preceding, the function c may of course be replaced by any larger function, and the best choice for $\Phi(r)$, $r \in (0, \infty)$, is the concave envelope of the function

$$\inf_{s>0} \left[\frac{n}{2} (\log s - 1) + \frac{r}{\pi e s} + c(s) \right]$$

(which may then be turned into a $\mathcal{C}^1$-function).

Proof Start from the Euclidean logarithmic Sobolev inequality (6.2.8) of Proposition 6.2.5, p. 284, which indicates that for any smooth compactly supported function g on $\mathbb{R}^n$ such that $\int_{\mathbb{R}^n} g^2 dx = 1$,

$$\int_{\mathbb{R}^n} g^2 \log g^2 dx \leq \frac{n}{2} \log \left(\frac{2}{n\pi e} \int_{\mathbb{R}^n} |\nabla g|^2 dx \right).$$

As for (7.1.2), by concavity of the logarithmic function, for every $s > 0$,

$$\int_{\mathbb{R}^n} g^2 \log g^2 dx \leq \frac{n}{2} \left(\log s - 1 + \frac{2}{n\pi e s} \int_{\mathbb{R}^n} |\nabla g|^2 dx \right).$$

Set now $g^2 = f^2 e^{-W}$. Then $\int_{\mathbb{R}^n} f^2 d\mu = 1$ and

$$\int_{\mathbb{R}^n} |\nabla g|^2 dx = \int_{\mathbb{R}^n} |\nabla f|^2 d\mu + \frac{1}{2} \int_{\mathbb{R}^n} f^2 \Delta W d\mu - \frac{1}{4} \int_{\mathbb{R}^n} f^2 |\nabla W|^2 d\mu.$$

7.3 Examples

It follows that, for every $s > 0$,

$$\int_{\mathbb{R}^n} f^2 \log f^2 d\mu \leq \frac{n}{2}\left(\log s - 1 + \frac{2}{n\pi es}\int_{\mathbb{R}^n}|\nabla f|^2 d\mu\right)$$
$$+ \int_{\mathbb{R}^n} f^2\left(\frac{1}{4\pi es}[2\Delta W - |\nabla W|^2] + W\right)d\mu$$
$$\leq \frac{n}{2}\left(\log s - 1 + \frac{2}{n\pi es}\int_{\mathbb{R}^n}|\nabla f|^2 d\mu\right) + c(s).$$

The conclusion then follows from the very definition of Φ. $\square$

The above proof is not specifically tied to the Euclidean case and shows how to go from an entropy-energy inequality $EE(\Phi)$ on (E, μ, Γ) to a new one for (E, ν, Γ), where $d\nu = e^{-W}d\mu$.

Proposition 7.3.1 may be applied in several instances of interest, including the model family μ_α of (7.0.1) described in the introduction of this chapter. Indeed, if $W(x) = (1 + |x|^2)^{\alpha/2}$ with $\alpha > 2$, it is easily checked that $c(s) \leq C s^{\alpha/(\alpha-2)}$, $s \geq 1$. Here and below, $C > 0$ depends on n and α, and may change from line to line. As a consequence, μ_α with $\alpha > 2$ satisfies an $EE(\Phi)$ inequality with

$$\Phi(r) = C\left(1 + r^{\alpha/(2\alpha-2)}\right), \quad r \in (0, \infty). \tag{7.3.1}$$

(It may be checked furthermore that whenever $\alpha < 2$, μ_α does not satisfy any $EE(\Phi)$ inequality, while μ_2 satisfies a logarithmic Sobolev inequality, that is $EE(\Phi)$ with $\Phi(r) = Cr$, $r \in (0, \infty)$.) According to Corollary 7.1.4, for $0 < t \leq 1$,

$$\|P_t\|_{1,\infty} \leq e^{Ct^{-\alpha/(\alpha-2)}}.$$

Conversely, this bound for some $t > 0$ shows that $EE(\Phi)$ holds with (7.3.1), so that the heat kernel bound has the right order of magnitude as $t \to 0$. Finally, Proposition 7.2.1 easily shows that $\mu_\alpha(|x| \geq r) \leq Ce^{-r^\alpha/C}$, $r > 0$, confirming again the exponents of Φ.

Along the same lines, if $W(x) = |x|^2[\log(1 + |x|^2)]^\alpha$, $\alpha > 1$, we may choose $c(s) = C(1 + e^{bs^{1/\alpha}})$, $s > 0$, and $\Phi(r) = C(1 + r\log^{-\alpha}(e+r))$, $r \in (0, \infty)$, from which

$$\|P_t\|_{1,\infty} \leq \exp\left(e^{Ct^{-1/(\alpha-1)}}\right), \quad 0 < t \leq 1.$$

In the same spirit, it may be checked that for $\alpha = 2$, the semigroup is not ultracontractive but immediately hypercontractive (cf. p. 351).

The same method applies on a compact interval of the real line. For example, on $(0, 1)$ with $W(x) = -m\log x$ for small $x > 0$ and $W(x) = -p\log(1-x)$ for x near to 1, $m, p > 2$, there exists an entropy-energy inequality $EE(\Phi)$ with

$$\Phi(r) \leq \frac{n}{2}(C + \log r)$$

when $r \to \infty$, where $n = \max(m+1, p+1)$. In that case $c(s) = \frac{C}{s} + \frac{\kappa}{2} \log s$ where $\kappa = \max(m, p)$. On the other hand, if $W(x) = C_1 x^{-a}$ for small $x > 0$ and $W(x) = C_2(1-x)^{-b}$ for x near 1, $a, b > 0$, one may choose

$$\Phi(r) \leq C\left(1 + r^{\kappa/(2\kappa+2)}\right), \quad r > 0,$$

with $\kappa = \max(a, b)$. This last inequality is easily seen to be compatible with boundedness of the diameter. These examples indicate that the behavior of the function Φ at infinity reflects the behavior of the measure at the boundaries of the interval, the smaller the weight around the edges, the bigger the function Φ at infinity.

7.4 Beyond Nash Inequalities

As described in Chap. 6 (Proposition 6.2.3, p. 281), Sobolev inequalities for a Markov Triple (E, μ, Γ) may be described equivalently via logarithmic entropy-energy inequalities, but also as families of Nash (and Gagliardo-Nirenberg) inequalities. This section investigates generalized Nash inequalities on the model of general entropy-energy inequalities of the preceding section through a growth function Φ. In a second step, these inequalities will also be extended so as to introduce weights. This further generalization carries many interesting features, such as bounds on the heat kernels when there is no a priori uniform bounds, or significant relationships with spectral properties of the generator, as already discussed in Sect. 4.10, p. 220.

7.4.1 Generalized Nash Inequalities

The generalized Nash inequalities involve a C^1 increasing concave and positive growth function $\Phi : (0, \infty) \to \mathbb{R}_+$. The function $\Phi(r)$, $r > 0$, then has a finite limit at $r = 0$, denoted $\Phi(0)$.

Definition 7.4.1 (Generalized Nash inequality) A Markov Triple (E, μ, Γ) is said to satisfy a (generalized) Nash inequality $N(\Phi)$ with respect to a growth function $\Phi : (0, \infty) \to \mathbb{R}_+$ if for every $f \in \mathcal{D}(\mathcal{E})$ with $\|f\|_1 = 1$ and $\mathcal{E}(f) > 0$,

$$\int_E f^2 d\mu \leq \Phi\big(\mathcal{E}(f)\big).$$

According to the discussion following Proposition 6.2.3, p. 281, the Euclidean Nash inequality in $\mathbb{R}^n$ corresponds to $\Phi(r) = C r^{n/(n+2)}$, $r \in (0, \infty)$ (for some constant $C > 0$). In the finite measure case, the tight Nash inequality corresponds to $\Phi(r) = (1 + Cr)^{n/(n+2)}$, $r \in (0, \infty)$.

Observe that due to the normalization $\|f\|_1 = 1$, necessarily $\lim_{r \to \infty} \Phi(r) = \infty$ (since if not $\|f\|_2 \leq C\|f\|_1$, which may only occur on discrete spaces). Furthermore, it may well be that the Nash inequality $N(\Phi)$ only holds for functions f such that $\|f\|_2^2 \geq c\|f\|_1$ for some $c > 0$. Then Φ may be replaced by

7.4 Beyond Nash Inequalities

$\max(\Phi, a_0)$. However, the latter function is no longer concave on $(0, \infty)$. Provided that $\Psi(r) = \Phi(r) - r\Phi'(r) \to \infty$ as $r \to \infty$ (which is usually the case), Φ may be compensated by a linear part near the origin in order to make it concave.

As for the entropy-energy inequality $EE(\Phi)$, *linearized versions* of the Nash inequality $N(\Phi)$ may be considered. Whenever Φ' is bijective (if it is not bijective, use its generalized inverse) and takes values in some interval (a, b), then the $N(\Phi)$ inequality is equivalent to the family of inequalities

$$\int_E f^2 d\mu \leq s\mathcal{E}(f) + \beta(s), \quad s \in (a, b), \tag{7.4.1}$$

for every $f \in \mathcal{D}(\mathcal{E})$, $\|f\|_1 = 1$, where $\beta(s) = \Phi(\Phi'^{-1}(s)) - s\Phi'^{-1}(s)$.

In this form, such inequalities have already been used in Sect. 4.10, p. 220, to describe the spectrum of the associated generator L, and in particular to provide criterions related to its discreteness. Generalized Nash inequalities will be used in these linearized versions in the investigation of measure-capacity inequalities in Chap. 8.

For probability measures μ, generalized Nash inequalities are closely related to Poincaré inequalities (Definition 4.2.1, p. 181).

Proposition 7.4.2 *Let (E, μ, Γ) be a Markov Triple with μ a probability measure. A Poincaré inequality $P(C)$ with constant $C > 0$ is equivalent to a Nash inequality $N(\Phi)$ with $\Phi(r) = 1 + Cr$, $r \in (0, \infty)$.*

Proof The proof is straightforward. Under $P(C)$, for every $f \in \mathcal{D}(\mathcal{E})$ with $\|f\|_1 = 1$,

$$\int_E f^2 d\mu \leq C\mathcal{E}(f) + \left(\int_E f d\mu\right)^2 \leq C\mathcal{E}(f) + 1.$$

Conversely, apply $N(\Phi)$ with $\Phi(r) = 1 + Cr$ to $1 + \varepsilon f$ where f is a bounded function in $\mathcal{D}(\mathcal{E})$. For $\varepsilon > 0$ small enough so that $1 + \varepsilon f \geq 0$, $N(\Phi)$ boils down to

$$1 + 2\varepsilon \int_E f d\mu + \varepsilon^2 \int_E f^2 d\mu \leq \varepsilon^2 C\mathcal{E}(f) + 1 + 2\varepsilon \int_E f d\mu + \varepsilon^2 \left(\int_E f d\mu\right)^2$$

from which the Poincaré inequality $P(C)$ follows as $\varepsilon \to 0$. □

Remark 7.4.3 When μ is a probability, tightness in $N(\Phi)$ corresponds to the fact that $\lim_{r\to 0} \Phi(r) = 1$, while the Poincaré inequality requires moreover that Φ has a bounded slope at the origin. However, as for entropy-energy inequalities, as soon as $\lim_{r\to 0} \Phi(r) = 1$, Corollary 7.5.7 below shows that a Poincaré inequality also holds, and therefore that the growth function Φ may then be replaced by another one satisfying $\Phi(r) \leq 1 + Cr$, $r > 0$.

Generalized Nash inequalities may be compared to the entropy-energy inequalities of Sect. 7.1, and are sometimes equivalent to the latter. However, the Nash family covers more general examples than the entropy-energy inequalities since they include Poincaré inequalities while the entropy-energy inequalities at least require stronger logarithmic Sobolev inequalities (together with tightness).

Proposition 7.4.4 *Let (E, μ, Γ) be a Markov Triple satisfying the entropy-energy inequality $EE(\Phi_1)$. Then it also satisfies the Nash inequality $N(\Phi_2)$ where $\Phi_2(r) = r[U(\log r)]^{-1}$ with U the inverse function of $\Phi_1 + \log r$, $r \in (0, \infty)$. In particular, if μ is a probability measure, the tight $EE(\Phi_1)$ inequality ($\Phi_1(0) = 0$) implies the tight $N(\Phi_2)$ inequality ($\Phi_2(0) = 1$).*

Proof Recall as in the proof of Proposition 6.2.3, p. 281, that the convex function $\phi : r \mapsto \log(\|f\|_{1/r})$, $r \in (0, 1]$, is such that $\phi'(\frac{1}{2}) = -\frac{\text{Ent}_\mu(f^2)}{\|f\|_2^2}$. Hence, by convexity,

$$\log\left(\frac{\|f\|_2^2}{\|f\|_1^2}\right) \leq \frac{\text{Ent}_\mu(f^2)}{\|f\|_2^2}.$$

Therefore, for $f \in \mathcal{D}(\mathcal{E})$ such that $\|f\|_1 = 1$, setting $a = \|f\|_2^2$ and $b = \mathcal{E}(f)$, $EE(\Phi_1)$ indicates that $\log a \leq \Phi_1(\frac{b}{a})$, or in other words,

$$\log b \leq \Phi_1\left(\frac{b}{a}\right) + \log\left(\frac{b}{a}\right).$$

Since Φ_1 is increasing, so is $\Phi_1(r) + \log r$, $r \in (0, \infty)$, and therefore $U(\log b) \leq \frac{b}{a}$, which immediately leads to the announced claim. □

As an example, a (defective) logarithmic Sobolev inequality $LS(C, D)$ (Definition 5.1.1, p. 236), thus with $\Phi_1(r) = 2Cr + D$, $r \in (0, \infty)$, leads to a Nash inequality $N(\Phi_2)$ with $\Phi_2(r) \sim C_1 \frac{r}{\log r}$ as $r \to \infty$. In the same way, $\Phi_1(r) = \frac{n}{2} \log Cr$, $r \in (0, \infty)$, leads to $\Phi_2(r) \sim C_1 r^{n/(n+2)}$ as $r \to \infty$ (classical Nash inequalities), while $\Phi_1(r) = Cr^\alpha$, $r \in (0, \infty)$, $\alpha > 0$, leads to $\Phi_2(r) \sim C_1 r(\log r)^{-1/\alpha}$ as $r \to \infty$.

At least when μ is a probability measure, the implication in the preceding proposition may easily be reversed to show that under a Nash inequality $N(\Phi_2)$, and provided $\Phi_2 \log(\Phi_2)$ is concave at infinity, an entropy-energy inequality $EE(\Phi_1)$ holds with $\Phi_1(r) = ae^{-r}\Phi_2^{-1}(e^r) + b$, $r \in (0, \infty)$, for constants $a, b > 0$. This is an easy consequence of Proposition 5.1.8, p. 241, and the linearization forms of both inequalities. Unfortunately, this conclusion is too rough for example to reach the dimensional Sobolev parameter from the classical Nash inequality and $\Phi_2(r) = C(1 + r)^{n/(n+2)}$ only yields $\Phi_1(r) \leq \frac{an}{2} \log(C_1(1 + r)) + b$, $r > 0$ (although $a = 1$ may actually be reached with some further analysis).

7.4.2 Heat Kernel Bounds and Tail Inequalities Under Generalized Nash Inequalities

As for the classical Nash inequalities in the proof of Theorem 6.3.1, p. 286, the generalized Nash inequalities $N(\Phi)$ may be used to obtain heat kernel bounds and ultracontractivity provided Φ does not grow too fast at infinity. While easier to obtain than under an entropy-energy inequality $EE(\Phi)$, the bounds are in general less precise (and do not reveal any information on the $\|\cdot\|_{p,q}$ norms). However, as described in the next section, the same method may provide non-uniform bounds on the heat kernel when the latter is not necessarily bounded. The next statement is a first illustration of this phenomenon.

Theorem 7.4.5 *Let (E, μ, Γ) be a Markov Triple satisfying a Nash inequality $N(\Phi)$ with growth function $\Phi : (0, \infty) \to \mathbb{R}_+$ such that $\int^\infty \Phi'(r)\frac{dr}{r} < \infty$. Let U be the (generalized) inverse of Φ, defined on $(\Phi(0), \infty)$, and define then $K(s)$, $s > 0$, to be the (decreasing) inverse function of $\int_{U(s)}^\infty \Phi'(r)\frac{dr}{r}$. Then, for every $t > 0$,*

$$\|P_t\|_{1,2}^2 \leq K(2t).$$

In particular, the density kernel $p_t(x, y)$, $t > 0$, $(x, y) \in E \times E$, of P_t with respect to the invariant measure μ is uniformly bounded from above by $K(t)$.

Observe from the hypotheses that since $U(\Phi(0)) = 0$, $\int_0^\infty \Phi'(r)\frac{dr}{r} = \infty$, so that the function K is a decreasing bijection from $(0, \infty)$ onto $(\Phi(0), \infty)$.

Proof The proof is very similar to that of the corresponding result for the classical Nash inequalities in Theorem 6.3.1, p. 286, so we shall only sketch it. For a positive function f in $\mathcal{D}(\mathcal{E})$ with $\int_E f d\mu = 1$, consider $\Lambda(t) = \int_E (P_t f)^2 d\mu$, $t \geq 0$. The $N(\Phi)$ inequality turns into the differential inequality $\Lambda \leq \Phi(-\frac{\Lambda'}{2})$. Fix now $t > 0$. Since $K(2t) \in [\Phi(0), \infty)$, if $\Lambda(t) \leq \Phi(0)$ there is nothing to prove. If not, since Λ is decreasing, $\Lambda(s) > \Phi(0)$ for $s \in (0, t)$. Therefore, by definition of the inverse function U, $\Lambda'(s) \leq -2U(\Lambda(s))$ for $s \in (0, t)$. Introducing the decreasing function

$$H(s) = \int_s^\infty \frac{du}{U(u)} = \int_{\Phi^{-1}(s)}^\infty \Phi'(r)\frac{dr}{r}, \quad s \in (\Phi(0), \infty),$$

it follows that $H(\Lambda(s))' \geq 2$ when $\Lambda(s) \geq \Phi(0)$. Therefore, $\Lambda(t) \leq \Phi(0)$ or $H(\Lambda(t)) \geq 2t + H(\Lambda(0)) \geq 2t$, $t > 0$. By definition of K, $\Lambda(t) \leq K(2t)$ and the upper bounds on the heat kernels are then deduced as usual. □

Remark 7.4.6 When the function Φ is such that $\int^\infty \Phi'(r)\frac{dr}{r} = \infty$, the power of the preceding strategy actually still yields useful, although weaker, results. Namely, choosing any real $a > 0$ and letting then $K(s)$, $s > 0$, be the (decreasing) inverse function of $H(s) = \int_{U(s)}^a \Phi'(r)\frac{dr}{r}$, for any f with $\|f\|_1 = 1$ and any $t > 0$,

$$\|P_t f\|_2^2 \leq K\big(H(\|f\|_2^2 + 2t)\big).$$

The latter is even equivalent to the generalized Nash inequality $N(\Phi)$ as $t \to 0$. The advantage of the previous form with respect of this last one is that under the hypothesis of Theorem 7.4.5, the bound on $\|P_t f\|_2^2$ no longer depends on $\|f\|_2^2$.

Generalized Nash inequalities may also be used to obtain tail estimates for Lipschitz functions. The following statement goes somewhat beyond Proposition 7.2.1 in case of entropy-energy inequalities.

Proposition 7.4.7 *Let (E, μ, Γ) be a Markov Triple, with μ a probability measure, satisfying a Nash inequality $N(\Phi)$, and let f be a 1-Lipschitz function on E. Let $r_0 > 0$ be such that $q_0 = \mu(f \geq r_0) \leq \frac{1}{2\Phi(0)}$. Then, for every $r \geq r_0$,*

$$\mu(f \geq r) \leq 2 F\big((r - r_0) \log 2\big)$$

where F is the inverse function of

$$G(v) = \int_{v/2}^{q_0} \left[u^3 \Phi^{-1}\left(\frac{1}{2u}\right) \right]^{-1/2} du, \quad v \in (0, q_0).$$

Whenever $\int_0^{\frac{1}{2\Phi(0)}} [u^3 \Phi^{-1}(\frac{1}{2u})]^{-1/2} du < \infty$, then f is bounded (μ-almost everywhere), and in particular the diameter of (E, μ, Γ) is finite.

In this statement, F is defined on $(0, A)$, where $A = \int_0^{q_0} [u^3 \Phi^{-1}(\frac{1}{2u})]^{-1/2} du$. When $A < \infty$, it is extended by 0 on $[A, \infty)$.

Proof For $r, s > 0$, apply $N(\Phi)$ to $g = \frac{1}{s}[(f - r)_+ \wedge s]$ where $f \in \mathcal{D}(\mathcal{E})$. Since $\mathbb{1}_{\{f \geq r+s\}} \leq g \leq \mathbb{1}_{\{f \geq r\}}$ and $\Gamma(g) \leq s^{-2} \mathbb{1}_{\{f \geq r\}}$, using that Φ is concave, and therefore that $u^{-1} \Phi(u)$ is decreasing, it follows that

$$q(r + s) \leq q^2(r) \, \Phi\left(\frac{1}{s^2 q(r)}\right)$$

where $q(u) = \mu(f \geq u)$, $u > 0$. Choose then

$$s = \left[q(r) \, \Phi^{-1}\left(\frac{1}{2q(r)}\right) \right]^{-1/2}$$

so that $q(r + s) \leq \frac{q(r)}{2}$. Now, with $r(q) = \inf\{u > 0 \,;\, q(u) \leq q\}$, $q > 0$, which plays the role of the inverse function of $q(r)$, the previous inequality indicates that

$$r\left(\frac{q}{2}\right) \leq r(q) + \left[q \, \Phi^{-1}\left(\frac{1}{2q}\right) \right]^{-1/2}.$$

7.4 Beyond Nash Inequalities

Recalling that $u^{-1}\Phi(u)$ is decreasing,

$$\left[q\Phi^{-1}\left(\frac{1}{2q}\right)\right]^{-1/2} = \left[q\Phi^{-1}\left(\frac{1}{2q}\right)\right]^{-1/2}\frac{1}{\log 2}\int_{q/2}^{q}\frac{du}{u}$$

$$\leq \frac{1}{\log 2}\int_{q/2}^{q}\left[u^3\Phi^{-1}\left(\frac{1}{2u}\right)\right]^{-1/2}du.$$

Applying the same procedure on $(\frac{q}{4}, \frac{q}{2})$ and iterating yields that

$$\left(r(2^{-k}q) - r(q)\right)\log 2 \leq \int_{2^{-k}q}^{q}\left[u^3\Phi^{-1}\left(\frac{1}{2u}\right)\right]^{-1/2}du$$

for any $k \geq 1$. Using that $r(q)$ is decreasing, it finally follows that for any $q_1 < q$

$$\left(r(q_1) - r(q)\right)\log 2 \leq \int_{q_1/2}^{q}\left[u^3\Phi^{-1}\left(\frac{1}{2u}\right)\right]^{-1/2}du$$

which yields the announced claim. □

To conclude this section, we illustrate some of the previous results again on the family μ_α defined in (7.0.1), in dimension one for simplicity. As mentioned earlier (p. 363), these measures μ_α with $\alpha \in [1, 2)$ on the real line do not satisfy any entropy-energy inequality (with respect to the standard carré du champ operator $\Gamma(f) = f'^2$). However an entropy-energy inequality $EE(\Phi_1)$ holds when $\alpha \geq 2$ with $\Phi_1(r) = C(1 + r^{\alpha/(2\alpha-2)})$, $r \in (0, \infty)$, and in turn a generalized Nash inequality $N(\Phi_2)$ with $\Phi_2 \sim r(\log r)^{-2(\alpha-1)/\alpha}$ at infinity (Proposition 7.4.4). For this growth function, the tail estimates produced by Proposition 7.4.7 provide the correct behavior of μ_α.

Using arguments close to those developed for the Muckenhoupt criterion (cf. Sect. 4.5.1, p. 194), it is not difficult to see that this Nash inequality actually extends to every $\alpha \in [1, 2)$ (and is optimal according to the tail estimates of Proposition 7.4.7). We only sketch the method, which may be used in other, similar, instances. In what follows, the constant $C > 0$ depends only on α and may vary from place to place. It is enough to deal with a smooth function f on $\mathbb{R}$ such that $f(0) = 0$ and $\int_\mathbb{R} f^2 d\mu_\alpha = 1$. We may moreover work separately on $[0, \infty)$ and $(-\infty, 0]$, say $[0, \infty)$ (thus assuming $\int_0^\infty f^2 d\mu_\alpha = 1$). If $p(x) = e^{-(1+x^2)^{\alpha/2}}$ is the density of μ_α (up to the normalizing constant), the only ingredient is actually the tail estimate $\mu_\alpha([x, \infty)) \leq Cx^{1-\alpha}p(x)$, $x > 0$. For some parameter $a > 0$, since $f(0) = 0$, write

$$\int_0^\infty f^2 d\mu_\alpha = \int_0^\infty f^2 \mathbb{1}_{\{|f|\leq a\}} d\mu_\alpha$$
$$+ 2\int_0^\infty f(x)f'(x)\left(\int_x^\infty \mathbb{1}_{\{|f|>a\}}d\mu_\alpha\right)dx. \quad (7.4.2)$$

The first integral is bounded above by $a\|f\|_1$, while for the second, use that

$$\int_x^\infty \mathbb{1}_{\{|f|>a\}} d\mu_\alpha \leq \min\bigl(\mu_\alpha([x,\infty)), a^{-2}\bigr) \leq C \min\bigl(p(x)x^{1-\alpha}, a^{-2}\bigr).$$

If $x_0 = x_0(a)$ is the point where $x^{1-\alpha} p(x) = a^{-2}$, then the right-hand side in the latter inequality is bounded from above by $\frac{Cp(x)}{a^2 p(x_0)}$ on $[0, x_0)$ and by $Cx_0^{1-\alpha} p(x)$ on $[x_0, \infty)$, hence bounded everywhere by $x_0^{1-\alpha} p(x)$. By the Cauchy-Schwarz inequality, the second integral on the right-hand side of (7.4.2) is then bounded from above by $Cx_0^{1-\alpha} \mathcal{E}(f)^{1/2}$, and therefore

$$1 = \int_0^\infty f^2 d\mu_\alpha \leq a\|f\|_1 + C x_0^{1-\alpha} \mathcal{E}(f)^{1/2}.$$

It remains to replace x_0 by its value in terms of a and to optimize in a to get that

$$\int_0^\infty f^2 d\mu_\alpha \leq \Phi\bigl(\mathcal{E}(f)\bigr)$$

for every $f \in \mathcal{D}(\mathcal{E})$ with $\|f\|_1 = 1$ and with $\Phi(r) \leq Cr(\log r)^{-2(\alpha-1)/\alpha}$ for large r. The announced claim follows.

7.4.3 Weighted Nash Inequalities

The main advantage of generalized Nash inequalities with respect to entropy-energy inequalities is that they may address measures which do not satisfy any logarithmic Sobolev inequality. However, as far as ultracontractive bounds are concerned, they do not offer much more up to now. But with some further mild extension, the same kind of ideas may lead to heat kernel upper bounds which are not uniform. In particular, non-ultracontractive settings such as $E = \mathbb{R}$, $\Gamma(f) = f'^2$ and $d\mu_\alpha = c_\alpha e^{-(1+x^2)^{\alpha/2}}$, $\alpha \in (1,2]$, on the line may be covered.

In this respect, this short section is concerned with an extension of the Nash inequalities by the introduction of a weight function which offers more flexibility and produces further heat kernel bounds. It deals with a Markov Triple (E, μ, Γ) where μ is a probability measure. Here, a *weight* function will be a positive (measurable) function $w : E \to \mathbb{R}_+$ such that $P_t w \leq L(t) w$, $t \geq 0$, for some increasing and positive function L on $\mathbb{R}_+$. The next definition of a generalized weighted Nash inequality is entirely similar to the generalized Nash inequality provided the $\mathbb{L}^1(\mu)$-norm is weighted by w. It involves similarly a $\mathcal{C}^1$ increasing and concave growth function $\Phi : (0, \infty) \to \mathbb{R}_+$.

Definition 7.4.8 (Weighted Nash inequality) A Markov Triple (E, μ, Γ), with μ a probability measure, is said to satisfy a (generalized) weighted Nash inequality

7.4 Beyond Nash Inequalities

$N(\Phi, w)$ with growth function $\Phi : (0, \infty) \to \mathbb{R}_+$ and weight $w : E \to \mathbb{R}_+$ if, for every $f \in \mathcal{D}(\mathcal{E})$ with $\int_E w|f|d\mu = 1$ and $\mathcal{E}(f) > 0$,

$$\int_E f^2 d\mu \leq \Phi(\mathcal{E}(f)).$$

Such inequalities were already considered earlier in Sect. 4.10 (cf. (4.10.4), p. 224) in connection with emptyness of the essential spectrum. The extra condition $P_t w \leq L(t) w$ on the weight, together with special properties of the growth function Φ, will actually allow for quantitative estimates on the (discrete) spectrum, as well as pointwise estimates on the density kernels $p_t(x, y)$.

The next statement is very similar to Theorem 7.4.5 and borrows from the latter the definition of $K(t)$, $t \geq 0$.

Proposition 7.4.9 *Let* (E, μ, Γ) *be a Markov Triple satisfying a weighted Nash inequality* $N(\Phi, w)$ *with growth function* $\Phi : (0, \infty) \to \mathbb{R}_+$ *such that* $r^{-1}\Phi'(r)$ *is integrable at infinity. Then* $(P_t)_{t \geq 0}$ *has density kernels* $p_t(x, y)$, $t > 0$, $(x, y) \in E \times E$, *such that, for every* $t > 0$ *and* $(x, y) \in E \times E$,

$$p_t(x, y) \leq K(t) L\left(\frac{t}{2}\right)^2 w(x) w(y), \qquad (7.4.3)$$

Proof The proof follows that of Theorem 7.4.5 and the method developed for the proof of Theorem 7.2.2. For f positive in $\mathcal{D}(\mathcal{E})$ with $\int_E f w d\mu = 1$ and $t > 0$, let $\Lambda(s) = L(t)^{-2} \int_E (P_s f)^2 d\mu$, $s \in [0, t]$. By symmetry and the hypothesis on the weight w, for every $s \leq t$, $\int_E w P_s f d\mu \leq L(t) \int_E w f d\mu$. Arguing then exactly as in the proof of Theorem 7.4.5 now under $N(\Phi, w)$ and using that $r \mapsto r^{-1} \Phi(r)$ is decreasing yields that

$$\|P_t f\|_2^2 \leq K(2t) L(t)^2$$

for every $t > 0$. Introduce then the new semigroup $Q_t f = \frac{1}{w} P_t(wf)$, $t \geq 0$, symmetric with respect to the weighted measure $dv = w^2 d\mu$. In terms of the semigroup $(Q_t)_{t \geq 0}$, and with $\mathbb{L}^p$-norms with respect to v, the preceding reads

$$\|Q_t f\|_2^2 \leq K(2t) L(t)^2$$

for every $t > 0$ and f such that $\|f\|_1 = 1$. This leads to an ultracontractive bound on Q_{2t}, $t > 0$, and therefore to the uniform bound $K(2t) L(t)^2$ on its kernel with respect to v. But if the semigroup $(Q_t)_{t \geq 0}$ has a kernel bounded from above by $\widetilde{K}$, then $(P_t)_{t \geq 0}$ has a density kernel, with respect to μ, which is bounded from above by $\widetilde{K} w(x) w(y)$. The conclusion then easily follows and Proposition 7.4.9 is established. □

Proposition 7.4.9 is most useful when $w \in \mathbb{L}^2(\mu)$, in which case the condition $P_t w \leq e^{ct} w$, $t \geq 0$, holds as soon as $w \in \mathcal{D}(L)$ and $Lw \leq cw$. The function w is thus similar to the Lyapunov functions of Sect. 4.6, p. 201. Furthermore, since $r^{-1}\Phi(r)$, $r > 0$, is decreasing (as Φ is concave and increasing), the condition $P_t w \leq L(t) w$, $t \geq 0$, may be replaced by $P_t w_1 \leq L_1(t) w_1$, $t \geq 0$, for some new weight $w_1 \geq w$. In particular, adding a constant, it may always be assumed that w is bounded from below by some strictly positive constant.

Corollary 7.4.10 *Under the hypotheses of Proposition 7.4.9, and assuming furthermore that $w \in \mathbb{L}^2(\mu)$, the semigroup $\mathbf{P} = (P_t)_{t \geq 0}$ is Hilbert-Schmidt, and if $(\lambda_k)_{k \in \mathbb{N}}$ denotes the sequence of eigenvalues of $-\mathrm{L}$,*

$$\sum_{k \in \mathbb{N}} e^{-\lambda_k t} \leq K(t) L\left(\frac{t}{2}\right)^2 \int_E w^2 d\mu$$

for every $t > 0$.

The proof is immediate. As $w \in \mathbb{L}^2(\mu)$, the upper bound (7.4.3) on $p_t(x, y)$ is in $\mathbb{L}^2(\mu \otimes \mu)$, and therefore P_t is Hilbert-Schmidt for every $t > 0$. The upper bound on the eigenvalues follows from the trace formula $\int_E p_t(x, x) d\mu(x) = \sum_{k \in \mathbb{N}} e^{-\lambda_k t}$ (see (1.7.4), p. 32).

Two-sided estimates may be obtained as in the case of standard Nash inequalities as soon as a Poincaré inequality holds. Recalling from the proof of Proposition 7.4.9 the semigroup $Q_t f = \frac{1}{w} P_t(wf)$, $t \geq 0$, symmetric with respect to the measure $d\nu = w^2 d\mu$, since $(Q_t)_{t \geq 0}$ and $(P_t)_{t \geq 0}$ share the same spectral properties, if a Poincaré inequality holds for $(P_t)_{t \geq 0}$ with constant $C > 0$, then $\|Q_t - Q_\infty\|_{2,2} \leq K e^{-t/C}$, $t \geq 0$, with $Q_\infty f = \frac{1}{w} \int_E wf d\mu$, the operator norm being understood in $\mathbb{L}^2(\nu)$. As soon as Q_T is bounded from $\mathbb{L}^1(\nu)$ into $\mathbb{L}^2(\nu)$ for some $T > 0$, then using the decomposition

$$Q_{t+T} - Q_\infty = Q_T (Q_t - Q_\infty) Q_T,$$

$\|Q_t - Q_\infty\|_{1,\infty} \leq K e^{-t/C}$ for large $t \geq T$. A uniform bound for the modulus of the density kernel of $Q_t - Q_\infty$ follows for large t, which in turn may be translated into a bound for the density kernel of P_t. The conclusion is summarized in the next statement with a proof similar to the one given for Proposition 6.3.4, p. 289.

Corollary 7.4.11 *Under the hypotheses of Proposition 7.4.9, assuming furthermore that (E, μ, Γ) satisfies a Poincaré inequality $P(C)$ for some $C > 0$, there exist a constant $K > 0$ and $T > 0$ such that, for $t \geq T$, the density kernel $p_t(x, y)$ of P_t, satisfies*

$$|p_t(x, y) - 1| \leq K e^{-C/t} w(x) w(y)$$

for all $(x, y) \in E \times E$.

Of course, since in general w is not bounded and goes to infinity at infinity, the lower bound is only useful (for a given $t > 0$) on a compact subset of $E \times E$, this subset enlarging to the full space when t goes to infinity.

7.5 Weak Poincaré Inequalities

We conclude this section with some examples. Starting from a classical Nash inequality for a given operator L, it is not hard to deduce weighted Nash inequalities for the operator $L - \Gamma(W, \cdot)$, that is to pass from a Nash inequality for some measure μ to a weighted Nash inequality for the (weighted) measure $e^{-W} d\mu$. The procedure is similar to the technique developed for entropy-energy inequalities in Sect. 7.3. We briefly illustrate it in $\mathbb{R}^n$ for the Lebesgue measure dx and the usual Nash inequality

$$\|f\|_2^{2+n} \leq C_n \|\nabla f\|_2^n \|f\|_1^2$$

from (6.2.4), p. 281 (the norms being understood with respect to dx). Considering $d\mu = e^{-W} dx$, choose $w = e^{W/2}$ as weight function. Changing then f into $fe^{-W/2}$ in the above Nash inequality, after integration by parts, it follows that for every smooth compactly supported function $f : \mathbb{R}^n \to \mathbb{R}$,

$$\left(\int_{\mathbb{R}^n} f^2 d\mu \right)^{(n+2)/n} \leq C_n \left(\int_{\mathbb{R}^n} |\nabla f|^2 d\mu + \int_{\mathbb{R}^n} f^2 R \, d\mu \right) \left(\int_{\mathbb{R}^n} |f| w \, d\mu \right)^{4/n}$$

(7.4.4)

where $R = \frac{1}{2} \Delta W - \frac{1}{4} |\nabla W|^2$. Provided then R is uniformly bounded, say by $c > 0$, a weighted Nash inequality for μ with weight $w = e^{W/2}$ and growth function $\Phi(r) = C(1+r)^{n/(n+2)}$, $r \in (0, \infty)$, holds, where $C > 0$ depends on n, C_n and c. There is no claim here that the weight w satisfies $P_t w \leq L(t) w$, which has to be studied separately (and is in general easy to check in practice). The weight w is never in $\mathbb{L}^2(\mu)$, so that the latter is on the other hand of little use for the control of the spectrum.

We do not expand more on weighted Nash inequalities. On the real line $\mathbb{R}$ with $\Gamma(f) = f'^2$ and $d\mu_\alpha(x) = p(x) dx$ with $p(x) = c_\alpha e^{-(1+x^2)^{\alpha/2}}$ (defined at (7.0.1)), one may obtain weighted Nash inequalities with $w(x) = p(x)^{1/2} (1+x^2)^{-\beta}$ for any $\beta > 0$. In this case, for constants $C > 0$ and $\delta \in (0, 1)$ depending on α and β, one may choose $\Phi(r) = C(1+r)^\delta$, $r \in (0, \infty)$. Thanks to the introduction of the weight, the function Φ may be chosen to be similar to the one used for the classical Nash inequalities, even in the case $\alpha > 2$ where ultracontractive bounds hold. But it also extends to non-ultracontractive settings corresponding to $1 < \alpha \leq 2$. When $\alpha = 1$ however, it is known that the spectrum of the associated operator is not discrete (cf. p. 189).

7.5 Weak Poincaré Inequalities

For simplicity of exposition, we deal here with a Markov Triple (E, μ, Γ) (Standard would be enough) where μ is a probability measure. Similar (useful) statements may be obtained in the sigma-finite case (in which case variances have to be replaced by $\mathbb{L}^2(\mu)$-norms).

Many functional inequalities encountered throughout this book may be used to control convergence to equilibrium. This property is related to tightness in the inequalities. As often used and described, tightness is a way to express how functions f converge to a constant when $\mathcal{E}(f) \to 0$. Up to now, we mainly used Poincaré inequalities to tighten the various functional inequalities under investigation (logarithmic Sobolev, Sobolev, entropy-energy and so on, cf. e.g. Proposition 5.1.3, p. 238, and Proposition 6.2.2, p. 280). The weak Poincaré inequalities developed in this section actually appear as the minimal requirement to ensure tightness. Also, by Proposition 4.4.2, p. 190, Poincaré inequalities require at least exponential decay on the tail behavior of the measure μ. When dealing with sub-exponential or polynomial decays, none of the inequalities investigated so far reveal any information about convergence to equilibrium. On the other hand, ergodicity, as presented in Definition 3.1.11, p. 135, is the minimal requirement that one may ask for in order to obtain convergence to equilibrium. But this property is purely qualitative. Weak Poincaré inequalities developed here are in some sense a way to quantify ergodicity, in the weakest possible sense. Finally, weak Poincaré inequalities essentially aim at capturing the behavior of Poincaré inequality constants on large compact sets (typically on large balls in Euclidean space). A Poincaré inequality holds when those constants do not diverge at infinity, and weak Poincaré inequalities describe the behavior of those constants when they diverge.

This section thus develops the tool of weak Poincaré inequalities, how they are used to tighten functional inequalities, and what type of information they entail.

7.5.1 Weak Poincaré Inequalities

At a technical level, weak Poincaré inequalities control how a sequence $(f_k)_{k \in \mathbb{N}}$ of (suitable) functions converge to a constant in the weakest sense (in measure) when $\mathcal{E}(f_k) \to 0$. Since $\mathcal{E}(f) = \mathcal{E}(f - a)$ for any $a \in \mathbb{R}$, the control of $\mathcal{E}(f_k)$ cannot say anything about the control of this constant, justifying the investigation of a tool such as weak Poincaré inequalities.

Before presenting the notion of weak Poincaré inequality itself, it is useful to study, as a preliminary investigation, some aspects of convergence in measure. Recall that a sequence $(f_k)_{k \in \mathbb{N}}$ of real-valued measurable functions on $(E, \mathcal{F}, \mu)$ converges in measure (or in probability, or in $\mathbb{L}^0(\mu)$) to a measurable function g, if, for any $\varepsilon > 0$, $\lim_{k \to \infty} \mu(|f_k - g| \geq \varepsilon) = 0$. Convergence in measure is metrizable, and $\int_E [|f - g| \wedge 1] d\mu$ is one example of a distance defining the $\mathbb{L}^0(\mu)$-topology.

The first statement tries to characterize the oscillation of a given function under convergence in measure. Dealing with functions which are a priori not integrable, we work instead with medians. Given f measurable on $(E, \mathcal{F})$, the set of $m \in \mathbb{R}$ such that $\mu(f \geq m) \geq \frac{1}{2}$ and $\mu(f \leq m) \geq \frac{1}{2}$ is a non-empty bounded interval (possibly reduced to one point). For simplicity, we agree to define the *median* $m(f)$ of f to be the middle point of this interval. On the other hand, for a given measurable

7.5 Weak Poincaré Inequalities

function $f : E \to \mathbb{R}$, set $\tau(f) = (f \wedge 1) \vee (-1)$, and introduce

$$T_\mu(f) = \mathrm{Var}_\mu\big(\tau(f)\big) \quad \text{and} \quad T^*_\mu(f) = \sup_{a \in \mathbb{R}} \mathrm{Var}_\mu\big(\tau(f-a)\big).$$

Obviously $T_\mu(f) \leq T^*_\mu(f) \leq 1$.

Proposition 7.5.1 *If a sequence $(f_k)_{k \in \mathbb{N}}$ of measurable functions on $(E, \mathcal{F})$ converges to 0 in measure, then $\lim_{k \to \infty} m(f_k) = 0$ and $\lim_{k \to \infty} T^*_\mu(f_k) = 0$. Conversely, if $\lim_{k \to \infty} m(f_k) = 0$ and $\lim_{k \to \infty} T_\mu(f_k) = 0$, then $(f_k)_{k \in \mathbb{N}}$ converges to 0 in measure. If the sequence $(f_k)_{k \in \mathbb{N}}$ is uniformly integrable, $m(f_k)$ may be replaced by $\int_E f_k d\mu$ in the latter implication.*

Recall that a sequence $(f_k)_{k \in \mathbb{N}}$ in $\mathbb{L}^1(\mu)$ (where μ is a probability measure) is uniformly integrable if $\lim_{c \to \infty} \sup_{k \in \mathbb{N}} \int_{\{|f_k| > c\}} |f_k| d\mu = 0$.

Proof Assuming that $f_k \to 0$ in measure, if $0 < \varepsilon \leq \frac{1}{3}$, for every k large enough, $\mu(|f_k| \geq \varepsilon) \leq \varepsilon \leq \frac{1}{3}$ and therefore $|m(f_k)| \leq \varepsilon$. Furthermore, since the map $r \mapsto \tau(r)$ is 1-Lipschitz, for every $a \in \mathbb{R}$, $|\tau(f_k - a) - \tau(-a)| \leq \varepsilon$ on a set of measure larger than $1 - \varepsilon$ and is bounded by 2 on its complement. It easily follows that for every k large enough, $\mathrm{Var}_\mu(\tau(f_k - a)) \leq 5\varepsilon$ uniformly in $a \in \mathbb{R}$, yielding the first claim.

For the converse implication, under the condition $\lim_{k \to \infty} \mathrm{Var}_\mu(\tau(f_k)) = 0$, by Markov's inequality, $\lim_{k \to \infty} \mu(|\tau(f_k) - \int_E \tau(f_k) d\mu| \geq \varepsilon) = 0$ for every $\varepsilon > 0$. Since the sequence $\int_E \tau(f_k) d\mu$, $k \in \mathbb{N}$, is bounded by 1, there exists a subsequence which converges to some $\alpha \in \mathbb{R}$. If $\alpha \neq 0$, then along a subsequence $\lim_{k \to \infty} \mu(|\tau(f_k) - \alpha| \geq \varepsilon) = 0$ contradicting $\lim_{k \to \infty} m(f_k) = 0$. Therefore $\lim_{k \to \infty} \int_E \tau(f_k) d\mu = 0$, and hence $\tau(f_k) \to 0$ in measure and thus also $f_k \to 0$. The last claim of the proposition follows in the same way. Namely, if along a subsequence $\lim_{k \to \infty} \int_E \tau(f_k) d\mu = 2\alpha \neq 0$, for example $\alpha > 0$, it would follow that $\lim_{k \to \infty} \mu(f_k \leq \alpha) = 0$, and then by uniform integrability that $\lim_{k \to \infty} \int_{\{f_k \leq \alpha\}} f_k d\mu = 0$. Then $\liminf_{k \to \infty} \int_E f_k d\mu \geq \alpha > 0$ contradicting that $\lim_{k \to \infty} \int_E f_k d\mu = 0$. □

Having set up the preliminaries, we introduce the notion of a *weak Poincaré inequality*, denoted by WP in what follows. Recall that throughout this section μ is a probability measure. Recall also that $T_\mu(f) = \mathrm{Var}_\mu(\tau(f))$ for a measurable function $f : E \to \mathbb{R}$. As in the preceding sections, weak Poincaré inequalities involve an increasing and concave (not necessarily $\mathcal{C}^1$ a priori) growth function $\Xi : (0, \infty) \to \mathbb{R}_+$.

Definition 7.5.2 (Weak Poincaré inequality) A Markov Triple (E, μ, Γ), with μ a probability measure, is said to satisfy a weak Poincaré inequality $WP(\Xi)$ with respect to a growth function $\Xi : (0, \infty) \to \mathbb{R}_+$ bounded by 1 and such that

$\lim_{r\to 0} \Xi(r) = \Xi(0) = 0$ if, for every $f \in \mathcal{D}(\mathcal{E})$,

$$T_\mu(f) \leq \Xi(\mathcal{E}(f)).$$

It is important to observe that the concavity assumption on Ξ in this definition is not necessary since the concave envelope of a bounded increasing function which converges to 0 at 0 shares the same properties. The restriction $\Xi \leq 1$ is obvious since $T_\mu(f) \leq 1$.

Weak Poincaré inequalities may be described in various equivalent forms. First note that since $\mathcal{E}(f - a) = \mathcal{E}(f)$ for every $a \in \mathbb{R}$, T_μ may be replaced by T_μ^* in Definition 7.5.2. By the definition of τ, Definition 7.5.2 is also equivalent to saying that

$$\operatorname{Var}_\mu(f) \leq \Xi(\mathcal{E}(f)) \tag{7.5.1}$$

for every $f \in \mathcal{D}(\mathcal{E})$ such that $\|f\|_\infty \leq 1$. It is then also equivalent for all bounded f's with either $\|f - \int_E f d\mu\|_\infty \leq 1$ or $\operatorname{osc}(f) = \sup f - \inf f \leq 1$. By linearization, it also amounts to the existence of a decreasing rate function $s \mapsto \gamma(s)$ on $\mathbb{R}_+$ such that

$$\operatorname{Var}_\mu(f) \leq s\,\mathcal{E}(f) + \gamma(s), \quad s \in \mathbb{R}_+, \tag{7.5.2}$$

for these respective classes of functions. For instance, for the class of bounded functions, since $\operatorname{Var}_\mu(f) \leq \|f\|_\infty^2$, it can be assumed that γ is furthermore bounded by 1. The important feature is the behavior of the rate function γ as $s \to \infty$ reflecting the behavior of $\Xi(r)$ near 0 in the weak Poincaré inequality, the fact that $\lim_{r\to 0} \Xi(r) = 0$ being equivalent to $\lim_{s\to\infty} \gamma(s) = 0$. This linearized form (7.5.2) will turn out to be particularly useful in the next chapter, Sect. 8.4, p. 403.

Weak Poincaré inequalities quantify convergence in measure under convergence of the Dirichlet form as illustrated by the following main claim.

Proposition 7.5.3 *A Markov Triple (E, μ, Γ) satisfies a weak Poincaré inequality $WP(\Xi)$ for some growth function Ξ (bounded by 1 and such that $\lim_{r\to 0} \Xi(r) = \Xi(0) = 0$) if and only if for any sequence $(f_k)_{k\in\mathbb{N}}$ in $\mathcal{D}(\mathcal{E})$ such that $\lim_{k\to\infty} \mathcal{E}(f_k) = 0$ together with $\lim_{k\to\infty} m(f_k) = 0$, $f_k \to 0$ in measure. The same is true if $\lim_{k\to\infty} m(f_k) = 0$ is replaced by $\sup_{k\in\mathbb{N}} \|f\|_\infty \leq 1$ and $\lim_{k\to\infty} \int_E f_k d\mu = 0$.*

Proof One direction is obvious thanks to Proposition 7.5.1. For the converse implication, consider

$$\Xi(r) = \sup\{T_\mu(f) ; \mathcal{E}(f) \leq r\}, \quad r \in (0, \infty).$$

Ξ is clearly increasing and bounded (by 1), and the fact that $\lim_{r\to 0} \Xi(r) = 0$ is a direct consequence of the hypothesis. Indeed, if not, there exist $\varepsilon > 0$ and a sequence $(f_k)_{k\in\mathbb{N}}$ in $\mathcal{D}(\mathcal{E})$ such that $\lim_{k\to\infty} \mathcal{E}(f_k) = 0$ and $T_\mu(f_k) \geq \varepsilon$ for every k. But now, each f_k may be translated so that $m(f_k) = 0$, thus contradicting the hypothesis.

7.5 Weak Poincaré Inequalities

It remains to change Ξ into its concave envelope as mentioned above. When the median $m(f_k)$ is replaced by $\int_E f_k d\mu$, the same proof holds changing Ξ into

$$\Xi(r) = \sup\{\mathrm{Var}_\mu(f); \|f\|_\infty \leq 1, \mathcal{E}(f) \leq r\}, \quad r \in (0, \infty).$$

The proof is complete. □

Corollary 7.5.4 *A tight Nash inequality $N(\Phi)$ implies a weak Poincaré inequality. The same holds with a tight entropy-energy inequality $EE(\Phi)$.*

The Nash case is an immediate consequence of the preceding proposition. The entropy-energy case follows from the fact that a tight $EE(\Phi)$ implies a tight $N(\Phi)$ (Corollary 7.4.4).

The relation with standard Poincaré inequalities is described in the following proposition.

Proposition 7.5.5 *A Markov Triple (E, μ, Γ) satisfies a Poincaré inequality $P(C)$ if and only if it satisfies a weak Poincaré inequality $WP(\Xi)$ with $\Xi(r) \leq Cr, r > 0$.*

Proof Under a Poincaré inequality $P(C)$, $T_\mu(f) \leq C\mathcal{E}(f)$ for every $f \in \mathcal{D}(\mathcal{E})$ and therefore $WP(\Xi)$ holds with $\Xi(r) = \min(Cr, 1), r \in (0, \infty)$. Conversely, it is enough by homogeneity to establish the Poincaré inequality for functions bounded by 1, so that the conclusion follows from the hypothesis $\Xi(r) \leq Cr, r > 0$. □

7.5.2 Tightening with Weak Poincaré Inequalities

Weak Poincaré inequalities are aimed at tightening functional inequalities. The following is an illustration of the principle in the context of (linearized) Nash inequalities. It may be applied similarly to most functional inequalities investigated in this monograph.

We start with a technical proposition describing the main tightening step.

Proposition 7.5.6 *Let (E, μ, Γ) be a Markov Triple such that, for constants $s_1, \beta_1 > 0$,*

$$\|f\|_2^2 \leq s_1 \mathcal{E}(f) + \beta_1 \|f\|_1^2 \quad (7.5.3)$$

for every $f \in \mathcal{D}(E)$. Assume furthermore that for some $s_2 > 0$ and $0 < \gamma_2 < \frac{1}{\beta_1 + 1}$, and any $f \in \mathcal{D}(\mathcal{E})$ bounded by 1,

$$\mathrm{Var}_\mu(f) \leq s_2 \mathcal{E}(f) + \gamma_2. \quad (7.5.4)$$

Then a Poincaré inequality $P(C)$ holds with constant

$$C = \frac{2(s_1 + s_2)}{1 - \sqrt{\gamma_2(\beta_1 + 1)}}.$$

On the basis of this result, a general tightening statement is summarized in the next corollary.

Corollary 7.5.7 (Tightening with a weak Poincaré inequality) *Assume that (E, μ, Γ) satisfies a generalized Nash inequality $N(\Phi)$ or an entropy-energy inequality $EE(\Phi)$ for some growth function Φ. Then, if a weak Poincaré inequality holds, (E, μ, Γ) satisfies a (true) Poincaré inequality.*

In particular, whenever a tight Nash inequality $N(\Phi_1)$ holds, or a tight entropy-energy inequality $EE(\Phi_2)$, then $\Phi_1(r)$ may be replaced by $\min(1 + Cr, \Phi_1(r))$ and $\Phi_2(r)$ by $\min(Cr, \Phi_2(r))$, $r \in (0, \infty)$, for some constant $C > 0$.

To deduce this corollary from Proposition 7.5.6, simply observe that by linearization, both a generalized Nash or entropy-energy inequality imply (7.5.3), while (7.5.4) holds under a weak Poincaré inequality since in its linearized form (7.5.2), $\lim_{s \to \infty} \gamma(s) = 0$. Concerning the second part of the statement, by Corollary 7.5.4, a weak Poincaré inequality holds under a tight Nash inequality $N(\Phi_1)$ or a tight $EE(\Phi_2)$ inequality, and therefore also a Poincaré inequality. Then Φ_1 may be chosen such that $\Phi_1(r) \leq Cr + 1$ (by Proposition 7.4.2) and $\Phi_2(r) \leq Cr$ (from the discussion p. 350), $r > 0$, for some constant $C > 0$.

Proof of Proposition 7.5.6 Let $f \in \mathcal{D}(\mathcal{E})$ be such that $\int_E f d\mu = 0$ and $\int_E f^2 d\mu = 1$. Set, for $R > 0$, $f_R = R\tau(\frac{f}{R})$, which coincides with f whenever $|f| \leq R$, and $d_R = f - f_R$. Since both f_R and d_R are contractions of f, $\mathcal{E}(f_R) \leq \mathcal{E}(f)$ and $\mathcal{E}(d_R) \leq \mathcal{E}(f)$. Since $\int_E f^2 d\mu = 1$, by the Cauchy-Schwarz and Markov inequalities,

$$\int_E |d_R| d\mu = \int_E |f| \mathbb{1}_{\{|f| \geq R\}} d\mu - R\mu(|f| \geq R)$$

$$\leq \sqrt{\mu(|f| \geq R)} - R\mu(|f| \geq R) \leq \frac{1}{4R}.$$

On the other hand, since $\int_E f d\mu = 0$, we also have that $|\int_E f_R d\mu| = |\int_E d_R d\mu| \leq \frac{1}{4R}$. Applying (7.5.4) to $R^{-1} f_R$ then yields

$$\int_E f_R^2 d\mu \leq \frac{1}{16R^2} + s_2 \mathcal{E}(f) + \gamma_2 R^2,$$

while, applying (7.5.3) to d_R produces

$$\int_E d_R^2 d\mu \leq \frac{\beta_1}{16R^2} + s_1 \mathcal{E}(f).$$

7.5 Weak Poincaré Inequalities

Now $f_R d_R = R d_R$ so that

$$1 = \int_E f^2 d\mu = \int_E f_R^2 d\mu + \int_E d_R^2 d\mu + 2R \int_E d_R d\mu$$

$$\leq \int_E f_R^2 d\mu + \int_E d_R^2 d\mu + \frac{1}{2}.$$

It follows from the preceding that

$$1 \leq (s_1 + s_2)\mathcal{E}(f) + \frac{\beta_1 + 1}{16 R^2} + \gamma_2 R^2 + \frac{1}{2}$$

$$\leq (s_1 + s_2)\mathcal{E}(f) + \frac{1}{2}\sqrt{\gamma_2(\beta_1 + 1)} + \frac{1}{2}$$

after optimizing in $R > 0$. Hence $1 - \sqrt{\gamma_2(\beta_1 + 1)} \leq 2(s_1 + s_2)\mathcal{E}(f)$ which amounts to the Poincaré inequality $P(C)$ since $\int_E f d\mu = 0$ and $\int_E f^2 d\mu = 1$. The proof is complete. □

In the following, we address the general question of how to obtain a weak Poincaré inequality. Clearly, weak Poincaré inequalities have much to do with Poincaré inequalities, and the results presented below rely on various tools and ideas developed for Poincaré inequalities in Chap. 4. The first statement describes how to obtain a weak Poincaré inequality from Poincaré inequalities on large sets.

Proposition 7.5.8 *Let (E, μ, Γ) be a Markov Triple such that for some measurable set $K \subset E$ with $\mu(K) > 0$, some constant $C > 0$ and any $f \in \mathcal{D}(\mathcal{E})$,*

$$\int_K f^2 d\mu \leq C \mathcal{E}(f) + \frac{1}{\mu(K)}\left(\int_K f d\mu\right)^2. \tag{7.5.5}$$

Then, for any bounded function $f \in \mathcal{D}(\mathcal{E})$ with mean zero,

$$\int_E f^2 d\mu \leq C \mathcal{E}(f) + \frac{\mu(K^c)}{\mu(K)} \|f\|_\infty^2.$$

Therefore, if (7.5.5) holds for a sequence $(K_\ell)_{\ell \in \mathbb{N}}$ of measurable subsets of E such that $\lim_{\ell \to \infty} \mu(K_\ell^c) = 0$, a weak Poincaré inequality holds.

Proof The proof is rather immediate. Write

$$\int_E f^2 d\mu = \int_K f^2 d\mu + \int_{K^c} f^2 d\mu$$

$$\leq C \mathcal{E}(f) + \frac{1}{\mu(K)}\left(\int_K f d\mu\right)^2 + \mu(K^c)\|f\|_\infty^2.$$

On the other hand, since $\int_E f d\mu = 0$, $|\int_K f d\mu| = |\int_{K^c} f d\mu| \leq \|f\|_\infty \mu(K^c)$, from which the claim follows. The conclusion then follows from the various characterizations of weak Poincaré inequalities. □

Remark 7.5.9 In many situations, a stronger inequality than (7.5.5) is considered, namely

$$\mathrm{Var}_\mu(f) \leq C \int_K \Gamma(f) d\mu + \frac{1}{\mu(K)} \left(\int_K f d\mu \right)^2$$

for some measurable set $K \subset E$ and every $f \in \mathcal{D}(\mathcal{E})$. In $\mathbb{R}^n$ or a smooth manifold, when K is a set with compact closure and regular boundary, this corresponds to the Poincaré inequality on K with Neumann boundary conditions. Indeed, it is not hard to see that any smooth function on K with zero normal derivative at the boundary may be extended to a smooth function with compact support in a neighborhood of K, and conversely that any smooth function f defined in a neighborhood of K may be approximated by a sequence $(f_k)_{k \in \mathbb{N}}$ of smooth functions satisfying Neumann conditions on the boundary of K such that all the quantities $\int_K f_k^2 d\mu$, $\int_K \Gamma(f_k) d\mu$ and $\int_K f_k d\mu$ converge to the respective quantities for f. It is therefore often useful to have such estimates on Poincaré inequalities on balls with Neumann boundary conditions, for example. These may be achieved under curvature-dimension hypotheses on balls whenever the set K is convex.

While many probability measures do satisfy a weak Poincaré inequality $WP(\Xi)$, the associated growth function Ξ may even be explicitly described. The idea, the same as for Poincaré or logarithmic Sobolev inequalities, consists of a suitable adaptation of the Muckenhoupt criterion to control Poincaré constants on balls. Results and classical examples will be illustrated in the next chapter in Proposition 8.4.4, p. 409.

7.5.3 Heat Kernel Bounds and Tail Inequalities Under Weak Poincaré Inequalities

To conclude this section, we briefly describe how weak Poincaré inequalities are related to heat kernel bounds, convergence to equilibrium and tail estimates. The strategy is entirely similar to the one developed under generalized or weighted Nash inequalities in the preceding sections. The procedure applies similarly to any functional inequality of the form

$$Q(f) \leq \Phi(\mathcal{E}(f))$$

for any $f \in \mathcal{D}(\mathcal{E})$ where Q is a positive functional which is closed under $(P_t)_{t \geq 0}$ and for which $Q(P_t f) \leq M(t) Q(f)$, $t \geq 0$, provided $\Phi(r)$ is increasing and $r^{-1} \Phi(r)$

7.5 Weak Poincaré Inequalities

decreasing, $r > 0$. According to the various descriptions of weak Poincaré inequalities, $Q(f)$ might be either $\|f\|_\infty^2$, or $\|f - \int f_E d\mu\|_\infty^2$ or $\text{osc}(f)^2$ for which $M(t) = 1$.

The following is then obtained by slightly modifying the proof of Theorem 7.4.5. For simplicity, the growth function Ξ in the weak Poincaré inequality $WP(\Xi)$ is assumed here to be $\mathcal{C}^1$.

Proposition 7.5.10 *Let (E, μ, Γ) be a Markov Triple satisfying a weak Poincaré inequality $WP(\Xi)$ for some growth function Ξ. Then, for any mean-zero bounded function $f \in \mathcal{D}(\mathcal{E})$ and any $t > 0$,*

$$\|P_t f\|_2^2 \leq K(2t) \|f\|_\infty^2$$

where $K(s)$, $s > 0$, is the inverse function of $\int_{U(s)}^\infty \Xi'(r) \frac{dr}{r}$ and U is the inverse function of Ξ defined on $[0, a]$ where $a \in (0, 1]$. Conversely, if for some $t > 0$, $\|P_t f\|_2^2 \leq K(t) \|f\|_\infty^2$ for any bounded mean-zero measurable function $f : E \to \mathbb{R}$, then, for any $f \in \mathcal{D}(\mathcal{E})$ with $\|f\|_\infty \leq 1$,

$$\text{Var}_\mu(f) \leq 4t \, \mathcal{E}(f) + 2K(t).$$

The converse part of the statement is a direct application of the $\mathbb{L}^2(\mu)$-bound (4.2.3), p. 184. This proposition tells us that a weak Poincaré inequality $WP(\Xi)$ holds if and only if $\text{Var}_\mu(P_t f) \leq M(t) \|f\|_\infty^2$ for any bounded mean-zero measurable function $f : E \to \mathbb{R}$ and any $t > 0$ with some function $M(t)$ such that $\lim_{t \to \infty} M(t) = 0$.

Even when there is no Poincaré inequality, it is possible to control convergence to equilibrium in $\mathbb{L}^2(\mu)$ provided that the initial function f is bounded. In general, this convergence will appear at a sub-exponential or polynomial rate, and this rate is related to the tail behavior of the invariant measure μ. For example, if $\Xi(r) \leq Cr^\alpha$ for $0 < r \leq 1$ with $\alpha \in (0, 1)$, then $K(t) \sim C_1 t^{-\alpha/(1-\alpha)}$ for large t, while if $K(t) \leq C t^{-\alpha/(1-\alpha)}$ for large t, a $WP(\Xi)$ inequality holds with $\Xi(r) \leq C_1 r^\alpha$, $r > 0$.

Finally, weak Poincaré inequalities also carry bounds on tails of Lipschitz functions following the usual procedure in this regard. The next statement is for example established by arguments similar to the proof of Proposition 7.4.7.

Proposition 7.5.11 *Let (E, μ, Γ) be a Markov Triple satisfying a weak Poincaré inequality $WP(\Xi)$, and let f be a 1-Lipschitz function on E. Choose $r_0 > 0$ such that $\mu(f \geq r_0) \leq \max(\frac{1}{4}, \sup_{r>0} \Xi(r)) = q_0$. Then, for any $r > r_0$,*

$$\mu(f \geq r) \leq 2F\big((r - r_0) \log 2\big)$$

where F is the inverse function of

$$G(v) = \int_{v/2}^{q_0} \left[u \, \Xi^{-1}\left(\frac{u}{4}\right) \right]^{-1/2} du, \quad v \in (0, q_0).$$

Proof Arguing as in Proposition 7.4.7, apply $WP(\Xi)$ to $g = \frac{1}{s}(((f-r) \vee 0) \wedge s)$, $r, s > 0$. With $q(u) = \mu(f \geq u)$, $u > 0$,

$$q(r+s) \leq q(r)^2 + \Xi\left(\frac{q(r)}{s^2}\right).$$

Choose r large enough so that $q(r) \leq \frac{1}{4}$ and then s such that $\Xi(\frac{q(r)}{s^2}) \leq \frac{q(r)}{4}$ which may be achieved for

$$s = \left(\frac{q(r)}{\Xi^{-1}(\frac{q(r)}{4})}\right)^{1/2}.$$

Hence $q(r+s) \leq \frac{q(r)}{2}$ and the conclusion follows as in Proposition 7.4.7. □

For measures with polynomial decay, this bound is optimal up to constants. For example, if $d\mu(x) = c_\alpha(1+x^2)^{-\alpha/2}dx$, $\alpha > 1$ then $\Xi(r) \leq Cr^{(\alpha-1)/(\alpha+1)}$, $0 < r < 1$, and for such Ξ, we get $\mu(f \geq r) \leq Cr^{(1-\alpha)}$ as $r \to \infty$, which is the correct order. For the family μ_α, $\alpha > 0$, of (7.0.1) on $\mathbb{R}$, $\Xi(r) \leq Cr(-\log r)^{2(1-\alpha)/\alpha}$, $r > 0$, from which $\mu_\alpha(f \geq r) \leq Ce^{-cr^\alpha}$ as $r \to \infty$ which again is of the correct order of magnitude up to constants.

7.6 Further Families of Functional Inequalities

This last section briefly surveys related families of functional inequalities which have been developed for different purposes. As for the inequalities investigated in the previous sections, each such family has its own interest and is more or less better suited to specific issues (convergence to equilibrium, heat kernel bounds or tail estimates). Only a few properties and illustrations are outlined here. In particular, the guideline is the study of the family of probability measures $d\mu_\alpha(x) = c_\alpha e^{-(1+x^2)^{\alpha/2}}dx$ defined in (7.0.1) on the real line for $\alpha \in [1, 2]$ which do satisfy a Poincaré inequality, but satisfy a logarithmic Sobolev inequality only for $\alpha = 2$ (cf. Theorem 4.5.1, p. 194, and Theorem 5.4.5, p. 256). This family has already been examined in the context of entropy-energy, generalized Nash or weighted Nash inequalities in Sects. 7.1 and 7.4. One feature of the various functional inequalities presented below is that they clearly distinguish the values of α, in particular in terms of tail behavior and convergence to equilibrium. A summary of the various properties of this family according to the parameter α will be presented in the next section. The exposition here is rather sketchy and we refer to the appropriate bibliography in the Notes and References for further details.

The various families of interest are presented in the usual general context of a Markov Triple (E, μ, Γ) where μ is a probability measure.

7.6 Further Families of Functional Inequalities

7.6.1 Φ-Entropy Inequality

The first family comprises the Φ-*entropy inequalities*. Such an inequality depends on some convex $\mathcal{C}^2$ function Φ defined on an open interval I of $(0, \infty)$. For every measurable function $f : E \to I$ in $\mathbb{L}^1(\mu)$ such that $\int_E |\Phi(f)|d\mu < \infty$, define the Φ-entropy of f as

$$\mathrm{Ent}_\mu^\Phi(f) = \int_E \Phi(f)d\mu - \Phi\left(\int_E f d\mu\right). \qquad (7.6.1)$$

The usual variance Var_μ is obtained with $\Phi(r) = r^2$ on $I = \mathbb{R}$ and the entropy Ent_μ ((5.1.1), p. 236) for $\Phi(r) = r\log r$ on $I = (0, \infty)$.

Given then a function $\Phi : I \to \mathbb{R}$ as above, a Markov Triple (E, μ, Γ) is said to satisfy a Φ-entropy inequality with constant $C > 0$ if

$$\mathrm{Ent}_\mu^\Phi(f) \leq \frac{C}{2} \int_E \Phi''(f) \Gamma(f) d\mu \qquad (7.6.2)$$

for every I-valued function $f \in \mathcal{D}(\mathcal{E})$. According to the previous examples of the function Φ, this definition appears as a direct extension of Poincaré and logarithmic Sobolev inequalities. To provide interesting consequences and applications, it is necessary to supplement the convexity assumption of Φ by the further hypothesis that Φ is $\mathcal{C}^4$ and that $\frac{1}{\Phi''}$ is concave on I. This is clearly the case for the examples of $\Phi(r) = r^2$ on $I = \mathbb{R}$ and $\Phi(r) = r \log r$ on $I = (0, \infty)$. In the following, only such admissible functions $\Phi : I \to \mathbb{R}$ on an interval $I \subset \mathbb{R}$ are considered.

The Φ-entropy inequalities satisfy the standard stability properties by bounded perturbation and by tensorization (which is less easy), analogous to the corresponding ones for Poincaré and logarithmic Sobolev inequalities. They may also be characterized by an exponential decay along the underlying semigroup, and hold under curvature conditions analogous to those for Poincaré and logarithmic Sobolev inequalities discussed in Chaps. 4 and 5. The next statement summarizes the latter two results.

Proposition 7.6.1 *Let (E, μ, Γ) be a Markov Triple with Markov semigroup $(P_t)_{t \geq 0}$ and let Φ be an admissible function on $I \subset \mathbb{R}$. The Φ-entropy inequality* (7.6.2) *with constant $C > 0$ is equivalent to the exponential decay*

$$\mathrm{Ent}_\mu^\Phi(P_t f) \leq e^{-t/C} \mathrm{Ent}_\mu^\Phi(f),$$

for every $t \geq 0$ and every I-valued function $f \in \mathbb{L}^1(\mu)$ such that $\int_E |\Phi(f)|d\mu < \infty$. Furthermore, under the curvature condition $CD(\rho, \infty)$ for some $\rho > 0$, the Triple (E, μ, Γ) satisfies a Φ-entropy inequality (7.6.2) *with constant $C = \frac{1}{\rho}$.*

The proof of the first part of the proposition is similar to that of Theorem 5.2.1, p. 244. The second part relies on the arguments put forward in the proof Proposition 5.5.3, p. 261, via the analysis of the function $\Lambda(s) = P_s(\Phi(P_{t-s}f))$, $s \in [0, t]$, $t > 0$, for $f : E \to I$ in $\mathcal{A}_0^{\mathrm{const}}$, and the use of the admissibility property of Φ.

By Theorem 4.2.5, p. 183, a Poincaré inequality for (E, μ, Γ) is equivalent to the exponential decay in $\mathbb{L}^2(\mu)$ of the associated Markov semigroup $(P_t)_{t \geq 0}$, while a logarithmic Sobolev inequality is equivalent, by Theorem 5.2.1, p. 244, to the exponential decay of the classical entropy. The main interest of Φ-entropy inequalities is their role in describing an exponential decay of intermediate norms between $\mathbb{L}^2(\mu)$ and $\mathbb{L}^1(\mu)$.

7.6.2 Beckner Inequality

One example of interest in the family of Φ-entropy inequalities concerns the choice of

$$\Phi(r) = \frac{r^p - 1}{p - 1}, \quad r \in (0, \infty),$$

which is admissible only if $p \in (1, 2]$, and includes the case $\Phi(r) = r \log r$ in the limit $p = 1$. For this family of functions Φ, the Φ-entropy inequality takes the form (after changing $f \geq 0$ into $f^{2/p}$ and letting $q = \frac{2}{p} \in [1, 2]$)

$$\frac{1}{2-q}\left[\int_E f^2 d\mu - \left(\int_E |f|^q d\mu\right)^{2/q}\right] \leq C \mathcal{E}(f) \tag{7.6.3}$$

for every $f \in \mathcal{D}(\mathcal{E})$, which is known as the *Beckner inequality* $B_q(C)$, $q \in [1, 2]$, with constant $C > 0$. According to the study of Sect. 7.4, $B_1(C)$ corresponds to the Poincaré inequality $P(C)$, while (in the limit) $B_2(C)$ amounts to the logarithmic Sobolev inequality $LS(C)$. The inequality $B_q(C)$ is also part of the family (6.8.7) of Remark 6.8.4, p. 312 (with $n = \infty$).

By convexity of the map $r \mapsto \log(\|f\|_{1/r})$, $r \in (0, 1]$, the expression

$$\frac{1}{\frac{1}{q} - \frac{1}{2}}\left[\int_E f^2 d\mu - \left(\int_E |f|^q d\mu\right)^{2/q}\right]$$

is increasing in $q \in [1, 2]$. In particular, if μ satisfies a logarithmic Sobolev inequality $LS(C)$, then it satisfies a Beckner inequality $B_q(\frac{2}{q}C)$ for every $q \in [1, 2)$. Conversely, by Jensen's inequality, a Poincaré inequality $P(C)$ implies a Beckner inequality $B_q(\frac{C}{2-q})$ for every $q \in [1, 2)$. Therefore, apart from the precise values of the constants, every member of this family of Beckner inequalities for $q \in [1, 2)$ is equivalent to a Poincaré inequality. The limiting case $q = 2$ corresponding to the logarithmic Sobolev inequality plays a special role. Note however that under the curvature condition $CD(\rho, \infty)$, $\rho > 0$, a logarithmic Sobolev inequality $LS(\frac{1}{\rho})$ holds (Proposition 5.7.1, p. 268) while Proposition 7.6.1 produces a Beckner inequality $B_q(\frac{1}{\rho})$ for every $q \in [1, 2]$, thus better than seen as consequence of $LS(\frac{1}{\rho})$.

7.6.3 Latała-Oleszkiewicz and Modified Logarithmic Sobolev Inequalities

According to the preceding discussion, in terms of tail behaviour, Beckner inequalities do not discriminate between members of the model family of probability measures μ_α of (7.0.1) on the real line for $\alpha \in [1, 2]$ (which do satisfy a Poincaré inequality but satisfy a logarithmic Sobolev inequality only for $\alpha = 2$). Two main options have been investigated in order to describe a family of intermediate inequalities between Poincaré and logarithmic Sobolev, which are in particular well-suited to the scale of the family μ_α, $\alpha \in [1, 2]$.

A first strategy consists of looking precisely at the behavior of the constants in Beckner inequalities when the parameter q approaches 2. Say, for instance, that a Markov Triple (E, μ, Γ) satisfies a *Latała-Oleszkiewicz inequality* $LO(a, C)$ with parameter $a \in [0, 1]$ and constant $C > 0$ if for every $f \in \mathcal{D}(\mathcal{E})$,

$$\sup_{q \in [1,2)} \frac{1}{(2-q)^a} \left[\int_E f^2 d\mu - \left(\int_E |f|^q d\mu \right)^{2/q} \right] \leq C \mathcal{E}(f). \quad (7.6.4)$$

Using as above Hölder's inequality, it is easily checked that a Poincaré inequality $P(C)$ coincides with $LO(0, C)$, that a logarithmic Sobolev inequality $LS(C)$ implies $LO(1, C)$, while conversely $LO(1, C)$ implies $LS(2C)$. Moreover, the family $LO(a, C)$ gets stronger (with the same C) as a increases. It may be shown (see Sect. 8.4, p. 403) that the measures μ_α satisfy a Latała-Oleszkiewicz inequality $LO(a, C)$ with parameter $a \in [0, 1]$ if and only if $\alpha \geq \frac{2}{2-a}$.

A second approach amounts to keeping the (classical) entropy and modifying the energy, yielding modified logarithmic Sobolev inequalities. For simplicity, we consider probability measures μ on $\mathbb{R}^n$, and say that μ satisfies a *modified logarithmic Sobolev inequality* with parameter $p \geq 1$ and constant $C > 0$ if for every smooth function $f : \mathbb{R}^n \to \mathbb{R}$,

$$\mathrm{Ent}_\mu(f^2) \leq C \int_{\mathbb{R}^n} \sum_{i=1}^n H_p\left(\frac{\partial_i f}{f}\right) f^2 d\mu \quad (7.6.5)$$

where $H_p(r) = r^2$ if $|r| \leq 1$ and $H_p(r) = |r|^{p^*}$ if $|r| > 1$, $\frac{1}{p} + \frac{1}{p^*} = 1$. Of course, for $p = 2$, we are left with the usual logarithmic Sobolev inequality. The measure μ_α (on $\mathbb{R}$ or $\mathbb{R}^n$) satisfies the modified logarithmic Sobolev inequality (7.6.5) if and only if $\alpha \geq p$.

The main interest in both the Latała-Oleszkiewicz and modified logarithmic Sobolev inequalities is that they are stable under products and capture the correct tail estimates of the given measure. That is, if the Markov Triple (E, μ, Γ) satisfies either a Latała-Oleszkiewicz inequality $LO(a, C)$ (for $a \in [0, 1]$), or a modified logarithmic Sobolev inequality with parameter $p \geq 1$ and constant $C > 0$, then for every Lipschitz function f on E, and every $r \geq 0$,

$$\mu\left(f \geq \int_E f d\mu + r\right) \leq C_1 e^{-r^p/C_1 \|f\|_{\mathrm{Lip}}^p}.$$

where $C_1 > 0$ only depends only on C and p (and $a = 2(1 - \frac{1}{p})$). The modified logarithmic Sobolev inequality implies the concentration result for all $p > 1$ while the Latała-Oleszkiewicz inequalities are valid only for $p \in [1, 2]$. Again, such tail estimates capture the correct behaviors of the model measures μ_α.

7.7 Summary for the Model Example μ_α

This brief section summarizes some of the conclusions of this and the preceding chapters in terms of functional inequalities, heat kernel bounds and tail behaviors for the model family (7.0.1) of probability measures

$$d\mu_\alpha(x) = c_\alpha \exp\bigl(-(1 + |x|^2)^{\alpha/2}\bigr) dx, \quad \alpha > 0,$$

on the Borel sets of $\mathbb{R}^n$. Recall that the density of μ_α is essentially $e^{-|x|^\alpha}$ but it is convenient to smooth out the potential to avoid regularity issues. As for the Markov Triple, the carré du champ operator is the usual one $\Gamma(f) = |\nabla f|^2$ on smooth functions f on $\mathbb{R}^n$. The associated Markov semigroup is denoted by $(P_t)_{t \geq 0}$.

According to (7.4.4), for all $\alpha > 0$, μ_α satisfies a weighted Nash inequality $N(\Phi)$ with growth function $\Phi(r) = r^{n/(n+2)}$, $r \in (0, \infty)$, and weight $w(x) = e^{-(1+|x|^2)^{\alpha/2}/2}$, $x \in \mathbb{R}^n$. In particular, following Proposition 7.4.9, the semigroup $(P_t)_{t \geq 0}$ has density kernels $p_t(x, y)$, $t > 0$, $(x, y) \in \mathbb{R}^n \times \mathbb{R}^n$, with respect to μ_α satisfying

$$p_t(x, y) \leq C(t) w(x) w(y)$$

with $C(t) > 0$. This inequality however does not yield any significant information about the spectrum since the weight is not in $\mathbb{L}^2(\mu_\alpha)$.

We next distinguish the various ranges of $\alpha > 0$, and for simplicity restrict to the one-dimensional case. Growth and rate functions Φ, Ξ, γ, β of the various inequalities are only given for small or large values of the variables. Constants $C > 0$ may change from line to line.

- $\alpha \in (0, 1)$. Only a weak Poincaré inequality is satisfied here with growth function $\Xi(r) = Cr(-\log r)^{2(1-\alpha)/\alpha}$ for small values of $r > 0$. Equivalently, in the linearized form (7.5.2), $\gamma(s) = \exp(-Cs^{\alpha/(2-2\alpha)})$ for large values of s. Poincaré, logarithmic Sobolev, Latała-Oleszkiewicz and modified logarithmic Sobolev inequalities are not satisfied. Moreover the essential spectrum is not empty (cf. Corollary 4.10.9, p. 229).
- $\alpha = 1$. The measure μ_1 behaves as the symmetric exponential measure. A Poincaré inequality $P(C)$ with constant $C > 0$ holds (cf. (4.4.3), p. 189), as well as a Latała-Oleszkiewicz inequality with $a = 0$ or a modified logarithmic Sobolev inequalities with $p = 1$. A weak Poincaré inequality holds with $\Xi(r) = Cr$, $r \in (0, \infty)$, or $\gamma(s) = 0$ for every $s \geq C$ (with C the Poincaré constant). The measure μ_1 does not satisfy any logarithmic Sobolev inequalities (see for example Theorem 5.4.5, p 256). Even if there is a spectral gap, the essential spectrum is not empty (cf. p. 229).

- $\alpha \in (1,2)$. A Poincaré inequality $P(C)$ holds (see Theorem 4.5.1, p. 194), and then also a weak Poincaré inequality with $\Xi(r) = Cr$, $r \in (0, \infty)$. Both the Latała-Oleszkiewicz inequality for any $0 \leq a \leq 2(1 - \frac{1}{\alpha})$, and the modified logarithmic Sobolev inequality for any $1 \leq p \leq \alpha$, hold. On the other hand, there is no logarithmic Sobolev inequality or entropy-energy inequality. However, a generalized Nash inequality $N(\Phi)$ holds with $\Phi(r) = Cr(\log r)^{-2(\alpha-1)/\alpha}$ for large $r > 0$, as well as, for any $\kappa \in (0, 1)$, a weighted Nash inequality $WN(\Phi)$ with $\Phi(r) = C(1+r)^\kappa$, $r \in (0, \infty)$, and weight

$$w(x) = \left(1 + x^2\right)^{-\sigma} e^{-(1+x^2)^{\alpha/2}/2}, \quad x \in \mathbb{R},$$

where $\sigma > 0$. For σ large enough, $w \in \mathbb{L}^2(\mu_\alpha)$ so that the operator P_t is Hilbert-Schmidt for any $t > 0$ (while not ultracontractive). In particular, the essential spectrum is empty (cf. the end of Sect. 7.4.3).

- $\alpha = 2$. The measure μ_2 behaves as the Gaussian measure and in particular satisfies a Poincaré and a logarithmic Sobolev inequality. All the properties of the case $\alpha \in (1, 2)$ extend to $\alpha = 2$, but due to the additional logarithmic Sobolev inequality, the semigroup $(P_t)_{t \geq 0}$ is hypercontractive (while still not ultracontractive), cf. Chap. 5.

- $\alpha > 2$. Both the Poincaré and the logarithmic Sobolev inequalities hold. As for $\alpha \in (1, 2]$, a weak Poincaré inequality holds with $\Xi(r) = Cr$, $r \in (0, \infty)$, where $C > 0$ is the Poincaré constant. The Latała-Oleszkiewicz inequality for any $a \in [0, 1]$ and the modified logarithmic Sobolev inequality for any $p \in [1, \alpha]$ hold. An entropy-energy inequality $EE(\Phi)$ holds with $\Phi(r) = C(1 + r^{\alpha/(2\alpha-2)})$, $r \in (0, \infty)$, (cf. Sect. 7.3) as well as a generalized Nash inequality $N(\Phi)$ with $\Phi(r) = Cr(\log r)^{2(1-\alpha)/\alpha}$ for large $r > 0$ (cf. p. 366). (Weighted Nash inequalities may be established for suitable weights but are not useful as compared to the generalized Nash inequality.) The semigroup $(P_t)_{t \geq 0}$ is ultracontractive, the density kernels are uniformly bounded for every $t > 0$, and hence the spectrum is discrete.

7.8 Notes and References

This chapter provides an introduction to and a brief exposition of some of the developments concerning functional inequalities in the framework of this monograph over the last 20 years. It is by no means a complete account of these fruitful activities.

Generalized functional inequalities in the form of entropy-energy or Nash inequalities emerged in various places in the literature. Among early sources are the book [144] by E. Davies and the lecture notes [26]. In particular, E. Davies describes in [144] the method based on families of logarithmic Sobolev inequalities (entropy-energy inequalities) to obtain on and off-diagonal heat kernel bounds, later refined in [26], on the basis of Gross' equivalence between logarithmic Sobolev inequality and hypercontractivity. The results of Sects. 7.1 and 7.2 along these lines mainly follow [26]. The optimal Euclidean bounds under the sharp Euclidean logarithmic Sobolev inequality presented in Corollary 7.1.6 and Proposition 7.2.4 are taken from [34]. The optimal heat kernel bounds (7.1.8) are part of a much deeper investigation of Gaussian kernels by E. Lieb [287]. More on heat kernel bounds in manifolds and metric spaces may be found in [217].

The examples of Sect. 7.3 are taken from [265] (where a more probabilistic viewpoint is emphasized) and [34]. Later and deeper refinements along these lines include in particular [46].

The ultracontractive bounds from the generalized Nash inequality (Theorem 7.4.5 in Sect. 7.4) are due to M. Tomisaki [411] and T. Coulhon [139]. The generalized Nash inequalities have also been considered in the equivalent linearized form (7.4.1) called *super Poincaré inequalities*, extensively developed by F.-Y. Wang (cf. [431] for a complete account as well as further bibliographical information in the context of probability theory). They are presented here in the equivalent language of generalized Nash inequalities which offer a somewhat more flexible treatment. The results of Sect. 7.4.1, in particular tail estimates, are mainly suitable translations of Wang's main contributions in this context. Weighted Nash inequalities were investigated in [430] and more recently in [32]. The results of Sect. 7.4.3 are essentially taken from these references. Other types of generalized Nash inequalities have been considered in the literature, in particular for infinite-dimensional models [63, 254, 445].

Weak Poincaré inequalities, and variations in the form of positivity improving properties, as presented in Sect. 7.5 were first used by S. Aida [4] and P. Mathieu [299] as a tool to tighten defective logarithmic Sobolev inequalities. The argument was further refined in for instance [361], and then extensively developed in the monograph [431] to which we refer for further information and complete references. Section 7.5 tries to offer a synthesized treatment to tighten arbitrary functional inequalities. The topic of Remark 7.5.9 has been studied extensively in the literature (cf. e.g. [431] and the references therein).

Section 7.6 collects various forms of functional inequalities which have been studied in recent years. Each form has its own advantages and privileged applications. The Φ-entropic inequalities were first considered in [36], and further developed in [84, 120, 238]. The Beckner inequality appeared in [55] for the Gaussian measure (cf. also in a manifold setting [66]) and the Latała-Oleszkiewicz inequality was introduced in [272]. Modified logarithmic Sobolev inequalities were considered in, among other references, [79] and [441], and later extended in [48, 199–201, 261]. See also [80] for further developments in connection with isoperimetric bounds. We refer to these references for the corresponding tensorization and tail properties, in particular with the measures μ_α as model examples.

To conclude, and in addition to the families of Sect. 7.6, we mention here the further family of *F-Sobolev inequalities*. It is established in [46] that for $\alpha \in (1, 2)$, the probability measure μ_α on the line satisfies

$$\int_{\mathbb{R}} f^2 \log^\beta(1+f^2) d\mu_\alpha - \int_{\mathbb{R}} f^2 d\mu_\alpha \log^\beta\left(1 + \int_{\mathbb{R}} f^2 d\mu_\alpha\right) \leq C \int_{\mathbb{R}} f'^2 d\mu_\alpha$$

for every smooth $f : \mathbb{R} \to \mathbb{R}$, where $\beta = 2(1 - \frac{1}{\alpha})$. The suitable general form for a Markov Triple (E, μ, Γ) appears as

$$\int_E f^2 F(f^2) d\mu \leq C \mathcal{E}(f) + D, \tag{7.8.1}$$

for every $f \in \mathcal{D}(\mathcal{E})$ with $\int_E f^2 d\mu = 1$, where $F : \mathbb{R}_+ \to \mathbb{R}$ is an increasing function satisfying $F(1) = 0$. Such inequalities entail useful regularity properties. For example, the Markov semigroup $(P_t)_{t \geq 0}$ associated with the model with invariant measure μ_α is not hypercontractive for $\alpha \in [1, 2)$. Nevertheless, the latter F-Sobolev inequality may be shown to imply intermediate hypercontractivity properties. F-Sobolev inequalities as in (7.8.1) go back to J. Rosen [365] and have recently been investigated in particular in [46, 436] in the analysis of hypercontractivity in intermediate Orlicz spaces. See also [360] for related Orlicz-Sobolev inequalities and sub-Gaussian bounds.

Chapter 8
Capacity and Isoperimetric-Type Inequalities

The capacity of a set is a way to measure its size from the point of view of potential theory. The theme of this chapter is inequalities comparing measure and capacity uniformly over a given class of sets as equivalent forms of functional inequalities. The passage from sets to functions is usually performed through the use of level sets of functions and co-area formulas. Inequalities on sets often offer more flexibility and a more transparent description of the hierarchy between functional inequalities. Moreover, measure-capacity inequalities allow for the description of more general forms of functional inequalities, dealing with Orlicz norms or with different measures on $\mathbb{L}^p$-spaces and Dirichlet forms, and it is in these contexts that they are most naturally studied. Furthermore, measure-capacity inequalities produce criteria to satisfy a functional inequality much in the spirit of the Muckenhoupt characterizations in dimension one (cf. Sect. 4.5.1, p. 194, and Sect. 5.4, p. 252), although these criteria might appear less useful for exhibiting examples in higher dimension.

Capacities, and measure-capacity inequalities, may be investigated (and have been considered in the literature) in rather large generality, but we only concentrate here on the so-called 2-capacities which capture the relevant Dirichlet form information on sets. A second form of interest are the 1-capacities which are related to boundary or surface measures. Measure-capacity inequalities in this case are then isoperimetric-type inequalities comparing the measure of a set with the measure of its boundary. Indeed, the first part of this chapter suitably adapts to the Markov Triple framework some of the basic tools and results of the theory of capacities, translating functional inequalities as measure-capacity inequalities. The second part of the chapter is devoted to the particular case of Gaussian isoperimetric-type inequalities as an illustration of the power of heat flow techniques.

The context of this chapter is the traditional one of a (Standard or Full) Markov Triple (E, μ, Γ) with Dirichlet form $\mathcal{E}$, generator L and semigroup $\mathbf{P} = (P_t)_{t \geq 0}$, as summarized in Sect. 3.4, p. 168. Full Markov Triples are actually only used in the study of the local heat kernel measures in the second part of this chapter, starting with Sect. 8.5, where curvature-dimension inequalities come into play. To translate, for comparison, the functional isoperimetric-type inequalities into more (classical) geometric statements, we will sometimes assume in addition that the intrinsic distance d defines a Polish topology on (E, μ, Γ).

As announced, the first part of this chapter mainly revisits from the viewpoint of sets some of the most important functional inequalities studied in the former chapters. The first section introduces the basic notions associated with capacities and the main technical tool to transfer (and back) functional inequalities into measure-capacity inequalities. This tool may be seen as the essence of the slicing method developed in previous chapters and is related to the famous co-area formulas in geometric measure theory. In Sect. 8.2, Sobolev-type inequalities are described via measure-capacity inequalities by a simple Orlicz space duality argument. Poincaré and logarithmic Sobolev inequalities are handled in Sect. 8.3, and Nash and weak Poincaré inequalities are examined in Sect. 8.4.

Replacing the 2-capacity by the boundary measure, measure-capacity inequalities turn into isoperimetric-type inequalities. In this direction, Sect. 8.5 investigates, with the heat kernel tools under a curvature condition, measure-capacity inequalities of isoperimetric-type, leading in particular to the Gaussian isoperimetric inequality as well as to comparison results under the curvature condition $CD(\rho, \infty)$. On the basis of these developments, Sect. 8.6 revisits the infinite-dimensional Harnack inequalities of Chap. 5, Sect. 5.6, p. 265, by providing a new approach via a reverse isoperimetric-type inequality along the semigroup. Combining the heat kernel isoperimetric inequality with its reverse form yields furthermore an isoperimetric-type Harnack inequality. Finally, Sect. 8.7 addresses the relationships between Poincaré and logarithmic Sobolev inequalities and the resulting concentration properties raised in Chaps. 4 and 5. While the latter are not enough in general to entail Poincaré and logarithmic Sobolev inequalities, they actually do under positive curvature bounds. More precise comparisons of the isoperimetric functions are moreover available in this context.

8.1 Capacity Inequalities and Co-area Formulas

This section introduces the notions of capacities and of measure-capacity inequalities, and the technical tool of co-area formulas for the transfer from set inequalities to function inequalities.

8.1.1 Measure-Capacity Inequalities

For a Markov Triple (E, μ, Γ) with associated Dirichlet form $\mathcal{E}(f) = \int_E \Gamma(f) d\mu$, $f \in \mathcal{D}(\mathcal{E})$, we first introduce the notion of the *capacity* of a measurable subset of E. Capacities related to $\mathbb{L}^p(\mu)$-norms of the gradient for any $p \geq 1$ are defined in the same way, but as announced, only the value $p = 2$ is considered here.

8.1 Capacity Inequalities and Co-area Formulas

Given a measurable subset $A \subset E$, set

$$\mathrm{Cap}_\mu(A) = \inf \int_E \Gamma(f) d\mu = \inf \mathcal{E}(f) \tag{8.1.1}$$

where the infimum is taken over all functions f in $\mathcal{D}(\mathcal{E})$ such that $\mathbb{1}_A \leq f \leq 1$ (the latter expression is $+\infty$ if the infimum is taken over the empty set. The quantity $\mathrm{Cap}_\mu(A)$ is usually referred to as the 2-*capacity* of A relative to the measure μ and the Dirichlet form $\mathcal{E}$ or carré du champ operator Γ.

For measurable subsets $A \subset B \subset E$, the capacity $\mathrm{Cap}_\mu(A, B)$ is defined by restricting E to B, and is referred to as the 2-*capacity* of A with respect to B relative to μ and $\mathcal{E}$ or Γ. In particular $\mathrm{Cap}_\mu(A, E) = \mathrm{Cap}_\mu(A)$. From the very definition, $\mathrm{Cap}_\mu(A, B)$ is increasing in A and decreasing in B.

If the constant function $\mathbb{1}$ belongs to $\mathcal{D}(\mathcal{E})$, as is the case for example when the measure μ is finite, then $\mathrm{Cap}_\mu(A) = 0$ for every measurable set $A \subset E$. In order to overcome this difficulty, introduce a modified capacity defined on all measurable sets A with $\mu(A) \leq \frac{1}{2}$ by

$$\mathrm{Cap}^*_\mu(A) = \inf\left\{\mathrm{Cap}_\mu(A, B) \,;\, A \subset B, \mu(B) \leq \frac{1}{2}\right\}. \tag{8.1.2}$$

For any A such that $\mu(A) \leq \frac{1}{2}$, $\mathrm{Cap}_\mu(A) \leq \mathrm{Cap}^*_\mu(A)$.

Usually, capacities are defined on topological spaces, and first for open sets and continuous functions. The preceding definition has the advantage of being considered in the context of (Standard) Markov Triples, but the price to pay is that such capacities might be quite hard to compute at this level. This is the case even in $\mathbb{R}^n$ for the Lebesgue measure and the Euclidean carré du champ operator (beyond the case where A and B are two concentric balls). Thus, in order to get some feeling about the definition of capacity, it might be useful to illustrate it on the example of the real line $E = \mathbb{R}$ with a measure μ admitting a strictly positive and smooth density p with respect to the Lebesgue measure and with carré du champ operator $\Gamma(f) = f'^2$ on smooth functions f. Then, for every measurable set $A \subset [b, \infty)$, $b \in \mathbb{R}$, such that $\mathrm{essinf}(A) > b$,

$$\mathrm{Cap}_\mu\big(A, [b, \infty)\big) = \left(\int_b^{\mathrm{essinf}(A)} \frac{dx}{p(x)}\right)^{-1}. \tag{8.1.3}$$

The critical point in this identity is the essential infimum which requires us to suitably approximate functions in the Dirichlet domain by smooth functions. To check (8.1.3), let $a = \mathrm{essinf}(A) > b$ and $C = \int_b^a \frac{dx}{p(x)}$. First, the estimate is attained for the choice of f such that $f(y) = 0$ for $y \leq b$, $f(y) = C^{-1} \int_b^y \frac{dx}{p(x)}$ if $y \in [b, a]$ and $f(y) = 1$ for $y \geq a$ (this function is not in the dense algebra $\mathcal{A}_0$ of smooth and compactly supported functions but may easily be replaced by a suitable smooth approximation). It remains to establish the lower bound. The difficulty, as announced, is to replace $f \in \mathcal{D}(\mathcal{E})$ so that $\mathbb{1}_A \leq f \leq \mathbb{1}_B$ by some smooth (or at least continuous) function with the same property. To this end, observe first that if

$(f_k)_{k \in \mathbb{N}}$ is a sequence of smooth compactly supported functions converging to f in $\mathcal{D}(\mathcal{E})$, then $(f'_k)_{k \in \mathbb{N}}$ converges in $\mathbb{L}^2(\mu)$ to some function g. Then, the function $\hat{f}(x) = \int_b^x g(u)du$ (with $\hat{f} = 0$ on $\{x < b\}$), which is the scalar product in $\mathbb{L}^2(\mu)$ of g and $p^{-1}\mathbb{1}_{[b,x]}$, is continuous and coincides with f almost everywhere on $\mathbb{R}$. Moreover $\mathcal{E}(f) = \int_{\mathbb{R}} g^2 d\mu$. By continuity, $\hat{f}(a) = 1$ since $f = 1$ almost everywhere on A. Now, $\bar{f}(x) = \int_b^{x \wedge a} g(t)dt$ (again with $\bar{f} = 0$ on $\{x < b\}$) clearly also satisfies $\mathbb{1}_A \leq \bar{f} \leq \mathbb{1}_B$ and $\int_b^a g^2 d\mu = \mathcal{E}(\bar{f}) \leq \mathcal{E}(f)$. By the Cauchy-Schwarz inequality,

$$1 = \int_b^a g(x)dx \leq \left(\int_b^a g^2 d\mu\right)^{1/2} \left(\int_b^a \frac{dx}{p(x)}\right)^{1/2}.$$

Therefore $\text{Cap}(A, [b, \infty)) \geq C^{-1}$ which completes the claim.

Even if no topology is used to define capacities, concrete estimates for a given set can thus only be reasonably achieved with topological arguments. In particular, it is not clear in the Markov Triple framework that the infimum in the definition of capacities may be restricted to functions in $\mathcal{A}_0$. Fortunately, capacities will only be used as an abstract tool below, without ever making use of a single explicit estimate.

Having defined capacities, we may introduce the notion of a *measure-capacity inequality*, or more simply a *capacity inequality*, with respect to some increasing growth function $\Phi : \mathbb{R}_+ \to \mathbb{R}_+$, in this context.

Definition 8.1.1 (Measure-capacity inequality) A Markov Triple (E, μ, Γ) is said to satisfy a measure-capacity inequality with growth function $\Phi : \mathbb{R}_+ \to \mathbb{R}_+$ if for every measurable set A in E such that $\mu(A) < \infty$,

$$\Phi(\mu(A)) \leq \text{Cap}_\mu(A). \tag{8.1.4}$$

In the finite measure case,

$$\Phi(\mu(A)) \leq \text{Cap}^*_\mu(A) \tag{8.1.5}$$

for all A with $\mu(A) \leq \frac{1}{2}$. (In particular, it is enough in this case to know Φ on $[0, \frac{1}{2}]$.)

One flexibility of measure-capacity inequalities is that one may consider entirely similar inequalities where the capacity of the set is evaluated under another measure ν on $(E, \mathcal{F})$. If ν has infinite mass, the inequality turns into

$$\Phi(\nu(A)) \leq \text{Cap}_\mu(A)$$

for every measurable set A in E (such that $\nu(A) < \infty$). If ν is finite and normalized into a probability measure, set

$$\text{Cap}^*_{\mu,\nu}(A) = \inf\left\{\text{Cap}_\mu(A, B) \, ; \, A \subset B, \, \nu(B) \leq \frac{1}{2}\right\}$$

8.1 Capacity Inequalities and Co-area Formulas

and the second part of the definition then takes the form

$$\Phi(v(A)) \leq \mathrm{Cap}^*_{\mu,v}(A)$$

for $v(A) \leq \frac{1}{2}$. There are several benefits of such generalizations in applications. Note that when μ is absolutely continuous with respect to v, this extension amounts to setting inequalities for v with a conformal change of metric (cf. Sect. 6.9, p. 313). Throughout this section, we deal for simplicity of exposition with the case of a single measure μ, but repeatedly point out the possible extensions to a pair of measures.

8.1.2 Co-area Formulas

The purpose of this chapter is to relate Definition 8.1.1 to the functional inequalities studied in the previous chapters. When μ has infinite mass, Sect. 8.2 studies how (8.1.4) of Definition 8.1.1 is an alternative description of a Sobolev inequality (for a suitable choice of the growth function Φ). When μ is a probability measure, (8.1.5) will be related, in Sects. 8.3 and 8.4, according again to the growth of Φ, to Poincaré, logarithmic Sobolev and others classical inequalities (such as generalized Nash or weak Poincaré inequalities).

The key to this transfer is a suitable description of the energy of a function in terms of capacities of its level sets, through versions of the so-called *co-area formula* for 2-capacities, and achieved at the technical level by the slicing method. The standard form of the co-area formula will be presented below in Theorem 8.5.1, but a first formula at the level of 2-capacities is stated next.

Theorem 8.1.2 (Co-area formula I) *For every function* $f \in \mathcal{D}(\mathcal{E})$,

$$\int_0^\infty \mathrm{Cap}_\mu(\mathcal{N}_r) \, r \, dr \leq 6\mathcal{E}(f) = 6 \int_E \Gamma(f) d\mu \qquad (8.1.6)$$

where $\mathcal{N}_r = \{x \in E, |f(x)| > r\}$, $r > 0$.

Proof We use the classical slicing principle on the basis of Proposition 3.1.17, p. 137. Letting $f \in \mathcal{D}(\mathcal{E})$, since the sets $\mathcal{N}_r$ are decreasing in r,

$$2 \int_0^\infty \mathrm{Cap}_\mu(\mathcal{N}_r) \, r \, dr = \sum_{k \in \mathbb{Z}} \int_{2^k}^{2^{k+1}} \mathrm{Cap}_\mu(\mathcal{N}_r) \, d(r^2)$$

$$\leq \sum_{k \in \mathbb{Z}} (2^{2k+2} - 2^{2k}) \, \mathrm{Cap}_\mu(\mathcal{N}_{2^k}) \qquad (8.1.7)$$

$$= 3 \sum_{k \in \mathbb{Z}} 2^{2k} \, \mathrm{Cap}_\mu(\mathcal{N}_{2^k}).$$

For every $k \in \mathbb{Z}$, let $f_k = \frac{1}{2^k}[(|f|-2^k)_+ \wedge 2^k]$. By construction,

$$\mathbb{1}_{\mathcal{N}_{2^k}} < f_{k-1} \leq \mathbb{1}_{\mathcal{N}_{2^{k-1}}}$$

and therefore, for every k,

$$\text{Cap}_\mu(\mathcal{N}_{2^k}) \leq \int_E \Gamma(f_{k-1})d\mu \leq 2^{-2k+2}\int_{\mathcal{N}_{2^{k-1}}\setminus \mathcal{N}_{2^k}} \Gamma(f)d\mu. \qquad (8.1.8)$$

By means of Proposition 3.1.17, p. 137, combining (8.1.7) and (8.1.8) yields

$$2\int_0^\infty \text{Cap}_\mu(\mathcal{N}_r) r\, dr \leq 12 \int_E \Gamma(f) d\mu$$

which is the claim. $\square$

The optimal constant in (8.1.6) is known to be 2 in the classical case of $\mathbb{R}^n$ (or an open set in $\mathbb{R}^n$) with the standard gradient and a measure absolutely continuous with respect to the Lebesgue measure, with however a different argument.

Theorem 8.1.2 can easily be generalized to functions with support $B \subset E$ by replacing $\text{Cap}_\mu(\mathcal{N}_r)$ by $\text{Cap}_\mu(\mathcal{N}_r, B)$. In particular, if the support of $f \in \mathcal{D}(\mathcal{E})$ is of μ-measure less than $\frac{1}{2}$, then

$$\int_0^\infty \text{Cap}_\mu^*(\mathcal{N}_r) r\, dr \leq 6 \mathcal{E}(f).$$

8.2 Capacity and Sobolev Inequalities

This section revisits some of the Sobolev inequalities of Chaps. 6 and 7 in terms of measure-capacity inequalities. As announced in the introduction, one of the advantages of capacities is their flexibility in describing functional inequalities directly on subsets of the state space. In particular, it is possible to deal with more general (Orlicz) spaces and with different measures on subsets and capacities respectively.

Dealing with Orlicz spaces, for $\pi : \mathbb{R}_+ \to \mathbb{R}_+$, assumed (for simplicity) to be strictly increasing and continuous, such that $\pi(0) = 0$, set

$$\Pi(r) = \int_0^r \pi(u) du, \quad r \in \mathbb{R}_+.$$

Setting $\upsilon = \pi^{-1}$ to be the inverse of π, define

$$\Upsilon(r) = \int_0^r \upsilon(v) dv, \quad r \in \mathbb{R}_+.$$

The two functions Π and Υ are convex increasing on $\mathbb{R}_+$ and dual with respect to the Legendre-Fenchel transform ($\Pi^{**}(r) = \Upsilon^*(r) = \sup_{s \in \mathbb{R}_+}[rs - \Pi(s)], r \in \mathbb{R}_+$),

8.2 Capacity and Sobolev Inequalities

and form a pair of *Orlicz functions*. If μ is a measure on $(E, \mathcal{F})$, the *Orlicz space* $\mathbb{L}^\Pi(\mu)$ is defined as the space of measurable functions $f : E \to \mathbb{R}$ such that

$$\|f\|_\Pi = \sup\left\{\int_E |f|g\, d\mu \,;\, g \geq 0,\, \int_E \Upsilon(g) d\mu \leq 1\right\} < \infty. \qquad (8.2.1)$$

It is easily checked, using convexity and Jensen's inequality, that for every measurable subset $A \subset E$ with $\mu(A) < \infty$,

$$\|\mathbb{1}_A\|_\Pi = \mu(A)\, \Upsilon^{-1}\left(\frac{1}{\mu(A)}\right). \qquad (8.2.2)$$

As a main example, if $1 < p < \infty$ and $\frac{1}{p} + \frac{1}{p^*} = 1$, for $\Upsilon(r) = r^{p^*}$, then $\Pi(r) = \frac{(p-1)^{p-1}}{p^p} r^p$, $r \in \mathbb{R}_+$, $\mathbb{L}^\Pi(\mu)$ is the usual Lebesgue space $\mathbb{L}^p(\mu)$ and $\|\cdot\|_\Pi = \|\cdot\|_p$. (The results below for the case $p = 1$ may simply be obtained by taking the limit $p \to 1$.) Observe also that when $\Upsilon(r) = e^r - 1$, $r \in \mathbb{R}_+$, comparison with the variational formula for entropy (5.1.3), p. 236, shows that for a probability measure μ and $f \geq 0$, $\mathrm{Ent}_\mu(f) \leq \|f\|_\Pi$, although these quantities are not equal in general.

The following proposition is a model example of the equivalence between capacity and functional inequalities.

Proposition 8.2.1 *Let (E, μ, Γ) be a Markov Triple and (Π, Υ) be a pair of dual Orlicz functions. The following assertions are equivalent.*

(i) *There is a constant $C_1 > 0$ such that for all measurable subsets $A \subset E$ with $\mu(A) < \infty$,*

$$\mu(A)\, \Upsilon^{-1}\left(\frac{1}{\mu(A)}\right) \leq C_1 \mathrm{Cap}_\mu(A). \qquad (8.2.3)$$

(ii) *There is a constant $C_2 > 0$ such that for all functions $f \in \mathcal{D}(\mathcal{E})$,*

$$\|f^2\|_\Pi \leq C_2 \mathcal{E}(f). \qquad (8.2.4)$$

Moreover the constants C_1 and C_2 satisfy $C_1 \leq C_2 \leq 12 C_1$.

The equivalence still holds whenever $\Pi(r) = r$, $r \in \mathbb{R}_+$, in which case $\Upsilon^{-1}(r) = 1$ and $\|f^2\|_\Pi$ is replaced by $\|f\|_2^2$.

As announced before, this statement, as well as most similar statements throughout this chapter, may be extended to a pair of measures (μ, ν) replacing μ on the left-hand side by some other measure ν on $(E, \mathcal{F})$. The modifications are minor and left to the reader.

Proposition 8.2.1 takes a particularly simple form in $\mathbb{L}^p(\mu)$-spaces, $1 < p < \infty$, as the equivalence between

$$\mu(A)^{1/p} \leq C_1\, \mathrm{Cap}_\mu(A)$$

for all measurable subsets $A \subset E$ and

$$\|f^2\|_p \leq C_2 \mathcal{E}(f)$$

for all functions $f \in \mathcal{D}(\mathcal{E})$, with moreover $C_1 \leq C_2 \leq 12 C_1$. The case $p = 1$ is included in the limit $p \to 1$, $p > 1$. Thus, in particular, the standard Sobolev inequality (6.1.1), p. 278, in Euclidean space

$$\|f\|_p^2 \leq C_n \int_{\mathbb{R}^n} |\nabla f|^2 dx$$

for all functions $f : \mathbb{R}^n \to \mathbb{R}$ in the associated Dirichlet domain and with $p = \frac{2n}{n-2}$, $n > 2$, is translated equivalently as a capacity inequality,

$$\mathrm{vol}_n(A)^{(n-2)/n} \leq C_n' \, \mathrm{Cap}_{dx}(A)$$

for all bounded measurable subsets $A \subset \mathbb{R}^n$. While not necessarily of practical value in order to reach the standard Sobolev inequality, this equivalence produces a somewhat different view of its meaning.

Proof of Proposition 8.2.1 Start with the implication from (i) to (ii). For any function $f \in \mathcal{D}(\mathcal{E})$, setting as above $\mathcal{N}_r = \{|f| > r\}$, $r > 0$,

$$\|f^2\|_\Pi = \sup \int_E f^2 g \, d\mu = 2 \sup \int_0^\infty \int_{\mathcal{N}_r} g \, d\mu \, r dr$$

$$\leq 2 \int_0^\infty \left(\sup \int_{\mathcal{N}_r} g \, d\mu \right) r dr,$$

where the supremum is taken over all positive measurable functions $g : E \to \mathbb{R}$ such that $\int_E \Upsilon(g) d\mu \leq 1$. Hence,

$$\|f^2\|_\Pi \leq 2 \int_0^\infty \|\mathbb{1}_{\mathcal{N}_r}\|_\Pi \, r dr$$

and therefore by (8.2.2),

$$\|f^2\|_\Pi \leq 2 \int_0^\infty \mu(\mathcal{N}_r) \Upsilon^{-1}\left(\frac{1}{\mu(\mathcal{N}_r)}\right) r dr.$$

The conclusion then follows from (8.2.3) and Theorem 8.1.2 with $C_2 = 12 C_1$. Turning to the converse implication, given $f \in \mathcal{D}(\mathcal{E})$ such that $\mathbb{1}_A \leq f \leq \mathbb{1}$ for $A \subset E$,

$$\|f^2\|_\Pi \geq \|\mathbb{1}_A\|_\Pi$$

from which (8.2.3) follows with $C_1 = C_2$ by the definition of the capacity and (8.2.2). Proposition 8.2.1 is proved. □

8.3 Capacity and Poincaré and Logarithmic Sobolev Inequalities 399

Remark 8.2.2 Replacing E by some measurable subset B, Proposition 8.2.1 can be generalized for functions with a fixed support. Indeed, given a measurable $B \subset E$, there is an equivalence between the existence of a constant $C_1 > 0$ such that for all measurable subsets $A \subset B$ with $\mu(A) < \infty$,

$$\mu(A) \Upsilon^{-1}\left(\frac{1}{\mu(A)}\right) \leq C_1 \operatorname{Cap}_\mu(A, B)$$

and the existence of a constant $C_2 > 0$ such that for all functions $f \in \mathcal{D}(\mathcal{E})$ with support in B,

$$\|f^2\|_\Pi \leq C_2 \mathcal{E}(f). \tag{8.2.5}$$

The optimal constants C_1 and C_2 similarly satisfy $C_1 \leq C_2 \leq 12 C_1$.

In the context of smooth (weighted) Riemannian manifolds, when $\Pi(r) = r$ and $B = \mathcal{O}$ is an open set, (8.2.5) turns into

$$\int_\mathcal{O} f^2 \leq C_2 \mathcal{E}(f),$$

for all functions $f \in \mathcal{D}(\mathcal{E})$ with compact support in $\mathcal{O}$. The optimal constant $C_2 > 0$ describes a lower bound of the spectrum of the restriction of the operator L on $\mathcal{O}$ with Dirichlet boundary conditions in the spirit of the Faber-Krahn inequalities (Remark 6.2.4, p. 284). More precisely, in a classical formulation,

$$\frac{1}{12} \operatorname{Cap}^\#(\mathcal{O}) \leq \lambda(\mathcal{O}) \leq \operatorname{Cap}^\#(\mathcal{O})$$

where $\operatorname{Cap}^\#(\mathcal{O}) = \inf_{A \subset \mathcal{O}} \frac{\operatorname{Cap}_\mu(A, \mathcal{O})}{\mu(A)}$ and $\lambda(\mathcal{O}) = \inf \frac{\mathcal{E}(f)}{\int_\mathcal{O} f^2 d\mu}$ where the infimum runs over all non-zero functions $f \in \mathcal{D}(\mathcal{E})$ with compact support in $\mathcal{O}$.

8.3 Capacity and Poincaré and Logarithmic Sobolev Inequalities

In the spirit of the previous equivalence between Sobolev-type inequalities and capacity inequalities, this section addresses similarly the capacity inequalities describing Poincaré and logarithmic Sobolev inequalities. The main difference between the preceding section on Sobolev inequalities is that both variance and entropy, which define Poincaré and logarithmic Sobolev inequalities, are not exactly Orlicz norms and must be compared with such norms for suitable Orlicz functions. Below, μ is a probability measure on $(E, \mathcal{F})$.

8.3.1 Capacity and Poincaré Inequalities

The first statement characterizes Poincaré inequalities in terms of capacities. Recall from Definition 4.2.1, p. 181, that a Markov Triple (E, μ, Γ) satisfies a Poincaré

inequality $P(C)$ for some constant $C > 0$ if
$$\mathrm{Var}_\mu(f) \leq C\, \mathcal{E}(f) = \int_E \Gamma(f) d\mu$$
for all $f \in \mathcal{D}(\mathcal{E})$.

Proposition 8.3.1 *Let (E, μ, Γ) be a Markov Triple. Assume that for some constant $C_1 > 0$,*
$$\mu(A) \leq C_1 \operatorname{Cap}^*_\mu(A) \tag{8.3.1}$$
for all subsets $A \subset E$ such that $\mu(A) \leq \frac{1}{2}$. Then μ satisfies a Poincaré inequality $P(C_2)$ with $C_2 = 12 C_1$. Conversely, under a Poincaré inequality $P(C_2)$, (8.3.1) holds with $C_1 = 2 C_2$.

Proof Let $f \in \mathcal{D}(\mathcal{E})$ and denote by m a median of f for μ (cf. p. 374). Setting $F_+ = (f - m)_+$ and $F_- = (f - m)_-$,
$$\mathrm{Var}_\mu(f) \leq \int_E (f - m)^2 d\mu = \int_E F_+^2 d\mu + \int_E F_-^2 d\mu.$$

Denoting by B_+ the support of F_+, (8.3.1) applied to B_+ (since $\mu(B_+) \leq \frac{1}{2}$) together with Proposition 8.2.1 and Remark 8.2.2 yield that
$$\int_E F_+^2 d\mu \leq 12\, C_1 \int_E \Gamma(F_+) d\mu.$$

With the same inequality for F_-, the announced Poincaré inequality $P(C_2)$ holds with $C_2 = 12 C_1$, using the fact, from (3.1.17), p. 129, that
$$\int_E \Gamma(F_+) d\mu + \int_E \Gamma(F_-) d\mu \leq \mathcal{E}(f). \tag{8.3.2}$$

Conversely, if μ satisfies a Poincaré inequality $P(C_2)$, let $B \subset E$ be such that $\mu(B) \leq \frac{1}{2}$ and $A \subset B$. If $f \in \mathcal{D}(\mathcal{E})$ satisfies $\mathbb{1}_A \leq f \leq \mathbb{1}_B$, the support of f is contained in B and thus, by the Cauchy-Schwarz inequality,
$$\left(\int_E f\, d\mu \right)^2 \leq \mu(B) \int_E f^2 d\mu.$$

Therefore
$$\mathrm{Var}_\mu(f) \geq \big(1 - \mu(B)\big) \int_E f^2 d\mu \geq \frac{1}{2} \mu(A)$$
and, under the Poincaré inequality $P(C_2)$, $2 C_2 \mathcal{E}(f) \geq \mu(A)$. Taking the infimum over all f's as above yields the announced claim. Proposition 8.3.1 is established. □

8.3 Capacity and Poincaré and Logarithmic Sobolev Inequalities

As initiated in the previous section, we may deal as easily with generalized Poincaré inequalities involving different measures for variance and energy of the type, for ν another probability measure on $(E, \mathcal{F})$,

$$\mathrm{Var}_\nu(f) \leq C \mathcal{E}(f) = C \int_E \Gamma(f) d\mu$$

for some $C > 0$ and every f in the Dirichlet domain $\mathcal{D}(\mathcal{E})$ of the Markov Triple (E, μ, Γ). This inequality is then characterized via Proposition 8.3.1 by

$$\nu(A) \leq C_1 \, \mathrm{Cap}^*_{\mu, \nu}(A)$$

for all sets $A \subset E$ such that $\nu(A) \leq \frac{1}{2}$. In particular, μ need not be finite in this formulation.

8.3.2 Capacity and Logarithmic Sobolev Inequalities

We turn to the corresponding statement for logarithmic Sobolev inequalities which is slightly more delicate, in particular due to the fact that the Orlicz norm related to the exponential function is not exactly the entropy. Recall from Definition 5.1.1, p. 236, that a Markov Triple (E, μ, Γ) satisfies a logarithmic Sobolev inequality $LS(C)$ for some constant $C > 0$ if

$$\mathrm{Ent}_\mu(f^2) \leq 2C \mathcal{E}(f) = \int_E \Gamma(f) d\mu$$

for all $f \in \mathcal{D}(\mathcal{E})$.

Proposition 8.3.2 *Let (E, μ, Γ) be a Markov Triple. Assume that for some constant $C_1 > 0$,*

$$\mu(A) \log\left(1 + \frac{e^2}{\mu(A)}\right) \leq C_1 \, \mathrm{Cap}^*_\mu(A) \qquad (8.3.3)$$

for all subsets $A \subset E$ such that $\mu(A) \leq \frac{1}{2}$. Then μ satisfies a logarithmic Sobolev inequality $LS(C_2)$ with $C_2 = 12 C_1$. Conversely, under a logarithmic Sobolev inequality $LS(C_2)$, (8.3.3) holds with $C_1 = 8 C_2$.

Again, there is a version of this statement for a pair of measures (μ, ν).

Proof The scheme of proof is rather similar to that of Proposition 8.3.1. One starting point is Lemma 5.1.4, p. 239, which indicates that for all $a \in \mathbb{R}$,

$$\mathrm{Ent}_\mu(f^2) \leq \mathrm{Ent}_\mu\big((f-a)^2\big) + 2 \int_E (f-a)^2 d\mu.$$

In particular for $a = m$ a median of f with respect to μ, the variational formula for entropy (5.1.3), p. 236, yields that

$$\operatorname{Ent}_\mu(f^2) \leq \sup \int_E (f-m)^2 g \, d\mu$$

where the supremum runs over all measurable functions g such that $\int_E e^g d\mu \leq e^2$. Since $\int_E (f-m)^2 g \, d\mu \leq \int_{\{g \geq 0\}} (f-m)^2 g \, d\mu$, setting $h = g \mathbb{1}_{\{g \geq 0\}}$,

$$\operatorname{Ent}_\mu(f^2) \leq \sup \int_E (f-m)^2 h \, d\mu \tag{8.3.4}$$

where the supremum now runs over the set $\mathcal{H}$ of all measurable positive functions h such that $\int_E e^h d\mu \leq e^2 + 1$.

Recall $F_+ = (f-m)_+$ and $F_- = (f-m)_-$ from the proof of Proposition 8.3.1 so that

$$\operatorname{Ent}_\mu(f^2) \leq \sup_{h \in \mathcal{H}_+} \int_{B_+} F_+^2 h \, d\mu + \sup_{h \in \mathcal{H}_-} \int_{B_-} F_-^2 h \, d\mu, \tag{8.3.5}$$

where B_+ (respectively B_-) is the support of F_+ (respectively F_-) and $\mathcal{H}_+$ (respectively $\mathcal{H}_-$) is the set of all measurable positive functions h such that $\int_{B_+} e^h d\mu \leq e^2 + 1$ (respectively $\int_{B_-} e^h d\mu \leq e^2 + 1$). A direct computation indicates that for any measurable set $A \subset B_+$

$$\sup_{h \in \mathcal{H}_+} \int_A h \, d\mu = \mu(A) \log\left(1 + \frac{e^2}{\mu(A)}\right)$$

(choose $h = \log(1 + \frac{e^2}{\mu(A)})\mathbb{1}_A$). Combining with (8.3.3), for all $A \subset B_+$ and $h \in \mathcal{H}_+$, $\int_A h \, d\mu \leq C_1 \operatorname{Cap}_\mu(A, B_+)$, from which, by Proposition 8.2.1 and Remark 8.2.2,

$$\int_{B_+} F_+^2 h \, d\mu \leq 12 C_1 \int_{B_+} \Gamma(F_+) d\mu.$$

Optimize then over $h \in \mathcal{H}_+$ and add the same bound for F_- to reach $LS(12C_1)$ due to (8.3.5) and (8.3.2).

Turning to the converse, let $B \subset E$ such that $\mu(B) \leq \frac{1}{2}$, $A \subset B$ and $f \in \mathcal{D}(\mathcal{E})$ such that $\mathbb{1}_B \geq f \geq \mathbb{1}_A$. Then (5.1.3), p. 236, implies that

$$\operatorname{Ent}_\mu(f^2) \geq \sup \int_A g \, d\mu = \mu(A) \log\left(1 + \frac{1-\mu(B)}{2\mu(A)}\right)$$

$$\geq \mu(A) \log\left(1 + \frac{1}{2\mu(A)}\right)$$

where the supremum may be taken over all positive measurable functions g such that $\int_B e^g d\mu \leq 1$. Now, for every $0 < \mu(A) \leq \frac{1}{2}$,

$$\frac{1}{4}\mu(A)\log\left(1 + \frac{e^2}{\mu(A)}\right) \leq \mu(A)\log\left(1 + \frac{1}{2\mu(A)}\right)$$

so that the claim immediately follows. The proposition is therefore established. □

One further advantage of measure-capacity inequalities is that they allow for some immediate comparison. Indeed, it is clear from Propositions 8.3.1 and 8.3.2 that a logarithmic Sobolev inequality is stronger than a Poincaré inequality. Sharp constants as in Proposition 5.1.3, p. 238, are however somewhat lost in this procedure.

8.4 Capacity and Further Functional Inequalities

There are many other functional inequalities which can be seen as a capacity inequalities with statements analogous to the preceding characterizations. This is the case, for example, for the entropy-energy inequalities of Sect. 7.1, p. 348, which may be handled as in Proposition 8.3.2. In this section, we focus on Nash, weak Poincaré and Latała-Oleszkiewicz inequalities as presented in Chap. 7.

Again, capacities allow us to deal as easily with pairs of measures (ν, μ) where (E, μ, Γ) is a Markov Triple and ν a measure on $(E, \mathcal{F})$, with μ possibly infinite (and ν finite). For simplicity, we only deal with the case where $\nu = \mu$ is a probability measure.

8.4.1 Capacity and Nash Inequalities

We start with the generalized Nash inequalities $N(\Phi)$ of Definition 7.4.1, p. 364, that is for a ($\mathcal{C}^1$ increasing and concave) growth function Φ on $(0, \infty)$,

$$\int_E f^2 d\mu \leq \Phi(\mathcal{E}(f))$$

for every $f \in \mathcal{D}(\mathcal{E})$ such that $\|f\|_1 = 1$. It will turn out to be more convenient to work with the linearized form (7.4.1), p. 365, $\int_E f^2 d\mu \leq s\mathcal{E}(f) + \beta(s)$ which holds for all functions $f \in \mathcal{D}(\mathcal{E})$ with $\|f\|_1 = 1$ where β is the corresponding decreasing rate function. Actually, capacity inequalities translate best for the (generalized) inverse function $\delta = \beta^{-1}$ of the decreasing rate function β. Hence, we consider here the (generalized) Nash inequality in the form

$$\int_E f^2 d\mu \leq \delta(u)\mathcal{E}(f) + u \qquad (8.4.1)$$

which holds for all functions $f \in \mathcal{D}(\mathcal{E})$ with $\|f\|_1 = 1$ with rate function $\delta : [1, \infty) \to (0, \infty)$ (since μ is a probability measure, only the values $u \geq 1$ are relevant).

The next statement is the announced equivalence between Nash and suitable capacity inequalities.

Proposition 8.4.1 *Let (E, μ, Γ) be a Markov Triple, with μ a probability measure, and let $\delta : [1, \infty) \to (0, \infty)$ be decreasing and such that $u \mapsto u\delta(u)$ is increasing. Assume that for all measurable subsets $A \subset E$ with $\mu(A) \leq \frac{1}{2}$,*

$$\frac{\mu(A)}{\delta\left(\frac{1}{\mu(A)}\right)} \leq \mathrm{Cap}_\mu^*(A). \tag{8.4.2}$$

Then μ satisfies a generalized Nash inequality (8.4.1) with associated rate function 12δ. Conversely, assume furthermore that there exists a $q \geq 4$ such that $q\delta(qu) \geq 4\delta(u)$ for all $u \geq 1$. Then, if μ satisfies a generalized Nash inequality (8.4.1) with rate function δ, then it satisfies (8.4.2) with rate function $u \mapsto 2q\delta(\frac{u}{2})$.

Observe that for the classical Euclidean Nash inequality (6.2.3), p. 281 (with $A = 0$, for which $\Phi(r) = Cr^{n/(n+2)}$ (although this case is formally excluded here since the statement is restricted for convenience to probability measures), it holds that $\mu(A)^{(n-2)/n} \leq C\,\mathrm{Cap}(A)$, which is of the same form as the inequality arising from Sobolev inequalities (Proposition 8.2.1). The conclusion here therefore recovers the equivalence between Sobolev and Nash inequalities studied in Sect. 6.2, p. 279.

Proof Given $f \in \mathcal{D}(\mathcal{E})$, let m be a median of f for μ. For every $u \geq 1$,

$$\int_E f^2 d\mu - u\left(\int_E |f| d\mu\right)^2 = \mathrm{Var}_\mu(|f|) - (u-1)\left(\int_E |f| d\mu\right)^2$$
$$\leq \int_E F^2 d\mu - (u-1)\left(\int_E |F| d\mu\right)^2$$

where $F = f - m$. Since $g \geq 0$,

$$\left(\int_E |F| d\mu\right)^2 = \inf\left\{\int_E F^2 g\, d\mu\,;\, g \geq 0,\, \int_E \frac{1}{g} d\mu \leq 1\right\},$$

for every $u \geq 1$,

$$\int_E F^2 d\mu - (u-1)\left(\int_E |F| d\mu\right)^2$$
$$= \sup\left\{\int_E F^2 h\, d\mu\,;\, h \leq 1,\, \int_E (1-h)^{-1} d\mu \leq \frac{1}{u-1}\right\}$$
$$\leq \sup\left\{\int_E F^2 g\, d\mu\,;\, g \in [0,1],\, \int_E (1-g)^{-1} d\mu \leq 1 + \frac{1}{u-1}\right\}.$$

8.4 Capacity and Further Functional Inequalities

It follows that

$$\int_E F^2 d\mu - (u-1)\left(\int_E |F| d\mu\right)^2$$

$$\leq \sup\left\{\int_{B_+} F_+^2 g\, d\mu\,;\, g \in [0,1],\, \int_E (1-g)^{-1} d\mu \leq 1 + \frac{1}{u-1}\right\}$$

$$+ \sup\left\{\int_{B_-} F_-^2 g\, d\mu\,;\, g \in [0,1],\, \int_E (1-g)^{-1} d\mu \leq 1 + \frac{1}{u-1}\right\},$$

where B_+ (respectively B_-) is the support of $F_+ = (f-m)_+$ (respectively $F_- = (f-m)_-$). For $u = 1$, the suprema run over all functions $g \in [0,1]$ without any assumption on the integral.

As for the logarithmic Sobolev inequalities in the previous section, by a direct computation, for any measurable subset $A \subset E$,

$$\sup\left\{\int_A g\, d\mu\,;\, g \in [0,1],\, \int_E (1-g)^{-1} d\mu \leq 1 + \frac{1}{u-1}\right\}$$
$$= \frac{\mu(A)}{1+(u-1)\mu(A)} \tag{8.4.3}$$

(this and similar formulas remain valid in the limit case $u = 1$). Now, for every $u \geq 1$ and $a \in (0, \frac{1}{2}]$, using the fact that $\delta(u)$ is decreasing and $u\delta(u)$ is increasing, it is easily checked according to whether $u \leq \frac{1}{a}$ or not that

$$\frac{a}{1+(u-1)a} \leq \frac{a\delta(u)}{\delta(\frac{1}{a})}.$$

Therefore, for every function $g \in [0,1]$ satisfying $\int_E (1-g)^{-1} d\mu \leq 1 + \frac{1}{u-1}$, since $\mu(B_+) \leq \frac{1}{2}$, the hypothesis (8.4.2) and (8.4.3) imply that, for every $A \subset B_+$ and $u \geq 1$,

$$\tilde{\mu}(A) = \int_A g\, d\mu \leq \delta(u)\, \mathrm{Cap}_\mu(A, B_+).$$

Proposition 8.2.1 and Remark 8.2.2 applied to μ and $\tilde{\mu}$ then yield that for every $u \geq 1$,

$$\int_{B_+} F_+^2 g\, d\mu \leq 12\, \delta(u) \int_{B_+} \Gamma(F_+) d\mu.$$

Optimizing over g together with the same inequality for F_- allows us to conclude the first assertion of the proposition.

Turning to the converse statement, let $A \subset B$ with $\mu(B) \leq \frac{1}{2}$ and let as usual f in $\mathcal{D}(\mathcal{E})$ be such that $\mathbb{1}_A \leq f \leq \mathbb{1}_B$. Define for every $k \in \mathbb{N}$, $f_k = (f - 2^{-k})_+ \wedge 2^{-k}$

and $\mathcal{N}_k = \{f > 2^{-k}\}$. For $k \in \mathbb{N}$, (8.4.1) applied to f_k shows that, for every $u \geq 1$,

$$\int_E f_k^2 d\mu \leq \delta(u) \int_E \Gamma(f_k) d\mu + u \left(\int_E |f_k| d\mu \right)^2.$$

Since $(\int_E |f_k| d\mu)^2 \leq \mu(\mathcal{N}_k) \int_E f_k^2 d\mu$, choosing $u = \frac{1}{2\mu(\mathcal{N}_k)}$ this inequality becomes

$$2^{-2k-1} \mu(\mathcal{N}_{k-1}) \leq \frac{1}{2} \int_E f_k^2 d\mu \leq \delta\left(\frac{1}{2\mu(\mathcal{N}_k)}\right) \mathcal{E}(f_k) \leq \delta\left(\frac{1}{2\mu(\mathcal{N}_k)}\right) \mathcal{E}(f)$$

for every $k \in \mathbb{N}$. Lemma 8.4.2 below for $k = 0$ shows that $\mu(\mathcal{N}_0) \leq 2q\delta(\frac{1}{2\mu(\mathcal{N}_0)})\mathcal{E}(f)$. Since $\mu(A) \leq \mu(\mathcal{N}_0)$ and $u \mapsto u\delta(u)$ is increasing, it follows that

$$\frac{\mu(A)}{2q\,\delta(\frac{1}{2\mu(A)})} \leq \mathcal{E}(f).$$

Optimizing over all f's as above then yields the converse statement of the proposition. Proposition 8.4.1 is therefore established. $\square$

For the proof to be complete, it remains to prove the following technical lemma.

Lemma 8.4.2 *Assume that $\delta : [1, \infty) \to \mathbb{R}_+$ is decreasing, that $u \mapsto u\delta(u)$ is increasing and that there exits a $q \geq 4$ such that $q\delta(qu) \geq 4\delta(u)$ for all $u \geq 1$. If for some increasing sequence $(a_k)_{k \in \mathbb{N}}$ of real numbers such that $0 < a_k \leq \frac{1}{2}, k \in \mathbb{N}$, and some constant $C > 0$ and every $k \in \mathbb{N}$,*

$$2^{-2k} a_{k-1} \leq C\,\delta\left(\frac{1}{2a_k}\right), \tag{8.4.4}$$

then, for every $k \in \mathbb{N}$,

$$2^{-2k} a_k \leq Cq\,\delta\left(\frac{1}{2a_k}\right). \tag{8.4.5}$$

Proof Define the sequence $\delta_k = \delta(\frac{1}{2a_k})$, $k \in \mathbb{N}$, which, by the hypotheses, is increasing and such that $\frac{\delta_k}{a_k}$, $k \in \mathbb{N}$, is decreasing. The aim is to prove that $\frac{a_k}{\delta_k} \leq Cq2^{2k}$ for every k. Since $a_k \leq \frac{1}{2}$ and $q \geq 4$, this is true if $2^{-2k} \leq 8C\delta_k$ and the latter clearly holds if k is large enough. It remains to see that if (8.4.5) holds for some $k\,(\geq 1)$, it also holds for $k - 1$. To this end, using (8.4.4) and the fact that $u \mapsto u\delta(u)$ is increasing, we have

$$\frac{a_{k-1}}{\delta_{k-1}} \leq \frac{C\,2^{2k} \delta_k}{\delta(\frac{1}{2C2^{2k}\delta_k})}.$$

8.4 Capacity and Further Functional Inequalities

By the recurrence hypothesis and since $u \mapsto \delta(u)$ is decreasing, it follows that

$$\frac{a_{k-1}}{\delta_{k-1}} \leq \frac{C\,2^{2k}\delta_k}{\delta(\frac{q}{2a_k})}.$$

The final conclusion is reached using that $\delta(qu) \geq \frac{4}{q}\delta(u)$. □

8.4.2 Capacity and Weak Poincaré Inequalities

We turn to the analogous discussion for the weak Poincaré inequalities of Definition 7.5.2, p. 375, that is (for μ a probability measure),

$$\mathrm{Var}_\mu(f) \leq \Xi\big(\mathcal{E}(f)\big)$$

for every bounded functions f in $\mathcal{D}(\mathcal{E})$ such that (for example) $\mathrm{osc}(f) = \sup f - \inf f = 1$, where $\Xi : (0, \infty) \to \mathbb{R}_+$ is a given ($\mathcal{C}^1$) concave increasing growth function bounded by 1 and such that $\lim_{r \to 0} \Xi(r) = \Xi(0) = 0$. Again, it will be better to work with the linearized version (7.5.2), p. 376, or rather, with the (generalized) inverse $\epsilon = \gamma^{-1}$ of the rate function γ there. Hence, we agree here that (E, μ, Γ) satisfies a weak Poincaré inequality if

$$\mathrm{Var}_\mu(f) \leq \epsilon(u)\mathcal{E}(f) + u, \quad u \in (0, 1), \tag{8.4.6}$$

for every bounded function f in $\mathcal{D}(\mathcal{E})$ such that $\mathrm{osc}(f) = 1$ where $\epsilon : (0, 1) \to (0, \infty)$ is decreasing.

The following statement is then an analogue of Proposition 8.4.1 above. Again, there is an analogous versions for a pair of measures.

Proposition 8.4.3 *Let (E, μ, Γ) be a Markov Triple, with μ a probability measure, and let $\epsilon : (0, 1) \to (0, \infty)$ be decreasing. Assume that for all measurable subsets $A \subset E$ with $\mu(A) \leq \frac{1}{2}$,*

$$\frac{\mu(A)}{\epsilon(\mu(A))} \leq \mathrm{Cap}^*_\mu(A). \tag{8.4.7}$$

Then μ satisfies a weak Poincaré inequality with rate function $12\,\epsilon$. Conversely, if μ satisfies a weak Poincaré inequality (8.4.6) with rate function ϵ, then it satisfies (8.4.7) with rate function $u \mapsto 4\epsilon(\frac{u}{4})$.

Proof The proof follows the same pattern as the previous ones. The second part of the proposition is rather easy. Namely, for $A \subset B \subset E$ such that $\mu(B) = \frac{1}{2}$, if

$\mathbb{1}_A \leq f \leq \mathbb{1}_B$, then $\mathrm{osc}(f) = 1$. Therefore, as for Poincaré inequalities in Proposition 8.3.1,

$$\sup_{u \in (0, \frac{1}{4})} \frac{1}{\epsilon(u)} \left(\frac{\mu(A)}{2} - u \right) \leq \mathrm{Cap}_\mu(A, B).$$

The choice of $u = \frac{\mu(A)}{4}$ then yields the claim.

Turning to the direct implication, assuming the capacity inequality (8.4.7), let $f \in \mathcal{D}(\mathcal{E})$ and let m be a median of f with respect to μ. As in previous similar proofs, set $F_+ = (f - m)_+$ and $F_- = (f - m)_-$ so that

$$\mathrm{Var}_\mu(f) \leq \int_E (f - m)^2 d\mu = \int_E F_+^2 d\mu + \int_E F_-^2 d\mu.$$

We handle the term $\int_E F_+^2 d\mu$, $\int_E F_-^2 d\mu$ being treated similarly. Denote by B_+ the support of F_+ thus satisfying $\mu(B_+) \leq \frac{1}{2}$. Fixing $u \in (0, 1)$, define

$$c = c(u) = \inf\{r \geq 0;\ \mu(F_+ > r) \leq u\}.$$

If $c = 0$, then $\int_E F_+^2 d\mu \leq u(\sup F_+)^2$ since $\mu(B_+) \leq u$. If $c > 0$, define for $a \in (0, 1)$ and $k \in \mathbb{N}$, $\mathcal{N}_k = \{F_+ > ca^k\}$ and $u_k = \mu(\mathcal{N}_k)$. Observe that

$$u_0 \leq u \leq u_1 \leq \cdots \leq u_k \leq \cdots \leq \mu(B_+) \leq \frac{1}{2}.$$

By the decomposition $\{F_+ > 0\} = \mathcal{N}_0 \cup \bigcup_{k \in \mathbb{N}} (\mathcal{N}_{k+1} \setminus \mathcal{N}_k)$,

$$\int_E F_+^2 d\mu = \int_{\mathcal{N}_0} F_+^2 d\mu + \sum_{k \geq 0} \int_{\mathcal{N}_{k+1} \setminus \mathcal{N}_k} F_+^2 d\mu$$

$$\leq u_0 (\sup F_+)^2 + \sum_{k \in \mathbb{N}} c^2 a^{2k} (u_{k+1} - u_k).$$

Since $u_0 \leq u$ and

$$\sum_{k \in \mathbb{N}} a^{2k} (u_{k+1} - u_k) = \frac{1 - a^2}{a^2} \sum_{k \geq 1} a^{2k} (u_k - u_0),$$

the latter inequality turns into

$$\int_E F_+^2 d\mu \leq u(\sup F_+)^2 + \frac{c^2(1 - a^2)}{a^2} \sum_{k \geq 1} a^{2k} (u_k - u_0).$$

Since $\epsilon : (0, 1) \to (0, \infty)$ is positive and decreasing, for any integer $k \geq 1$, the map $\theta \mapsto \frac{u_k - u_0 + \theta u_0}{\epsilon(u + \theta(u_k - u))}$ is increasing on $(0, 1)$. Hence

$$\frac{u_k - u_0}{\epsilon(u)} \leq \frac{u_k}{\epsilon(u_k)} \leq \mathrm{Cap}_\mu^*(\mathcal{N}_k) \leq \mathrm{Cap}_\mu(\mathcal{N}_k, \mathcal{N}_{k+1}),$$

8.4 Capacity and Further Functional Inequalities

and
$$u_k - u_0 \leq \epsilon(u)\,\mathrm{Cap}_\mu(\mathcal{N}_k, \mathcal{N}_{k+1}).$$

Define then, for every $k \in \mathbb{N}$,
$$f_k = \frac{1}{c a^k(1-a)}\left[\left(F_+ - c a^{k+1}\right)_+ \wedge \left(c a^k(1-a)\right)\right].$$

Then $\mathbb{1}_{\mathcal{N}_k} \leq f_k \leq \mathbb{1}_{\mathcal{N}_{k+1}}$ and, by definition,
$$\mathrm{Cap}_\mu(\mathcal{N}_k, \mathcal{N}_{k+1}) \leq \int_{\mathcal{N}_{k+1}\setminus \mathcal{N}_k} \Gamma(f_k)d\mu$$
$$\leq \frac{1}{c^2 a^{2k}(1-a)^2} \int_{\mathcal{N}_{k+1}\setminus \mathcal{N}_k} \Gamma(F_+)d\mu.$$

Collecting the various estimates, it finally follows that
$$\int_E F_+^2 d\mu \leq \frac{1+a}{a^2(1-a)} \epsilon(u) \int_{B_+} \Gamma(F_+)d\mu + u(\sup F_+)^2.$$

After optimizing on $a \in (0, 1)$, the inequality becomes
$$\int_E F_+^2 d\mu \leq 12\,\epsilon(u) \int_{B_+} \Gamma(F_+)d\mu + u\,(\sup F_+)^2.$$

Together with the corresponding bound for F_-, it follows by (8.3.2) that
$$\mathrm{Var}_\mu(f^2) \leq 12\,\epsilon(u)\mathcal{E}(f) + u\left[(\sup f - m)^2 + (\inf f - m)^2\right].$$

The conclusion then follows since if $p = \sup f$ and $q = \inf f$, then for any $m \in [p, q]$, $(p-m)^2 + (q-m)^2 \leq (p-q)^2$. The proof is complete. $\square$

As mentioned in Proposition 7.5.8, p. 379, any reasonable probability measure satisfies a weak Poincaré inequality. However, to describe the associated growth function Ξ, or rate function ϵ of (8.4.6), is not an easy task in general. In dimension one, however, since capacities can be made explicit, the description of Ξ or ϵ is often reduced to a Muckenhoupt criterion as for Poincaré or logarithmic Sobolev inequalities (cf. Theorem 4.5.1, p. 194, and Theorem 5.4.5, p. 256). The following result is essentially established in the same way and there is also a version of this characterization for a pair of measures.

Proposition 8.4.4 *Let μ be a probability measure on the Borel sets of $\mathbb{R}$ with strictly positive density p with respect to the Lebesgue measure and median m. Denote by $C > 0$ the optimal constant of the weak Poincaré inequality in its linearized form*
$$\mathrm{Var}_\mu(f) \leq C\,\epsilon(u) \int_\mathbb{R} f'^2 d\mu + u, \quad u \in (0, 1),$$

for every bounded function $f \in \mathcal{D}(\mathcal{E})$ such that $\operatorname{osc}(f) = 1$. Then, there exist universal constants $c, c' > 0$ such that

$$c \max(b_-, b_+) \leq C \leq c' \max(B_-, B_+),$$

where

$$b_+ = \sup_{x \geq m} \mu\big([x, +\infty)\big) \frac{1}{\epsilon(\mu([x, +\infty))/4)} \int_m^x \frac{1}{p(t)} dt,$$

$$B_+ = \sup_{x \geq m} \mu\big([x, +\infty)\big) \frac{1}{\epsilon(\mu([x, +\infty)))} \int_m^x \frac{1}{p(t)} dt,$$

and similarly,

$$b_- = \sup_{x \leq m} \mu\big((-\infty, x]\big) \frac{1}{\epsilon(\mu((-\infty, x])/4)} \int_x^m \frac{1}{p(t)} dt,$$

and

$$B_- = \sup_{x \leq m} \mu\big((-\infty, x]\big) \frac{1}{\epsilon(\mu((-\infty, x]))} \int_x^m \frac{1}{p(t)} dt.$$

What this proposition tells us is that in order to determine a rate function ϵ, it is necessary that B_+ and B_- are finite, which amounts to the comparison as $x \to \infty$ of $\mu([x, \infty))$ and $\int_x^\infty \frac{1}{p(t)} dt$ together with a similar estimate as $x \to -\infty$. As an illustration, the model example $d\mu_\alpha(x) = c_\alpha e^{-(1+x^2)^{\alpha/2}} dx$ on the line (summarized in Sect. 7.7, p. 386) satisfies a weak Poincaré inequality for $0 < \alpha < 1$ with rate function $\epsilon(u) = C(-\log u)^{2(1-\alpha)/\alpha}$, $u \in (0, 1)$. Alternatively, the growth function Ξ is given by $\Xi(r) = (Cr(-\log r)^{2(1-\alpha)/\alpha}) \wedge 1$, $r \in (0, \infty)$, for some other constant $C > 0$. When $\alpha = 1$, since μ_1 satisfies a Poincaré inequality $P(C)$, $\epsilon(u) = C$ for all $u \in (0, 1)$. As a further example, if $d\mu(x) = c_\gamma (1 + x^2)^{-\gamma/2} dx$ with $\gamma > 1$, a weak Poincaré inequality holds with rate function $\epsilon(u) = Cu^{2/(1-\gamma)}$, $u \in (0, 1)$, or equivalently $\Xi(r) = (Cr^{(\gamma-1)/(\gamma+1)}) \wedge 1$, $r \in (0, \infty)$.

8.4.3 Capacity and Latała-Oleszkiewicz Inequalities

In the last part of this section, we analogously briefly translate Latała-Oleszkiewicz-type inequalities in terms of capacity inequalities. Recall that, according to (7.6.4), p. 385, a Markov Triple (E, μ, Γ) with μ a probability measure is said to satisfy a Latała-Oleszkiewicz inequality $LO(a, C)$ with parameter $a \in [0, 1]$ and constant $C > 0$ if

$$\sup_{q \in [1,2)} \frac{1}{(2-q)^a} \left[\int_E f^2 d\mu - \left(\int_E |f|^q d\mu \right)^{2/q} \right] \leq C \mathcal{E}(f). \tag{8.4.8}$$

for every $f \in \mathcal{D}(\mathcal{E})$. The following statement is in the same spirit as those for generalized Nash or weak Poincaré inequalities. We omit the proof.

Proposition 8.4.5 *Let (E, μ, Γ) be a Markov Triple, with μ a probability measure. Assume that for some constant $C_1 > 0$ and $a \in [0, 1]$,*

$$\mu(A)\left[\log\left(1 + \frac{1}{\mu(A)}\right)\right]^a \leq C_1 \operatorname{Cap}^*_\mu(A) \tag{8.4.9}$$

for all measurable sets $A \subset E$ such that $\mu(A) \leq \frac{1}{2}$. Then μ satisfies a Latała-Oleszkiewicz inequality $LO(a, cC_1)$ where $c > 0$ is a numerical constant. Conversely, under the Latała-Oleszkiewicz inequality $LO(a, C_2)$ of (8.4.8) with $C_2 > 0$, the inequality (8.4.9) holds with constant $C_1 = c'C_2$ where $c' > 0$ is numerical.

This result recovers the fact that Latała-Oleszkiewicz inequalities are an interpolation between Poincaré and logarithmic Sobolev inequalities according to the parameter $a \in [0, 1]$. Moreover, following Proposition 8.4.1, it provides a link with generalized Nash inequalities. Namely, if μ satisfies $LO(a, C)$, $a \in [0, 1]$, $C > 0$, then it satisfies a generalized Nash inequality in the linearized form (8.4.1) with rate function

$$\delta(u) = c\, C\big[\log(1 + u)\big]^{-a}, \quad u \in (1, \infty),$$

where $c > 0$ is a numerical constant. With the tools developed above, Latała-Oleszkiewicz inequalities may be shown to be characterized similarly through a Muckenhoupt-type criterion. In particular, the model family μ_α on the real line satisfies $LO(a, C)$ if and only if $\alpha \geq \frac{2}{2-a}$.

8.5 Gaussian Isoperimetric-Type Inequalities Under a Curvature Condition

In the second part of this chapter, starting with this section, we consider another form of measure-capacity inequalities: the isoperimetric-type inequalities. The $p = 1$ definition of capacity may be viewed as a boundary or surface measure so that a capacity inequality compares in this case measures and surface measures, which is the essence of isoperimetric inequalities.

To address these issues, the first sub-section collects general properties on boundary measures and co-area formulas. Our main focus is then Gaussian-type isoperimetric inequalities for heat kernel measures under curvature condition, providing in particular a semigroup proof of the classical Gaussian isoperimetric inequality.

The setting of this and the following sections is that of a Full Markov Triple (E, μ, Γ). As in all this work, the emphasis is on functional inequalities, and the main results here actually describe functional forms of isoperimetric properties (for heat kernel and invariant measures).

However, it is of interest to relate these functional inequalities to the more geometric and traditional statements. To this end, and only for the matter of comparison, it is convenient to supplement the Markov Triple framework with convenient topological features. As we know, a Markov Triple (E, μ, Γ) includes the notion of intrinsic distance d (Sect. 3.3.7, p. 166). In general, d is not always a true metric. When it will be useful and necessary to suitably describe the geometric conclusions from the functional ones, we will therefore assume that the intrinsic distance d defines a true Polish metric on E, and that μ is a non-atomic Borel measure on (E, d). We furthermore assume that the functions in $\mathcal{A}_0$ are continuous (in the metric d) and that for functions f in $\mathcal{A}$, $\sqrt{\Gamma(f)}$ may be identified with a modulus of gradient

$$\sqrt{\Gamma(f)}(x) = \limsup_{d(x,y) \to 0} \frac{|f(x) - f(y)|}{d(x, y)} \qquad (8.5.1)$$

(for every $x \in E$). In this setting, and when μ is a probability, Lipschitz means that $|f(x) - f(y)| \leq C d(x, y)$ for some $C > 0$ on the support of $\mu \otimes \mu$, and it will be assumed in addition that bounded Lipschitz functions (in this sense) are in the Dirichlet domain $\mathcal{D}(\mathcal{E})$.

This description suitably covers the examples of weighted Riemannian manifolds with $\Gamma(f)$ the usual Riemannian length squared of the gradient of a smooth function f (cf. Sect. C.4, p. 509). These assumptions will be implicit as soon as geometric statements on sets (involving isoperimetric neighborhood and surface measure) are considered below, and they are in particular in force in the first sub-section. It should be pointed out that these hypotheses might rule out natural infinite-dimensional examples such as Wiener spaces (cf. Sect. 2.7.2, p. 108). Finite-dimensional approximations may nevertheless be developed to cover such instances on the basis of the dimension-free inequalities described here (for example for Gaussian measures).

8.5.1 Isoperimetric-Type Inequalities

As explained in the previous introduction, an isoperimetric-type inequality is a measure-capacity inequality where the 2-capacity is replaced by a boundary measure identified with a 1-capacity.

In the (true) intrinsic metric d, and together with (8.5.1), whenever f approximates the characteristic function of a closed set $A \subset E$ (for example $f_\varepsilon = (1 - \frac{1}{\varepsilon} d(\cdot, A))_+$ and $\varepsilon \to 0$), the integral $\int_E \sqrt{\Gamma(f)} d\mu$ approaches the so-called Minkowski (exterior) boundary measure of the set A defined by

$$\mu^+(A) = \liminf_{\varepsilon \to 0} \frac{1}{\varepsilon} \big[\mu(A_\varepsilon) - \mu(A)\big] \qquad (8.5.2)$$

where $A_\varepsilon = \{x \in E \,;\, d(x, A) < \varepsilon\}$, $\varepsilon > 0$, is the ε-(open) neighborhood of $A \subset E$. The quantity $\mu^+(A)$ describes the *surface measure* of A.

8.5 Gaussian Isoperimetric-Type Inequalities Under a Curvature Condition

The measure μ is then said to satisfy an *isoperimetric-type inequality* if there exists a (non-trivial) positive function $\mathcal{J}$ on $\mathbb{R}_+$, called an *isoperimetric function* for μ, such that

$$\mathcal{J}\big(\mu(A)\big) \le \mu^+(A) \tag{8.5.3}$$

for every measurable (closed) subset A of E (of finite measure). Equivalently, define the *isoperimetric profile* $\mathcal{I}_\mu$ of μ on (E, d) (or of (E, μ, Γ)) as the pointwise maximal function $\mathcal{J}$ in (8.5.3), so that an isoperimetric-type inequality bounds from below the isoperimetric profile $\mathcal{I}_\mu$. When equality is achieved in (8.5.3) for some non-trivial sub-family of sets A, called the extremal sets, determining $\mathcal{I}_\mu$, the inequality (8.5.3) expresses a true isoperimetric statement, comparing the surface measure of a set to that of an extremal set with the same measure.

For a given measure on the Borel sets of a metric space (E, d), not that many isoperimetric profiles are known. One may mention the standard isoperimetric inequality for the Lebesgue measure on $\mathbb{R}^n$ given by the isoperimetric function $n\widetilde{\omega}_n^{1/n} v^{(n-1)/n}$, $v \in \mathbb{R}_+$, where $\widetilde{\omega}_n$ is the volume of the Euclidean unit ball, and for which balls are the extremal sets. Similarly, (geodesic) balls are the extremal sets of the isoperimetric problem on the sphere $\mathbb{S}^n \subset \mathbb{R}^{n+1}$ with respect to the uniform measure and the isoperimetric profile is again described by its value on these extremal sets. The next sub-section will feature the further example of the isoperimetric profile of Gaussian measures (for which half-spaces are the extremal sets). Another example is that of the exponential measure $d\mu(x) = e^{-x}dx$ on $\mathbb{R}_+$ for which $\mathcal{I}_\mu(v) = \min(v, 1-v)$, $v \in [0, 1]$.

The theorem below presents a co-area formula involving the surface measure μ^+ instead of the 2-capacity of Theorem 8.1.2. Both the statement itself and its proof are purely metric.

Theorem 8.5.1 (Co-area formula II) *In the metric space setting (E, d) induced by the Markov Triple (E, μ, Γ) as described above, for every Lipschitz function f on E,*

$$\int_{-\infty}^{+\infty} \mu^+(\mathcal{N}_r) dr \le \int_E \sqrt{\Gamma(f)}\, d\mu \tag{8.5.4}$$

where $\mathcal{N}_r = \{x \in E,\ f(x) > r\}$, $r \in \mathbb{R}$.

In the context of a general Markov Triple, there is no reason why (8.5.4) should be an equality. Equality is known in specific instances such as $\mathbb{R}^n$ with a measure μ which is absolutely continuous with respect to the Lebesgue measure, in which case (8.5.4) is classically referred to as a *co-area formula*.

Proof Assume first that $f : E \to \mathbb{R}$ is bounded and moreover, without loss of generality, positive. For each $\varepsilon > 0$, set $f_\varepsilon(x) = \sup_{d(x,y) < \varepsilon} f(y)$, $x \in E$. Since f is Lipschitz, f_ε is finite and lower semi-continuous. From (8.5.1), for every $x \in E$,

$$\limsup_{\varepsilon \to 0} \frac{f_\varepsilon(x) - f(x)}{\varepsilon} \le \sqrt{\Gamma(f)(x)}.$$

The dominated convergence Theorem then implies that

$$\limsup_{\varepsilon \to 0} \frac{1}{\varepsilon} \int_E (f_\varepsilon - f) d\mu \leq \int_E \sqrt{\Gamma(f)} \, d\mu.$$

On the other hand, for all $r > 0$,

$$\{x \in E \, ; \, f_\varepsilon(x) > r\} = \left(\{x \in E \, ; \, f(x) > r\}\right)_\varepsilon = (\mathcal{N}_r)_\varepsilon,$$

the ε-neighborhood of $\mathcal{N}_r$. Since $\int_E f d\mu = \int_0^\infty \mu(\mathcal{N}_r) dr$ and $\int_E f_\varepsilon d\mu = \int_0^\infty \mu((\mathcal{N}_r)_\varepsilon) dr$,

$$\frac{1}{\varepsilon} \int_E (f_\varepsilon - f) d\mu = \frac{1}{\varepsilon} \int_0^\infty \left[\mu((\mathcal{N}_r)_\varepsilon) - \mu(\mathcal{N}_r)\right] dr.$$

By Fatou's Lemma,

$$\liminf_{\varepsilon \to 0} \frac{1}{\varepsilon} \int_0^\infty \left[\mu((\mathcal{N}_r)_\varepsilon) - \mu(\mathcal{N}_r)\right] dr \geq \int_0^\infty \liminf_{\varepsilon \to 0} \frac{\mu((\mathcal{N}_r)_\varepsilon) - \mu(\mathcal{N}_r)}{\varepsilon} dr$$

$$= \int_0^\infty \mu^+(\mathcal{N}_r) dr.$$

The desired inequality follows. Whenever f is unbounded, apply the preceding to the sequence of Lipschitz functions $f_k = (f \wedge a_k) \vee (-a_k)$, $a_k \in \mathbb{N}$ (where the sequence $(a_k)_{k \in \mathbb{N}}$ is chosen such that $\lim_{k \to \infty} a_k = \infty$ and $\mu(f = a_k) = 0$) and let $k \to \infty$. Theorem 8.5.1 is established. $\square$

Theorem 8.5.1 is typically used to provide evidence that a lower bound on the isoperimetric profile $\mathcal{I}_\mu$ may imply standard functional inequalities for μ. A classical example is that of Lebesgue measure in $\mathbb{R}^n$, $n > 1$. Namely, if $f : \mathbb{R}^n \to \mathbb{R}$ is smooth and compactly supported, then by the co-area formula, denoting respectively by vol_n and vol_n^+ the volume and surface measures,

$$\int_0^\infty \text{vol}_n^+\left(|f| > r\right) dr \leq \int_{\mathbb{R}^n} |\nabla f| dx.$$

By the isoperimetric inequality, for every $r > 0$,

$$n \widetilde{\omega}_n^{1/n} \text{vol}_n\left(|f| > r\right)^{(n-1)/n} \leq \text{vol}_n^+\left(|f| > r\right)$$

from which (after integration in r and a little bit of work)

$$n \widetilde{\omega}_n^{1/n} \|f\|_p \leq \int_{\mathbb{R}^n} |\nabla f| dx$$

8.5 Gaussian Isoperimetric-Type Inequalities Under a Curvature Condition

with $p = \frac{n}{n-1}$. Changing f into f^2 then yields the standard Sobolev inequality (6.1.1), p. 278, in $\mathbb{R}^n$ (however not with its best constant).

Some more care has to be taken for finite measures. One classical example in this case is the following statement expressing that whenever $\mathcal{I}_\mu$ may be compared to the isoperimetric profile of the exponential measure, then it satisfies a Poincaré inequality. Similar statements may be obtained for different behaviors of the isoperimetric function and different functional inequalities. For example, as will be used below in Theorem 8.7.2, comparison with the Gaussian isoperimetric profile yields logarithmic Sobolev inequalities.

Proposition 8.5.2 *In the preceding context with μ a probability measure, assume that there exists an isoperimetric function $\mathcal{J}$ for μ such that, for some $c > 0$*

$$\mathcal{J}(v) \geq c \min(v, 1-v), \quad v \in [0,1].$$

Then μ satisfies a Poincaré inequality $P(\frac{4}{c^2})$.

Proof By Theorem 8.5.1 and the hypothesis, for any positive and Lipschitz g,

$$c \int_0^\infty \min\bigl(\mu(g > r), 1 - \mu(g > r)\bigr) dr \leq \int_E \sqrt{\Gamma(g)}\, d\mu. \tag{8.5.5}$$

Let now $f : E \to \mathbb{R}$ be Lipschitz and denote by m a median of f for μ. Set as usual $F_+ = (f - m)_+$ and $F_- = (f - m)_-$ so that $f - m = F_+ - F_-$. By the definition of the median, for every $r > 0$,

$$\mu\bigl((F_+)^2 > r\bigr) \leq \frac{1}{2} \quad \text{and} \quad \mu\bigl((F_-)^2 > r\bigr) \leq \frac{1}{2}.$$

Hence, (8.5.5) applied to $g = (F_+)^2$ and $g = (F_-)^2$ yields

$$\begin{aligned}
c \int_E |f - m|^2 d\mu &= c \int_E (F_+)^2 d\mu + c \int_E (F_-)^2 d\mu \\
&= c \int_0^\infty \mu\bigl((F_+)^2 \geq r\bigr) dr + c \int_0^\infty \mu\bigl((F_-)^2 \geq r\bigr) dr \quad (8.5.6) \\
&\leq \int_E \sqrt{\Gamma\bigl((F_+)^2\bigr)}\, d\mu + \int_E \sqrt{\Gamma\bigl((F_-)^2\bigr)}\, d\mu.
\end{aligned}$$

By the Cauchy-Schwarz inequality,

$$\int_E \sqrt{\Gamma\bigl((F_+)^2\bigr)}\, d\mu = 2 \int_E F_+ \sqrt{\Gamma(F_+)}\, d\mu$$

$$\leq 2 \left(\int_E F_+^2 d\mu \right)^{1/2} \left(\int_E \Gamma(F_+) d\mu \right)^{1/2}$$

and similarly for F_-. By (8.3.2), the right-hand side of (8.5.6) is less than or equal to

$$2\left(\int_E |f-m|^2 d\mu\right)^{1/2} \left(\int_E \Gamma(f) d\mu\right)^{1/2}.$$

Therefore, for every median m of f,

$$\frac{c^2}{4} \int_E |f-m|^2 d\mu \leq \int_E \Gamma(f) d\mu.$$

Since the variance of f is the infimum of $\int_E |f-a|^2 d\mu$ over $a \in \mathbb{R}$, it follows that μ satisfies a Poincaré inequality $P(\frac{4}{c^2})$. Since in the considered setting any function f in $\mathcal{A}_0$ is Lipschitz, the claim follows. □

8.5.2 Gaussian Isoperimetric-Type Inequalities Under $CD(\rho, \infty)$

While it is in general a difficult issue to determine extremal sets for an isoperimetric problem such as (8.5.3), however one might expect to find comparison theorems involving a given isoperimetric function which might not be the optimal one. The theme of this section is comparison with the Gaussian isoperimetric profile under curvature conditions. Accordingly, we will speak of a *Gaussian isoperimetric-type inequality* for a Markov Triple (E, μ, Γ) if (8.5.3) holds for $\mathcal{J}$ the isoperimetric profile of the (standard) Gaussian measure (see below).

The standard Gaussian measure on the Borel sets of $\mathbb{R}^n$ satisfies, with respect to the standard carré du champ operator $\Gamma(f) = |\nabla f|^2$ on smooth functions, numerous functional inequalities of interest, in particular the logarithmic Sobolev inequality (Proposition 5.5.1, p. 258). The powerful curvature tools allowed in Sects. 5.5 and 5.7, p. 257 and p. 268, for the extension of these inequalities to Markov Triples (E, μ, Γ) under the curvature condition $CD(\rho, \infty)$ with $\rho > 0$. This covers in particular the case of probability measures $d\mu = e^{-W} dx$ on $\mathbb{R}^n$ where the smooth potential W is such that uniformly $\nabla\nabla W \geq \rho \, \text{Id} > 0$, $\rho > 0$ (as symmetric matrices).

The main conclusion of the investigation here is a stronger property in the form of the isoperimetric inequality for the standard Gaussian measure and its extension, as a comparison property, to positively curved diffusion operators. The key will again be provided by the Γ-calculus on heat kernel measures via a suitable functional description of isoperimetry. The heat flow proof which will be developed towards this goal therefore also entails a proof of the isoperimetric inequality for the Gaussian measure itself.

To state the main result, we introduce the isoperimetric profile of Gaussian measures. Let $\Phi(r) = \int_{-\infty}^r e^{-x^2/2} \frac{dx}{\sqrt{2\pi}}$, $r \in \mathbb{R}$, be the distribution function of the standard Gaussian measure on the real line and $\varphi = \Phi'$ its density with respect to the Lebesgue measure. Then, as will be justified below,

$$\mathcal{I} = \varphi \circ \Phi^{-1} : [0, 1] \to \left[0, \frac{1}{\sqrt{2\pi}}\right] \quad (8.5.7)$$

8.5 Gaussian Isoperimetric-Type Inequalities Under a Curvature Condition

defines the *Gaussian isoperimetric profile* (in any dimension). Note that the function $\mathcal{I}$ is concave continuous, symmetric with respect to the vertical line going through $\frac{1}{2}$, and such that $\mathcal{I}(0) = \mathcal{I}(1) = 0$, and satisfies the fundamental differential equation $\mathcal{I}\mathcal{I}'' = -1$.

The main result of this section is that $\sqrt{\rho}\,\mathcal{I}$ is an isoperimetric function for the invariant probability measure μ of a Markov Triple (E,μ,Γ) satisfying the curvature condition $CD(\rho,\infty)$ for some $\rho > 0$. To reach this goal, we use an ad-hoc functional description of isoperimetry, known as a *Bobkov inequality*, and, as for Poincaré and logarithmic Sobolev inequalities, we establish such Bobkov inequalities for heat kernel measures under curvature conditions.

Theorem 8.5.3 (Local Bobkov inequalities) *Let (E,μ,Γ) be a Markov Triple with semigroup $\mathbf{P} = (P_t)_{t \geq 0}$. The following assertions are equivalent.*

(i) *The curvature condition $CD(\rho,\infty)$ holds for some $\rho \in \mathbb{R}$.*

(ii) *For every function f in $\mathcal{A}_0$ with values in $[0,1]$, every (or some) $\alpha \geq 0$ and every $t \geq 0$,*

$$\sqrt{\mathcal{I}^2(P_t f) + \alpha\,\Gamma(P_t f)} \leq P_t\left(\sqrt{\mathcal{I}^2(f) + c_\alpha(t)\Gamma(f)}\right) \tag{8.5.8}$$

where

$$c_\alpha(t) = \frac{1 - e^{-2\rho t}}{\rho} + \alpha\,e^{-2\rho t}, \quad t \geq 0$$

(with $c_\alpha(t) = 2t + \alpha$ whenever $\rho = 0$).

Before turning to the proof of this result, let us illustrate its isoperimetric content in the form of the following corollary, which follows from the choice of $\alpha = \frac{1}{\rho}$ with $\rho > 0$ (so that $c_\alpha(t) = \frac{1}{\rho}$ for every $t \geq 0$) and letting $t \to \infty$ by ergodicity. In this context, μ is a probability measure.

Corollary 8.5.4 (Bobkov inequality under $CD(\rho,\infty)$) *Let (E,μ,Γ) be a Markov Triple, with μ a probability measure, satisfying the curvature condition $CD(\rho,\infty)$ for some $\rho > 0$. Then, for every f in $\mathcal{D}(\mathcal{E})$ with values in $[0,1]$,*

$$\sqrt{\rho}\,\mathcal{I}\left(\int_E f\,d\mu\right) \leq \int_E \sqrt{\rho\,\mathcal{I}^2(f) + \Gamma(f)}\,d\mu. \tag{8.5.9}$$

In a metric context, applying this corollary to functions approximating the characteristic function of a measurable (closed) subset A in E yields that (using $\mathcal{I}(0) = \mathcal{I}(1) = 0$),

$$\sqrt{\rho}\,\mathcal{I}\big(\mu(A)\big) \leq \mu^+(A) \tag{8.5.10}$$

where we recall the surface measure $\mu^+(A)$ as defined by the Minkowski content formula (8.5.2). In an alternative integrated form, for any measurable (closed) subset

A in E and any $r > 0$,

$$\Phi^{-1} \circ \mu(A_r) \geq \Phi^{-1} \circ \mu(A) + \sqrt{\rho}\, r \qquad (8.5.11)$$

(since $\frac{d}{dr}\Phi^{-1} \circ \mu(A_r)|_{r=0} = \frac{\mu^+(A)}{\mathcal{I}(\mu(A))}$).

Note that, for the Gaussian measure in Euclidean space itself, integrating (8.5.10) along the level sets of a function yields conversely, due to the dimension-free character of the Gaussian isoperimetric profile $\mathcal{I}$, the functional form (8.5.9) which therefore fully describes Gaussian isoperimetry.

The preceding conclusion may be summarized as follows.

Corollary 8.5.5 (Lévy-Gromov isoperimetric comparison Theorem) *In the preceding metric context, under the curvature condition $CD(\rho, \infty)$ for some $\rho > 0$, the isoperimetric profile $\mathcal{I}_\mu$ of (E, μ, Γ) is bounded from below by $\sqrt{\rho}\,\mathcal{I}$ where $\mathcal{I}$ is the Gaussian isoperimetric profile.*

The corollary applies in particular to measures $d\mu = e^{-W}dx$ on $\mathbb{R}^n$ such that $\nabla\nabla W \geq \rho\, \mathrm{Id}$, $\rho > 0$. The result actually covers the optimal isoperimetric inequality for the standard Gaussian measure $d\mu(x) = (2\pi)^{-n/2}e^{-|x|^2/2}dx$ on $\mathbb{R}^n$ for which $\rho = 1$ and the half-spaces $H = \{x \in \mathbb{R}^n ; x \cdot u \leq r\}$, $r \in \mathbb{R}$, u unit vector in $\mathbb{R}^n$, achieve equality in (8.5.10) (since $\mu(H) = \Phi(r)$ and $\mu^+(H) = \varphi(r)$), providing therefore a purely functional proof of the Gaussian isoperimetric inequality itself. Note that since the measures of half-spaces are one-dimensional, the isoperimetric statement is dimension-free. The previous isoperimetric application may be developed similarly at the level of the heat kernel measures in Theorem 8.5.3. In particular, for the Euclidean semigroup in $\mathbb{R}^n$, we recover in the same way the Gaussian isoperimetric inequality.

Furthermore, it should be pointed out that the Bobkov inequality (8.5.9) in Corollary 8.5.4 may actually be studied in its own right, as for Poincaré or logarithmic Sobolev inequalities. In particular, it may be shown to be stable under products, justifying its relevance towards dimension-free isoperimetric properties. It also improves upon the logarithmic Sobolev inequality, preserving constants, when applied to εf^2 with a Taylor expansion as $\varepsilon \to 0$ using the asymptotics $\mathcal{I}(v) \sim v\sqrt{2\log\frac{1}{v}}$ as $v \to 0$. In this way, Corollary 8.5.4 implies the logarithmic Sobolev inequality of Proposition 5.7.1, p. 268, in the same context (and similarly Theorem 8.5.3 improves upon Theorem 5.5.2, p. 259).

We next turn to the semigroup properties of Theorem 8.5.3 and to its proof. First observe that, as for the Poincaré and logarithmic Sobolev inequalities for heat kernel measures, since $c_{c_\alpha(s)}(t) = c_\alpha(s+t)$, (8.5.8) is stable under $(P_t)_{t\geq 0}$. With the same Taylor expansion as the one used above under the invariant measure, (ii) improves upon the Poincaré and logarithmic Sobolev inequalities of Theorem 4.7.2, p. 209, and Theorem 5.5.2, p. 259, and therefore (8.5.8), which holds for every f in $\mathcal{A}_0$ with values in $[0, 1]$, every $t > 0$ and some fixed $\alpha \geq 0$, ensures conversely the curvature condition $CD(\rho, \infty)$. This may also be checked directly by a suitable (although somewhat tedious) Taylor expansion at $t = 0$ together with the further asymptotics

8.5 Gaussian Isoperimetric-Type Inequalities Under a Curvature Condition

$f = \frac{1}{2} + \varepsilon g$ as $\varepsilon \to 0$. In addition, whenever $\alpha \to \infty$, we recover the fundamental strong gradient bound of Theorem 3.3.18, p. 163,

$$\sqrt{\Gamma(P_t f)} \leq e^{-\rho t} P_t\left(\sqrt{\Gamma(f)}\right)$$

for any function $f \in \mathcal{A}_0$ and any $t \geq 0$. In particular, these various arguments settle the implication from (ii) to (i).

We now address the converse implication in the proof of Theorem 8.5.3. The argument is very similar to the semigroup monotonicity proofs of the local Poincaré and logarithmic Sobolev inequalities. Basic use will be made of the differential equation $\mathcal{II}'' = -1$. It will be convenient to record the following technical lemma.

Lemma 8.5.6 *Let Ψ be smooth on $\mathbb{R}^3$ (or some open set in $\mathbb{R}^3$), $f \in \mathcal{A}_0$ and $t > 0$ be fixed. Then*

$$\frac{d}{ds} P_s\big(\Psi\big(s, P_{t-s}f, \Gamma(P_{t-s}f)\big)\big) = P_s(K)$$

with

$$K = \partial_1 \Psi + 2\partial_3 \Psi\, \Gamma_2(g) + \partial_2^2 \Psi\, \Gamma(g) + 2\partial_2\partial_3 \Psi\, \Gamma\big(g, \Gamma(g)\big) + \partial_3^2 \Psi\, \Gamma\big(\Gamma(g)\big)$$

where we wrote on the right-hand side $g = P_{t-s}f$ and Ψ for $\Psi(s, g, \Gamma(g))$.

Proof We have

$$\frac{d}{ds} P_s\big(\Psi\big(s, P_{t-s}f, \Gamma(P_{t-s}f)\big)\big)$$
$$= P_s\left(L\Psi\big(s, P_{t-s}f, \Gamma(P_{t-s}f)\big) + \frac{d}{ds}\Psi\big(s, P_{t-s}f, \Gamma(P_{t-s}f)\big)\right).$$

By the diffusion property and using the notation of the statement,

$$L\Psi\big(s, P_{t-s}f, \Gamma(P_{t-s}f)\big) + \frac{d}{ds}\Psi\big(s, P_{t-s}f, \Gamma(P_{t-s}f)\big)$$
$$= \partial_2 \Psi\, Lg + \partial_3 \Psi\, L\Gamma(g) + \partial_2^2 \Psi\, \Gamma(g) + 2\partial_2\partial_3 \Psi\, \Gamma\big(g, \Gamma(g)\big)$$
$$+ \partial_3^2 \Psi\, \Gamma\big(\Gamma(g)\big) + \partial_1 \Psi - \partial_2 \Psi\, Lg - 2\partial_3 \Psi\, \Gamma(g, Lg).$$

The lemma then follows by the very definition of the Γ_2 operator. □

Proof of Theorem 8.5.3 Assume the curvature $CD(\rho, \infty)$ condition for some $\rho \in \mathbb{R}$. Let $f \in \mathcal{A}_0^{\text{const}\,1}$ with values in $[0, 1]$ and $t > 0$ be fixed. For every $s \in [0, t]$, set

$$\Lambda(s) = P_s\left(\sqrt{\mathcal{I}^2(P_{t-s}f) + c_\alpha(s)\Gamma(P_{t-s}f)}\right).$$

(Strictly speaking, it would be necessary to replace $\Gamma(P_{t-s}f)$ by $\Gamma(P_{t-s}f) + \varepsilon$ for some $\varepsilon > 0$.) Since $c_\alpha(0) = \alpha$, it will be enough to show that Λ is increasing. Apply Lemma 8.5.6 with

$$\Psi(s, u, v) = \sqrt{\mathcal{I}^2(u) + c_\alpha(s)v}, \quad (s, u, v) \in (0, t) \times (0, 1) \times (0, \infty).$$

It is immediate that, with $\mathcal{I} = \mathcal{I}(u)$, $\mathcal{I}' = \mathcal{I}'(u)$,

$$\Psi \partial_1 \Psi = \frac{c_\alpha'}{2} v, \qquad \Psi \partial_2 \Psi = \mathcal{I}\mathcal{I}', \qquad \Psi \partial_3 \Psi = \frac{c_\alpha}{2},$$

and (using $\mathcal{I}\mathcal{I}'' = -1$)

$$\Psi^3 \partial_2^2 \Psi = -\mathcal{I}^2 \mathcal{I}'^2 + \Psi^2(\mathcal{I}'^2 - 1), \qquad \Psi^3 \partial_2 \partial_3 \Psi = -\frac{c_\alpha}{2}\mathcal{I}\mathcal{I}',$$

$$\Psi^3 \partial_3^2 \Psi = -\frac{c_\alpha^2}{4}.$$

Therefore, K of Lemma 8.5.6 satisfies

$$\Psi^3 K = \Psi^2 \frac{c_\alpha'(s)}{2}\Gamma(g) + \Psi^2 c_\alpha(s)\Gamma_2(g)$$
$$- \mathcal{I}^2(g)\mathcal{I}'^2(g)\Gamma(g) + \Psi^2(\mathcal{I}'^2(g) - 1)\Gamma(g)$$
$$- c_\alpha(s)\mathcal{I}(g)\mathcal{I}'(g)\Gamma(g, \Gamma(g)) - \frac{c_\alpha(s)^2}{4}\Gamma(\Gamma(g)),$$

where $\Psi = \Psi(s, g, \Gamma(g))$ (and $g = P_{t-s}f$). Since

$$\Psi = \Psi(s, g, \Gamma(g)) = \sqrt{\mathcal{I}^2(g) + c_\alpha(s)\Gamma(g)},$$

it follows that

$$\Psi^3 K = [\mathcal{I}^2(g) + c_\alpha(s)\Gamma(g)]\frac{c_\alpha'(s)}{2}\Gamma(g)$$
$$+ [\mathcal{I}^2(g) + c_\alpha(s)\Gamma(g)]c_\alpha(s)\Gamma_2(g)$$
$$- \mathcal{I}^2(g)I'^2(g)\Gamma(g) + [\mathcal{I}^2(g) + c_\alpha(s)\Gamma(g)](\mathcal{I}'^2(g) - 1)\Gamma(g)$$
$$- c_\alpha(s)\mathcal{I}(g)\mathcal{I}'(g)\Gamma(g, \Gamma(g)) - \frac{c_\alpha(s)^2}{4}\Gamma(\Gamma(g)).$$

Therefore, after some algebra,

$$\Psi^3 K = c_\alpha(s)\Gamma(g)\left[c_\alpha(s)\Gamma_2(g) - \left(1 - \frac{c'_\alpha(s)}{2}\right)\Gamma(g)\right]$$
$$- \frac{c_\alpha(s)^2}{4}\Gamma(\Gamma(g)) + \mathcal{I}^2(g)\left[c_\alpha(s)\Gamma_2(g) - \left(1 - \frac{c'_\alpha(s)}{2}\right)\Gamma(g)\right]$$
$$- c_\alpha(s)\mathcal{I}'(g)\mathcal{I}(g)\Gamma(g,\Gamma(g)) + c_\alpha(s)\mathcal{I}'^2(g)\Gamma(g)^2.$$

By the very definition of $c_\alpha(s)$, $1 - \frac{c'_\alpha(s)}{2} = \rho\, c_\alpha(s)$ so that

$$\Psi^3 K = c_\alpha(s)^2\left(\Gamma(g)[\Gamma_2(g) - \rho\,\Gamma(g)] - \frac{1}{4}\Gamma(\Gamma(g))\right)$$
$$+ c_\alpha(s)\left(\mathcal{I}^2(g)[\Gamma_2(g) - \rho\,\Gamma(g)] - \mathcal{I}'(g)\mathcal{I}(g)\Gamma(g,\Gamma(g)) + \mathcal{I}'^2(g)\Gamma(g)^2\right).$$

Now, as a consequence of the reinforced curvature condition (3.3.6), p. 159,

$$4\,\Gamma(f)\bigl[\Gamma_2(f) - \rho\,\Gamma(f)\bigr] \geq \Gamma(\Gamma(f)), \quad f \in \mathcal{A},$$

applied twice,

$$\Psi^3 K \geq c_\alpha(s)\left(\mathcal{I}^2(g)\frac{\Gamma(\Gamma(g))}{4\Gamma(g)} - \mathcal{I}'(g)\mathcal{I}(g)\Gamma(g,\Gamma(g)) + \mathcal{I}'^2(g)\Gamma(g)^2\right).$$

The right-hand-side of this inequality is a quadratic form in $\mathcal{I}(g)$ and $\mathcal{I}'(g)$ which is positive since, as $\Gamma(f) \geq 0$ for every f,

$$\Gamma\bigl(f,\Gamma(f)\bigr)^2 \leq \Gamma(f)\Gamma\bigl(\Gamma(f)\bigr).$$

Hence $K \geq 0$ and thus $\Lambda'(s) = P_s(K) \geq 0$ for every $s \in [0,t]$. This completes the proof of Theorem 8.5.3. □

The natural challenge after Theorem 8.5.3 would be to investigate similar conclusions under the curvature-dimension condition $CD(\rho, n)$ (for some finite n). The aim would be to reach in this way the isoperimetric inequality on the n-sphere, and the corresponding comparison statement for manifolds with dimension n and with a strictly positive lower bound on the Ricci curvature. However, this program is mainly open since in particular a functional description of isoperimetry on the sphere suitable for the Γ-calculus is still missing. The analysis should probably also involve the fast diffusion equation rather than the heat flow as in Sect. 6.11, p. 330.

8.6 Harnack Inequalities Revisited

Section 5.6, p. 265, investigated Harnack inequalities for infinite-dimensional Markov generators with curvature $CD(\rho, \infty)$. In particular, it was observed there (see (5.6.4), p. 267) that the reverse forms of logarithmic Sobolev inequalities for heat kernel measures may be used to get quite close to the optimal statement. Together with the refined Gaussian isoperimetric function $\mathcal{I}$, we show in this section how to obtain these Harnack inequalities by means of a suitable reverse form of Theorem 8.5.3. In addition, the conjunction of Theorem 8.5.3 together with its reverse form will lead to an isoperimetric-type Harnack inequality.

The following statement is therefore a kind of reverse Bobkov inequality along the semigroup. Recall $c_0(t) = \frac{1-e^{-2\rho t}}{\rho}$ ($= 2t$ if $\rho = 0$) from Theorem 8.5.3.

Proposition 8.6.1 *Let (E, μ, Γ) be a Markov Triple satisfying the curvature condition $CD(\rho, \infty)$ for some $\rho \in \mathbb{R}$. Then, for every f in $\mathcal{A}_0$ and every $t \geq 0$,*

$$\left[\mathcal{I}(P_t f)\right]^2 - \left[P_t(\mathcal{I}(f))\right]^2 \geq e^{2\rho t} c_0(t) \Gamma(P_t f). \tag{8.6.1}$$

By the usual extension procedure, the inequality (for $t > 0$) applies to any measurable function f with values in $[0, 1]$. As a consequence, for every bounded measurable function f and every $t > 0$,

$$\Gamma(P_t f) \leq \frac{e^{-2\rho t}}{2\pi c_0(t)} \|f\|_\infty^2. \tag{8.6.2}$$

Proof The proof relies on the standard heat flow interpolation argument. For $f \in \mathcal{A}_0^{\text{const}+}$ with values in $[0, 1]$ and $t > 0$, write

$$\left[\mathcal{I}(P_t f)\right]^2 - \left[P_t(\mathcal{I}(f))\right]^2 = -\int_0^t \frac{d}{ds}\left[P_s(\mathcal{I}(P_{t-s} f))\right]^2 ds.$$

By the change of variables formula for the diffusion operator L,

$$-\frac{d}{ds}\left[P_s(\mathcal{I}(P_{t-s} f))\right]^2$$
$$= -2 P_s(\mathcal{I}(P_{t-s} f)) P_s(L\mathcal{I}(P_{t-s} f) - \mathcal{I}'(P_{t-s} f) LP_{t-s} f)$$
$$= -2 P_s(\mathcal{I}(P_{t-s} f)) P_s(\mathcal{I}''(P_{t-s} f) \Gamma(P_{t-s} f))$$
$$= 2 P_s(\mathcal{I}(P_{t-s} f)) P_s\left(\frac{\Gamma(P_{t-s} f)}{\mathcal{I}(P_{t-s} f)}\right)$$

where we used that $\mathcal{I}\mathcal{I}'' = -1$ in the last step. Since P_s is given by a kernel, it satisfies a Cauchy-Schwarz inequality, and hence

$$P_s(B) P_s\left(\frac{A^2}{B}\right) \geq \left[P_s(A)\right]^2, \quad A, B \geq 0.$$

Therefore, with $A = \sqrt{\Gamma(P_{t-s} f)}$ and $B = \mathcal{I}(P_{t-s} f)$,

$$\left[\mathcal{I}(P_t f)\right]^2 - \left[P_t(\mathcal{I}(f))\right]^2 \geq 2 \int_0^t \left[P_s\left(\sqrt{\Gamma(P_{t-s} f)}\right)\right]^2 ds.$$

8.6 Harnack Inequalities Revisited

By the strong gradient bounds of Theorem 3.3.18, p. 163, $P_s(\sqrt{\Gamma(g)}) \geq e^{\rho s}\sqrt{\Gamma(P_s g)}$. With $g = P_{t-s}f$, it follows that

$$[\mathcal{I}(P_t f)]^2 - [P_t(\mathcal{I}(f))]^2 \geq 2\int_0^t e^{2\rho s} ds\, \Gamma(P_t f)$$

which is the result. The gradient bound (8.6.2) is sharp on the example of the Ornstein-Uhlenbeck semigroup (2.7.3), p. 104, for f the characteristic function of a half-space. □

It may be shown that as $t \to 0$ (8.6.1) implies in return the curvature condition $CD(\rho, \infty)$. Also, by applying it to εf and letting $\varepsilon \to 0$, one recovers the reverse logarithmic Sobolev inequality (5.5.6) of Theorem 5.5.2, p. 259.

On the basis of Proposition 8.6.1, the aim is now to fully recover the Harnack inequalities of Theorem 5.6.1, p. 265. Indeed, in terms of gradient bounds, since $(\Phi^{-1})' = \frac{1}{\mathcal{I}}$, for every $f \in \mathcal{A}_0$ with values in $[0, 1]$ and $t > 0$,

$$\Gamma(\Phi^{-1} \circ P_t f) = \frac{\Gamma(P_t f)}{\mathcal{I}(P_t f)^2} \leq \frac{1}{e^{2\rho t} c_0(t)}$$

as a consequence of (8.6.1). In particular, for $x, y \in E$ and $d(x, y)$ the intrinsic distance between x and y,

$$\Phi^{-1} \circ P_t f(x) \leq \Phi^{-1} \circ P_t f(y) + \frac{d(x, y)}{e^{\rho t}\sqrt{c_0(t)}} \quad (8.6.3)$$

which extends to any measurable function f with values in $[0, 1]$. This Lipschitz property actually entails the announced Harnack inequalities. Set $\delta = e^{-\rho t} c_0(t)^{-1/2} d(x, y)$, so that (8.6.3) reads as

$$P_t f(x) \leq \Phi(\Phi^{-1} \circ P_t f(y) + \delta).$$

Now apply this inequality with f replaced by $\mathbb{1}_{\{f \geq r\}}$, $r \geq 0$, for a positive measurable function f on E. Denoting by ν the distribution of f under P_t at the point y (that is $\nu(B) = P_t(\mathbb{1}_{\{f \in B\}})(y)$ for every Borel set B in $\mathbb{R}$),

$$P_t(\mathbb{1}_{\{f \geq r\}})(x) \leq \Phi(\Phi^{-1}(\nu([r, \infty))) + \delta).$$

Integrating in $r \geq 0$ and using Fubini's Theorem,

$$P_t f(x) \leq \int_0^\infty \int_{-\infty}^{\Phi^{-1}(\nu([r,\infty)))+\delta} e^{-u^2/2}\frac{du}{\sqrt{2\pi}}\, dr$$

$$= \int_{-\infty}^\infty \left(\int_0^\infty \mathbb{1}_{\{u \leq \Phi^{-1}(\nu([r,\infty)))+\delta\}} dr\right) e^{-u^2/2}\frac{du}{\sqrt{2\pi}}.$$

Change u into $u + \delta$ to get

$$P_t f(x) \leq e^{-\delta^2/2} \int_{-\infty}^\infty e^{-\delta u}\left(\int_0^\infty \mathbb{1}_{\{\Phi(u) \leq \nu([r,\infty))\}} dr\right) e^{-u^2/2}\frac{du}{\sqrt{2\pi}}.$$

Changing u into $-u$ and denoting by F the distribution function of v, it follows that

$$P_t f(x) \le e^{-\delta^2/2} \int_{-\infty}^{\infty} e^{\delta u} \left(\int_0^{\infty} \mathbb{1}_{\{F(r) \le \Phi(u)\}} dr \right) e^{-u^2/2} \frac{du}{\sqrt{2\pi}}.$$

After the further change of variable $v = \Phi(u)$,

$$P_t f(x) \le e^{-\delta^2/2} \int_0^1 e^{\delta \Phi^{-1}(v)} \left(\int_0^{\infty} \mathbb{1}_{\{F(r) \le v\}} dr \right) dv.$$

The next statement summarizes the conclusions reached so far.

Theorem 8.6.2 *Let* (E, μ, Γ) *be a Markov Triple satisfying the curvature condition* $CD(\rho, \infty)$ *for some* $\rho \in \mathbb{R}$. *For every positive measurable function* f *on* E, *every* $t > 0$ *and every* $x, y \in E$,

$$P_t f(x) \le e^{-\delta^2/2} \int_0^{\infty} e^{\delta \Phi^{-1} \circ F(s)} s \, dF(s)$$

where $\delta = e^{-\rho t} c_0(t)^{-1/2} d(x, y)$ *and* F *is the distribution function of* f *under* P_t *at the point* y.

Although not expressed in a very tractable form, Theorem 8.6.2 may be viewed as the root of the various Harnack inequalities in this context. For example, by the Cauchy-Schwarz inequality,

$$\int_0^{\infty} e^{\delta \Phi^{-1} \circ F(s)} s \, dF(s) \le \left(\int_0^{\infty} e^{2\delta \Phi^{-1} \circ F(s)} dF(s) \right)^{1/2} \left(\int_0^{\infty} s^2 \, dF(s) \right)^{1/2}$$

$$\le e^{\delta^2} \left(P_t(f^2)(y) \right)^{1/2}$$

since

$$\int_0^{\infty} e^{2\delta \Phi^{-1} \circ F(s)} dF(s) = \int_0^1 e^{2\delta \Phi^{-1}(v)} dv = \int_{-\infty}^{\infty} e^{2\delta u - u^2/2} \frac{du}{\sqrt{2\pi}} = e^{2\delta^2}.$$

The preceding therefore exactly recovers Wang's Harnack inequality of Theorem 5.6.1, p. 265, for $\alpha = 2$,

$$(P_t f)^2(x) \le P_t(f^2)(y) \exp\left(\frac{d(x, y)^2}{e^{2\rho t} c_0(t)} \right).$$

By Hölder's inequality rather than the Cauchy-Schwarz inequality, one obtains the whole family of inequalities with $\alpha > 1$. Using the entropic inequality (5.1.2), p. 236, yields similarly the log-Harnack inequality (5.6.2) of Remark 5.6.2, p. 266. With respect to the original argument of Theorem 5.6.1, the proof here avoids interpolation along geodesics and holds true in the general context of Markov Triples.

8.7 From Concentration to Isoperimetry

As announced, Theorem 8.6.2 may be coupled with Theorem 8.5.3 to reach a kind of isoperimetric Harnack inequality. We only sketch the argument which assumes the necessary metric structure.

Theorem 8.6.3 (Isoperimetric Harnack inequality) *Let (E, μ, Γ) be a Markov Triple satisfying the curvature condition $CD(\rho, \infty)$ for some $\rho \in \mathbb{R}$. For every measurable (closed) set A in E, every $t \geq 0$ and every $x, y \in E$ with $d(x, y) > 0$,*

$$P_t(\mathbb{1}_A)(x) \leq P_t(\mathbb{1}_{A_{d_t}})(y)$$

where $d_t = e^{-\rho t} d(x, y)$. In particular, when $\rho = 0$,

$$P_t(\mathbb{1}_A)(x) \leq P_t(\mathbb{1}_{A_{d(x,y)}})(y).$$

Proof In its integrated form (cf. (8.5.11)), Theorem 8.5.3 implies that for any $y \in E$, any measurable set $A \subset E$, any $t > 0$ and any $r > 0$,

$$\Phi^{-1} \circ P_t(\mathbb{1}_{A_r})(y) \geq \Phi^{-1} \circ P_t(\mathbb{1}_A)(y) + \frac{r}{\sqrt{c_0(t)}}$$

where A_r is the r-neighborhood of A with respect to the distance d. On the other hand, the Lipschitz property (8.6.3) applied to $f = \mathbb{1}_A$ ensures that

$$\Phi^{-1} \circ P_t(\mathbb{1}_A)(x) \leq \Phi^{-1} \circ P_t(\mathbb{1}_A)(y) + \delta$$

where $\delta = e^{-\rho t} c_0(t)^{-1/2} d(x, y)$. The combination of these two inequalities immediately yields the result. □

8.7 From Concentration to Isoperimetry

In this last section, we address the questions raised in Remarks 4.4.5, p. 193, and 5.4.4, p. 255, concerning possible converses from concentration properties towards Poincaré or logarithmic Sobolev inequalities under curvature hypotheses. Actually, the investigation will provide a complete picture of the links between measure concentration and isoperimetric-type inequalities for a Markov Triple satisfying the curvature condition $CD(0, \infty)$. Since we will be dealing at some point with isoperimetric enlargements, it is convenient to further examine the metric setting emphasized in the introduction of Sect. 8.5. A typical illustrative setting is the case of a weighted Riemannian manifold with $d\mu = e^{-W} d\mu_{\mathfrak{g}}$ and $\nabla \nabla W \geq 0$ (in particular log-concave probability measures on $\mathbb{R}^n$).

We saw in Chap. 4, Proposition 4.4.2, p. 190, and (4.4.6), p. 192, that under a Poincaré inequality $P(C)$, the probability μ entails exponential concentration in the sense that any Lipschitz function $f : E \to \mathbb{R}$ is integrable and satisfies

$$\mu\left(f \geq \int_E f d\mu + r\right) \leq 3 e^{-r/\sqrt{C} \, \|f\|_{\mathrm{Lip}}}, \quad r \geq 0.$$

426   8 Capacity and Isoperimetric-Type Inequalities

This section investigates the converse implication. While it may not be expected to hold in general, as is clear for example from the characterization of Poincaré inequalities on the line (cf. Theorem 4.5.1, p. 194), surprisingly a converse does hold under suitable curvature assumptions.

In order to properly state the result, we say that (E, μ, Γ) has *exponential concentration* if there are constants $C, c > 0$ such that for every integrable 1-Lipschitz function f on E,

$$\mu\left(f \geq \int_E f d\mu + r\right) \leq C e^{-cr}, \quad r \geq 0. \tag{8.7.1}$$

It should be mentioned that the relevant constant in this property is c which controls the exponential decay. The first constant C is usually easily handled.

The following is then the announced converse implication. Recall the isoperimetric profile $\mathcal{I}_\mu$ of a measure μ.

Theorem 8.7.1 (Milman's Theorem I) *Let (E, μ, Γ) be a Markov Triple, with μ a probability measure, satisfying the curvature condition $CD(0, \infty)$. If (E, μ, Γ) has exponential concentration with constants $C, c > 0$, then*

$$\mathcal{I}_\mu(v) \geq c' \min(v, 1-v), \quad v \in [0, 1], \tag{8.7.2}$$

where $c' > 0$ only depends on C, c. (In other words, the isoperimetric profile of (E, μ, Γ) is bounded from below, up to the constant c', by the isoperimetric profile of the exponential measure). In particular, (E, μ, Γ) satisfies a Poincaré inequality $P(C')$ where $C' > 0$ only depends on C, c.

By Theorem 4.6.3, p. 203, a log-concave probability measure on $\mathbb{R}^n$ satisfies a Poincaré inequality with a constant depending on the dimension. Theorem 8.7.1 indicates that under exponential concentration, the Poincaré constant only depends on the concentration parameters.

There is a corresponding statement for Gaussian concentration. Say that (E, μ, Γ) has *Gaussian concentration* if there are $C, c > 0$ such that for every integrable 1-Lipschitz function $f : E \to \mathbb{R}$,

$$\mu\left(f \geq \int_E f d\mu + r\right) \leq C e^{-cr^2}, \quad r \geq 0. \tag{8.7.3}$$

As shown in Proposition 5.4.1, p. 252, under a logarithmic Sobolev inequality $LS(C')$, (E, μ, Γ) has Gaussian concentration with $C = 2$ and $c = \frac{1}{2C'}$. Again, there is a converse under the curvature condition $CD(0, \infty)$.

Theorem 8.7.2 (Milman's Theorem II) *Let (E, μ, Γ) be a Markov Triple, with μ a probability measure, satisfying the curvature condition $CD(0, \infty)$. If (E, μ, Γ) has Gaussian concentration with constants $C, c > 0$, then*

$$\mathcal{I}_\mu \geq c' \mathcal{I},$$

8.7 From Concentration to Isoperimetry

where $c' > 0$ only depends on C, c and $\mathcal{I}$ is the Gaussian isoperimetric function defined at (8.5.7). (In other words, (E, μ, Γ) satisfies a Gaussian isoperimetric-type inequality.) In particular, (E, μ, Γ) satisfies a logarithmic Sobolev inequality $LS(C')$ where $C' > 0$ only depends on C, c.

Both Theorems 8.7.1 and 8.7.2 bound the isoperimetric functions $\mathcal{I}_\mu$ for the Markov Triple (E, μ, Γ) from below by the isoperimetric functions of exponential and (standard) Gaussian measure. It is not too difficult to verify, either directly or via the subsequent Poincaré or logarithmic Sobolev inequalities, that conversely such lower bounds ensure the exponential, respectively Gaussian, concentration properties (8.7.1) and (8.7.3). As such, Theorems 8.7.1 and 8.7.2 provide a complete connection between concentration properties and isoperimetric-type inequalities under positive curvature bounds.

The schemes of the proofs of Theorems 8.7.1 and 8.7.2 are rather similar. For simplicity, we only concentrate on the first one on exponential concentration.

Proof of Theorem 8.7.1 By Proposition 8.5.2, the Poincaré inequality is a consequence of the lower bound (8.7.2) on the isoperimetric function. It is therefore enough to prove the latter.

Let f in $\mathcal{A}_0^{\text{const}+}$. For every $t \geq 0$, by the usual heat flow interpolation and de Bruijn's identity (5.2.3), p. 245,

$$\int_E f \log f \, d\mu - \int_E P_t f \log P_t f \, d\mu = -\int_0^t \frac{d}{ds}\left(\int_E P_s f \log P_s f \, d\mu\right) ds$$

$$= \int_0^t \int_E \frac{\Gamma(P_s f)}{P_s f} d\mu \, ds.$$

By the reverse local logarithmic Sobolev inequality (5.5.5) of Theorem 5.5.2, p. 259, whenever $0 < \varepsilon \leq f \leq 1$ where $0 < \varepsilon < 1$, for every $s > 0$,

$$\sqrt{\Gamma(P_s f)} \leq \sqrt{\frac{1}{s} \log \frac{1}{\varepsilon}} \, P_s f.$$

Hence

$$\int_E \frac{\Gamma(P_s f)}{P_s f} d\mu \leq \sqrt{\frac{1}{s} \log \frac{1}{\varepsilon}} \int_E \sqrt{\Gamma(P_s f)} \, d\mu$$

$$\leq \sqrt{\frac{1}{s} \log \frac{1}{\varepsilon}} \int_E \sqrt{\Gamma(f)} \, d\mu$$

where the last inequality follows from the fact that, under $CD(0, \infty)$, the strong gradient bound $\sqrt{\Gamma(P_s f)} \leq P_s(\sqrt{\Gamma(f)})$ holds (cf. Theorem 3.3.18, p. 163). It follows from the preceding that for every $t > 0$ and $0 < \varepsilon < 1$, and every function f in

$\mathcal{A}_0^{\mathrm{const}+}$ such that $\varepsilon \leq f \leq 1$,

$$\int_E f \log f \, d\mu - \int_E P_t f \log P_t f \, d\mu \leq 2\sqrt{t \log \frac{1}{\varepsilon}} \int_E \sqrt{\Gamma(f)} \, d\mu. \tag{8.7.4}$$

Recall next (5.6.4), p. 267, which tells us that $-\psi = -\sqrt{\log \frac{2}{P_t f}}$ is $\frac{1}{2\sqrt{t}}$-Lipschitz. (Note that Proposition 8.6.1 may also be used for this task.) By the exponential concentration hypothesis (8.7.1), for every $r \geq 0$,

$$\mu(\psi \leq m - r) \leq C e^{-2cr\sqrt{t}}$$

where $m = \int_E \psi \, d\mu$. By convexity of the map $u \mapsto \sqrt{\log \frac{2}{u}}$ on $(0, 1]$, $m \geq \sqrt{\log \frac{2}{\int_E f d\mu}}$ so that, for every $r > 0$ and $t > 0$,

$$\mu\left(\psi \leq \sqrt{\log \frac{2}{\int_E f d\mu}} - r\right) \leq C e^{-2cr\sqrt{t}}.$$

This implies that, for every $0 \leq r \leq \sqrt{\log \frac{2}{\int_E f d\mu}}$ and $t > 0$,

$$\mu\left(P_t f \geq \sqrt{2 \int_E f d\mu \, e^{r^2}}\right) \leq C e^{-2cr\sqrt{t}}.$$

Therefore, whenever $0 < \delta \leq 1$ is such that $\delta \geq \sqrt{2 \int_E f d\mu \, e^{r^2}}$,

$$\int_E P_t f \log \frac{1}{P_t f} \, d\mu \geq \log \frac{1}{\delta} \int_{\{P_t f \leq \delta\}} P_t f \, d\mu$$

$$= \left[\int_E f \, d\mu - \int_{\{P_t f > \delta\}} P_t f \, d\mu\right] \log \frac{1}{\delta} \tag{8.7.5}$$

$$\geq \left[\int_E f \, d\mu - C e^{-2cr\sqrt{t}}\right] \log \frac{1}{\delta}.$$

Let $\delta = [2(\varepsilon + (1-\varepsilon)\mu(A))]^{1/2} e^{r^2}$, which we assume to be less than 1. The previous inequalities (8.7.4) and (8.7.5) extend to $\mathcal{D}(\mathcal{E})$. Now, for a closed set $A \subset E$, apply these to $f = \max(\mathbb{1}_A, \varepsilon)$, $0 < \varepsilon < 1$, actually first to some suitable Lipschitz approximations $f_\eta = \max(\varepsilon, (1 - \frac{1}{\eta} d(\cdot, A))_+)$ with $\eta \to 0$, $\eta > 0$ (which belong to $\mathcal{D}(\mathcal{E})$ by hypothesis). For this choice of f, (8.7.4) and (8.7.5) (together with (8.5.2)) imply that

$$-\varepsilon \log \frac{1}{\varepsilon} + \left[(1-\varepsilon)\mu(A) - C e^{-2cr\sqrt{t}}\right] \log \frac{1}{\delta} \leq 2\sqrt{t \log \frac{1}{\varepsilon}} \, \mu^+(A).$$

It remains to suitably optimize the various parameters. The simple choices $\varepsilon = \mu(A)^2$ and $r^2 = \frac{1}{8} \log \frac{1}{\varepsilon} = \frac{1}{4} \log \frac{1}{\mu(A)}$ satisfy $\delta \leq 2\mu(A)^{1/4} \leq 1$, for $\mu(A) \leq \frac{1}{16}$, and $r \leq \sqrt{\log \frac{2}{\int_E f d\mu}}$. Moreover, for every $t > 0$ and $\mu(A) \leq \frac{1}{16}$

$$-2\mu(A)^2 \log \frac{1}{\mu(A)} + \frac{1}{4}\left[\frac{\mu(A)}{2} - Ce^{-2cr\sqrt{t}}\right] \log \frac{1}{16\mu(A)} \leq 4\sqrt{2}\, r\sqrt{t}\, \mu^+(A).$$

As a consequence, for $t > 0$ well chosen of the order of $\log \frac{1}{\mu(A)}$ (for instance satisfying $r\sqrt{t} = \frac{1}{2c} \log \frac{4C}{\mu(A)}$), there exists a $c' > 0$ depending only on $C, c > 0$ such that for every set A with $0 < \mu(A) \leq c'$,

$$c'\mu(A) \leq \mu^+(A). \tag{8.7.6}$$

Similar, and actually easier, arguments show that (8.7.6) may be extended to every set A with $0 < \mu(A) \leq \frac{1}{2}$. Taking the complement yields (8.7.2) and therefore Theorem 8.7.1. □

8.8 Notes and References

The notion of capacity has been developed in the second part of the 20th century in various parts of mathematics, including functional analysis, potential theory, harmonic analysis and geometric measure theory. A comprehensive account on the notion of measure-capacity inequalities and the various developments in analysis, together with bibliographical references, is the monograph [303] by V. G. Maz'ya. Early developments in the context of geometric measure theory are considered by H. Federer in [184]. See also [97, 98].

The first co-area formula of Theorem 8.1.2 may be found in [303, Chap. 2].

The topic of Sect. 8.2 linking classical Sobolev inequalities and capacity inequalities is also discussed in [216, 303], where in particular the general Proposition 8.2.1 on capacity inequalities and pairs of Orlicz functions is emphasized. The classical Faber-Krahn inequality compares the first eigenvalue of the Laplacian on a bounded open domain in $\mathbb{R}^n$ with Dirichlet boundary conditions to that of the ball with the same volume (cf. e.g. [121, 303]). As alluded to in Remark 8.2.2, such inequalities may be studied similarly in the context of measure-capacity inequalities to formulate equivalently functional inequalities [35, 114, 216, 303].

The corresponding statements for Poincaré and logarithmic Sobolev inequalities in Sect. 8.3 have been developed more recently in the works of F. Barthe, P. Cattiaux and C. Roberto [46, 49]. Their results may be understood as a generalization of the one-dimensional characterizations of B. Muckenhoupt [321] and S. Bobkov and

F. Götze [77] discussed respectively in Chap. 4, Sect. 4.5, p. 193, and Chap. 5, Sect. 5.4, p. 252.

Generalized Nash inequalities from the point of view of capacity inequalities are studied in [47, 207, 450]. For the corresponding results for weak Poincaré inequalities, see [45]. Characterizations in dimension one of Latała-Oleszkiewicz inequalities are considered in [49].

Extensions of the method have been studied for further families of functional inequalities including F-Sobolev inequalities (cf. [46]), weak logarithmic Sobolev inequalities (cf. [116]) or $\mathbb{L}^q$-logarithmic Sobolev inequalities (cf. [162]).

The co-area formula of Theorem 8.5.1 in Sect. 8.5 is a standard statement which may be found in numerous references including [97, 122, 184, 303]. See also [78, 295, 313]. We refer to these references for an account of the historical developments. Proposition 8.5.2 linking (exponential) isoperimetric-type bounds with Poincaré inequalities is an observation going back to J. Cheeger [124] in Riemannian geometry (see also [99, 442]). It has been the source of many extensions on the basis of the same principle (cf. for instance [78, 312]). The monograph [78] by S. Bobkov and C. Houdré studies in particular a variety of statements along these lines connecting isoperimetric-type inequalities and functional inequalities, including those for Gaussian measures of Sect. 8.5.2 (see below). Moreover, the metric framework of Sect. 8.5 (surface measure, co-area formulas etc.) is carefully described there and we refer the reader to it for all the necessary technical details. General introductions to isoperimetric and geometric inequalities include [98, 122, 195, 338].

The isoperimetric inequality for Gaussian measures (Sect. 8.5.2) is due to C. Borell [87] and V. Sudakov and B. Cirel'son [401]. The functional form of the Gaussian isoperimetric inequality illustrated in Corollary 8.5.4 was introduced by S. Bobkov [73] who established it first on the two-point space and then in the limit for the Gaussian measure by the central limit Theorem (following the original approach of L. Gross [224] in his proof of the logarithmic Sobolev inequality, cf. the Notes and References in Chap. 5). On the basis of the Bobkov inequality, the local inequalities of Theorem 8.5.3 were established in [38]. Corollary 8.5.5 may be considered as the infinite-dimensional extension of a famous result of P. Lévy and M. Gromov [220, 281, 316] comparing the isoperimetric profile of a Riemannian manifold with a strictly positive lower bound on the Ricci curvature to that of the sphere with the same (constant) curvature and dimension (see also [60, 193, 221, 315]). A purely Markov operator proof of this statement under the curvature-dimension condition $CD(\rho, n)$, $\rho > 0$, $n < \infty$, is yet to be found. Alternative proofs of Corollary 8.5.5 for a probability measure $d\mu = e^{-W} dx$ on $\mathbb{R}^n$ with $\nabla\nabla W \geq \rho\, \text{Id}$, $\rho > 0$, have been provided in [103] by mass transportation methods (more precisely using Theorem 9.3.4, p. 447, in the next chapter), in [75] via the localization method of [293] and in [318] with a geometric derivation. See [315] for a recent complete geometric picture of isoperimetric comparison theorems with families of one-dimensional models under curvature-dimension conditions and diameter bounds.

8.8 Notes and References

Section 8.6 is taken from [37]. Proposition 8.6.1 already appeared in [38] (in an alternative proof of Gaussian isoperimetry).

Theorems 8.7.1 and 8.7.2 of Sect. 8.7 connecting isoperimetric and concentration properties in spaces with positive curvature are due to E. Milman [312, 314] in a (weighted) Riemannian manifold setting. His results go far beyond the statements presented here and cover a large spectrum of isoperimetric regimes and functional inequalities. Preliminary contributions in this context go back to [38, 99, 273] (see also [279]). The semigroup proof of Theorems 8.7.1 and 8.7.2 presented here is taken from [280]. It should be pointed out that, in a (weighted) Riemannian manifold framework with (extended) positive Ricci curvature, the isoperimetric profile of the (weighted) Riemannian measure is always concave as established by V. Bayle [52] (see [312, 314]). Therefore, in this case, the weakest concentration rate actually implies its comparison with that of the exponential measure (and thus exponential concentration). Earlier steps in the relationships between (Gaussian) measure concentration and logarithmic Sobolev inequalities are due to F.-Y. Wang [428] on the basis of his Theorem 5.6.1, p. 265 (see also [431], [14, Chap. 7] and [80]).

Chapter 9
Optimal Transportation and Functional Inequalities

This chapter is a brief investigation of the links between optimal transportation methods and functional inequalities of Poincaré, logarithmic Sobolev or Sobolev-type. Optimal transportation tools and ideas have arisen from the geometric analysis of partial differential equations, and the study of gradient flows in Wasserstein spaces and of interpolation along the geodesics of optimal transport have been particularly useful in establishing some of the functional and geometric inequalities recorded in this monograph.

Along these lines, optimal transportation will be used here in particular to present an alternative approach to sharp Sobolev or Gagliardo-Nirenberg inequalities in Euclidean space. Transportation cost inequalities comparing relative entropy and Wasserstein distances between probability measures are further investigated and we study their relationships to logarithmic Sobolev inequalities, hypercontractivity and measure concentration.

This chapter is not a complete exposition of optimal transportation (in particular most proofs of general results are omitted) or of its applications to functional inequalities. It only aims at giving a flavor of some of the results and methods in this context, and mainly concentrates on the links with functional inequalities.

In contrast to the rest of this work, this chapter will be presented in somewhat specific frameworks adapted to the various results. More precisely, the general theory of optimal transportation, as well as several transportation cost inequalities, may (and will) be presented in the classical context of topological Polish spaces. On the other hand, smooth (complete connected, weighted) Riemannian manifolds, often even Euclidean spaces, form a convenient framework in which we can easily deal with transportation maps and basic Hamilton-Jacobi equations and their (viscosity) solutions by Hopf-Lax formulas. Recent developments (see Sects. 9.8 and 9.9) have addressed extensions to general metric measure spaces involving refined non-smooth analysis. Nevertheless, to better highlight the main results, we restrict here to the smooth Riemannian setting as soon as these analytic tools are exploited.

Finally, some statements may be given in the (Full) Markov Triple (E, μ, Γ) framework of this monograph. While the general measurable Markov Triple structure might be pushed further to cover most of the results in this chapter (see again the Notes and References for recent progress in this direction), for the sake of clarity, the exposition is restricted to the specific smooth frameworks, indicating when such extensions are possible.

The first section introduces the basic definitions and general results (without proofs) of optimal transportation as well as the basic tool of the Brenier map. The main topic of transportation cost inequalities and first examples, in particular for Gaussian measures, are discussed in Sect. 9.2. Section 9.3 develops the tool of optimal transportation to establish logarithmic Sobolev and Sobolev inequalities (in Euclidean space) with sharp constants. Non-linear Hamilton-Jacobi equations are briefly presented in Sect. 9.4, while the subsequent section emphasizes a hypercontractivity property of solutions of Hamilton-Jacobi equations analogous to the one for linear heat equations. The preceding results are then applied in Sect. 9.6 to investigate the relationships between (quadratic) transportation cost inequalities and logarithmic Sobolev inequalities. Section 9.7 investigates contraction properties in Wasserstein space along the heat semigroup under a curvature condition by means of commutation between the heat and Hopf-Lax semigroups. Section 9.8 is a very brief overview of recent developments towards a notion of Ricci curvature lower bounds based on optimal transportation.

9.1 Optimal Transportation

This section presents some of the basic definitions and fundamental theorems (without proofs) concerning optimal transportation costs and couplings.

9.1.1 Optimal Transportation

Although we will mostly deal with Euclidean or Riemannian spaces in the various illustrations, in order to develop some of the preliminaries of optimal transportation, a natural topological framework is that of a Polish space (complete separable metric space) (E, d) equipped with its Borel σ-field $\mathcal{F}$. The product space $E \times E$ is equipped with the product Borel σ-field. $\mathcal{P}(E)$ denotes the set of probability measures on $(E, \mathcal{F})$.

Definition 9.1.1 (Optimal transportation cost) Let $c : E \times E \to [0, \infty]$ be a lower semi-continuous function and let $\mu, \nu \in \mathcal{P}(E)$. The optimal transportation cost for transporting μ onto ν with cost c is defined by

$$\mathcal{T}_c(\mu, \nu) = \inf \int_{E \times E} c(x, y) d\pi(x, y) \in [0, \infty] \qquad (9.1.1)$$

where the infimum runs over the set of probability measures $\pi \in \mathcal{P}(E \times E)$ with respective marginals μ and ν. That is, for all bounded measurable functions u and

9.1 Optimal Transportation

v on E,

$$\int_{E \times E} [u(x) + v(y)] d\pi(x, y) = \int_E u \, d\mu + \int_E v \, dv.$$

The function c is called a cost function and a measure $\pi \in \mathcal{P}(E \times E)$ with marginals μ and v is called a coupling of (μ, v).

Note that the set over which we take the infimum in (9.1.1) is not empty, the product measure $\mu \otimes v$ being a coupling of μ and v. The first result concerns the existence of such an optimal coupling.

Proposition 9.1.2 (Optimal coupling) *Given $\mu, v \in \mathcal{P}(E)$, there exists an optimal coupling π of (μ, v) such that*

$$\mathcal{T}_c(\mu, v) = \int_{E \times E} c \, d\pi.$$

The optimal transportation cost admits a powerful dual representation.

Theorem 9.1.3 (Kantorovich's Theorem) *Let $c : E \times E \to [0, \infty]$ be a cost function and let $\mu, v \in \mathcal{P}(E)$. Then*

$$\mathcal{T}_c(\mu, v) = \sup \left(\int_E u \, d\mu - \int_E v \, dv \right) \in [0, \infty] \qquad (9.1.2)$$

where the supremum runs over all bounded continuous functions u and v (or in $\mathbb{L}^1(\mu)$ and $\mathbb{L}^1(v)$ respectively) satisfying, for all $(x, y) \in E \times E$,

$$u(x) - v(y) \leq c(x, y). \qquad (9.1.3)$$

If the cost function is given by a distance function $\bar{d}(x, y) = c(x, y)$ on $E \times E$ then Kantorovich's Theorem takes an another formulation.

Theorem 9.1.4 (Kantorovich-Rubinstein Theorem) *Let $\bar{d}$ be a distance on E, lower semi-continuous with respect to d (not necessarily the initial distance of (E, d)). Then*

$$\mathcal{T}_{\bar{d}}(\mu, v) = \sup \left(\int_E u \, d\mu - \int_E u \, dv \right)$$

where the supremum runs over all 1-Lipschitz (with respect to $\bar{d}$) functions u. (The functions u may be assumed furthermore to be bounded.)

As already indicated by the preceding theorem, in this short exposition of optimal transportation we will not be interested in the most general cost functions c, and mainly consider $c = d^p$ where d is the metric on E and $p \geq 1$ (actually, we even restrict ourselves below to $p = 1$ or 2). Denote by $\mathcal{P}_p(E)$ the space of probability measures μ on E with finite p-th moment (i.e. such that $\int_E d(x,x_0)^p d\mu(x) < \infty$ for some $x_0 \in E$). Therefore, for the choice of $c = d^p$, the optimal transportation cost gives rise to the so-called *Wasserstein distance*

$$W_p(\mu, \nu) = \inf\left(\int_{E\times E} d(x,y)^p d\pi(x,y)\right)^{1/p} = \mathcal{T}_{d^p}(\mu,\nu)^{1/p}, \qquad (9.1.4)$$

where as above the infimum runs over all couplings π of (μ, ν) such that $\mu, \nu \in \mathcal{P}_p(E)$. Note, as is easily seen, that if $\mu \in \mathcal{P}_p(E)$ and $W_p(\mu, \nu) < \infty$, then necessarily $\nu \in \mathcal{P}_p(E)$. From a more probabilistic point of view,

$$W_p(\mu, \nu) = \inf\bigl(\mathbb{E}\bigl(d(X,Y)^p\bigr)\bigr)^{1/p}$$

where the infimum is over all random variables X and Y with respective laws μ and ν. On $\mathcal{P}_p(E)$, the Wasserstein distance W_p metrizes the weak convergence topology together with convergence of the respective p-th moments, defining the corresponding *Wasserstein space*.

The Kantorovich-Rubinstein Theorem 9.1.4 is of particular interest in two special cases. First, choose $\bar{d} = d$ where d is the distance on the Polish space E. This choice gives rise to the Wasserstein distance W_1. Another choice is the trivial metric $\bar{d}(x,y) = \mathbb{1}_{x\neq y}$, $(x,y) \in E \times E$, which yields the total variation distance between μ and ν,

$$\mathcal{T}_{\bar{d}}(\mu,\nu) = \frac{1}{2}\int_E \left|1 - \frac{d\nu}{d\mu}\right| d\mu = \|\mu - \nu\|_{\mathrm{TV}} = \sup_{A\in\mathcal{F}} |\mu(A) - \nu(A)|. \qquad (9.1.5)$$

9.1.2 The Brenier Map

One main question in optimal transportation is to find an explicit coupling π of (μ, ν) optimizing (9.1.1) or equivalently, in the dual formulation, to find optimal functions u and v such that (9.1.2) is an equality. An even more stringent requirement, known as *Monge's problem*, is to find a (measurable) map $T : E \to E$ pushing forward μ onto ν and such that $\pi = (\mathrm{Id}, T)\#\mu$ is an optimal coupling. By definition of the push-forward $\#$,

$$\int_{E\times E} h(x,y) d\pi(x,y) = \int_E h(x, T(x)) d\mu(x)$$

for all, say, bounded measurable functions $h : E \times E \to \mathbb{R}$. The following theorem, emphasizing the *Brenier map*, answers this question in Euclidean (or Riemannian) spaces for the quadratic cost $c(x,y) = |x-y|^2$, $(x,y) \in \mathbb{R}^n \times \mathbb{R}^n$.

Theorem 9.1.5 (Brenier's map) *Let μ and ν be two probability measures on the Borel sets of $\mathbb{R}^n$ with μ absolutely continuous with respect to the Lebesgue measure*

9.1 Optimal Transportation

and such that $W_2(\mu, \nu) < \infty$. Then there exists a convex function $\phi : \mathbb{R}^n \to \mathbb{R}$ such that $T = \nabla \phi$ maps μ onto ν (denoted $T\#\mu = \nu$), where, here, $\nabla \phi$ is considered as a map $\mathbb{R}^n \to \mathbb{R}^n$. In other words, for every bounded measurable function h on $\mathbb{R}^n$,

$$\int_{\mathbb{R}^n} h \, d\nu = \int_{\mathbb{R}^n} h(\nabla \phi) d\mu. \tag{9.1.6}$$

Moreover, $T = \nabla \phi$ is uniquely determined μ-almost everywhere and if μ and ν have a finite second moment, $T = \nabla \phi$ is optimal in the sense that

$$W_2(\mu, \nu)^2 = \int_{\mathbb{R}^n} \left| x - T(x) \right|^2 d\mu(x). \tag{9.1.7}$$

Brenier's Theorem may be suitably extended in a Riemannian manifold for the cost given by the square of the Riemannian distance. The Brenier map $T = \nabla \phi$ then has to be replaced by the geodesic exponential map generated by the gradient of a function satisfying a suitable convexity property (which would be $\frac{|x|^2}{2} - \phi$ in the Euclidean case).

If μ and ν have densities f and g with respect to the Lebesgue measure on $\mathbb{R}^n$, according to (9.1.6), for every bounded measurable map $h : \mathbb{R}^n \to \mathbb{R}$,

$$\int_{\mathbb{R}^n} h(y) g(y) dy = \int_{\mathbb{R}^n} h\bigl(\nabla \phi(x)\bigr) f(x) dx.$$

Whenever the change of variable $y = \nabla \phi(x)$ is licit, the preceding leads to the so-called *Monge-Ampère equation*

$$f = g(\nabla \phi) \det(\nabla \nabla \phi) \tag{9.1.8}$$

(where $\nabla \nabla \phi$ is the matrix of the second derivatives of ϕ). However, the map $T = \nabla \phi$ exists only almost everywhere, and may not be differentiable in any usual sense. The Monge-Ampère equation may then be understood in a generalized sense. Actually it can be proved that ϕ is locally Lipschitz on the interior of its domain and that the Monge-Ampère equation is valid fdx-almost everywhere with $\nabla \nabla \phi$ being understood as the Hessian of ϕ in the sense of Aleksandrov (the absolutely continuous part of the distributional Hessian of ϕ). Alternatively, suitable assumptions on f and g ensure its validity as in the next statement.

In the following, we say that a function f defined on an open set $\mathcal{O}$ in $\mathbb{R}^n$ belongs to $\mathcal{C}^{k,\alpha}(\mathcal{O})$ for some $k \in \mathbb{N}$ if $f \in \mathcal{C}^k(\mathcal{O})$ and all its derivatives up to order k are locally Hölder continuous with exponent $\alpha \in (0, 1)$.

Theorem 9.1.6 (Caffarelli's regularity properties) *Let μ and ν be probability measures on the Borel set of $\mathbb{R}^n$ with respective densities f and g defined on bounded open sets $\mathcal{O}_\mu$ and $\mathcal{O}_\nu$ in $\mathbb{R}^n$ such that $\varepsilon \leq f, g \leq \frac{1}{\varepsilon}$ for some $\varepsilon > 0$. Then, if $f \in \mathcal{C}^{1,\alpha}(\mathcal{O}_\mu)$ and $g \in \mathcal{C}^{1,\alpha}(\mathcal{O}_\nu)$ with $\alpha \in (0, 1)$, and if $\mathcal{O}_\mu$ is a convex set, the Brenier map ϕ is $\mathcal{C}^{2,\alpha}(\mathcal{O}_\mu)$ and the Monge-Ampère equation (9.1.8) holds on $\mathcal{O}_\mu$. Moreover, if f and g are $\mathcal{C}^\infty$, then ϕ is also $\mathcal{C}^\infty$.*

9.2 Transportation Cost Inequalities

This section addresses transportation cost inequalities which are built on the cost functions discussed in the preceding section. More precisely, transportation cost inequalities compare a transportation cost distance to a fluctuation distance expressed by (relative) entropy.

The first example of a transportation cost inequality is the Pinsker-Csizsár-Kullback inequality (5.2.2), p. 244,

$$\|\mu - \nu\|_{TV}^2 \leq \frac{1}{2} H(\nu \mid \mu),$$

where we recall that, for probability measures μ, ν on a metric space (E, d),

$$H(\nu \mid \mu) = \int_E \log \frac{d\nu}{d\mu} \, d\nu$$

if $\nu \ll \mu$ and $H(\nu \mid \mu) = \infty$ if not, is the relative entropy of ν with respect to μ (cf. (5.2.1), p. 244). Now, as discussed in (9.1.5), the total variation distance may be interpreted as a Wasserstein distance W_1 with respect to the trivial metric $\bar{d}(x, y) = \mathbb{1}_{x \neq y}$ on $E \times E$.

On the basis of this example, one may more generally consider *transportation cost inequalities* $T_1(C)$ for a given $\mu \in \mathcal{P}_1(E)$ as

$$W_1(\mu, \nu)^2 \leq 2C \, H(\nu \mid \mu) \tag{9.2.1}$$

for some $C > 0$ and all $\nu \in \mathcal{P}(E)$. (Note that under this inequality, $H(\nu \mid \mu) = \infty$ whenever $\nu \notin \mathcal{P}_1(E)$.) This family of transportation cost inequalities will in particular be studied below in connection with concentration inequalities.

However, the connections between transportation cost inequalities, partial differential equations and functional inequalities as discussed in this work are actually of most interest for a quadratic cost and the associated W_2 Wasserstein distance (with respect to the Euclidean or Riemannian structures). The first basic example in this setting is the following quadratic transportation cost inequality for the Gaussian measure in $\mathbb{R}^n$. Recall that $\mathcal{P}_2(\mathbb{R}^n)$ denotes the set of probability measures on $\mathbb{R}^n$ with a second moment.

Theorem 9.2.1 (Talagrand's inequality) *Let $d\mu(x) = (2\pi)^{-n/2} e^{-|x|^2/2} dx$ be the standard Gaussian measure on the Borel sets of $\mathbb{R}^n$. Then, for any $\nu \in \mathcal{P}(\mathbb{R}^n)$,*

$$W_2(\mu, \nu)^2 \leq 2 H(\nu \mid \mu).$$

It is instructive to first present a one-dimensional proof this theorem.

Proof Let $n = 1$. Let $f \geq 0$ be such that $\int_\mathbb{R} f d\mu = 1$ and $W_2(\mu, \nu) < \infty$ where $d\nu = f d\mu$. In this one-dimensional setting, the idea is to describe the explicit map

9.2 Transportation Cost Inequalities

pushing forward μ onto ν. To this end, by standard approximation arguments, assume that $f > 0$ everywhere. We may therefore uniquely define $T : \mathbb{R} \to \mathbb{R}$ by

$$\nu\big((-\infty, T(x)]\big) = \mu\big((-\infty, x]\big), \quad x \in \mathbb{R},$$

so that ν is the image measure of μ under the strictly increasing differentiable map T. Since then, for every bounded measurable $h : \mathbb{R} \to \mathbb{R}$,

$$\int_{\mathbb{R}} h(y) f(y) e^{-y^2/2} dy = \int_{\mathbb{R}} h(T(x)) e^{-x^2/2} dx,$$

by the change of variables formula $y = T(x)$,

$$f(T(x)) e^{-T(x)^2/2} T'(x) = e^{-x^2/2}, \quad x \in \mathbb{R}.$$

This identity is the Monge-Ampère equation (9.1.8) in this case. Hence, taking logarithms, for every x,

$$\log f(T(x)) + \log T'(x) - \frac{1}{2} T(x)^2 = -\frac{x^2}{2}.$$

Integrating with respect to μ, and using that $T\#\mu = \nu$,

$$\int_{\mathbb{R}} \log f \, d\nu = \frac{1}{2} \int_{\mathbb{R}} [T(x)^2 - x^2] d\mu - \int_{\mathbb{R}} \log T' d\mu.$$

Integrating by parts with respect to the Gaussian measure μ classically yields (cf. (2.7.7), p. 107) that $\int_{\mathbb{R}} x(T-x) d\mu = \int_{\mathbb{R}} (T'-1) d\mu$ so that

$$\mathrm{H}(\nu \mid \mu) = \int_{\mathbb{R}} \log f \, d\nu = \frac{1}{2} \int_{\mathbb{R}} |x - T(x)|^2 d\mu + \int_{\mathbb{R}} [T' - 1 - \log T'] d\mu$$

$$\geq \frac{1}{2} \int_{\mathbb{R}} |x - T(x)|^2 d\mu$$

where we used that $r - 1 - \log r \geq 0$, $r > 0$. Now, since ν is the image of μ under T, the image measure π of μ under the map $x \mapsto (x, T(x))$ has marginals μ and ν respectively so that

$$\int_{\mathbb{R}} |x - T(x)|^2 d\mu = \int_{\mathbb{R}\times\mathbb{R}} |x - y|^2 d\pi \geq W_2(\mu, \nu)^2.$$

The theorem is therefore established (in the one-dimensional case). □

The extension to arbitrary dimension n raises interesting issues, and different options. The first option is to tensorize the Talagrand inequality of Theorem 9.2.1 as was done in the earlier chapters on Poincaré and logarithmic Sobolev inequalities. This may indeed be performed, either directly or on a dual formulation as a consequence of the Kantorovich duality. We follow this route below, before studying another option in the next section. To suitably present the arguments, let us introduce the general definition of a quadratic transportation cost inequality (covering Theorem 9.2.1 for Gaussian measures). As before, (E, d) denotes a Polish space. Recall the space $\mathcal{P}_2(E)$ of probability measures on the Borel sets of E with a finite second moment.

Definition 9.2.2 (Quadratic transportation cost inequality) We say that $\mu \in \mathcal{P}_2(E)$ satisfies a quadratic transportation cost inequality $T_2(C)$ with constant $C > 0$ if for every $\nu \in \mathcal{P}(E)$,

$$W_2(\mu, \nu)^2 \leq 2C\, \mathrm{H}(\nu \,|\, \mu).$$

Talagrand's inequality of Theorem 9.2.1 tells us that the standard Gaussian measure on $\mathbb{R}^n$ satisfies $T_2(1)$ (justifying again the choice in the normalization of the constant C).

The following equivalent dual formulation of Definition 9.2.2 will turn out to be important and useful. With this task in mind, we introduce, for any continuous function $f : E \to \mathbb{R}$, the infimum-convolutions

$$Q_t f(x) = Q_t(f)(x) = \inf_{y \in E} \left\{ f(y) + \frac{1}{2t} d(x, y)^2 \right\}, \quad t > 0, \; x \in E. \quad (9.2.2)$$

Note the homogeneity $Q_t f = \frac{1}{t} Q_1(tf)$, $t > 0$. The operators Q_t, $t > 0$, will be identified in Sect. 9.4 as the fundamental solutions of Hamilton-Jacobi equations (in smooth settings). Their relevance at this point is expressed by the Kantorovich duality Theorem 9.1.3 which indicates that

$$\frac{1}{2} W_2(\mu, \nu)^2 = \sup \left(\int_E Q_1 f \, d\nu - \int_E f \, d\mu \right) \quad (9.2.3)$$

where the supremum runs over all bounded continuous functions $f : E \to \mathbb{R}$.

Proposition 9.2.3 (Dual formulation of the quadratic transportation cost inequality) *A probability measure μ in $\mathcal{P}_2(E)$ satisfies the quadratic transportation cost inequality $T_2(C)$ for some $C > 0$ if and only if for every bounded continuous function $f : E \to \mathbb{R}$,*

$$\int_E e^{Q_C f} d\mu \leq e^{\int_E f d\mu}. \quad (9.2.4)$$

Proof We first prove (9.2.4) from $T_2(C)$. According to (9.2.3), if $\frac{d\nu}{d\mu} = g$, the quadratic transportation cost inequality $T_2(C)$ expresses that

$$\int_E Q_1 f \, g \, d\mu - \int_E f \, d\mu \leq C \, \mathrm{H}(\nu \,|\, \mu) = C \int_E g \log g \, d\mu.$$

Choosing

$$g = \frac{e^{(Q_1 f)/C}}{\int_E e^{(Q_1 f)/C} d\mu}$$

9.2 Transportation Cost Inequalities

in the latter shows that

$$\log \int_E e^{(Q_1 f)/C} d\mu \leq \frac{1}{C} \int_E f \, d\mu.$$

The conclusion follows since by homogeneity $\frac{1}{C} Q_1 f = Q_C(\frac{f}{C})$. Conversely, (9.2.4) implies that $\int_E e^h d\mu \leq 1$ with $h = Q_C f - \int_E f \, d\mu$. If $g = \frac{d\nu}{d\mu}$ where $\nu \in \mathcal{P}(E)$, the variational formula (5.1.3), p. 236, yields that

$$\int_E Q_C f \, g \, d\mu - \int_E f \, d\mu = \int_E g h \, d\mu \leq \int_E g \log g \, d\mu = \mathrm{H}(\nu \mid \mu).$$

Again by means of (9.2.3), the $T_2(C)$ inequality follows after changing f into $\frac{f}{C}$. □

Before turning to the tensorization issues, note that a quadratic transportation cost inequality $T_2(C)$ (for $\mu \in \mathcal{P}_2(E)$) implies an inequality $T_1(C)$ of the form (9.2.1), with the same constant $C > 0$. Moreover, following the scheme of proof of the above Proposition 9.2.3, such an inequality $T_1(C)$ may be translated to the equivalent

$$\int_E e^{sf} d\mu \leq e^{s \int_E f d\mu + C s^2/2} \tag{9.2.5}$$

for every $s \in \mathbb{R}$ and every (bounded) 1-Lipschiz function $f : E \to \mathbb{R}$. Such T_1 inequalities appear therefore as consequences of logarithmic Sobolev inequalities by the Herbst argument of Proposition 5.4.1, p. 252. In particular, they equivalently describe Gaussian concentration properties in the sense of (8.7.3), p. 426. Indeed, as already illustrated in Sect. 5.4, p. 252, (9.2.5) implies by Markov's inequality that

$$\mu\left(f \geq \int_E f d\mu + r\right) \leq e^{-r^2/2C}, \quad r \geq 0, \tag{9.2.6}$$

for every integrable 1-Lipschitz function $f : E \to \mathbb{R}$. After integration, these concentration properties are equivalent to transportation cost inequalities T_1 in the form (9.2.5), up to the numerical constants.

On the basis of the characterization of Proposition 9.2.3, the stability of quadratic transportation cost inequalities under products is easily obtained. It may be worthwhile to observe that T_1 inequalities, being equivalent to (9.2.5), do not tensorize independently of the dimension of the product.

Proposition 9.2.4 (Stability under product) *Let μ_1 and μ_2 be probability measures on metric spaces (E_1, d_1) and (E_2, d_2) satisfying respectively $T_2(C_1)$ and $T_2(C_2)$ for constants $C_1 > 0$ and $C_2 > 0$. Then $\mu_1 \otimes \mu_2$ on $(E_1 \times E_2, d_1 \oplus d_2)$ where $d_1 \oplus d_2 = (d_1^2 + d_2^2)^{1/2}$ satisfies $T_2(C)$ with $C = \max(C_1, C_2)$.*

Proof We work accordingly with the dual formulation of Proposition 9.2.3. Let $f : E_1 \times E_2 \to \mathbb{R}$ be bounded and continuous. For $C = \max(C_1, C_2)$ and

$(x_1, x_2) \in E_1 \times E_2$,

$$Q_C f(x_1, x_2)$$
$$\leq \inf_{(y_1, y_2) \in E_1 \times E_2} \left\{ f(y_1, y_2) + \frac{1}{2C_1} d_1(x_1, y_1)^2 + \frac{1}{2C_2} d_2(x_2, y_2)^2 \right\}.$$

For every $x_2 \in E_2$, define $z_1 \in E_1 \mapsto \widetilde{Q}_{C_2} f(z_1, x_2)$ as

$$\widetilde{Q}_{C_2} f(z_1, x_2) = Q_{C_2} f(z_1, \cdot)(x_2) = \inf_{y_2 \in E_2} \left\{ f(z_1, y_2) + \frac{1}{2C_2} d_2(x_2, y_2)^2 \right\}$$

so that

$$Q_C f(x_1, x_2) \leq Q_{C_1}\big(\widetilde{Q}_{C_2} f(\cdot, x_2)\big)(x_1).$$

By the dual transportation cost inequality (9.2.4) along the x_1 variable,

$$\int_{E_1} e^{Q_C f(x_1, x_2)} d\mu_1(x_1) \leq \int_{E_1} e^{Q_{C_1}(\widetilde{Q}_{C_2}(f(\cdot, x_2))) (x_1)} d\mu_1(x_1)$$
$$\leq \exp\left(\int_{E_1} \widetilde{Q}_{C_2} f(x_1, x_2) d\mu_1(x_1) \right).$$

Now, since $\widetilde{Q}_{C_2}$ is an infimum,

$$\int_{E_1} \widetilde{Q}_{C_2} f(x_1, x_2) d\mu_1(x_1) \leq Q_{C_2}\left(\int_{E_1} f(x_1, \cdot) d\mu_1(x_1) \right)(x_2),$$

so that integrating in $d\mu_2(x_2)$ and applying Proposition 9.2.3 again, this time along the variable x_2, yields the conclusion. □

Together with Proposition 9.2.4, we may then conclude the proof of Talagrand's inequality of Theorem 9.2.1 in any dimension. Now, another way to obtain this theorem is to try to perform a transportation proof directly in dimension n with the help of the Brenier transportation map. This is the approach followed in the next section, which will prove useful not only for transportation cost inequalities but also for other functional inequalities such as logarithmic Sobolev and Sobolev inequalities.

9.3 Transportation Proofs of Functional Inequalities

As announced, this section develops the mass transportation method in order to reach the preceding transportation cost inequality for Gaussian measures as well as, in an alternative way, some of the functional inequalities considered so far such as logarithmic Sobolev and Sobolev inequalities (in Euclidean or Riemannian settings). The key idea is the use of the Brenier map $T = \nabla \phi$ and of the associated

9.3 Transportation Proofs of Functional Inequalities

Monge-Ampère equation (9.1.8) together with the convexity properties of ϕ. In particular, log-concavity of the map $Q \mapsto \det(Q)$ or concavity of $Q \mapsto \det(Q)^{1/n}$ on the set of symmetric positive $n \times n$ matrices Q will provide the technical step towards the various conclusions. The precise arguments usually require some refined regularity analysis, which we do not necessarily enter into here. To some extent, these may be handled either by Theorem 9.1.6 or by using the Hessian in the sense of Aleksandrov.

9.3.1 HWI Inequality

This section introduces a new inequality which encompasses several inequalities of interest such as logarithmic Sobolev and transportation cost inequalities. The following statement is a general formulation from which the various cases will be easily drawn.

For simplicity we deal here with measures on $\mathbb{R}^n$, but the methods and results may be extended to a (weighted) Riemannian setting. Let $d\mu = e^{-W} dx$ be a probability measure on the Borel sets of $\mathbb{R}^n$ where $W : \mathbb{R}^n \to \mathbb{R}$ is a smooth (of class $\mathcal{C}^2$) potential such that $\nabla\nabla W \geq \rho \, \mathrm{Id}$ as symmetric matrices for some $\rho \in \mathbb{R}$ (equivalently, $W(x) - \frac{\rho|x|^2}{2}$ is convex). In other words, $\mathbb{R}^n$ with the measure $d\mu = e^{-W} dx$ and the standard carré du champ operator $\Gamma(f) = |\nabla f|^2$ of the diffusion operator $L = \Delta - \nabla W \cdot \nabla$ invariant with respect to μ is a Markov Triple (as a weighted Riemannian manifold) of curvature $CD(\rho, \infty)$. The standard Gaussian measure corresponds as usual to $\rho = 1$. The following theorem is established in the same way as Theorem 9.2.1 using the Brenier map in place of the explicit one-dimensional transport. If $f : \mathbb{R}^n \to \mathbb{R}_+$ is measurable, for simplicity we denote by $f\mu$ the measure with density f with respect to μ.

Theorem 9.3.1 *Let $d\mu = e^{-W} dx$ be a probability measure on the Borel sets of $\mathbb{R}^n$ where $W : \mathbb{R}^n \to \mathbb{R}$ is a smooth potential such that $\nabla\nabla W \geq \rho \, \mathrm{Id}$ for some $\rho \in \mathbb{R}$. Furthermore, let $f, g : \mathbb{R}^n \to \mathbb{R}_+$ be probability densities with respect to μ of class $\mathcal{C}^1$ and compactly supported. If $T(x) = x + \nabla\theta(x)$ is the Brenier map pushing $f\mu$ onto $g\mu$, then*

$$\mathrm{H}(g\mu \mid \mu) \geq \mathrm{H}(f\mu \mid \mu) + \int_{\mathbb{R}^n} \nabla f \cdot \nabla \theta \, d\mu + \frac{\rho}{2} W_2(f\mu, g\mu)^2.$$

Proof We only sketch the argument, without justifying several of the regularity issues involved in the proof. Since T pushes $f\mu$ onto $g\mu$, the corresponding Monge-Ampère equation (9.1.8) reads

$$f(x) e^{-W(x)} = g(T(x)) e^{-W(x+\nabla\theta(x))} \det(\mathrm{Id} + \nabla\nabla\theta(x)), \quad x \in \mathbb{R}^n$$

(under the suitable regularity assumptions or in a generalized sense). As in the proof of Theorem 9.2.1, taking logarithms,

$$\log g(T(x)) = \log f(x) + W(x + \nabla\theta(x)) - W(x) - \log\det(\mathrm{Id} + \nabla\nabla\theta(x)).$$

Note that the matrix $\mathrm{Id} + \nabla\nabla\theta$ is positive at any point since the Brenier map ϕ is convex. Hence

$$\log\det(\mathrm{Id} + \nabla\nabla\theta(x)) \leq \Delta\theta(x).$$

Furthermore, by the convexity assumption on W, for every $x \in \mathbb{R}^n$,

$$W(x + \nabla\theta(x)) - W(x) \geq \nabla W(x) \cdot \nabla\theta(x) + \frac{\rho}{2}|\nabla\theta(x)|^2.$$

Therefore,

$$\log g(T(x)) \geq \log f(x) + \nabla W(x) \cdot \nabla\theta(x) - \Delta\theta(x) + \frac{\rho}{2}|\nabla\theta(x)|^2.$$

After integration with respect to $f\mu$,

$$\int_{\mathbb{R}^n} g \log g \, d\mu \geq \int_{\mathbb{R}^n} f \log f \, d\mu$$
$$- \int_{\mathbb{R}^n} [\Delta\theta - \nabla W \cdot \nabla\theta] f \, d\mu + \frac{\rho}{2} W_2(f\mu, g\mu)^2$$

where we used that T is the Brenier map sending $f\mu$ onto $g\mu$ so that, by (9.1.7), $\int_{\mathbb{R}^n} |\nabla\theta|^2 f \, d\mu = W_2(f\mu, g\mu)^2$. Finally, by integration by parts for the operator $L = \Delta - \nabla W \cdot \nabla$ with invariant measure μ,

$$\int_{\mathbb{R}^n} [\Delta\theta - \nabla W \cdot \nabla\theta] f \, d\mu = -\int_{\mathbb{R}^n} \nabla\theta \cdot \nabla f \, d\mu$$

which yields the claim. The proof of Theorem 9.3.1 is complete. □

Theorem 9.3.1 covers a variety of interesting cases according to the respective choices of f and g. First choosing f so that it approaches the constant function $\mathbb{1}$ smoothly, we obtain a generalization of Talagrand's inequality (Theorem 9.2.1) under the curvature condition $CD(\rho, \infty)$ for some $\rho > 0$.

Corollary 9.3.2 *Let $d\mu = e^{-W} dx$ be a probability measure on the Borel sets of $\mathbb{R}^n$ where $W : \mathbb{R}^n \to R$ is a smooth potential such that $\nabla\nabla W \geq \rho\,\mathrm{Id}$ for some $\rho > 0$. Then μ satisfies the quadratic transportation cost inequality $T_2(\frac{1}{\rho})$.*

Another instance of interest is the case where g approaches $\mathbb{1}$. We then end up with

$$\mathrm{H}(f\mu \mid \mu) \leq -\int_{\mathbb{R}^n} \nabla f \cdot \nabla\theta \, d\mu - \frac{\rho}{2} W_2(f\mu, \mu)^2.$$

9.3 Transportation Proofs of Functional Inequalities

By the Cauchy-Schwarz inequality, $-\int_{\mathbb{R}^n} \nabla f \cdot \nabla \theta \, d\mu$ is bounded from above by

$$\left(\int_{\mathbb{R}^n} |\nabla \theta|^2 f \, d\mu\right)^{1/2} \left(\int_{\mathbb{R}^n} \frac{|\nabla f|^2}{f} \, d\mu\right)^{1/2} = W_2(f\mu, \mu) I_\mu(f)^{1/2}$$

where

$$I_\mu(f) = \int_{\mathbb{R}^n} \frac{|\nabla f|^2}{f} \, d\mu$$

is the Fisher information (5.1.6), p. 237, of f (or $f\mu$) with respect to μ. (Recall that according to Remark 5.1.2, p. 238, this Fisher information $I_\mu(f)$ has to be understood as a suitable limit of strictly positive functions decreasing to f.) The resulting inequality is known as the HWI inequality (involving entropy H, Wasserstein distance W and Fisher information I_μ) and is exhibited in the next corollary.

Corollary 9.3.3 (HWI inequality) *Let $d\mu = e^{-W} dx$ be a probability measure on the Borel sets of $\mathbb{R}^n$ with a finite second moment where $W : \mathbb{R}^n \to \mathbb{R}$ is a smooth potential such that $\nabla\nabla W \geq \rho \, \mathrm{Id}$ for some $\rho \in \mathbb{R}$. For every probability density $f : \mathbb{R}^n \to \mathbb{R}_+$ with respect to μ such that $W_2(f\mu, \mu) < \infty$,*

$$H(f\mu \mid \mu) \leq W_2(f\mu, \mu) I_\mu(f)^{1/2} - \frac{\rho}{2} W_2(f\mu, \mu)^2. \qquad (9.3.1)$$

The HWI inequality actually admits an alternative proof which relies on the semigroup tools developed in this work, allowing in particular with the material presented in Sect. 8.6, p. 421, for the extension to the setting of a (Full) Markov Triple (E, μ, Γ) under the curvature condition $CD(\rho, \infty)$. We briefly outline the argument keeping for simplicity the Euclidean notation, all the arguments holding true in the same way in a Markov Triple.

Alternative proof of Corollary 9.3.3 Denote by $\mathbf{P} = (P_t)_{t \geq 0}$ the Markov semigroup with generator $L = \Delta - \nabla W \cdot \nabla$. Let $T > 0$ and let f be a smooth bounded and strictly positive function on $\mathbb{R}^n$ such that $\int_{\mathbb{R}^n} f \, d\mu = 1$. Following de Bruijn's identity (5.7.3), p. 269,

$$\mathrm{Ent}_\mu(f) = \int_0^T I_\mu(P_t f) \, dt + \mathrm{Ent}_\mu(P_T f).$$

Since $\nabla\nabla W \geq \rho \, \mathrm{Id}$, and thus the $CD(\rho, \infty)$ criterion holds, it follows from the exponential decay (5.7.4), p. 269, of Fisher information that

$$\mathrm{Ent}_\mu(f) \leq \alpha(T) I_\mu(f) + \mathrm{Ent}_\mu(P_T f) \qquad (9.3.2)$$

where $\alpha(T) = \frac{1 - e^{-2\rho T}}{2\rho}$ $(= T$ if $\rho = 0)$. The idea is now to control $\mathrm{Ent}_\mu(P_T f)$ by a mass transportation factor with the help of the log-Harnack inequality (5.6.2),

p. 266. The latter applied to $P_T f$ yields, for $x, y \in \mathbb{R}^n$,

$$P_T(\log P_T f)(x) \leq \log P_{2T} f(y) + \frac{|x-y|^2}{2\beta(T)}$$

where $\beta(T) = \frac{e^{2\rho T}-1}{\rho}$ ($= 2T$ if $\rho = 0$). In other words, taking the infimum over $y \in \mathbb{R}^n$, using the infimum-convolution notation (9.2.2),

$$P_T(\log P_T f) \leq Q_{\beta(T)}(\log P_{2T} f).$$

Therefore, by time reversibility,

$$\begin{aligned}
\mathrm{Ent}_\mu(P_T f) &= \int_{\mathbb{R}^n} f\, P_T(\log P_T f) d\mu \\
&\leq \int_{\mathbb{R}^n} f\, Q_{\beta(T)}(\log P_{2T} f) d\mu \\
&\leq \frac{1}{\beta(T)} \int_{\mathbb{R}^n} f\, Q_1\bigl(\beta(T) \log P_{2T} f\bigr) d\mu.
\end{aligned}$$

Since furthermore, by Jensen's inequality, $\int_{\mathbb{R}^n} \log P_{2T} f d\mu \leq \log(\int_{\mathbb{R}^n} P_{2T} f d\mu) = 0$, the Kantorovich duality in the form (9.2.3) ensures that

$$\mathrm{Ent}_\mu(P_T f) \leq \frac{1}{2\beta(T)} W_2(f\mu, \mu)^2.$$

Since $\frac{1}{\beta(T)} = \frac{1}{2\alpha(T)} - \rho$, combining this estimate with (9.3.2) and optimizing in $T > 0$ yields the conclusion. $\square$

The HWI inequality is a bridge between quadratic transportation cost inequalities and logarithmic Sobolev inequalities. When $\rho > 0$, using the fact that for real numbers $rs \leq \frac{\rho r^2}{2} + \frac{s^2}{2\rho}$, the HWI inequality implies that

$$\mathrm{H}(f\mu \mid \mu) \leq \frac{1}{2\rho} \mathrm{I}_\mu(f),$$

which is the logarithmic Sobolev inequality for the measure $d\mu = e^{-W} dx$ (as deduced in Corollary 5.7.2, p. 268, from the curvature condition $CD(\rho, \infty)$ for some $\rho > 0$). When only $\rho \geq 0$, the quadratic transportation cost inequality $T_2(C)$ still implies a logarithmic Sobolev inequality, but with a weaker constant, namely $LS(4C)$ (the implication actually still holds as soon as $\rho > -\frac{1}{C}$). With a less precise dependence on the constants, this implication actually holds if μ only satisfies the $T_1(C)$ transportation cost inequality (9.2.1) as a consequence of Theorem 8.7.2, p. 426. In general, however, a quadratic transportation cost inequality is not enough to ensure the validity of a logarithmic Sobolev inequality (see Sect. 9.6 below and the Notes and References).

9.3 Transportation Proofs of Functional Inequalities

To conclude this section, and from a somewhat different perspective, it is worth mentioning (without proof) the following result which easily allows us to deduce quadratic transportation cost, logarithmic Sobolev and even isoperimetric-type inequalities from their counterparts for Gaussian measures.

Theorem 9.3.4 (Caffarelli's contraction Theorem) *Let $d\mu = e^{-W} dx$ be a probability measure on the Borel sets of $\mathbb{R}^n$ where $W : \mathbb{R}^n \to \mathbb{R}$ is a smooth potential such that $\nabla\nabla W \geq \rho \, \text{Id}$ for some $\rho > 0$. Then the Brenier transportation map $T = \nabla\phi : \mathbb{R}^n \to \mathbb{R}^n$ pushing the standard Gaussian measure on $\mathbb{R}^n$ onto μ is Lipschitz, with Lipschitz coefficient $\frac{1}{\sqrt{\rho}}$.*

9.3.2 Sobolev Inequality in $\mathbb{R}^n$

The transportation method developed in the preceding section to recover in particular logarithmic Sobolev inequalities may be used similarly to directly obtain the standard Sobolev inequality on $\mathbb{R}^n$ with optimal constants, thus providing another approach. One further interesting feature of the method is that it allows us to similarly deal with arbitrary norms on $\mathbb{R}^n$ (with optimal constants), something which is not allowed by the semigroup approach restricted to Euclidean metrics. We therefore present here the result and the transportation proof of Sobolev inequalities in this framework. Although not discussed in detail here, the transportation proof leads to a unique description of the extremal functions.

Consider therefore an arbitrary norm $\|\cdot\|$ on $\mathbb{R}^n$. Denote by $\|\cdot\|_*$ its dual norm defined by

$$\|x\|_* = \sup_{y \in \mathbb{R}^n, \|y\| \leq 1} x \cdot y, \quad x \in \mathbb{R}^n,$$

where $x \cdot y$ is as usual the Euclidean scalar product in $\mathbb{R}^n$. The proof below will make use of Young's inequality

$$x \cdot y \leq \frac{\|x\|^p}{p} + \frac{\|y\|_*^{p^*}}{p^*} \tag{9.3.3}$$

for $x, y \in \mathbb{R}^n$, $1 \leq p \leq \infty$, $\frac{1}{p} + \frac{1}{p^*} = 1$, and of Hölder's inequality for (suitable) functions $F, G : \mathbb{R}^n \to \mathbb{R}^n$,

$$\int_{\mathbb{R}^n} F \cdot G \, dx \leq \left(\int_{\mathbb{R}^n} \|F\|^p \, dx \right)^{1/p} \left(\int_{\mathbb{R}^n} \|G\|_*^{p^*} \, dx \right)^{1/p^*}. \tag{9.3.4}$$

The next statement is the announced version of the Sobolev inequality (6.1.1), p. 278, on $\mathbb{R}^n$ ($n > 2$) with respect to such an arbitrary norm. We refer to Remark 6.9.5, p. 319, for the class of functions satisfying the Sobolev inequality in Euclidean space.

Theorem 9.3.5 (Sobolev inequality) *Let $n > 2$ and $p = \frac{2n}{n-2}$. For every smooth function $f : \mathbb{R}^n \to \mathbb{R}$ with $f \in \mathbb{L}^p(dx)$ and $\|\nabla f\|_* \in \mathbb{L}^2(dx)$,*

$$\|f\|_p \leq C_n \big\| \|\nabla f\|_* \big\|_2$$

where the norms are understood to be with respect to the Lebesgue measure. The sharp constant $C_n > 0$ is achieved on the extremal functions $f_{\sigma,b,x_0}(x) = (\sigma^2 + b \|x - x_0\|^2)^{(2-n)/2}$, $x \in \mathbb{R}^n$, $\sigma > 0$, $b > 0$ and $x_0 \in \mathbb{R}^n$.

Proof Before starting the proof, note that the regularity properties of Theorem 9.1.6 will not be enough since the various densities should be smooth and compactly supported and bounded from below by a strictly positive constant. A careful proof requires the use of distributional gradients, Laplacians and Hessians in the sense of Aleksandrov. As in the previous sub-section, this will be mostly implicit below and the proof therefore only emphasizes the main ideas and arguments.

It is enough to establish the theorem for positive functions. Let u, v be smooth compactly supported probability densities with respect to the Lebesgue measure. According to Theorem 9.1.5, there is a convex function $\phi : \mathbb{R}^n \to \mathbb{R}$ such that $T = \nabla \phi$, the Brenier map, is the optimal map from $u\,dx$ to $v\,dx$. The associated Monge-Ampère equation (9.1.8) reads

$$u = v(\nabla \phi) \det(\nabla \nabla \phi).$$

Hence,

$$v^{-1/n}(\nabla \phi) = u^{-1/n} \det(\nabla \nabla \phi)^{1/n}.$$

The arithmetic-geometric inequality (for the determinant of symmetric positive matrices) indicates that

$$\det(\nabla \nabla \phi)^{1/n} \leq \frac{1}{n} \Delta \phi$$

since $\Delta \phi$ is the trace of $\nabla \nabla \phi$ (which is at the heart of the $CD(0,n)$ condition). Therefore, after integration with respect to the measure $u\,dx$,

$$\int_{\mathbb{R}^n} v^{-1/n}(\nabla \phi) u\, dx \leq \frac{1}{n} \int_{\mathbb{R}^n} u^{1-(1/n)} \Delta \phi\, dx. \qquad (9.3.5)$$

Since $T = \nabla \phi$ maps $u\,dx$ onto $v\,dx$,

$$\int_{\mathbb{R}^n} v^{-1/n}(\nabla \phi) u\, dx = \int_{\mathbb{R}^n} v^{1-(1/n)} dx.$$

Integrating the right-hand side of (9.3.5) by parts therefore yields that

$$\int_{\mathbb{R}^n} v^{1-(1/n)} dx \leq -\frac{1}{n} \int_{\mathbb{R}^n} \nabla\big(u^{1-(1/n)}\big) \cdot \nabla \phi\, dx. \qquad (9.3.6)$$

Note that since u and v are compactly supported, $\nabla \phi$ is bounded.

9.3 Transportation Proofs of Functional Inequalities

Let now f and g be two positive smooth compactly supported functions on $\mathbb{R}^n$ such that $f^p = u$ and $g^p = v$ ($p = \frac{2n}{n-2}$) are probability densities. Then (9.3.6) reads as

$$\int_{\mathbb{R}^n} g^{2(n-1)/(n-2)} dx \leq -\frac{2(n-1)}{n(n-2)} \int_{\mathbb{R}^n} f^{n/(n-2)} \nabla f \cdot \nabla \phi \, dx.$$

The Cauchy-Schwarz inequality (cf. (9.3.4)) implies that

$$\int_{\mathbb{R}^n} g^{2(n-1)/(n-2)} dx \leq \frac{2(n-1)}{n(n-2)} \left(\int_{\mathbb{R}^n} \|\nabla f\|_*^2 dx \right)^{1/2} \left(\int_{\mathbb{R}^n} f^p \|\nabla \phi\|^2 dx \right)^{1/2}$$

and thus, by construction of $T = \nabla \phi$,

$$\int_{\mathbb{R}^n} g^{2(n-1)/(n-2)} dx \leq \frac{2(n-1)}{n(n-2)} \left(\int_{\mathbb{R}^n} \|\nabla f\|_*^2 dx \right)^{1/2} \left(\int_{\mathbb{R}^n} g^p \|x\|^2 dx \right)^{1/2}.$$

Hence, removing the normalization condition on f, it follows after a density argument that for every positive smooth function f and every positive function g such that g^p is a probability density (both compactly supported),

$$\|f\|_p \leq \frac{2(n-1)}{n(n-2)} \left(\int_{\mathbb{R}^n} \|\nabla f\|_*^2 dx \right)^{1/2}$$
$$\times \left(\int_{\mathbb{R}^n} g^p \|x\|^2 dx \right)^{1/2} \left(\int_{\mathbb{R}^n} g^{2(n-1)/(n-2)} dx \right)^{-1}. \tag{9.3.7}$$

Now, it may be observed that whenever $f = g = f_{\sigma,b,x_0}$ for suitable choices of σ and b in order that g^p is a probability density, $\nabla \phi = x$. Then, even though these densities are not compactly supported, the preceding computations are actually still valid for these specific functions and (9.3.5) as well as the subsequent Cauchy-Schwarz inequality are equalities in this case. Therefore, the choice of $g = f_{\sigma,b,x_0}$ in the preceding yields the Sobolev inequality with its optimal constant. Moreover, following this reasoning, it may be shown that the functions f_{σ,b,x_0} are the only extremal functions (see the Notes and References). □

9.3.3 Sobolev Trace Inequality in $\mathbb{R}^n$

The same strategy may be used to derive an (optimal) *Sobolev trace inequality* on the half-space, illustrating the power of the transportation argument. Set below $\mathbb{R}_+^n = \mathbb{R}^{n-1} \times [0, \infty)$ and $\partial \mathbb{R}_+^n = \mathbb{R}^{n-1} \times \{0\}$ and consider as before $\|\cdot\|$ a norm on $\mathbb{R}^n$. Denote a generic point $x \in \mathbb{R}_+^n$ by $x = (y, s)$ so that $y \in \mathbb{R}^{n-1}$ is identified with $(y, 0) \in \partial \mathbb{R}_+^n$.

Theorem 9.3.6 (Sobolev trace inequality) *Let $n > 2$ and $p = \frac{2n}{n-2}$, $\tilde{p} = \frac{2(n-1)}{n-2}$. For every smooth function $f : \mathbb{R}^n_+ \to \mathbb{R}$ with $f \in L^p(\mathbb{R}^n_+, dx)$ and $\|\nabla f\|_* \in L^2(\mathbb{R}^n_+, dx)$,*

$$\|f\|_{L^{\tilde{p}}(\partial \mathbb{R}^n_+)} \leq C_n \|\|\nabla f\|_*\|_{L^2(\mathbb{R}^n_+)}. \tag{9.3.8}$$

The sharp constant $C_n > 0$ is attained for the extremal functions $h(x) = \|x - e\|^{2-n}$, $x \in \mathbb{R}^n_+$, with $e = (0, \ldots, 0, -1) \in \mathbb{R}^n$.

Proof The starting point is similar to the proof of Theorem 9.3.5 for the Sobolev inequality. We again make use of the Monge-Ampère equation (9.1.8) in the sense of Aleksandrov without any further notice.

Let u and v be two smooth probability densities with respect to the Lebesgue measure on $\mathbb{R}^n_+$ with compact support, and denote by $T = \nabla \phi$ the Brenier map from $u\,dx$ onto $v\,dx$. As in (9.3.5), integrating with respect to the measure $u\,dx$,

$$\int_{\mathbb{R}^n_+} v^{1-(1/n)} dx \leq \frac{1}{n} \int_{\mathbb{R}^n_+} u^{1-(1/n)} \Delta \phi \, dx. \tag{9.3.9}$$

To obtain an inequality at the boundary, let $\psi = \phi - e \cdot x$. Since $\Delta \psi = \Delta \phi$, an integration by parts on the right-hand side yields

$$\begin{aligned}\int_{\mathbb{R}^n_+} v^{1-(1/n)} dx \leq &-\frac{1}{n} \int_{\mathbb{R}^n_+} \nabla\big(u^{1-(1/n)}\big) \cdot \nabla(\phi - e \cdot x) dx \\ &+ \frac{1}{n} \int_{\partial \mathbb{R}^n_+} u^{1-(1/n)} (\nabla \phi - e) \cdot e \, dx.\end{aligned} \tag{9.3.10}$$

Again the integration by parts is justified since u and v are compactly supported, and the map ϕ may be extended to the full space as a convex function so that $\nabla \phi$ has a bounded variation at the boundary. Since $T = \nabla \phi$ is the optimal map between $u\,dx$ and $v\,dx$, $\nabla \phi \in \mathbb{R}^n_+$ and then $\nabla \phi \cdot e \leq 0$ on $\partial \mathbb{R}^n_+$, so that the previous inequality yields

$$\begin{aligned}\int_{\mathbb{R}^n_+} v^{1-(1/n)} dx \leq &-\frac{1}{n} \int_{\mathbb{R}^n_+} \nabla\big(u^{1-(1/n)}\big) \cdot \nabla(\phi - e \cdot x) dx \\ &- \frac{1}{n} \int_{\partial \mathbb{R}^n_+} u^{1-(1/n)} dx.\end{aligned} \tag{9.3.11}$$

Let now f and g be two positive smooth compactly supported functions such that $u = f^p$ and $v = g^p$ ($p = \frac{2n}{n-2}$) are probability densities. The Cauchy-Schwarz inequality and the properties of the transportation map $T = \nabla \phi$ then imply, as in the proof of Theorem 9.3.5, that

$$\int_{\partial \mathbb{R}^n_+} f^{\tilde{p}} dx \leq \tilde{p} \bigg(\int_{\mathbb{R}^n_+} g^p \|x - e\|^2 dx\bigg)^{1/2} \|\|\nabla f\|_*\|_{L^2(\mathbb{R}^n_+)} - n \int_{\mathbb{R}^n_+} g^{\tilde{p}} dx.$$

9.4 Hamilton-Jacobi Equations

Therefore, removing the normalization on f, for any pair of smooth positive functions f and g such that $\int_{\mathbb{R}^n_+} g^p dx = 1$, and with $\kappa = \|\|\nabla f\|_*\|_{\mathbb{L}^2(\mathbb{R}^n_+)}/\|f\|_{\mathbb{L}^p(\mathbb{R}^n_+)}$,

$$\frac{\|f\|^{\tilde{p}}_{\mathbb{L}^{\tilde{p}}(\partial\mathbb{R}^n_+)}}{\|\|\nabla f\|_*\|^{\tilde{p}}_{\mathbb{L}^2(\mathbb{R}^n_+)}} \leq \kappa^{-\tilde{p}}\big[\kappa A(g) - B(g)\big]$$

where

$$A(g) = \tilde{p}\left(\int_{\mathbb{R}^n_+} g^p \|x - e\|^2 dx\right)^{1/2} \quad \text{and} \quad B(g) = n\int_{\mathbb{R}^n_+} g^{\tilde{p}} dx.$$

The worse case is when $\kappa = \frac{2(n-1)}{n}\frac{B(g)}{A(g)}$, for which it follows that for all positive g such that g^p is a probability density,

$$\|f\|_{\mathbb{L}^{\tilde{p}}(\partial\mathbb{R}^n_+)} \leq C_n(g)\|\|\nabla f\|_*\|_{\mathbb{L}^2(\mathbb{R}^n_+)}$$

where $C_n(g) > 0$. The optimal Sobolev trace inequality follows with the choice of $g = ch$ for a suitable normalization constant $c > 0$ since when $f = g = ch$, all the previous inequalities turn into equalities. As in the proof of Theorem 9.3.5, even if h is not compactly supported, all the arguments are justified since in this case $\nabla\phi = x$. The proof of Theorem 9.3.6 is then easily completed along these lines. $\square$

Remark 9.3.7 The preceding transportation method may be developed similarly to reach the general family of optimal Sobolev inequalities in $\mathbb{R}^n$ with $\mathbb{L}^q$-norms of the gradient, $1 \leq q < n$, and arbitrary norms on $\mathbb{R}^n$,

$$\|f\|_p \leq C_{n,q}\|\|\nabla f\|_*\|_q$$

where now $p = \frac{qn}{n-q}$. Similarly, the Sobolev trace inequality

$$\|f\|_{\mathbb{L}^{\tilde{p}}(\partial\mathbb{R}^n_+)} \leq C_{n,q}\|\|\nabla f\|_*\|_{\mathbb{L}^q(\mathbb{R}^n_+)}$$

now with $\tilde{p} = \frac{q(n-1)}{n-q}$ is established in the same way. The strategy also includes the family of Gagliardo-Nirenberg inequalities of Theorem 6.10.4, p. 326, again with $\mathbb{L}^q$-norms of gradients and arbitrary norms on $\mathbb{R}^n$.

9.4 Hamilton-Jacobi Equations

This section presents the basic material on Hamilton-Jacobi equations and their relationships to optimal transport at the level of the dual formulation expressed by Theorem 9.1.3. The main aspects may be presented in the context of a (smooth complete connected) Riemannian manifold $(M, \mathfrak{g})$, $\mathbb{R}^n$ for example, with d the Riemannian or Euclidean distance.

9.4.1 Hamilton-Jacobi Equations and Hopf-Lax Formulas

In its general form, a Hamilton-Jacobi equation is a non-linear partial differential equation of the form

$$\partial_t u + H(t, x, u, \nabla u) = 0, \qquad (9.4.1)$$

$u = u(t, x)$, $t > 0$, $x \in M$, where the function H is called a *Hamiltonian*. We will only be interested here in the very particular case when $H : \mathbb{R}_+ \to \mathbb{R}_+$ only depends on the length $|\nabla u|$ of the gradient of u. That is, given the initial condition in the form of a continuous and bounded function f on M, the equation takes the form

$$\begin{cases} \partial_t u + H(|\nabla u|) = 0 & \text{in } (0, \infty) \times M, \\ u = f & \text{on } \{t = 0\} \times M. \end{cases} \qquad (9.4.2)$$

Existence and properties of the solutions are summarized in the next proposition.

Proposition 9.4.1 (Hopf-Lax formula) *Assume that $H : \mathbb{R}_+ \to \mathbb{R}_+$ is a $\mathcal{C}^2$ increasing and convex function satisfying $H(0) = 0$ and*

$$\lim_{r \to \infty} \frac{H(r)}{r} = \infty.$$

Let $f : M \to \mathbb{R}$ be bounded and continuous and, for $t > 0$ and $x \in M$, set

$$\begin{cases} u(t, x) = \inf_{y \in M} \{ f(y) + t H^*(\frac{d(x,y)}{t}) \} & \text{for } (t, x) \in (0, \infty) \times M, \\ u(0, x) = f(x) & \text{for } x \in M, \end{cases} \qquad (9.4.3)$$

where H^* is the Legendre-Fenchel transform of H, $H^*(r) = \sup_{s \in \mathbb{R}_+} [rs - H(s)]$, $r \in \mathbb{R}_+$. Then for all $t > 0$, $x \mapsto u(t, x)$ is Lipschitz and, almost everywhere in the variable $x \in M$, u satisfies the Hamilton-Jacobi equation (9.4.2) starting from f.

If the initial condition f is a $\mathcal{C}^2$ Lipschitz map such that $\|f\|_{\text{Lip}} \leq (H^*)'(\infty)$, then it may be shown moreover that u satisfies the Hamilton-Jacobi equation down to $t = 0$.

For any (bounded and continuous) initial condition f, the solution given by u presented in Proposition 9.4.1 is called the *Hopf-Lax infimum-convolution representation* of the solutions of the Hamilton-Jacobi equation (9.4.2). The Hopf-Lax formula actually defines a non-linear semigroup starting from f, and we denote, by similarity with Markov semigroups,

$$Q_t^{(H)} f(x) = \inf_{y \in M} \left\{ f(y) + t H^*\left(\frac{d(x, y)}{t}\right) \right\}, \quad t > 0, \ x \in M,$$

together with $Q_0^{(H)} f = f$. The semigroup property is then expressed by the fact that $Q_t^{(H)} \circ Q_s^{(H)} = Q_{t+s}^{(H)}$, $t, s \geq 0$. We will not be interested in the most general

9.4 Hamilton-Jacobi Equations

Hamiltonians, and actually only concentrate below (as for Wasserstein distances in the preceding section) on the quadratic example $H(r) = \frac{r^2}{2}$ for which $H^* = H$. In this case, $Q_t^{(H)}$ will be denoted more simply by Q_t, so that

$$Q_t f(x) = \inf_{y \in M} \left\{ f(y) + \frac{1}{2t} d(x, y)^2 \right\}, \quad t > 0, \ x \in M. \tag{9.4.4}$$

As already emphasized in (9.2.3), the link between the Hopf-Lax formula and optimal transportation is made clear on the dual Kantorovich problem. Indeed, for $c(x, y) = H(d(x, y))$, $(x, y) \in M \times M$, Theorem 9.1.3 implies that for all probability measures $\mu, \nu \in \mathcal{P}(M)$,

$$\mathcal{T}_c(\mu, \nu) = \sup \left(\int_M Q_1^{(H)} f \, d\mu - \int_M f \, d\nu \right) \tag{9.4.5}$$

where the supremum is taken over all bounded continuous functions $f : M \to \mathbb{R}$.

9.4.2 Vanishing Viscosity

In order to get a flavor of the connections between Hamilton-Jacobi and heat equations, and therefore between transportation inequalities and functional inequalities related to heat kernels as developed earlier in this monograph, consider the example of a diffusion semigroup $(P_t)_{t \geq 0}$ with generator $\mathrm{L} = \Delta_{\mathfrak{g}} - \nabla W \cdot \nabla$ on a Riemannian manifold $(M, \mathfrak{g})$ with carré du champ operator $\Gamma(f) = |\nabla f|^2$. The Hamilton-Jacobi equation, with $H(r) = \frac{r^2}{2}$, $r \in \mathbb{R}_+$, with a bounded continuous initial condition $f : M \to \mathbb{R}$ takes here the form

$$\partial_t u + \frac{1}{2} |\nabla u|^2 = 0, \quad u(0, \cdot) = f(\cdot). \tag{9.4.6}$$

Consider now, for every $\varepsilon > 0$,

$$u^\varepsilon = -2\varepsilon \log P_{\varepsilon t}\left(e^{-f/2\varepsilon}\right), \tag{9.4.7}$$

for which, by the heat equation and the change of variables formula,

$$\partial_t u^\varepsilon = \varepsilon \, \mathrm{L} u^\varepsilon - \frac{1}{2} |\nabla u^\varepsilon|^2, \quad u^\varepsilon(0, \cdot) = f.$$

As $\varepsilon \to 0$, it is expected that u^ε will approach the solution u of (9.4.6) in a reasonable sense. This approximation is classically known as the *vanishing viscosity method*. Note that the limiting solution u given by the infimum-convolution $Q_t f$ is independent of the potential W and thus of the invariant measure $d\mu = e^{-W} d\mu_{\mathfrak{g}}$ of the diffusion operator $\mathrm{L} = \Delta_{\mathfrak{g}} - \nabla W \cdot \nabla$. In particular, this asymptotics is explicit on the basic Brownian and Ornstein-Uhlenbeck examples.

In a more probabilistic interpretation, there is a clear picture of the preceding asymptotics $\varepsilon \to 0$ via the Laplace-Varadhan asymptotics with rate precisely described by the infimum convolution of f with the quadratic large deviation rate function for the heat semigroup according to (6.7.13), p. 303,

$$\lim_{t \to 0} 4t \log p_t(x, y) = -d^2(x, y),$$

where $p_t(x, y)$, $t > 0$, $(x, y) \in M \times M$, are the density kernels with respect to the reference measure μ. Under this asymptotics, u^ε in (9.4.7) takes the form

$$u^\varepsilon(x) = -\log \left\| \exp\left(-f - \frac{d^2(x, \cdot)}{2t} + o(1)\right) \right\|_{1/2\varepsilon}, \quad x \in M$$

(where the norm is understood to be with respect to μ). Using the fact that $\mathbb{L}^p$-norms converge to $\mathbb{L}^\infty$-norms as $p \to \infty$, it turns out in the end that u^ε converges when $\varepsilon \to 0$ to the infimum-convolution $Q_t f$ of (9.4.4) which recovers the Hopf-Lax formulation. Of course, this should only be considered as an heuristic, the precise approximation having to be set up more carefully (see Sect. 9.5 below).

9.5 Hypercontractivity of Solutions of Hamilton-Jacobi Equations

This section addresses smoothing properties of the solutions of Hamilton-Jacobi equations (9.4.2) in the form of the Hopf-Lax semigroup $(Q_t)_{t>0}$ similar to the hypercontractivity property of linear semigroups. As presented in Sect. 5.2.2, p. 246, a logarithmic Sobolev inequality for the invariant measure of a Markov semigroup $(P_t)_{t \geq 0}$ may be translated equivalently into hypercontractive bounds on the latter. We discuss here the analogous description for $(Q_t)_{t>0}$. In continuation of the previous section, we work below on a (smooth complete connected) Riemannian manifold $(M, \mathfrak{g})$ with Riemannian measure $\mu_\mathfrak{g}$ and distance d.

For simplicity, we only deal with the Hamilton-Jacobi equation for the quadratic cost, leading by Proposition 9.4.1 to the Hopf-Lax representation

$$Q_t f(x) = \inf_{y \in M} \left\{ f(y) + \frac{1}{2t} d(x, y)^2 \right\}, \quad t > 0, \ x \in M,$$

of (9.4.4). Following the naive approximation of the Hamilton-Jacobi equation described in Sect. 9.4, the results described below should not look too surprising.

Recall from Definition 5.1.1, p. 236, that, after a change of functions, a probability measure μ on the Borel sets of M satisfies a logarithmic Sobolev inequality $LS(C)$ with constant $C > 0$ if for every smooth function f on M,

$$\mathrm{Ent}_\mu(e^f) \leq \frac{C}{2} \int_M |\nabla f|^2 e^f d\mu.$$

9.5 Hypercontractivity of Solutions of Hamilton-Jacobi Equations

The following statement is then the analogue of the (linear) hypercontractivity Theorem 5.2.3, p. 246.

Theorem 9.5.1 (Hypercontractivity of Solutions of Hamilton-Jacobi Equations) *Let μ be a probability measure on $(M, \mathfrak{g})$ (absolutely continuous with respect to the Riemannian measure $\mu_\mathfrak{g}$). If μ satisfies the logarithmic Sobolev inequality $LS(C)$, then for all $t > 0$, all $a > 0$, and all bounded continuous functions f on M,*

$$\|e^{Q_t f}\|_{a+\frac{t}{C}} \leq \|e^f\|_a. \tag{9.5.1}$$

Conversely, if (9.5.1) holds for some $a > 0$ and all $t > 0$ and all f bounded and continuous, then μ satisfies $LS(C)$.

Proof The proof is similar to the proof of Theorem 5.2.3, p. 246. The derivative along P_t is replaced by the derivative along Q_t, which together with the Hamilton-Jacobi equation (9.4.2) yields that

$$\partial_t \int_M e^{q\, Q_t f}\, d\mu = -\frac{q}{2} \int_M |\nabla Q_t f|^2\, e^{q\, Q_t f}\, d\mu.$$

Then, setting

$$\Lambda = \Lambda(t, q) = \int_M e^{q\, Q_t f}\, d\mu, \quad t \geq 0,\ q > 0,$$

the $LS(C)$ inequality applied to $q\, Q_t f$ expresses that

$$q\, \partial_q \Lambda - \Lambda \log \Lambda \leq -Cq\, \partial_t \Lambda.$$

For the choice of $q = q(t) = a + \frac{t}{C}$, it is then immediately checked from this inequality that $H(t) = \frac{1}{q(t)} \log(\Lambda(t, q(t)))$ is decreasing in $t > 0$. Inequality (9.5.1) is then simply $H(t) \leq H(0)$.

Conversely, assuming that, for some $a > 0$, (9.5.1) holds for any $t > 0$ and any initial condition f, the time derivative at $t = 0$ yields

$$\mathrm{Ent}_\mu\left(e^{af}\right) \leq \frac{Ca^2}{2} \int_M |\nabla f|^2\, e^{af}\, d\mu,$$

which amounts (as $a > 0$) to the logarithmic Sobolev inequality $LS(C)$. Theorem 9.5.1 is therefore established. (Note that the absolute continuity of μ is implicitly used in the Hamilton-Jacobi equation, which only holds almost everywhere. More refined analyses developed in works mentioned in the Notes and References actually allow us to suitably extend the conclusion without this assumption.) □

While Theorem 9.5.1 describes, under a logarithmic Sobolev inequality for μ, a hypercontractivity property for the semigroup $(Q_t)_{t>0}$ of the classical Hamilton-Jacobi equation similar to hypercontractivity of the semigroup $(P_t)_{t\geq 0}$ with generator L and invariant measure μ, it is actually possible to derive hypercontractivity of $(Q_t)_{t>0}$ directly from that for $(P_t)_{t\geq 0}$ using the vanishing viscosity method outlined in Sect. 9.4. Indeed, under the logarithmic Sobolev inequality $LS(C)$, apply the reverse hypercontractivity property described in Remark 5.2.4, p. 247, stating that

$$\|P_t h\|_q \geq \|h\|_p$$

for any strictly positive $h : M \to \mathbb{R}$ and any $-\infty < q < p < 1$ such that $e^{2t/C} = \frac{q-1}{p-1}$. For $\varepsilon > 0$, let $0 > p = -2\varepsilon a > q = -2\varepsilon b$ and $t > 0$ so that

$$e^{2\varepsilon t/C} = \frac{1+2\varepsilon b}{1+2\varepsilon a}.$$

It follows that, with $u^\varepsilon = -2\varepsilon \log P_{\varepsilon t}(e^{-f/2\varepsilon})$ from (9.4.7),

$$\|e^{u^\varepsilon}\|_b \leq \|e^f\|_a.$$

As $\varepsilon \to 0$, $u^\varepsilon \to Q_t f$ where $t > 0$ is such that $b = a + \frac{t}{C}$, and we recover Theorem 9.5.1 in the limit $\varepsilon \to 0$.

The preceding hypercontractive property of the Hopf-Lax solutions $(Q_t)_{t>0}$ of the classical Hamilton-Jacobi equation may be developed similarly at the level of the Euclidean logarithmic Sobolev inequality (6.2.8), p. 284, providing analogues for Q_t of the optimal heat kernel bounds in Theorem 6.3.1, p. 286. We work below in $\mathbb{R}^n$ with the Lebesgue measure dx. Recall thus the Euclidean logarithmic Sobolev inequality from Proposition 6.2.5, p. 284,

$$\mathrm{Ent}_{dx}(f^2) \leq \frac{n}{2} \log\left(\frac{2}{n\pi e}\int_{\mathbb{R}^n} |\nabla f|^2 dx\right) \qquad (9.5.2)$$

which holds for f on $\mathbb{R}^n$ in the associated Dirichlet domain with $\int_{\mathbb{R}^n} f^2 dx = 1$. Arguing as in the proof of Theorem 9.5.1, one similarly concludes the following statement.

Theorem 9.5.2 *For every bounded continuous function $f : \mathbb{R}^n \to \mathbb{R}$, any $0 < \alpha \leq \beta$ and any $t > 0$,*

$$\|e^{Q_t f}\|_\beta \leq \left(\frac{\beta-\alpha}{2\pi t}\right)^{\frac{n}{2}\frac{\beta-\alpha}{\beta\alpha}} \left(\frac{\alpha}{\beta}\right)^{\frac{n}{2}\frac{\alpha+\beta}{\alpha\beta}} \|e^f\|_\alpha \qquad (9.5.3)$$

(where the norms are understood here with respect to the Lebesgue measure). Furthermore, as $\beta \to \infty$ (with $\alpha = 1$ for example), for every $x \in \mathbb{R}^n$,

$$Q_t f(x) \leq \log \int_{\mathbb{R}^n} e^f dx + \frac{n}{2}\log\left(\frac{1}{2\pi t}\right).$$

Proof By scaling, it is enough to assume that $t = 1$. By concavity of the logarithmic function, the Euclidean logarithmic Sobolev inequality (9.5.2) is as usual linearized

9.5 Hypercontractivity of Solutions of Hamilton-Jacobi Equations

as

$$\mathrm{Ent}_{dx}(e^f) \leq \frac{n}{2r}\int_{\mathbb{R}^n}|\nabla f|^2 e^f\,dx + \frac{n}{2}\log\left(\frac{r}{2\pi e^2 n}\right)\int_{\mathbb{R}^n} e^f\,dx, \quad r>0.$$

Setting $\Lambda(t) = \|e^{Q_t f}\|_{q(t)}$ with $q(t) = \frac{\alpha\beta}{(\alpha-\beta)t+\beta}$, $0 < \alpha \leq \beta$, this family of logarithmic Sobolev inequalities with $r = nq'(t)$ implies as in the proof of Theorem 9.5.1 that

$$\Lambda'(t) \leq \Lambda(t)\frac{nq'(t)}{2q^2(t)}\log\left(\frac{q'(t)}{2\pi e^2}\right).$$

The conclusion follows after integration over $[0, 1]$. Note that the limiting case $\beta = \infty$ may also be established directly since by definition of $Q_t f(x)$,

$$e^{Q_t f(x) - |x-y|^2/2t} \leq e^{f(y)}$$

for every $x, y \in \mathbb{R}^n$ and $t > 0$, so that integration in y yields the claim. □

It should be noted that (9.5.3) of Theorem 9.5.2, which holds for every $t \geq 0$ and $\beta \geq \alpha$ for a fixed $\alpha > 0$, is formally equivalent to the Euclidean logarithmic Sobolev inequality.

To conclude this section we briefly describe a local version of the main hypercontractivity Theorem 9.5.1 for solutions of the Hamilton-Jacobi equations in the spirit of local hypercontractivity for the linear semigroup $(P_t)_{t\geq 0}$ of Theorem 5.5.5, p. 262. The result in particular produces a characterization of the curvature condition $CD(\rho, \infty)$. We still work in the context of a (weighted) Riemannian manifold $(M, \mathfrak{g})$.

Theorem 9.5.3 (Local hypercontractivity) *Consider the Markov Triple consisting of a smooth complete connected Riemannian manifold $(M, \mathfrak{g})$ and the diffusion operator $\mathrm{L} = \Delta_{\mathfrak{g}} - \nabla W \cdot \nabla$, where W is a smooth potential, with invariant measure $d\mu = e^{-W}d\mu_{\mathfrak{g}}$ and Markov semigroup $(P_t)_{t\geq 0}$. The following assertions are equivalent.*

(i) *The curvature condition $CD(\rho, \infty)$ holds for some $\rho \in \mathbb{R}$.*
(ii) *For all $q > p > 0$ and $u, t > 0$ satisfying $q = p + \frac{\rho t}{1-e^{-2\rho u}}$,*

$$P_u(e^{q Q_t f})^{1/q} \leq P_u(e^{pf})^{1/p}, \tag{9.5.4}$$

for all bounded continuous functions f on M. When $\rho = 0$ the condition has to be replaced by $q = p + \frac{t}{2u}$.

Proof We only sketch the argument following the proofs of Theorem 5.5.5, p. 262, and Theorem 9.5.1. Theorem 5.5.2, p. 259, indicates that (i) is equivalent to the local logarithmic Sobolev inequality (5.5.5), p. 260. Fixing $u > 0$, let $\Lambda(s) = P_u(e^{q(s)Q_s f})^{1/q(s)}$, $s \in [0, t]$, where q is affine and satisfies $q'(s) = \frac{\rho}{1-e^{-2\rho u}}$.

Arguing as in the proof of Theorem 9.5.1, the local logarithmic Sobolev inequality (5.5.5), p. 260, at time u readily implies that $\Lambda'(s) \leq 0$ on $[0, t]$, proving (ii). Conversely, a simple asymptotics at $t = 0$ of (9.5.4) implies (5.5.5), p. 260, and thus (i). $\square$

9.6 Transportation Cost and Logarithmic Sobolev Inequalities

On the basis of the preceding developments, this section examines the connections between logarithmic Sobolev inequalities and (quadratic) transportation cost inequalities, as well as stability properties of transportation cost inequalities. We again deal with the setting of a (smooth complete connected) Riemannian manifold $(M, \mathfrak{g})$.

9.6.1 From Logarithmic Sobolev to Transportation Cost Inequalities

The first main result is that a logarithmic Sobolev inequality is stronger than a quadratic transportation cost inequality.

Theorem 9.6.1 (Otto-Villani Theorem) *Let μ be a probability measure on M. If μ satisfies a logarithmic Sobolev inequality $LS(C)$ for some constant $C > 0$, then it satisfies a quadratic transportation cost inequality $T_2(C)$ with the same constant C.*

The proof is actually an immediate consequence of the hypercontractivity Theorem 9.5.1 (assuming in addition that μ is absolutely continuous with respect to $\mu_{\mathfrak{g}}$). Indeed, if μ satisfies $LS(C)$, we may let $a \to 0$ and $t = C$ therein to obtain that for every bounded continuous function f on M,

$$\int_M e^{Q_C f} d\mu \leq e^{\int_M f d\mu}.$$

But this is exactly the dual formulation of the $T_2(C)$ inequality as expressed in Proposition 9.2.3.

We will present below an alternative proof of the Otto-Villani Theorem. However, before we begin, it should be pointed out that the converse implication in Theorem 9.5.1 only works when $a > 0$, that is a transportation cost inequality $T_2(C)$ does not in general imply in return a logarithmic Sobolev inequality (and explicit counterexamples may be given on the real line). Nevertheless, a $T_2(C)$ inequality implies a Poincaré inequality $P(C)$.

Proposition 9.6.2 *Let μ be a probability measure on M. If μ satisfies a quadratic transportation cost inequality $T_2(C)$ for some constant $C > 0$, then it satisfies a Poincaré inequality $P(C)$ with the same constant C.*

9.6 Transportation Cost and Logarithmic Sobolev Inequalities

The proof again relies on the dual description of $T_2(C)$ which yields that for every bounded continuous function f on M, and every $t > 0$,

$$\int_M e^{t\, Q_{Ct} f} d\mu \leq e^{t \int_M f d\mu}.$$

A second order Taylor expansion at $t = 0$ together with the Hamilton-Jacobi equation immediately yields the claim.

While the converse of Theorem 9.5.1 does not hold in general, a partial result in this direction is of use in the study of perturbation properties of transportation cost inequalities, which we briefly describe without proof. Namely, whenever $a = 0$, that is starting from $T_2(C)$, for f bounded and continuous and $t > 0$,

$$\int_M e^{t\, Q_t f/C} d\mu \leq e^{(t/C) \int_M f d\mu} \leq \left(\int_M e^f d\mu \right)^{t/C}$$

where we used Jensen's inequality in the last inequality. By the entropic inequality (5.1.2), p. 236,

$$\frac{t}{C} \int_M e^f Q_t f\, d\mu \leq \mathrm{Ent}_\mu(e^f) + \int_M e^f d\mu \, \log\!\left(\int_M e^{t\, Q_t f/C} d\mu \right).$$

By the preceding, after some algebra, it follows that

$$\mathrm{Ent}_\mu(e^f) \leq \frac{t^2}{t-C} \int_M \frac{1}{t}[f - Q_t f] e^f d\mu \qquad (9.6.1)$$

for every $t > C$. While we cannot let $t \to 0$ in (9.6.1) to reach a logarithmic Sobolev inequality, it turns out that, after some work, (9.6.1) implies in return a transportation cost inequality $T_2(C_1)$ for some $C_1 > 0$ only depending on C. In other words, (9.6.1) is a kind of ersatz of the logarithmic Sobolev inequality replacing the generator L by $\frac{1}{t}(\mathrm{Id} - Q_t)$ which is equivalent to $T_2(C)$. The argument relies on a refined analysis of the Herbst argument of Sect. 5.4, p. 252, but we skip the details here (see Sect. 9.9).

Now, due to Lemma 5.1.7, p. 240, and since $f - Q_t f \geq 0$, it is clear that (9.6.1) is stable under bounded perturbations of μ. As a consequence, similarly to Poincaré and logarithmic Sobolev inequalities, quadratic transportation cost inequalities T_2 are stable under bounded perturbations.

Proposition 9.6.3 (Bounded perturbation) *Assume that a probability measure μ on M satisfies a quadratic transportation cost inequality $T_2(C)$ for some $C > 0$. Let μ_1 be a probability measure with density h with respect to μ such that $\frac{1}{b} \leq h \leq b$ for some constant $b > 0$. Then μ_1 satisfies $T_2(C_1)$ where $C_1 > 0$ only depends on C and b.*

As for Poincaré and logarithmic Sobolev inequalities, if $h = e^k$, the dependence on b may be replaced by dependence on $\mathrm{osc}(k) = \sup k - \inf k$.

9.6.2 Dimension-Free Transportation Inequalities

Towards an alternative approach to the Otto-Villani Theorem 9.6.1, we next investigate the relationships between the T_1 and T_2 transportation cost inequalities of independent interest. The main statement is the following result which clarifies the difference between T_1 and T_2 inequalities under tensorization. As mentioned before Proposition 9.2.4 establishing the stability under products of T_2 inequalities, T_1 inequalities do not tensorize independently of the dimension. If (E, d) is a metric (Polish) space, the k-fold product space E^k is assumed to be equipped with the ℓ^2-metric $(d^2 \oplus \cdots \oplus d^2)^{1/2}$.

Theorem 9.6.4 (Gozlan's Theorem) *Let μ be a probability measure on the Borel sets of (E, d) such that $\mu^{\otimes k}$ satisfies the transportation cost inequality $T_1(C)$ of (9.2.1) on E^k uniformly in $k \geq 1$. Then μ satisfies the quadratic transportation cost inequality $T_2(C)$.*

The proof shows that the conclusion actually holds under a (uniform) Gaussian measure concentration property (8.7.3), p. 426 (essentially equivalent to $T_1(C)$).

Proof Let $f : E \to \mathbb{R}$ be positive (bounded measurable) and let $\varepsilon > 0$. Set

$$A = \left\{ x \in E \,;\, f(x) \leq (1 + \varepsilon) \int_E f d\mu \right\}.$$

Then $1 - \mu(A) \leq \frac{1}{1+\varepsilon}$ and, for every $t > 0$ and every $x \in E$,

$$Q_t f(x) \leq \inf_{y \in A} \left\{ f(y) + \frac{1}{2t} d(x, y)^2 \right\} \leq (1 + \varepsilon) \int_E f d\mu + \frac{1}{2t} d(x, A)^2$$

where $d(x, A)$ is the distance from the point x to the set A. For every $r > 0$, the map $h(x) = \min(d(x, A), r)$, $x \in E$, is 1-Lipschitz and

$$\int_E h d\mu \leq r \big[1 - \mu(A)\big] \leq \frac{r}{1+\varepsilon}.$$

Hence, if $s = \frac{\varepsilon r}{1+\varepsilon}$,

$$\mu\big(d(\cdot, A) \geq r\big) = \mu(h \geq r) = \mu\left(h \geq \frac{r}{1+\varepsilon} + s\right) \leq \mu\left(h \geq \int_E h d\mu + s\right),$$

so that by the dual formulation (9.2.5) of the T_1 inequality and its application (9.2.6) via Markov's inequality,

$$\mu\big(d(\cdot, A) \geq r\big) \leq e^{-s^2/2C} = e^{-\varepsilon^2 r^2 / 2C(1+\varepsilon)^2}$$

9.6 Transportation Cost and Logarithmic Sobolev Inequalities

for every $r > 0$. On the basis of this tail estimate, we may actually control $\int_E h\, d\mu$ in a finer way as

$$\int_E h\, d\mu = \int_0^r \mu\bigl(d(\cdot, A) \geq u\bigr) du$$

$$\leq \int_0^r e^{-\varepsilon^2 u^2 / 2C(1+\varepsilon)^2} du \leq \sqrt{\frac{2}{\pi}} \, \frac{\sqrt{C}(1+\varepsilon)}{\varepsilon} = r_0(\varepsilon)$$

to get similarly that, for every $r \geq r_0(\varepsilon)$,

$$\mu\bigl(d(\cdot, A) \geq r\bigr) = \mu(h \geq r) \leq \mu\left(h \geq \int_E h\, d\mu + r - r_0(\varepsilon)\right)$$

$$\leq e^{-(r - r_0(\varepsilon))^2 / 2C}.$$

In particular, for every $t > C$,

$$\int_E e^{d(\cdot, A)^2 / 2t}\, d\mu \leq K$$

where $K = K(C, t, \varepsilon)$. Therefore, in conclusion, for $t > C$,

$$\int_E e^{Q_t f}\, d\mu \leq K\, e^{(1+\varepsilon) \int_E f d\mu}.$$

By the hypothesis, apply the latter on $(E^k, \mu^{\otimes k})$, with respect to the ℓ^2-metric (thus with K uniform in $k \geq 1$), to

$$f(x_1, \ldots, x_k) = g(x_1) + \cdots + g(x_k), \quad (x_1, \ldots, x_k) \in E^k,$$

for a given positive (bounded measurable) function $g : E \to \mathbb{R}$. Since

$$Q_t f(x_1, \ldots, x_k) = Q_t g(x_1) + \cdots + Q_t g(x_k),$$

it follows that

$$\int_E e^{Q_t g}\, d\mu \leq K^{1/k}\, e^{(1+\varepsilon) \int_E g d\mu}.$$

Since $K = K(C, t, \varepsilon)$ is independent of k, as $k \to \infty$, and then $\varepsilon \to 0$, $t \to C$, the inequality $T_2(C)$ in its dual formulation (Proposition 9.2.3) follows. The proof is complete. □

As announced, the preceding Theorem 9.6.4 may be used to provide an alternative proof of Theorem 9.6.1 linking logarithmic Sobolev inequalities and quadratic transportation inequalities, in particular going beyond smooth settings. Indeed, under a logarithmic Sobolev inequality for μ, by the tensorization Proposition 5.2.7, p. 249, and the Herbst argument, Proposition 5.4.1, p. 252, $\mu^{\otimes k}$ satisfies a T_1 inequality independently of k, and thus a T_2 inequality by Theorem 9.6.4 (with the same constant). In this form, Theorem 9.6.1 may be extended to the Markov Triple structure (E, μ, Γ) (provided the intrinsic distance d defines a Polish topology on E).

9.7 Heat Flow Contraction in Wasserstein Space

This section is concerned with contraction properties in Wasserstein distance along the heat flow $(P_t)_{t \geq 0}$ of a diffusion operator with some curvature. We work again in the context on a (smooth complete connected) Riemannian manifold $(M, \mathfrak{g})$ with a diffusion operator $L = \Delta_\mathfrak{g} - \nabla W \cdot \nabla$ where W is a smooth potential on M, which is the infinitesimal generator of a Markov semigroup $(P_t)_{t \geq 0}$. As will be pointed out, the results here may be developed in the usual setup of a (Full) Markov Triple but by homogeneity with the rest of this chapter, we stick to the preceding framework.

The first step towards contractions is a basic commutation property between the heat and Hamilton-Jacobi semigroups under curvature conditions.

Proposition 9.7.1 (Commutation property) *In the preceding weighted Riemannian setting, the following assertions are equivalent.*

(i) *The curvature condition $CD(\rho, \infty)$ holds for some $\rho \in \mathbb{R}$.*
(ii) *For every bounded continuous function $f : M \to \mathbb{R}$ and every $t, s \geq 0$,*

$$P_t(Q_s f) \leq Q_{e^{2\rho t} s}(P_t f). \tag{9.7.1}$$

Proof For simplicity we only prove the result for $\rho = 0$, the general result being obtained in the same way. The most important part is the proof of (ii) under the curvature condition. Assume for simplicity that $f \geq 0$. To this end, use the log-Harnack inequality (5.6.2), p. 266, which applied to $e^{u Q_1 f}$ yields

$$P_t(Q_1 f) \leq \frac{1}{u} Q_{2t}\left(\log P_t\left(e^{u Q_1 f}\right)\right)$$

for every $t, u > 0$. Now, the local hypercontractivity inequality (9.5.4) as $p \to 0$ implies that

$$\log P_t\left(e^{\frac{1}{2t} Q_1 f}\right) \leq \frac{1}{2t} P_t f$$

for every $t > 0$. Combining the inequalities with $u = \frac{1}{2t}$ yields (i) for $s = 1$. Arbitrary values of s are obtained by homogeneity. Conversely, under (ii), the s derivative in (9.7.1) at $s = 0$ implies in return, by the Hamilton-Jacobi equation, the gradient bound $|\nabla P_t f|^2 \leq P_t(|\nabla f|^2)$ which is equivalent to the curvature dimension $CD(0, \infty)$. The proof is complete. □

Under the curvature condition $CD(\rho, \infty)$, the commutation (9.7.1) actually extends to the (Full) Markov Triple setting (E, μ, Γ) on the basis of the developments

9.7 Heat Flow Contraction in Wasserstein Space

in Sect. 8.6, p. 421. Indeed (assuming in addition that the intrinsic distance defines a Polish topology on E), apply the isoperimetric Harnack inequality of Theorem 8.6.3, p. 425, to $A = \{Q_1 f \geq u\}$, $u \geq 0$, for $f : E \to \mathbb{R}$ bounded continuous and positive. If $z \in A_\varepsilon$ (where $\varepsilon > 0$), there exists an $a \in A$ such that $d(z, a) < \varepsilon$ so that, by definition of the infimum-convolution Q_1,

$$f(z) + \frac{\varepsilon^2}{2} \geq f(z) + \frac{d(z,a)^2}{2} \geq Q_1 f(a) \geq u.$$

Hence

$$(\{Q_1 f \geq u\})_{d(x,y)} \leq \left\{ f \geq u + \frac{d(x,y)^2}{2} \right\}$$

for all $(x, y) \in E \times E$ with $d(x, y) > 0$. Again taking $\rho = 0$ for simplicity, Theorem 8.6.3 after integration with respect to $u \geq 0$ then yields

$$P_t(Q_1 f)(x) \leq P_t f(y) + \frac{d(x,y)^2}{2}$$

for every x and y in E. In other words, $P_t(Q_1 f) \leq Q_1(P_t f)$ which amounts to (ii).

On the basis of the commutation property (9.7.1) between the heat and Hopf-Lax semigroups of the last proposition, we next obtain contraction in Wasserstein distance, which is again shown to be equivalent to the curvature condition. Recall that $f\mu$ denotes for simplicity the measure with density f with respect to μ.

Theorem 9.7.2 (Heat flow contraction in Wasserstein space) *Consider the Markov Triple consisting of a smooth complete connected Riemannian manifold $(M, \mathfrak{g})$ and of a diffusion operator $L = \Delta_\mathfrak{g} - \nabla W \cdot \nabla$, where W is a smooth potential, with invariant measure $d\mu = e^{-W} d\mu_\mathfrak{g}$ and Markov semigroup $(P_t)_{t\geq 0}$. The following assertions are equivalent.*

(i) *The curvature condition $CD(\rho, \infty)$ holds for some $\rho \in \mathbb{R}$.*
(ii) *For every $t \geq 0$ and all probability densities f and g with respect to μ with a finite second moment,*

$$W_2(P_t f \mu, P_t g \mu) \leq e^{-\rho t} W_2(f \mu, g \mu). \tag{9.7.2}$$

Proof We start with (i). For any bounded continuous $h : M \to \mathbb{R}$, by time reversibility and (9.7.1)

$$\int_M Q_1 h\, P_t f\, d\mu - \int_M h\, P_t g\, d\mu = \int_M P_t(Q_1 h) f\, d\mu - \int_M P_t h\, g\, d\mu$$

$$\leq \int_M Q_{e^{2\rho t}}(P_t h) f\, d\mu - \int_M P_t h\, g\, d\mu.$$

Since by homogeneity $Q_{e^{2\rho t}}(P_t h) = e^{-2\rho t} Q_1(e^{2\rho t} P_t h)$, the Kantorovich dual description of the Wasserstein distance (Theorem 9.1.3) yields the contraction property (9.7.2).

Assume conversely that (ii) holds, that is, again by Kantorovich duality and symmetry,

$$\int_M P_t(Q_1 h) f \, d\mu - \int_M P_t h \, g \, d\mu \leq e^{-\rho t} \, W_2(f\mu, g\mu)$$

for any bounded continuous $h : M \to \mathbb{R}$. Given $x, y \in M$, by a suitable approximation, let $f\mu$ and $g\mu$ approach respectively Dirac masses at x and y so that the latter inequality turns into

$$P_t(Q_1 h)(x) - P_t h(y) \leq e^{-2\rho t} \frac{d(x,y)^2}{2}$$

(alternatively take $v_1 = \delta_x$ and $v_2 = \delta_y$ in (9.7.4) below). Taking the infimum over y then exactly expresses the commutation property (9.7.1), which is equivalent to the $CD(\rho, \infty)$ condition. The theorem is established. □

With similar arguments, it may be shown that the curvature-dimension condition $CD(0, n)$ is equivalent to the contraction bound

$$W_2^2(P_t f\mu, P_s g\mu) \leq W_2^2(f\mu, g\mu) + n(\sqrt{t} - \sqrt{s})^2, \qquad (9.7.3)$$

for any $t, s \geq 0$ and any probability densities f and g with respect to μ.

Remark 9.7.3 Theorem 9.7.2 expresses equivalently that, under the curvature condition $CD(\rho, \infty)$,

$$W_2(P_t^* v_1, P_t^* v_2) \leq e^{-\rho t} \, W_2(v_1, v_2) \qquad (9.7.4)$$

for all probability measures v_1 and v_2 in $\mathcal{P}_2(M)$, where according to (1.1.2), p. 9, $P_t^* v_1$, respectively $P_t^* v_2$, is the law at time $t \geq 0$ of the underlying Markov process $\{X_t^x ; t \geq 0, x \in M\}$ with initial data v_1, respectively v_2. It should also be observed that the preceding inequalities extend, via the commutation property, to the whole family of Wasserstein distances W_p, $1 \leq p < \infty$, and even to more general costs.

9.8 Curvature of Metric Measure Spaces

This last section briefly outlines recent developments on notions of Ricci curvature lower bounds based on optimal transport. These notions, providing an alternative approach to Ricci curvature in Riemannian spaces, may actually be extended to the context of metric measure spaces, justifying their interest.

To start with, let $M = \mathbb{R}^n$ and $d\mu = e^{-W} dx$ be a probability measure on the Borel sets of $\mathbb{R}^n$ for a smooth potential $W : \mathbb{R}^n \to \mathbb{R}$. If μ_0 and μ_1 are absolutely continuous probability measures on $\mathbb{R}^n$, recall $T = \nabla \phi$, the Brenier map pushing μ_0 onto μ_1. For any $\theta \in [0, 1]$, interpolate between μ_0 and μ_1 by letting

$$\mu_\theta = \big((1-\theta)\,\mathrm{Id} + \theta \, T\big) \# \mu_0.$$

9.8 Curvature of Metric Measure Spaces

One then speaks of the geodesics of optimal transportation joining μ_0 to μ_1. Namely, since the Brenier map achieves optimality in the Wasserstein distance W_2,

$$W_2(\mu_0, \mu_\theta) + W_2(\mu_\theta, \mu_1) = W_2(\mu_0, \mu_1),$$

justifying the terminology (see below).

As we have seen, under suitable regularity assumptions, μ_θ has a density f_θ (with respect to the Lebesgue measure) satisfying

$$f_0(x) = f_\theta\big((1-\theta)x + \theta\nabla\phi(x)\big)\det\big((1-\theta)\operatorname{Id} + \theta\nabla\nabla\phi(x)\big)$$

$f_0(x)dx$-almost everywhere. Arguing then as in the proof of Theorem 9.3.1, it may be shown that whenever $\nabla\nabla W \geq 0$, the relative entropy $H(\cdot\,|\,\mu)$ with respect to μ is convex along the geodesics of optimal transportation in the sense that

$$H(\mu_\theta \,|\, \mu) \leq (1-\theta)H(\mu_0 \,|\, \mu) + \theta H(\mu_1 \,|\, \mu) \qquad (9.8.1)$$

for every $\theta \in [0, 1]$.

The preceding convexity property (9.8.1) has taken the name of *displacement convexity* (convexity along the geodesics of optimal transportation). Together with the suitable Riemannian extension of the Brenier map, (9.8.1) extends to (weighted) Riemannian manifolds $(M, \mathfrak{g})$ under the convexity property $\operatorname{Ric} + \nabla\nabla W \geq 0$. Actually, under $\operatorname{Ric} + \nabla\nabla W \geq \rho\,\mathfrak{g}$, $\rho \in \mathbb{R}$, the displacement convexity of order ρ has to be understood with respect to the Wasserstein distance W_2 as

$$H(\mu_\theta \,|\, \mu) \leq (1-\theta)H(\mu_0 \,|\, \mu) + \theta H(\mu_1 \,|\, \mu)$$
$$- \frac{\rho\theta(1-\theta)}{2} W_2(\mu_0, \mu_1)^2, \quad \theta \in [0, 1]. \qquad (9.8.2)$$

Now, these displacement convexity inequalities may be formulated in purely metric terms, opening the extension to abstract metric measure spaces. A key observation in this regard is that the displacement inequality (9.8.2) in Riemannian manifolds characterizes in return the Ricci curvature lower bound $\operatorname{Ric} + \nabla\nabla W \geq \rho\,\mathfrak{g}$. Therefore, (9.8.2) provides an equivalent (synthetic, as opposed to the more local definition of Riemannian curvature) definition of Ricci curvature lower bounds. This interpolation along the geodesics of optimal transport moreover shares some similarities with the semigroup interpolation along the heat flow. The optimal transportation formalism actually induces a formal Riemannian structure on the Wasserstein space $\mathcal{P}_2(E)$ for which convexity of entropy provides the corresponding curvature lower bounds.

To formulate the extension to metric measure spaces, it is necessary to clarify a few definitions from metric geometry. In a metric (Polish) space (E, d), a curve joining two points $x, y \in E$ is a continuous map $\gamma : [0, 1] \to E$ such that $\gamma(0) = x$ and $\gamma(1) = y$. The length of γ is defined by $L(\gamma) = \sup \sum_{i=1}^{k} d(\gamma(s_{i-1}), \gamma(s_i))$ where the supremum runs over all subdivisions $0 = s_0 < s_1 < \cdots < s_k = 1$. Such a curve γ is called a geodesic if $L(\gamma) = d(\gamma(0), \gamma(1))$. The metric space (E, d)

is called a length space if, for all $(x, y) \in E \times E$, $d(x, y) = \inf L(\gamma)$ where the infimum is taken over all curves γ joining x to y. The space (E, d) is called geodesic if all points $x, y \in E$ are connected by a geodesic. In a geodesic space, there exist middle points, or θ-middle points ($\theta \in [0, 1]$), z between $x, y \in E$, i.e. $d(x, z) = \theta d(x, y)$, $d(z, y) = (1 - \theta) d(x, y)$, and such a point z is necessarily on a geodesic joining x to y. The Wasserstein space $\mathcal{P}_2(E)$ of probability measures on the Borel sets of E with a finite second moment equipped with the Wasserstein distance W_2 is a length space if (E, d) is a length space.

In the following, μ is a fixed reference probability measure on (E, d), so that (E, d, μ) is a *metric measure space*. According to the preceding developments in a Riemannian context, we say that (E, d, μ) is of Ricci curvature bounded from below by ρ, $\rho \in \mathbb{R}$, if for every pair (μ_0, μ_1) in $\mathcal{P}_2(E)$ (with supports contained in the support of μ), there exists a geodesic $(\mu_\theta)_{\theta \in [0,1]}$ in $\mathcal{P}_2(E)$ joining μ_0 and μ_1 such that

$$H(\mu_\theta \mid \mu) \leq (1 - \theta) H(\mu_0 \mid \mu) + \theta H(\mu_1 \mid \mu) - \frac{\rho \theta (1 - \theta)}{2} W_2(\mu_0, \mu_1)^2 \quad (9.8.3)$$

for every $\theta \in [0, 1]$.

As already mentioned above, this definition is thus equivalent to the Ricci curvature lower bound $\mathrm{Ric} + \nabla \nabla W \geq \rho \, \mathfrak{g}$ in Riemannian manifolds. Among the most noticeable aspects of this synthetic definition, not discussed here, it may be shown to be stable under the so-called Gromov-Hausdorff convergence (a weak notion of convergence of metric measure spaces for which a limit of Riemannian manifolds is not necessarily a Riemannian manifold anymore). The preceding definition of Ricci curvature lower bound also allows for Poincaré or logarithmic Sobolev inequalities as investigated in this monograph. Further definitions may also be used to include curvature and dimension of a metric measure space as the $CD(\rho, n)$ condition (see the Notes and References).

9.9 Notes and References

This chapter and the notes and references here are only a rough introduction to mass transportation and transportation cost inequalities. General references on mass transportation are the comprehensive books by S. T. Rachev [351], S. T. Rachev and L. Rüschendorf [352, 353], and C. Villani [424, 426]. The first references are more oriented towards probabilistic issues. The monographs by C. Villani present a complete and deep investigation of modern mass transportation theory and its rich interplay with partial differential equations, probability theory and geometry. A recent account of transportation cost inequalities with bibliographical references is the survey [210] by N. Gozlan and C. Léonard.

The reader may find in the preceding references the basic Kantorovich Theorem 9.1.3 and Kantorovich-Rubinstein Theorem 9.1.4 of Sect. 9.1, presented in greater generality. The existence of a transport map as the gradient of a convex function (Theorem 9.1.5) goes back to M. Knott and C. Smith [267], L. Rüschendorf and S. T. Rachev [373] and Y. Brenier [94], the latter having emphasized the geometric relevance to partial differential equations. The manifold case is due to R. McCann [304]. The associated Monge-Ampère equation (9.1.8) with the Hessian of ϕ in the sense of Aleksandrov is proved in [305], on the basis of results in [180] and the regularity properties of the Brenier map as expressed in Theorem 9.1.6 due to L. Caffarelli [101, 102] (cf. [424, 426] for details and numerous further contributions).

The quadratic transportation cost inequality for the Gaussian measure (Theorem 9.2.1) is due to M. Talagrand [404] with the one-dimensional proof together with tensorization. A multi-dimensional transportation proof, further extended to strictly log-concave measures, appeared in [71, 133] (see also below). The dual formulation of the quadratic transportation cost inequality T_2 (Proposition 9.2.3) was introduced by S. Bobkov and F. Götze in [77]. The corresponding statement for T_1 may also be found there. The link between the latter and measure concentration goes back to K. Marton [298] (cf. [278, 426]). See [210, 213] for further developments on transportation cost inequalities.

The new HWI inequality of Corollary 9.3.3 was introduced in the seminal paper [340] by F. Otto and C. Villani. In this work, deep links between mass transportation and functional inequalities were identified on the basis of the Otto differential calculus in the Wasserstein space of probability measures [259, 339]. The proof of Theorem 9.3.1 given here, and its consequence for logarithmic Sobolev inequalities and quadratic transportation cost inequalities, is due to D. Cordero-Erausquin [133]. The alternative heat flow proof of the HWI inequality comes from [76]. Caffarelli's contraction Theorem 9.3.4 is proved in [103] (see also [424]). On the basis of the transportation proof of the HWI and logarithmic Sobolev inequalities, the optimal transportation method to obtain classical Sobolev inequalities with their extremal functions was developed by D. Cordero-Erausquin, B. Nazaret and C. Villani [137] (where in particular more details on the integration by parts formula (9.3.6) may be found). The optimal Sobolev trace inequality of Theorem 9.3.6 is due independently to J. F. Escobar and W. Beckner [56, 177]. The proof presented in the monograph and its generalization in $\mathbb{L}^p$ (as presented in Remark 9.3.7) is due to B. Nazaret [324]. The transportation mass method has been further developed to investigate a wide variety of inequalities between entropy functionals and energy in a partial differential equation context (see e.g. [3, 112, 113, 134, 296], and cf. [424, 426]). It should be noted that, as far as functional inequalities such as Sobolev-type inequalities are concerned, alternative maps may be considered in the transportation proofs. In particular, the earlier triangular Knothe map [266] can be used to this end, as M. Gromov did in a proof of the standard isoperimetric inequality in Euclidean space [316] (see also [137, 426]). The optimal Brenier map becomes relevant as soon as more (Riemannian) geometric features enter into the picture (see [426] for much more in this regard).

Classical properties of Hamilton-Jacobi equations as briefly outlined in Sect. 9.4 such as the Hopf-Lax infimum-convolution formula and the vanishing viscosity method are presented, for example, in [44, 104, 179] (see also [424, 426] or [291] in the context of length spaces). For recent developments in metric structures, see [11, 12, 43, 212]).

Hypercontractivity of solutions of Hamilton-Jacobi equations in Sect. 9.5 have been emphasized in [76] towards a new proof of the Otto-Villani Theorem. Theorem 9.5.1 appears there whereas ultracontractivity of the Hamilton-Jacobi solutions is examined in [197] and in a more general context in [188, 198]. In the latter papers, generalizations of the Euclidean logarithmic Sobolev inequality in $\mathbb{L}^p(dx)$, originally introduced in [148], are also considered. Local hypercontractivity for Hamilton-Jacobi solutions, Theorem 9.5.3, is investigated in [31].

A first contraction result in Wasserstein distance (along the Boltzmann equation) goes back to H. Tanaka [407, 408]. The Wasserstein contraction property put forward in Theorem 9.7.2 may be traced back to the investigation [339] of the heat flow as a gradient flow in the Wasserstein space, further developed in this spirit in [112, 113, 341] and [9, 175] in connection with the Evolutionary Variational Inequality. For coupling arguments, see [128, 129, 427, 431]. The crucial equivalence with curvature bounds as expressed by Theorem 9.7.2 is due to M. von Renesse and K.-T. Sturm [427]. See [424, 426] for more on Wasserstein contraction properties. The proof given here, based of the commutation property between the heat and Hopf-Lax semigroup (Proposition 9.7.1) is due to K. Kuwada [270] (see also [12, 37]). Contraction under any curvature-dimension is studied, among other recent developments, in [85, 176, 271].

The major link between quadratic transportation inequality and logarithmic Sobolev inequality, at the source of many developments at the interface between partial differential equations, probability theory and geometry, is due to F. Otto and C. Villani [340]. The proof presented here is taken from [76]. That logarithmic Sobolev inequalities are actually strictly stronger that quadratic transportation cost inequalities is due to P. Cattiaux and A. Guillin [117] (see [208] for a direct characterization). The perturbation Proposition 9.6.3 is due to N. Gozlan, C. Roberto and P.-M. Samson [211], and was part of the early motivation of [340]. Theorem 9.6.4 is due to N. Gozlan [206], following the preliminary investigation [209] (cf. [210]).

Section 9.8 is a very short overview of recent deep investigations involving a notion of Ricci curvature lower bounds in metric measure spaces due to J. Lott and C. Villani [292] and K.-T. Sturm [399, 400]. Preliminary steps consisted of the displacement convexity (9.8.1) by R. McCann [304], the heat flow as gradient flow of entropy with respect to the Wasserstein metric by R. Jordan, D. Kinderlehrer and F. Otto [259] and the Otto calculus [339], the links with functional inequalities by F. Otto and C. Villani [340], the Borell-Brascamp-Lieb inequalities by D. Cordero-Erausquin, R. McCann and M. Schmuckenschläger [135, 136] and the equivalence with the Riemannian Ricci curvature lower bounds by M. von Renesse and K.-T. Sturm [427]. This has been the starting point of many fruitful developments in recent years. The monumental monograph [426] by C. Villani is a complete exposition to which the reader is warmly referred for both mathematical developments and historical background. See also [9].

In connection with the Markov operator curvature-dimension viewpoint emphasized in this book, it should be mentioned that recent developments by L. Ambrosio,

9.9 Notes and References

N. Gigli and G. Savaré [10, 11] actually conclude that in a suitable sub-class of metric measure spaces for which the gradient energy is Hilbertian, the curvature as convexity of relative entropy (9.8.3) is equivalent to the Markov operator curvature condition $CD(\rho, \infty)$, thus providing an even deeper link between Markov semigroups and optimal transportation. The approach relies on the identification of the heat flow as gradient flow of either Dirichlet energy or entropy in curved spaces. An improved version of contraction in Wasserstein distance in the form of the Evolutionary Variational Inequality combining the heat flow interpolation along the geodesics of optimal transportation provides one key step in this direction. This investigation partially relies on the extension to a non-smooth framework of some of the basic semigroup tools described in this monograph by means of refined non-smooth analysis in metric measure spaces. The case of curvature-dimension $CD(\rho, n)$ is investigated by M. Erbar, K. Kuwada and K.-T. Sturm [176, 271]. We refer to these and the previously mentioned references for a complete account of these deep developments.

Appendices

The following appendices are devoted to the three basic aspects of this monograph, the analytic aspect of semigroups of operators, the probabilistic aspect of diffusion processes and their generators, and the geometric aspect of Laplace-type operators and their curvature properties. These appendices only briefly feature a few central objects and properties, in perspective with the main developments in the core of the book. In particular, the analysis of Markov semigroups and their infinitesimal generators may be found there. In this interplay between these different viewpoints, the aim here is to help the reader to access these topics in a balanced way. It is of course strongly advised to supplement these short surveys with more complete and detailed references. A few such references are indicated at the end of each appendix.

The first appendix is thus devoted to some basics of the analysis of semigroups of bounded operators on a Banach space, the second is a brief introduction to stochastic calculus while the third presents some basic notions in differential and Riemannian geometry. Note that the last two sections of the third appendix describe in this geometric context some features of the Γ_2 operator, in particular the reinforced curvature condition, which are critically used in the core of the book.

Appendix A
Semigroups of Bounded Operators on a Banach Space

This appendix presents some of the basics of the analysis of semigroups of operators on a Banach space. It provides in particular the necessary material for the investigation of Markov semigroups and their infinitesimal generators as developed in the core of the book. The appendix surveys the Hille-Yosida theory, symmetric and self-adjoint operators and their Friedrichs extensions, spectral decompositions and Hilbert-Schmidt operators. Some general references on these topics are provided at the end of the appendix.

A.1 The Hille-Yosida Theory

The *Hille-Yosida theory* is devoted to the study of semigroups of bounded operators on Banach spaces. The application to Markov semigroups mainly deals with $\mathbb{L}^p$-spaces, and moreover with contraction semigroups (as is the case for Markov semigroups when the reference measure is invariant). For convenience, we restrict to separable Banach spaces in what follows, although this assumption may be removed in many places.

Thus let $\mathcal{B}$ be a (real separable) Banach space with norm $\|\cdot\|$.

Definition A.1.1 (Semigroup) A family $(P_t)_{t \geq 0}$ of linear operators $P_t : \mathcal{B} \to \mathcal{B}$, $t \geq 0$, is called a *contraction semigroup* provided the following holds:

(i) For all $t \geq 0$ and all $x \in \mathcal{B}$, $\|P_t x\| \leq \|x\|$.
(ii) For all $t, s \geq 0$, $P_t \circ P_s = P_{t+s}$.
(iii) For all $x \in \mathcal{B}$, $\lim_{t \to 0} P_t x = x \; (= P_0 x)$.

For simplicity here (and throughout the book) $P_t x = P_t(x)$. Note that linearity together with the contraction property imply that for any $x \in \mathcal{B}$, $t \mapsto P_t x$ is continuous in $\mathcal{B}$ since for every $s \geq 0$,

$$\|P_{t+s} x - P_t x\| = \|P_t(P_s x - x)\| \leq \|P_s x - x\|,$$

and similarly $\|P_{t-s} x - P_t x\| \le \|x - P_s x\|$ for $0 \le s \le t$.

Given such a contraction semigroup $(P_t)_{t \ge 0}$, the Hille-Yosida theory ensures the existence of a dense linear subspace of $\mathcal{B}$ on which the function $t \mapsto P_t x$ has a bounded derivative at $t = 0$ (and therefore at any point $t > 0$ by the semigroup property). The derivative at $t = 0$ is a linear operator L, often unbounded. It is called the infinitesimal generator of the semigroup $\mathbf{P} = (P_t)_{t \ge 0}$ and the dense domain on which this generator is defined is called the domain of the generator L. Formally, P_t may be described as $e^{t\mathrm{L}}$.

To briefly understand these results, it is convenient to introduce new operators acting on $\mathcal{B}$. Given a bounded measure ν on the Borel sets of $\mathbb{R}_+$, define the operator P_ν as

$$P_\nu = \int_0^\infty P_t \, d\nu(t).$$

Hence $P_\nu x = \int_0^\infty P_t x \, d\nu(t)$, $x \in \mathcal{B}$. The latter quantity has to be understood as a limit of Riemann sums

$$\sum_i P_{t_i} x \, \nu([a_i, a_{i+1}))$$

along a sequence of partitions (a_i) which approximate the measure ν. When ν is a finite measure, this operator is bounded on $\mathcal{B}$ with norm bounded from above by $|\nu|([0, \infty))$. In this extended framework, the semigroup property (ii) may then be translated into

$$P_\nu P_{\nu'} = P_{\nu'} P_\nu = \int_0^\infty \int_0^\infty P_{t+s} \, d\nu(t) d\nu'(s) \tag{A.1.1}$$

for bounded measures ν, ν' on $\mathbb{R}_+$. A special case is the so-called *resolvent operator*, defined for $\lambda > 0$ as

$$R_\lambda = \int_0^\infty P_t e^{-\lambda t} dt \tag{A.1.2}$$

(thus corresponding to the choice of $d\nu(t) = e^{-\lambda t} dt$). Anticipating the next results, note that formally $R_\lambda = (\lambda \operatorname{Id} - \mathrm{L})^{-1}$. Indeed, for any $t \ge 0$ and $x \in \mathcal{B}$,

$$R_\lambda P_t x = e^{\lambda t} \int_t^\infty P_s x \, e^{-\lambda s} ds.$$

The time derivative at $t = 0$ then yields $\mathrm{L}(R_\lambda x) = \lambda R_\lambda x - x$.

Now (A.1.1) and (A.1.2) yield the resolvent equation, that is, for $\lambda, \lambda' > 0$,

$$R_\lambda - R_{\lambda'} = (\lambda' - \lambda) R_\lambda R_{\lambda'} = (\lambda' - \lambda) R_{\lambda'} R_\lambda. \tag{A.1.3}$$

In other words, $R_\lambda = R_{\lambda'}(\operatorname{Id} + (\lambda' - \lambda) R_\lambda)$. This identity indicates that the image of R_λ in $\mathcal{B}$ is included in the image of $R_{\lambda'}$, and exchanging the roles of λ and λ', that this image is independent of λ. Call this image $\mathcal{D}$.

Since $P_t x \to x$ as $t \to 0$ by (i) and $\lambda R_\lambda x = \int_0^\infty P_{s/\lambda} x\, e^{-s} ds$, it follows that for every $x \in \mathcal{B}$,

$$\lim_{\lambda \to \infty} \lambda R_\lambda x = x.$$

Therefore the image $\mathcal{D}$ of R_λ is dense in $\mathcal{B}$. Fix $\lambda > 0$, and define L on $\mathcal{D}$ by setting, for $x = R_\lambda y$,

$$Lx = \lambda x - y.$$

Then Lx may indeed be seen as the derivative of $P_t x$ at $t = 0$. Indeed,

$$R_\lambda P_t x = e^{\lambda t} \int_t^\infty P_s x\, e^{-\lambda s} ds,$$

so that the time derivative at $t = 0$ gives $\partial_t R_\lambda P_t x |_{t=0} = \lambda R_\lambda x - x$. Therefore for any $t > 0$,

$$\partial_t P_t x = L P_t x = P_t L x.$$

A bit more is available. $\mathcal{D}$ is, in fact, precisely the set of points x in $\mathcal{B}$ for which $P_t x$ admits a derivative at $t = 0$. Furthermore, R_λ is a (bi-) continuous bijection between $\mathcal{B}$ and the domain $\mathcal{D}$ when the latter is given the domain topology defined from the norm by

$$\|x\|_\mathcal{D} = \|x\| + \|Lx\|, \quad x \in \mathcal{D}. \tag{A.1.4}$$

The space $\mathcal{D}$ is then described as the *domain* of the operator L and denoted by $\mathcal{D}(L)$. Moreover, the knowledge of $\mathcal{D}(L)$ and of the action of L on it completely determines the semigroup $\mathbf{P} = (P_t)_{t \geq 0}$ as the unique solution of the (parabolic) heat equation (with respect to L)

$$\partial_t P_t x = L P_t x \tag{A.1.5}$$

for $x \in \mathcal{D}(L)$, and subsequently for $x \in \mathcal{B}$. The operator L is called accordingly the *infinitesimal generator* of the semigroup $\mathbf{P}$ with domain $\mathcal{D}(L)$.

A.2 Symmetric Operators

In the special case when $\mathcal{B}$ is a (real) Hilbert space $\mathcal{H}$ with scalar product $\langle \cdot, \cdot \rangle$, one may consider semigroups of bounded symmetric operators. Their generators are then (unbounded) self-adjoint operators on $\mathcal{H}$. In this book, $\mathcal{H}$ will mainly appear as an $\mathbb{L}^2(\mu)$-space for some (positive) measure μ on a measurable space $(E, \mathcal{F})$, but what follows has nothing to do with this particular case. (Unbounded) self-adjoint operators have special features, such as, for example, spectral decompositions (cf. Sect. A.4).

The first point we need to pay attention to is the difference between symmetry and self-adjointness of unbounded operators. A linear operator A defined on a dense subspace $\mathcal{D}(A)$ of $\mathcal{H}$ is said to be *symmetric* if for all x, y in $\mathcal{D}(A)$,

$$\langle Ax, y \rangle = \langle x, Ay \rangle. \tag{A.2.1}$$

For a symmetric operator A, its adjoint operator A^* is defined on the domain $\mathcal{D}(A^*)$ consisting of the points $x \in \mathcal{H}$ for which there exists a finite constant $C(x)$ such that for any $y \in \mathcal{D}(A)$, $|\langle x, Ay \rangle| \leq C(x) \|y\|$. Since A is symmetric, $\mathcal{D}(A) \subset \mathcal{D}(A^*)$. On $\mathcal{D}(A^*)$, A^* is then defined by duality

$$\langle A^*x, y \rangle = \langle x, Ay \rangle.$$

More precisely, the map $y \mapsto \langle x, Ay \rangle$, which is defined on the dense subspace $\mathcal{D}(A)$, may be uniquely extended into a continuous linear form on $\mathcal{H}$, and thus may be represented as the scalar product with some element $A^*x \in \mathcal{H}$. By the symmetry assumption, $A^*x = Ax$ for every $x \in \mathcal{D}(A)$ and A^* is therefore an extension of A. Moreover, A^* is a closed operator on its domain, meaning that whenever $(x_k)_{k \in \mathbb{N}}$ is a sequence in $\mathcal{D}(A^*)$ converging to x such that $(A^*x_k)_{k \in \mathbb{N}}$ converges to y, then $x \in \mathcal{D}(A^*)$ and $A^*x = y$. Thus, in particular, densely defined symmetric operators are closable (they admit a closed extension).

The following definition makes precise the main difference between symmetry and *self-adjointness*.

Definition A.2.1 (Self-adjoint operator) In the preceding setting, a symmetric operator A is self-adjoint if $\mathcal{D}(A^*) = \mathcal{D}(A)$.

A self-adjoint operator therefore has no symmetric extension other than itself. Except for bounded operators (bounded symmetric operators are self-adjoint), there is a huge difference between symmetric and self-adjoint operators. Every symmetric operator considered in this book admits at least one self-adjoint extension (in the sense that its domain may be extended until this extension is self-adjoint), but this extension may not be unique. Therefore, the description of an operator L on a domain which is too small (even if dense in $\mathcal{H}$) may not ensure the complete description of the semigroup $\mathbf{P} = (P_t)_{t \geq 0}$ with infinitesimal generator L.

A semigroup $(P_t)_{t \geq 0}$ is said to be symmetric if the operators P_t, $t \geq 0$, are symmetric. In any case, the generator L of a symmetric semigroup $(P_t)_{t \geq 0}$ of contractions on $\mathcal{H}$, defined on its domain $\mathcal{D}(L)$, is always self-adjoint. This is a consequence of the fact that the operators P_t, $t \geq 0$, themselves are self-adjoint, and of the very definition of the domain. The true problem is that in general the full domain of L is unknown. What is usually given is the description of L on a dense subspace $\mathcal{D}_0$ of $\mathcal{H}$, and the question is to determine whether $\mathcal{D}_0$ is dense in $\mathcal{D}(L)$ for the topology of the domain. Such a dense subspace $\mathcal{D}_0$ is called a *core*. A core completely determines a unique self-adjoint extension, and therefore a unique semigroup $\mathbf{P} = (P_t)_{t \geq 0}$ (consisting of symmetric bounded operators) with infinitesimal generator L. When a symmetric linear operator L defined on a dense subset $\mathcal{D}_0$ of $\mathcal{H}$ has a unique self-adjoint extension to some larger domain $\mathcal{D}(L)$, then L is said to be essentially self-adjoint (this notion refers both to L and $\mathcal{D}_0$). Determining if a given operator L defined on some domain $\mathcal{D}_0$ is essentially self-adjoint is investigated in Sect. A.5.

A.3 Friedrichs Extension of Positive Operators

The generators L of symmetric Markov semigroups considered in this book share some further positivity property. To stick with the standard notations, set here $A = -L$. The operators A are then positive in the following sense.

Definition A.3.1 (Positive operator) Let $\mathcal{H}$ be a (real) Hilbert space, and $\mathcal{D}_0$ be a dense linear subspace of $\mathcal{H}$ on which is given a symmetric linear operator A. The operator A is said to be positive if, for any $x \in \mathcal{D}_0$, $\langle x, Ax \rangle \geq 0$.

Symmetric positive operators A may be extended to self-adjoint operators in a minimal way (which does not mean that the domain of the extension is minimal, since in general the domains of the various self-adjoint extensions of A are not comparable). Indeed, define $\overline{\mathcal{D}_0}$ as the set of points $x \in \mathcal{H}$ which are limits of sequences $(x_k)_{k \in \mathbb{N}}$ in $\mathcal{D}_0$ for which the sequence $(Ax_k)_{k \in \mathbb{N}}$ converges. It is easily checked, using the symmetry of A, that this limit does not depend on the sequence $(x_k)_{k \in \mathbb{N}}$. This procedure thus defines an extension $\bar{A}$ of the operator A on $\overline{\mathcal{D}_0}$, which is closed with respect to the topology of the domain given by the norm (A.1.4). The operator $\bar{A}$ is a closed extension of A, however it is not self-adjoint in general so that A has to be further extended.

With this aim in mind, on the linear space $\overline{\mathcal{D}_0}$, consider the norm

$$\langle x, x \rangle_A = \|x\|_A^2 = \langle x, x \rangle + \langle x, \bar{A}x \rangle$$

and then take the completion of $\overline{\mathcal{D}_0}$ under this norm. Since the norm $\|\cdot\|_A$ is larger than the norm in $\mathcal{H}$, this completion naturally imbeds into $\mathcal{H}$. The resulting completion thus defines a new Hilbert space $\mathcal{H}_1$ which is such that $\overline{\mathcal{D}_0} \subset \mathcal{H}_1 \subset \mathcal{H}$. Define then the domain $\mathcal{D}(A)$ as the set of points $x \in \mathcal{H}_1$ for which the map $\ell(x) : y \mapsto \langle x, y \rangle_A$ is bounded for the $\mathcal{H}$-topology (that is such that $|\ell(x)(y)| \leq C(x)\|y\|$ for some $C(x) < \infty$). For these points x, $\ell(x)(y)$ may be represented as $\langle T(x), y \rangle$ for some linear operator T defined on $\mathcal{D}(A)$ which is easily seen to be self-adjoint by construction. The self-adjoint extension of A is then given by $T - \mathrm{Id}$, and the domain $\mathcal{D}(A)$ of A satisfies $\mathcal{D}_0 \subset \mathcal{D}(A) \subset \mathcal{H}_1$. This self-adjoint extension is called the *Friedrichs extension* of A.

The Friedrichs extension is not in general the unique self-adjoint extension of A, and two different self-adjoint extensions may have domains which cannot be compared. However, the Friedrichs extension is minimal in some other sense, namely in terms of Dirichlet domains. The construction of the Hilbert space $\mathcal{H}_1$ above may actually be performed for any symmetric (and therefore any self-adjoint) extension of A and is called the domain of the *Dirichlet form*. It is this domain of the Dirichlet form which is minimal for the Friedrichs extension. When dealing with operators on a bounded domain, with $\mathcal{D}_0$ the set of smooth compactly supported functions on it, this corresponds to the Dirichlet boundary condition, the Neumann boundary condition corresponding to the maximal corresponding $\mathcal{H}_1$ space. The space $\mathcal{D}_0$ is a core as soon as these minimal and maximal extensions coincide.

A.4 Spectral Decompositions

When A is a linear operator defined on a dense domain $\mathcal{D}(A)$ of a Banach space $\mathcal{B}$, the spectrum of A is defined according to the following definition. It is necessary here to deal with complex Banach spaces, for which the developments of the preceding sections are similar.

Definition A.4.1 (Spectrum) Given a linear operator A on $\mathcal{B}$, *the resolvent set* $\rho(A)$ of A is the set of all complex values $\lambda \in \mathbb{C}$ such that the range of $\lambda \operatorname{Id} - A$ is dense in $\mathcal{B}$ and such that $\lambda \operatorname{Id} - A$ has a bounded inverse $R_\lambda = (\lambda \operatorname{Id} - A)^{-1}$. The spectrum $\sigma(A)$ of A is $\mathbb{C} \setminus \rho(A)$.

When the operator $(A, \mathcal{D}(A))$ is closed (for the domain topology), if $\lambda \in \rho(A)$, then $R_\lambda(\lambda \operatorname{Id} - A) = \mathcal{B}$. Therefore the inverse $(\lambda \operatorname{Id} - A)^{-1}$ is everywhere defined. Also, the resolvent set $\rho(A)$ is always an open set in $\mathbb{C}$.

A complex number λ may be in the spectrum $\sigma(A)$ of A for three different reasons. If there is a non-zero solution $x \in \mathcal{B}$ to $Ax = \lambda x$, then λ is called an *eigenvalue* (of A) and such an $x \in \mathcal{B}$ ($x \neq 0$) is an *eigenvector* (associated to the eigenvalue λ). The set of eigenvalues forms the so-called *point spectrum*. In case the range of $\lambda \operatorname{Id} - A$ is dense in $\mathcal{B}$ but the inverse is not bounded, the set of those λ's forms the *continuous spectrum*. Finally, it may happen that the range of $\lambda \operatorname{Id} - A$ is not dense, but λ is not an eigenvalue. The set of those λ's forms the *residual spectrum*.

On real Banach spaces, as is the case throughout this book, one has to look at the complexification of the space. Fortunately, this is irrelevant for self-adjoint operators on separable Hilbert spaces for which the spectrum is always real, and the residual spectrum is empty.

Self-adjoint operators are quite similar to symmetric matrices in finite dimension which may be diagonalized in orthonormal bases of eigenvectors associated to real eigenvalues. Things are of course more intricate in infinite dimension where there may not exist any eigenvectors. Think for example of the operator $Af = -f''$ on the line $\mathbb{R}$, which is self-adjoint on $\mathbb{L}^2(dx)$ for the Lebesgue measure dx (the space of smooth compactly supported functions being a core), and for which eigenvectors are not in $\mathbb{L}^2(dx)$. Diagonalization then has to be replaced by what is called a spectral decomposition.

We restrict ourselves here to positive self-adjoint operators A on a (real, separable) Hilbert space $\mathcal{H}$ as described in the previous section.

A *spectral decomposition* is then an increasing family $(\mathcal{H}_\lambda)_{\lambda \geq 0}$ of closed linear subspaces of $\mathcal{H}$, right-continuous in the sense that $\bigcap_{\lambda' > \lambda} \mathcal{H}_{\lambda'} = \mathcal{H}_\lambda$. It is furthermore required that $\bigcup_{\lambda \geq 0} \mathcal{H}_\lambda$ is dense in $\mathcal{H}$. Consider then, for every $\lambda \geq 0$, the

A.4 Spectral Decompositions

orthogonal projection E_λ onto $\mathcal{H}_\lambda$. For any $x \in \mathcal{H}$, the map $\lambda \mapsto E_\lambda x$ is then right-continuous and converges to x when λ converges to ∞. Furthermore, the map

$$\lambda \mapsto \|E_\lambda x\|^2 = \langle E_\lambda x, x \rangle$$

is bounded and increasing (and positive). Therefore, for any pair of points $(x, y) \in \mathcal{H} \times \mathcal{H}$, the map $\lambda \mapsto \langle E_\lambda x, y \rangle$ is right-continuous and of bounded variations as a difference of bounded increasing functions

$$\langle E_\lambda x, y \rangle = \frac{1}{2} \big[\langle E_\lambda(x+y), (x+y) \rangle - \langle E_\lambda x, x \rangle - \langle E_\lambda y, y \rangle \big].$$

Then given any bounded measurable function $\psi : \mathbb{R}_+ \to \mathbb{R}$, and any $x, y \in \mathcal{H}$, one may define the Stieltjes integral

$$\int_0^\infty \psi(\lambda) d \langle E_\lambda x, y \rangle.$$

By convention, we assume that $E_\lambda = \{0\}$ for $\lambda < 0$, so that the previous integral has to be understood, extending ψ by 0 on $(-\infty, 0)$, as $\int_{-\infty}^\infty \psi(\lambda) d \langle E_\lambda x, y \rangle$, taking into account the possible gap at $\lambda = 0$.

This construction defines by duality a bounded symmetric linear operator which may be written symbolically as

$$\Psi = \int_0^\infty \psi(\lambda) \, dE_\lambda,$$

therefore satisfying $\langle \Psi x, y \rangle = \int_0^\infty \psi(\lambda) d \langle E_\lambda x, y \rangle$ for every $x, y \in \mathcal{H}$. When ψ is unbounded, the operator $\Psi = \int_0^\infty \psi(\lambda) \, dE_\lambda$ is unbounded on the domain

$$\left\{ x \in \mathcal{H} \,;\, \int_0^\infty \psi(\lambda)^2 d \langle E_\lambda x, x \rangle < \infty \right\}. \quad (A.4.1)$$

The main existence result is the following statement.

Theorem A.4.2 (Spectral decomposition) *Given a positive self-adjoint operator A on $\mathcal{H}$, there exists a spectral decomposition $(E_\lambda)_{\lambda \geq 0}$ such that*

$$A = \int_0^\infty \lambda \, dE_\lambda.$$

For any measurable function ψ on $\mathbb{R}_+$, one may then define the operator $\psi(A)$ as

$$\psi(A) = \int_0^\infty \psi(\lambda) \, dE_\lambda$$

on the domain (A.4.1) on which it satisfies $\|\psi(A)x\|^2 = \int_0^\infty \psi(\lambda)^2 \, d \langle E_\lambda x, x \rangle$.

When A is a positive and self-adjoint operator, the semigroup $P_t = e^{-tA}$, $t \geq 0$, with infinitesimal generator $-A$ is defined via the spectral resolution $(E_\lambda)_{\lambda \geq 0}$ by

$$P_t = \int_0^\infty e^{-\lambda t} dE_\lambda, \quad t \geq 0.$$

In the context of this monograph, it is convenient to feature a few norms associated with a Markov semigroup $\mathbf{P} = (P_t)_{t \geq 0}$ on $(E, \mathcal{F}, \mu)$ with invariant and symmetric measure μ, and its generator $A = -\mathrm{L}$ on $\mathbb{L}^2(\mu)$ in terms of the preceding spectral decomposition. (See Chaps. 1 and 3 for the full description of Markov semigroups and their infinitesimal generators.) Some of the following identities and inequalities are presented and used repeatedly throughout the various chapters. For example, on suitable classes of functions $f : E \to \mathbb{R}$,

$$\int_E (-\mathrm{L}f)^2 d\mu = \int_0^\infty \lambda^2 d\langle E_\lambda f, f \rangle,$$

and also

$$\mathcal{E}(f, f) = \int_E f(-\mathrm{L}f) d\mu = \int_0^\infty \lambda d\langle E_\lambda f, f \rangle = \int_E \left((-\mathrm{L})^{1/2} f\right)^2 d\mu.$$

On the semigroup $P_t = e^{t\mathrm{L}}$, $t \geq 0$, itself, for every $t \geq 0$,

$$\|(P_t - \mathrm{Id}) f\|^2 \leq t^2 \|\mathrm{L}f\|^2,$$

and using that $(1 - e^{-\lambda t})^2 \leq 2\lambda t$,

$$\|(P_t - \mathrm{Id}) f\|^2 \leq 2t \, \mathcal{E}(f, f).$$

The relationship of the spectral decomposition Theorem A.4.2 with the spectrum $\sigma(A)$ of A as described in Definition A.4.1 is the following. For a positive real number λ, $\lambda \in \sigma(A)$ if and only if, for any $\varepsilon > 0$, the dimension of $\mathcal{H}_{\lambda+\varepsilon}$ is strictly larger than the dimension of $\mathcal{H}_{\lambda-\varepsilon}$. Furthermore, λ belongs to the point spectrum if and only if the closed space $\mathcal{H}_\lambda^-$ spanned by $\bigcup_{\lambda' < \lambda} \mathcal{H}_{\lambda'}$ is such that $\mathcal{H}_\lambda^- \neq \mathcal{H}_\lambda$. In that case, any point $x \in \mathcal{H}_\lambda$ which is orthogonal to $\mathcal{H}_\lambda^-$ is an eigenvector of A, with eigenvalue λ.

The spectrum $\sigma(A)$ of a positive self-adjoint operator A may also be decomposed in another way, much more useful in practice.

Definition A.4.3 (Essential spectrum) Let A be a positive self-adjoint operator on $\mathcal{H}$. A number $\lambda \in \sigma(A)$ belongs to the essential spectrum $\sigma_{ess}(A)$ if, for any $\varepsilon > 0$, the dimension of the orthogonal of $\mathcal{H}_{\lambda-\varepsilon}$ in $\mathcal{H}_{\lambda+\varepsilon}$ is infinite. The complement $\sigma_d(A) = \sigma(A) \setminus \sigma_{ess}(A)$ is called the discrete spectrum.

According to this definition, a point lies in the discrete spectrum if it is isolated in the point spectrum and if the corresponding eigenspace is finite-dimensional. If

$\lambda \in \sigma_d(A)$, then the set of solutions x of the equation $Ax = \lambda x$ is non-zero and finite-dimensional. The following useful criterion characterizes the elements of the spectrum and of the essential spectrum.

Theorem A.4.4 (Weyl's criterion) *Let A be a positive self-adjoint operator with domain $\mathcal{D}(A) \subset \mathcal{H}$. Then $\lambda \in \sigma(A)$ if and only if for any $\varepsilon > 0$, there exists an $x \in \mathcal{D}(A)$ with $\|x\| = 1$ such that $\|Ax - \lambda x\| < \varepsilon$. An element $\lambda \in \sigma_{ess}(A)$ if and only if, for any $\varepsilon > 0$, there exists an orthonormal sequence $(x_k)_{k \in \mathbb{N}}$ in $\mathcal{D}(A) \subset \mathcal{H}$ such that $\|Ax_k - \lambda x_k\| < \varepsilon$ for every $k \in \mathbb{N}$.*

As an illustration, let $\mathcal{H} = \mathbb{L}^2(dx)$, and let $A = -\Delta$ where Δ is the Laplace operator (more precisely the closure of this operator when defined on smooth functions with compact support). The operator $A = -\Delta$ is self-adjoint and its spectral decomposition is given by the family $(\mathcal{H}_\lambda)_{\lambda > 0}$ where, for each $\lambda > 0$, $\mathcal{H}_\lambda$ is the set of functions in $\mathbb{L}^2(dx)$ with Fourier transform compactly supported on $[-\sqrt{\lambda}, \sqrt{\lambda}]$. The point spectrum is empty and $\sigma_{ess} = [0, \infty)$.

As a consequence of Theorem A.4.4, when $\sigma_{ess}(A)$ is empty, the operator A is diagonalizable. That is, there exist a sequence $(\lambda_k)_{k \in \mathbb{N}}$ in $\sigma(A)$ and an orthonormal basis $(e_k)_{k \in \mathbb{N}}$ of $\mathcal{D}(A) \subset \mathcal{H}$ such that $Ae_k = \lambda_k e_k$ for every k. Since A is positive, $\lambda_k \geq 0$, and it may be assumed for simplicity that the λ_k's are ordered increasingly $0 \leq \lambda_0 \leq \cdots \leq \lambda_k \leq \cdots$. It is actually more convenient to distinguish the different eigenvalues, that is $0 \leq \lambda_0 < \cdots < \lambda_k < \cdots$ and to consider for any $k \in \mathbb{N}$, the vector space F_k spanned by the eigenvectors associated to the eigenvalues $\lambda_0, \ldots, \lambda_k$. In this case, the spectral decomposition of A may be described in the following simplified way. Namely, E_k is the orthogonal projection onto F_k and $E_\lambda = E_k$ on $[\lambda_k, \lambda_{k+1})$ and is constant between two jumps, which occur at the values λ_k. If $\psi : \mathbb{R}_+ \to \mathbb{R}$, $\psi(A)$ is the operator which maps e_k into $\psi(\lambda_k)e_k$, $k \in \mathbb{N}$, which is the natural way of thinking of $\psi(A)$, for example when ψ is a polynomial function.

A.5 Essentially Self-adjoint Operators

As mentioned earlier, given a symmetric operator L on some dense linear space $\mathcal{D}_0$ of a Hilbert space $\mathcal{H}$, a central question is whether the knowledge of L on $\mathcal{D}_0$ is enough to determine a unique self-adjoint extension of L (or equivalently a unique symmetric semigroup $\mathbf{P} = (P_t)_{t \geq 0}$ with generator L)? In other words, is $\mathcal{D}_0$ a core? The discussion below concentrates on positive symmetric operators A ($A = -$L in the case of Markov generators in the framework of this book).

In this setting, one may always consider the closure of $\mathcal{D}_0$ with respect to the domain topology given by the norm (A.1.4) (already used when dealing with the Friedrichs extension). This closure represents the *minimal extension* of A one may think of. The adjoint operator A^* of A and its domain may therefore be defined as before and A^* is always an extension of A. This leads to the following definition.

Definition A.5.1 (Essentially self-adjoint operator) In the preceding setting, a positive symmetric operator A is said to be essentially self-adjoint on $\mathcal{D}_0$ if the minimal extension is self-adjoint.

Now let A be an unbounded positive symmetric operator defined on some dense domain $\mathcal{D}(A)$ in a complex Hilbert space $\mathcal{H}$. If A is symmetric, $\dim(\mathrm{Ker}(\lambda\,\mathrm{Id} - A^*))$, as a function $\mathbb{C} \to \mathbb{R}$, is constant on $\Im(\lambda) > 0$ and $\Im(\lambda) < 0$. Furthermore, $\mathrm{Ker}(\lambda\,\mathrm{Id} - A^*)$ is the orthogonal of $\mathrm{Im}(\bar{\lambda}\,\mathrm{Id} - A)$. In general, choose two particular complex numbers, for example $\lambda = \pm i$, and consider the kernels $\mathcal{K}_+$ and $\mathcal{K}_-$ of $\pm i\,\mathrm{Id} - A^*$. One fundamental (although quite easy) result is the following.

Theorem A.5.2 (Von Neumann's Theorem) *For a closed symmetric operator A defined on $\mathcal{D}(A)$, the domain $\mathcal{D}(A^*)$ is $\mathcal{D}(A) + \mathcal{K}_+ + \mathcal{K}_-$. In particular, A is self-adjoint if and only if the two linear spaces $\mathcal{K}_+$ and $\mathcal{K}_-$ are 0.*

A useful criterion for an operator A defined on $\mathcal{D}_0$ to be essentially self-adjoint is then given by the following property.

Proposition A.5.3 *Let A be a positive symmetric operator defined on some dense linear subspace $\mathcal{D}_0$. Then A is essentially self-adjoint as soon as there exists a real number λ such that $\lambda\,\mathrm{Id} - A^*$ is injective.*

Indeed, since the resolvent set is an open subset in $\mathbb{C}$, $\lambda\,\mathrm{Id} - A^*$ is still injective for some imaginary complex number close to λ with positive or negative imaginary part. Therefore, both $\mathcal{K}_+$ and $\mathcal{K}_-$ are reduced to 0 and the conclusion follows from Theorem A.5.2. In particular, in order to ensure self-adjointness, it is enough to exhibit some real number λ for which the equation $\lambda x = A^* x$ admits no non-zero solutions (in other words, if x is such that $\lambda \langle y, x \rangle = \langle Ay, x \rangle$ and $|\langle y, x \rangle| \leq C \|y\|$ for all $y \in \mathcal{D}_0$, then x must be 0).

We conclude this section with a classical illustration. A lot of attention has been paid over the last century to the class of so-called Schrödinger operators on $\mathbb{R}^n$, that is, operators of the form $Af = -\Delta f + Vf$, where Δ is the standard Laplacian and V is some $\mathbb{L}^1_{\mathrm{loc}}$-function on $\mathbb{R}^n$. The main results on this family of operators are summarized in the next statement.

Theorem A.5.4 *Assume that $V \geq 0$ and that $V \in \mathbb{L}^2_{\mathrm{loc}}$. Then $A = -\Delta + V$ is symmetric in $\mathbb{L}^2$ with respect to the Lebesgue measure and essentially self-adjoint on the class of smooth functions with compact support.*

The statement may be further completed with a description of the bottom of the essential spectrum. That is, as soon as $V \in \mathbb{L}^p_{\mathrm{loc}}$ with $p > \frac{n}{2}$ and

$$\lim_{r \to \infty} \mathrm{essinf}_{|x|>r} V(x) > -\infty,$$

then

$$\inf \sigma_{ess}(A) = \sup_{K \subset \mathbb{R}^n} \inf_{f \in C_c^\infty(K^c), f \neq 0} \frac{\int_{\mathbb{R}^n} fAf dx}{\int_{\mathbb{R}^n} f^2 dx}$$

where K ranges among all compact subsets of $\mathbb{R}^n$ and where $C_c^\infty(K^c)$ denotes the class of smooth functions compactly supported in $K^c = \mathbb{R}^n \setminus K$. Operators A on $\mathbb{R}^n$ sharing this property are called Persson operators in the literature and aspects of their analysis are discussed in Sect. 4.10.3, p. 227. In particular, for such V, the spectrum is discrete as soon as V goes to ∞ at infinity.

A.6 Compact and Hilbert-Schmidt Operators

When $\mathcal{B}_1$ and $\mathcal{B}_2$ are Banach spaces, a linear operator $A : \mathcal{B}_1 \to \mathcal{B}_2$ is said to be compact if the image of the unit ball (of $\mathcal{B}_1$) is relatively compact (in $\mathcal{B}_2$). If an operator A may be written as QK or KQ, with Q bounded and K compact, then it is compact. In addition, a strong limit of compact operators is compact. In other words, if $(A_k)_{k \in \mathbb{N}}$ is a sequence of compact operators and if $\lim_{k \to \infty} \|A_k - A\| = 0$ (for the operator norm), then A is compact. Furthermore, if the image of an operator is finite-dimensional (a finite range operator), this operator is compact. Actually, on a Hilbert space a compact operator is always a strong limit of finite range operators.

In the context of this book, the following statement will be relevant.

Theorem A.6.1 *For a compact symmetric operator on a Hilbert space, the spectrum consists of a sequence of eigenvalues $(\lambda_k)_{k \in \mathbb{N}}$, the only limit point possibly not in the discrete spectrum being 0. The non-zero eigenvalues form a (possibly finite) sequence converging to 0 at infinity, have finite-dimensional eigenspaces, and there is an orthonormal basis of eigenvectors. Conversely, an operator with a sequence of eigenvalues converging to 0 is compact.*

Hilbert-Schmidt operators are significant examples of compact symmetric operators.

Definition A.6.2 (Hilbert-Schmidt operator) A symmetric (bounded) operator K defined on a separable Hilbert space $\mathcal{H}$ is Hilbert-Schmidt if, for some orthonormal basis $(e_k)_{k \in \mathbb{N}}$ of $\mathcal{H}$,

$$\sum_{k \in \mathbb{N}} \|Ke_k\|^2 < \infty.$$

This property is independent of the chosen basis.

Proposition A.6.3 *Hilbert-Schmidt operators are compact.*

Hilbert-Schmidt operators are thus operators for which the sequence $(\lambda_k)_{k \in \mathbb{N}}$ of eigenvalues satisfy $\sum_{k \in \mathbb{N}} \lambda_k^2 < \infty$. When the Hilbert space is $\mathbb{L}^2(\mu)$ over a measure space $(E, \mathcal{F}, \mu)$, a Hilbert-Schmidt operator K may be represented by a kernel $k(x, y) \in \mathbb{L}^2(\mu \otimes \mu)$ such that

$$Kf(x) = \int_E k(x, y) f(y) d\mu(y), \quad x \in E,$$

for every $f \in \mathbb{L}^2(\mu)$. Indeed, if $(e_\ell)_{\ell \in \mathbb{N}}$ is a sequence of eigenvectors associated with the eigenvalues $(\lambda_\ell)_{\ell \in \mathbb{N}}$, then $k(x, y) = \sum_{\ell \in \mathbb{N}} \lambda_\ell e_\ell(x) e_\ell(y)$, and the fact that K is Hilbert-Schmidt is equivalent to the fact that this series converges in $\mathbb{L}^2(\mu \otimes \mu)$. Conversely, such operators K represented by $\mathbb{L}^2(\mu \otimes \mu)$-kernels are compact. Indeed, given a sequence $(f_m)_{m \in \mathbb{N}}$ bounded in $\mathbb{L}^2(\mu)$, there exists a subsequence $(f_{m_p})_{p \in \mathbb{N}}$ which converges weakly in $\mathbb{L}^2(\mu)$. Then $Kf_{m_p}(x)$ converges as $p \to \infty$ for every $x \in E$ such that $\int_E k^2(x, y) d\mu(y) < \infty$ and

$$|Kf_{m_p}(x)| \leq C \left(\int_E k^2(x, y) d\mu(y) \right)^{1/2}$$

where $C = \sup_{m \in N} \|f_m\|_2 < \infty$. Since $k(x, y) \in \mathbb{L}^2(\mu \otimes \mu)$, it follows that Kf_{m_p} does indeed converge in $\mathrm{L}^2(\mu)$. By Theorem A.6.1, there exists an orthonormal basis of eigenvectors of $\mathbb{L}^2(\mu)$ with corresponding sequence of eigenvalues $(\lambda_\ell)_{\ell \in \mathbb{N}}$. Using standard Hilbertian tools, it is easily seen that

$$\sum_{\ell \in \mathbb{N}} \lambda_\ell^2 = \int_E \int_E k^2(x, y) d\mu(x) d\mu(y)$$

from which the claim follows.

These compact operators should be distinguished from those defined by some $\mathbb{L}^2(\mu)$-kernel as described for example in Definition 1.2.4, p. 14, where it is only required that for μ-almost every $x \in E$, $\int_E k^2(x, y) d\mu(y) < \infty$.

In the context of this book, the generators L of Markov semigroups $\mathbf{P} = (P_t)_{t \geq 0}$ are in general not Hilbert-Schmidt (they are not even bounded) whereas, for $t > 0$, the semigroup P_t may be Hilbert-Schmidt and therefore compact. The following statement describes conditions for P_t to be compact.

Theorem A.6.4 *For a symmetric semigroup* $\mathbf{P} = (P_t)_{t \geq 0}$ *with (unbounded) infinitesimal generator* L *such that* $-$L *is positive with domain* $\mathcal{D}(\mathrm{L})$, *the following are equivalent*:

(i) $\sigma_{ess}(-\mathrm{L}) = \emptyset$.
(ii) *For all, or some,* $t > 0$, P_t *is compact.*
(iii) *For all, or some,* $\lambda > 0$, *the resolvent* $(\lambda \operatorname{Id} -\mathrm{L})^{-1}$ *is compact.*

At the level of the spectral decomposition, the equivalent conditions in Theorem A.6.4 amount to the fact that the eigenvalues of $-$L go to infinity. In particular, when P_t is compact for some $t > 0$, the spectrum of $-$L is discrete and forms a sequence of eigenvalues $(\lambda_k)_{k \in \mathbb{N}}$ going to infinity, associated to eigenspaces E_k with finite dimension n_k, $k \in \mathbb{N}$. Then, P_t is Hilbert-Schmidt if and only if $\sum_{k \in \mathbb{N}} n_k e^{-2\lambda_k t} < \infty$.

A.7 Notes and References

As mentioned earlier, the material briefly presented here forms a classical part of functional analysis. General references on these topics, used during the preparation of this appendix, are [95, 96, 100, 143, 154, 168, 169, 189, 190, 205, 241, 264, 344, 355, 356, 359, 444]. More recent references include, among many others, [142, 174, 244, 419].

Somewhat more specifically, a standard reference on the Hille-Yosida theory is the monograph [444] by K. Yosida. Most of the material on symmetric and self-adjoint operators may be found in the volumes [355] by M. Reed and B. Simon or [169] by N. Dunford and J.T. Schwartz. The definition of essential spectrum seems to go back to H. Weyl.

Appendix B
Elements of Stochastic Calculus

Stochastic calculus is a wide topic. For the reader not familiar with advanced probability theory, we briefly present in this appendix some basic notions adapted to the calculus related to Brownian motion. This short exposition surveys stochastic integrals with respect to Brownian motion, Itô's formula, stochastic differential equations and diffusion processes. In particular, these elements are aimed at a better understanding of the links with diffusion operators, semigroups and heat kernels as informally described in Chap. 1. The Notes and References include pointers to the literature and more complete accounts.

B.1 Brownian Motion and Stochastic Integrals

A real-valued *Brownian motion* defined on some probability space $(\Omega, \Sigma, \mathbb{P})$ is a family $(B_t)_{t \geq 0}$ of real-valued random variables (i.e. measurable maps) $B_t : \Omega \to \mathbb{R}$, $t \geq 0$, with the properties that $B_0 = 0$, for almost all $\omega \in \Omega$ the map $t \mapsto B_t(\omega)$ is continuous, and for any $0 < t_1 < \cdots < t_k$, the random variables $B_{t_1}, B_{t_2} - B_{t_1}, \ldots, B_{t_k} - B_{t_{k-1}}$ are independent centered Gaussian random variables with respective variances $t_1, t_2 - t_1, \ldots, t_k - t_{k-1}$. That is, the distribution of the random vector $(B_{t_1}, B_{t_2} - B_{t_1}, \ldots B_{t_k} - B_{t_{k-1}})$ has density

$$\frac{1}{(2\pi)^{k/2}[t_1(t_2 - t_1) \cdots (t_k - t_{k-1})]^{1/2}} \exp\left(-\frac{1}{2}\left(\frac{x_1^2}{t_1} + \cdots + \frac{x_k^2}{t_k - t_{k-1}}\right)\right)$$

with respect to the Lebesgue measure on $\mathbb{R}^k$.

Brownian motion $(B_t)_{t \geq 0}$ with values in $\mathbb{R}^n$ is constructed similarly, the random vectors $B_{t_1}, B_{t_2} - B_{t_1}, \ldots B_{t_k} - B_{t_{k-1}}$ now being independent centered Gaussian vectors with respective covariance matrices

$$t_1 \operatorname{Id}, (t_2 - t_1) \operatorname{Id}, \ldots, (t_k - t_{k-1}) \operatorname{Id}$$

where Id is the identity matrix of dimension n. Equivalently, such a Brownian motion $(B_t)_{t\geq 0}$ in $\mathbb{R}^n$ may be described as $B_t = (B_t^1, \ldots, B_t^n)$, $t \geq 0$, where $(B_t^i)_{t\geq 0}$, $1 \leq i \leq n$, is a collection of n independent one-dimensional Brownian motions.

It is not obvious that such a process $(B_t)_{t\geq 0}$ exists, in particular with respect to continuity of the trajectories. From a functional analysis point of view, we may as well consider that such a stochastic process $(B_t)_{t\geq 0}$ induces a probability measure on the space $\mathcal{C}([0, \infty))$ of continuous functions on $[0, \infty)$ with values in $\mathbb{R}$ or $\mathbb{R}^n$ (that is, we may always choose $\Omega = \mathcal{C}([0, \infty))$ and set $B_t(\omega)$ to be $\omega(t)$). This Gaussian measure on the Borel sets of $\mathcal{C}([0, \infty))$ is known as the *Wiener measure*.

Although continuous, the functions $t \mapsto B_t(\omega)$ are very irregular. With probability 1, they do not have a bounded variation on any interval. In particular, expressions like $\frac{dB_t}{dt}$ do not make sense. Nevertheless, there is a way to define *stochastic integrals*, that is expressions like

$$\int_0^t H_s(\omega) dB_s(\omega)$$

for some stochastic processes $H_s(\omega)$ provided they "depend only on the past". To understand this notion, it is necessary to introduce the so-called filtration $\mathcal{F}_t$, $t \geq 0$, of Brownian motion consisting of the sub-σ-fields of Σ defined as $\mathcal{F}_t = \sigma(B_s\,;\, 0 \leq s \leq t)$ with, by convention, $\mathcal{F}_\infty = \sigma(B_s\,;\, s \geq 0)$. In other words, $\mathcal{F}_t$ is the smallest σ-field for which all the random variables B_s for $s \leq t$ are measurable, and thus contains all information up to time t (as far as observations on the process $(B_s)_{s\geq 0}$ are concerned). The collection $(\mathcal{F}_t)_{t\geq 0}$ defines an increasing family of σ-fields and, for technical reasons, it is usually enlarged so to contain all the elements of $\mathcal{F}_\infty$ which have 0 probability. Sometimes, the σ-field $\mathcal{F}_t$ for every t is also enlarged in order to include non-constant variables in $\mathcal{F}_0$, which are in general independent of $\mathcal{F}_\infty$, and this is systematically the case when solving stochastic differential equations with non-constant initial data.

Definition B.1.1 (Martingale) A *martingale* (relative to the filtration $(\mathcal{F}_t)_{t\geq 0}$) is a real-valued process $(M_t)_{t\geq 0}$ defined on $(\Omega, \Sigma, \mathbb{P})$, such that for any $t \geq 0$, M_t is $\mathbb{P}$-integrable, M_t is $\mathcal{F}_t$-measurable (the process $(M_t)_{t\geq 0}$ is then said to be adapted to the filtration $(\mathcal{F}_t)_{t\geq 0}$) and for all $s < t$,

$$M_s = \mathbb{E}(M_t \mid \mathcal{F}_s).$$

Here $\mathbb{E}(M_t \mid \mathcal{F}_s)$ denotes the conditional expectation of the integrable random variable M_t with respect to the σ-field $\mathcal{F}_s$ (defined almost surely). In particular (and this is an important feature of the martingale property), for a martingale $(M_t)_{t\geq 0}$, $\mathbb{E}(M_t)$ is constant and $\mathbb{E}(M_t) = \mathbb{E}(M_0)$ for every $t > 0$.

The Brownian motion itself is a martingale. Martingales may be constructed in many ways. For example, if X is an integrable random variable on $(\Omega, \Sigma, \mathbb{P})$, then $M_t = \mathbb{E}(X \mid \mathcal{F}_t)$, $t \geq 0$, is a martingale (with respect to $(\mathcal{F}_t)_{t\geq 0}$). Many martingales may be represented in this way. In fact, if for some $p > 1$, $\sup_t \mathbb{E}(|M_t|^p) < \infty$, then there exists a random variable $M_\infty \in \mathbb{L}^p(\mathbb{P})$ such that $M_t = \mathbb{E}(M_\infty \mid \mathcal{F}_t)$ for

B.1 Brownian Motion and Stochastic Integrals

every $t \geq 0$, and moreover M_t converges to M_∞ as $t \to \infty$ (almost surely and in $\mathbb{L}^p(\mathbb{P})$). A specific property of the filtration $(\mathcal{F}_t)_{t \geq 0}$ of Brownian motion is that any martingale relative to it is continuous, which makes the analysis of processes constructed with Brownian motion much more pleasant than the general theory.

One further way to construct martingales is via the stochastic integrals (B.1.1). Such integrals are defined for any processes $H_s(\omega)$, $s \geq 0$, $\omega \in \Omega$, assumed to be continuous (more general conditions are available), adapted to the filtration $(\mathcal{F}_t)_{t \geq 0}$ in the sense that $\omega \mapsto H_s$ is $\mathcal{F}_s$-measurable for any s, and square integrable

$$\mathbb{E}\left(\int_0^t H_s^2 ds\right) < \infty$$

(for every $t \geq 0$). Given such a process $H = (H_s)_{s \geq 0}$, the stochastic integrals

$$X_t = \int_0^t H_s dB_s, \quad t \geq 0, \tag{B.1.1}$$

define a new stochastic process, again with continuous paths (even if H itself is not continuous). Although the integral X_t does not make sense for any particular choice of ω since the map $t \mapsto B_t(\omega)$ has no bounded variation on any interval, the usual way to define X_t is to consider it as a suitable limit of Riemann sums. Indeed, start with $H_s(\omega)$ piecewise constant in the sense that

$$H_s(\omega) = \sum_{i=0}^{p-1} K_i(\omega) \mathbb{1}_{(t_i, t_{i+1}]}(s)$$

for some finite sequence $0 = t_0 \leq t_1 < \cdots < t_p$ and random variables $K_1, \ldots, K_{p-1}$. In order for this process to be adapted, assume that K_i is $\mathcal{F}_{t_i}$-measurable for every i, and of course in $\mathbb{L}^2(\mathbb{P})$. In this case, one directly defines

$$\int_0^\infty H_s dB_s = \sum_{i=0}^{p-1} K_i (B_{t_{i+1}} - B_{t_i})$$

and

$$X_t = \int_0^\infty H_s \mathbb{1}_{[0,t]}(s) dB_s, \quad t \geq 0.$$

It is easily checked that the process $t \mapsto X_t$ is a martingale and that

$$\mathbb{E}(X_t^2) = \int_0^t \mathbb{E}(H_s^2) ds$$

for every $t \geq 0$. This identity thus defines an isometry between the piecewise constant processes H and $\mathbb{L}^2(\mathbb{P})$ random variables, which may then be extended by $\mathbb{L}^2(\mathbb{P})$-continuity to general (continuous) processes H.

Via this construction, the resulting process $(X_t)_{t\geq 0}$ is then a continuous square integrable martingale (with respect to the filtration $(\mathcal{F}_t)_{t\geq 0}$). In general $X_0 = 0$. It is however convenient to add some $\mathcal{F}_0$-measurable random variable X_0 and to consider the process

$$X_t = X_0 + \int_0^t H_s dB_s, \quad t \geq 0,$$

with initial value X_0. These stochastic integrals are in general as irregular (in the time variable) as the original Brownian motion $(B_t)_{t\geq 0}$ itself.

A simple special case is obtained with the choice of non-random functions $H_s = h_s$, $s \geq 0$. In this case, the random variable $X(h) = \int_0^\infty h_s dB_s$, known as a *Wiener integral*, is a Gaussian variable with mean 0 and variance $\int_0^\infty h_s^2 ds$ (provided it is finite). The covariance of two such variables $X(h)$ and $X(k)$ is given similarly by

$$\mathbb{E}(X(h)X(k)) = \int_0^\infty h_s k_s \, ds.$$

This identity thus yields an embedding of $\mathbb{L}^2([0,\infty), dt)$ into the $\mathbb{L}^2(\mathbb{P})$-space of random variables defined on $(\Omega, \Sigma, \mathbb{P})$ giving rise to the so-called *reproducing kernel Hilbert space* or *Cameron-Martin Hilbert space* associated to Brownian motion. More on the Cameron-Martin Hilbert space in the context of this book may be found in Sect. 2.7.2, p. 108.

In general, it is not enough to construct stochastic integrals only for square integrable processes H and it is required to "localize" this procedure. This localization procedure is performed with the notion of *stopping time*.

Definition B.1.2 (Stopping time) A stopping time, with respect to the filtration $(\mathcal{F}_t)_{t\geq 0}$, is a random variable $T : \Omega \to [0, \infty]$ such that, for any $t \geq 0$, the set $\{T \leq t\} = \{\omega \in \Omega ; T(\omega) \leq t\}$ is $\mathcal{F}_t$-measurable.

It is important to allow stopping times to take the value $+\infty$. A typical stopping time is

$$T(\omega) = \inf\{t \geq 0 ; X_t(\omega) \in A\}, \quad \omega \in \Omega,$$

where $(X_t)_{t\geq 0}$ is a continuous process (with values in $\mathbb{R}$ or $\mathbb{R}^n$, for example) adapted to the filtration $(\mathcal{F}_t)_{t\geq 0}$ and A any Borel set in $\mathbb{R}$ or $\mathbb{R}^n$ (with the usual convention that $\inf\{\emptyset\} = +\infty$).

Associated with a stopping time T, one introduces the σ-field $\mathcal{F}_T$ of events which take place before T, defined as

$$\mathcal{F}_T = \{A \in \mathcal{F}_\infty ; A \cap \{T \leq t\} \in \mathcal{F}_t \text{ for every } t \geq 0\}.$$

Clearly T is $\mathcal{F}_T$-measurable.

When $(X_t)_{t\geq 0}$ is an adapted stochastic process (with respect to $(\mathcal{F}_t)_{t\geq 0}$), this process may be stopped at any stopping time T (for $(\mathcal{F}_t)_{t\geq 0}$) by setting

$$X_{t \wedge T}(\omega) = X(\omega)_{\min(t, T(\omega))}, \quad t \geq 0, \ \omega \in \Omega.$$

B.2 The Itô Formula

Set $X_T = X(\omega)_{T(\omega)}$, being careful with this expression when $T(\omega) = \infty$ (in general assume that some X_∞ random variable is given). This procedure thus produces a new adapted stochastic process $(X_{t \wedge T})_{t \geq 0}$, and it is a basic result, known as *Doob's stopping time Theorem*, that if $(X_t)_{t \geq 0}$ is a continuous martingale, so is the new process $(X_{t \wedge T})_{t \geq 0}$ (relative to the underlying filtration $(\mathcal{F}_t)_{t \geq 0}$). Moreover, when $X_t = \mathbb{E}(X_\infty \mid \mathcal{F}_t)$, $t \geq 0$, for some integrable random variable X_∞, then $X_T = \mathbb{E}(X_\infty \mid \mathcal{F}_T)$.

Local martingales are then defined as adapted processes $(X_t)_{t \geq 0}$ for which there exists an increasing sequence $(T_k)_{k \in \mathbb{N}}$ of stopping times (relative to $(\mathcal{F}_t)_{t \geq 0}$) which converges to ∞ and for which the processes $(X_{t \wedge T_k})_{t \geq 0}$, $k \in \mathbb{N}$, are martingales. Stochastic integrals of the type (B.1.1)

$$\int_0^t H_s dX_s, \quad t \geq 0,$$

with respect to such a local martingale $(X_t)_{t \geq 0}$ are then defined for (continuous) adapted processes $(H_s)_{s \geq 0}$ provided that

$$\mathbb{E}\left(\int_0^{T_k} H_s^2 ds\right) < \infty$$

for any k. The resulting stochastic integral is then again a local martingale. This construction yields much more freedom to define stochastic integrals (and later solutions of stochastic differential equations) although one can no longer be sure that a local martingale $(X_t)_{t \geq 0}$ has a constant expectation. To reach this conclusion usually requires further integrability properties, such as $\sup_{t \geq 0} \mathbb{E}(|X_t|^p) < \infty$ for some $p > 1$.

B.2 The Itô Formula

The fundamental property of the stochastic calculus is the famous Itô formula which expresses a change of variables adapted to stochastic integration.

Theorem B.2.1 (Itô's formula) *Given stochastic integrals*

$$X_t = X_0 + \int_0^t H_s dB_s, \quad t \geq 0,$$

as constructed before, if $f : \mathbb{R} \to \mathbb{R}$ is a C^2 function, then

$$f(X_t) = f(X_0) + M_t + A_t, \quad t \geq 0,$$

where

$$M_t = \int_0^t f'(X_s) H_s dB_s \quad \text{and} \quad A_t = \frac{1}{2} \int_0^t f''(X_s) H_s^2 ds.$$

In particular, $(M_t)_{t \geq 0}$ is a local martingale and $(A_t)_{t \geq 0}$ has locally bounded variations.

In other terms, using the differential notation $dX_t = H_t dB_t$, and introducing the "bracket"

$$d\langle X, X \rangle_t = H_t^2 dt,$$

Itô's formula indicates that for a C^2 function $f : \mathbb{R} \to \mathbb{R}$,

$$df(X_t) = f'(X_t)dX_t + \frac{1}{2} f''(X_t)d\langle X, X \rangle_t.$$

The quadratic notation $d\langle X, X \rangle$ is natural under the interpretation

$$d\langle X, X \rangle_t = \langle dX, dX \rangle_t = \langle HdB, HdB \rangle_t = H_t^2 dt$$

(in particular $\langle dB, dB \rangle_t = dt$).

There is a more intrinsic definition of the increasing process $\langle X, X \rangle_t = \int_0^t d\langle X, X \rangle_s$, $t \geq 0$, as the limit under subdivisions $0 = t_0 < t_1 < \cdots < t_k = t$ of the interval $[0, t]$ of

$$\sum_{i=0}^{k-1} (X_{t_{i+1}} - X_{t_i})^2$$

when the mesh of the subdivision goes to 0. It is easily seen that such a quantity is 0 as soon as $(X_t)_{t \geq 0}$ is a continuous bounded variation process (and therefore this bracket extracts the purely martingale part of the process $(X_t)_{t \geq 0}$ in the same way as the carré du champ operator Γ of Definition 1.4.2, p. 20, extracts the second order part of a differential operator L).

The change of variables formula expressed by Itô's formula is valid for the larger class of so-called semi-martingales. A process $(X_t)_{t \geq 0}$ is called a *semi-martingale* (with respect to a given filtration $(\mathcal{F}_t)_{t \geq 0}$) if it may be decomposed into a sum

$$X_t = X_0 + M_t + A_t, \quad t \geq 0,$$

where $(M_t)_{t \geq 0}$ is a local martingale and $(A_t)_{t \geq 0}$ has locally bounded variations. For example, as illustrated by Itô's formula, a process $(X_t)_{t \geq 0}$ given by

$$X_t = X_0 + \int_0^t H_s dB_s + \int_0^t K_s ds, \quad t \geq 0, \tag{B.2.1}$$

for suitable stochastic processes $(H_s)_{s \geq 0}$ and $(K_s)_{s \geq 0}$ such that $\int_0^t H_s dB_s$, $t \geq 0$, is a local martingale and $\int_0^t K_s ds$, $t \geq 0$, has bounded variations, is a semi-martingale. Replacing in Itô's formula $dX_t = H_t dB_t$ by $dX_t = dM_t + dA_t$ shows that the class of semi-martingales is stable under the action of C^2 functions.

B.3 Stochastic Differential Equations

One may also consider stochastic integrals driven by semi-martingales, for example in the framework of example (B.2.1), set, for an adapted process $(R_s)_{s \geq 0}$,

$$(R \cdot X)_t = \int_0^t R_s dX_t = \int_0^t R_s H_s dB_s + \int_0^t R_s K_s ds, \quad t \geq 0.$$

The differential rules then take the form

$$d(R \cdot X)_t = R_t dX_t, \qquad d\langle R \cdot X, R \cdot X \rangle_t = R_t^2 d\langle X, X \rangle_t = R_t^2 H_t^2 dt.$$

The Itô chain rule formula is thus quite different than the formula for differentiable processes for which the second order term vanishes. The full strength of stochastic calculus relies on this specific feature, which deeply connects stochastic integration to second order differential operators.

When dealing with multivariate processes, introduce (by polarization) for a pair $(X_t)_{t \geq 0}$, $(Y_t)_{t \geq 0}$ of semi-martingales, the bounded variation process

$$d\langle X, Y \rangle_t = \frac{1}{2}\big[d\langle X+Y, X+Y \rangle_t - d\langle X, X \rangle_t - d\langle Y, Y \rangle_t\big], \quad t \geq 0.$$

The multi-variable form of Itô's chain rule formula for a vector $X_t = (X_t^1, \ldots, X_t^n)$, $t \geq 0$, of semi-martingales and $f : \mathbb{R}^k \to \mathbb{R}$ a $\mathcal{C}^2$ function is then

$$df(X_t) = \sum_{i=1}^k \partial_i f(X_t) dY_t^i + \frac{1}{2} \sum_{i,j=1}^k \partial_{ij}^2 f(X_t) d\langle X^i, X^j \rangle_t.$$

This formula is of course closely related to the change of variables formula for diffusion processes (1.11.4), p. 43, in Chap. 1.

B.3 Stochastic Differential Equations

Stochastic integrals lead to the solution of *stochastic differential equations*. Stochastic differential equations provide a natural framework for the probabilistic analysis of diffusions processes and their Markov generators as further developed in Sect. B.4 (see also Sect. 1.10, p. 38).

In dimension one, the simplest stochastic differential equation appears in the form

$$dX_t = \sigma(X_t)dB_t + b(X_t)dt, \quad X_0 = x,$$

where σ and b are smooth functions on the real line (random coefficients σ may also be considered although they are not addressed here). In other words,

$$X_t = X_0 + \int_0^t \sigma(X_s)dB_s + \int_0^t b(X_s)ds, \quad t \geq 0.$$

Stochastic calculus and stochastic differential equations related to the study of diffusion semigroups deal mostly with vector-valued Brownian motion $B_t = (B_t^1, \ldots, B_t^p)$, $t \geq 0$, thus consisting of p independent copies of the standard Brownian motion. Stochastic integrals with respect to those p independent Brownian motions may be considered simultaneously and the corresponding bracket rule is $d\langle B^i, B^j \rangle_t = \delta^{ij} dt$, $1 \leq i, j \leq p$. The prototype stochastic differential equation driven by such a multi-dimensional Brownian motion $(B_t)_{t \geq 0}$ is then described as a vector-valued process $X_t = (X_t^1, \ldots, X_t^n)$, $t \geq 0$, which is a solution of

$$dX_t^j = \sum_{i=1}^{p} \sigma_i^j(X_t) dB_t^i + b^j(X_t) dt, \quad X_0^j = x^j, \ 1 \leq j \leq n.$$

Here $\sigma = \sigma(x) = (\sigma_i^j(x))_{1 \leq i, j \leq n}$ is an $n \times p$ matrix depending on $x \in \mathbb{R}^n$ and $b = b(x) = (b^j(x))_{1 \leq j \leq n}$ is a vector depending on $x \in \mathbb{R}^n$ satisfying suitable growth and smoothness assumptions (see Theorem B.3.1 below). The preceding equation is then summarized more simply in the form

$$dX_t = \sigma(X_t) dB_t + b(X_t) dt, \quad X_0 = x, \tag{B.3.1}$$

and the solution is the vector-valued process denoted by $\{X_t^x; t \geq 0, x \in \mathbb{R}^n\}$ to describe dependence on the initial condition x (there should be no confusion with the j-th coordinate of X_t).

There are many ways in which existence and uniqueness of such stochastic differential equations may be discussed. The existence may be thought of as the existence of a Brownian motion $(B_t)_{t \geq 0}$ on some probability space $(\Omega, \Sigma, \mathbb{P})$ such that this holds. The uniqueness may also be regarded in terms of the uniqueness of the laws of the processes $\{X_t^x; t \geq 0, x \in \mathbb{R}^n\}$. The discussion here is restricted to the simplest notions. Say that $\{X_t^x; t \geq 0, x \in \mathbb{R}^n\}$ is a solution of (B.3.1) if it satisfies the equation path-wise (that is for almost every $\omega \in \Omega$ in the probability space on which the Brownian motion is defined). Moreover, say that the solution is (path-wise) unique if for any initial value x, two solutions X and X' of (B.3.1) are such that, outside a set of probability 0, the maps $t \mapsto X_t$ and $t \mapsto X_t'$ coincide. In this context, the main conclusions may be described by the following statement. There, $\|\cdot\|$ denotes any norm on points, matrices and vectors which are compatible with the usual topology.

Theorem B.3.1 (Existence and uniqueness of solutions of stochastic differential equations) *Assume that the maps $x \mapsto \sigma(x)$ and $x \mapsto b(x)$ on $\mathbb{R}^n$ have a linear growth*

$$\|\sigma(x)\| + \|b(x)\| \leq D(1 + \|x\|) \tag{B.3.2}$$

for some finite constant D and all $x \in \mathbb{R}^n$ and are locally Lipschitz in the sense that for any compact set $K \subset \mathbb{R}^n$, there exists a constant $C_K > 0$ such that

$$\|\sigma(x) - \sigma(y)\| + \|b(x) - b(y)\| \leq C_K \|x - y\| \tag{B.3.3}$$

for all $x, y \in K$. Then, there exists a unique solution $\{X_t^x; t \geq 0, x \in \mathbb{R}^n\}$ of (B.3.1) defined on the time interval $[0, \infty)$.

Solutions $\{X_t^x ; t \geq 0, x \in \mathbb{R}^n\}$ of stochastic differential equations (B.3.1) satisfy many properties, in particular the *strong Markov property*. This property indicates that for any finite stopping time T of the given filtration $(\mathcal{F}_t)_{t\geq 0}$, the law of the process $(Y_t = X_{T+t})_{t\geq 0}$ conditioned on $\mathcal{F}_T$ only depends on X_T itself, and this law conditioned to $\{X_T = x\}$ is the law of $(X_t^x)_{t\geq 0}$. The process $(Y_t)_{t\geq 0}$ is thus a solution of the same stochastic differential equation (B.3.1) but with initial value $Y_0 = X_T$ (with of course some other Brownian motion driving the equation, namely $(B_{T+t} - B_T)_{t\geq 0}$). In other words, the process starts afresh at any stopping time. This strong Markov property also allows for the definition of solutions of such stochastic differential equations up to a given stopping time. As will be developed below, this is an important tool in the investigation of stochastic differential equations on manifolds where there is no global chart.

With the latter task in mind, Theorem B.3.1 is not always enough. For example, it is often necessary to work on some open subset $\mathcal{O}$ in $\mathbb{R}^n$ and to define the solution up to the stopping time T which is the time at which the process leaves the set $\mathcal{O}$. In this case, only the Lipschitz property (B.3.3) is necessary, and uniqueness and existence then only hold up to this stopping time which is called the *explosion time* of the process. When $\mathcal{O}$ is $\mathbb{R}^n$ itself, this explosion time is the time at which the process reaches infinity, which could be finite whenever (B.3.2) is not satisfied.

The preceding processes solutions of stochastic differential equations may also be constructed on manifolds, not only on open sets in $\mathbb{R}^n$, replacing σ and b by a system of vector fields. This is usually achieved in local charts via localization as briefly outlined below in Sect. B.4.

B.4 Diffusion Processes

The link between stochastic differential equations and diffusion processes is one of the main features of this monograph. Solutions of stochastic differential equations with smooth coefficients define stochastic processes driven by second order differential operators. Solutions of the associated heat equation give rise to semigroups of operators and heat kernels.

Let $\{X_t^x ; t \geq 0, x \in \mathbb{R}^n\}$ be a solution of the stochastic differential equation (B.3.1) with coefficients σ and b and driven by a p-dimensional Brownian motion $(B_t)_{t\geq 0}$. For every smooth function $f : \mathbb{R}^n \to \mathbb{R}$, by Itô's formula,

$$df(X_t) = \sum_{i=1}^{p} \sum_{j=1}^{n} \sigma_i^j(X_t) \partial_j f(X_t) dB_t^i + \mathrm{L} f(X_t) dt, \quad X_0 = x,$$

where L is the second order operator

$$\mathrm{L} f = \frac{1}{2} \sum_{i,j=1}^{n} \sum_{k=1}^{p} \sigma_k^i \sigma_k^j \partial_{ij}^2 f + \sum_{i=1}^{n} b^i \partial_i f.$$

Now, the first term in the expression of $df(X_t)$ gives rise to a local martingale $(M_t^f)_{t\geq 0}$ so that

$$f(X_t^x) = M_t^f + \int_0^t \mathrm{L}f(X_s^x)ds, \quad t \geq 0, \ x \in \mathbb{R}^n,$$

and these processes are therefore always semi-martingales. The process $\{X_t^x ; t \geq 0, x \in \mathbb{R}^n\}$ is then said to solve the *martingale problem* associated with the operator L. As soon as we know that the local martingale $(M_t^f)_{t\geq 0}$ is indeed a true martingale, for example as soon as f and $\mathrm{L}f$ are bounded functions, it follows by taking expectations that, for every $t \geq 0$,

$$\mathbb{E}(f(X_t^x)) = f(x) + \frac{1}{2}\int_0^t \mathbb{E}(\mathrm{L}f(X_s^x))ds.$$

This formula tells us that the process $\{X_t^x ; t \geq 0, x \in \mathbb{R}^n\}$ is a so-called *diffusion process* with generator $\frac{1}{2}\mathrm{L}$. In other words, setting

$$P_t f(x) = \mathbb{E}(f(X_t^x)) = \mathbb{E}(f(X_t) \mid X_0 = x), \quad t \geq 0, \ x \in \mathbb{R}^n,$$

it holds that

$$P_t f(x) = f(x) + \frac{1}{2}\int_0^t P_s \mathrm{L}f(x)ds, \quad t \geq 0, \ x \in \mathbb{R}^n,$$

and therefore $P_t f$ the solution of the (parabolic) *heat equation* $\partial_t = \frac{1}{2}\mathrm{L}$ starting from f. This construction thus produces a process whose law represents the heat kernels associated with the generator $\frac{1}{2}\mathrm{L}$. It should be mentioned that throughout this monograph, in spite of this probabilistic representation, we work with the generator L rather than $\frac{1}{2}\mathrm{L}$.

The *Stratonovich integral* is a way to get simpler and easier expressions in the preceding when dealing with operators given as a sum of squares of vector fields (in Hörmander form, cf. (1.10.2), p. 41). The associated Stratonovich notation ∘ is used when both the integrator $(Y_t)_{t\geq 0}$ and the integrand $(H_t)_{t\geq 0}$ are semi-martingales in which case the Stratonovich integral reads

$$X_t = X_0 + \int_0^t H_s \circ dY_s = X_0 + \int_0^t H_s dY_s + \frac{1}{2}\int_0^t d\langle H, Y\rangle_s, \quad t \geq 0.$$

Theorems on the existence and uniqueness of solutions of stochastic differential equations under the Stratonovich integral are exactly the same as those under the Itô form. Concerning the change of variables formula in the Stratonovich formalism, if for example the real-valued diffusion process $(X_t)_{t\geq 0}$ solves $dX_t = \sigma(X_t) \circ dB_t$, then for any smooth f,

$$df(X_t) = f'(X_t)\sigma(X_t)dB_t + \frac{1}{2}\big[f'(X_t)\sigma(X_t)\sigma'(X_t) + f''(X_t)\sigma^2(X_t)\big]dt.$$

B.4 Diffusion Processes

As mentioned above, the Stratonovich formulation is a much more convenient way to represent the operator L in the Hörmander form

$$L = \frac{1}{2}\sum_{j=1}^{d} Z_j^2 + Z_0$$

where, for $j = 0, 1, \ldots, d$, Z_j is a vector field (in other words, the vector $Z_j(x) = (Z_j^1(x), \ldots, Z_j^n(x))$ is identified to the first order differential operator $Z_j f = \sum_{i=1}^{n} Z_j^i \partial_i f$). Indeed, in this case, as soon as

$$dX_t = \sum_{j=1}^{d} Z_j(X_t) \circ dB_t^j + Z_0(X_t) dt$$

where $(B_t^j)_{t \geq 0}$, $j = 1, \ldots, d$, are independent Brownian motions, the Itô formula immediately yields that

$$f(X_t) = M_t^f + \int_0^t Lf(X_s) ds$$

where $(M_t^f)_{t \geq 0}$ is the local martingale

$$dM_t^f = \sum_{j=1}^{d} \partial_j f(X_t) Z_j(X_t) dB_t^j.$$

The process $(X_t)_{t \geq 0}$ thus similarly solves the martingale problem with respect to the operator L.

When dealing with stochastic processes and stochastic differential equations on a smooth manifold, one usually uses the localization procedure. In some open set $\mathcal{O}$ on which there exists a local chart, one solves the stochastic differential equation up to the stopping time $T = \inf\{t \geq 0; X_t \notin \mathcal{O}\}$ and then changes the local chart. The change of variable laws according to Itô's formula assert that the result is independent of the choice of coordinate system. Then, the strong Markov property at time T is used to start the process again in the new chart with initial value X_T. In such a way, the diffusion process is well defined as long as it does not reach the boundary of the manifold (or infinity), and then specific arguments are required to analyze whether or not the process reaches this boundary in finite time or not. (However, the global growth condition of Theorem B.3.1 provides a useful criterion towards non-explosion.) When dealing with Laplace operators on compact or complete Riemannian manifolds, there exist some more global procedures (embedding the manifold in some Euclidean space with larger dimensions, lifting the process to the frame fiber bundle etc.), each of them having its specific advantages.

As already mentioned, solutions of stochastic differential equations are very irregular in the time variable t (as irregular as the Brownian paths). However, they depend smoothly on the initial condition x. In fact, solving for example the stochastic differential equation

$$dX_t = \sum_{j=1}^{d} Z_j(X_t) \circ dB_t^j + Z_0(X_t)dt, \quad X_0 = x,$$

for vector fields $Z_0, Z_1, \ldots, Z_d$, the solution $\{X_t^x ; t \geq 0, x \in \mathbb{R}^n\}$ is as smooth in the variable x as the vector fields Z_j are. For example, if every Z_j is C^k for some $k \geq 2$, so is the process $(X_t^x)_{t \geq 0}$, as long as the solution exists on some time interval $[0, T]$ (which may be random when T is a stopping time). This explains why the laws of the random variables X_t^x, described by Markov kernels $p_t(x, dy)$ (cf. Sect. 1.2, p. 9), are in general as smooth as the vector fields Z_j themselves. On the other hand, to study densities $p_t(x, y)$ with respect to some fixed reference measure $dm(y)$, the regularity in the y variable requires more hypotheses on the vector fields Z_j, for example ellipticity (or hypo-ellipticity, cf. Sect. 1.12, p. 49). The probabilistic analysis of those densities in terms of the variable y (especially the mere existence of those densities) requires more sophisticated tools, for example Malliavin calculus, or stochastic calculus of variation, which go far beyond this elementary introduction.

B.5 Notes and References

The material presented in this short appendix is rather standard and may be found in any classic reference on stochastic calculus, including [132, 153, 252, 255, 263, 336, 350, 358, 362, 363] etc. For accounts of the stochastic calculus of variations (Malliavin calculus), see [142, 297, 334].

Appendix C
Basic Notions in Differential and Riemannian Geometry

This appendix presents some basic notions of differential geometry which are used in many parts of this book. Even for the reader only interested in diffusions in $\mathbb{R}^n$ or open sets in $\mathbb{R}^n$, it may sometimes be useful to consider the case of manifolds. For example, it is much easier to obtain the optimal Sobolev inequality on spheres than on Euclidean spaces, and then to carry those optimal inequalities from spheres to $\mathbb{R}^n$ by conformal transformations as described in Sect. 6.9, p. 313. It is also true that the analysis of Markov semigroups on compact Riemannian manifolds without boundaries (like spheres, toruses and so on) is in many respects much easier than the analysis of the same objects on $\mathbb{R}^n$ or open sets of $\mathbb{R}^n$. (This is somewhat similar to the difference in the analysis of Markov chains on a finite set or on an infinite set.) Unfortunately, on manifolds there is no way to locate a point through a unique chart, thus explaining why it is necessary to develop an (apparently) somewhat heavy machinery. Finally, there are some computation rules which have a natural interpretation in terms of differential geometry, even if only the case of $\mathbb{R}^n$ is considered (such as connections, curvature tensors, Γ_2 operators etc.).

This appendix thus briefly presents the language of differentiable and Riemannian manifolds, and some of the classical rules. It also introduces the notion of Ricci curvature and its connection with the curvature-dimension conditions extensively used throughout this work. It should be mentioned in particular that the last two sections outline arguments and results on the Γ_2 operator and the reinforced curvature-dimension condition, which are central to this book. As for the previous appendices, this appendix is far from a complete and formal exposition of differentiable and Riemannian manifolds and only emphasizes some basic elements necessary for the further developments. The reader is referred to standard textbooks on the subject for complete details, some of which are listed in the Notes and References.

In several places in this appendix, as well as in the corresponding parts of this book, index conventions are used according to the covariant or contravariant rules of differential geometry. (In particular, coordinates of vectors are indicated with upper indices. For simplicity, in the core of the book, lower indices are usually used.) The Kronecker symbol δ_i^j is equal to 1 if $i = j$, and 0 otherwise. The Einstein notation summarizes a summation over an index when it appears twice, one time up, one time down (usually from 1 to the dimension of the underlying structure). For example, $Z^i e_i$ is a shorthand for $\sum_{i=1}^{n} Z^i e_i$. In the same spirit

$$g^{ij} e_i e_j = \sum_{i,j=1}^{n} g^{ij} e_i e_j.$$

C.1 Differentiable Manifolds

An n-dimensional differentiable manifold is a topological space in which a point is located through coordinates $x \in \mathbb{R}^n$, called a chart. It is not always possible to specify such a localization globally so sometimes points might be located in two different charts. For example, the unit sphere $\mathbb{S}^n \subset \mathbb{R}^{n+1}$ is an n-differentiable manifold. A point $x = (x^1, \ldots, x^{n+1}) \in \mathbb{S}^n$ such that $x^{n+1} > 0$ and $x^1 > 0$ may be located through its projection on the hyperplane $\{x^{n+1} = 0\}$, but also through its projection on $\{x^1 = 0\}$. Those two projections are in fact points in the n-dimensional open unit ball (in $\mathbb{R}^n$). It is then necessary to establish a correspondence between these two points to assert that they represent the same point on $\mathbb{S}^n$. In this example, a point $(y^1, \ldots, y^n)$ in the unit ball corresponds to

$$\left(y^1, \ldots, y^n, \sqrt{1 - (y^1)^2 - \cdots - (y^n)^2}\right) \in \mathbb{S}^n$$

in the first projection, and to

$$\left(\sqrt{1 - (y^1)^2 - \cdots - (y^n)^2}, y^1, \ldots, y^n\right) \in \mathbb{S}^n$$

in the second one. Therefore, this correspondence defines a map ψ from the intersection of the unit ball with $\{y^1 > 0\}$ into the intersection of the unit ball with $\{y^n > 0\}$ given by

$$\psi(y^1, \ldots, y^n) = \left(y^2, \ldots, y^n, \sqrt{1 - (y^1)^2 - \cdots - (y^n)^2}\right).$$

The key point in differential geometry is to be able to perform computations which are independent of the chosen chart. Therefore, objects from one chart to another have to be identified, and some basic rules when changing charts have to be prescribed. The conclusions should then be independent of such changes of coordinates. For example, when solving stochastic differential equations, one has to make sure that the solution (or at least its distribution) does not depend on the chosen chart (cf. Sect. B.4). According to these requirements, one definition of differentiable manifolds is the following.

Definition C.1.1 (Differentiable manifold) A C^k ($k \geq 0$) n-dimensional differentiable manifold M is a topological space endowed with a countable family of open

C.1 Differentiable Manifolds

sets $(\mathcal{O}_i)$, such that each $\mathcal{O}_i$ is homeomorphic to some open set in $\mathbb{R}^n$ via a homeomorphism ψ_i. Every point $x \in M$ belongs to at least one $\mathcal{O}_i$, and the changes of coordinates

$$\psi_j \circ \psi_i^{-1} : \psi_i(\mathcal{O}_i \cap \mathcal{O}_j) \to \psi_j(\mathcal{O}_i \cap \mathcal{O}_j)$$

are $\mathcal{C}^k$-diffeomorphisms in $\mathbb{R}^n$.

A function from one manifold to another (in particular curves from $\mathbb{R}$ into a manifold) is $\mathcal{C}^k$ if it is $\mathcal{C}^k$ in the given charts. In the following and throughout the monograph, mostly $\mathcal{C}^\infty$ connected manifolds are considered ($\mathcal{C}^k$ manifolds for every k) without further mention. "Smooth" is usually understood as $\mathcal{C}^\infty$ below (and throughout the book).

The first objects of importance in this context are *tangent vectors*. A $\mathcal{C}^1$ curve on a differentiable manifold M is a $\mathcal{C}^1$ map from $\mathbb{R}$, or an open interval $I \subset \mathbb{R}$ containing 0, into M. It is given in a local chart by coordinates $\gamma(t) = (x^i(t))_{1 \le i \le n}$, $t \in I$. The derivative at $t = 0$ is the tangent vector $(Z^i)_{1 \le i \le n}$ of the curve at $\gamma(0) = x_0$. Through a change of coordinates given by $y = y(x)$ in a neighborhood of x_0, this tangent vector will have coordinates $(V^j)_{1 \le j \le n}$ at the point $y_0 = y(x_0)$. Then (recall the Einstein convention),

$$V^j = J_i^j Z^i, \quad 1 \le j \le n,$$

where $(J_i^j) = \left(\frac{\partial y^j}{\partial x^i}\right)_{x=x_0}$ is the Jacobian matrix at the point x_0.

The set of all tangent vectors to all curves passing through a given point x_0 is an n-dimensional vector space which is called the tangent space of M at x_0 and is denoted $T_{x_0}(M)$. Its dual vector space (the linear forms on it) is spanned by the differentials of functions $f : M \to \mathbb{R}$. Actually, given such a smooth function f and a curve $\gamma(t)$ onto M with derivative $Z = (Z^i)_{1 \le i \le n}$ at $t = 0$, the derivative at $t = 0$ of the function $f(\gamma(t))$ is nothing else than

$$Z^i \frac{\partial f}{\partial x^i} = Z^i \partial_i f.$$

The form (more precisely the 1-form) $w = df$ with coordinates $w_i = \partial_i f$, $1 \le i \le n$, in the local coordinate system x follows a different rule in the change of variables from x to $y = y(x)$. Namely, in the coordinate system y, df has components η_j, $1 \le j \le n$, with

$$\eta_j J_i^j = w_i, \quad 1 \le i \le n,$$

which is precisely the inverse of the change of coordinates rule for vectors.

A *vector field* Z is for any $x \in M$ a tangent vector $Z(x) \in T_x(M)$ depending in general smoothly on the point x, meaning that in a local system of coordinates, its components are smooth functions. To any vector field Z is associated a first order differential operator $f \mapsto Zf$ given in a local system of coordinates by $Zf = Z^i \partial_i f$. Thanks to the change of variables rules for vectors and forms, this expression is independent of the system of coordinates in which it is considered.

In this monograph, more complicated objects than vectors and 1-forms will be considered, namely tensors with many indices, some up (like vectors), some down (like 1-forms). Tensors are required in particular to introduce Riemannian metrics, which are simply Euclidean structures on the tangent space $T_x(M)$ moving smoothly with the point $x \in M$. The next section describes these objects first in the classical Euclidean framework.

C.2 Some Elementary Euclidean Geometry

This section is devoted to some basic notions concerning tensor calculus in Euclidean spaces. On $\mathbb{R}^n$, or any n-dimensional vector space E, a Euclidean structure is a strictly positive (positive-definite in the standard terminology) symmetric bilinear form $G(Z, Y)$ which to any pair (Z, Y) of vectors in E associates a real number $G(Z, Y)$ such that $G(Z, Z) \geq 0$ for any $Z \in E$ and $G(Z, Z) = 0$ only for $Z = 0$.

Given any basis $e = (e_i)_{1 \leq i \leq n}$ in E, such a strictly positive symmetric bilinear form G is represented by the positive-definite symmetric matrix $g_{ij} = G(e_i, e_j)$, $1 \leq i, j \leq n$, so that if a vector Z is $Z = Z^i e_i$ (that is $(Z^i)_{1 \leq i \leq n}$ are the coordinates of Z in the basis e), then

$$G(Z, Z) = Z^i Z^j g_{ij}.$$

Such a non-degenerate bilinear form G provides an isomorphism Ψ_G between E and its dual space E^* through $\Psi_G(Z)(Y) = G(Z, Y)$, $Z, Y \in E$. The linear map Ψ_G is represented by the matrix G in the dual basis e^*. The Euclidean structure G on E may then be transferred to a Euclidean structure G^* on E^* by

$$G^*(V, W) = G\big(\Psi_G^{-1}(V), \Psi_G^{-1}(W)\big), \quad (V, W) \in E^* \times E^*.$$

It turns out that if G has matrix $(g_{ij})_{1 \leq i,j \leq n}$ in a basis e, the matrix of G^* in the dual basis e^* is the inverse matrix of G, denoted $(g^{ij})_{1 \leq i,j \leq n} = ((g_{ij})_{1 \leq i,j \leq n})^{-1}$. When V and W have respective coordinates $(V_i)_{1 \leq i \leq n}$ and $(W_i)_{1 \leq i \leq n}$ in the dual basis e^*, then $G^*(V, W) = g^{ij} V_i W_j$ and the correspondence $\Psi_G(Z) = V$ is such that $V_i = g_{ij} Z^j$, $Z^i = g^{ij} V_j$, $1 \leq i \leq n$. This operation of identification between E and E^* is called the *lifting* or *lowering* of indices. Observe also that since $g^{ij} g_{jk} = \delta^i_k$, the lifting and lowering operators are inverse to each other.

Multi-linear forms acting both on E and E^* may be investigated similarly. For example, a trilinear map $T(Z, Y, V)$ on $E \times E \times E^*$ would have coordinates $T_{ij}{}^k = T(e_i, e_j, e^{*k})$ such that

$$T(Z, Y, V) = T_{ij}{}^k Z^i Y^j V_k.$$

Quite often, it is not important to precisely locate the position of the upper and lower indices. When it is implicit or not necessary to specify, this aspect will simply be omitted, as is usually the case. Such an object is called a tensor (here a 3-tensor). It may have indices up or down, depending on whether it acts for some component on

C.2 Some Elementary Euclidean Geometry

E or E^*. The spaces of multi-linear tensors may be regarded as tensor products of copies of E and E^* (in the previous example $E^* \otimes E^* \otimes E$, with basis $e^* \otimes e^* \otimes e$, and the coefficients $T_{ij}{}^k$ are just the coordinates on T in this basis). Via the identification between E and E^* one may always raise or lower any component of such a tensor. In the previous example, the components of $S(Z, Y, X) = T(Z, Y, \Psi_G(X))$ would be $S_{ijk} = T_{ij}{}^\ell g_{\ell k}$. With this operation in mind, we may always assume that a given tensor has all its indices up or down.

Tensors may be multiplied (in fact through tensor products). For example, a 3-tensor $T(Z, Y, X)$ may be multiplied by a 2-tensor $S(Z', Y')$ to give rise to a 5-tensor $R = T \otimes S$ via $R(Z, Y, X, Z', Y') = T(Z, Y, X) S(Z', Y')$. The coordinates of the new tensor are just the term by term products of the coordinates of the two initial tensors. Tensors may also be contracted. The simplest example of contraction is the action of a linear form V on a vector Z, which may be written in coordinates as $V(Z) = V_i Z^i$. A tensor $T = T_{ij}^k$ may be contracted as T_{ki}^i, or T_{ik}^i (they are not the same in general). For tensors of the form M_i^j which correspond to linear transformations from E to E, or E^* to E^*, the contraction operation is just the trace.

The quadratic forms G and G^* defined on E and E^* may be used to transfer the Euclidean metric onto some new Euclidean metric on the vector space of the tensors of any form, which means for example that if e is an orthonormal basis of E, then $e \otimes \cdots \otimes e$ is a basis of $E \otimes \cdots \otimes E$, which is again orthonormal (of course, there is a more intrinsic definition for this Euclidean metric). To compute the norm of a tensor T in any basis, one has to apply the metric G as many times as there are indices. For example, if $T = T_{ij}{}^k$ is defined in $E \otimes E \otimes E^*$, then its norm is given by

$$|T|^2 = T_{ij}{}^k T_{\ell p}{}^q g^{i\ell} g^{jp} g_{kq}.$$

(The notation $|\cdot|$ is used throughout the monograph to denote the Euclidean norm of vectors and tensors.) It is easier to use the operation of lifting or lowering indices to obtain simpler expressions such as $|T|^2 = T_{ijk} T^{ijk}$, or $T_{ij}{}^k T^{ij}{}_k$ (the multiplication by the matrix g^{ij} or the matrix g_{ij} being hidden in the operation of lifting or lowering indices).

Recall finally that the Lebesgue measure is in general defined on a vector space up to some scaling constant (as the unique Radon measure which is invariant under translation). It is well-defined on any Euclidean space, so that for example, that the measure of the cube C constructed on an orthonormal basis, that is $C = \{t^i e_i; 0 \le t_i \le 1\}$, has measure 1. This property is independent of the chosen orthonormal basis (due to the fact that the Lebesgue measure in $\mathbb{R}^n$ is invariant under orthogonal transformations). In a non-orthonormal basis, with the metric given by the matrix $G = (g_{ij})_{1 \le i, j \le n}$, the Lebesgue measure is given by $\det(G)^{1/2} dx^1 \cdots dx^n$.

In Euclidean spaces, it is thus easier to develop the associated calculus in orthonormal bases. But in Riemannian geometry, it is impossible in general to choose a system of variables for which the tangent spaces, spanned by the vectors ∂_i, are orthonormal everywhere in the neighborhood of any point. This is why the calculus has to be developed in non-orthonormal bases.

C.3 Basic Notions in Riemannian Geometry

In Riemannian geometry, the preceding notions from Euclidean geometry evolve as the metric G on the tangent space $T_x(M)$ now depends (smoothly) on the point x in the manifold M. More precisely, a *Riemannian metric* on a n-dimensional differentiable manifold M is a Euclidean metric $G(x)$ given on any tangent space $T_x(M)$, $x \in M$. In a local system of coordinates, it is given by a positive-definite symmetric matrix $G(x) = (g_{ij}(x))_{1 \leq i,j \leq n}$, such that if $Z \in T_x(M)$ has coordinates $(Z^i(x))_{1 \leq i \leq n}$, then

$$|Z|^2 = G(Z, Z) = g_{ij} Z^i Z^j$$

(for simplicity, dependence upon x is omitted here). The entries of the matrix $G(x)$ are supposed to be as smooth in the parameter $x \in M$ as the manifold is, C^∞ for simplicity in this appendix (and in most parts of the book). By ordinary computation, this is then also the case for the inverse matrix $G^*(x) = (g^{ij}(x))_{1 \leq i,j \leq n}$ denoted below by $\mathfrak{g} = G^*$ and called the *co-metric*. Outside the Euclidean case, it is always possible to choose a coordinate system such that, at a given point x, $G(x)$ is the identity matrix, but it will in general not be possible to make it the identity everywhere in a neighborhood of any given point. Nevertheless, when computing quantities such as traces or norms of tensors at a given point $x \in M$, it is often convenient to assume that $\mathfrak{g}(x) = \mathrm{Id}$.

Definition C.3.1 (Riemannian manifold) A pair (M, G) comprising a smooth (connected) n-dimensional differentiable manifold M and a (smooth) Riemannian metric $G = G(x)$, $x \in M$ is called a *Riemannian manifold* (with dimension n).

In the framework of this monograph, working with functions rather than points, the relevant Riemannian object is the co-metric $\mathfrak{g}$ (thus with upper indices) rather than the metric G. It is indeed the co-metric which naturally enters into the description of the Markov generators and their carré du champ operators (cf. Sects. 1.10 and 1.11, p. 38 and p. 42.) We thus prefer to write $(M, \mathfrak{g})$ for the basic Riemannian structure.

Due to the dependence on $x \in M$, it is important to be able to follow various quantities in the tangent spaces, such as metrics and tensors, in different coordinate systems. In a given coordinate system, a tensor $T = T(x)$ is represented by its coordinates $T^{i_1 \cdots i_\ell}_{j_1 \cdots j_m}(x)$, with as many functions of $x \in M$ as there are possible values for the multi-indices $i_1 \cdots i_\ell, j_1 \cdots j_m$, each of them varying between 1 and the dimension n (this range for the various indices is not always indicated in similar expressions below). For example, a vector field $Z(x) = (Z^i(x))_{1 \leq i \leq n}$ has one index, and represents a first order differential operator. A matrix field like $G(x) = (g_{ij}(x))_{1 \leq i,j \leq n}$ has two indices. Recall that tensors with more indices may be considered similarly and appear naturally under the operation of tensor products such as $(Z^i g_{k\ell})$ (corresponding to the tensor product $Z \otimes G$ of the vector field Z and the metric G). Here and below, dependence on $x \in M$ is often omitted.

C.3 Basic Notions in Riemannian Geometry

Under a change of coordinates $y(x)$, the coordinates of a tensor transform using the Jacobian matrix $J = \frac{\partial y}{\partial x} = (J_j^i) = (\frac{\partial y^i}{\partial x^j})$, $1 \leq i, j \leq n$, with the rule of multiplication by J for any index down and by its inverse for any index up. For example, denoting by $\bar{J}$ the inverse matrix of J (which is $\frac{\partial x}{\partial y}$), a vector field Z with coordinates $(Z^i)_{1 \leq i \leq n}$ in the coordinates x, would have coordinates $\widehat{Z}^j = Z^i J_i^j$, $1 \leq j \leq n$, in the new system y such that

$$Z^i \frac{\partial f}{\partial x^i} = Z^i \frac{\partial y^j}{\partial x^i} \frac{\partial f}{\partial y^j} = \widehat{Z}^j \frac{\partial f}{\partial y^j}.$$

Similarly, a 2-tensor $T = (T_{ij})_{1 \leq i, j \leq n}$ would be changed into $T_{k\ell} \bar{J}_i^k \bar{J}_j^\ell$, $1 \leq i, j \leq n$. A simple reminder is the change of variables formula

$$\frac{\partial f}{\partial y^i} = \frac{\partial f}{\partial x^j} \frac{\partial x^j}{\partial y^i}$$

which amounts to the change of variables formula for tensors with one index down. Upper indices are called *covariant* and lower indices are called *contravariant*.

Since the Riemannian metric $G(x) = (g_{ij}(x))_{1 \leq i, j \leq n}$ is non-degenerate, the inverse matrix $\mathfrak{g}(x) = (g^{ij}(x))_{1 \leq i, j \leq n}$ corresponds to the metric on the dual of the space of 1-forms. Recall that, as vector fields correspond to first order differential operators, a vector field Z in a given coordinate system may be written as $Zf = Z^i \partial_i f$ (in short $Z = Z^i \partial_i$) and the operators ∂_i, $1 \leq i \leq n$, form a linear basis of this space of vector fields (or tangent vectors) in this system of coordinates. In the dual space, a basis is given by the forms dx^i, $1 \leq i \leq n$, and

$$df = \partial_i f dx^i, \quad \langle Z, df \rangle = Z^i \partial_i f.$$

Therefore, the duality action between a vector field Z and a 1-form df is what was denoted by Zf. Note that not every 1-form is of the form df. For example, when g and f are smooth functions, gdf is a 1-form, but may not in general be written as dh.

Recall that the indices for vectors are up and for forms are down. As in the Euclidean case, to compute the length of a 1-form $w = (w_i)_{1 \leq i \leq n}$, one uses the inverse matrix $\mathfrak{g} = (g^{ij}(x))_{1 \leq i, j \leq n}$

$$|w|^2 = g^{ij} w_i w_j.$$

In particular, this formula defines the length $|df|^2$. As mentioned above, in the context of this monograph, while working with second order differential operators, we will be dealing precisely with the co-metric $\mathfrak{g}$ via the formula

$$\Gamma(f, f) = |df|^2 = g^{ij} \partial_i f \partial_j f$$

for the carré du champ operator Γ. This corresponds to a Riemannian metric only when the underlying operator is elliptic. Some of the formulas defined in Riemannian geometry however still make sense for degenerate operators, although the computations are much easier in the language of Riemannian geometry (cf. Sect. C.5 below).

As already discussed, the metric and its inverse may be used to lift or lower indices. For example, one may lift in this way the form df to yield the *gradient* vector denoted by ∇f with coordinates

$$\nabla^i f = g^{ij}\partial_j f, \quad 1 \leq i \leq n.$$

In particular $\nabla^i f \partial_i g = \Gamma(f,g)$. Since, as in Euclidean geometry, lifting or lowering indices do not change the norms of tensors, it holds that $|\nabla f|^2 = |df|^2$. It is however of importance to clearly distinguish df and ∇f.

The partial derivative of a function is a tensor. But in general, the partial derivative of a tensor is not a tensor as can be seen for example in the change of variable formula for $\partial^2_{ij} f$ which involves the second derivative of $y(x)$. To compensate this effect, it is necessary to introduce the notion of a *connection*. On vector fields $Z = (Z^j)_{1 \leq j \leq n}$ for example, $\partial_i Z^j$ has to be replaced by the more complicated object $\nabla_i Z^j$ defined by

$$\nabla_i Z^j = \partial_i Z^j + \gamma^j_{ik} Z^k, \quad 1 \leq i, j \leq n,$$

where the coefficients γ^k_{ij} will satisfy a specific change of variables formula, different from the one for tensors, in order for the resulting operation $\nabla_i Z^i$ to behave as a tensor. Those coefficients $\gamma^k_{ij} = \gamma_{ij}{}^k$, $1 \leq i, j, k \leq n$, are called the *Christoffel symbols* of the connection. With tensors $w = (w_j)_{1 \leq j \leq n}$ with one index down, signs have to be changed. For example,

$$\nabla_i w_j = \partial_i w_j - \gamma^k_{ij} w_k, \quad 1 \leq i, j \leq n.$$

This yields in particular the following rule, for Z a vector field and w a 1-form,

$$\nabla(Z \cdot w) = (\nabla Z) \cdot w + Z \cdot \nabla w,$$

meaning in coordinates

$$\partial_i (Z^j w_j) = (\nabla_i Z^j) \cdot w_j + Z^j (\nabla_i w_j), \quad 1 \leq i \leq n.$$

For tensors T with many indices, the rule may be applied on any index. For example,

$$\nabla_i T^{jk} = \partial_i T^{jk} + \gamma^j_{i\ell} T^{\ell k} + \gamma^k_{i\ell} T^{j\ell}.$$

The rule (changing sign for indices down) is applied as many times as there are indices in the tensor. In this way,

$$\nabla(T \otimes U) = (\nabla T) \otimes U + T \otimes (\nabla U),$$

being especially careful with the position of the indices in the resulting tensor.

The first fundamental theorem in Riemannian geometry is the existence of a particular connection.

C.3 Basic Notions in Riemannian Geometry

Theorem C.3.2 (Levi-Civita connection) *On a Riemannian manifold* $(M, \mathfrak{g})$, *there exists a unique connection* (*the Levi-Civita connection*) *such that the derivative of the metric is zero, that is* $\nabla(g^{ij}) = 0$ (*or equivalently* $\nabla(g_{ij}) = 0$), *and for any smooth function* $f : M \to \mathbb{R}$, $\nabla_i \nabla_j f = \nabla_j \nabla_i f = \nabla \nabla_{ij} f$ (*the connection is said to be torsion-free*). *In this case, the resulting operation* $(\nabla \nabla^{ij} f = \nabla^i \nabla^j f)_{1 \leq i, j \leq n}$ *is also symmetric and is called* (*in this book*) *the Hessian* $\nabla \nabla f$ *of* f (*sometimes also denoted by* Hess(f)). *The symbols of this connection are given by*

$$\gamma_{ij}^k = \frac{1}{2} g^{kp} (\partial_i g_{jp} + \partial_j g_{ip} - \partial_p g_{ij}), \quad 1 \leq i, j, k \leq n.$$

The derivative of any quantity according to this connection rule is called the *covariant derivative*.

Of course, when the metric is constant (that is, on Euclidean spaces), the connection is just the usual derivative, as long as the system of coordinates is such that (g^{ij}) is constant. But even in this case, in some other coordinate system (for example polar coordinates in $\mathbb{R}^n$), the usual derivatives differ from the covariant ones. The connection rules may be used to compute what corresponds to the usual derivative under a change of coordinates without coming back to the standard system of coordinates. The only necessary information is the matrix $\mathfrak{g}(x) = (g^{ij}(x))$.

There is often no need to explicitly compute the Christoffel symbols. This connection is nevertheless a powerful tool when working with and differentiating tensors. Thanks to the rules of the Riemannian connection, the operation ∇ commutes with lifting and lowering of indices, and also with the contraction of indices.

The Levi-Civita connection is chosen to be torsion-free, meaning that second derivatives of functions (0-tensors) are symmetric 2-tensors. This is no longer the case with derivatives of higher order tensors. For example, for a vector field Z, $\nabla_i \nabla_j Z^k$ may differ from $\nabla_j \nabla_i Z^k$. Similarly, for a smooth function f, $\nabla_i \nabla_j \nabla_k f$ is in general not symmetric with respect to (i, j, k) as is the case in Euclidean spaces. Actually, the symmetry defect $(\nabla_i \nabla_j - \nabla_j \nabla_i) Z^k$ of a vector field Z gives rise to a 4-tensor $R = (R_{ij}{}^k{}_\ell)_{1 \leq i, j, k, \ell \leq n}$, independent of the vector field Z, such that

$$(\nabla_i \nabla_j - \nabla_j \nabla_i) Z^k = R_{ij}{}^k{}_\ell Z^\ell.$$

This tensor R is the *Riemann curvature tensor*. It may be expressed (in a complicated way) in terms of the metric (g_{ij}) and its first and second derivatives in a coordinate system.

A fundamental theorem of geometry asserts that the Riemann tensor R entirely characterizes the metric. In particular, when it vanishes, the metric is Euclidean, which means that there exists (locally) a change of coordinates in which the metric is constant. It is also true that when it is constant (strictly) positive or constant (strictly) negative, the metric is spherical or hyperbolic (provided the notion of constant tensor is properly described). The Riemann curvature tensor R has a lot of symmetries. In particular, the tensor $(R_{ijk\ell})$ obtained by lowering the index k is anti-symmetric in (i, j), anti-symmetric in (k, ℓ) and symmetric under the exchange of the pairs (i, j)

and (k, ℓ). There are many other identities involving the covariant derivatives of R, for example the so-called Bianchi identities, not described here.

Most of the tools and results in this book do not use the full Riemann tensor R, but a simpler one, known as the *Ricci 2-tensor*, which is the trace of R defined as

$$\text{Ric}_{ij} = R_{ki}{}^k{}_j, \quad 1 \le i, j \le n. \tag{C.3.1}$$

The Ricci tensor is a symmetric tensor. While defined here with lower indices, depending on the context it may be convenient to use Ric^{ij} for which we retain the same notation Ric without risk of confusion. In low dimension $n \le 3$, due to the many symmetries of the Riemann tensor R, it also characterizes the metric. To say that the Ricci tensor is constant (equal to some constant ρ) amounts to saying that $\text{Ric}_{ij} = \rho \, g_{ij}$. Therefore, in dimension less than or equal to 3, a constant Ricci tensor indicates that the metric is a metric of a sphere when $\rho > 0$ and a metric of hyperbolic space when $\rho < 0$ ($\rho = 0$ corresponding to the flat Euclidean space). The Ricci tensor of $(M, \mathfrak{g})$ is usually denoted by $\text{Ric}_\mathfrak{g}$ to emphasize the underlying (co-) metric, or more simply Ric (to avoid confusion with indices).

To say that the Ricci tensor is bounded below by $\rho \in \mathbb{R}$ means that the tensor $\text{Ric} - \rho \, \mathfrak{g}$ is positive. In other words, the lowest eigenvalue of the 2-tensor Ric in a system of coordinates where at a point x, $\mathfrak{g}(x)$ is the identity, is bounded from below by ρ. This is usually denoted as $\text{Ric} \ge \rho \, \mathfrak{g}$. In this definition, ρ may be a function, although in most parts of this book only constant ρ is considered.

A further trace operation on this Ricci tensor produces the *scalar curvature*, that is

$$\text{sc}_\mathfrak{g} = g^{ij} \text{Ric}_{ij} = \text{Ric}^i_i. \tag{C.3.2}$$

The function $\text{sc}_\mathfrak{g}(x)$, $x \in M$, plays an important role in conformal invariance of Sobolev inequalities (Chap. 6). In dimension 2, it is also enough to characterize the metric, as for the Riemann tensor in general and the Ricci tensor in dimension 3. (In dimension one, there is only one metric after a change of variables, and it is the usual, Euclidean, flat metric.)

This Ricci tensor appears crucially in the analysis of Markov semigroups through the Bochner-Lichnerowicz formula. This formula connects the Laplace operator with the Ricci curvature. In the same way as in $\mathbb{R}^n$ where it is given on a smooth function f as the trace of the symmetric matrix (Hessian) $(\partial^2_{ij} f)_{1 \le i, j \le n}$, the *Laplace-Beltrami operator* (or Laplacian) $\Delta_\mathfrak{g}$ on a Riemannian manifold $(M, \mathfrak{g})$ is defined on smooth functions $f : M \to \mathbb{R}$ as

$$\Delta_\mathfrak{g} f = g^{ij} \nabla_i \nabla_j f = \nabla^i \nabla_i f = \nabla_i \nabla^i f. \tag{C.3.3}$$

This definition is coordinate-free and one would obtain the same operator in any coordinate system, using the change of variables rules for tensors.

The Laplace-Beltrami operator $\Delta_\mathfrak{g}$ is presented in Sect. 1.11, p. 42, in a different way as the differential operator whose carré du champ operator is $\Gamma(f, f) = |df|^2$ and which is invariant with respect to the Riemannian measure. In a local system of coordinates $(x^i)_{1\leq i\leq n}$, the *Riemannian measure* has density $\det(\mathfrak{g})^{-1/2}$ with respect to the Lebesgue measure $dx^1 \cdots dx^n$ where $\det(\mathfrak{g})$ is the determinant of the matrix $\mathfrak{g} = (g^{ij})$. This is coherent with the representation of the Lebesgue measure in Euclidean space when the coordinate system is not orthogonal. Then, the Riemannian measure, often denoted by $\mu_\mathfrak{g}$ in this monograph, corresponds to the Lebesgue measure on the tangent space $T_x(M)$ equipped with the Euclidean metric associated with $\mathfrak{g}$. These two definitions of the Laplacian $\Delta_\mathfrak{g}$ are equivalent.

The following theorem introduces the Bochner-Lichnerowicz formula, which inspires many of the developments in this book. Indeed, it is the basis for the geometric and functional analysis of curvature developed in the context of the Γ and Γ_2 operators in Sect. C.6 below.

Theorem C.3.3 (Bochner-Lichnerowicz formula) *For any smooth function f on the Riemannian manifold $(M, \mathfrak{g})$,*

$$\frac{1}{2}\Delta_\mathfrak{g}(|\nabla f|^2) = \nabla f \cdot \nabla(\Delta_\mathfrak{g} f) + |\nabla\nabla f|^2 + \mathrm{Ric}_\mathfrak{g}(\nabla f, \nabla f).$$

C.4 Riemannian Distance

Riemannian manifolds admit a natural structure as metric spaces. Let $(M, \mathfrak{g})$ be a Riemannian manifold. Given the metric $G = (g_{ij})_{1\leq i,j\leq n}$, one may compute at any point the length of a tangent vector. Given a smooth curve $t \mapsto c(t)$ with values in M, if $\dot{c}(t) = \frac{dc(t)}{dt}$ denotes the tangent vector at time t, its length computed with respect to the metric at the point $c(t)$ is

$$|\dot{c}(t)|_{c(t)} = \left(g_{ij}\dot{c}^i\dot{c}^j\right)^{1/2}.$$

By comparison with the Euclidean case, the *length* of the curve c between 0 and t may be defined as

$$\int_0^t |\dot{c}(s)|\, ds.$$

Curves of minimal length are called *geodesics*. In a given system of coordinates, the point $c(t) = (c^i(t))_{1\leq i\leq n}$ along a geodesic satisfies the differential equation (Euler-Lagrange equation)

$$\frac{d^2 c^i(t)}{dt^2} + \gamma^i_{jk}\frac{dc^j}{dt}\frac{dc^k}{dt} = 0, \quad 1\leq i\leq n,$$

where γ^i_{jk} are the Christoffel symbols. Geodesics are also curves of minimal energy, minimizing $\int_0^t |\dot{c}(s)|^2 ds$.

It is not always the case that there is a geodesic going from one point to another. For example, in the Euclidean plane, if one removes the point 0, there is no geodesic connecting x to $-x$ since such a geodesic would be a straight line through 0 which has been removed. Also, if there exist geodesics between two points, it is not always the case that they are unique. For example on a sphere, geodesics are meridian lines (big circles centered at the center of the sphere, or intersections of the sphere with hyperplanes containing the center of the sphere), and there are infinitely many geodesics joining two antipodal points (x and $-x$). On the other hand, when two points x and y are sufficiently close to each other, there always exists a unique geodesic connecting them.

The (*geodesic*) *distance* between two points x and y in a Riemannian manifold $(M, \mathfrak{g})$ is defined as the infimum of the lengths of all curves joining x to y. This provides the manifold M with a distance (a metric structure) $d(x, y)$, $(x, y) \in M \times M$, called the *Riemannian distance*. The manifold is complete when it is complete for this distance. When the manifold is complete, there always exists a geodesic (but not necessarily a unique one, as in the example of the sphere) joining two points.

When two different geodesics starting from x meet at the same point y, the latter point y is said to be in the *cut-locus* of x (and this is a symmetric relation between x and y). The cut-locus of any point has measure 0 for the Riemannian measure, and may therefore often be forgotten (however unfortunately not always). Given a point $x \in M$, the map $d_x : M \to R$ defined by $d_x(y) = d(x, y)$, $y \in M$, is not smooth in general. It is not smooth even in the Euclidean space $\mathbb{R}^n$ at $y = x$. But, in $\mathbb{R}^n$, d_x^2 is smooth, in contrast to what happens in general in manifolds. On a manifold, this latter map is smooth, except at the cut-locus of x, where it is not even $\mathcal{C}^2$. The map d_x is nevertheless continuous and Lipschitz, and satisfies $|\nabla d_x| \leq 1$, with equality everywhere except possibly at the cut-locus.

In the context of this book, as long as diffusion operators are concerned, there is a dual description of the Riemannian distance as

$$d(x, y) = \sup[f(x) - f(y)], \quad (x, y) \in M \times M,$$

where the supremum is taken over all smooth functions $f : M \to \mathbb{R}$ such that $|\nabla f| \leq 1$. This is reminiscent of the Rademacher Theorem describing Lipschitz functions (in Euclidean space) as functions with (almost everywhere) bounded gradients. Functions which approximate this infimum are in fact smooth approximations of the distance $d(x, y)$. This dual description is much more convenient when considering many aspects of functional inequalities. Furthermore, the latter definition makes sense for any diffusion operator, even for non-elliptic ones (hypo-ellipticity is however necessary to effectively define a finite distance between points), and this formulation is the one used throughout this monograph to introduce a natural pseudo-distance associated with any generator of a diffusion semigroup (see (3.3.9), p. 166).

Completeness of a Riemannian manifold $(M, \mathfrak{g})$ with respect to the Riemannian distance d may be described similarly according to the following proposition. This characterization will prove most useful in the analysis of self-adjointness or gradient bounds as developed in Sect. 3.2.2, p. 141.

Proposition C.4.1 *A smooth Riemannian manifold $(M, \mathfrak{g})$ is complete if and only if there exists an increasing sequence $(\zeta_k)_{k \in \mathbb{N}}$ of smooth compactly supported functions on M such that $|\nabla \zeta_k| \leq \frac{1}{k}$ for every k (≥ 1) and $\lim_{k \to \infty} \zeta_k = 1$ (pointwise).*

C.5 The Riemannian Γ and Γ_2 Operators

This section and the next one go beyond the purpose of this appendix and deal with the central objects of this monograph, namely diffusion operators and their associated carré du champ and iterated carré du champ operators Γ and Γ_2. In particular, we refer to the core of the book for the proper definitions and discussions of diffusion operators and their Γ and Γ_2 operators (cf. Chaps. 1 and 3). This section however aims to describe these objects in the present Riemannian setting, connecting to and justifying their geometric origin and interpretation.

Most of this book deals with diffusion operators, more precisely second order differential operators with no constant terms. These are operators on open sets $\mathcal{O}$ in $\mathbb{R}^n$, or on a manifold, which may be represented in the form (in Einstein's notation)

$$\mathrm{L} f = g^{ij} \partial^2_{ij} f + b^i \partial_i f$$

where $(g^{ij}(x))_{1 \leq i, j \leq n}$ is a symmetric positive matrix and $(b^i(x))_{1 \leq i \leq n}$ is a vector, both depending smoothly on $x \in \mathcal{O}$. If the matrix $(g^{ij}(x))$ is everywhere non-degenerate so that the operator L is *elliptic* in the sense of (1.12.2), p. 50, the matrix $\mathfrak{g} = \mathfrak{g}(x) = (g^{ij}(x))_{1 \leq i, j \leq n}$ gives rise to a Riemannian structure (co-metric) $\mathfrak{g}$ on this open set. Then L may be written as

$$\mathrm{L} = \Delta_\mathfrak{g} + Z$$

where $\Delta_\mathfrak{g}$ is the Laplace-Beltrami operator associated with $\mathfrak{g}$. The difference between L and $\Delta_\mathfrak{g}$ is thus a first order differential operator (vector field) Z with no 0-order term. Note that this canonical decomposition into first and second order parts is more difficult to set up when dealing with the Hörmander form (see e.g. Sect. B.4) $\sum_{j=1}^{d} Z_j^2 + Z_0$.

For operators as above which are moreover symmetric with respect to some measure with smooth strictly positive density w with respect to the Riemannian measure $\mu_\mathfrak{g}$, this decomposition reads

$$\mathrm{L} = \Delta_\mathfrak{g} + \nabla(\log w).$$

Actually, when the operator L is of the form $\Delta_\mathfrak{g} + Z$, and provided the coefficients of Z are smooth, whenever it is symmetric with respect to a measure μ, this measure has a smooth density w with respect to $\mu_\mathfrak{g}$. Setting $w = e^{-W}$, the vector field Z is the gradient field $-\nabla W$ and in particular ∇Z is a symmetric tensor. This necessary condition on Z in order for L be symmetric in $\mathbb{L}^2(\mu)$ is also sufficient on simply connected domains. A Riemannian manifold $(M, \mathfrak{g})$ equipped with a measure $d\mu = e^{-W} d\mu_\mathfrak{g}$ is referred to as a *weighted Riemannian manifold*.

The preceding decomposition of elliptic operators thus amounts to the investigation of operators of the form $L = \Delta_{\mathfrak{g}} + Z$ on a manifold $(M, \mathfrak{g})$. For such an operator L, its *carré du champ operator* Γ is then defined on smooth functions $f : M \to \mathbb{R}$ as

$$\Gamma(f, f) = |df|^2 = |\nabla f|^2. \tag{C.5.1}$$

The *iterated carré du champ operator* Γ_2 is introduced in (1.16.1), p. 71, as

$$\Gamma_2(f, f) = \frac{1}{2}\big[L\Gamma(f, f) - 2\Gamma(f, Lf)\big] \tag{C.5.2}$$

on smooth functions $f : M \to \mathbb{R}$. To ease the notation, we often use $\Gamma(f, f) = \Gamma(f)$ and $\Gamma_2(f, f) = \Gamma_2(f)$. Both Γ and Γ_2 are extended by polarization to define bilinear forms $\Gamma(f, g)$ and $\Gamma_2(f, g)$ on pairs (f, g) of smooth functions. In the local system of coordinates $\mathfrak{g}$,

$$\Gamma(f, g) = g^{ij} \partial_i f \partial_j g.$$

The Bochner-Lichnerowicz formula (Theorem C.3.3) shows (and actually motivates the definition of Γ_2) that

$$\Gamma_2(f) = |\nabla\nabla f|^2 + (\mathrm{Ric}_{\mathfrak{g}} - \nabla_S Z)(\nabla f, \nabla f) \tag{C.5.3}$$

where $\mathrm{Ric}_{\mathfrak{g}}$ denotes the Ricci tensor and

$$(\nabla_S Z)_{ij} = \frac{1}{2}(\nabla_i Z_j + \nabla_j Z_i), \quad 1 \leq i, j \leq n,$$

is the symmetric part of the (non-symmetric) tensor ∇Z. The symmetric tensor $\mathrm{Ric}_{\mathfrak{g}} - \nabla_S Z$ on the right-hand side of (C.5.3) is denoted by

$$\mathrm{Ric}(L) = \mathrm{Ric}_{\mathfrak{g}} - \nabla_S Z \tag{C.5.4}$$

and is often called the *generalized Ricci tensor* (of $L = \Delta_{\mathfrak{g}} + Z$).

Remark C.5.1 (Hessian) When working with diffusion operators L and their carré du champ operators Γ, there is no need to compute the Christoffel symbols for formulas on functions. For instance, to compute the Hessian of a smooth function f acting on the gradient of smooth functions g and h on M,

$$\nabla\nabla^{ij} f \partial_i g \partial_j h = \frac{1}{2}\big[\Gamma\big(g, \Gamma(f, h)\big) + \Gamma\big(h, \Gamma(f, g)\big) - \Gamma\big(f, \Gamma(g, h)\big)\big]. \tag{C.5.5}$$

This formula for Hessians is widely used in the construction of Γ_2 operators on the domain of L (in particular in Sect. 3.3, p. 151). The right-hand side makes sense even in non-elliptic settings, when the (co-) metric is degenerate.

C.6 Curvature-Dimension Conditions

Operators defined on an open set in $\mathbb{R}^n$ or on a manifold often satisfy uniform ellipticity conditions in the sense that, in a given system of coordinates, the metric (g^{ij}) is bounded from below by $c\,\mathrm{Id}$ for some $c > 0$. This condition is not invariant under a coordinate change (and is thus not intrinsic). The intrinsic condition which is the closest to uniform ellipticity could in many cases be replaced by a uniform (possibly negative) lower bound on the Ricci curvature. The curvature conditions presented below cover these instances. This section is more precisely devoted to the Riemannian description of the curvature-dimension condition of a diffusion operator, which is extensively investigated throughout this work.

Start with the Laplace-Beltrami operator $\mathrm{L} = \Delta_\mathfrak{g}$ on an n-dimensional Riemannian manifold $(M, \mathfrak{g})$ with Ricci tensor $\mathrm{Ric} = \mathrm{Ric}_\mathfrak{g}$. In this case, thanks to the Bochner-Lichnerowicz formula (Theorem C.3.3) and (C.5.3), the Γ_2 operator takes the form, on smooth functions $f : M \to \mathbb{R}$,

$$\Gamma_2(f) = |\nabla\nabla f|^2 + \mathrm{Ric}(\nabla f, \nabla f).$$

In this form, the second and first order terms are well separated, and it is easily seen that

$$\Gamma_2(f) \geq \rho\, \Gamma(f) = \rho\, |\nabla f|^2 \qquad (C.6.1)$$

for all f's if and only if

$$\mathrm{Ric}(\nabla f, \nabla f) \geq \rho\, |\nabla f|^2$$

for all f's. In other words, (C.6.1) holds if and only if the Ricci tensor at every point is bounded from below by ρ.

Here ρ could be a function, satisfying, for any smooth function f, $\mathrm{Ric}(\nabla f, \nabla f) \geq \rho(x)\Gamma(f, f)$, which we denote by $\mathrm{Ric} \geq \rho\, \mathfrak{g}$. The best $\rho(x)$ may be seen as the lowest eigenvalue of the matrix Ric^i_j. The inequality (C.6.1), equivalent to the lower bound $\mathrm{Ric} \geq \rho\, \mathfrak{g}$ on the Ricci curvature, is called the *curvature condition* $CD(\rho, \infty)$ throughout this book, and ρ will usually be a (real) constant.

However, there is a second parameter buried in the Γ_2 operator. Namely, recall that, on a given function f,

$$\Delta_\mathfrak{g} f = \nabla^i \nabla_i f = \mathrm{Tr}(\nabla\nabla f)$$

(where Tr is the trace). In dimension n, and in an orthonormal basis, the Hilbert-Schmidt norm $|N|^2$ of the symmetric matrix $N = \nabla\nabla f$ is the sum $\sum_{ij=1}^n N_{ij}^2$, while $\mathrm{Tr}(N) = \sum_{i=1}^n N_{ii}$. Therefore, by the Cauchy-Schwarz inequality,

$$\mathrm{Tr}(N) \leq n \sum_{i=1}^n N_{ii}^2 \leq n\, |N|^2. \qquad (C.6.2)$$

Note that this inequality may actually be seen in a more intrinsic way. Indeed, in the Euclidean space of symmetric matrices N endowed with the Hilbert-Schmidt norm $|N|^2$, $\mathrm{Tr}(N)$ is the scalar product with the matrix Id (or equivalently with the metric $\mathfrak{g}$). This identity has norm $|\mathrm{Id}|^2 = n$, and the inequality $(\mathrm{Tr}(N))^2 \leq n|N|^2$ is nothing else than the Cauchy-Schwarz inequality in this Euclidean space.

On the basis of (C.6.2), the Γ_2 operator of the Laplace-Beltrami operator $\Delta_\mathfrak{g}$ is bounded from below, on every smooth function $f : M \to \mathbb{R}$, by

$$\Gamma_2(f) \geq \rho\, \Gamma(f) + \frac{1}{n}(\Delta_\mathfrak{g} f)^2,$$

where n is the dimension of the manifold and ρ the lowest eigenvalue of the Ricci tensor. This inequality is called a *curvature-dimension condition* or *condition $CD(\rho, n)$* throughout this book. Note that in this example, the best possible choice for (ρ, n) in the $CD(\rho, n)$ condition is the lower bound ρ on the Ricci curvature and the dimension n of the manifold.

The preceding curvature-dimension conditions may be addressed similarly for general operators of the form $\mathrm{L} = \Delta_\mathfrak{g} + Z$ where Z is a vector field on an n-dimensional manifold $(M, \mathfrak{g})$. From (C.5.3),

$$\Gamma_2(f) \geq \rho\, \Gamma(f) + \frac{1}{m}(\mathrm{L}f)^2$$

for every smooth $f : M \to \mathbb{R}$ if and only if $m \geq n$ and, as symmetric tensors,

$$\mathrm{Ric}(\mathrm{L}) = \mathrm{Ric}_\mathfrak{g} - \nabla_S Z \geq \rho\, \mathfrak{g} + \frac{1}{m-n}\, Z \otimes Z. \tag{C.6.3}$$

Here, there is no longer a best optimal choice both for m and ρ, except in particular cases. In particular, the parameter m may be equal to the dimension of the manifold only for Laplace-Beltrami operators. In this sense, among elliptic operators on a manifold, the family of Laplace operators plays a particular role.

The condition (C.6.3) above takes a simpler form for symmetric operators $\mathrm{L} = \Delta_\mathfrak{g} - \nabla W \cdot \nabla$ for the invariant (reversible) measure with density e^{-W} with respect to the Riemannian measure $\mu_\mathfrak{g}$. Namely, setting $e^{-W} = w_1^{m-n}$, (C.6.3) then turns into

$$\mathrm{Ric}_m(\mathrm{L}) = \mathrm{Ric}_\mathfrak{g} - (m-n)\frac{1}{w_1}\nabla\nabla w_1 \geq \rho\, \mathfrak{g}.$$

It is worth mentioning that this tensor $\mathrm{Ric}_m(\mathrm{L})$ has a simple geometric interpretation when the dimension m is an integer. Indeed, set $p = m - n$, and start from the n-dimensional manifold $(M, \mathfrak{g})$ together with a smooth function $w_1 : M \to (0, \infty)$. Consider then an auxiliary p-dimensional manifold $(M_1, \mathfrak{g}_1)$. Equip the product $\widehat{M} = M \times M_1$ with the Riemannian metric

$$\widehat{G}(x, y) = G(x) + w_1^2(x) G_1(y)$$

(such metrics are often called *wrapped products*). In terms of the carré du champ operator, for a function $f(x, y) : M \times M_1 \to \mathbb{R}$ in a local system of coordinates

C.6 Curvature-Dimension Conditions

(x^i, y^j),

$$\widehat{\Gamma}(f, f) = g^{ij}(x) \partial_{x^i} f \partial_{x^j} f + \frac{1}{w_1^2(x)} g_1^{k\ell}(y) \partial_{y^k} f \partial_{y^\ell} f.$$

The Riemannian measure on $M \times M_1$ is given by $w_1^p(x) d\mu_{\mathfrak{g}}(x) d\mu_{\mathfrak{g}_1}(y)$. In the same way, the new Laplace operator is

$$\widehat{\Delta} f = \Delta_{\mathfrak{g}} f + p \Gamma(\log w_1, f) + \frac{1}{w_1^2} \Delta_{\mathfrak{g}_1} f.$$

The Ricci operator for this new structure on $M \times M_1$ splits into two parts, that is

$$\widehat{\mathrm{Ric}} = \mathrm{Ric}^0 + \mathrm{Ric}^1$$

where the action of Ric^0 only depends on $(\partial_{x^i} f)$ and that of Ric^1 only on $(\partial_{y^j} f)$. Using the techniques described in Sect. 6.9, p. 313, it may be shown that

$$\mathrm{Ric}^0 = \mathrm{Ric}_{\mathfrak{g}} - \frac{p}{w_1} \nabla \nabla w_1,$$

$$\mathrm{Ric}^1 = \frac{1}{w_1^4} \mathrm{Ric}_{\mathfrak{g}_1} - \big(\Delta_{\mathfrak{g}}(\log w_1) + p \Gamma_{\mathfrak{g}}(\log w_1)\big) \frac{1}{w_1^2} \Gamma_{\mathfrak{g}_1}.$$

In particular, when acting on functions f depending only on the first variable $x \in M$, the action of the Laplacian on the product space $M \times M_1$ is a function depending only on $x \in M$ and is the weighted Laplace operator $\Delta_{\mathfrak{g}} f + p \Gamma(\log w_1, f)$. Furthermore,

$$\widehat{\mathrm{Ric}}(\nabla f, \nabla f) = \mathrm{Ric}^0(\nabla f, \nabla f) = \Big(\mathrm{Ric}_{\mathfrak{g}} - \frac{m-n}{w_1} \nabla \nabla w_1\Big)(\nabla f, \nabla f).$$

Observe then that if $W = -p \log w_1$,

$$\mathrm{Ric}_{\mathfrak{g}} - \frac{m-n}{w_1} \nabla \nabla w_1 = \mathrm{Ric}_{\mathfrak{g}} + \nabla \nabla W - \frac{1}{m-n} \nabla W \otimes \nabla W = \mathrm{Ric}_m(L).$$

In some places throughout this book, the curvature-dimension conditions $CD(\rho, \infty)$ or $CD(\rho, n)$ are used in a reinforced version. For example, the $CD(\rho, \infty)$ condition actually implies that, for all smooth functions $f : M \to \mathbb{R}$,

$$4 \Gamma(f) \big[\Gamma_2(f) - \rho \Gamma(f)\big] \geq \Gamma\big(\Gamma(f)\big). \tag{C.6.4}$$

This reinforcement is rather easy to understand in this differential geometry context. Namely, the $CD(\rho, \infty)$ condition amounts to $\mathrm{Ric}(L) \geq \rho \mathfrak{g}$. The reinforced inequality expresses that

$$4 |\nabla f|^2 \big[|\nabla \nabla f|^2 + \mathrm{Ric}(L)(\nabla f, \nabla f) - \rho |\nabla f|^2\big] \geq \big|\nabla(|\nabla f|^2)\big|^2.$$

Now, $\nabla(|\nabla f|^2)$ is a vector whose scalar product with any vector V is $2\nabla\nabla f(\nabla f, V)$, or in coordinates $2\nabla^i \nabla^j f \nabla_j f$. Its norm is bounded above by $2|\nabla\nabla f||\nabla f|$ from which the reinforced inequality immediately follows.

Much of the spirit of this book is actually to address a reinforced inequality such as (C.6.4) not by differential geometry tools, but rather by differential calculus. This is in particular necessary (for the $CD(\rho, \infty)$ condition) in infinite dimension where the usual Riemannian geometry is not available. The proof then relies on the change of variables formula for the Γ_2 operator. Indeed, in the geometric description (C.5.3) of the Γ_2 operator, the standard differentiation rules

$$\nabla\nabla\psi(f) = \psi'(f)\nabla\nabla f + \psi''(f)\nabla f \otimes \nabla f$$

and

$$T(\psi(f), \phi(g)) = \psi'(f)\phi'(g)T(f, g),$$

for smooth functions $\psi, \phi : \mathbb{R} \to \mathbb{R}$, smooth functions $f, g : M \to \mathbb{R}$ and 2-tensor T, lead to, for smooth functions $\Psi : \mathbb{R}^k \to \mathbb{R}$ and vectors $f = (f_1, \ldots, f_k)$ of smooth functions on M,

$$\Gamma_2(\Psi(f)) = \sum_{i,j=1}^{k} X_i X_j \Gamma_2(f_i, f_j) + 2 \sum_{i,j,\ell=1}^{k} X_i Y_{j\ell} H(f_i)(f_i, f_\ell) \\ + \sum_{i,j,\ell,m=1}^{k} Y_{ij} Y_{\ell m} \Gamma(f_i, f_\ell) \Gamma(f_j, f_m) \quad \text{(C.6.5)}$$

where

$$X_i = \partial_i \Psi(f_1, \ldots, f_k), \qquad Y_{ij} = \partial^2_{ij} \Psi(f_1, \ldots, f_k)$$

and

$$H(f)(g, h) = \frac{1}{2}\Big[\Gamma\big(g, \Gamma(f, h)\big) + \Gamma\big(h, \Gamma(f, g)\big) - \Gamma(f, \Gamma(g, h))\Big]. \quad \text{(C.6.6)}$$

This last expression is directly related to the representation of the Hessian (C.5.5). A similar formula applies to $\Gamma_2 - \rho\Gamma$ instead of Γ_2.

Note that (C.6.5) of course takes a simpler form when dealing with a single function $\psi : M \to \mathbb{R}$,

$$\Gamma_2(\psi(f)) = \psi'(f)^2 \Gamma_2(f) + \psi'(f)\psi''(f) \Gamma\big(f, \Gamma(f)\big) + \psi''(f)^2 \Gamma(f)^2 \quad \text{(C.6.7)}$$

(recall the notation $\Gamma(f) = \Gamma(f, f)$ and $\Gamma_2(f) = \Gamma_2(f, f)$), a formula which will be used extensively in Chaps. 5 and 6.

While of a standard differential calculus nature, it should be pointed out that a more intrinsic derivation of these change of variables formulas may be developed on the basis of the diffusion property of the generator L as presented in Sect. 1.11,

C.6 Curvature-Dimension Conditions

p. 42. This point of view will be systematically emphasized throughout this work in the form of the Γ-calculus.

Let us now describe how the reinforced inequality (C.6.4) may be obtained from the standard curvature condition $CD(\rho, \infty)$ by differential calculus and the change of variables formula (C.6.5). Indeed, given functions $(f_1, \ldots, f_k)$ and at any point x, the function Ψ may be chosen in such a way that the coefficients X_i and Y_{ij} take any particular value, provided the symmetries $Y_{ij} = Y_{ji}$ are respected (for example just letting Ψ vary among second degree polynomials). The curvature condition $CD(\rho, \infty)$ therefore yields a positive quadratic form in the variables (X_i, Y_{ij}). It is for example a quadratic form in the nine variables $X_i, Y_{i,j}, i, j = 1, \ldots, 3, i < j$, in case of three functions f_1, f_2, f_3. Illustrating the argument in this sample example, restrict this quadratic form to the set where all the variables are 0 except X_1 and Y_{23}. Its determinant

$$\left[\Gamma_2(f_1) - \rho\,\Gamma(f_1)\right]\left[\Gamma(f_2, f_3)^2 + \Gamma(f_2)\Gamma(f_3)\right] - 2H(f_1)(f_2, f_3)^2$$

is positive, which in turn yields

$$H(f_1)(f_2, f_3)^2 \leq \left[\Gamma_2(f_1) - \rho\,\Gamma(f_1)\right]\Gamma(f_2)\Gamma(f_3). \tag{C.6.8}$$

Now,

$$H(f_1)(f_1, f_2) = \frac{1}{2}\Gamma\bigl(f_2, \Gamma(f_1)\bigr),$$

so that choosing $f_2 = \Gamma(f_1)$, it follows that

$$\Gamma(f_2)^2 \leq 4\left[\Gamma_2(f_1) - \rho\,\Gamma(f_1)\right]\Gamma(f_1)\Gamma(f_2).$$

The announced reinforced inequality (C.6.4) is thus obtained.

The preceding strategy may look strange. However, it allows for an efficient geometric treatment of operators which are not Laplacians, somewhat far away from the Riemannian geometry context. Even in a Riemannian setting, it often provides a useful form of the curvature conditions. The technique (with variations) moreover often provides optimal inequalities in numerous functional inequalities of interest. This Γ-calculus, and its numerous applications, lies at the heart of this monograph.

Finally, and to conclude this section, we observe that in several places in this work we consider operators L given in Hörmander form

$$\mathrm{L} = \sum_{j=1}^{d} Z_j^2 + Z_0$$

where $Z_0, Z_1, \ldots, Z_d$ are vector fields, that is first order differential operators with no zero-order terms. Here, the carré du champ operator is given, on smooth functions, by $\Gamma(f) = \sum_{j=1}^{d}(Z_j f)^2$. In order to have a tractable form for the Γ_2 operator, it is useful to introduce the symmetric second order derivatives associated to this decomposition, that is $D_{ij} f = \frac{1}{2}(Z_i Z_j f + Z_j Z_i f)$, $0 \leq i, j \leq d$, and to denote the usual commutators $[Z_i, Z_j]$ by Z_{ij}. Introduce moreover $D_{k,ij} = \frac{1}{2}(Z_k Z_{ij} + Z_{ij} Z_k)$. Then

$$\Gamma_2(f) = \sum_{i,j=1}^d \left(D_{ij}f + \frac{1}{2} Z_{ij}f \right)^2 + 2 \sum_{i,j=1}^d Z_i f D_{j,ji} f + \sum_{j=1}^d Z_j Z_{0j} f.$$

To get a useful formula, as is done in the elliptic case with the use of the language of differential geometry, it is necessary to separate the first and second order terms. This is possible for example whenever, for $0 \leq i, j \leq d$,

$$Z_{ij} = \sum_{k=1}^d \alpha_{ij}^k Z_k$$

for some (smooth) functions α_{ij}^k. In this case,

$$\Gamma_2(f) = \sum_{i,j=1}^d \left(D_{ij}f + \sum_{k=1}^d \alpha_{jk}^i Z_k f \right)^2 + R(f, f),$$

where $R(f, f)$ (which plays the role of the Ricci tensor) is given by

$$R(f, f) = \sum_{i,j=1}^d \left[\frac{1}{4} \left(\sum_{k=1}^d \alpha_{ij}^k Z_k f \right)^2 - \left(\sum_{k=1}^d \alpha_{jk}^i Z_k f \right)^2 \right.$$
$$\left. + Z_i f Z_j f \left(\alpha_{0i}^j + \sum_{k=1}^d Z_k(\alpha_{ki}^j) \right) \right].$$

C.7 Notes and References

Standard references on differentiable manifolds include [68, 159, 166, 167]. Basics on Riemannian geometry and Ricci curvature may be found, for example, in [62, 123, 126, 194, 260, 346] where in particular the construction of the Laplace-Beltrami operator and the Bochner-Lichnerowicz formula (Theorem C.3.3) are emphasized. Comparison methods based on Ricci curvature bounds are surveyed in [123, 126, 448]. See also [61] for an overview of modern Riemannian geometry.

The characterization of completeness of Proposition C.4.1 is due to R. Strichartz [390] (see also [23] in the context of this work).

The Γ and Γ_2-calculus of Sects. C.5 and C.6 was introduced in [36] in the context of logarithmic Sobolev inequalities and in [22, 24, 25] in the study of Riesz transforms. The definition of the curvature-dimension condition $CD(\rho, n)$ (also called the Γ_2 criterion), which for Laplace-Beltrami operators boils down to the Bochner

C.7 Notes and References

formula in the context of harmonic maps between manifolds, may be found there. For more on Bochner-Weitzenböck formulas, see [92]. These ideas have been further developed in the early lecture notes [26] (see also [27, 28, 277]). Some geometric properties of the tensor Ric(L) are examined in [290].

Afterword

Chicken "Gaston Gérard"

After all, the content of this book is nothing but a sequence of recipes.

Let us thus conclude this monograph with the famous recipe of the chicken *à la façon Gaston Gérard*. This is a creation of the Mayor of Dijon's wife, Madame Reine Geneviève Gérard. After reading this book, with a white Burgundy wine, such as Chablis premier cru Montée de Tonnerre or Chassagne Montrachet, this recipe is perfect for a nice dinner with friends.

Preparation: 60 minutes.
Cooking time: 45 minutes.

- 1 Bresse chicken
- 50 g of butter
- 2 glasses of white wine (Jura wine, preferably Savagnin, is perfect)
- 150 g of Comté cheese, plus 50 g for gratin (cheese topping)
- 400 g of single cream
- 1 teaspoon of paprika
- 1 tablespoon of Dijon mustard
- 1 tablespoon of breadcrumbs
- salt and pepper

(i) Put the pieces of chicken in a cooking pot with the butter, the paprika, salt and pepper. Turn them over periodically and let the preparation cook for 30 minutes.
(ii) In a separate pan, mix slowly while heating the wine, cheese, mustard and cream. Do not let the mixture boil.
(iii) Put the chicken in a gratin dish with the above mixture. Cover the preparation with the spare grated cheese and breadcrumbs.
(iv) Place the dish in a hot oven (240 °C) for 15 minutes.

What about the heat equation in the oven: is this some practical application of the theory developed here, or just an exercise?

Notation and List of Symbols

$\langle \cdot, \cdot \rangle$	Scalar product in Euclidean and Hilbert space	xv
$\|\cdot\|$	Euclidean norm of vectors and tensors	xv, 503
$\mathbb{1}_A$	Characteristic function of the set A	xv
Id	Identity operator or Identity matrix	9
$\mathcal{O}$	Open set	50
E	State space	120, 168
$\mathcal{F}$	σ-field	120
μ	Positive measure	xv
dx, vol_n	Lebesgue measure on $\mathbb{R}^n$	xvi
$\mathbb{L}^p(E, \mu)$, $\mathbb{L}^p(\mu)$	Space of pth-power integrable functions on $(E, \mathcal{F}, \mu)$	10
$\mathbb{L}^2_0(\mu)$	Functions in $\mathbb{L}^2(\mu)$ with mean zero	70
$\|\cdot\|_p$, $\|\cdot\|_{\mathbb{L}^p(\mu)}$	$\mathbb{L}^p(\mu)$-norm	10
$\|\cdot\|_{p,q}$	Operator norm from $\mathbb{L}^p(\mu)$ into $\mathbb{L}^q(\mu)$	286
$\|\cdot\|_{\mathrm{TV}}$	Total variation distance	244
(E, μ, Γ)	Diffusion, Standard, Full Markov Triple	132, 136, 167, 168
$\mathbf{P} = (P_t)_{t \geq 0}$	Markov semigroup	9, 168
$p_t(x, dy)$	Markov probability kernel	8, 168
$p_t(x, y)$	Markov density kernel	14
P_t^*	Dual Markov semigroup	9
$(P_t^D)_{t \geq 0}$	Semigroup with Dirichlet boundary conditions	93
$(P_t^N)_{t \geq 0}$	Semigroup with Neumann boundary conditions	93
$\mathcal{E}$	Dirichlet form	30, 168
$\mathcal{D}(\mathcal{E})$	Domain of the Dirichlet form $\mathcal{E}$	125, 168
$\|\cdot\|_\mathcal{E}$	Dirichlet norm	125, 168

L	Infinitesimal generator/Markov generator	18, 168
$\mathcal{D}(L)$	Domain of the generator L	18, 168
$\|\cdot\|_{\mathcal{D}(L)}$	Domain norm on $\mathcal{D}(L)$	133, 168
L^*	Adjoint operator	24, 168
Γ	Carré du champ operator	20, 168
Γ_2	Iterated carré du champ operator	71, 168
$\mathcal{A}_0$	Standard algebra of functions	120, 168
$\mathcal{A}_0^{\text{const}}$	Adding a constant to functions in $\mathcal{A}_0$	154, 168
$\mathcal{A}_0^{\text{const}+}$	Adding a strictly positive constant to positive functions in $\mathcal{A}_0$	154, 168
$\mathcal{A}$	Extended algebra of functions	151, 168
ESA	Essentially self-adjointness property	134, 168
$\|f\|_{\text{Lip}}$	Lipschitz coefficient	166, 168
σ_{ess}	Essential spectrum	480
σ_d	Discrete spectrum	480
$(E_\lambda)_{\lambda\in\mathbb{R}}$	Spectral decomposition	479
$(\Omega, \Sigma, \mathbb{P})$	Probability space	7
$\mathbb{E}(\cdot), \mathbb{E}(\cdot\,\vert\,\cdot)$	Expectation, conditional expectation	8
$(\mathcal{F}_t)_{t\geq 0}$	Filtration	8
$(B_t)_{t\geq 0}$	Standard Brownian motion	487
$(X_t^x)_{t\geq 0}$	Stochastic process starting at x	8
$\mathcal{C}^k(\mathcal{O})$	Space of k-times differentiable functions on $\mathcal{O}$	42
$\mathcal{C}^\infty$	Space of infinitely differentiable functions	138
$\mathcal{C}_c^\infty$	Space of infinitely differentiable compactly supported functions	95
$\mathbb{S}^n \subset \mathbb{R}^{n+1}$	n-dimensional sphere	81
σ_n	Uniform probability measure on $\mathbb{S}^n$	81
Δ	Laplacian on $\mathbb{R}^n$	4
$(M, \mathfrak{g})$	Riemannian manifold	504
$\mathfrak{g}$	Co-metric	504
∇	Gradient	5
∇_S	Symmetric derivative	73
$\nabla\nabla f$	Hessian of f	507
$H(f)(g,h)$	Hessian of f evaluated at $(\nabla g, \nabla h)$	158
$\Delta_\mathfrak{g}$	Laplace-Beltrami operator in $(M, \mathfrak{g})$	504
$\mu_\mathfrak{g}$	Riemannian measure on $(M, \mathfrak{g})$	509
$\text{Ric}_\mathfrak{g}$	Ricci tensor of the (co)-metric $\mathfrak{g}$	508
$\text{Ric}(L)$	Ricci tensor of L	514
$\text{sc}_\mathfrak{g}$	Scalar curvature of $(M, \mathfrak{g})$	508
d	Intrinsic or Riemannian distance	166, 509, 168
d^H	Harmonic distance	360

Notation and List of Symbols

Var_μ	Variance	181
Ent_μ	Entropy	236
Ent_μ^Φ	Φ-entropy	383
$H(\cdot \mid \cdot)$	Relative entropy	244
I_μ	Fisher information	237
$I_{\mu, F}$	Generalized Fisher information	333
$P(C)$	Poincaré inequality	181
$LS(C, D), LS(C)$	Logarithmic Sobolev inequality	236
$S(p; A, C), S_n(A, C)$	Sobolev inequality	279
$GN_n(s, q; A, C)$	Gagliardo-Nirenberg inequality	324
$EE(\Phi)$	Entropy-energy inequality	349
$N(\Phi)$	Generalized Nash inequality	364
$N(\Phi, w)$	Weighted Nash inequality	370
$WP(C)$	Weak Poincaré inequality	375
$B_q(C)$	Beckner inequality	384
$LO(C)$	Latała-Oleszkiewicz inequality	385
$\text{Cap}_\mu(A)$	Capacity of the set A	393
$\text{Cap}_\mu^*(A)$	Modified capacity of the set A	394
$\mathcal{I}_\mu$	Isoperimetric profile of μ	413
$\mathcal{I}$	Gaussian isoperimetric profile	416
$\mathcal{P}(E)$	Space of probability measures on $(E, \mathcal{F})$	434
$\mathcal{P}_p(E)$	Space of probability measures on $(E, \mathcal{F})$ with a finite p-th moment	436
$T \# \mu$	Image measure of μ by T	437
$\mathcal{T}_c$	Optimal transportation cost	434
W_p	Wasserstein distance	436

Bibliography

1. R.A. Adams, *Sobolev Spaces*. Pure and Applied Mathematics, vol. 65 (Academic Press, New York, 1975)
2. R.A. Adams, J.J.F. Fournier, *Sobolev Spaces*, 2nd edn. Pure and Applied Mathematics, vol. 140 (Elsevier/Academic Press, Amsterdam, 2003)
3. M. Agueh, N. Ghoussoub, X. Kang, Geometric inequalities via a general comparison principle for interacting gases. Geom. Funct. Anal. **14**(1), 215–244 (2004)
4. S. Aida, Uniform positivity improving property, Sobolev inequalities, and spectral gaps. J. Funct. Anal. **158**(1), 152–185 (1998)
5. S. Aida, K.D. Elworthy, Differential calculus on path and loop spaces. I. Logarithmic Sobolev inequalities on path spaces. C. R. Math. Acad. Sci. Paris, Sér. I **321**(1), 97–102 (1995)
6. S. Aida, T. Masuda, I. Shigekawa, Logarithmic Sobolev inequalities and exponential integrability. J. Funct. Anal. **126**(1), 83–101 (1994)
7. S. Aida, D.W. Stroock, Moment estimates derived from Poincaré and logarithmic Sobolev inequalities. Math. Res. Lett. **1**(1), 75–86 (1994)
8. G. Allaire, A la recherche de l'inégalité perdue. Matapli **98**, 52–64 (2012)
9. L. Ambrosio, N. Gigli, G. Savaré, *Gradient Flows in Metric Spaces and in the Space of Probability Measures*, 2nd edn. Lectures in Mathematics ETH Zürich (Birkhäuser, Basel, 2008)
10. L. Ambrosio, N. Gigli, G. Savaré, Bakry-Emery curvature-dimension condition and Riemannian Ricci curvature bounds. Preprint, 2012
11. L. Ambrosio, N. Gigli, G. Savaré, Metric measure spaces with Riemannian Ricci curvature bounded from below. Preprint, 2012
12. L. Ambrosio, N. Gigli, G. Savaré, Calculus and heat flow in metric measure spaces and applications to spaces with Ricci bounds from below. Invent. Math. (2013). doi:10.1007/s00222-013-0456-1
13. L. Ambrosio, P. Tilli, *Topics on Analysis in Metric Spaces*. Oxford Lecture Series in Mathematics and Its Applications, vol. 25 (Oxford University Press, Oxford, 2004)
14. C. Ané, S. Blachère, D. Chafaï, P. Fougères, I. Gentil, F. Malrieu, C. Roberto, G. Scheffer, *Sur les Inégalités de Sobolev Logarithmiques*. Panoramas et Synthèses, vol. 10 (Société Mathématique de France, Paris, 2000)
15. D. Applebaum, *Lévy Processes and Stochastic Calculus*, 2nd edn. Cambridge Studies in Advanced Mathematics, vol. 116 (Cambridge University Press, Cambridge, 2009)
16. A. Arnold, P. Markowich, G. Toscani, A. Unterreiter, On convex Sobolev inequalities and the rate of convergence to equilibrium for Fokker-Planck type equations. Commun. Partial Differ. Equ. **26**(1–2), 43–100 (2001)
17. T. Aubin, Équations différentielles non linéaires et problème de Yamabe concernant la courbure scalaire. J. Math. Pures Appl. **55**(3), 269–296 (1976)

18. T. Aubin, Espaces de Sobolev sur les variétés Riemanniennes. Bull. Sci. Math. **100**(2), 149–173 (1976)
19. T. Aubin, Problèmes isopérimétriques et espaces de Sobolev. J. Differ. Geom. **11**(4), 573–598 (1976)
20. T. Aubin, *Nonlinear Analysis on Manifolds. Monge-Ampère Equations*. Grundlehren der mathematischen Wissenschaften [Fundamental Principles of Mathematical Sciences], vol. 252 (Springer, New York, 1982)
21. T. Aubin, *Some Nonlinear Problems in Riemannian Geometry*. Springer Monographs in Mathematics (Springer, Berlin, 1998)
22. D. Bakry, Transformations de Riesz pour les semi-groupes symétriques. II. Étude sous la condition $\Gamma_2 \geq 0$, in *Séminaire de Probabilités, XIX*, 1983/84. Lecture Notes in Math., vol. 1123 (Springer, Berlin, 1985), pp. 145–174
23. D. Bakry, Un critère de non-explosion pour certaines diffusions sur une variété Riemannienne complète. C. R. Math. Acad. Sci. Paris, Sér. I **303**(1), 23–26 (1986)
24. D. Bakry, Étude des transformations de Riesz dans les variétés Riemanniennes à courbure de Ricci minorée, in *Séminaire de Probabilités, XXI*. Lecture Notes in Math., vol. 1247 (Springer, Berlin, 1987), pp. 137–172
25. D. Bakry, The Riesz transforms associated with second order differential operators, in *Seminar on Stochastic Processes*, Gainesville, FL, 1988. Progr. Probab., vol. 17 (Birkhäuser, Boston, 1989), pp. 1–43
26. D. Bakry, L'hypercontractivité et son utilisation en théorie des semigroupes, in *Lectures on Probability Theory*, Saint-Flour, 1992. Lecture Notes in Math., vol. 1581 (Springer, Berlin, 1994), pp. 1–114
27. D. Bakry, On Sobolev and logarithmic Sobolev inequalities for Markov semigroups, in *New Trends in Stochastic Analysis*, Charingworth, 1994 (World Sci., River Edge, 1997), pp. 43–75
28. D. Bakry, Functional inequalities for Markov semigroups, in *Probability Measures on Groups: Recent Directions and Trends* (Tata Inst. Fund. Res, Mumbai, 2006), pp. 91–147
29. D. Bakry, F. Barthe, P. Cattiaux, A. Guillin, A simple proof of the Poincaré inequality for a large class of probability measures including the log-concave case. Electron. Commun. Probab. **13**, 60–66 (2008)
30. D. Bakry, F. Baudoin, M. Bonnefont, D. Chafaï, On gradient bounds for the heat kernel on the Heisenberg group. J. Funct. Anal. **255**(8), 1905–1938 (2008)
31. D. Bakry, F. Bolley, I. Gentil, Dimension dependent hypercontractivity for Gaussian kernels. Probab. Theory Relat. Fields **154**(3), 845–874 (2012)
32. D. Bakry, F. Bolley, I. Gentil, P. Maheux, Weighed Nash inequalities. Rev. Mat. Iberoam. **28**(3), 879–906 (2012)
33. D. Bakry, P. Cattiaux, A. Guillin, Rate of convergence for ergodic continuous Markov processes: Lyapunov versus Poincaré. J. Funct. Anal. **254**(3), 727–759 (2008)
34. D. Bakry, D. Concordet, M. Ledoux, Optimal heat kernel bounds under logarithmic Sobolev inequalities. ESAIM Probab. Stat. **1**, 391–407 (1995/97) (electronic)
35. D. Bakry, T. Coulhon, M. Ledoux, L. Saloff-Coste, Sobolev inequalities in disguise. Indiana Univ. Math. J. **44**(4), 1033–1074 (1995)
36. D. Bakry, M. Émery, Diffusions hypercontractives, in *Séminaire de Probabilités, XIX*, 1983/1984. Lecture Notes in Math., vol. 1123 (Springer, Berlin, 1985), pp. 177–206
37. D. Bakry, I. Gentil, M. Ledoux, On Harnack inequalities and optimal transportation. Ann. Sc. Norm. Sup. Pisa (2012). doi:10.2422/2036-2145.201210_007
38. D. Bakry, M. Ledoux, Lévy-Gromov's isoperimetric inequality for an infinite-dimensional diffusion generator. Invent. Math. **123**(2), 259–281 (1996)
39. D. Bakry, M. Ledoux, Sobolev inequalities and Myers's diameter theorem for an abstract Markov generator. Duke Math. J. **85**(1), 253–270 (1996)
40. D. Bakry, M. Ledoux, A logarithmic Sobolev form of the Li-Yau parabolic inequality. Rev. Mat. Iberoam. **22**(2), 683–702 (2006)

41. D. Bakry, S.Y. Orevkov, M. Zani, Orthogonal polynomials and diffusion operators. Preprint, 2013
42. D. Bakry, Z.M. Qian, Some new results on eigenvectors via dimension, diameter, and Ricci curvature. Adv. Math. **155**(1), 98–153 (2000)
43. Z.M. Balogh, A. Engulatov, L. Hunziker, O.E. Maasalo, Functional inequalities and Hamilton–Jacobi equations in geodesic spaces. Potential Anal. **36**(2), 317–337 (2012)
44. G. Barles, *Solutions de Viscosité des Équations de Hamilton-Jacobi*. Mathématiques & Applications (Berlin) [Mathematics & Applications], vol. 17 (Springer, Paris, 1994)
45. F. Barthe, P. Cattiaux, C. Roberto, Concentration for independent random variables with heavy tails. Appl. Math. Res. Express **2**, 39–60 (2005)
46. F. Barthe, P. Cattiaux, C. Roberto, Interpolated inequalities between exponential and Gaussian, Orlicz hypercontractivity and isoperimetry. Rev. Mat. Iberoam. **22**(3), 993–1067 (2006)
47. F. Barthe, P. Cattiaux, C. Roberto, Isoperimetry between exponential and Gaussian. Electron. J. Probab. **12**(44), 1212–1237 (2007) (electronic)
48. F. Barthe, A.V. Kolesnikov, Mass transport and variants of the logarithmic Sobolev inequality. J. Geom. Anal. **18**(4), 921–979 (2008)
49. F. Barthe, C. Roberto, Sobolev inequalities for probability measures on the real line. Stud. Math. **159**(3), 481–497 (2003). Dedicated to Professor Aleksander Pełczyński on the occasion of his 70th birthday
50. R.F. Bass, *Diffusions and Elliptic Operators*. Probability and Its Applications (New York) (Springer, New York, 1998)
51. F. Baudoin, N. Garofalo, Curvature-dimension inequalities and Ricci lower bounds for sub-Riemannian manifolds with transverse symmetries. Preprint, 2012
52. V. Bayle, A differential inequality for the isoperimetric profile. Int. Math. Res. Not. **7**, 311–342 (2004)
53. M. Bebendorf, A note on the Poincaré inequality for convex domains. Z. Anal. Anwend. **22**(4), 751–756 (2003)
54. W. Beckner, Inequalities in Fourier analysis. Ann. Math. (2) **102**(1), 159–182 (1975)
55. W. Beckner, A generalized Poincaré inequality for Gaussian measures. Proc. Am. Math. Soc. **105**(2), 397–400 (1989)
56. W. Beckner, Sharp Sobolev inequalities on the sphere and the Moser-Trudinger inequality. Ann. Math. (2) **138**(1), 213–242 (1993)
57. W. Beckner, Geometric asymptotics and the logarithmic Sobolev inequality. Forum Math. **11**(1), 105–137 (1999)
58. W. Beckner, M. Pearson, On sharp Sobolev embedding and the logarithmic Sobolev inequality. Bull. Lond. Math. Soc. **30**(1), 80–84 (1998)
59. J. Bennett, A. Carbery, M. Christ, T. Tao, The Brascamp-Lieb inequalities: finiteness, structure and extremals. Geom. Funct. Anal. **17**(5), 1343–1415 (2008)
60. P. Bérard, G. Besson, S. Gallot, Sur une inégalité isopérimétrique qui généralise celle de Paul Lévy-Gromov. Invent. Math. **80**(2), 295–308 (1985)
61. M. Berger, *A Panoramic View of Riemannian Geometry* (Springer, Berlin, 2003)
62. M. Berger, P. Gauduchon, E. Mazet, *Le Spectre d'Une Variété Riemannienne*. Lecture Notes in Mathematics, vol. 194 (Springer, Berlin, 1971)
63. L. Bertini, B. Zegarliński, Coercive inequalities for Gibbs measures. J. Funct. Anal. **162**(2), 257–286 (1999)
64. J. Bertoin, *Lévy Processes*. Cambridge Tracts in Mathematics, vol. 121 (Cambridge University Press, Cambridge, 1996)
65. J. Bertoin, Subordinators: examples and applications, in *Lectures on Probability Theory and Statistics*, Saint-Flour, 1997. Lecture Notes in Math., vol. 1717 (Springer, Berlin, 1999), pp. 1–91
66. M.-F. Bidaut-Véron, L. Véron, Nonlinear elliptic equations on compact Riemannian manifolds and asymptotics of Emden equations. Invent. Math. **106**(3), 489–539 (1991)

67. P. Billingsley, *Probability and Measure*. Wiley Series in Probability and Statistics (Wiley, Hoboken, 2012). Anniversary edition [of MR1324786], with a foreword by Steve Lalley and a brief biography of Billingsley by Steve Koppes
68. R.L. Bishop, R.J. Crittenden, *Geometry of Manifolds*. Pure and Applied Mathematics, vol. XV (Academic Press, New York, 1964)
69. J.-M. Bismut, *Large Deviations and the Malliavin Calculus*. Progress in Mathematics, vol. 45 (Birkhäuser, Boston, 1984)
70. W. Blaschke, *Kreis und Kugel* (Chelsea, New York, 1949)
71. G. Blower, The Gaussian isoperimetric inequality and transportation. Positivity **7**(3), 203–224 (2003)
72. R.M. Blumenthal, R.K. Getoor, *Markov Processes and Potential Theory*. Pure and Applied Mathematics, vol. 29 (Academic Press, New York, 1968)
73. S.G. Bobkov, An isoperimetric inequality on the discrete cube, and an elementary proof of the isoperimetric inequality in Gauss space. Ann. Probab. **25**(1), 206–214 (1997)
74. S.G. Bobkov, Isoperimetric and analytic inequalities for log-concave probability measures. Ann. Probab. **27**(4), 1903–1921 (1999)
75. S.G. Bobkov, A localized proof of the isoperimetric Bakry-Ledoux inequality and some applications. Teor. Veroâtn. Ee Primen. **47**(2), 340–346 (2002)
76. S.G. Bobkov, I. Gentil, M. Ledoux, Hypercontractivity of Hamilton-Jacobi equations. J. Math. Pures Appl. **80**(7), 669–696 (2001)
77. S.G. Bobkov, F. Götze, Exponential integrability and transportation cost related to logarithmic Sobolev inequalities. J. Funct. Anal. **163**(1), 1–28 (1999)
78. S.G. Bobkov, C. Houdré, Some connections between isoperimetric and Sobolev-type inequalities. Mem. Am. Math. Soc. **129**, 616 (1997), pp. viii+111
79. S.G. Bobkov, M. Ledoux, Poincaré's inequalities and Talagrand's concentration phenomenon for the exponential distribution. Probab. Theory Relat. Fields **107**(3), 383–400 (1997)
80. S.G. Bobkov, B. Zegarliński, Entropy bounds and isoperimetry. Mem. Am. Math. Soc. **176**, 829 (2005)
81. T. Bodineau, B. Zegarliński, Hypercontractivity via spectral theory. Infin. Dimens. Anal. Quantum Probab. Relat. Top. **3**(1), 15–31 (2000)
82. V.I. Bogachev, *Gaussian Measures*. Mathematical Surveys and Monographs, vol. 62 (American Mathematical Society, Providence, 1998)
83. V.I. Bogachev, *Measure Theory*, vol. I, II (Springer, Berlin, 2007)
84. F. Bolley, I. Gentil, Phi-entropy inequalities for diffusion semigroups. J. Math. Pures Appl. **93**(5), 449–473 (2010)
85. F. Bolley, I. Gentil, A. Guillin, Dimensional contraction via Markov transportation distance. Preprint, 2013
86. L. Boltzmann, *Lectures on Gas Theory* (University of California Press, Berkeley, 1964). Translated by Stephen G. Brush
87. C. Borell, The Brunn-Minkowski inequality in Gauss space. Invent. Math. **30**(2), 207–216 (1975)
88. C. Borell, Positivity improving operators and hypercontractivity. Math. Z. **180**(2), 225–234 (1982)
89. A.A. Borovkov, S.A. Utev, An inequality and a characterization of the normal distribution connected with it. Teor. Veroâtn. Ee Primen. **28**(2), 209–218 (1983)
90. S. Boucheron, G. Lugosi, P. Massart, *Concentration Inequalities: A Nonasymptotic Theory of Independence* (Oxford University Press, Oxford, 2013)
91. N. Bouleau, F. Hirsch, *Dirichlet Forms and Analysis on Wiener Space*. de Gruyter Studies in Mathematics, vol. 14 (Walter de Gruyter, Berlin, 1991)
92. J.-P. Bourguignon, The "magic" of Weitzenböck formulas, in *Variational Methods*, Paris, 1988. Progr. Nonlinear Differential Equations Appl., vol. 4 (Birkhäuser, Boston, 1990), pp. 251–271

93. H.J. Brascamp, E.H. Lieb, On extensions of the Brunn-Minkowski and Prékopa-Leindler theorems, including inequalities for log concave functions, and with an application to the diffusion equation. J. Funct. Anal. **22**(4), 366–389 (1976)
94. Y. Brenier, Polar factorization and monotone rearrangement of vector-valued functions. Commun. Pure Appl. Math. **44**(4), 375–417 (1991)
95. H. Brézis, *Opérateurs Maximaux Monotones et Semi-Groupes de Contractions dans les Espaces de Hilbert*. North-Holland Mathematics Studies, No. 5. Notas de Matemática (50) (North-Holland, Amsterdam, 1973)
96. H. Brézis, *Functional Analysis, Sobolev Spaces and Partial Differential Equations* (Springer, New York, 2011)
97. D. Burago, Y.D. Burago, S. Ivanov, *A Course in Metric Geometry*. Graduate Studies in Mathematics, vol. 33 (American Mathematical Society, Providence, 2001)
98. Y.D. Burago, V.A. Zalgaller, *Geometric Inequalities*. Grundlehren der mathematischen Wissenschaften [Fundamental Principles of Mathematical Sciences], vol. 285 (Springer, Berlin, 1988). Translated from the Russian by A.B. Sosinskiĭ, Springer Series in Soviet Mathematics
99. P. Buser, Notes: a geometric approach to invariant subspaces of orthogonal matrices. Am. Math. Mon. **89**(10), 751 (1982)
100. P.L. Butzer, H. Berens, *Semi-Groups of Operators and Approximation*. Die Grundlehren der mathematischen Wissenschaften, vol. 145 (Springer, New York, 1967)
101. L.A. Caffarelli, Boundary regularity of maps with convex potentials. II. Ann. Math. (2) **144**(3), 453–496 (1996)
102. L.A. Caffarelli, A priori estimates and the geometry of the Monge Ampère equation, in *Nonlinear Partial Differential Equations in Differential Geometry*, Park City, UT, 1992. IAS/Park City Math. Ser., vol. 2. (Amer. Math. Soc., Providence, 1996), pp. 5–63
103. L.A. Caffarelli, Monotonicity properties of optimal transportation and the FKG and related inequalities. Commun. Math. Phys. **214**(3), 547–563 (2000)
104. P. Cannarsa, C. Sinestrari, *Semiconcave Functions, Hamilton-Jacobi Equations, and Optimal Control*. Progress in Nonlinear Differential Equations and Their Applications, vol. 58 (Birkhäuser, Boston, 2004)
105. M. Capitaine, E.P. Hsu, M. Ledoux, Martingale representation and a simple proof of logarithmic Sobolev inequalities on path spaces. Electron. Commun. Probab. **2**, 71–81 (1997) (electronic)
106. E.A. Carlen, Superadditivity of Fisher's information and logarithmic Sobolev inequalities. J. Funct. Anal. **101**(1), 194–211 (1991)
107. E.A. Carlen, J.A. Carrillo, M. Loss, Hardy-Littlewood-Sobolev inequalities via fast diffusion flows. Proc. Natl. Acad. Sci. USA **107**(46), 19696–19701 (2010)
108. E.A. Carlen, S. Kusuoka, D.W. Stroock, Upper bounds for symmetric Markov transition functions. Ann. Inst. Henri Poincaré Probab. Stat. **23**(2), 245–287 (1987)
109. E.A. Carlen, E.H. Lieb, M. Loss, A sharp analog of Young's inequality on S^N and related entropy inequalities. J. Geom. Anal. **14**(3), 487–520 (2004)
110. E.A. Carlen, M. Loss, Sharp constant in Nash's inequality. Int. Math. Res. Not. **7**, 213–215 (1993)
111. J.A. Carrillo, A. Jüngel, P.A. Markowich, G. Toscani, A. Unterreiter, Entropy dissipation methods for degenerate parabolic problems and generalized Sobolev inequalities. Monatshefte Math. **133**(1), 1–82 (2001)
112. J.A. Carrillo, R.J. McCann, C. Villani, Kinetic equilibration rates for granular media and related equations: entropy dissipation and mass transportation estimates. Rev. Mat. Iberoam. **19**(3), 971–1018 (2003)
113. J.A. Carrillo, R.J. McCann, C. Villani, Contractions in the 2-Wasserstein length space and thermalization of granular media. Arch. Ration. Mech. Anal. **179**, 217–263 (2006)
114. G. Carron, Inégalités isopérimétriques de Faber-Krahn et conséquences, in *Actes de la Table Ronde de Géométrie Différentielle*, Luminy, 1992. Sémin. Congr., vol. 1 (Soc. Math. France, Paris, 1996), pp. 205–232

115. P. Cattiaux, A pathwise approach of some classical inequalities. Potential Anal. **20**(4), 361–394 (2004)
116. P. Cattiaux, I. Gentil, A. Guillin, Weak logarithmic Sobolev inequalities and entropic convergence. Probab. Theory Relat. Fields **139**(3–4), 563–603 (2007)
117. P. Cattiaux, A. Guillin, On quadratic transportation cost inequalities. J. Math. Pures Appl. **86**(4), 341–361 (2006)
118. P. Cattiaux, A. Guillin, *Long Time Behavior of Markov process and Functional Inequalities*. Forthcoming Monograph, 2014
119. P. Cattiaux, A. Guillin, F.-Y. Wang, L. Wu, Lyapunov conditions for super Poincaré inequalities. J. Funct. Anal. **256**(6), 1821–1841 (2009)
120. D. Chafaï, Entropies, convexity, and functional inequalities: on Φ-entropies and Φ-Sobolev inequalities. J. Math. Kyoto Univ. **44**(2), 325–363 (2004)
121. I. Chavel, *Eigenvalues in Riemannian Geometry*. Pure and Applied Mathematics, vol. 115 (Academic Press, Orlando, 1984). Including a chapter by Burton Randol, with an appendix by Jozef Dodziuk
122. I. Chavel, *Isoperimetric Inequalities*. Cambridge Tracts in Mathematics, vol. 145 (Cambridge University Press, Cambridge, 2001). Differential geometric and analytic perspectives
123. I. Chavel, A modern introduction, in *Riemannian Geometry*. Cambridge Studies in Advanced Mathematics, vol. 98, 2nd edn. (Cambridge University Press, Cambridge, 2006)
124. J. Cheeger, A lower bound for the smallest eigenvalue of the Laplacian, in *Problems in Analysis (Papers Dedicated to Salomon Bochner, 1969)* (Princeton Univ. Press, Princeton, 1970), pp. 195–199
125. J. Cheeger, Differentiability of Lipschitz functions on metric measure spaces. Geom. Funct. Anal. **9**(3), 428–517 (1999)
126. J. Cheeger, D.G. Ebin, *Comparison Theorems in Riemannian Geometry* (AMS Chelsea, Providence, 2008). Revised reprint of the 1975 original
127. L.H.Y. Chen, An inequality for the multivariate normal distribution. J. Multivar. Anal. **12**(2), 306–315 (1982)
128. M.-F. Chen, Trilogy of couplings and general formulas for lower bound of spectral gap, in *Probability Towards 2000*, New York, 1995. Lecture Notes in Statist., vol. 128 (Springer, New York, 1998), pp. 123–136
129. M.-F. Chen, F.-Y. Wang, Application of coupling method to the first eigenvalue on manifold. Prog. Nat. Sci. **5**(2), 227–229 (1995)
130. H. Chernoff, A note on an inequality involving the normal distribution. Ann. Probab. **9**(3), 533–535 (1981)
131. T.S. Chihara, *An Introduction to Orthogonal Polynomials*. Mathematics and Its Applications, vol. 13 (Gordon and Breach, New York, 1978)
132. K.L. Chung, *Lectures from Markov Processes to Brownian Motion*. Grundlehren der mathematischen Wissenschaften [Fundamental Principles of Mathematical Science], vol. 249 (Springer, New York, 1982)
133. D. Cordero-Erausquin, Some applications of mass transport to Gaussian-type inequalities. Arch. Ration. Mech. Anal. **161**(3), 257–269 (2002)
134. D. Cordero-Erausquin, W. Gangbo, C. Houdré, Inequalities for generalized entropy and optimal transportation, in *Recent Advances in the Theory and Applications of Mass Transport*. Contemp. Math., vol. 353 (Am. Math. Soc., Providence, 2004), pp. 73–94
135. D. Cordero-Erausquin, R.J. McCann, M. Schmuckenschläger, A Riemannian interpolation inequality à la Borell, Brascamp and Lieb. Invent. Math. **146**(2), 219–257 (2001)
136. D. Cordero-Erausquin, R.J. McCann, M. Schmuckenschläger, Prékopa-Leindler type inequalities on Riemannian manifolds, Jacobi fields, and optimal transport. Ann. Fac. Sci. Toulouse **15**(4), 613–635 (2006)
137. D. Cordero-Erausquin, B. Nazaret, C. Villani, A mass-transportation approach to sharp Sobolev and Gagliardo-Nirenberg inequalities. Adv. Math. **182**(2), 307–332 (2004)
138. T. Coulhon, Inégalités de Gagliardo-Nirenberg pour les semi-groupes d'opérateurs et applications. Potential Anal. **1**(4), 343–353 (1992)

139. T. Coulhon, Ultracontractivity and Nash type inequalities. J. Funct. Anal. **141**(2), 510–539 (1996)
140. R. Courant, D. Hilbert, *Methoden der mathematischen Physik*, vol. 2 (1937) (German)
141. T.M. Cover, J.A. Thomas, *Elements of Information Theory*, 2nd edn. (Wiley-Interscience, Hoboken, 2006)
142. G. Da Prato, *An Introduction to Infinite-Dimensional Analysis*. Universitext (Springer, Berlin, 2006). Revised and extended from the 2001 original by Da Prato
143. E.B. Davies, *One-Parameter Semigroups*. London Mathematical Society Monographs, vol. 15 (Academic Press, London, 1980)
144. E.B. Davies, *Heat Kernels and Spectral Theory*. Cambridge Tracts in Mathematics, vol. 92 (Cambridge University Press, Cambridge, 1989)
145. E.B. Davies, L. Gross, B. Simon, Hypercontractivity: a bibliographic review, in *Ideas and Methods in Quantum and Statistical Physics*, Oslo, 1988 (Cambridge Univ. Press, Cambridge, 1992), pp. 370–389
146. E.B. Davies, B. Simon, Ultracontractivity and the heat kernel for Schrödinger operators and Dirichlet Laplacians. J. Funct. Anal. **59**(2), 335–395 (1984)
147. M. Del Pino, J. Dolbeault, Best constants for Gagliardo-Nirenberg inequalities and applications to nonlinear diffusions. J. Math. Pures Appl. **81**(9), 847–875 (2002)
148. M. Del Pino, J. Dolbeault, The optimal Euclidean L^p-Sobolev logarithmic inequality. J. Funct. Anal. **197**(1), 151–161 (2003)
149. H. Delin, A proof of the equivalence between Nash and Sobolev inequalities. Bull. Sci. Math. **120**(4), 405–411 (1996)
150. C. Dellacherie, *Capacités et Processus Stochastiques*. Ergebnisse der Mathematik und ihrer Grenzgebiete, vol. 67 (Springer, Berlin, 1972)
151. C. Dellacherie, P.-A. Meyer, *Probabilities and Potential*. North-Holland Mathematics Studies, vol. 29 (North-Holland, Amsterdam, 1978)
152. C. Dellacherie, P.-A. Meyer, Theory of martingales, in *Probabilities and Potential. B*. North-Holland Mathematics Studies, vol. 72 (North-Holland, Amsterdam, 1982). Translated from the French by J.P. Wilson
153. C. Dellacherie, P.-A. Meyer, Théorie du potentiel associée à une résolvante. Théorie des processus de Markov, in *Probabilités et Potentiel*. 2nd edn. Publications de l'Institut de Mathématiques de l'Université de Strasbourg (Hermann, Paris, 1987). Chapitres XII–XVI
154. C. Dellacherie, P.-A. Meyer, Potential theory for discrete and continuous semigroups, in *Probabilities and Potential. C*. North-Holland Mathematics Studies, vol. 151 (North-Holland, Amsterdam, 1988). Translated from the French by J.R. Norris
155. J. Demange, Porous media equation and Sobolev inequalities under negative curvature. Bull. Sci. Math. **129**(10), 804–830 (2005)
156. J. Demange, Improved Gagliardo-Nirenberg-Sobolev inequalities on manifolds with positive curvature. J. Funct. Anal. **254**(3), 593–611 (2008)
157. A. Dembo, T.M. Cover, J.A. Thomas, Information-theoretic inequalities. IEEE Trans. Inf. Theory **37**(6), 1501–1518 (1991)
158. J.-D. Deuschel, D.W. Stroock, *Large Deviations*. Pure and Applied Mathematics, vol. 137 (Academic Press, Boston, 1989)
159. M.P. do Carmo, *Riemannian Geometry*. Mathematics: Theory & Applications (Birkhäuser, Boston, 1992). Translated from the second Portuguese edition by Francis Flaherty
160. J. Dolbeault, M.J. Esteban, M. Kowalczyk, M. Loss, Sharp interpolation inequalities on the sphere: new methods and consequences. Chin. Ann. Math., Ser. B **34**(1), 99–112 (2013)
161. J. Dolbeault, M. Esteban, M. Loss, Nonlinear flows and rigidity results on compact manifolds. Preprint, 2013
162. J. Dolbeault, I. Gentil, A. Guillin, F.-Y. Wang, L^q-functional inequalities and weighted porous media equations. Potential Anal. **28**(1), 35–59 (2008)
163. H. Donnelly, Exhaustion functions and the spectrum of Riemannian manifolds. Indiana Univ. Math. J. **46**(2), 505–527 (1997)

164. J.L. Doob, *Classical Potential Theory and Its Probabilistic Counterpart*. Grundlehren der mathematischen Wissenschaften [Fundamental Principles of Mathematical Sciences], vol. 262 (Springer, New York, 1984)
165. O. Druet, E. Hebey, The AB program in geometric analysis: sharp Sobolev inequalities and related problems. Mem. Am. Math. Soc. **160**, 761 (2002), pp. viii+98
166. B.A. Dubrovin, A.T. Fomenko, S.P. Novikov, The geometry and topology of manifolds, in *Modern Geometry—Methods and Applications. Part II*. Graduate Texts in Mathematics, vol. 104 (Springer, New York, 1985). Translated from the Russian by Robert G. Burns
167. B.A. Dubrovin, A.T. Fomenko, S.P. Novikov, The geometry of surfaces, transformation groups, and fields, in *Modern Geometry—Methods and Applications. Part I*, 2nd edn. Graduate Texts in Mathematics, vol. 93 (Springer, New York, 1992). Translated from the Russian by Robert G. Burns
168. N. Dunford, J.T. Schwartz, *Linear Operators. I. General Theory*. Pure and Applied Mathematics, vol. 7 (Interscience, New York, 1958). With the assistance of W.G. Bade and R.G. Bartle
169. N. Dunford, J.T. Schwartz, *Linear Operators. Part II: Spectral Theory. Self Adjoint Operators in Hilbert Space* (Interscience, New York, 1963). With the assistance of William G. Bade, and Robert G. Bartle
170. E.B. Dynkin, *Markov Processes. Vols. I, II*. Die Grundlehren der mathematischen Wissenschaften, vols. 121, 122 (Academic Press, New York, 1965). Translated with the authorization and assistance of the author by J. Fabius, V. Greenberg, A. Maitra, G. Majone
171. K.D. Elworthy, *Stochastic Differential Equations on Manifolds*. London Mathematical Society Lecture Note Series, vol. 70 (Cambridge University Press, Cambridge, 1982)
172. K.D. Elworthy, Geometric aspects of diffusions on manifolds, in *École d'Été de Probabilités de Saint-Flour XV–XVII, 1988–87*. Lecture Notes in Math., vol. 1362 (Springer, Berlin, 1985), pp. 277–425
173. M. Émery, J.E. Yukich, A simple proof of the logarithmic Sobolev inequality on the circle, in *Séminaire de Probabilités, XXI*. Lecture Notes in Math., vol. 1247 (Springer, Berlin, 1987), pp. 173–175
174. K.-J. Engel, R. Nagel, *One-Parameter Semigroups for Linear Evolution Equations*. Graduate Texts in Mathematics, vol. 194 (Springer, New York, 2000). With contributions by S. Brendle, M. Campiti, T. Hahn, G. Metafune, G. Nickel, D. Pallara, C. Perazzoli, A. Rhandi, S. Romanelli and R. Schnaubelt
175. M. Erbar, The heat equation on manifolds as a gradient flow in the Wasserstein space. Ann. Inst. Henri Poincaré Probab. Stat. **46**(1), 1–23 (2010)
176. M. Erbar, K. Kuwada, K.-T. Sturm, On the equivalence of the entropy curvature-dimension condition and Bochner's inequality on metric measure spaces. Preprint, 2013
177. J.F. Escobar, Sharp constant in a Sobolev trace inequality. Indiana Univ. Math. J. **37**(3), 687–698 (1988)
178. S.N. Ethier, T.G. Kurtz, Characterization and convergence, in *Markov Processes*. Wiley Series in Probability and Mathematical Statistics: Probability and Mathematical Statistics (Wiley, New York, 1986)
179. L.C. Evans, *Partial Differential Equations*, 2nd edn. Graduate Studies in Mathematics, vol. 19 (American Mathematical Society, Providence, 2010)
180. L.C. Evans, R.F. Gariepy, *Measure Theory and Fine Properties of Functions*. Studies in Advanced Mathematics (CRC Press, Boca Raton, 1992)
181. E.B. Fabes, D.W. Stroock, A new proof of Moser's parabolic Harnack inequality using the old ideas of Nash. Arch. Ration. Mech. Anal. **96**(4), 327–338 (1986)
182. S. Fang, Inégalité du type de Poincaré sur l'espace des chemins Riemanniens. C. R. Math. Acad. Sci. Paris, Sér. I **318**(3), 257–260 (1994)
183. P. Federbush, A partially alternate derivation of a result of Nelson. J. Math. Phys. **10**(1), 50–52 (1969)
184. H. Federer, *Geometric Measure Theory*. Die Grundlehren der mathematischen Wissenschaften, vol. 153 (Springer, New York, 1969)

185. É. Fontenas, Sur les constantes de Sobolev des variétés riemanniennes compactes et les fonctions extrémales des sphères. Bull. Sci. Math. **121**(2), 71–96 (1997)
186. P. Fougères, Spectral gap for log-concave probability measures on the real line, in *Séminaire de Probabilités XXXVIII*. Lecture Notes in Math., vol. 1857 (Springer, Berlin, 2005), pp. 95–123
187. J. Franchi, Y. Le Jan, *Hyperbolic Dynamics and Brownian Motions: An Introduction* (Oxford University Press, Oxford, 2012)
188. Y. Fujita, An optimal logarithmic Sobolev inequality with Lipschitz constants. J. Funct. Anal. **261**(5), 1133–1144 (2011)
189. M. Fukushima, *Dirichlet Forms and Markov Processes*. North-Holland Mathematical Library, vol. 23 (North-Holland, Amsterdam, 1980)
190. M. Fukushima, Y. Oshima, M. Takeda, *Dirichlet Forms and Symmetric Markov Processes*. de Gruyter Studies in Mathematics, vol. 19 (Walter de Gruyter, Berlin, 2011), extended edn.
191. M.P. Gaffney, The conservation property of the heat equation on Riemannian manifolds. Commun. Pure Appl. Math. **12**, 1–11 (1959)
192. E. Gagliardo, Proprietà di alcune classi di funzioni in più variabili. Ric. Mat. **7**, 102–137 (1958)
193. S. Gallot, Inégalités isopérimétriques et analytiques sur les variétés Riemanniennes. Astérisque (1988), no. 163–164, pp. 5–6, 31–91, 281 (1989). On the geometry of differentiable manifolds (Rome, 1986)
194. S. Gallot, D. Hulin, J. Lafontaine, *Riemannian Geometry*, 3nd edn. Universitext (Springer, Berlin, 2004)
195. R.J. Gardner, The Brunn-Minkowski inequality. Bull. Am. Math. Soc. (N.S.) **39**(3), 355–405 (2002)
196. A.M. Garsia, *Martingale Inequalities: Seminar Notes on Recent Progress*. Mathematics Lecture Notes Series (Benjamin, Reading, 1973)
197. I. Gentil, Ultracontractive bounds on Hamilton-Jacobi solutions. Bull. Sci. Math. **126**(6), 507–524 (2002)
198. I. Gentil, The general optimal L^p-Euclidean logarithmic Sobolev inequality by Hamilton-Jacobi equations. J. Funct. Anal. **202**(2), 591–599 (2003)
199. I. Gentil, From the Prékopa-Leindler inequality to modified logarithmic Sobolev inequality. Ann. Fac. Sci. Toulouse **17**(2), 291–308 (2008)
200. I. Gentil, A. Guillin, L. Miclo, Modified logarithmic Sobolev inequalities and transportation inequalities. Probab. Theory Relat. Fields **133**(3), 409–436 (2005)
201. I. Gentil, A. Guillin, L. Miclo, Modified logarithmic Sobolev inequalities in null curvature. Rev. Mat. Iberoam. **23**(1), 235–258 (2007)
202. B. Gidas, J. Spruck, Global and local behavior of positive solutions of nonlinear elliptic equations. Commun. Pure Appl. Math. **34**(4), 525–598 (1981)
203. D. Gilbarg, N.S. Trudinger, *Elliptic Partial Differential Equations of Second Order*. Classics in Mathematics (Springer, Berlin, 2001). Reprint of the 1998 edition
204. J. Glimm, Boson fields with nonlinear selfinteraction in two dimensions. Commun. Math. Phys. **8**, 12–25 (1968) (English)
205. J.A. Goldstein, *Semigroups of Linear Operators and Applications*. Oxford Mathematical Monographs (The Clarendon Press, New York, 1985)
206. N. Gozlan, A characterization of dimension free concentration in terms of transportation inequalities. Ann. Probab. **37**(6), 2480–2498 (2009)
207. N. Gozlan, Poincaré inequalities and dimension free concentration of measure. Ann. Inst. Henri Poincaré Probab. Stat. **46**(3), 708–739 (2010)
208. N. Gozlan, Transport-entropy inequalities on the line. Electron. J. Probab. **17**(49), 1–18 (2012)
209. N. Gozlan, C. Léonard, A large deviation approach to some transportation cost inequalities. Probab. Theory Relat. Fields **139**(1–2), 235–283 (2007)
210. N. Gozlan, C. Léonard, Transport inequalities. A survey. Markov Process. Relat. Fields **16**, 635–736 (2010)

211. N. Gozlan, C. Roberto, P.-M. Samson, From concentration to logarithmic Sobolev and Poincaré inequalities. J. Funct. Anal. **260**(5), 1491–1522 (2011)
212. N. Gozlan, C. Roberto, P.-M. Samson, Hamilton Jacobi equations on metric spaces and transport entropy inequalities. Rev. Mat. Iberoam. (2013, to appear)
213. N. Gozlan, C. Roberto, P.-M. Samson, Characterization of Talagrand's transport-entropy inequality in metric spaces. Ann. Probab. **41**(5), 3112–3139 (2013)
214. A. Grigor'yan, The heat equation on noncompact Riemannian manifolds. Mat. Sb. **182**(1), 55–87 (1991)
215. A. Grigor'yan, Analytic and geometric background of recurrence and non-explosion of the Brownian motion on Riemannian manifolds. Bull. Am. Math. Soc. (N.S.) **36**(2), 135–249 (1999)
216. A. Grigor'yan, Isoperimetric inequalities and capacities on Riemannian manifolds, in *The Maz'ya Anniversary Collection, Vol. 1*, Rostock, 1998. Oper. Theory Adv. Appl., vol. 109 (Birkhäuser, Basel, 1999), pp. 139–153
217. A. Grigor'yan, *Heat Kernel and Analysis on Manifolds*. AMS/IP Studies in Advanced Mathematics, vol. 47 (American Mathematical Society, Providence, 2009)
218. A. Grigor'yan, Yau's work on heat kernels, in *Geometry and Analysis. No. 1*. Adv. Lect. Math. (ALM), vol. 17 (Int. Press, Somerville, 2011), pp. 113–117
219. G. Grillo, On Persson's theorem in local Dirichlet spaces. Z. Anal. Anwend. **17**(2), 329–338 (1998)
220. M. Gromov, Paul Lévy's isoperimetric inequality. Preprint, 1980
221. M. Gromov, *Metric Structures for Riemannian and Non-Riemannian Spaces*. Progress in Mathematics, vol. 152 (Birkhäuser, Boston, 1999). Based on the 1981 French original. With appendices by M. Katz, P. Pansu and S. Semmes, Translated from the French by Sean Michael Bates
222. M. Gromov, V.D. Milman, A topological application of the isoperimetric inequality. Am. J. Math. **105**(4), 843–854 (1983)
223. L. Gross, Existence and uniqueness of physical ground states. J. Funct. Anal. **10**, 52–109 (1972)
224. L. Gross, Logarithmic Sobolev inequalities. Am. J. Math. **97**(4), 1061–1083 (1975)
225. L. Gross, Logarithmic Sobolev inequalities and contractivity properties of semigroups, in *Dirichlet Forms*, Varenna, 1992. Lecture Notes in Math., vol. 1563 (Springer, Berlin, 1993), pp. 54–88
226. L. Gross, Hypercontractivity, logarithmic Sobolev inequalities, and applications: a survey of surveys, in *Diffusion, Quantum Theory, and Radically Elementary Mathematics*. Math. Notes, vol. 47 (Princeton Univ. Press, Princeton, 2006), pp. 45–73
227. L. Gross, O.S. Rothaus, Herbst inequalities for supercontractive semigroups. J. Math. Kyoto Univ. **38**(2), 295–318 (1998)
228. O. Guédon, Concentration phenomena in high dimensional geometry. ESAIM Proc. (2013, to appear)
229. A. Guionnet, B. Zegarliński, Lectures on logarithmic Sobolev inequalities, in *Séminaire de Probabilités, XXXVI*. Lecture Notes in Math., vol. 1801 (Springer, Berlin, 2003), pp. 1–134
230. P. Hajłasz, P. Koskela, Sobolev met Poincaré. Mem. Am. Math. Soc. **145**, 688 (2000), p. x+101
231. G.H. Hardy, Note on a theorem of Hilbert. Math. Z. **6**(3–4), 314–317 (1920)
232. G.H. Hardy, J.E. Littlewood, G. Pólya, *Inequalities*, 2nd edn. (Cambridge University Press, Cambridge, 1952)
233. E. Hebey, *Sobolev Spaces on Riemannian Manifolds*. Lecture Notes in Mathematics, vol. 1635 (Springer, Berlin, 1996)
234. E. Hebey, *Nonlinear Analysis on Manifolds: Sobolev Spaces and Inequalities*. Courant Lecture Notes in Mathematics, vol. 5 (New York University Courant Institute of Mathematical Sciences, New York, 1999)
235. E. Hebey, M. Vaugon, The best constant problem in the Sobolev embedding theorem for complete Riemannian manifolds. Duke Math. J. **79**(1), 235–279 (1995)

236. J. Heinonen, *Lectures on Analysis on Metric Spaces*. Universitext (Springer, New York, 2001)
237. B. Helffer, Remarks on decay of correlations and Witten Laplacians, Brascamp-Lieb inequalities and semiclassical limit. J. Funct. Anal. **155**(2), 571–586 (1998)
238. B. Helffer, *Semiclassical Analysis, Witten Laplacians, and Statistical Mechanics*. Series in Partial Differential Equations and Applications, vol. 1 (World Scientific, River Edge, 2002)
239. B. Helffer, J. Sjöstrand, On the correlation for Kac-like models in the convex case. J. Stat. Phys. **74**(1–2), 349–409 (1994)
240. S. Helgason, Integral geometry, invariant differential operators, and spherical functions, in *Groups and Geometric Analysis*. Mathematical Surveys and Monographs, vol. 83 (American Mathematical Society, Providence, 2000). Corrected reprint of the 1984 original
241. E. Hille, R.S. Phillips, *Functional Analysis and Semi-Groups*. American Mathematical Society Colloquium Publications, vol. 31 (American Mathematical Society, Providence, 1957), rev. ed.
242. M. Hino, On short time asymptotic behavior of some symmetric diffusions on general state spaces. Potential Anal. **16**(3), 249–264 (2002)
243. F. Hirsch, Intrinsic metrics and Lipschitz functions. J. Evol. Equ. **3**(1), 11–25 (2003). Dedicated to Philippe Bénilan
244. P.D. Hislop, I.M. Sigal, *Introduction to Spectral Theory*. Applied Mathematical Sciences, vol. 113 (Springer, New York, 1996). With applications to Schrödinger operators
245. R. Holley, D.W. Stroock, Logarithmic Sobolev inequalities and stochastic Ising models. J. Stat. Phys. **46**, 1159–1194 (1987)
246. L. Hörmander, L^2 estimates and existence theorems for the $\bar{\partial}$ operator. Acta Math. **113**, 89–152 (1965)
247. L. Hörmander, Differential operators with constant coefficients, in *The Analysis of Linear Partial Differential Operators. II*. Grundlehren der mathematischen Wissenschaften [Fundamental Principles of Mathematical Sciences], vol. 257 (Springer, Berlin, 1983)
248. L. Hörmander, Distribution theory and Fourier analysis, in *The Analysis of Linear Partial Differential Operators. I*, 2nd edn. (Springer, Berlin, 1990). Springer Study Edition
249. L. Hörmander, *An Introduction to Complex Analysis in Several Variables*, 3nd edn. North-Holland Mathematical Library, vol. 7 (North-Holland, Amsterdam, 1990)
250. E.P. Hsu, Logarithmic Sobolev inequalities on path spaces over Riemannian manifolds. Commun. Math. Phys. **189**(1), 9–16 (1997)
251. E.P. Hsu, *Stochastic Analysis on Manifolds*. Graduate Studies in Mathematics, vol. 38 (American Mathematical Society, Providence, 2002)
252. N. Ikeda, S. Watanabe, *Stochastic Differential Equations and Diffusion Processes*, 2nd edn. North-Holland Mathematical Library, vol. 24 (North-Holland, Amsterdam, 1989)
253. S. Ilias, Constantes explicites pour les inégalités de Sobolev sur les variétés Riemanniennes compactes. Ann. Inst. Fourier (Grenoble) **33**(2), 151–165 (1983)
254. J. Inglis, M. Neklyudov, B. Zegarliński, Ergodicity for infinite particle systems with locally conserved quantities. Infin. Dimens. Anal. Quantum Probab. Relat. Top. **15**(1), 1250005 (2012), 28
255. K. Itô, H.P. McKean, *Diffusion Processes and Their Sample Paths*. Die Grundlehren der mathematischen Wissenschaften, vol. 125 (Springer, Berlin, 1974). Second printing, corrected
256. J. Jacod, *Calcul Stochastique et Problèmes de Martingales*. Lecture Notes in Mathematics, vol. 714 (Springer, Berlin, 1979)
257. S. Janson, *Gaussian Hilbert Spaces*. Cambridge Tracts in Mathematics, vol. 129 (Cambridge University Press, Cambridge, 1997)
258. F. John, *Partial Differential Equations*, 4th edn. Applied Mathematical Sciences, vol. 1 (Springer, New York, 1982)
259. R. Jordan, D. Kinderlehrer, F. Otto, The variational formulation of the Fokker-Planck equation. SIAM J. Math. Anal. **29**(1), 1–17 (1998)

260. J. Jost, *Riemannian Geometry and Geometric Analysis*, 5th edn. Universitext (Springer, Berlin, 2008)
261. A. Joulin, N. Privault, Functional inequalities for discrete gradients and application to the geometric distribution. ESAIM Probab. Stat. **8**, 87–101 (2004)
262. R. Kannan, L. Lovász, M. Simonovits, Isoperimetric problems for convex bodies and a localization lemma. Discrete Comput. Geom. **13**(3–4), 541–559 (1995)
263. I. Karatzas, S.E. Shreve, *Brownian Motion and Stochastic Calculus*, 2nd edn. Graduate Texts in Mathematics, vol. 113 (Springer, New York, 1991)
264. T. Kato, *Perturbation Theory for Linear Operators*, 2nd edn. Grundlehren der mathematischen Wissenschaften, vol. 132 (Springer, Berlin, 1976)
265. O. Kavian, G. Kerkyacharian, B. Roynette, Quelques remarques sur l'ultracontractivité. J. Funct. Anal. **111**(1), 155–196 (1993)
266. H. Knothe, Contributions to the theory of convex bodies. Mich. Math. J. **4**, 39–52 (1957)
267. M. Knott, C.S. Smith, On the optimal mapping of distributions. J. Optim. Theory Appl. **43**(1), 39–49 (1984)
268. R. Koekoek, P.A. Lesky, R.F. Swarttouw, *Hypergeometric Orthogonal Polynomials and Their q-Analogues*. Springer Monographs in Mathematics (Springer, Berlin, 2010). With a foreword by Tom H. Koornwinder
269. A. Kufner, L.-E. Persson, *Weighted Inequalities of Hardy Type* (World Scientific, River Edge, 2003)
270. K. Kuwada, Duality on gradient estimates and Wasserstein controls. J. Funct. Anal. **258**(11), 3758–3774 (2010)
271. K. Kuwada, Space-time Wasserstein controls and Bakry-Ledoux type gradient estimates. Preprint, 2013
272. R. Latała, K. Oleszkiewicz, Between Sobolev and Poincaré, in *Geometric Aspects of Functional Analysis*. Lecture Notes in Math., vol. 1745 (Springer, Berlin, 2000), pp. 147–168
273. M. Ledoux, A simple analytic proof of an inequality by P. Buser. Proc. Am. Math. Soc. **121**(3), 951–959 (1994)
274. M. Ledoux, L'algèbre de Lie des gradients itérés d'un générateur markovien—développements de moyennes et entropies. Ann. Sci. Éc. Norm. Super. **28**(4), 435–460 (1995)
275. M. Ledoux, Remarks on logarithmic Sobolev constants, exponential integrability and bounds on the diameter. J. Math. Kyoto Univ. **35**(2), 211–220 (1995)
276. M. Ledoux, Concentration of measure and logarithmic Sobolev inequalities, in *Séminaire de Probabilités, XXXIII*. Lecture Notes in Math., vol. 1709 (Springer, Berlin, 1999), pp. 120–216
277. M. Ledoux, The geometry of Markov diffusion generators. Ann. Fac. Sci. Toulouse **9**(2), 305–366 (2000). Probability theory
278. M. Ledoux, *The Concentration of Measure Phenomenon*. Mathematical Surveys and Monographs, vol. 89 (American Mathematical Society, Providence, 2001)
279. M. Ledoux, Spectral gap, logarithmic Sobolev constant, and geometric bounds, in *Surveys in Differential Geometry*, vol. IX (Int. Press, Somerville, 2004), pp. 219–240
280. M. Ledoux, From concentration to isoperimetry: semigroup proofs, in *Concentration, Functional Inequalities and Isoperimetry*. Contemp. Math., vol. 545 (Am. Math. Soc., Providence, 2011), pp. 155–166
281. P. Lévy, *Problèmes Concrets d'Analyse Fonctionnelle*, 2nd edn. Des Leçons d'Analyse Fonctionnelle (Gauthier-Villars, Paris, 1951) (French). Avec un complément par F. Pellegrino
282. H.-Q. Li, Estimation optimale du gradient du semi-groupe de la chaleur sur le groupe de Heisenberg. J. Funct. Anal. **236**(2), 369–394 (2006)
283. H.-Q. Li, Estimations asymptotiques du noyau de la chaleur sur les groupes de Heisenberg. C. R. Math. Acad. Sci. Paris **344**(8), 497–502 (2007)
284. P. Li, *Geometric Analysis* (Cambridge University Press, Cambridge, 2012)
285. P. Li, S.-T. Yau, On the parabolic kernel of the Schrödinger operator. Acta Math. **156**(3–4), 153–201 (1986)

286. A. Lichnerowicz, *Géométrie des Groupes de Transformations*. Travaux et Recherches Mathématiques, vol. III (Dunod, Paris, 1958)
287. E.H. Lieb, Gaussian kernels have only Gaussian maximizers. Invent. Math. **102**(1), 179–208 (1990)
288. E.H. Lieb, M. Loss, *Analysis*, 2nd edn. Graduate Studies in Mathematics, vol. 14 (American Mathematical Society, Providence, 2001)
289. R.S. Liptser, A.N. Shiryayev, *Theory of Martingales*. Mathematics and Its Applications (Soviet Series), vol. 49 (Kluwer Academic, Dordrecht, 1989). Translated from the Russian by K. Dzjaparidze [Kacha Dzhaparidze]
290. J. Lott, Some geometric properties of the Bakry-Émery-Ricci tensor. Comment. Math. Helv. **78**(4), 865–883 (2003)
291. J. Lott, C. Villani, Hamilton-Jacobi semigroup on length spaces and applications. J. Math. Pures Appl. **88**(3), 219–229 (2007)
292. J. Lott, C. Villani, Ricci curvature for metric-measure spaces via optimal transport. Ann. Math. (2) **169**(3), 903–991 (2009)
293. L. Lovász, M. Simonovits, Random walks in a convex body and an improved volume algorithm. Random Struct. Algorithms **4**(4), 359–412 (1993)
294. Z.M. Ma, M. Röckner, *Introduction to the Theory of (Nonsymmetric) Dirichlet Forms*. Universitext (Springer, Berlin, 1992)
295. F. Maggi, *Sets of Finite Perimeter and Geometric Variational Problems* (Cambridge University Press, Cambridge, 2012)
296. F. Maggi, C. Villani, Balls have the worst best Sobolev inequalities. II. Variants and extensions. Calc. Var. Partial Differ. Equ. **31**(1), 47–74 (2008)
297. P. Malliavin, *Stochastic Analysis*. Grundlehren der mathematischen Wissenschaften [Fundamental Principles of Mathematical Sciences], vol. 313 (Springer, Berlin, 1997)
298. K. Marton, Bounding $\bar{d}$-distance by informational divergence: a method to prove measure concentration. Ann. Probab. **24**(2), 857–866 (1996)
299. P. Mathieu, Quand l'inégalité log-Sobolev implique l'inégalité de trou spectral, in *Séminaire de Probabilités, XXXII*. Lecture Notes in Math., vol. 1686 (Springer, Berlin, 1998), pp. 30–35
300. J. Mawhin, Henri Poincaré and partial differential equations. Nieuw Arch. Wiskd. **13**(3), 159–169 (2012)
301. O. Mazet, Classification des semi-groupes de diffusion sur **R** associés à une famille de polynômes orthogonaux, in *Séminaire de Probabilités, XXXI*. Lecture Notes in Math., vol. 1655 (Springer, Berlin, 1997), pp. 40–53
302. V.G. Maz'ja, *Sobolev Spaces*. Springer Series in Soviet Mathematics (Springer, Berlin, 1985). Translated from the Russian by T.O. Shaposhnikova
303. V.G. Maz'ya, *Sobolev Spaces with Applications to Elliptic Partial Differential Equations*. Grundlehren der mathematischen Wissenschaften [Fundamental Principles of Mathematical Sciences], vol. 342 (Springer, Heidelberg, 2011). augmented ed.
304. R.J. McCann, Existence and uniqueness of monotone measure-preserving maps. Duke Math. J. **80**(2), 309–323 (1995)
305. R.J. McCann, A convexity principle for interacting gases. Adv. Math. **128**(1), 153–179 (1997)
306. H.P. McKean, An upper bound to the spectrum of Δ on a manifold of negative curvature. J. Differ. Geom. **4**, 359–366 (1970)
307. P.-A. Meyer, *Probability and Potentials* (Blaisdell, Waltham, 1966)
308. P.-A. Meyer, Note sur les processus d'Ornstein-Uhlenbeck, in *Seminar on Probability, XVI*. Lecture Notes in Math., vol. 920 (Springer, Berlin, 1982), pp. 95–133
309. S. Meyn, R.L. Tweedie, *Markov Chains and Stochastic Stability*, 2nd edn. (Cambridge University Press, Cambridge, 2009). With a prologue by Peter W. Glynn
310. L. Miclo, Quand est-ce que des bornes de Hardy permettent de calculer une constante de Poincaré exacte sur la droite? Ann. Fac. Sci. Toulouse **17**(1), 121–192 (2008)
311. L. Miclo, On hyperboundedness and spectrum of Markov operators. Preprint, 2013

312. E. Milman, On the role of convexity in isoperimetry, spectral gap and concentration. Invent. Math. **177**(1), 1–43 (2009)
313. E. Milman, A converse to the Maz'ya inequality for capacities under curvature lower bound, in *Around the Research of Vladimir Maz'ya. I*. Int. Math. Ser. (N. Y.), vol. 11 (Springer, New York, 2010), pp. 321–348
314. E. Milman, Isoperimetric and concentration inequalities: equivalence under curvature lower bound. Duke Math. J. **154**(2), 207–239 (2010)
315. E. Milman, Sharp isoperimetric inequalities and model spaces for curvature-dimension-diameter condition. Preprint, 2012
316. V.D. Milman, G. Schechtman, *Asymptotic Theory of Finite-Dimensional Normed Spaces*. Lecture Notes in Mathematics, vol. 1200 (Springer, Berlin, 1986). With an appendix by M. Gromov
317. D.S. Mitrinović, J.E. Pečarić, A.M. Fink, *Inequalities Involving Functions and Their Integrals and Derivatives*. Mathematics and Its Applications (East European Series), vol. 53 (Kluwer Academic, Dordrecht, 1991)
318. F. Morgan, Manifolds with density. Not. Am. Math. Soc. **52**(8), 853–858 (2005)
319. J. Moser, A Harnack inequality for parabolic differential equations. Commun. Pure Appl. Math. **17**, 101–134 (1964)
320. J. Moser, A sharp form of an inequality by N. Trudinger. Indiana Univ. Math. J. **20**, 1077–1092 (1970/71)
321. B. Muckenhoupt, Hardy's inequality with weights. Stud. Math. **44**, 31–38 (1972). Collection of articles honoring the completion by Antoni Zygmund of 50 years of scientific activity, I
322. C.E. Mueller, F.B. Weissler, Hypercontractivity for the heat semigroup for ultraspherical polynomials and on the n-sphere. J. Funct. Anal. **48**(2), 252–283 (1982)
323. J. Nash, Continuity of solutions of parabolic and elliptic equations. Am. J. Math. **80**, 931–954 (1958)
324. B. Nazaret, Best constant in Sobolev trace inequalities on the half-space. Nonlinear Anal. **65**(10), 1977–1985 (2006)
325. E. Nelson, A quartic interaction in two dimensions, in *Mathematical Theory of Elementary Particles*, Dedham, MA, 1965. Proc. Conf. (MIT Press, Cambridge, 1966), pp. 69–73
326. E. Nelson, *Dynamical Theories of Brownian Motion* (Princeton University Press, Princeton, 1967)
327. E. Nelson, The free Markoff field. J. Funct. Anal. **12**, 211–227 (1973)
328. E. Nelson, Quantum fields and Markoff fields, in *Partial Differential Equations*, Berkeley, CA, 1971. Proc. Sympos. Pure Math., vol. XXIII (Am. Math. Soc., Providence, 1973), pp. 413–420
329. J. Neveu, Sur l'espérance conditionnelle par rapport à un mouvement brownien. Ann. Inst. Henri Poincaré, Sect. B (N. S.) **12**(2), 105–109 (1976)
330. V.H. Nguyen, Sharp weighted Sobolev and Gagliardo-Nirenberg inequalities on half-spaces via mass transport and consequences. Preprint, 2013
331. L. Nirenberg, On elliptic partial differential equations. Ann. Sc. Norm. Super. Pisa, Cl. Sci. (3) **13**, 115–162 (1959)
332. J.R. Norris, Heat kernel asymptotics and the distance function in Lipschitz Riemannian manifolds. Acta Math. **179**(1), 79–103 (1997)
333. J.R. Norris, *Markov Chains*. Cambridge Series in Statistical and Probabilistic Mathematics, vol. 2 (Cambridge University Press, Cambridge, 1998). Reprint of 1997 original
334. D. Nualart, *The Malliavin Calculus and Related Topics*, 2nd edn. Probability and Its Applications (New York) (Springer, Berlin, 2006)
335. M. Obata, Certain conditions for a Riemannian manifold to be isometric with a sphere. J. Math. Soc. Jpn. **14**, 333–340 (1962)
336. B. Øksendal, *Stochastic Differential Equations*, 6th edn. Universitext (Springer, Berlin, 2003). An introduction with applications
337. E. Onofri, On the positivity of the effective action in a theory of random surfaces. Commun. Math. Phys. **86**(3), 321–326 (1982)

338. R. Osserman, The isoperimetric inequality. Bull. Am. Math. Soc. **84**(6), 1182–1238 (1978)
339. F. Otto, The geometry of dissipative evolution equations: the porous medium equation. Commun. Partial Differ. Equ. **26**(1–2), 101–174 (2001)
340. F. Otto, C. Villani, Generalization of an inequality by Talagrand and links with the logarithmic Sobolev inequality. J. Funct. Anal. **173**(2), 361–400 (2000)
341. F. Otto, M. Westdickenberg, Eulerian calculus for the contraction in the Wasserstein distance. SIAM J. Math. Anal. **37**(4), 1227–1255 (2005) (electronic)
342. E.M. Ouhabaz, *Analysis of Heat Equations on Domains*. London Mathematical Society Monographs Series, vol. 31 (Princeton University Press, Princeton, 2005)
343. L.E. Payne, H.F. Weinberger, An optimal Poincaré inequality for convex domains. Arch. Ration. Mech. Anal. **5**, 286–292 (1960)
344. A. Pazy, *Semigroups of Linear Operators and Applications to Partial Differential Equations*. Applied Mathematical Sciences, vol. 44 (Springer, New York, 1983)
345. A. Persson, Bounds for the discrete part of the spectrum of a semi-bounded Schrödinger operator. Math. Scand. **8**, 143–153 (1960)
346. P. Petersen, *Riemannian Geometry*. Graduate Texts in Mathematics, vol. 171 (Springer, New York, 1998)
347. R.G. Pinsky, *Positive Harmonic Functions and Diffusion*. Cambridge Studies in Advanced Mathematics, vol. 45 (Cambridge University Press, Cambridge, 1995)
348. H. Poincaré, Sur la théorie analytique de la chaleur. C. R. Séances Acad. Sci. **104**, 1753–1759 (1887)
349. H. Poincaré, Sur les equations aux derivees partielles de la physique mathematique. Am. J. Math. **12**(3), 211–294 (1890)
350. P.E. Protter, *Stochastic Integration and Differential Equations*, 2nd edn. Stochastic Modelling and Applied Probability, vol. 21 (Springer, Berlin, 2005). Version 2.1, Corrected third printing
351. S.T. Rachev, *Probability Metrics and the Stability of Stochastic Models*. Wiley Series in Probability and Mathematical Statistics: Applied Probability and Statistics (Wiley, Chichester, 1991)
352. S.T. Rachev, L. Rüschendorf, *Mass Transportation Problems. Vol. I*. Probability and Its Applications (New York) (Springer, New York, 1998). Theory
353. S.T. Rachev, L. Rüschendorf, *Mass Transportation Problems. Vol. II*. Probability and Its Applications (New York) (Springer, New York, 1998). Applications
354. J.G. Ratcliffe, *Foundations of Hyperbolic Manifolds*, 2nd edn. Graduate Texts in Mathematics, vol. 149 (Springer, New York, 2006)
355. M. Reed, B. Simon, *Methods of Modern Mathematical Physics. II. Fourier Analysis, Self-Adjointness* (Academic Press, New York, 1975)
356. M. Reed, B. Simon, *Methods of Modern Mathematical Physics. I*, 2nd edn. (Academic Press, New York, 1980). Functional analysis
357. D. Revuz, *Markov Chains*, 2nd edn. North-Holland Mathematical Library, vol. 11 (North-Holland, Amsterdam, 1984)
358. D. Revuz, M. Yor, *Continuous Martingales and Brownian Motion*, 3rd edn. Grundlehren der mathematischen Wissenschaften [Fundamental Principles of Mathematical Sciences], vol. 293 (Springer, Berlin, 1999)
359. F. Riesz, B.Sz. Nagy, *Functional Analysis* (Frederick Ungar, New York, 1955). Translated by Leo F. Boron
360. C. Roberto, B. Zegarliński, Orlicz-Sobolev inequalities for sub-Gaussian measures and ergodicity of Markov semi-groups. J. Funct. Anal. **243**(1), 28–66 (2007)
361. M. Röckner, F.-Y. Wang, Weak Poincaré inequalities and L^2-convergence rates of Markov semigroups. J. Funct. Anal. **185**(2), 564–603 (2001)
362. L.C.G. Rogers, D. Williams, Itô calculus, in *Diffusions, Markov Processes, and Martingales. Vol. 2*. Wiley Series in Probability and Mathematical Statistics: Probability and Mathematical Statistics (Wiley, New York, 1987)

363. L.C.G. Rogers, D. Williams, Foundations, in *Diffusions, Markov Processes, and Martingales. Vol. 1*. Wiley Series in Probability and Mathematical Statistics: Probability and Mathematical Statistics, 2nd edn. (Wiley, Chichester, 1994)
364. G. Rosen, Minimum value for c in the Sobolev inequality $\|\phi^3\| \leq c \|\nabla \phi\|^3$. SIAM J. Appl. Math. **21**, 30–32 (1971)
365. J. Rosen, Sobolev inequalities for weight spaces and supercontractivity. Trans. Am. Math. Soc. **222**, 367–376 (1976)
366. J.-P. Roth, Opérateurs dissipatifs et semi-groupes dans les espaces de fonctions continues. Ann. Inst. Fourier (Grenoble) **26**, 1–97 (1976)
367. O.S. Rothaus, Logarithmic Sobolev inequalities and the spectrum of Sturm-Liouville operators. J. Funct. Anal. **39**(1), 42–56 (1980)
368. O.S. Rothaus, Diffusion on compact Riemannian manifolds and logarithmic Sobolev inequalities. J. Funct. Anal. **42**(1), 102–109 (1981)
369. O.S. Rothaus, Logarithmic Sobolev inequalities and the spectrum of Schrödinger operators. J. Funct. Anal. **42**(1), 110–120 (1981)
370. O.S. Rothaus, Analytic inequalities, isoperimetric inequalities and logarithmic Sobolev inequalities. J. Funct. Anal. **64**(2), 296–313 (1985)
371. O.S. Rothaus, Hypercontractivity and the Bakry-Emery criterion for compact Lie groups. J. Funct. Anal. **65**(3), 358–367 (1986)
372. G. Royer, *An Initiation to Logarithmic Sobolev Inequalities*. SMF/AMS Texts and Monographs, vol. 14 (American Mathematical Society, Providence, 2007). Translated from the 1999 French original by Donald Babbitt
373. L. Rüschendorf, S.T. Rachev, A characterization of random variables with minimum L^2-distance. J. Multivar. Anal. **32**(1), 48–54 (1990)
374. L. Saloff-Coste, A note on Poincaré, Sobolev, and Harnack inequalities. Int. Math. Res. Not. **2**, 27–38 (1992)
375. L. Saloff-Coste, Convergence to equilibrium and logarithmic Sobolev constant on manifolds with Ricci curvature bounded below. Colloq. Math. **67**(1), 109–121 (1994)
376. L. Saloff-Coste, *Aspects of Sobolev-Type Inequalities*. London Mathematical Society Lecture Note Series, vol. 289 (Cambridge University Press, Cambridge, 2002)
377. L. Saloff-Coste, Sobolev inequalities in familiar and unfamiliar settings, in *Sobolev Spaces in Mathematics. I*. Int. Math. Ser. (N. Y.), vol. 8 (Springer, New York, 2009), pp. 299–343
378. L. Saloff-Coste, The heat kernel and its estimates, in *Probabilistic Approach to Geometry*. Adv. Stud. Pure Math., vol. 57 (Math. Soc. Japan, Tokyo, 2010), pp. 405–436
379. K.-i. Sato, *Lévy Processes and Infinitely Divisible Distributions*. Cambridge Studies in Advanced Mathematics, vol. 68 (Cambridge University Press, Cambridge, 1999). Translated from the 1990 Japanese original, Revised by the author
380. R. Schoen, Conformal deformation of a Riemannian metric to constant scalar curvature. J. Differ. Geom. **20**(2), 479–495 (1984)
381. R. Schoen, Variational theory for the total scalar curvature functional for Riemannian metrics and related topics, in *Topics in Calculus of Variations*, Montecatini Terme, 1987. Lecture Notes in Math., vol. 1365 (Springer, Berlin, 1989), pp. 120–154
382. R. Schoen, S.-T. Yau, Conformally flat manifolds, Kleinian groups and scalar curvature. Invent. Math. **92**(1), 47–71 (1988)
383. C.E. Shannon, A mathematical theory of communication. Bell Syst. Tech. J. **27**, 379–423, 623–656 (1948)
384. B. Simon, R. Høegh-Krohn, Hypercontractive semigroups and two dimensional self-coupled Bose fields. J. Funct. Anal. **9**, 121–180 (1972)
385. J. Sjöstrand, Correlation asymptotics and Witten Laplacians. Algebra Anal. **8**(1), 160–191 (1996)
386. S. Sobolev, Sur un théorème d'analyse fonctionnelle. Mat. Sb. (N.S.) **4**, 471–497 (1938). (Russian)
387. A.J. Stam, Some inequalities satisfied by the quantities of information of Fisher and Shannon. Inf. Control **2**, 101–112 (1959)

388. E.M. Stein, *Singular Integrals and Differentiability Properties of Functions*. Princeton Mathematical Series, vol. 30 (Princeton University Press, Princeton, 1970)
389. E.M. Stein, G. Weiss, *Introduction to Fourier Analysis on Euclidean Spaces*. Princeton Mathematical Series, vol. 32 (Princeton University Press, Princeton, 1971)
390. R.S. Strichartz, Analysis of the Laplacian on the complete Riemannian manifold. J. Funct. Anal. **52**(1), 48–79 (1983)
391. D.W. Stroock, *An Introduction to the Analysis of Paths on a Riemannian Manifold*. Mathematical Surveys and Monographs, vol. 74 (American Mathematical Society, Providence, 2000)
392. D.W. Stroock, *Partial Differential Equations for Probabilists*. Cambridge Studies in Advanced Mathematics, vol. 112 (Cambridge University Press, Cambridge, 2008)
393. D.W. Stroock, S.R.S. Varadhan, *Multidimensional Diffusion Processes*. Classics in Mathematics (Springer, Berlin, 2006). Reprint of the 1997 edition
394. D.W. Stroock, O. Zeitouni, Variations on a theme by Bismut. Astérisque **236**, 291–301 (1996). Hommage à P. A. Meyer et J. Neveu
395. K.-T. Sturm, Analysis on local Dirichlet spaces. I. Recurrence, conservativeness and L^p-Liouville properties. J. Reine Angew. Math. **456**, 173–196 (1994)
396. K.-T. Sturm, Analysis on local Dirichlet spaces. II. Upper Gaussian estimates for the fundamental solutions of parabolic equations. Osaka J. Math. **32**(2), 275–312 (1995)
397. K.-T. Sturm, Diffusion processes and heat kernels on metric spaces. Ann. Probab. **26**(1), 1–55 (1998)
398. K.-T. Sturm, The geometric aspect of Dirichlet forms, in *New Directions in Dirichlet Forms*. AMS/IP Stud. Adv. Math., vol. 8 (Amer. Math. Soc., Providence, 1998), pp. 233–277
399. K.-T. Sturm, On the geometry of metric measure spaces. I. Acta Math. **196**(1), 65–131 (2006)
400. K.-T. Sturm, On the geometry of metric measure spaces. II. Acta Math. **196**(1), 133–177 (2006)
401. V.N. Sudakov, B.S. Cirel'son, Extremal properties of half-spaces for spherically invariant measures. Zap. Nauč. Semin. LOMI **41**, 14–24, 165 (1974). Problems in the theory of probability distributions, II
402. G. Szegő, *Orthogonal Polynomials*, 4th edn. American Mathematical Society Colloquium Publications, vol. XXIII (American Mathematical Society, Providence, 1975)
403. M. Talagrand, A new isoperimetric inequality and the concentration of measure phenomenon, in *Geometric Aspects of Functional Analysis (1989–90)*. Lecture Notes in Math., vol. 1469 (Springer, Berlin, 1991), pp. 94–124
404. M. Talagrand, Transportation cost for Gaussian and other product measures. Geom. Funct. Anal. **6**(3), 587–600 (1996)
405. G. Talenti, Osservazioni sopra una classe di disuguaglianze. Rend. Semin. Mat. Fis. Milano **39**, 171–185 (1969)
406. G. Talenti, Best constant in Sobolev inequality. Ann. Mat. Pura Appl. (4) **110**, 353–372 (1976)
407. H. Tanaka, An inequality for a functional of probability distributions and its application to Kac's one-dimensional model of a Maxwellian gas. Z. Wahrscheinlichkeitstheor. Verw. Geb. **27**, 47–52 (1973)
408. H. Tanaka, Probabilistic treatment of the Boltzmann equation of Maxwellian molecules. Z. Wahrscheinlichkeitstheor. Verw. Geb. **46**(1), 67–105 (1978/79)
409. A. Terras, *Harmonic Analysis on Symmetric Spaces—Euclidean Space, the Sphere, and the Poincaré Upper Half-Plane* (Springer, Berlin, 2013)
410. G. Tomaselli, A class of inequalities. Boll. Unione Mat. Ital. **2**, 622–631 (1969)
411. M. Tomisaki, Comparison theorems on Dirichlet norms and their applications. Forum Math. **2**(3), 277–295 (1990)
412. G. Toscani, Entropy production and the rate of convergence to equilibrium for the Fokker-Planck equation. Q. Appl. Math. **57**(3), 521–541 (1999)

413. H. Triebel, *Theory of Function Spaces*. Monographs in Mathematics, vol. 78 (Birkhäuser, Basel, 1983)
414. H. Triebel, *Theory of Function Spaces. II*. Monographs in Mathematics, vol. 84 (Birkhäuser, Basel, 1992)
415. H. Triebel, *Interpolation Theory, Function Spaces, Differential Operators*, 2nd edn. (Johann Ambrosius Barth, Heidelberg, 1995)
416. N.S. Trudinger, On imbeddings into Orlicz spaces and some applications. J. Math. Mech. **17**, 473–483 (1967)
417. A.S. Üstünel, *An Introduction to Analysis on Wiener Space*. Lecture Notes in Mathematics, vol. 1610 (Springer, Berlin, 1995)
418. J.A. van Casteren, *Markov Processes, Feller Semigroups and Evolution Equations*. Series on Concrete and Applicable Mathematics, vol. 12 (World Scientific, Hackensack, 2011)
419. J. van Neerven, *The Asymptotic Behaviour of Semigroups of Linear Operators*. Operator Theory: Advances and Applications, vol. 88 (Birkhäuser, Basel, 1996)
420. S.R.S. Varadhan, On the behavior of the fundamental solution of the heat equation with variable coefficients. Commun. Pure Appl. Math. **20**, 431–455 (1967)
421. N.T. Varopoulos, Hardy-Littlewood theory for semigroups. J. Funct. Anal. **63**(2), 240–260 (1985)
422. N.T. Varopoulos, L. Saloff-Coste, T. Coulhon, *Analysis and Geometry on Groups*. Cambridge Tracts in Mathematics, vol. 100 (Cambridge University Press, Cambridge, 1992)
423. J.L. Vázquez, *The Porous Medium Equation*. Oxford Mathematical Monographs (The Clarendon Press, Oxford, 2007). Mathematical theory
424. C. Villani, *Topics in Optimal Transportation*. Graduate Studies in Mathematics, vol. 58 (American Mathematical Society, Providence, 2003)
425. C. Villani, Hypocoercivity. Mem. Am. Math. Soc. **202**, 950 (2009), pp. iv+141
426. C. Villani, *Optimal Transport, old and new*. Grundlehren der mathematischen Wissenschaften [Fundamental Principles of Mathematical Sciences], vol. 338 (Springer, Berlin, 2009)
427. M.-K. von Renesse, K.-T. Sturm, Transport inequalities, gradient estimates, entropy, and Ricci curvature. Commun. Pure Appl. Math. **58**(7), 923–940 (2005)
428. F.-Y. Wang, Logarithmic Sobolev inequalities on noncompact Riemannian manifolds. Probab. Theory Relat. Fields **109**(3), 417–424 (1997)
429. F.-Y. Wang, Functional inequalities for empty essential spectrum. J. Funct. Anal. **170**(1), 219–245 (2000)
430. F.-Y. Wang, Functional inequalities and spectrum estimates: the infinite measure case. J. Funct. Anal. **194**(2), 288–310 (2002)
431. F.-Y. Wang, *Functional Inequalities, Markov Processes, and Spectral Theory* (Science Press, Beijing, 2004)
432. F.-Y. Wang, Spectral gap for hyperbounded operators. Proc. Am. Math. Soc. **132**(9), 2629–2638 (2004)
433. F.-Y. Wang, Dimension-free Harnack inequality and its applications. Front. Math. China **1**(1), 53–72 (2006)
434. F.-Y. Wang, Harnack inequalities on manifolds with boundary and applications. J. Math. Pures Appl. **94**(3), 304–321 (2010)
435. F.-Y. Wang, Equivalent semigroup properties for the curvature-dimension condition. Bull. Sci. Math. **135**(6–7), 803–815 (2011)
436. F. Wang, Y. Zhang, F-Sobolev inequality for general symmetric forms. Northeast. Math. J. **19**(2), 133–138 (2003)
437. F.B. Weissler, Logarithmic Sobolev inequalities for the heat-diffusion semigroup. Trans. Am. Math. Soc. **237**, 255–269 (1978)
438. F.B. Weissler, Two-point inequalities, the Hermite semigroup, and the Gauss-Weierstrass semigroup. J. Funct. Anal. **32**(1), 102–121 (1979)
439. F.B. Weissler, Logarithmic Sobolev inequalities and hypercontractive estimates on the circle. J. Funct. Anal. **37**(2), 218–234 (1980)

440. D. Williams, *Probability with Martingales*, Cambridge Mathematical Textbooks (Cambridge University Press, Cambridge, 1991)
441. L. Wu, A new modified logarithmic Sobolev inequality for Poisson point processes and several applications. Probab. Theory Relat. Fields **118**(3), 427–438 (2000)
442. S.-T. Yau, Isoperimetric constants and the first eigenvalue of a compact Riemannian manifold. Ann. Sci. Éc. Norm. Super. **8**(4), 487–507 (1975)
443. S.-T. Yau, On the heat kernel of a complete Riemannian manifold. J. Math. Pures Appl. **57**(2), 191–201 (1978)
444. K. Yosida, *Functional Analysis*, 6th edn. Grundlehren der mathematischen Wissenschaften [Fundamental Principles of Mathematical Sciences], vol. 123 (Springer, Berlin, 1980)
445. B. Zegarliński, Entropy bounds for Gibbs measures with non-Gaussian tails. J. Funct. Anal. **187**(2), 368–395 (2001)
446. B. Zegarliński, Analysis on extended Heisenberg group. Ann. Fac. Sci. Toulouse **20**(2), 379–405 (2011)
447. A. Zettl, *Sturm-Liouville Theory*. Mathematical Surveys and Monographs, vol. 121 (American Mathematical Society, Providence, 2005)
448. S. Zhu, The comparison geometry of Ricci curvature, in *Comparison Geometry*, Berkeley, CA, 1993–94. Math. Sci. Res. Inst. Publ., vol. 30 (Cambridge Univ. Press, Cambridge, 1997), pp. 221–262
449. W.P. Ziemer, Sobolev spaces and functions of bounded variation, in *Weakly Differentiable Functions*. Graduate Texts in Mathematics, vol. 120 (Springer, New York, 1989)
450. P.-A. Zitt, Super Poincaré inequalities, Orlicz norms and essential spectrum. Potential Anal. **35**(1), 51–66 (2011)

Index

A

Adjoint operator, 45, 48, 133, 141, 170

B

Beckner inequality, 312, 384, 388
Bessel
    operator, 95
    process, 94
    semigroup, 61, 94
Beta distribution, 114
Bobkov inequality, 417, 430
    local, 417
    reverse, 422
Bochner-Lichnerowicz formula, 509, 519
Boundary conditions
    Dirichlet, 93, 97, 99, 200, 221
    Neumann, 93, 97, 100, 199, 200, 273
    periodic, 200, 273
Bounded perturbation
    Φ-entropy inequality, 383
    logarithmic Sobolev inequality, 240, 274
    Poincaré inequality, 185, 231
    transportation cost inequality, 459, 468
Brascamp-Lieb inequality, 215, 232
Brenier's map, 436, 447, 467
Brownian motion, 487
    Euclidean, 78
    hyperbolic, 89
    killed, 94
    reflected, 94
    spherical, 85
Brownian semigroup
    Euclidean, 79
    hyperbolic, 89
    spherical, 81, 85

C

Caffarelli's
    contraction Theorem, 447
    regularity properties, 437, 467
Cameron-Martin Hilbert space, 110, 490
Capacity, 392, 429
Capacity inequality, 394, 397, 400, 401, 404, 407, 411, 429
Carré du champ operator Γ, 20, 34, 37, 42, 55, 74, 78, 82, 84, 89, 91, 98, 103, 111, 114, 120, 140, 169, 512
Cattiaux's inequality, 248, 353
Cauchy kernel, 80
Change of coordinates, 56
Change of variables formula, 43, 124, 169, 171, 493, 516
Chapman-Kolmogorov equation, 16, 28, 264
Chart, 500
Christoffel symbols, 506
Closable operator, 476
Co-area formula, 395, 413, 429, 430
Co-metric, 47, 82, 504, 505
Commutation property, 144–146, 173, 259
    Hamilton-Jacobi, 462, 468
Compact operator, 483
Completeness, 141, 157, 172, 510
Concentration
    exponential, 426, 431, 467
    Gaussian, 426, 431, 441, 460, 467
Concentration under
    logarithmic Sobolev inequality, 255, 274
    Nash inequality, 368
    Poincaré inequality, 192, 231
    weak Poincaré inequality, 381
Conformal
    invariance, 314, 342
    invariant, 315, 321

map, 84, 92
Connection, 506
    Levi-Civita, 507
Connexity, 33, 140, 156, 171
Continuity property, 11, 54
Contraction in Wasserstein space, 463, 468
Contraction principle
    logarithmic Sobolev inequality, 254
    Poincaré inequality, 192
Contraction semigroup, 10, 18, 473
Convergence to equilibrium, 33
Core, 19, 52, 101, 134, 142, 477
Cost function, 434
Coupling, 434
Covariant derivative, 507
Curvature condition $CD(\rho, \infty)$, 72, 104, 111, 144, 209, 259, 416, 443, 457, 462, 463, 513
    reinforced, 146, 172, 421, 515
Curvature-dimension condition $CD(\rho, n)$, 72, 75, 79, 87, 92, 104, 115, 159, 172, 211, 232, 268, 275, 298, 305, 337, 514
Cut-locus, 510

D
De Bruijn's identity, 245, 274, 333
Del Pino-Dolbeault Theorem, 326, 342
Density kernel, 14, 25, 286, 356, 371
Diameter, 143, 166, 173, 293, 341
Differentiable manifold, 500
Diffusion
    operator, 43
    process, 38, 496
    property, 56, 122
Dirichlet
    domain, 30, 55, 125, 155, 169
    form, 30, 55, 125, 169, 477
    norm, 125, 169
Displacement convexity, 465, 468
Distance
    geodesic, 510
    harmonic, 360
    intrinsic, 166, 173
    Riemannian, 166, 510
    total variation, 244, 436, 438
    Wasserstein, 436, 445
Doob's stopping time Theorem, 491
Dual semigroup, 9, 464
Duhamel's formula, 130, 144, 145, 207, 214, 232

E
Eigenvalue, 478
Eigenvector, 478
Ellipticity, 40, 49–51, 511
    hypo-, 51
    semi-, 40, 43, 49, 50
    uniform, 50
    weak hypo-, 141, 157, 172
Entropic inequality, 236
Entropy, 236, 273, 348
    relative, 237, 244, 438, 445
Entropy-energy inequality, 281, 349, 387
Ergodic semigroup, 33
Ergodicity, 33, 70, 135, 170
ESA property, 134, 142, 157, 170
Essentially self-adjoint, 95, 101, 134, 170, 477, 482
Explosion time, 495
Exponential decay
    Φ-entropy, 383
    entropy, 244, 274
    variance, 183, 231
Exponential integrability
    eigenvector, 250, 274
    logarithmic Sobolev inequality, 252, 274
    Poincaré inequality, 190, 231
Exponential measure, 188, 231
Extended algebra, 152
Extremal function, 259, 285, 320, 448, 467

F
F-Sobolev inequality, 388, 430
Faber-Krahn inequality, 183, 284, 399, 429
Fast diffusion equation, 331, 342
Feynman-Kac formula, 64
Filtration, 488
Fisher information, 237, 333, 445
Fokker-Planck equation, 24, 28, 332
Friedrichs extension, 128, 477

G
Γ_2 operator, 75, 79, 104, 144, 157, 172, 173, 275, 512, 519
Gagliardo-Nirenberg inequality, 324, 342, 451
Gamma distribution, 111
Gaussian measure, 103, 109, 296
Generalized inverse, 58, 350
Geodesic, 509
    optimal transportation, 465
Girsanov transformation, 62
Good measurable space, 7, 53, 169
Gozlan's Theorem, 460, 468
Gradient, 506
    vector field, 47, 65

Index   549

Gradient bound (strong), 105, 144–146, 163, 173, 209, 259, 298, 416
Gross' hypercontractivity Theorem, 246, 274, 387
Ground state transformation, 64, 107
Growth function, 225, 348, 375

H
h-transform, 66, 95
Hamilton-Jacobi equation, 452, 454, 467
Harmonic
    distance, 360
    extension, 81, 86, 116
    function, 66, 92
    oscillator, 107
Harnack inequality, 232, 302, 341, 421, 431
    isoperimetric, 425, 431
    Wang, 265, 275, 424
Heat equation, 4, 19, 24, 28, 55, 127, 496, 521
Heat flow, 19, 55, 127
    monotonicity, 130, 144, 207, 214, 257, 261, 269, 275, 299, 343, 419, 422
Heat kernel bound, 264, 302, 341, 350, 388
Heat semigroup, 19, 55, 127
    Euclidean, 79
    hyperbolic, 89
    spherical, 81, 85
Herbst's argument, 252, 274, 294, 356
Hermite polynomials, 105, 179
Hessian, 146, 158, 172, 507, 512
Hilbert-Schmidt operator, 29, 290, 355, 483
Hille-Yosida theory, 18, 473, 485
Hopf-Lax formula, 452, 467
Hörmander form, 41, 43, 51, 497, 517
HWI inequality, 443, 467
Hyperbolic space, 89
Hypercontractivity, 246, 273, 389
    Hamilton-Jacobi, 454, 468
    immediate, 351
    local, 262, 275, 342, 457
    reverse, 274

I
Indices
    contravariant, 505
    covariant, 505
Infimum-convolution, 467
Infinitesimal generator, 18, 55, 169, 475, 496
    domain, 18, 52, 55, 99, 126, 155, 170, 475
Integration by parts, 27, 49, 55, 132, 169, 171
Intermediate value Theorem, 136, 173
Invariant measure, 10, 20, 25, 33, 35, 37, 45, 54, 78, 82, 89, 103, 109, 111, 114, 123

Inversion, 91
Isoperimetric
    function, 413
    profile, 413
        Gaussian, 417
Isoperimetric inequality, 413, 430
    Euclidean, 413
    exponential, 413
    Gaussian, 411, 417
    spherical, 413
Isoperimetric-type inequality, 412, 426
    Gaussian, 416, 426
    local, 418
Itô's formula, 491

J
Jacobi
    operator, 113
    polynomials, 114
    semigroup, 115
    spectrum, 114

K
Kannan-Lovász-Simonovits-Bobkov Theorem, 203, 232
Kantorovich-Rubinstein Theorem, 435, 466
Kantorovich's Theorem, 435, 466
Kernel
    density, 78, 90, 97, 101, 104, 116, 118
    representation, 13, 104

L
Laguerre
    operator, 111
    polynomials, 112
    semigroup, 113
    spectrum, 112
Laplace-Beltrami operator, 46, 81, 146, 508, 514
Laplace-Varadhan asymptotics, 303, 453
Laplacian, 46, 508
    Euclidean, 78
    hyperbolic, 89, 92
    spherical, 81, 82, 84, 85, 115
Latała-Oleszkiewicz inequality, 385, 388, 410, 430
Legendre-Fenchel transform, 397, 452
Lévy-Gromov Theorem, 342, 418, 430
Li-Yau inequality, 301, 341
Lichnerowicz' Theorem, 213, 215, 232
Linearized form
    entropy-energy, 349
    Nash, 365, 403
    weak Poincaré, 376, 407

Lipschitz function, 166, 173
Local
  Bobkov inequality, 417
  hypercontractivity, 262, 275, 304, 342, 457
  isoperimetric-type inequality, 418
  logarithmic Sobolev inequality, 259, 275, 299
  Poincaré inequality, 209, 232
Local inequalities, 206
Log-concave measure, 203, 425
Log-Harnack inequality, 266
Logarithmic entropy-energy inequality, 281, 306
Logarithmic Sobolev constant, 237, 259
Logarithmic Sobolev inequality, 236, 273, 401, 429, 446, 458
  defective, 236, 282, 349
  dimensional, 296
  Euclidean, 284, 328, 341, 354, 456
  Gaussian, 258, 275
  local, 259, 275
  modified, 385, 388
  reverse, 259
  segment, 273, 275
  under $CD(\rho, \infty)$, 268
  under $CD(\rho, n)$, 270, 275
Lyapunov
  criterion, 202, 231
  function, 202, 372

M

Manifold
  differentiable, 500
  Riemannian, 137, 504
Markov
  chain, 33
  generator, 18
  kernel, 12, 54
  operator, 10, 15
  process, 8, 17, 37, 54
  semigroup, 12, 54, 128, 170
    symmetric, 25, 55
Markov property (strong), 8, 495
Markov Triple, 28, 120, 168, 177, 235, 277, 347, 392
  Compact, 167, 171
  Diffusion, 132, 169
  Full, 167, 171, 177, 235, 277, 347, 392
  Standard, 136, 170, 177, 235, 277, 347, 392
Martingale, 488
  local, 491
  problem, 22, 39, 496
Mass conservation, 10, 54, 135, 150, 165, 170, 173

Measure decomposition theorem, 7
Measure-capacity inequality, 394, 397, 400, 401, 404, 407, 411, 429
Median
  function, 375
  measure, 194, 256
Metric measure space, 466
Milman's Theorems, 426, 431
Minimal extension, 134, 221, 481
Monge-Ampère equation, 437, 439, 448
Muckenhoupt's criterion, 231, 274, 429
  Latała-Oleszkiewicz inequality, 411
  logarithmic Sobolev inequality, 256
  Poincaré inequality, 194
  weak Poincaré inequality, 409

N

Nash inequality, 281, 403
  generalized, 364, 387, 430
  weighted, 224, 370, 388
Non-explosion, 135, 165

O

Obata's Theorem, 214, 232, 313
Onofri's inequality, 342
Opposite generator, 31
Optimal transportation, 433, 466
Orlicz
  function, 396
  space, 389, 396
Ornstein-Uhlenbeck
  operator, 103
  process, 104
  semigroup, 103, 178
    in infinite dimension, 110
  spectrum, 107, 111
Orthogonal polynomials, 102
Oscillation, 185, 240, 376, 459
Otto-Villani Theorem, 458, 468

P

Φ-entropy inequality, 383, 388
Perron-Frobenius eigenvector, 36, 67, 222
Persson operator, 227, 233
Pinsker-Csiszár-Kullback inequality, 244, 274, 438
Poincaré constant, 181, 190, 200
Poincaré inequality, 181, 365, 377, 400, 429, 458
  ball, 204
  convex set, 204
  exponential measure, 189
  Gaussian, 180, 231
  hyperbolic space, 222, 233
  local, 209, 232

Index 551

log-concave measure, 203
segment, 199, 231
Sturm-Liouville operator, 198
super, 388
under $CD(\rho,\infty)$, 212
under $CD(\rho,n)$, 215, 232
weak, 375, 388, 430
Poisson formula, 86, 116
Polish space, 7, 53, 412, 434
Porous medium equation, 331, 342
Positive operator, 477
Positivity preserving, 10, 54, 127
Potential, 63

R

Rellich-Kondrachov Theorem, 229, 291, 341
Representation
open ball, 91
orthogonal projection, 81
stereographic projection, 83, 327
upper half-space, 89
Reproducing kernel Hilbert space, 110, 490
Resolvent
operator, 68, 289, 312, 474
set, 478
Reversible measure, 25, 35, 37, 46, 55, 78, 82, 89, 103, 111, 114, 123
Ricci
curvature, 72, 73, 82, 144, 146, 314, 317, 464, 508, 513
tensor, 71, 82, 508, 512
Riemann curvature tensor, 507
Riemannian
distance, 166, 510
measure, 46, 509
metric, 504
Riemannian manifold, 137, 504
weighted, 47, 73, 137, 213, 269, 412, 433, 511
Riesz potential, 69
Riesz-Thorin Theorem, 287, 289, 353
Rota's Lemma, 26, 75
Rothaus' Lemma, 239, 274

S

Scalar curvature, 314, 508
Second order differential operator, 40, 42, 139
Self-adjoint operator, 476
Semi-martingale, 492
Semigroup, 18, 473
Markov, 12, 54, 128, 170
product, 59
quotient, 60

Semigroup interpolation, 130, 144, 207, 214, 232, 257, 261, 269, 275, 299, 419, 422
Semigroup property, 10, 54
Slicing, 137, 241, 283, 341, 395
Smooth, 501
function, 138, 167
Sobolev constant, 280, 318
Sobolev inequality, 279, 340, 397, 429
Euclidean, 278, 318, 341, 448, 467
hyperbolic, 279, 319, 341
spherical, 278, 318, 341
trace, 449, 467
under $CD(\rho,n)$, 308, 342
Spectral
decomposition, 31, 33, 131, 478, 479
gap, 182
Spectrum, 32, 221, 478
continuous, 478
discrete, 224, 228, 290, 355, 480
essential, 223, 228, 480, 485
point, 478
residual, 478
Stability under product
Φ-entropy inequality, 383
logarithmic Sobolev inequality, 249, 274
Poincaré inequality, 185, 231
quadratic transportation cost inequality, 441
Sobolev inequality, 292
State space, 7, 53
Stationary measure, 10, 11, 123
Stochastic
differential equation, 38, 493
integral, 488
Stochastic completeness, 135, 165
Stopping time, 490
Stratonovich integral, 41, 496
Sturm-Liouville operator, 97, 198, 199, 273
Sub-Markov semigroup, 12, 63
Subordination, subordinator, 67, 86
Super Poincaré inequality, 388
Surface measure, 412
Symmetric operator, 476

T

Talagrand's inequality, 438, 440, 444, 467
Tangent vector, 501
Tensor, 502
Tight
inequality, 182
logarithmic Sobolev inequality, 236
Sobolev inequality, 280

Tightening, 238, 274, 280, 388
Time change, 58
Total variation distance, 244, 436, 438
Trace formula, 32
Transportation cost inequality, 466
    quadratic T_2, 438, 440, 458, 467
    T_1, 438

U
Ultracontractivity, 286, 341, 353, 367, 456
Uniqueness extension, 134

V
Vanishing viscosity, 453, 455
Variance, 181
Variational formula of entropy, 236
Vector field, 41, 62, 501
    gradient, 47, 65

W
Wang's Harnack inequality, 265, 275, 424
Wasserstein
    contraction, 463, 468
    distance, 436, 445, 465
    space, 436, 463, 465, 466
Weak hypo-ellipticity, 141, 157, 172
Weak Poincaré inequality, 375, 388, 407, 430
Weight function, 370
Weyl's criterion, 223, 481
Wiener
    integral, 110, 490
    measure, 109, 488
Wirtinger's inequality, 200, 231, 232

Z
Zonal function, 115